AF477379

BIODIVERSITY, ECOSYSTEMS, AND CONSERVATION
IN NORTHERN MEXICO

The editors acknowledge the generous sponsorship of this book by the following institutions:

Arizona-Sonora Desert Museum
Centro de Investigación en Alimentación y Desarrollo (CIAD), A.C.
Centro de Investigaciones Biológicas del Noroeste
Cesar Kleberg Wildlife Research Institute
Cooperativa La Cruz Azul S.C.L.
EDM International, Inc.
JM Kaplan Fund
Museum of Southwestern Biology, University of New Mexico
New Mexico Museum of Natural History
New Mexico State University
T & E, Inc.
Universidad Nacional Autónoma de México
U.S. Fish and Wildlife Service
Wallace Research Foundation

BIODIVERSITY, ECOSYSTEMS, AND CONSERVATION IN NORTHERN MEXICO

Edited by
Jean-Luc E. Cartron
Gerardo Ceballos
Richard Stephen Felger

UNIVERSITY PRESS

2005

OXFORD

UNIVERSITY PRESS

Oxford University Press, Inc., publishes works that further
Oxford University's objective of excellence
in research, scholarship, and education.

Oxford   New York
Auckland   Cape Town   Dar es Salaam   Hong Kong   Karachi
Kuala Lumpur   Madrid   Melbourne   Mexico City   Nairobi
New Delhi   Shanghai   Taipei   Toronto

With offices in
Argentina   Austria   Brazil   Chile   Czech Republic   France   Greece
Guatemala   Hungary   Italy   Japan   Poland   Portugal   Singapore
South Korea   Switzerland   Thailand   Turkey   Ukraine   Vietnam

Published by Oxford University Press, Inc.
198 Madison Avenue, New York, New York 10016

www.oup.com

Oxford is a registered trademark of Oxford University Press

Library of Congress Cataloging-in-Publication Data
Biodiversity, ecosystems, and conservation in northern Mexico / [edited by] Jean-Luc E.
Cartron, Gerardo Ceballos, and Richard Stephen Felger.
p. cm.
Includes index.
ISBN-13 978-0-19-515672-0
ISBN 0-19-515672-2
1. Biological diversity—Mexico, North. 2. Ecology—Mexico, North. I. Cartron, Jean-Luc E.
II. Ceballos, Gerardo. III. Felger, Richard Stephen.
QH107.B525 2004
333.95'16'0972—dc22   2004012002

9 8 7 6 5 4 3 2 1

Printed in the United States of America
on acid-free paper

For Dominique, Matthieu, Melanie, and Olivia. Also to the memory of my grandfather Marcel Chichery, geologist, naturalist, and teacher, killed during WWII.
—Jean-Luc Cartron

To Pupa, Pablo, and Regina, the light of my life.
—Gerardo Ceballos

For Silke especially and Grayce and the rest of the world's children.
—Richard Felger

Foreword

To the average informed layperson, the word "megadiversity" probably elicits images of steamy tropical rainforests and perhaps also of coral reefs. If both occur together, like in Indonesia, so much the better because the stereotype is reinforced. It is seldom appreciated, however, that outside the wet tropics there are regions of the world that harbor extremely high numbers of species, high levels of endemism, or fascinating ecosystems. For most Mexicans, and I daresay also for many Americans, news that the arid and desert zones of North America represent one such hot spot of biodiversity would be surprising. Even for professional biologists from other parts of the world, the tremendous biological richness of the deserts of North America may remain unappreciated. I witnessed this while serving in one of the scientific advisory bodies of the United Nations: the perceptions and points of view of the expert on the sub-Saharan deserts were totally at odds with my own view on deserts. Our personal experiences regarding arid lands were based on starkly contrasting systems: mine was based mainly on the Sonoran and Tehuacan Deserts, with their astounding biological richness, and hers mainly on the degraded lands of many Sahelian countries.

Even a country like Mexico, which is fortunate enough to harbor both tropical rainforests and deserts among its biomes, may have more species or a higher percentage of endemics in its arid lands than in its very rich but much less extensive wet tropics. Unfortunately, the pace of destruction that for many years was focused on the tropical forests is now shifting toward the arid lands. The combined pressure of demographics and development, including the extensive introduction of alien species (e.g., buffelgrass, *Pennisetum ciliare* or tamarisk, *Tamarix* spp.), is causing widespread deterioration of precious ecosystems, which until recently were still relatively well preserved over large areas.

Within this dual context of high biological diversity and accelerated pace of anthropogenic impacts, the publication of *Biodiversity, Ecosystems and Conservation in Northern Mexico* is a welcome addition to the armory of ecologists, nongovernmental organizations, and consultants interested in the conservation and sustainable management of our arid lands. The thematic scope of the book is quite ambitious: it includes chapters on terrestrial, marine, and freshwater species; on the geology, biology, human history, and even environmental laws of the region, which includes the whole Baja California peninsula, the state of Sonora, and the Chihuahuan Desert all the way south to San Luis Potosí. The emphasis of the volume is mostly on plants and vertebrates. However, one chapter focuses on scorpions, and two others include coverage of marine and intertidal macroinvertebrates. One chapter addresses ecosystem integrity and functioning in the state of Baja California. In general, even those chapters with a more narrow focus have broad applications and should be of interest to all ecologists regardless of their area of expertise.

An interesting and welcome feature of the book is its balanced national authorship: about three-quarters of the chapters are the product of collaboration between authors on both sides of the U.S.–Mexico border or were written entirely by Mexicans. Not very long ago a book of this kind would have had 90% or more American authors. Increased activity and expertise on the Mexican side, along with a sustained tradition of international collaboration, are transforming the approach to dealing with the manifold problems of our shared ecosystems.

The personnel at the National Commission on Biodiversity of Mexico (CONABIO) was very happy to be able to collaborate in the production of this volume, which adds another title to the growing list of books dealing with the megadiversity of our country. I hope that this book will have the wide distribution it deserves. I also hope that together with other recent works on northern Mexico, once regarded as good only to raise cattle in a permanent fight against harsh environments, the present volume will be used as an important tool to help preserve the beautiful and unique ecosystems of the region.

Jorge Soberón Mainero

Contents

**Part III: Natural Resource Impacts and Conservation
at a Population, Species, and Landscape Level**

Contributors

EDGAR S. AMADOR-SILVA
Centro de Investigaciones Biológicas del Noroeste, S. C., La Paz, Baja California Sur, Mexico

DANIEL W. ANDERSON
Department of Wildlife, Fish, and Conservation Biology, University of California, Davis, CA, USA

MA. CARMEN BLÁZQUEZ
Centro de Investigaciones Biológicas del Noroeste, La Paz, Baja California Sur, Mexico

RICHARD C. BRUSCA
Arizona-Sonora Desert Museum, Tucson, AZ, USA

ALBERTO BÚRQUEZ
Instituto de Ecología, Universidad Nacional Autónoma de México, Hermosillo, Sonora, Mexico

WILLIAM A. CALDER
Department of Ecology and Evolutionary Biology, University of Arizona, Tucson, AZ, USA

DOMINIQUE CARTRON
Daniel B. Stephens and Associates, Albuquerque, NM, USA

JEAN-LUC E. CARTRON
Department of Biology, University of New Mexico, Albuquerque, NM, USA

GERARDO CEBALLOS
Instituto de Ecología, Universidad Nacional Autónoma de México, México D.F., Mexico

MARTHA J. DESMOND
Department of Fishery and Wildlife Sciences, New Mexico State University, Las Cruces, NM, USA

MARK A. DIMMITT
Arizona-Sonora Desert Museum, Tucson, AZ, USA

DIANA DOAN-CRIDER
Caesar Kleberg Wildlife Research Institute, Texas A&M University-Kingsville, Kingsville, TX, USA

HUGH DRUMMOND
Instituto de Ecología, Universidad Nacional Autónoma de México, México D.F., Mexico

HÉCTOR ESPINOSA PÉREZ
Colección Nacional de Peces, Instituto de Biología, Universidad Nacional Autónoma de México, México D.F., Mexico

RICHARD S. FELGER
Drylands Institute, Tucson, AZ, USA

ISMAEL FERRUSQUÍA-VILLAFRANCA
Instituto de Geología, Universidad
Nacional Autónoma de México, México
D.F., Mexico

LLOYD T. FINDLEY
Centro de Investigación en Alimentación y
Desarrollo, A.C.-Unidad Guaymas, Guaymas,
Sonora, Mexico

ERNESTO FRANCO-VIZCAÍNO
Departamento de Ecología, Centro de
Investigación Científica y de Educación
Superior de Ensenada (CICESE), Ensenada,
Baja California, Mexico

TIMOTHY E. FULBRIGHT
Caesar Kleberg Wildlife Research Institute,
Texas A&M University-Kingsville, Kingsville,
TX, USA

FELIPE GALVÁN-MAGAÑA
Centro Interdisciplinario de Ciencias Marinas, La Paz, Baja California Sur, Mexico

FRANCISCO J. GARCÍA DE LEÓN
Laboratorio de Biología Integrativa,
Instituto Tecnológico de Ciudad Victoria,
Tamaulipas, Mexico

CARLOS GÓMEZ-HINOSTROSA
Instituto de Biología, Universidad
Nacional Autónoma de México, México
D.F., Mexico

LAURA I. GONZÁLEZ GUZMAN
Section of Integrative Biology, University of
Texas at Austin, Austin, TX, USA

PATRICIA GONZÁLEZ-ZAMORANO
Centro de Investigaciones Biológicas del
Noroeste, S. C., La Paz, Baja California Sur,
Mexico

L. LEE GRISMER
Department of Biology, La Sierra University,
Riverside, CA, USA

MERCEDES GUERRERO-RUÍZ
Departamento de Biología Marina,
Universidad Autónoma de Baja California
Sur, La Paz, Baja California Sur, Mexico

DELLADIRA GUTIÉRREZ TIRADO
Laboratorio de Biología Integrativa, Instituto
Tecnológico de Ciudad Victoria, Tamaulipas,
Mexico

DAVID J. HAFNER
New Mexico Museum of Natural History,
Albuquerque, NM, USA

RICHARD E. HARNESS
EDM International, Inc., Fort Collins, CO,
USA

PHILLIP A. HASTINGS
Scripps Institution of Oceanography,
University of California at San Diego,
La Jolla, CA, USA

DEAN A. HENDRICKSON
Texas Memorial Museum and Section of
Integrative Biology, University of Texas,
Austin, TX, USA

MICHEL E. HENDRICKX
Estación Mazatlán, ICML- Universidad
Nacional Autónoma de México, Mazatlán,
Sinaloa, Mexico

BRENT E. HENDRIXSON
Department of Life, Earth, and Environmen-
tal Sciences, West Texas A&M University,
Canyon, TX, USA

HÉCTOR M. HERNÁNDEZ
Instituto de Biología, Universidad Nacional
Autónoma de México, México D.F., Mexico

DAVID G. HEWITT
Caesar Kleberg Wildlife Research Institute,
Texas A&M University-Kingsville, Kingsville,
TX, USA

GINA HOLGUIN
Centro de Investigaciones Biológicas del
Noroeste, S. C., La Paz, Baja California Sur,
Mexico

ARMANDO JARAMILLO-LEGORRETA
Programa Nacional de Mamíferos Marinos,
Instituto Nacional de Ecología/ Centro de
Investigación Científica y de Educación
Superior de Ensenada (CICESE), Ensenada,
Baja California, Mexico

KAREN KREBBS
Arizona-Sonora Desert Museum, Tucson,
AZ, USA

ALBERTO LAFÓN TERRAZAS
Facultad Zootecnia, Universidad Autónoma
de Chihuahua, Chihuahua, Mexico

JOSÉ LUIS LEÓN DE LA LUZ
Centro de Investigaciones Biológicas del
Noroeste, S. C., La Paz, Baja California Sur,
Mexico

RURIK LIST
Instituto de Ecología, Universidad
Nacional Autónoma de México, México
D.F., Mexico

JUAN MANUEL LOBATO G.
Centro de Investigaciones Biológicas del
Noroeste, S. C., La Paz, Baja California Sur,
Mexico

CARLOS MANTEROLA
Unidos para la Conservación, A. C., México
D.F., Mexico.

PATRICIA MANZANO-FISCHER
Instituto de Ecología, Universidad Nacional
Autónoma de México, México D.F., Mexico

LUIS OCTAVIO MARTÍNEZ MORALES
Despacho Székely y Asociados, México D.F.,
Mexico

CAROLE C. MCIVOR
USGS-Biological Resources Division, c/o
Center for Coastal Studies, St. Petersburg,
FL, USA

RODRIGO A. MEDELLÍN
Instituto de Ecología, Universidad Nacional
Autónoma de México, México D.F., Mexico

ERIC MELLINK
Centro de Investigación Científica y de
Educación Superior de Ensenada (CICESE),
Ensenada, Baja California, Mexico

RENATO MENDOZA-SALGADO
Centro de Investigaciones Biológicas del
Noroeste, S. C., La Paz, Baja California Sur,
Mexico

RICHARD A. MINNICH
Department of Earth Sciences, University of
California, Riverside, CA, USA

WALLACE J. NICHOLS
Blue Ocean Institute, Davenport, CA, USA

JESÚS PACHECO
Instituto de Ecología, Universidad Nacional
Autónoma de México, México D.F., Mexico

EDUARDO PALACIOS
Centro de Investigación Científica y de
Educación Superior de Ensenada (CICESE),
Ensenada, Baja California, Mexico

CAROL RAISH
U.S. Forest Service, Rocky Mountain Research Station, Albuquerque, NM, USA

FANNY REBÓN GALLARDO
Museo de Zoología, Facultad de Ciencias,
Universidad Nacional Autónoma de México,
México D.F., Mexico

CHARLOTTE G. REEDER
University of Arizona Herbarium, Tucson,
AZ, USA

JOHN R. REEDER
University of Arizona Herbarium, Tucson,
AZ, USA

ANA LILIA REINA G.
Arizona-Sonora Desert Museum, Tucson,
AZ, USA

BRETT R. RIDDLE
Department of Biological Sciences, University
of Nevada Las Vegas, Las Vegas, NV, USA

CRISTINA RODRÍGUEZ
Instituto de Ecología, Universidad Nacional
Autónoma de México, México D.F., Mexico

RICARDO RODRÍGUEZ-ESTRELLA
Centro de Investigaciones Biológicas del
Noroeste, S. C., La Paz, Baja California Sur,
Mexico

ROBERT C. ROGERS
Biology Department, University of New
Mexico, Albuquerque, NM, USA

LORENZO ROJAS-BRACHO
Programa Nacional de Mamíferos Marinos,
Instituto Nacional de Ecología/ Centro de
Investigación Científica y de Educación
Superior de Ensenada (CICESE), Ensenada,
Baja California, Mexico

MARIO ROYO
Instituto de Ecología, Universidad Nacional
Autónoma de México, México D.F., Mexico

GEORGINA SANTOS
Instituto de Ecología, Universidad Nacional
Autónoma de México, México D.F., Mexico

JEFFREY A. SEMINOFF
Southwest Fisheries Science Center, NOAA-
National Marine Fisheries Service, La Jolla,
CA, USA

W. DAVID SISSOM
Department of Life, Earth, and Environmen-
tal Sciences, West Texas A&M University,
Canyon, TX, USA

MARK J. SPALDING
Graduate School of International Relations
and Pacific Studies, University of California
at San Diego, La Jolla, CA, USA

SCOTT H. STOLESON
U.S. Forest Service, Northeastern Research
Station, Irvine, PA, USA

ALBERTO SZÉKELY
Despacho Székely y Asociados, México D.F.,
Mexico

BRUCE C. THOMPSON
U.S.G.S.-Biological Resources, New Mexico
Cooperative Fish and Wildlife Research Unit,
Las Cruces, NM, USA

JORGE TORRE COSIO
Comunidad y Biodiversidad, A.C., Guaymas,
Sonora, Mexico

JORGE URBÁN R.
Departamento de Biología Marina,
Universidad Autónoma de Baja California
Sur, La Paz, Baja California Sur, Mexico

MANUEL VALDÉZ
Unidos para la Conservación, A. C., México
D.F., Mexico

RAUL VALDEZ
Department of Fishery and Wildlife Sciences,
New Mexico State University, Las Cruces,
NM, USA

ALBERT M. VAN DER HEIDEN
Centro de Investigación en Alimentación y
Desarrollo, A.C.-Unidad Mazatlán, Sinaloa,
Mexico

THOMAS R. VAN DEVENDER
Arizona-Sonora Desert Museum, Tucson,
AZ, USA

ENRIQUETA VELARDE
Centro de Ecología y Pesquerías, Universidad
Veracruzana, Xalapa, Veracruz, Mexico

ROBERT C. WHITMORE
West Virginia University, Division of Forestry, Morgantown, WV, USA

JOHN F. WIENS
Arizona-Sonora Desert Museum, Tucson,
AZ, USA

MICHAEL F. WILSON
Drylands Institute, Tucson, AZ, USA

KENDAL E. YOUNG
U.S.G.S.-Biological Resources, New Mexico
Cooperative Fish and Wildlife Research Unit,
Las Cruces, NM, USA

BIODIVERSITY, ECOSYSTEMS, AND CONSERVATION
IN NORTHERN MEXICO

Biodiversity, Ecosystems, and Conservation: Prospects for Northern Mexico

JEAN-LUC E. CARTRON

GERARDO CEBALLOS

RICHARD S. FELGER

Conservation biology is a crisis discipline rooted in the realization that the world's biodiversity is rapidly disappearing. Environmental problems (including loss of species and ecosystems) are mainly the result of explosive human population growth, with an ever-growing demand on natural resources and a steady increase in waste products. Minimizing the negative effects of human activities on the environment requires a conservation-minded culture, an adequate legal framework that is implemented and enforced, properly trained professionals, and funding. Just as important, minimizing environmental damage requires solid scientific information on the biodiversity and functioning of every ecosystem and on conservation threats at a local, regional, and global scale. This knowledge becomes a crucial tool for guiding where and when the protection of nature is most needed and for designing sound management strategies. According to the Latin proverb, *Scientia est potentia* (knowledge is power), and the seventeenth-century philosopher Francis Bacon was one of the first to argue the importance of gathering scientific information on the natural world. Bacon's reasoning was that by learning more about nature we could control or harness it. Today, this same type of knowledge needs to be applied instead to protect ecosystems and species.

This book is first and foremost about conservation in northern Mexico. The information it contains is intended primarily for Mexican government agencies, environmental organizations, and research institutions, but it also has applications at the scale of North America. The stakes are high in Mexico, one of the world's 17 megadiversity countries (Mittermeier et al. 1997). Among all nations, Mexico may support the highest number of reptile species (Flores-Villela and Geréz 1994). It also has one of the highest numbers of vascular plants and amphibians (Dirzo and Gómez 1996; Mittermeier et al. 1997), along with 12% of the world's mammals (Ceballos et al. 2002). As many as 1076 bird species occur in Mexico, a total 30% greater than that for the United States and Canada combined (Ceballos and Marquéz 2000). It is in Mexico that the Neotropic and Nearctic regions merge, producing unique communities of mixed biogeographic affinities. The percentage of endemic species is surprisingly high for a country that is part of a continent (Ceballos and Marquéz 2000).

Much of Mexico's biodiversity is in the arid, northern part of the country (loosely defined in this volume) and in waters along this region's extensive coastline. To use examples from this volume, the Baja California peninsula harbors the highest known density of scorpion species in the world (chapter 6). The Huizache area in the state of San Luis Potosí has 75 cactus species, the highest concentration of Cactaceae recorded anywhere to date (chapter 13). With 186 taxa (183 species) documented and 38 (37 species) more expected, the Municipio (County) de Yécora (3300 km^2) in eastern Sonora has one of the highest grass diversities in Mexico (chapter 5). With 44 succulent plant taxa occurring over its 10-km² area, Cerro Colorado in Baja California Sur epitomizes the biological wealth of the Sonoran Desert. The largest remaining prairie-dog town complex in North America and its surrounding area in northwestern Chihuahua support about 21% of the mammals of Mexico (chapter 21). The Gulf of California has a macrofauna composed of at least 5969 nominal species and subspecies, including 891 fish taxa, 31 (37%) of the world's cetaceans, and 5 of the 7 sea turtles, with thousands of additional taxa remaining to be described (chapters 9, 14, and 20). As a means of comparison, the Mediterranean Sea covers an area 10 times larger than that of the Gulf of California, yet harbors a total of only 664 fish species (Whitehead et al. 1986; Quignard and Tomasini 2000) and 21 cetaceans (Notarbartolo di Sciara 2002).

As in other parts of the world, northern Mexico's biodiversity is now threatened. Rapid loss of habitat is occurring throughout the region, threatening mangroves (chapter 15), deserts (chapters 11 and 13), grasslands (chapter 21), forests (chapter 3), and oases (chapter 16). Anthropogenic impacts on species include the collapse of northwestern Mexico's sea turtle populations (chapter 20), endangerment of the vaquita in the Gulf of California (chapter 14), rapid decline of the Janos–Casas Grandes prairie dog town complex (chapter 21), and a high incidence of bird electrocutions on concrete power poles in Chihuahua (chapter 17). Exotic species have become the scourge of many ecosystems and wildlife populations, including seabirds (chapters 3 and 23) and freshwater fish (chapter 7).

The present volume begins with four introductory chapters that provide background information and frame many of the issues developed later in the book. In the introductory chapter describing the recently developed legal framework for the conservation of species and ecosystems in Mexico, Szekely et al. (chapter 4) are critical of what they and others view as government inaction. Certainly, lack of implementation or enforcement of Mexico's laws highly contributes to some of the most pressing conservation issues in the northern part of the country (see chapter 20). At the same time, some conservation efforts on the part of individuals and environmental organizations have been truly heroic, and there are success stories to report (chapters 19 and 23).

Among some of the other factors complicating conservation are the social ills of Mexico, which is dogged by poverty and population growth. On the other hand, northern Mexico still has vast expanses of pristine habitat, which cannot be said for many other parts of the world (e.g., Mittermeier et al. 2003). For a long time, some Mexicans have expressed interest in and concern over environmental issues in general and conservation in particular. Since about 1990, a growing number of national and international NGOs (nongovernmental organizations) and government research organizations have become actively involved in conservation issues. Among the most important are World Wildlife Fund Mexico, The Nature Conservancy-Mexico, Pronatura, Instituto de Ecologia A.C., Instituto de Ecologia UNAM, Centro de Investigación Cientifica y de Educación Superior de Ensenada (CICESE), and the state universities of northern states, including Baja California, Baja California Sur, Sonora, Chihuahua, Nuevo León, and Tamaulipas.

The federal government of Mexico relies heavily on the creation of Natural Protected Areas. As of June 2003, it has decreed a total of 33 Biosphere Reserves, 65 National Parks, 4 National Monuments, 2 Areas for the Protection of Natural Resources, 26 Areas for the Protection of Flora and Fauna, and 17 Sanctuaries. As a result, there are many vast federally recognized reserves in the northern part of Mexico (tables I.1 and I.2), though they often lack sufficient infrastructure and funding to be effectively protected. Sierra de San Pedro Mártir in the state of Baja California is a National Park (see chapter 18). The Cuatro Ciénegas basin (see chapter 6) in Coahuila is an example of Area for the Protection of Flora and Fauna. There are also additional reserves created by individual states, and Biosphere Reserves created under UNESCO's Man and the Biosphere (MAB) Program, some of which do not have the status of Biosphere Reserves under Mexican law. For example, El Cielo in Tamaulipas is a MAB Biosphere Reserve, but under Mexican law it is only a State-decreed protected area. Another MAB Biosphere Reserve, Islas del Golfo de California, is protected under federal law but has the status only of Area for the Protection of Flora and Fauna.

With international financial support, the existence of all these reserves in northern Mexico can become a powerful tool for conserving ecosystems in northern Mexico. Some of them could be incorporated into a network of protected areas connecting vast tracts of land on both sides of the U.S.–Mexican border. Felger et al. (2004) proposed an *Escalera Ecológica* in place of the *Escalera Náutica* (an

Table I.1. Biosphere Reserves and National Parks established under federal Mexican law in the northern states of Baja California, Baja California Sur, Chihuahua, Coahuila, Durango, Nuevo León, Sinaloa, Sonora, and Tamaulipas.

Protected Areas	Location	Date of Creation	Size (ha)
Biosphere Reserves			
Alto Golfo de California y Delta del Río Colorado[a]	Baja California and Sonora	June 10, 1993	934,756
El Vizcaíno[a,b]	Baja California Sur	November 30, 1988	2,546,790
Sierra La Laguna	Baja California Sur	June 6, 1994	112,437
Mapimí[a]	Chihuahua, Coahuila, and Durango	November 27, 2000	342,388
La Michilía[a]	Durango	July 18, 1979	9,325
El Pinacate y Gran Desierto de Altar[a]	Sonora	June 10, 1993	714,557
Isla San Pedro Mártir	Sonora	June 13, 2002	30,165
National Parks			
Constitución de 1857	Baja California	April 27, 1962	5,009
Sierra de San Pedro Mártir	Baja California	April 26, 1947	72,911
Bahía de Loreto	Baja California Sur	July 19, 1996	206,581
Cabo Pulmo	Baja California Sur	June 6, 1995	7,111
Cascada de Bassaseachic	Chihuahua	February 2, 1981	5,803
Cumbres de Majalca	Chihuahua	September 1, 1939	4,772
Los Novillos	Coahuila	June 18, 1940	42
Cumbres de Monterrey	Nuevo León	November 17, 2000	177,396
El Sabinal	Nuevo León	August 25, 1938	8

Natural Protected Areas are created upon presidential decrees. Adapted from CONANP (2003).

[a]Also designated as a Biosphere Reserve under UNESCO's Man and the Biosphere Program.
[b]Includes the Laguna Ojo de Liebre complex and Laguna San Ignacio. These lagoons have been designated as Whale Refuges by presidential decrees.

Table I.2. Natural Monuments, Areas for the Protection of Flora and Fauna, and Sanctuaries established under federal Mexican law in the northern states of Baja California, Baja California Sur, Chihuahua, Coahuila, Durango, Nuevo León, Sinaloa, Sonora, and Tamaulipas.

Protected Areas	Location
Natural Monuments	
Cerro de la Silla	Nuevo León
Areas for the Protection of Flora and Fauna	
Valle de los Cirios	Baja California
Cabo San Lucas	Baja California Sur
Islas del Golfo de California	Baja California, Baja California Sur, Sinaloa, Sonora
Cañon de Santa Elena	Chihuahua
Tutuaca	Chihuahua
Campo Verde	Chihuahua
Papigochic	Chihuahua
Cuatro Ciénegas	Coahuila
Maderas del Carmen	Coahuila
Meseta de Cacaxtla	Sinaloa
Sierra de Álamos-Río Cuchujaqui	Sonora
Sanctuaries	
Playa Ceuta	Sinaloa
Playa el Verde Camacho	Sinaloa
Playa de Rancho Nuevo	Tamaulipas

Natural Protected Areas are created upon presidential decrees. There is no Area for the Protection of Natural Resources in northern Mexico. The three listed Sanctuaries are mainly for the protection of sea turtles. Adapted from CONANP (2003).

ambitious project calling for the development of marinas, waterways, and hotels along the Pacific and Gulf of California coasts, discussed in several chapters). The steps include:

- On the U.S. side, the creation of a Sonoran Desert National Park with existing federal lands (Organ Pipe Cactus National Monument, Cabeza Prieta National Wildlife Refuge, and Tinajas Altas) to fulfill conservation promises to Mexico and to parallel Mexico's commitment.
- The establishment of the contiguous Sonoran Desert Peace Parks (Parques Hermanos, *un corredor ecológico fronterizo*). This protected area would consist of the Sonoran Desert National Park, Reserva de Biosfera El Pinacate y El Gran Desierto de Altar, and Reserva de la Biosfera Alto Golfo de California y Delta del Rio Colorado. Teamed with neighboring areas that include the 1.8 million-acre Gold-water Military Range and the 500,000-acre Sonoran Desert National Monument, the Sonoran Desert Peace Parks would encompass 7.5 million acres of contiguous federal lands on both sides of the border.
- Linking the Reserva Alto Golfo contiguously across desert and mountains to conifer forests of the Sierra Juárez.
- The creation of an *Escalera Ecológica* of diverse reserves—for example, Sonoran Desert Peace Parks, plus an additional 7 million acres of Mexican federal lands comprising Reserva de la Biosfera El Vizcaíno, Valle de Cirios, Parque Nacional Sierra de San Pedro Mártir, proposed Parque Nacional Bahía de los Angeles, Parque Nacional Bahía de Loreto, Reserva de la Biosfera Sierra La Laguna, Área de Protección Cajón del Diablo, Área de Protección de Flora y Fauna Islas del Golfo de California, and Área Protegida Isla Guadalupe.

Similar projects have been proposed or are suitable for other areas in northern Mexico: the grasslands along the Chihuahua–New Mexico border (chapter 21); the scrub and temperate forests in the Big Bend (Texas) and Maderas del Carmen (Coahuila) reserves; and the Laguna Atascosa National Wildlife Refuge (Texas) and the proposed Laguna Madre Biosphere Reserve (Tamaulipas).

What is the outlook for northern Mexico? Impacts described in this volume may continue or even accelerate. The *Escalera Náutica* project, which is currently in its early stage of implementation, may well transform the regional landscape and result in wholesale destruction of coastal ecosystems. More formidable threats are looming on the horizon. First, there are the global climate and water issues, 2 paramount challenges of the twenty-first century. Arid northern Mexico is particularly at risk of water shortage, but water-related impacts on wildlife and ecosystems may result equally from human actions aimed at solving the issue. Another conservation threat is the spread of viruses and other pathogens around the globe, resulting from factors such as globalization, climate change, or habitat fragmentation. Currently, the West Nile virus is a source of grave concern, as it has the potential to decimate wild populations of birds throughout North, Central, and South America (Malakoff 2002; Ananthaswamy 2003).

For northern Mexico, as for most regions on earth, the next few decades will decide the ultimate fate of many ecosystems. The pessimistic outlook for the region is that the vast expanses of coastal habitat, grassland, and forest will largely disappear, as meeting the needs of growing human populations for more space and natural resources (chiefly water) completely overrides conservation. The optimistic scenario is that public awareness and education along with funding will continue to grow, natural protected areas can be incorporated into networks such as the ones presented above, and northern Mexico will remain the magnificent and biologically wealthy land it is today. It is our hope that this book will contribute toward realizing this optimistic scenario.

Literature Cited

Ananthaswamy, A. 2003. Death in the sun. New Scientist 26 July 2003. Vol. 179, no. 2405: 12–13.

Ceballos, G. and L. Marquéz. 2000. Las Aves de México en Peligro de Extinción. CONABIO–UNAM–Fondo de Cultura Economica, México D.F.

Ceballos, G., J. Arroyo-Cabrales, and R. A. Medellín. 2002. The mammals of México: composition, distribution, and status. Occasional Papers, Texas Tech University 218: 1–27.

Comisión Nacional de Áreas Naturales Protegidas (CONANP). 2003. Áreas Naturales Protegidas. [Available: www.conanp.gob.mx/]

Dirzo, R., and G. Gomez. 1996. Temporal patterns of taxonomic research on Mexican vascular plants and estimates of the number of known species. Annals of the Missouri Botanical Garden 83: 396–403.

Felger, R., W. J. Nichols, A. Aquirre, F. Arcas, L. Bourillón-Moreno, R. Brusca, J. Compoy, A. Cantú, J.-L. Cartron, G. D. Danemann, H. D'Antoni, P. K. Dayton, S. A. Diehn, L. Gerber, J. L. León de la Luz, C. Mártinez del Rio, R. A. Medellín, F. Molina Freaner, E. Palacious-Castro, and E. Sala. 2004. Escalera Ecológica Initiative: A staircase of reserves through the Gulf of California Corridor. P. 71 *in* Proceedings of the Gulf of California Conference 2004, Tucson.

Flores-Villela, O., and P. Geréz. 1994. Biodiversidad y Conservación en México: Vertebrados, Vegetación y Uso del Suelo. CONABIO and UNAM, Mexico D.F.

Malakoff, D. 2002. Bird advocates fear that West Nile virus could silence the spring. Science 297: 1989.

Mittermeier, R. A., P. Robles Gil, and C. Goettsch Mittermeier. 1997. Megadiversity: Earth's Biologically Wealthiest Nations. CEMEX, México D.F.

Mittermeier, R. A., P. Robles Gil, and C. Goettsch Mittermeier. 2003. The Last Wilderness. CEMEX, México D.F.

Notarbartolo di Sciara G. 2002. Cetacean species occurring in the Mediterranean and Black Seas. *In* G. Notarbartolo di Sciara (ed.), Cetaceans of the Mediterranean and Black Seas: State of Knowledge and Conservation Strategies. A report to the ACCOBAMS Secretariat, Monaco, February 2002, section 3.

Quignard, J. P., and J. A. Tomasini. 2000. Mediterranean fish biodiversity. Biologia Marina Mediterranea 7(3): 1–66.

Whitehead, P. J. P., M.-L. Bauchot, J.-C. Hureau, J. Nielsen, and E. Tortonese (eds.). 1986. Fishes of the North-Eastern Atlantic and the Mediterranean, 3 vols. UNESCO, Paris.

I

HISTORICAL, GEOGRAPHICAL, AND LEGAL SETTING

1

Northern Mexico's Landscape, Part I: The Physical Setting and Constraints on Modeling Biotic Evolution

ISMAEL FERRUSQUÍA-VILLAFRANCA

LAURA I. GONZÁLEZ GUZMÁN

JEAN-LUC E. CARTRON

A proper understanding of the biodiversity of any area should include adequate knowledge of the regional geologic and biotic evolution. In northern Mexico (and many other regions as well), the history of biotic evolution is poorly known. As a result, the studies that model this evolution are largely speculative, yielding not only different but sometimes contradictory findings. The Cenozoic terrestrial vertebrate record is small, but it is sufficient to give us a tantalizing view of the fauna that inhabited this region. In contrast, the Cenozoic angiosperm/gymnosperm record is practically nonexistent; thus, knowledge of the floristic evolution of northern Mexico is filled with inferences or extrapolations from (a) fossil floras from outside the region, (b) this region's Recent floristic composition and/ or biogeographic relationships, or (c) a combination of both approaches. The same applies to the terrestrial invertebrate record. Under these circumstances, it is necessary to have means (factual constraints, for instance) to help generate or choose the most parsimonious model.

The solution to this problem would be hopeless were it not for the fact that in any given region, geologic history is intimately linked to the evolution of the biota. Geologic forces continuously mold and change the land upon which the biota develops through time. In this chapter, we describe and discuss the geographic/geologic framework of northern Mexico, sketching how its present-day landscape developed and how the geologic processes/events and the geographic features involved most likely influenced the Cenozoic biota, thus setting constraints on modeling its development. Three examples show what is meant.

(1) The latitudinal position of northern Mexico largely determines the amount of solar energy the region receives and the manner in which this energy is distributed throughout the year. Therefore, to understand the regional history of biotic development, it is essential to establish whether the position of northern Mexico has changed in the geologic past. If it has changed, what was the magnitude of the latitudinal and/or longitudinal displacement, and when did it occur? The answer will greatly influence the model used for describing biotic development.

(2) Were northern Mexico's major geographic features, such as its cordilleras, generated concurrently or at different times? Regardless of the answer, the origin and development of geographic features must have produced discontinuities in the distribution of the then-existing biota, promoting some degree of differentiation. Are the regional biotas now present in northern Mexico the result of such process or processes? When did this all happen? These biotic questions could be addressed in part through an analysis of the geologic (lithic, structural, and fossil) record, which again will set constraints on models attempting to

delineate the region's biotic development or evolution.

(3) It is known that geographic diversity (largely physiographic) enhances habitat diversity, which in turn greatly influences species richness (i.e., biodiversity). Northern Mexico's present-day landscape is physiographically quite diverse and sustains a highly diverse biota. Therefore, knowledge of when and how this region's high geographic diversity came about will also help us understand how and when its biota originated and became diversified.

Each of these issues is addressed in the following sections of this chapter.

As defined in this chapter, northern Mexico corresponds to the territory located from the Trans-Mexican Volcanic Belt north to the Mexico–U.S. border (fig. 1.1). It covers approximately 1,245,900 km², or nearly two-thirds of the country's total area, and includes 19 states. It lies between 20°30'–32°30' N and 96°30'–117°15' W. Oceans bound northern Mexico both to the east and west, and it also includes a narrow northwest-southeast trending sea, the Gulf of California. Northern Mexico's littoral is approximately 6,400 km long (four-fifths of it is along the Pacific Ocean–Gulf of California). Two of Mexico's largest cordilleras lie in this territory, which shows a wide array of landforms, climates, and biomes. Figures 1.2–1.6 depict the region's major geographic factors greatly influencing species distribution.

Given the fact that the physical geographic features (e.g., relief, landforms) reflect their geologic makeup, structure, and history, it follows that in a large territory it is possible to discriminate zones that have geomorphic and geologic/tectonic features distinctive enough to differentiate them from neighboring ones. Such zones are the morphotectonic provinces. Once distinguishing criteria have been defined, the recognition of the provinces is objective. Morphotectonic provinces allow an orderly description of vast territories. The description of northern Mexico presented below follows the morphotectonic province division presented elsewhere (Ferrusquía-Villafranca 1993) and represents the only such classification, where the provinces are precisely defined and characterized. Cli-

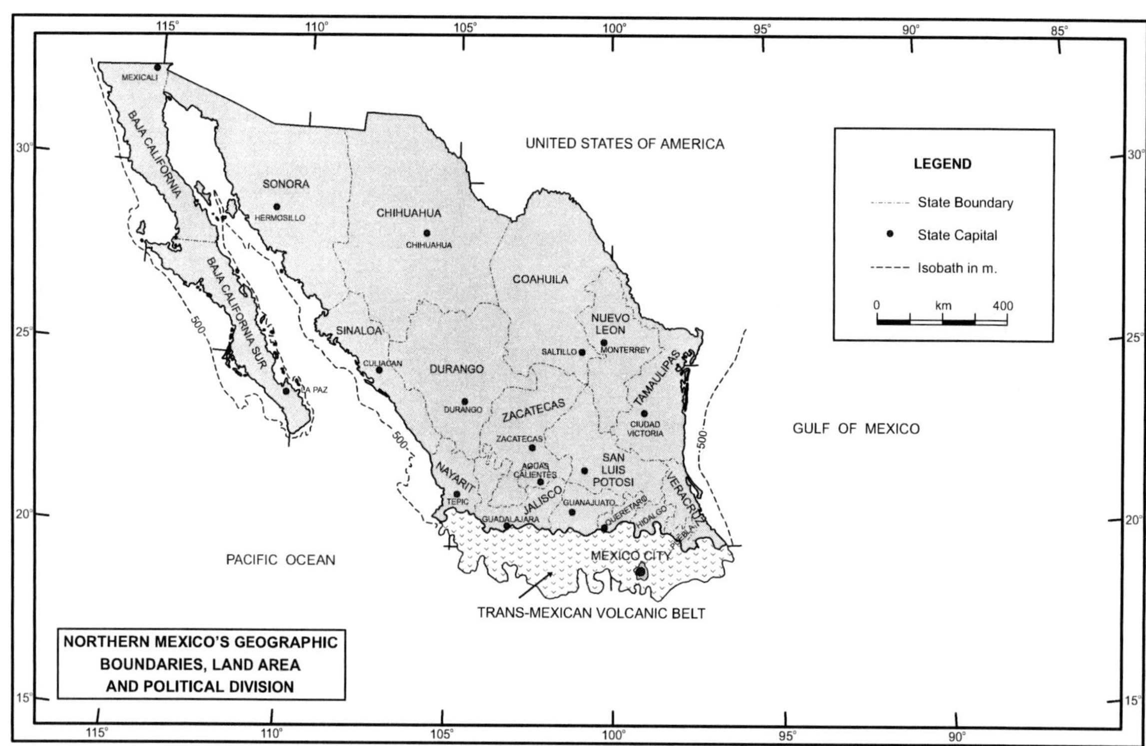

Figure 1.1. Geographic boundaries, land area, and political division of northern Mexico.

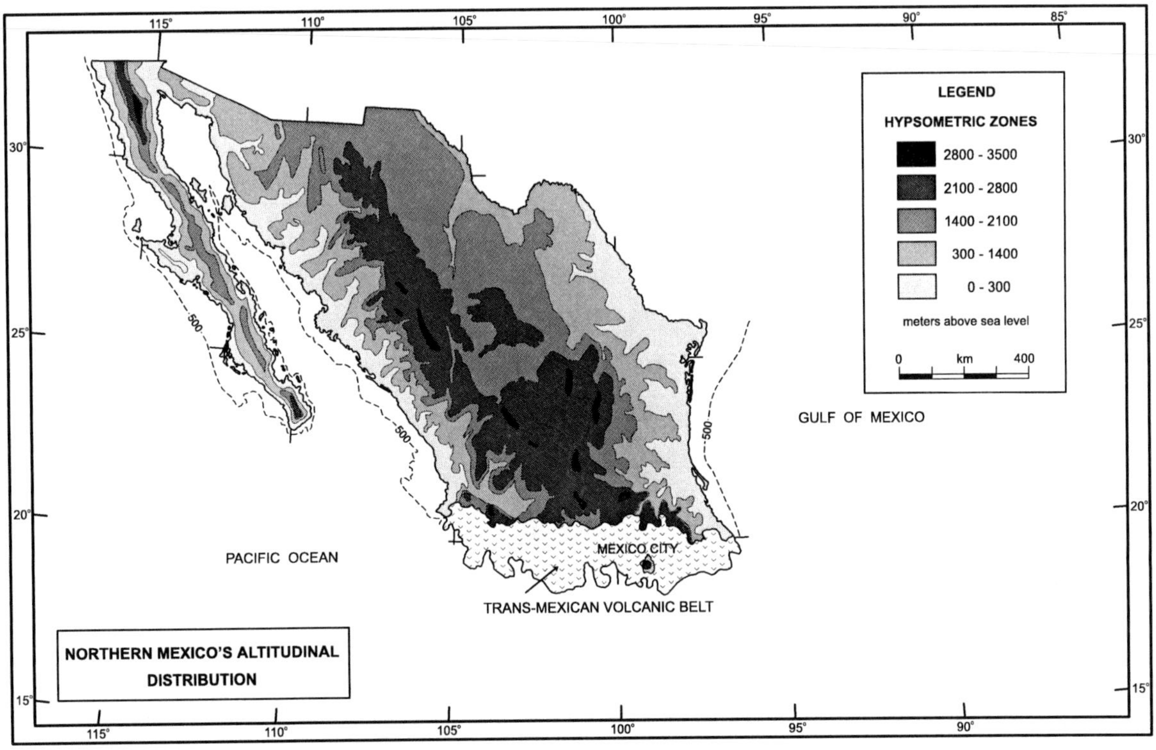

Figure 1.2. Northern Mexico's altitudinal distribution map (modified from Perez-Villegas 1990).

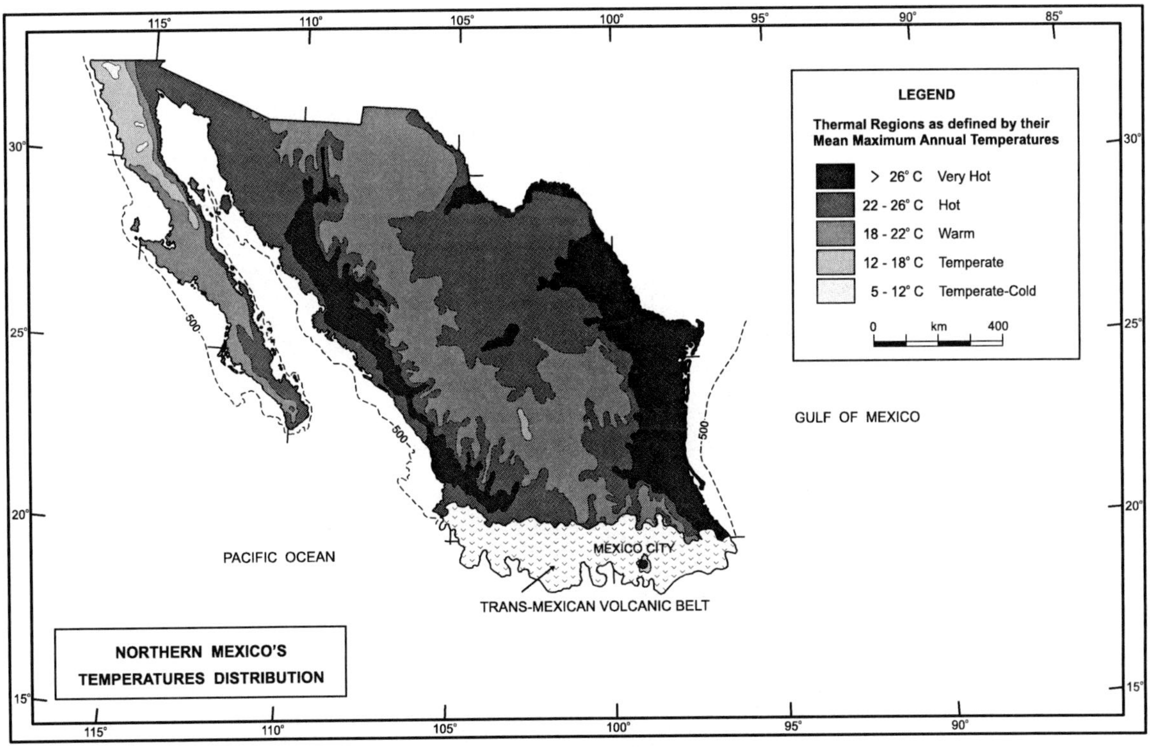

Figure 1.3. Northern Mexico's temperature distribution map (modified from Vidal-Zepeda 1990a).

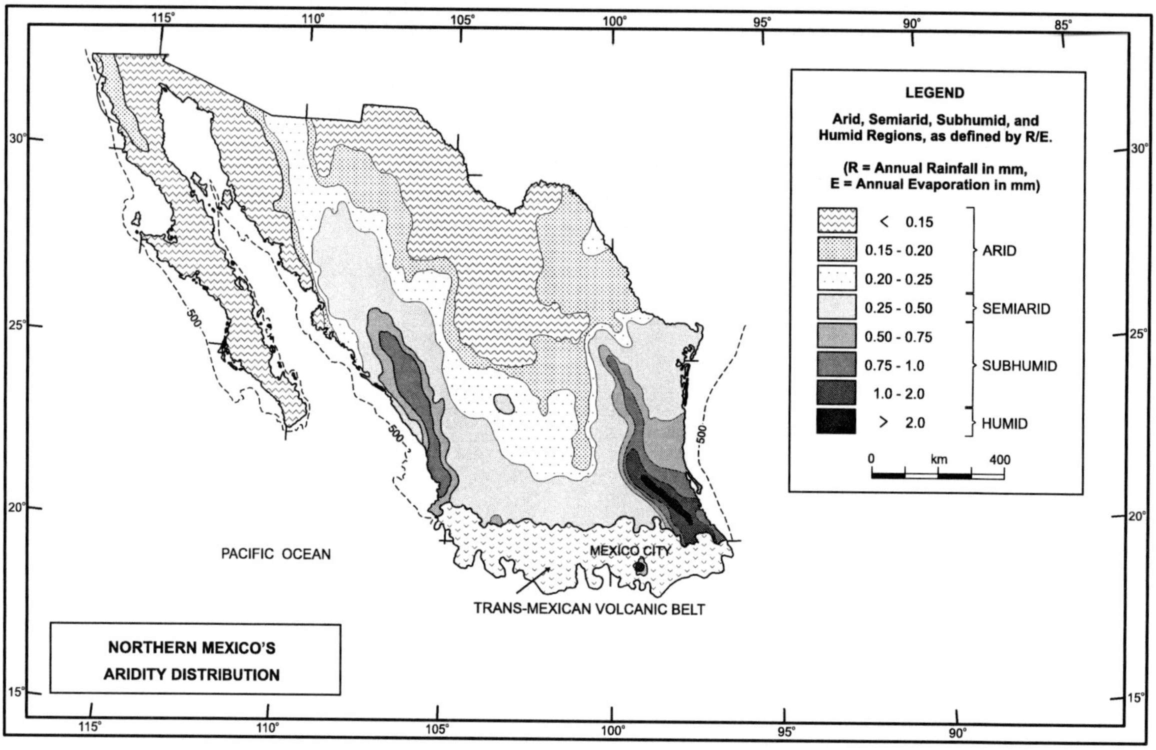

Figure 1.4. Northern Mexico's aridity distribution map (modified from Hernandez 1990).

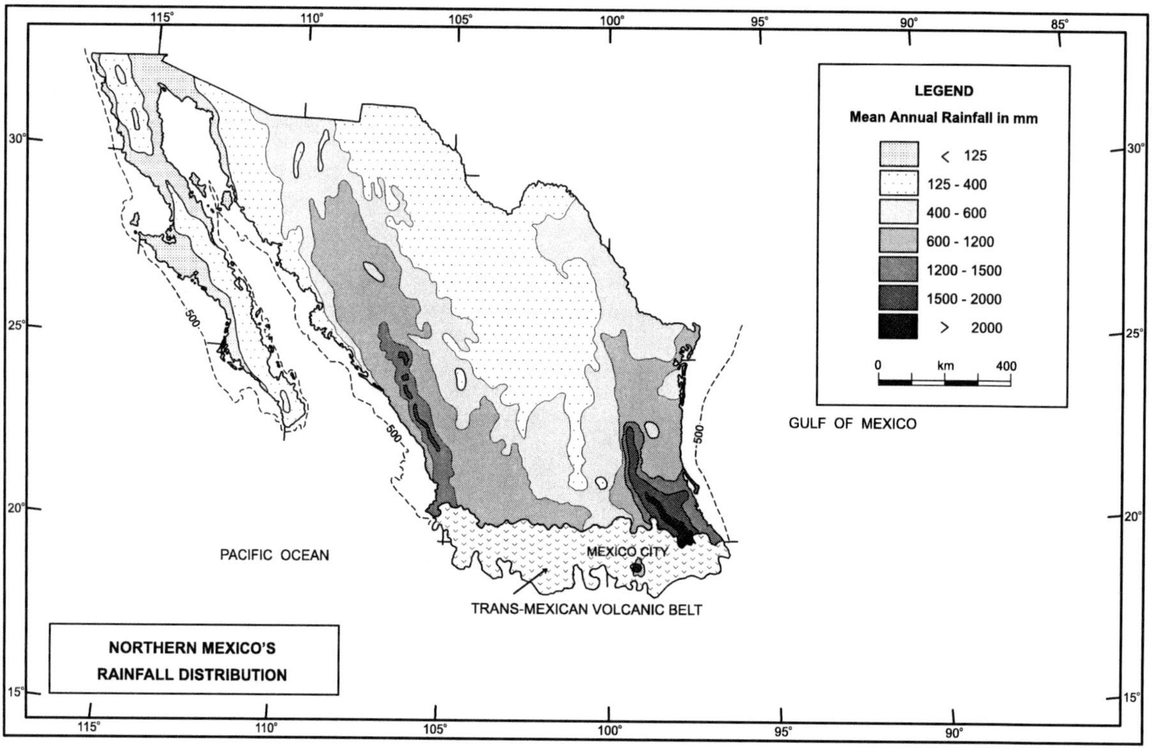

Figure 1.5. Northern Mexico's rainfall distribution map (adapted from Vidal-Zepeda 1990b).

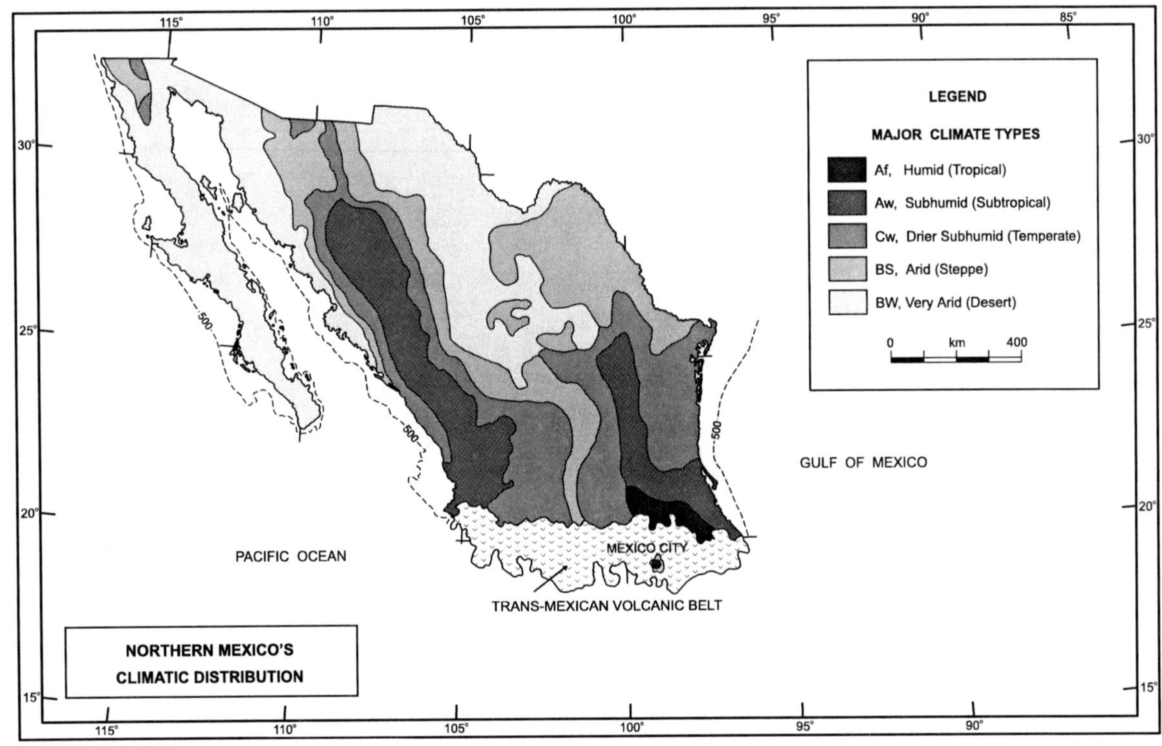

Figure 1.6. Northern Mexico's climatic regions map (adapted from García 1990).

mate description is based on García's (1990) and Köppen-Trewartha's (Trewartha 1968; adapted by García 1988) classifications.

Physical Setting: Northern Mexico's Morphotectonic Provinces

Northern Mexico has 7 morphotectonic provinces: (1) Baja California Peninsula, (2) Northwestern Plains and Sierras, (3) Sierra Madre Occidental, (4) Chihuahua-Coahuila Plateaus and Ranges, (5) Sierra Madre Oriental, (6) Gulf Coastal Plain, and (7) Central Plateau. Their boundaries are defined in figure 1.7, and their main geographic features and climates, designated by three-letter abbreviations, are given in table 1.1. The geology of the morphotectonic provinces can be visualized with the aid of figure 1.8, a generalized geologic map of northern Mexico. To complement the information presented, some selected references to comprehensive or recent works dealing with specific or controversial views are provided following the geologic description of each province.

1. Baja California Peninsula Morphotectonic Province

Geographic Aspect

This province is located between 23°00'–32°30' N and 109°30'–117°15' W. It is about 1,200 km long, with an average width of 95 km and an area of 144,000 km²; the dominant climate is BWh (dry desert group, hot tropical-subtropical type; see table 1.1), cooling northward. Elevation varies between 0 and 2130 m above sea level (masl); most of the mountainous country lies below 1000 masl; the northern half of the province is more mountainous. The peninsula has some 13 rather short, largely seasonal rivers that discharge into the Pacific Ocean. None of these rivers is large (the Río Colorado, the river marking the boundary between the states of Baja California and Sonora, lies east of the Baja California Peninsula Morphotectonic Province, in the Northwestern Plains and Sierras Morphotectonic Province [see fig. 1.7]). There are no lakes.

The Baja California peninsula includes a series of sierras, collectively known as Peninsular Ranges.

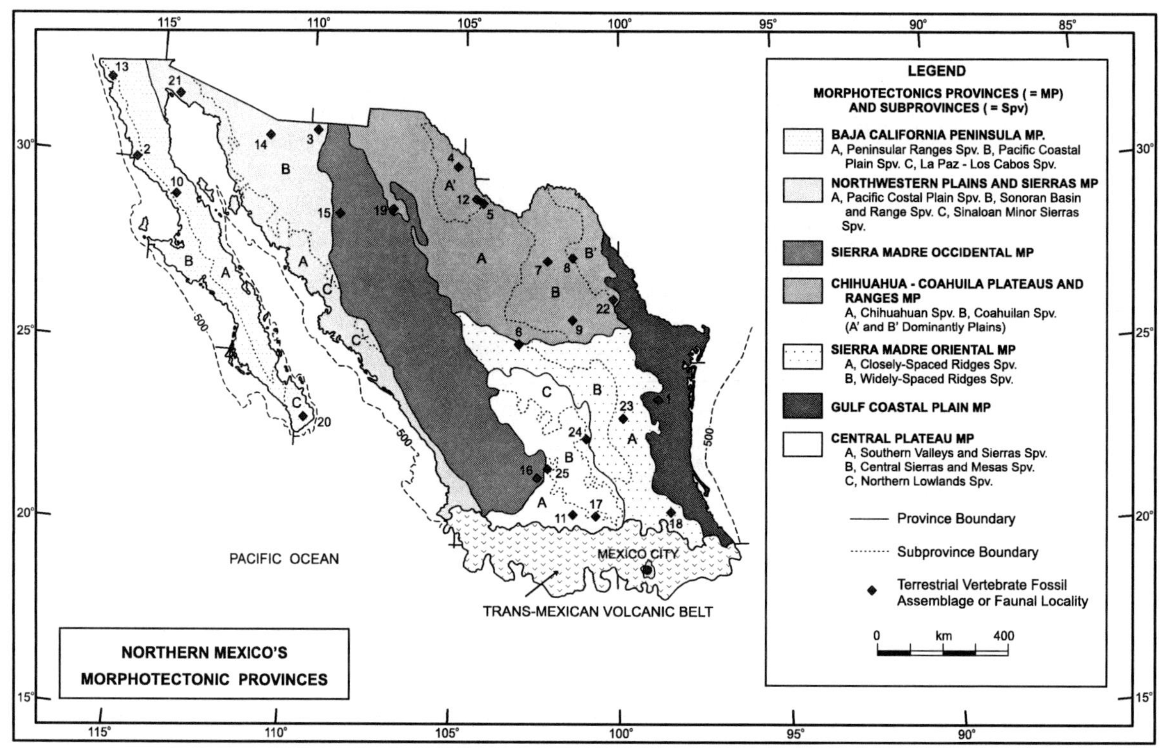

Figure 1.7. Northern Mexico's morphotectonic provinces map (adapted from Ferrusquía-Villafranca 1993, figs. 1.1, 1.2, 1.3, 1.4, 1.5, and 1.8). The chief terrestrial vertebrate fossil localities are also shown (black diamonds). Early Jurassic: 1, Cañón de Huizachal, Tams. Late Cretaceous: 2, El Rosario, B. C. N.; 3, Cuenca de Cabullona, Son; 4, Sierra Mojada, Chih; 5, Ojinaga, Chih; 6, Torreón, Coah; 7, Hipólito, Coah; 8, Palau-Múzquiz-Nueva Rosita area, Coah; 9, Rincón Colorado-Cerro del Pueblo area, Coah. Eocene: Early (Wasatchian): 10 Punta Prieta (Lomas Las Tetitas de Cabra, also known as Occidental Butes), B. C. N. Middle (Bridgerian-Uintan): 11, Marfil, Gto. Late (Chadronian): 12, Ojinaga (Rancho Gaitán), Chih. Miocene, Early (Hemingfordian): 13, La Misión, B. C. N.; 14, Tubutama, Son; 15, Yécora, Son; 16, Zoyatal, Ags. Late (Hemphillian): 17, San Miguel de Allende area, Gto; 18, Tehuichila-Zacualtipan area, Hgo; 19, Yepómera, Chih. Late Pliocene-earliest Pleistocene (Blancan): 17, 19 and 20, Las Tunas, B. C. S. Pleistocene, Middle (Irvingtonian): 21, El Golfo, Son. Late (Rancholabrean): 22, Bustamante Cave, N. L; 23, San Josecito Cave, N. L./Tams. Border; 24, El Cedral, S. L. P.; and 25, Arroyo El Cedazo, Ags.

These span nearly the entire length of the peninsula. In Baja California Sur the ranges are narrower, allowing enough space westward for a broad lowland, the Pacific Coastal Plain (<200 masl), which is widest where it forms the Vizcaíno Desert. The Peninsular Ranges and Pacific Coastal Plain make up the first two subprovinces of the peninsula (fig. 1.7). The Peninsular Ranges end by Bahía de La Paz, giving way southward to the La Paz–Los Cabos (Range and Upland) Subprovince.

Geologic Aspect

The geologic features of the northern and southern halves of the peninsula are quite different, so we describe them separately (fig. 1.8).

Northern Baja California. The Peninsular Ranges include some small portions of Paleozoic crystalline rocks, but largely they consist of Mesozoic (mainly Cretaceous) metamorphic and granitoid plutonic rock bodies. Clastic marine and conti-

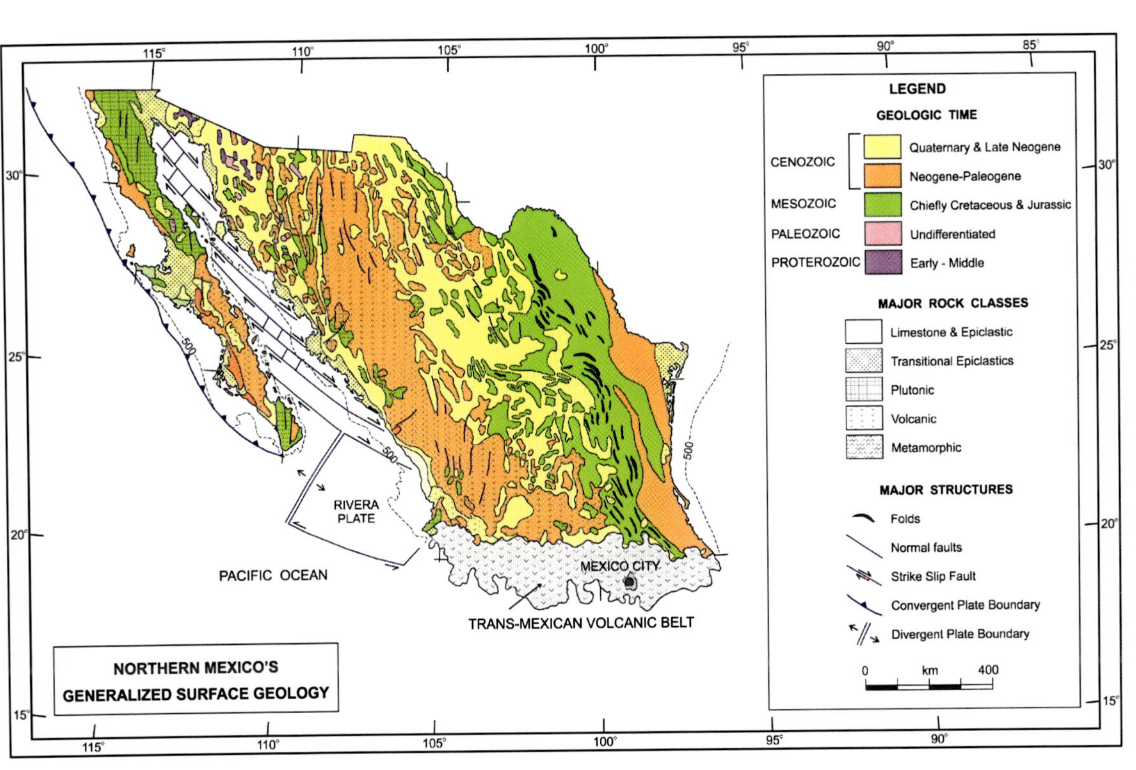

Figure 1.8. Northern Mexico's generalized geologic map (adapted from De Cserna 1990; Ortega-Gutiérrez et al. 1992; Ferrusquía-Villafranca 1993, figs. 1.3, 1.5, 1.7, and 1.9).

Table 1.1. Location and main features of northern Mexico's morphotectonic provinces.

Province[a]	Location	Surface Area (km²)	Elevational Ranges (m)	Climate[d]	Chief Land Form
1	Northwestern Mexico 109°30'–117°00' W, 23°00'–32°30' N	144,000 (7.3%)[b]	0–2130 (0–1,000)[c]	BWh, BShs, Csa	Sierras and plains
2	Northwestern Mexico 107°00'–116°00' W, 23°00'–32°30' N	236,800 (12%)	0–2200 (200–1000)	BWh, BSh	Sierras and plains
3	Western and northwestern Mexico 102°20'–109°40' W, 20°30'–31°20' N	289,000 (14.7%)	200–3000 (2000–3000)	Cfb, Aw	Sierras and plateaus
4	Northern Mexico 101°31'–110°31' W, 26°00'–31°45' N	255,900 (12.5%)	200–2000 (800–1200)	BShw, BWh, BSk	Sierras and plateaus
5	Northeastern and northcentral Mexico Transverse Sector 100°00'–105°00' W, 24°30'–26°00' N Eastern Sector 97°30'–101°20' W, 19°40'–26°00' N	145,500 (7.5%)	200–3000 (1000–2000)	BWh, BSk Cfa, Cwa, BSh	Sierras
6	Eastern Mexico/Northern Sector 95°30'–100°20' W, 20°00'–26°00' N	87,200 (4.4%)	0–200	Aw', Cw, Cx'w'	Plains
7	Central Mexico 100°00'–104°00' W, 21°00'–24°00' N	85,300 (4.3%)	1000–3300 (2000–3000)	BSh	Plateaus
TOTAL		1,245,900 (62.9%)			

[a]Province names: 1 = Baja California; 2 = Northwestern Plains and Sierras; 3 = Sierra Madre Occidental; 4 = Chihuahuan-Coahuilan Plateaus and Ranges; 5 = Sierra Madre Oriental; 6 = Gulf Coastal Plain; 7 = Central Plateau.
[b]Percent area of Mexico.
[c]Dominant elevational range.
[d]Climate types: BWh = desertlike, mean annual temperature (MAT) > 18°C; BShs Csa = temperate with dry winter; BSh = dry, MAT > 18°C; Cfb = temperate humid with no dry season; BShw = steppelike, winter dry season, MAT > 18°C; BSk = steppelike, MAT > 18°C; Cfa = temperate, no defined dry season; Cwa = temperate with dry winter; Aw = subtropical with dry winter and warm rainy summer; Aw' = tropical with dry winter and rainy summer; Cw = temperate with dry winter; Cx'w' = temperate with little rain throughout the year. The key to the letter symbology is: A = warm humid and subhumid climate group (lack of a well-defined dry season); m = rainy season restricted to the summer; w = dry winter and warm season from April to September; w' = less rainy summer with a short dry season. B = warm to cold and very arid to semiarid climate group; BS = warm to semicold and arid to semiarid climate subgroup; BW = warm to semicold and very arid climate subgroup; h = semiwarm with cool winter; k = temperate with a warm summer; s = rainy winter; C = temperate to semicold and humid to semihumid climate group; a = warm summer; b = cool and long summer; x = rainfall. Source: García 1988.

From Ferrusquía-Villafranca (1993).

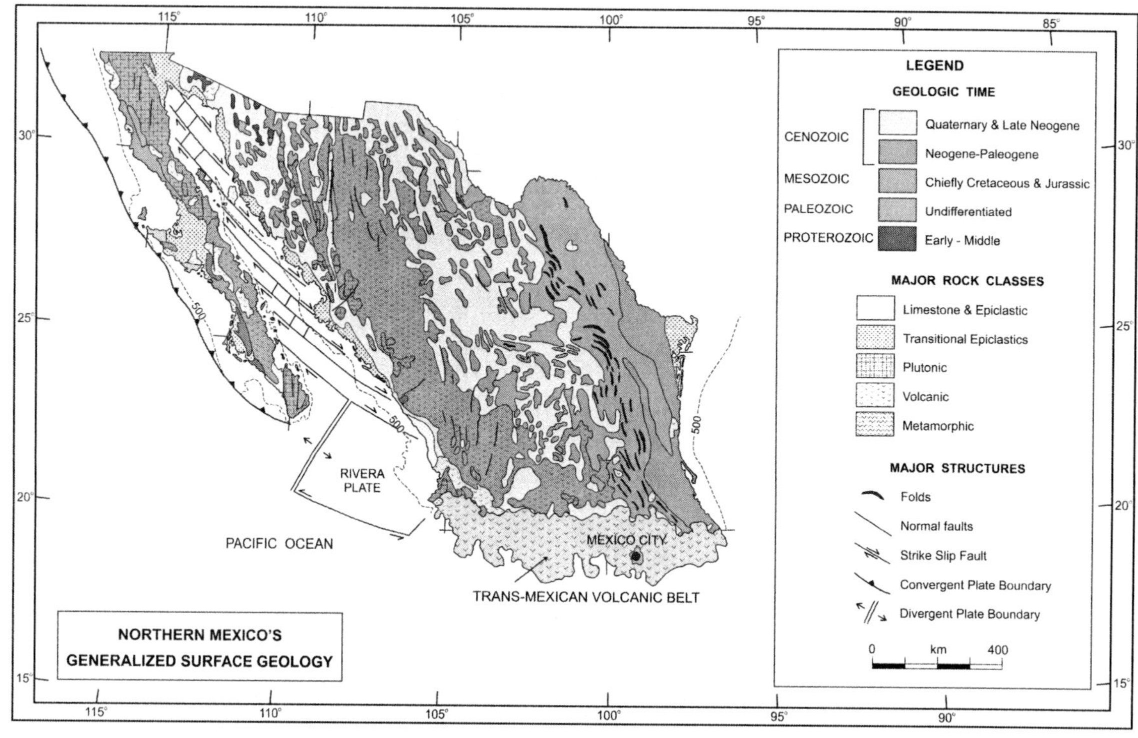

Figure 1.8. Northern Mexico's generalized geologic map (adapted from De Cserna 1990; Ortega-Gutiérrez et al. 1992; Ferrusquía-Villafranca 1993, figs. 1.3, 1.5, 1.7, and 1.9).

nental Cretaceous units occupy part of the narrow Pacific Plain. Dinosaurs and mammals have been found in it (fig. 1.7, location 2; Morris 1967, 1973, 1981; Lillegraven 1972, 1976; Molnar 1974; Clemens 1994; Montellano-Ballesteros 2002). Early Tertiary shallow marine and continental units occur in the south. An Early Eocene (Wasatchian) land mammal fauna, the oldest Cenozoic one from Mexico, was collected from a site near Punta Prieta (fig. 1.7, location 10; Novacek et al. 1991). Late Tertiary volcanics and Plio-Quaternary beach deposits occur in the rest of the plain. Faults deformed the Pre-Quaternary units (fig. 1.8).

Southern Baja California. The largest feature is the Sierra de la Giganta, consisting of Early Neogene volcanics (mafic lava flows and silicic pyroclastic sheets), and fluvio-lacustrine epiclastics and volcaniclastics ("Comondu Formation"). West of La Giganta, the Pacific Plain widens and largely consists of Late Tertiary-Quaternary, fine-grained, shallow marine to beach deposits. Older clastic units occur to the east. Some have yielded Late Oligocene

and Miocene marine vertebrates and dicotyledon (legume) wood (Weber and Cevallos-Ferriz 1994). Baja California Sur has a Mesozoic core of sedimentary marine rock bodies, ophiolites, and basaltic volcanics, all indicative of an ocean floor origin.

The peninsular southern end is set apart by a major north–south trending fault and largely consists of a Cretaceous granitoid pluton, unconformably covered by Late Cenozoic fine-grained marine and continental clastics (fig. 1.8); a large Late Pliocene terrestrial vertebrate fauna was recovered from these clastics (fig. 1.7, location 20; Miller 1980).

Gulf of California. The Gulf includes islands formed by volcanic, sedimentary, and/or metamorphic bodies similar to those of adjacent lands. The islands represent uplifted or detached blocks of the submerged continental platform. The oceanic (basaltic) floor is present only in the southern half; it is about 4 ma old and is affected by left lateral faults (arranged *en echelon*), ultimately related to the San Andres Fault System and to the opening of the Gulf (fig. 1.8).

Late Oligocene and Early Miocene marine fossils (e.g., mammals, sharks, mollusks) found in the Gulf margin of Baja California Sur (San Telmo-Bahía de La Paz, ~24–26° N; Vanderhoof 1942; Durham 1950; Smith 1991; Cruz-Marín and Barnes 1996; Barnes 2002) reveal the presence of a marine environment, the "Protogulf," long before the oceanization of its floor (ca. 4 ma).

Selected references. The whole peninsula: Beal 1948; Gastil et al. 1981; Frizzell 1984; Hausback 1984. Northern Baja California: Gastil et al. 1975. Southern Baja California and the Gulf Region: Henry 1989; Ortlieb 1991a,b; Fletcher et al. 2003; Ledezma-Vázquez and Johnson 2003; Oskin and Martín-Barajas 2003.

2. Northwestern Plains and Sierras Morphotectonic Province

Geographic Aspect

The Northwestern Plains and Sierras Province lies between 23°00'–32°30' N and 116°00'–117°00' W; its shape is that of an elongated, obtuse triangle (the southern two-thirds is quite narrow). The base of the triangle is an approximately 640-km long segment of the U.S.–Mexico boundary. Westward, the province is bounded by the 1,870-km long Sonora-Sinaloa-Nayarit Pacific margin, and to the east it shares its 1,280-km long boundary with the Sierra Madre Occidental (fig. 1.7).

The area is about 236,800 km². Elevation ranges from 0 to 2,200 masl, but the 200–1,000 masl hypsometric zone is dominant. The climate is desertic (BWh group; see table 1.1) in the northwest, and dry (Bsh group) elsewhere. This province has 7 major rivers, all of which drain to the Gulf of California. Dams have been built along some of these rivers, at least 2 of them where natural lakes existed.

Three main geomorphic zones are present and are recognized as distinct subprovinces: (1) The Pacific Costal Plain, defined by the 200 masl contour line; its southern Sonora–Sinaloan littoral is especially diverse, including delta complexes, estuaries, lagoons and sand bars. (2) The Sonoran Basin and Range (an extension of the U.S. namesake province) consists of narrow, closely spaced, north-northwest–south-southeast trending block mountains, separated by corresponding basins. (3) The Sinaloan Minor Sierras form the smallest geomorphic zone, and show no basin-and-range pattern (fig. 1.7).

Geologic Aspect

Numerous north-northwest–south-southeast trending faults divide the larger part of the province into narrow, uplifted blocks (the sierras or ranges) and downthrown blocks (the basins), where late Middle to Late Tertiary clastic deposits have accumulated; in Tubutama, some have yielded Early Miocene camels (see below). The upthrown blocks exhibit different parts of the geologic column (from Precambrian to Mesozoic), depending on their actual individual uplift and erosion (fig. 1.8).

The Pre-Cambrian consists of high-grade metamorphics, intruded by granitic plutons. The Paleozoic is formed by marine carbonate and siliciclastic units. The Late Triassic-Early Jurassic units are largely continental, both volcanic and clastic. The Cretaceous includes volcanic, plutonic, and marine carbonate units.

The Cenozoic includes Early Tertiary granitoid plutons (genetically related to copper deposits), Middle Tertiary silicic, mesa-forming pyroclastics, epiclastics, and Late Tertiary-Quaternary basaltic flows and clastic units. It appears that coeval to this volcanic activity, extensive parallel block-faulting tectonically overprinted all preexisting rock bodies, giving this province its characteristic basin and range structural pattern. Neogene continental deposits were preserved in the basins thus formed; in one such deposit at Tubutama, a Miocene, highly specialized camel, *Stenomylus tubutamensis*, was recovered and described (fig. 1.7, location 14; Ferrusquía-Villafranca 1990a). The Stenomylini had gazellelike proportions, a short face, and highly hypsodont, very elongated molars; they tended to be smaller than contemporary camels, left no descendants, and geographically were restricted to West Texas, southern New Mexico–Arizona, and, of course, northern Sonora (Ferrusquia-Villafranca 1990a; Honey et al. 1998).

In the southern part of the province, the basin and range pattern does not exist. The Paleozoic, Creataceous, and Early Tertiary units are largely crystalline (high-grade metamorphics, granitoid plutons, and rhyolitic to basaltic volcanics). Middle Tertiary pyroclastics occur near the Sierra Madre Occidental (fig. 1.8).

The Quaternary in the Pacific Coastal Plain consists of tabular, marine-continental bodies developed by prograding delta complex systems, whose sediments were redistributed by waves, tides, and along-shoreline currents. A large Middle Pleistocene terrestrial vertebrate fauna is known from the northwestern end of this subprovince (fig. 1.7, location 21; Shaw 1981; Arroyo-Cabrales et al. 2002).

Selected references. Cook and Bally 1975; Roldan-Quintana 1984; Henry and Aranda-Gómez 1992; Bortolini et al. 1995; McDowell et al. 1997; Roldan-Quintana et al. 2003.

3. Sierra Madre Occidental Morphotectonic Province

Geographic Aspect

The Sierra Madre Occidental Province lies between 20°30'–31°20' N and 102°20'–109°40' W; it includes parts of Sonora, Chihuahua, Durango, Sinaloa, Zacatecas, Nayarit, and Jalisco. It is Mexico's largest morphotectonic province (~289,000 km^2), and it has a rectangular shape (length 1300 km, average width 190 km). Elevation varies from 200 to 3000 masl, dominating in the 2000–3000 masl hypsometric zone. Elevation exerts a greater influence on the climate than does the 10° latitudinal spread. The Cfb temperate humid climate (see table 1.1) dominates at the higher elevations, while lower terrain is characterized by its humid subtropical, winter-dry climate (Cwa). Near the coast, the climate type is tropical wet-and-dry (Aw).

The Sierra Madre's western slope, being more humid, has more rivers (11) than the eastern slope (5). Along the western slope, those rivers in the northern half of the Sierra Madre Occidental discharge into the Gulf of California, while the other rivers drain into the Pacific Ocean. The rivers of the eastern slope drain into the Rio Grande (Río Bravo) or in "La Laguna," and in one endorheic basin located in northwest Durango–southeast Coahuila. Lago Santiaguillo, about 70 km north of Durango City, is the only lake in this province.

Geomorphologically, the province consists of closely spaced volcanic sierras and plateaus that coalesce to form larger ranges. River systems draining largely into the Gulf of California separate the individual sierras and plateaus (fig. 1.7).

Geologic Aspect

The volcanic bodies that make up the Sierra Madre Occidental Province are arranged in 2 complexes. The Lower Complex chiefly consists of andesitic lavas and pyroclastic sheets (some are rhyolitic) dating 100–45 ma; it is broadly arcuate and intensely faulted; its basement is largely unknown, but some Cretaceous carbonates have already been detected. The Upper Complex unconformably overlies the Lower Complex and consists of an extensive silicic ignimbrite succession up to 1000 m thick that spans the province. The ignimbrites issued largely through calderas (200–400), some very large (diameter 40 km), dating 54–34 ma (dominating the younger ones). The Upper Volcanic Complex is largely horizontal (fig. 1.8).

Locally, Late Cenozoic sedimentary clastic bodies occupy structural or topographic depressions; a Middle Miocene leporid and some large Quaternary mammals have been recovered from these clastic bodies (fig. 1.7, location 15; Ferrusquía-Villafranca 1990a, unpubl. data). In the northwestern part of the province, near Yépomera, Chihuahua, a large latest Miocene-earliest Pliocene mammal fauna has been known for a long time (fig. 1.7, location 19; Lance 1950; Ferrusquía-Villafranca 1978; Lindsay and Jacobs 1981; Lindsay 1984). In the southeast, Miocene mammals are also known from Aguascalientes (fig. 1.7, location 16; Dalquest and Mooser 1974; Ferrusquía-Villafranca 1990a, 2003).

Selected references. McDowell and Clabaugh 1979; Roldan-Quintana 1984; Moore et al. 1994; Nieto-Samaniego et al. 1999; Aranda-Gómez et al. 2003; Roldan-Quintana et al. 2003.

4. Chihuahua-Coahuila Plateaus and Ranges Morphotectonic Province

Geographic Aspect

The Chihuahua–Coahuila Province lies between 26°00'–31°45' N and 101°31'–110°31' W; it is bounded to the west and south by the Sierras Madre Occidental and Oriental, respectively, and to the north by the Rio Grande; to the east, it grades into the Gulf Coastal Plain (the 200 masl contour line is the limit). The area is about 255,900 km^2. Elevation ranges from 200 masl to a little higher than 2000 masl. The western half is higher (>1200 m); in the eastern half, eastern Coahuila gradually slopes down to the Coastal Plain. The climate is very arid in the west (Bw group; see table 1.1), grading to less arid to the northeast (steppelike; Bs group).

The Rio Grande and its tributary the Río Conchos are the only major rivers in this province. Smaller rivers discharge into endorheic basins; some basins have lakes, such as Laguna Guzman, Laguna

Santa María, and Laguna Patos in northern Chihuahua, and Lago Toronto and Lago Palomas in northwestern Durango. Coahuila has only 1 lake (along the Río Salado, in the northeast, close to the Coahuila–Nuevo León border), which has been dammed and named Venustiano Carranza Dam.

Geomorphologically, the province includes low, northwest–southeast trending block-folded ranges and block-mesas, separated by flat-lying basins and plateaus. In the northwest, the ranges are fairly narrow and largely consist of Cretaceous limestone and Neogene volcanics; in the north, large mesas are present; everywhere else mesas and wider ranges dominate, formed also by Cretaceous limestone, while volcanics are scarce. Lowlands occur in the southwest (Bolson de Mapimi) and south (La Laguna area); the latter area separates this province from the Sierra Madre Oriental (fig. 1.7; small-scale maps do not allow us to plot these last features).

Geologic Aspect

The Late Precambrian and Paleozoic units have a very small total outcrop area (Villa Aldama-Placer de Guadalupe in northern Chihuahua, and Delicias in southwestern Coahuila). Jurassic units also are little exposed, chiefly occurring in the Placer de Guadalupe area, so that the Mesozoic sequence is mainly composed of Cretaceous units. The Cretaceous units largely consist of marine carbonates and clastics; this character and the stratigraphic succession are similar to those of the Sierra Madre Oriental. The major difference between the 2 provinces pertains to structural style. The folds of the Chihuahua-Coahuila Plateaus and Ranges are somewhat broader, axially trending northwest-southeast, and form 2 discontinuous or distinct sets. The northwestern set of folds is parallel to the Rio Grande in the north of Chihuahua and seems to correspond to a pull-apart basin, the Chihuahua Trough (Haenggi 2002). The other set of folds originates in the Big Bend area of Texas, ending north of Monterrey, but the folds do not coalesce with those of the Sierra Madre Oriental. Elsewhere in Coahuila, the Cretaceous units form block-faulted, somewhat tilted, mesas (fig. 1.8). Late Cretaceous dinosaurs have been recovered from scattered localities both in Chihuahua and Coahuila (fig. 1.7, locations 4–9; Rodríguez de la Rosa and Cevallos-Ferriz 1998; Ferrusquía-Villafranca, unpubl. data).

The Cenozoic is continental, unconformably overlies Cretaceous units, and in Chihuahua includes silicic volcanics of Oligocene-Miocene age that form northwest–southeast trending sierras, frequently bounded by faults. Small silicic plutons and mafic lava flows complete the igneous rock succession in Chihuahua. In Coahuila, this succession is less well represented. The intermontane basins and lowlands are frequently grabens, where volcaniclastic and epiclastic fluviolacustrine Late Paleogene or Neogene deposits occur (fig. 1.8). Near Ojinaga, Chihuahua, a Late Eocene mammal fauna was collected from one such deposit, the Prietos Formation (fig. 1.7, location 12; Ferrusquía-Villafranca 1969; Ferrusquía-Villafranca and Wood 1969; Ferrusquía-Villafranca et al. 1997). Thin, flat-lying Quaternary deposits floor the basins and lowlands.

Selected references. Imlay 1936, 1938, 1943, 1944; Bridges 1965; Bridges and DeFord 1965; Navarro and Tovar 1974; Cook and Bally 1975; Charleston 1981; Eguiluz 1984; Brown and Dyer 1987; Haenggi 2001, 2002.

5. Sierra Madre Oriental Morphotectonic Province

Geographic Aspect

The Sierra Madre Oriental Province is divided into the Closely-Spaced Ridges and Widely-Spaced Ridges Subprovinces (fig. 1.7), which largely differ in structural features (see below). Another more informal but quite practical geographic subdivision, into Eastern and Transverse Sectors, is first used to describe this province. The name and position of these sectors are descriptive enough, so that they were not discriminated in figure 1.7.

The Eastern Sector lies between 19°40'– 26°00' N and 97°30'–101°20' W; its area is about 77,000 km², it is 550 km long and 140 km wide (average) and extends from the Trans-Mexican Volcanic Belt in the south to the Monterrey area, Nuevo León, in the north. The Transverse Sector is the western extension of the Eastern Sector, reaching as far as the Sierra Madre Occidental. This sector then chiefly lies east–west (i. e., transverse to the Eastern Sector [hence the name]), between 24°30'-26°00' N and 100°00'-105°00' W; its area is about 68,000 km², and it is 400 km long and 120 km wide (average).

Elevation varies from 200 to slightly higher than 3000 masl, but with an uneven distribution; the 1000–2000 masl hypsometric zone is dominant. The Eastern Sector's climate varies from Cfa temperate humid to Cwa temperate dry (see table 1.1), whereas the Transverse Sector's climate is dominantly BWh desertic.

The humid eastern slope of the Sierra Madre's Eastern Sector is the one that has rivers. All 10 of them drain to the Gulf of Mexico; most have carved deep gorges and canyons. The Ríos Santa María and Moctezuma, tributaries of the Río Pánuco, are the only rivers that have cut through the entire width of this sector. There are no lakes in the Sierra Madre Oriental. Karstification is extensive.

Geomorphologically, this province consists of folded ridges and intermontane, elongated valleys, and plateaus. It is the spacing and other features of the ridges that allow the distinction between the 2 subprovinces.

Geologic Aspect

The province consists largely of Early Jurassic continental and shallow marine (siliciclastics and evaporates) units and Cretaceous marine carbonate units that change upward to fine clastic, shallow marine to transitional units. The exposed area is much less for the Jurassic than for the Cretaceous. This latter system is dominated in extent and thickness by the Lower Cretaceous (Ortega-Gutiérrez et al. 1992); it represents the transgression apex of the epicontinental sea that covered much of northern Mexico at that time (fig. 1.8).

Late Cretaceous dinosaurs have been collected from several localities in the Eastern Sector (fig. 1.7, location 6; Ferrusquía-Villafranca, unpubl. data), and in southern Coahuila, near this sector; in 1 locality, scarce angiosperm remains were also found (fig. 1.7, location 9; Rodriguez de la Rosa and Cevallos-Ferriz 1998). The only Jurassic (Early) vertebrate fauna of this province comes from the Huizachal Canyon in Tamaulipas (fig. 1.7, location 1; Clark et al. 1994, 1998; Fastovsky et al. 1995; Reynoso-Rosales 1996; Montellano-Ballesteros 2002).

The Mesozoic units are folded into anticlinoria and synclinoria, whose spacing diminishes from west to east; around Monterrey, Nuevo León, they bend westward, changing their axial orientation from north-northwest–south-southeast to nearly east–west. These structures are also cut by faults. The resulting fold-ranges are separated by long and narrow intermontane depressions, where Late Paleogene-Neogene terrigenous fluviocustrine clastic units have been formed (fig. 1.8). This extensive regional uplift and structural deformation is part of the Laramide Orogeny, which occurred during the Middle and Late Eocene (~50–40 Ma).

Quaternary karstification of the limestone fold-ranges in this province and in the neighboring Chihuahua-Coahuila Plateaus and Ranges has produced numerous and complex cave systems, where Pleistocene vertebrates were trapped, thus generating fossil assemblages (fig. 1.7, locations 22 and 23). The best known such assemblage comes from the San Josecito Cave along the Nuevo León/Tamaulipas border (Stock 1943; Ferrusquía-Villafranca 1978; Arroyo-Cabrales et al. 2002; McDonald 2002).

Selected references. Imlay 1936, 1938, 1943, 1944; De Cserna 1956, 1960, 1989; Carrillo-Bravo 1961; McBride et al. 1974; Cook and Bally 1975; Sutter 1980, 1984, 1987; Padilla-Sanchez 1985.

6. Gulf Coastal Plain Morphotectonic Province

Geographic Aspect

The Gulf Coastal Plain Province consists of the lowlands bordering the Gulf of Mexico exclusive of the Yucatán Platform. The Teziutlan Massif (regarded by some as part of the Trans-Mexican Volcanic Belt) and associated volcanic hills nearly divide the province into 2 sectors, Northern and Eastern, leaving a narrow strip or corridor that communicates between them. The Northern Sector falls within northern Mexico's boundaries and is described here (fig. 1.7). This sector lies between 20°00'–26°00' N and 96°30'–100°20' W; it covers an area of about 87,200 km^2 and is 100 km wide (average), becoming wider northward. Elevation varies from 0–200 masl. The climate is Aw' tropical in the southern third, becoming temperate and drier (Cw to Cx'w) to the north.

The Gulf Coastal Plain is traversed by the rivers that drain the Sierra Madre Oriental. Lakes are present only at and near the mouth of the Río Pánuco; they include Laguna de la Culebra, Laguna de la Tortuga, and Laguna de Altamira in southern Tamaulipas, and Laguna del Pueblo Viejo in northern Veracruz.

The Gulf Coastal Plain littoral is affected by prograding fluvial sedimentation related to the development of offshore islands, lagoons, and estuaries. The flats are the dominant geomorph and include floodplains, alluvial fans, marine terraces, and beach lands (both sandy and muddy).

Geologic Aspect

This province is well known due to its oil reaches. Numerous units have been defined, and their outcrops show a regional pattern of bandlike zones successively younger toward the gulf. The units consist of marine limestones and fine-grained clastics that make up tabular bodies gently dipping eastward. The Tertiary sequence spans the whole period and unconformably rests on Cretaceous carbonates. The different-age bodies are wider in the northern part, where they were deposited in a large paleobay, the Rio Grande Embayment (fig. 1.8). A large Miocene palynoflora is known from the Coastal Plain in southern Veracruz (Paraje Solo Formation; Graham 1978, 1993).

Subsurface information discloses the presence of some large sedimentary thickenings, which correspond to paleobasins, where subsidence occurred at a greater rate than in the rest of the province. The Quaternary sedimentary bodies are typically transitional, deposited in shore and fluvial environments. Quaternary and Tertiary sedimentary bodies are broader in the north.

The province's Cenozoic geologic record depicts the gulfward marine regression of the epicontinental sea that once covered northern Mexico, concomitantly increasing its territory.

Selected references. Murray 1961; Viniegra-Osorio 1965; Barker and Blow 1975; Cook and Bally 1975; Wilson 1987; Galloway and Blow 1989; Galloway et al. 1991; Coleman et al. 1991.

7. The Central Plateau Morphotectonic Province

Geographic Aspect

The Central Plateau Province is bounded by the Sierras Madres Occidental and Oriental and by the Trans-Mexican Volcanic Belt. It lies between 21°00'–24°00' N and 100°00'–104°00' W. It is parallelogram-shaped, 450 km long and 280 km wide (average); the area is about 85,300 km². Elevation ranges from 1000 to 3300 masl, but the 2000–3000 masl hypsometric zone (Sierra de Guanajuato-Zacatecas) is dominant. The climate is arid hot to semiarid temperate (BSh group; see table 1.1). There are no major rivers in this province, and only a single lake, Laguna de Yuriria in southern Guanajuato.

Geomorphologically, the Central Plateau consists of 3 zones or subprovinces (fig. 1.7): The Southern Valleys and Sierras, which largely corresponds to the Río Lerma-Santiago basin; the Central Sierras and Mesas, which corresponds to the Sierra de Guanajuato-Zacatecas; and the Northern Lowlands, formed by rolling lands and isolated hills.

Geologic Aspect

The Central Plateau Province is especially complex (fig. 1.8). The Pre-Cenozoic rock bodies largely consist of Late Paleozoic metamorphic units that crop out in Zacatecas; Late Triassic marine carbonate and siliciclastic units exposed both in Zacatecas and in San Luis Potosí; and Middle and Late Jurassic marine and Cretaceous marine and low-grade metamorphic bodies chiefly exposed in Zacatecas, San Luis Potosí, and Guanajuato (largely the Central Sierras and Mesas Subprovince).

The Cenozoic unconformably rests on the older rock bodies; it includes sedimentary clastic, volcanic (both lavic and pyroclastic), and plutonic units. The sedimentary units are of Paleogene and Late Neogene age, such as the Guanajuato Conglomerate (Middle Eocene) and the San Miguel de Allende Formation (latest Miocene), which have both yielded mammal faunas. Early Neogene units are scarce, but one such unit in Aguascalientes yielded a mammal fauna.

Volcanics overlie the Paleogene clastics; they are lava flows and pyroclastic sheets of andesitic to rhyolitic composition that form mesas and occasional peaks. Their age is 35 ma (similar to that of the Sierra Madre Occidental's Upper Volcanic Complex). Small- to medium-size plutons of granitic to dioritic or even gabbroic composition intrude into the Tertiary sequence; their emplacement is related to the genesis of this province's mineral deposits. Locally, other younger (Neogene) volcanic flows or fine clastics cover the mesas or floor the adjacent lowlands. Thin, flat-lying Quaternary deposits crop out in the lowlands.

The red Guanajuato Conglomerate has yielded at Marfil the only known Middle Eocene terrestrial vertebrate fauna from Mexico (also the southernmost known fauna from North America), with rodents related to the ancestral stock of the South American Caviomorpha (fig. 1.7, location 11; Edwards 1955; Fries et al. 1955; Black and Stephens 1973; Ferrusquía-Villafranca 1987, 1989; Ferrusquía-Villafranca et al. 2002). The Early Miocene Zoyatal Tuff of Aguascalientes yielded the namesake mammal fauna, which includes perissodactyls and artiodactyls that belong to extant families, except the extinct Merycoidodontidae (fig. 1.7, location 16; Dalquest and Mooser 1974; Ferrusquía-Villafranca 1990a, 2003; the site lies near the Central Plateau–Sierra Madre Occidental boundary, within the latter province).

Eastern Guanajuato has yielded a large and highly diverse mammal assemblage of latest Miocene–Early Pleistocene age (fig. 1.7, location 17; Carranza-Castañeda and Ferrusquía-Villafranca 1978, 1979; Dalquest and Mooser 1980; Carranza-Castañeda et al. 1981, 1994; Miller and Carranza-Castañeda 1984, 2002; Carranza-Castañeda and Miller 2002). This assemblage includes the second earliest record of South American immigrants in North America (*Glossotherium*, fission-track dated as 4.1 ± 0.3 ma and 3.9 ± 0.3 ma old; Kowallis et al. 1998). These ages indicate an earlier date for the beginning of the Great American Faunal Interchange (GAFI) than usually thought. However, the presence of large North American mammals in the Amazonian Peru, dated about 9 ma, indicates even a far earlier date for this event (Campbell et al. 2000; Ferrusquía-Villafranca 2003).

Aguascalientes and San Luis Potosí also carry large Pleistocene vertebrate assemblages (fig. 1.7, locations 24 and 25; Mooser and Dalquest 1975; Ferrusquía-Villafranca 1978; Arroyo-Cabrales et al. 2002; McDonald 2002).

The sedimentary Mesozoic units are deformed into folds of varied extent, amplitude, and axis orientation. They are also affected by faulting, which appears to have 2 dominant directions, north-northwest–south-southeast and east–west. There are several grabens largely in the southern part, whose structural orientation coincides with the fault directions; their genesis must have occurred during the same faulting episode.

Selected references. Echegoyen et al. 1970; Cook and Bally 1975; Labarthe-Hernández et al. 1982; Martínez-Reyes 1992; Barboza-Gudiño et al. 1999; Ferrari et al. 1999; Cerca-Martínez et al. 2000; Aranda-Gómez et al. 2003.

Constraints on Modeling: Correlation of Historical, Geologic, Geographic, and Biotic Processes, Events, and Features

This correlation is presented as a series of integrative considerations and remarks aimed at establishing the relevance of geologic processes and geographic features in influencing the development of northern Mexico's biota throughout the Cenozoic (see figs 1.1–1.9). The timing and duration of the processes, events, and features discussed below are plotted in appendix 1.1; their supporting evidence has already been given above and will be supplemented as needed. We provide the most parsimonious interpretations. Inferences are as objective as possible and kept at a minimum.

(1) Tectonically, northern Mexico's territory has been part of the North American Plate at least since the Cretaceous, and its latitudinal position has remained stable since that time (Ross and Scotese 1988; De Cserna 1989; Sedlock et al. 1993). Significant tectonic rearrangement during the Cenozoic has occurred only on the adjacent Pacific Ocean floor, involving subduction (of the Farallon Plate) and detachment of the Baja California peninsula by deep transform faults, thereby originating the Gulf of California (Ortlieb 1991a,b; Ferrari 1995; Henry and Aranda-Gómez 2000; Fletcher et al. 2003).

(2) This extensional deformation seems to be genetically related to the magmatism that originated the Sierra Madre Occidental volcanic field and to the associated plutonic and volcanic bodies, which chiefly occurred in the Late Paleogene (ca. 35–30 ma). The deformation is related as well to the basin and range faulting of western Sonora and the adjacent southwestern United States, which took place in the Late Neogene (McDowell and Clabaugh 1979; Roldan-Quintana 1984; Henry 1989; Henry and Aranda-Gómez 1992, 2000; McDowell and Mauger 1994).

(3) The Sierra Madre Oriental and much of the Chihuahua-Coahuila Plateaus and Ranges have a different and more complicated origin that largely involves folding (generated by compression), faulting (of various kinds), and regional uplift; preexisting structures partly controlled the behavior of tectonic forces. This complex process is known as the Laramide Orogeny; its major episodes took place during the Middle and Late Eocene (De Cserna 1956, 1960, 1989; Sutter 1980, 1984, 1987; Haenggi 2002).

(4) As a result of the regional uplift, the epicontinental sea that covered much of northern Mexico during a good part of the Mesozoic at last retrograded (gulfward) in the Cenozoic, leaving a record of its gradual regression in the Gulf Coast Plain (and elsewhere in Coahuila), and forming a low and narrow land-corridor located between the Gulf of Mexico and the increasingly higher Sierra Madre Oriental ranges (Murray 1961; Galloway et al. 1991).

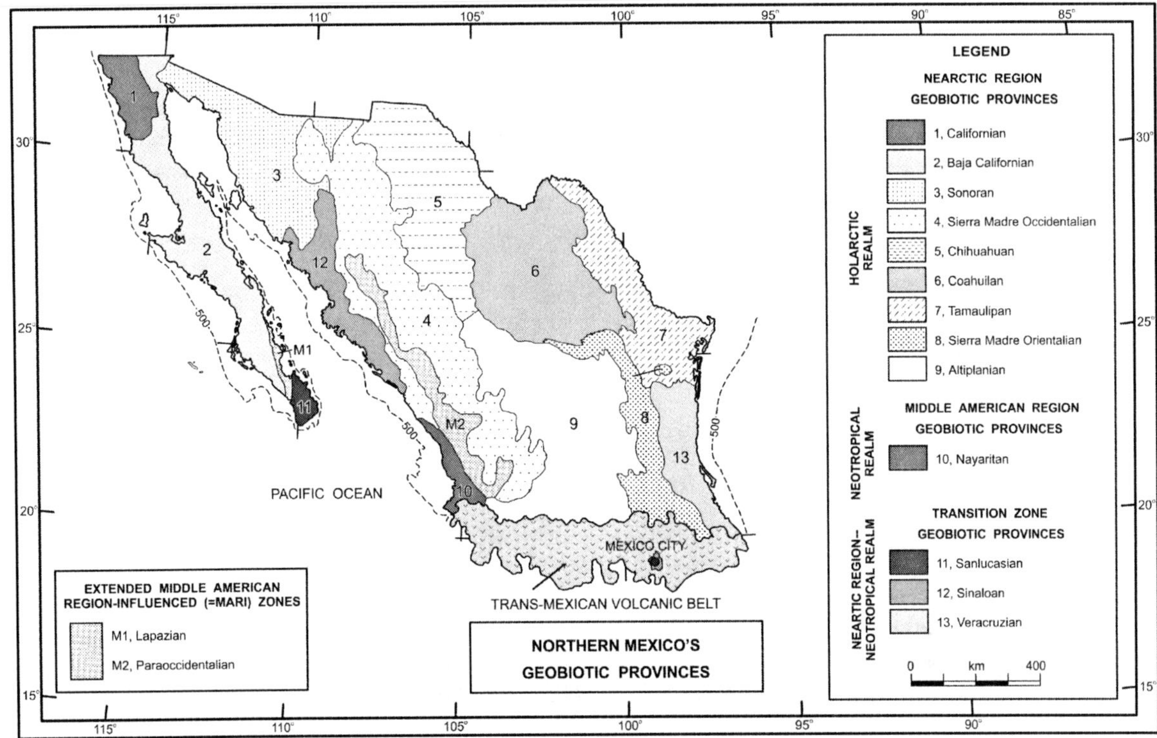

Figure 1.9. Geobiotic provinces of northern Mexico. Adapted from Ferrusquía-Villafranca (1990b). Although largely congruent with biotic province systems, the geobiotic province delineation scheme is conceptually distinct. It is based on multiple parameters including morphotectonic and paleontologic information. Each geobiotic province was presumably subjected to a distinct paleoenvironment during much of its history during the Cenozoic. The separation between Neotropic and Holarctic Realms is set at the Trans-Mexican Volcanic Belt.

(5) Additional tectonic and magmatic activity has caused the emplacement of Middle and Late Tertiary volcanic and plutonic bodies in Chihuahua and in the Central Plateau, as well as the faulting that generated the horsts (upthrown blocks) and grabens (downthrown blocks) of these regions. The horsts and grabens frequently correspond to the topographic highs (uplands, mesas, etc.) and lows (basins, flats, lowlands, etc.) seen there. Some horsts consist of Pre-Cenozoic rocks that became exposed during subsequent erosion in the Late Cenozoic (Bridges and DeFord 1965; De Cserna 1989; Haenggi 2001, 2002).

(6) The Trans-Mexican Volcanic Belt (TMVB) started to develop during the Late Oligocene–Early Miocene with the emplacement of silicic to andesitic lava flows and pyroclastic sheets in the western half that mainly formed high plateaus and a few peaks. The magmatic activity migrated eastward, producing in the Late Neogene–Quaternary large stacks of andesitic to basaltic flows and pyroclas-

tics that formed plateaus, sierras, and spectacular peaks (Demant 1984; Nieto-Obregon et al. 1985; Pasquare et al. 1986, 1987a,b; Ferrari et al. 1995, 1999; García-Palomo et al. 2002).

It should be noted that in the Gulf Coastal Plain, the Late Neogene–Quaternary volcanics were nonconformably emplaced upon the emerged Paleogene–Early Neogene sedimentary units. Such emplacement locally reduced the plain to a narrow land strip or corridor, thereby generating the present-day subdivision of this province into its Northern and Eastern Sectors.

During Late Neogene–Quaternary time, fluvio-lacustrine units were laid down in the lowlands throughout the TMVB; many sites have yielded large mammal faunas from such units (Ferrusquia-Villafranca 1978; Miller and Carranza-Castañeda 1984, 1996, 2002; Arroyo Cabrales et al. 2002).

(7) As a result of this complex geologic makeup and evolution, northern Mexico eventually acquired

a very diverse geographic layout. The region covers about 1,245,900 km² of rugged territory and spans 12° of latitude (from 20°30' to 32°30' N). It includes a large variety of geomorphs such as flats (of diverse size and altitudinal position: from near sea level to ~2000 masl), mountains (isolated or forming sierras or cordilleras), plateaus, valleys, and basins (fig. 1.2; Raisz 1964; Espinasa-Pereña 1990; Lugo-Hubp 1990). The climate is also diverse, changing from very arid, hot desert in the west and north to tropical, very humid with no defined dry season in the southeast (figs. 1.3–1.6; García 1990). It is befitting, then, that this geographically diverse territory be populated by a very diverse biota.

(8) The detailed Cenozoic paleogeographic evolution of northern Mexico, and particularly of its terrestrial landscape, is not well known (Morán-Zénteno 1984; De Cserna 1989). Worldwide, there were important climatic changes during the Cenozoic. However, the region's latitudinal position has changed little since the Late Cretaceous–Paleocene (Ross and Scotese 1988). Thus, climatic factors strongly influenced by latitudinal position (i.e., general insolation, light and temperature spatial distribution, and major wind patterns) remained largely (elevation did vary; see below) similar to those of the Recent during the Cenozoic. The increasing complexity of the landscape developed during the Cenozoic must have greatly influenced the climate (and related processes such as soil development) because, among other factors, high cordilleras force the wind to discharge rain on the oceanside slopes. On the landward side, the wind is depleted of its moisture, and as a result, there is less rain throughout the year. Given that the Sierras Madres and the Chihuahua-Coahuila Plateaus and Ranges were present in northern Mexico at least by the Late Paleogene–Early Neogene (ca. 30–25 ma), it is reasonable to postulate that the present-day major climatic regions of northern Mexico have existed since then and that increasing aridity in the north developed over time in a spatially uniform manner. The persistence of such different climatic regions in this vast territory for an interval nearly half as long as the Cenozoic must have had a profound effect on soil development and, of course, on the resident biota.

It should be noted that northern Mexico's present-day climate is influenced and partly determined by factors or phenomena occurring elsewhere, such as the El Niño/Southern Oscillation, polar air-mass invasions, and hurricanes, which in turn are controlled by physical or astronomical causes of worldwide reach. This must have been so in the past, too. Thus, among other factors influencing the climate of northern Mexico, the actual temperatures, humidity, and rainfall distribution and dynamics must have reflected the influence of worldwide climatic trends and changes that took place in the Cenozoic (see chapter 2 for a discussion of this topic).

The influence of elevation likely played an increasingly important role as northern Mexico's tall physiographic features developed. Southward range extensions of northern biotas with high-elevation affinities (or "acrobionts") would have been promoted when and where mountain ranges developed. Conversely, mountain ranges could have blocked the northward range extension of tropical, low-elevation species and communities.

(9) The continental glacier advance and retreat episodes that occurred in North America during the Middle and Late Pleistocene must have caused successive contractions and expansions of the climatic belts, thus producing major changes (fluctuations?) in the fluviolacustrine sedimentary regime, as well as important north–south and/or south–north shifts or displacements of the then-existing ecological associations (communities, ecosystems, and biomes) and species distributions. These changes probably increased in intensity northward.

However, the detailed comprehensive study and mapping of such climatic, sedimentary-stratigraphic, and biotic records of shifts for northern Mexico as a whole during the Pleistocene remains to be made. Even in the best-known areas, such as the Mexican Basin or the Valley of Puebla, nothing like an integrative geologic/paleontologic/paleoclimatic study has been published. There are, however, several works that deal with particular aspects of the record, such as those by Axelrod (1979), Van Devender (1990), and Arroyo-Cabrales (1994), to name just a few.

(10) Biogeographically, the role of the major geographic/geologic features of the region remains to be precisely assessed. The available geologic/geographic information, parsimoniously interpreted, allows one to recognize the following sequence of biogeographic events:

- The Sierra Madre Oriental and much of the Chihuahua-Coahuila Plateaus and Ranges were the first Cenozoic cordilleras to develop in northern Mexico. Their presence generated geographic diversity, which in turn led to habitat diversity. It also created barriers that restrained dispersal of the species making up the biota then inhabiting northern Mexico. In short, these cordilleras promoted regional biotic differentiation as early as early Late Eocene time, as they had fully developed by that time. The role of these cordilleras as upland platforms that allowed southward range extension of acrobiont communities, such as pine forest or pine-oak forest, merits consideration in its own right.

- At present, the Sierra Madre Oriental Eastern Sector ends by the parallel 20° N, where it meets the TMVB. However, the striking similarity in makeup and structure of the Sierra Madre Oriental Eastern Sector and the eastern cordillera just south of the TMVB (a fold-belt in southeastern Veracruz, northern Oaxaca, and adjacent Puebla; Ortega-Gutiérrez et al. 1992) strongly suggests physical continuity between them (De Cserna 1960, 1989; Murray 1961; Sedlock et al. 1993). If in fact there was such a continuity, it follows that during the Late Eocene–Early Neogene, both cordilleras would have formed a much longer fold-mountain belt, or "Mega-Eastern Sector," later severed (buried) by the TMVB. Should this be so, by Late Eocene–Early Neogene time the Mega- Eastern Sector functioned as a highland corridor that penetrated into Middle America (at least down to 17°30' N), allowing significant dispersal of acrobiont communities much earlier than in western Mexico. This effect was enhanced by the presence of the Sierra de Zongolica, a block mountain range formed by uplifted Mesozoic metamorphic basement rocks, which lies just west of the eastern cordillera mentioned above (Ortega-Gutiérrez et al. 1992). The eastern development of the TMVB during the Late Neogene–Quaternary maintained this upland platform.

- The gradually widening Gulf Coastal Plain (an event related to the Sierra Madre Oriental development and to the regional uplift) probably afforded during the Paleogene–Early Neogene a land corridor that allowed biotic exchange between the northern (which became Nearctic) and southern (which became Neotropical) regions. This corridor became restricted to a narrow land strip by Late Neogene–Quaternary time, as the TMVB became a more effective barrier, due to the emplacement of vol-

canics in its eastern part reaching close to the Gulf coast.

- The Sierra Madre Occidental could have acted as a biotic provincial barrier within the Nearctic Region no earlier than Late Oligocene time, when it was fully developed. From this upland platform, acrobiont communities, such as pine forest and pine–oak forest, could have expanded their range southward, as highlands in the Sierra Madre del Sur and in the TMVB were or became available.

On the ocean side, the Sierra Madre Occidental promoted the development of moist-tropical conditions, which contributed to generate a corridor that allowed northward dispersal of tropical (to transitional) communities, enabling them to become established farther north than the normal tropical latitude.

- The TMVB could have functioned as an effective Nearctic/Neotropical–Middle American regional barrier no earlier than the Middle to early Late Miocene, at which time it was essentially complete. At an earlier time, the territory corresponding to the TMVB probably functioned as a broad transitional zone between the northern (which became Nearctic) and southern (which became Neotropical) biotas.

- The Baja California peninsula's biota seems to have had little exchange with mainland northern Mexico for a long time. The Gulf of California and its predecessor, the "Protogulf," as well as the developing Sierra Madre Occidental, probably played a significant role, becoming barriers that effectively forestalled biotic exchange for many lineages (see chapter 11).

Conclusions

There is practically no specific factual information on the Cenozoic biotic development of northern Mexico. However, some of this information gap can be filled using knowledge of the region's geologic/geographic history, which allows us to trace the major features of such development. The geologic/geographic history also introduces constraints on models aimed at describing the development of this vast territory's biota.

Northern Mexico's present-day geographic position largely determines the amount of solar energy it receives and the way this energy is distributed throughout the year. Because northern Mexico's position has changed little since at least the Late Cretaceous, spatial patterns for atmospheric factors directly related to solar energy input (e. g., temperature, humidity, rainfall, and wind circulation) could have become increasingly locked in place with the development of this territory's present-day major geomorphic features during the Cenozoic. However, the climate varied a great deal during this era (contrasting significantly with that of the Cretaceous), responding to worldwide changes and trends, which themselves were fine-tuned by evolving particular features of the landscape.

The broad landscape and climatic uniformity of northern Mexico during the Early Paleogene became episodically complicated through the Tertiary. By Middle–Late Eocene time, extensive tectonic activity generated the Sierra Madre Oriental, the fold-ranges and high plateaus of Chihuahua and Coahuila, the gulfward regression of the epicontinental sea that covered much of northern Mexico, and the development of the Coastal Plain of the Gulf of Mexico. By Early Oligocene time, extensive magmatic activity in the west generated the Sierra Madre Occidental. Such activity is related to extensional deformation ultimately responsible for the detachment of the Baja California peninsula and for producing the basin and range structural pattern, so characteristic of southwestern North America. These last 2 processes occurred during the Late Neogene.

By Middle Miocene time, extensive magmatic activity across central Mexico generated the Trans-Mexican Volcanic Belt, thus completing the major geographic features included in or bounding northern Mexico's landscape. The species that lived in northern Mexico during the Late Paleogene–Early Neogene adapted to varying local/regional conditions, and in due time became differentiated into several regional biotas. The whole spectrum of environmental factors must have played a role in this differentiation process, but to trace it precisely requires information that is not available at present. Nevertheless, it remains clear that in broad terms, the current biotic differentiation of northern Mexico could not be older than Middle Miocene time (ca. 15 ma).

The Pleistocene glaciations that affected the Northern Hemisphere must have exerted an important effect on climate and on biotic distribution in northern Mexico; however, a detailed, region-wide study of the influence of glaciations remains to be conducted.

Acknowledgments

I. F.-V. thanks Gerardo Ceballos Gonzalez, Instituto de Ecología, University of Mexico, for his kind invitation to participate in this important project. I. F.-V. also expresses his gratitude to Zoltan De Cserna (Instituto de Geología, University of Mexico), Jose Luis Macías (Instituto de Geofísica, University of Mexico), and Richard Felger (director, Drylands Institute) for their careful review of the manuscript; their comments and questions helped much to improve this chapter. Discussions with colleagues both from the University of Mexico and from other institutions provided important information or ideas, among them Shelton P. Applegate and Oscar Carranza-Castañeda (paleontologic issues) and Jaime Roldan-Quintana and Jose Luis Macías (geologic issues). The maps were drafted by Jose Ruiz, and the manuscript prepared by Karla and Ismael Ferrusquía Jr.; we thank them for their kind help. The Cooperativa La Cruz Azul S.C.L., México, D.F. graciously covered the additional cost of producing figure 1.8 in color.

Appendix 1.1: Proposed Holistic Geobiologic Integration on Northern Mexico's Biotic Development

The approximate timing and duration of the major Cenozoic geologic/geographic factors that shaped this vast territory are shown on the following page and correlated with key climatic, paleontologic, and biotic events and features; their relationships are briefly worked out.

In the heading, the chronostratigraphic terms are separated by a slash from the corresponding geochronologic term. The geologic time (AB. or Geol.) is expressed in megaanni (ma, millions of years). The chronostratigraphic/geochronologic framework is adapted from Remane et al 2001. A continuous, numbered vertical line in the chart's body represents the known recorded duration or time span of a given process, event, or feature; a discontinuous line represents its inferred or probable duration. The small inset square in the lower right corner of the chart is developed in the large inset in the lower left corner.

General Abbreviations

BCEF, biotic and climatic processes, events, or features; Cretac., Cretaceous; E, Early; ER, era; ET, erathema; G, regional advance/retreat of the continental icecap (glacier); H, humidity; HO, Holocene (= Recent); I, inferior; lower; L, Late. LOREF, Laramide Orogeny related processes, events, and features. M, Middle. NOREF, Middle and Late Cenozoic processes, events, and features generated by extensional deformation and magmatism, ulti-

mately related to plate tectonic dynamics of southwestern North America (northwest Mexico included) and the adjacent Pacific crust (= the Nevada Orogeny). OGEF, other geologic events and features; Oligoc., Oligocene; Paleoc., Paleocene; Pleist., Pleistocene; Pli., Pliocene; S, mean sea level; Syst., system; T, temperature; U, upper.

Number Symbology and Remarks

1. The Geographic position of the Mexican territory has remained more or less constant since the Late Cretaceous.

LOREF: 2. Major folding, faulting, magmatic activity, and regional uplift (essence of the Laramide Orogeny). **3.** Genesis and development of the Sierra Madre Oriental (SMOR). Recorded existence of this cordillera as a major physiographic feature of northern Mexico. **4.** Genesis and development of the fold-ranges and high plateaus of the Chihuahua-Coahuila Morphotectonic Province. **5.** Genesis of structural lowlands (grabens) and intermontane basins in 2 and 3. The co-occurrence of events 2, 3, and 4 promoted the extensive development of rugged-relief territory in northern Mexico, increasing significantly its physiographic diversity. **6.** Regional marine regression. **7.** Development of the Coastal Plain of the Gulf of Mexico.

NOREF: 8. Magmatic activity (chiefly volcanic) generated the Sierra Madre Occidental (SMOC). **9.**

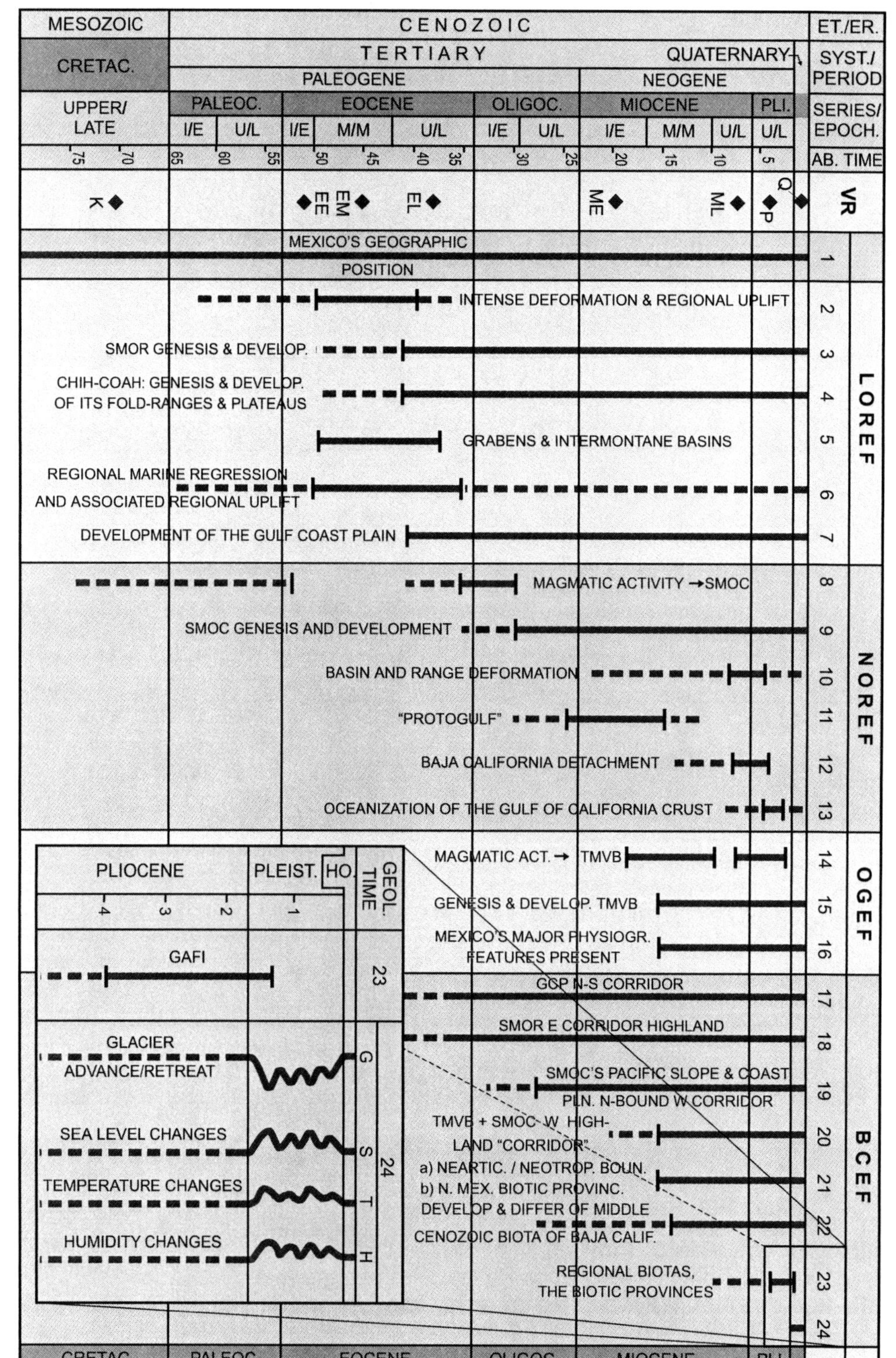

MESOZOIC
CENOZOIC
ET./ER.
CRETAC.
TERTIARY
QUATERNARY
SYST./ PERIOD
PALEOGENE
NEOGENE
UPPER/ LATE
PALEOC.
EOCENE
OLIGOC.
MIOCENE
PLI.
SERIES/ EPOCH.
I/E
U/L
I/E
M/M
U/L
I/E
U/L
I/E
M/M
U/L
U/L
75
70
65
60
55
50
45
40
35
30
25
20
15
10
5
AB. TIME
K
EE
EM
EL
ME
ML
Q
P
VR
MEXICO'S GEOGRAPHIC
POSITION
1
INTENSE DEFORMATION & REGIONAL UPLIFT
2
SMOR GENESIS & DEVELOP.
3
CHIH-COAH: GENESIS & DEVELOP.
OF ITS FOLD-RANGES & PLATEAUS
4
GRABENS & INTERMONTANE BASINS
5
REGIONAL MARINE REGRESSION
AND ASSOCIATED REGIONAL UPLIFT
6
DEVELOPMENT OF THE GULF COAST PLAIN
7
LOREF
MAGMATIC ACTIVITY → SMOC
8
SMOC GENESIS AND DEVELOPMENT
9
BASIN AND RANGE DEFORMATION
10
"PROTOGULF"
11
BAJA CALIFORNIA DETACHMENT
12
OCEANIZATION OF THE GULF OF CALIFORNIA CRUST
13
NOREF
MAGMATIC ACT. → TMVB
14
GENESIS & DEVELOP. TMVB
15
MEXICO'S MAJOR PHYSIOGR. FEATURES PRESENT
16
OGEF
PLIOCENE
PLEIST.
HO.
GEOL. TIME
GCP N-S CORRIDOR
17
4
3
2
1
SMOR E CORRIDOR HIGHLAND
18
GAFI
23
SMOC'S PACIFIC SLOPE & COAST
19
GLACIER ADVANCE/RETREAT
G
PLN. N-BOUND W CORRIDOR
TMVB + SMOC W HIGH-
LAND "CORRIDOR"
20
SEA LEVEL CHANGES
S
24
a) NEARTIC. / NEOTROP. BOUN.
b) N. MEX. BIOTIC PROVINC.
21
TEMPERATURE CHANGES
T
DEVELOP & DIFFER OF MIDDLE
CENOZOIC BIOTA OF BAJA CALIF.
22
HUMIDITY CHANGES
H
REGIONAL BIOTAS,
THE BIOTIC PROVINCES
23
24
BCEF
CRETAC.
PALEOC.
EOCENE
OLIGOC.
MIOCENE
PLI.

Genesis and development of SMOC. Recorded existence of this cordillera as a major physiographic feature of northern Mexico. 10. Genesis and development of the Basin and Range structural deformation best recorded in the Northwestern Plains and Sierras Morphotectonic Province. 11. Presence of the Protogulf, a narrow lowland transgressed by a shallow sea. 12. Baja California detachment from mainland Mexico. 13. Oceanization of the Gulf of California crust (completed only in the southern half).

OGEF: 14. Magmatic and tectonic activity across central Mexico generated the Trans-Mexican Volcanic Belt (TMVB). **15.** Genesis and development of the TMVB. Recorded existence of this cordillera as a major physiographic feature of Mexico. **16.** All the major physiographic features included in or bounding northern Mexico are now present. The regional landscape acquires at last a modern look, similar to that seen today.

BCEF: 17. The Coastal Plain of the Gulf of Mexico functioned as a north–south trending corridor for temperate and tropical biotic elements. **18.** The SMOR functioned as a highland corridor that allowed southward dispersal of acrobiont communities. The Zongolica Range enhanced this effect. **19.** The SMOC's Pacific slope and associated coastal plain functioned as a western corridor that fostered northward dispersal of tropical biotic elements. **20.** The TMVB's western part and the adjacent SMOC functioned as a western highland corridor that allowed southward dispersal of acrobiont communities.

21. Establishment of: (a) The Nearctic Realm/Neotropical Realm boundary; (b) the biotic provincialization of continental northern Mexico. (a) Before the presence of the TMVB, the temperate and tropical communities that inhabited Mexico had a broad zone of contact and interaction. The geologically rapid development of the TMVB put an end to this broad zone of contact. (b) The different morphotectonic provinces of continental northern Mexico, which resulted from this region's complex geologic history, were all in existence at least by Middle Miocene time. The biota that inhabited northern Mexico during the Cenozoic must have followed suit, becoming differentiated into regional biotas best adapted to local conditions. This process most likely was completed also by Middle Miocene time, producing distinctive regional biotas. The differentiation of regional biotas occurred in geobiotic provinces, each of them characterized by a distinct paleoenvironment (fig. 1.9; see also Ferrusquía-Villafranca 1990b).

22. Development and differentiation of the Middle Cenozoic biota of the Baja California peninsula into regional or zonal biotas distinctive enough to be recognized as different biotic provinces. The striking differences between the biota of Baja California and that of continental northern Mexico duly attest the prolonged near-isolation of the former. **23.** The Great American Biotic Interchange (GAFI) was made possible by the Central American land connection. GAFI may have started earlier than 4 ma, and it had a greater impact in southern than in northern Mexico.

24. The Quaternary Glaciation events. Four major continental ice cap advance/retreat episodes occurred in the Pleistocene (curve 24G). They had a profound effect on sea level, making it fall (curve 24S) or rise. The climate was deeply affected, so that for places at a given latitude, the weather became colder and drier during glacier advances (curves 24 T and H), and warmer and more humid during the retreats. Climatic zones were displaced, contracted, or expanded. Biogeographic distributions of floras, faunas, and/or biotic elements were severely affected.

Notes on the Terrestrial Vertebrate Record

The presentation that follows is arranged from older to younger (see figure 1.7 for the geographic position of the localities). The Late Cretaceous Assemblage is based on records from sites in Baja California, Sonora, Chihuahua, and Coahuila. Dinosaurs occur in all sites, mammals only in Baja California Norte. The Early Eocene (Wasatchian) fauna is from Punta Prieta, Baja California Norte. It mainly consists of archaic mammals. The Middle Eocene (Bridgerian) fauna is from Marfil, Guanajuato. It includes rodents related to the ancestral stock of the South American Caviomorpha. The Late Eocene (Chadronian) fauna is from Ojinaga, Chihuahua. It largely consists of rodents and large herbivores; most belong to extinct families worldwide. The Early Miocene (Heminfordian) faunas are from Baja California Norte, Sonora, and Aguascalientes. They chiefly include herbivores that belong to extant modern families. The Late Miocene (Hemphillian) faunas are from Chihuahua, Guanajuato, and Hidalgo. They are diverse and include both herbivores and carnivores that belong to modern families, now largely extinct in North America (including northern Mexico). The Late Pliocene

(Blancan) faunas are from Baja California Sur, Chihuahua, and Guanajuato. Some include early records of South American immigrants (xenarthrans and caviomorph rodents). The Middle and Late Pleistocene (Irvingtonian and Rancholabrean) faunas are from many sites in northern Mexico. Large assemblages are known from Nuevo León, San Luis Potosí, Aguascalientes, and Guanajuato. The record is biased toward large, now extinct mammals such as proboscidians, artiodactyls, and perissodactyls. Some related taxa remain extant elsewhere (Asia, Africa, and South America).

Literature Cited

Aranda-Gómez, J. J., M. M. Godchaux, G. J. Aguirre-Díaz, B. Bonnichsen, and J. Martínez-Reyes. 2003. Three superimposed volcanic arcs in the southern Cordillera: A record of tectono-magmatic activity from the Early Cretaceous to the Miocene, Guanajuato, Mexico. Pp. 123–168 *in* Geologic Transects across Cordilleran Mexico, Guidebook of the Fieldtrips for the 99th Annual Meeting of the Geological Society of America, Cordilleran Section, Puerto Vallarta, Jalisco, Mexico. Publicación Especial 1. Instituto de Geología, Universidad Nacional Autónoma de México, México D.F.

Arroyo-Cabrales, J. 1994. Taphonomy and paleoecology of the San Josecito Cave, Nuevo León, Mexico. Ph.D. dissertation, Texas Tech University, Lubbock.

Arroyo-Cabrales, J., O. J. Polaco, and E. Johnson. 2002. La mastofauna del Cuaternario Tardío en México. Pp. 103–123 *in* M. Montellano-Ballesteros and J. Arroyo-Cabrales (coords.), Avances en los Estudios Paleomastozoológicos en México. Seria Arqueología del Instituto Nacional Antropología e Historia (INAH), México D.F.

Axelrod, D. I. 1979. Age and origin of Sonoran Desert vegetation. California Academy of Sciences Occasional Paper 132.

Barboza-Gudiño, J. R., M. Tristán-González, and J. R. Torres-Hernández. 1999. Tectonic setting of pre-Oxfordian units from central and northeastern Mexico: a review. Geological Society of America, Special Paper 340: 197–210.

Barker, R. W., and W.H. Blow. 1975. Biostratigraphy of some Tertiary formations in the Tampico-Misantla Embayment, Mexico. Journal of Foraminiferal Research 6: 39–58.

Barnes, L. G. 2002. Evolutionary history of the fossil marine mammals of Mexico. Pp. 125–225 *in* M. Montellano-Ballesteros and J. Arroyo-Cabrales (coords.), Avances en los Estudios Paleomastozoológicos en México. Seria Arqueología del Instituto Nacional Antropología e Historia (INAH), México D.F.

Beal, C. H. 1948. Reconnaissance of the geology and oil possibilities of Baja California, Mexico. Geological Society of America Memoir 31.

Black, C. C., and J. J. Stephens, III. 1973. Rodents from the Paleogene of Guanajuato, Mexico. Texas Tech University Museum, Occasional Paper 14.

Bortolini, C., P. E. Damon, M. Shafiqullah, and M. M. Morales. 1995. Geochronologic contributions to the Tertiary sedimentary-volcanic sequences ("Baucarit Formation") in Sonora, Mexico. Geofísica Internacional 34: 67–77.

Bridges, L. W. 1965. Geología del área de Plomosas, Chihuahua. Universidad Nacional Autónoma de México, Instituto de Geología Boletín 74.

Bridges, L. W., and R. K. DeFord. 1965. Precarboniferous Paleozoic rocks in Central Chihuahua, Mexico. American Association of Petroleum Geologists, Bulletin 45: 98–104.

Brown, M. L., and R. Dyer. 1987. Mesozoic geology of northwestern Chihuahua, Mexico. *In* Mesozoic Rocks of Southern Arizona and Adjacent Areas. Arizona Geological Society Digest 18: 381–394.

Campbell, K. C., M. Frailey, L. Romero-Pittman, S. B. Ingement, and D. R. Prothero. 2000. Late Miocene dynamics of the Great American Faunal Interchange: waifs are out. Journal of Vertebrate Paleontology 20 (suppl. 3): 33A.

Carranza-Castañeda, O., and I. Ferrusquía-Villafranca. 1978. Nuevas investigaciones sobre la fauna del Rancho El Ocote, Plioceno medio de Guanajuato, México. Informe Preliminar. Universidad Nacional Autónoma de México Revista 3: 29–38.

Carranza-Castañeda, O., and I. Ferrusquía-Villafranca. 1979. El género *Neohipparion* (Mammalia-Perissodactyla) de la fauna local Rancho El Ocote, Plioceno Medio, de Guanajuato, México. Universidad Nacional Autónoma de México, Instituto de Geología Revista 3: 29–38.

Carranza-Castañeda, O., I. Ferrusquía-Villafranca, and W. E. Miller. 1981. Roedores caviomorfos pliocenicos de la región Central de México. II Congreso Latinoamericano de Paleontología, Porto Alegre, Brasil Annals II: 721–729.

Carranza-Castañeda, O., and W. E. Miller. 2002. Inmigrantes sudamericanos en las faunas del

Terciario Tardío del Centro de México. Pp. 69–81 *in* M. Montellano-Ballesteros and J. Arroyo-Cabrales (coords.), Avances en los Estudios Paleomastozoológicos en México. Seria Arqueología del Instituto Nacional Antropología e Historia (INAH), México D.F.

Carranza-Castañeda, O., M. S. Petersen, and W. E. Miller. 1994. Preliminary investigation of the geology of the northern San Miguel Allende area, northeastern Guanajuato, Mexico. Brigham Young University Geology Studies 40: 1–9.

Carrillo-Bravo, J. 1961. Geología del Anticlinorio Huizachal-Peregrina al Noroeste de Ciudad Victoria, Tamaulipas, Asociación Mexicana de Geólogos Petroleros Boletín 13: 1–198.

Cerca-Martínez, M., G. J. Aguirre-Díaz, and M. López-Martínez. 2000. The geologic evolution of southern Sierra de Guanajuato, Mexico: a documented example of the transition from the Sierra Madre Occidental to the Mexican Volcanic Belt. International Geology Review 12: 131–151.

Charleston, S. 1981. A summary of the structural geology and tectonics of the State of Coahuila, Mexico. Pp. 28–36 *in* S. Charleston, C. I. Smith, and J. B. Brown (eds.), Lower Cretaceous Stratigraphy and Structure, Northern Mexico. Fieldtrip Guide. West Texas Geological Society Publication 81–74.

Clark, J., M. Montellano, J. Hopson, R. Hernández, and D. Fastovsky. 1994. An Early or Middle Jurassic tetrapod assemblage from the La Boca Formation, northeastern Mexico. Pp. 295–302 *in* N. Fraser and H. D. Sues (eds.), In the Shadow of the Dinosaurs: Early Mesozoic Tetrapods. Cambridge University Press, New York.

Clark, J., M. Montellano, J. Hopson, R. Hernández, and V. H. Reynoso. 1998. The Jurassic vertebrates of Huizachal Canyon, Tamaulipas. Pp. 1–17 *in* Avances en Investigación-Paleontológica de Vertebrados. Instituto de Investigaciones en Ciencias de la Tierra, Publicación Especial 1. Universidad Autónoma del Estado de Hidalgo Pachuca, Hidalgo, Mexico.

Clemens, W. A. 1994. *Gallolestes pachymandibularis* (Theria, incertae sedis; Mammalia) from Late Cretaceous deposits in Baja California Norte, Mexico. PaleoBios 33: 1–10.

Coleman, J. M., H. H. Robert, and W. R. Bryant. 1991. Late Quaternary Sedimentation. Pp. 325–352 *in* A. Salvador (ed.), The Geology of North America, vol. J, The Gulf of Mexico Basin. Geological Society of America, Boulder, Colorado.

Cook, T. D., and A. W. Bally (eds.). 1975. Stratigraphic Atlas of North and Central America. Princeton University Press, Princeton, New Jersey.

Cruz-Marín, A., and L. G. Barnes. 1996. A new agorophiid from San Juan de la Costa, Baja California Sur; The oldest fossil Odontocete from Mexico. Sixth North American Paleontological Convention, Abstracts of Papers. Paleontological Society Special Publication 8: 91.

Dalquest, W. W., and O. Mooser. 1974. Miocene vertebrates of Aguascalientes, Central Mexico. University of Texas-Austin, Texas Memorial Museum Pearce-Sellards Series 21.

Dalquest, W. W., and O. Mooser 1980. Late Hemphillian mammals of the Ocote local fauna, Guanajuato, Mexico. University of Texas-Austin, Texas Memorial Museum Pearce-Sellards Series 32.

De Cserna, Z. 1956. Tectónica de la Sierra Madre Oriental de México entre Torreón y Monterrey. Twentieth International Geological Congress, México D.F.

De Cserna, Z. 1960. Orogenesis in time and space in Mexico. Geologische Rundschau 50: 595–605.

De Cserna, Z. 1989. An outline of the Geology of Mexico. Pp. 233–264 *in* A. W. Bally and A. R. Palmer (eds.), The Geology of North America, vol. A, The Geology of North America, an Overview. Geological Society of America, Boulder, Colorado.

De Cserna, Z. 1990. Tectónica, Map IV.2.1; scale 1: 4,000,000. Atlas Nacional de México. Instituto de Geografía, Universidad Autónoma de México, México D.F.

Demant, A., 1984. Interpretación geodinámica del vulcanismo del Eje Neovolcánico Transmexicano. Universidad Nacional Autónoma de México, Instituto de Geología Revista 5: 217–222.

Durham, J. W. 1950. Megascopic paleontology and marine stratigraphy, Part 2 of the "E. W. Scripps" cruise to the Gulf of California. Geological Society of America Memoir 43.

Echegoyen, S. J., S. M. Romero, and S. Velásquez S. 1970. Geología y yacimientos minerales de la parte central del Distrito Minero de Guanajuato. Consejo de Recursos Naturales No Renovables Boletín 75: 1–36.

Edwards, J. D. 1955. Studies of some Early Tertiary red conglomerates of Central Mexico, U. S. Geological Service Professional Paper 264: 153–185.

Eguiluz, A. S. 1984. Tectónica cenozoica del norte de Mexico. Asociasión Mexicana de Geólogos Petroleros Boletín 36: 43–62.

Espinasa-Pereña, R. 1990. Geomorfología I, Map IV.3.4; scale 1: 4,000,000. Atlas Nacional de

México. Instituto de Geografía, Universidad Autónoma de México, México D.F.

Fastovsky, D. E., J. M. Clark, N. Strater, M Montellano, R. Hernández, and J. A. Hopson. 1995. Depositional environments of a Middle Jurassic terrestrial vertebrate assemblage, Huizachal Canyon, Mexico. Journal of Vertebrate Paleontology 15: 561–575.

Ferrari, L. 1995. Miocene shearing along the northern boundary of the Jalisco block and the opening of the southern Gulf of California. Geology 23: 751–754.

Ferrari, L., H. V. Garduño, G. Pasquare, and A. Tibaldi. 1995. Volcanic and tectonic evolution of Central Mexico: Oligocene to present. Geofísica Internacional 33: 91–105.

Ferrari, L., M. López-Martínez, G. J. Aguirre-Díaz, and G. Carrasco-Núñez. 1999. Space-time patterns of Cenozoic arc volcanism in Central Mexico, from the Sierra Madre Occidental to the Mexican Volcanic Belt. Geology 27: 303–306.

Ferrusquía-Villafranca, I. 1969. Rancho Gaitán local fauna, Early Chadronian, northeastern Chihuahua. Boletín de la Sociedad Geológica Mexicana 30: 99–138.

Ferrusquía-Villafranca, I. 1978. Distribution of Cenozoic vertebrate faunas in Middle America and problems in migration between North and South America. Pp. 193–329 *in* I. Ferrusquía-Villafranca (ed.), Conexiones Terrestres entre Norte y Sudamérica. Universidad Nacional Autónoma de México Boletín 101.

Ferrusquía-Villafranca, I. 1987. Reubicación geocronológica del Conglomerado Guanajuato basado en nuevos mamíferos. Pp. 21–22 *in* Simposio sobre la Geología de la Región de la Sierra de Guanajuato. Programa Resumenes y Guía de Excursiones, Instituto de Geología, Universidad Nacional Autónoma de México, México D.F.

Ferrusquía-Villafranca, I. 1989. A new rodent genus from Central Mexico and its bearing on the origin of the Caviomorpha. Pp. 91–117 *in* C. C. Black and M. R. Dawson (eds.), Papers on Fossil Rodents in Honor of Albert Elmer Wood. Natural History Museum of Los Angeles County, Science Series 33: 91–117.

Ferrusquía-Villafranca, I. 1990a. Biostratigraphy of the Mexican Continental Miocene. Universidad Nacional Autónoma de México, Paleontología Mexicana 56: 1–149.

Ferrusquía-Villafranca, I. 1990b. Regionalización Biogeográfica, Map IV.4.10; scale 1: 4,000,000. Atlas Nacional de México. Instituto de Geografía, Universidad Nacional Autónoma de México, México D.F.

Ferrusquía-Villafranca, I. 1993. Geology of Mexico: A Synopsis. Pp. 3–107 *in* T. P. Ramamoorthy, R. Bye, A. Lot, and J. Fa (eds.), Biological Diversity of Mexico, Origins and Distribution. Oxford University Press, New York.

Ferrusquía-Villafranca, I. 2003. Mexico´s Middle Miocene mammalian assemblages: an overview. American Museum of Natural History Bulletin 245: 321–347.

Ferrusquía-Villafranca, I., C. Galindo-Hernández, and H. Barrios-Rivera. 1997. Los mamíferos oligocénicos de México: revisión y adición a la fauna local Rancho Gaitán, Formación Prietos, Chadroniano de Chihuahua Nororiental. Pp. 97–134 *in* J. Arroyo-Cabrales and O. J. Polaco (coords.), Homenaje al Profesor Ticul Alvarez. Colecciones Cientificas, Instituto Nacional de Antropología e Historia, México D.F.

Ferrusquía-Villafranca, I., E. Jiménez-Hidalgo, J. A. Ortiz-Mendieta, and V. M. Bravo-Cuevas. 2002. El registro paleogénico de mamíferos de México y su significación geológico-paleontológica. Pp. 25–45 *in* M. Montellano-Ballesteros and J. Arroyo-Cabrales (coords.), Avances en los Estudios Paleomastozoológicos en México. Seria Arqueología del Instituto Nacional Antropología e Historia (INAH), México D.F.

Ferrusquía-Villafranca, I., and A. E. Wood. 1969. New fossil rodents from the Early Oligocene Rancho Gaitán local fauna, northeastern Chihuahua, Mexico. University of Texas-Austin, Texas Memorial Museum Pearce-Sellards Series 16.

Fletcher, J. M., J. A. Pérez-Venzor, G. González-Barba, and J. J. Aranda-Gómez. 2003. Ridge-trench interactions and the ongoing capture of the Baja California microplate: New insights from the southern Gulf Extensional Province. Pp. 13–31 *in* Geologic Transects across Cordilleran Mexico, Guidebook of the Fieldtrips for the 99th Annual Meeting of the Geological Society of America, Cordilleran Section, Puerto Vallarta, Jalisco, Mexico. Publicación Especial 1. Instituto de Geología, Universidad Nacional Autónoma de México, México D.F.

Fries, C., Jr., W. C. Hibbard, and D. H. Dunkle. 1955. Early Cenozoic vertebrates in the red conglomerate at Guanajuato, Mexico. Smithsonian Miscellaneous Collections 123.

Frizzell, V. A., Jr., ed. 1984. Geology of the Baja California Peninsula. Society of Economic Paleontologists and Mineralogists, Pacific Section, Los Angeles.

Galloway, W. E., and W. H. Blow. 1989. Genetic

stratigraphic sequence in basin analysis II. Application to North-west Gulf of Mexico Cenozoic Basin. American Association of Petroleum Geologists Bulletin 73: 143–154.

Galloway, W. E., D. G. Bebout, W. L. Fisher, J. B. Dunlap, R. Cabrera-Castro, J. E. Lugo-Rivera, and T. M. Scott. 1991. Cenozoic. Pp. 325–352 *in* A. Salvador (ed.), The Geology of North America, vol. J, The Gulf of Mexico Basin. Geological Society of America, Boulder, Colorado.

García, E. 1988. Modificaciones al Sistema de Clasificación Climática de Copen (Para Adaptarlo a las Condiciones de la República Mexicana). Instituto de Geografía, Universidad Nacional Autónoma de México, México D.F.

García, E. 1990. Climas, Map. IV.4.10, scale 1:4,000,000. Atlas Nacional de México, vol. II. Instituto de Geografía, Universidad Nacional Autónoma de México, México D.F.

García-Palomo, A., J. L. Macías, G. Tolson, G. Valdez, and J. C. Mora. 2002. Volcanic stratigraphy and geological evolution of the Apan region, east-central sector of the Trans-Mexican Volcanic Belt. Geofísica Internacional 41: 133–150.

Gastil, G., G. Morgan, and D. Krummenacher. 1981. The tectonic history of Peninsular California. Pp. 285–305 *in* W. G. Ernst (ed.), The Geotectonic Development of California, vol. 1. Prentice-Hall, Englewood Ciffs, New Jersey.

Gastil, G., R. P. Phillips, and E. C. Allison. 1975. Reconnaissance geology of the State of Baja California Norte. Geological Society of America Memoir 140.

Graham, A. 1978. Distribution and migration of Cenozic floras in Mesoamérica. Pp. 166–192 *in* I. Ferrusquía-Villafranca (ed.), Conexiones Terrestres entre Norte y Sudamerica. Universidad Nacional Autónoma de México, Instituto de Geología Boletín 101: 166–192.

Graham, A. 1993. Historical factors and biological diversity. Pp. 109–127 *in* T. P. Ramamoorthy, R. Bye, A. Lot, and J. Fa (eds.), Biological Diversity of Mexico, Origins and Distribution. Oxford University Press, New York.

Haenggi, W. T. 2001. Tectonic history of the Chihuahua Trough, Mexico and adjacent USA, Part I: The pre-Mesozoic setting. Boletín de la Sociedad Geológica Mexicana 54: 28–66.

Haenggi, W. T. 2002. Tectonic history of the Chihuahua Trough, Mexico and adjacent USA, Part II: Mesozoic and Cenozoic. Boletín de la Sociedad Geológica Mexicana 55: 38–94.

Hausback, B. P. 1984. Cenozoic volcanic and tectonic evolution of Baja California Sur, Mexico. Pp. 219–236 *in* V. A. Frizzell (ed.), Geology of the Baja Peninsula. Society of Economic Paleontologists and Mineralogists, Pacific Section, Los Angeles.

Henry, C. D. 1989. Late Cenozoic Basin and Range structure in western Mexico adjacent to the Gulf of California. Geological Society of America Bulletin 101: 1147–1156.

Henry, C. D., and J. J. Aranda-Gómez. 1992. The real Southern Basin and Range mid-to Late Cenozoic extension in Mexico. Geology 20: 701–704.

Henry, C. D., and J. J. Aranda-Gómez. 2000. Plate interactions control Middle-Late Miocene, Proto-Gulf and Basin and Range extension in the southern Basin and Range. Tectonophysics 318: 1–26.

Hernandez, M. E. 1990. Medidas de Aridez, Map IV.4.9D, scale 1: 8,000,000. Atlas Nacional de México. Instituto de Geografía, Universidad Autónoma de México, México D.F.

Honey, J. G., J. A. Harrison, D. R. Prothero, and M. S. Stevens. 1998. Camelidae. Pp. 439–462 *in* C. M. Janis, K. M. Scott, and L. L. Jacobs (eds.), Evolution of Tertiary Mammals of North America, vol. 1: Terrestrial Carnivores, Ungulates and Ungulate-like Mammals. Cambridge University Press, New York.

Imlay, R. W. 1936. Evolution of the Coahuila Peninsula, Mexico. Part. IV, Geology of the Western part of the Sierra de Parras. Geological Society of America Bulletin 47: 1091–1152.

Imlay, R. W. 1938. Studies of the Mexican Geosyncline. Geological Society of America Bulletin 49: 1651–1694.

Imlay, R. W. 1943. Evidence for Upper Jurassic landmass in Eastern Mexico. American Association of Petroleum Geologists Bulletin 27: 524–549.

Imlay, R. W. 1944. Correlations of the Cretaceous formations of the Greater Antilles. Central America and Mexico. Geological Society of America Bulletin 55: 1005–1045.

Kowallis, B. J., C. C. Swisher, O. Carranza-Castañeda, W. E. Miller, and G. D. Tingey. 1998. Preliminary radiometric dates in selected Late Tertiary vertebrate faunas from Mexico. Universidad Autónoma del Estado de Hidalgo Pachuca, Hidalgo, México. Instituto de Investigaciones en Ciencias de la Tierra, Publicación Especial 1, Paleontología de Vetebrados: 103–108.

Labarthe-Hernández, G., M. Tristán-González, and J. J. Aranda-Gómez. 1982. Revisión estratigráfica del Cenozoico de la parte central del Estado de San Luis Potosí.

Universidad Autónoma de San Luis Potosí, Instituto de Geología, Folleto Técnico 85: 1–208.

Lance, J. F. 1950. Paleontología y Estratigrafía del Plioceno de Yepómera, Estado de Chihuahua 1. Equidos, excepto *Neohipparion*. Universidad Nacional Autónoma de México, Instituto de Geología Boletín 54: 1–81.

Ledezma-Vázquez, J., and M. E. Johnson. 2003. Pliocene sedimentary units and tectonic evolution at the Santa Rosalía-Loreto Region, Gulf of California. Pp. 1–12 *in* Geologic Transects across Cordilleran Mexico, Guidebook of the Fieldtrips for the 99th Annual Meeting of the Geological Society of America, Cordilleran Section, Puerto Vallarta, Jalisco, Mexico. Publicación Especial 1. Instituto de Geología, Universidad Nacional Autónoma de México, México D.F.

Lillegraven, J. A. 1972. Preliminary report on Late Cretaceous mammals from the El Gallo Formation, Baja California Norte, Mexico. Contributions in Sciences 232.

Lillegraven, J. A. 1976. A new genus of therian mammals from the Late Cretaceous, El Gallo Formation Baja California, Mexico. Journal of Paleontology 50: 437–444.

Lindsay, H. E. 1984. Late Cenozoic mammals from northwestern, Mexico. Journal of Vertebrate Paleontology 4: 208–215.

Lindsay, H. E., and L. L. Jacobs. 1981. Pliocene small mammals from Chihuahua, Mexico. Instituto de Geología, Universidad Nacional Autónoma de México, Paleontología Mexicana 51.

Lugo-Hupb, J. 1990. Geomorfología 2, Maps IV.3.4A and C, scale 1:8,000,000. Atlas Nacional de México, vol. II. Instituto de Geografía, Universidad Nacional Autónoma de México, México D.F.

Martínez-Reyes, J. 1992. Mapa geológico de la Sierra de Guanajuato con resumen de la geología de la Sierra de Guanajuato. Cartas Geológicas y Mineras. Map scale 1:100,000. Instituto de Geología, Universidad Nacional Autónoma de México, México D.F.

McBride, E. F., A. E. Weidie, J. A. Wolleben, and R. C. Laudon. 1974. Stratigraphy and structure of the Parras and La Popa Basinis, Northeastern Mexico. Geological Society of America Bulletin 84: 1603–1622.

McDonald, H. G. 2002. Fossil Xenarthra of Mexico: a review. Pp. 227–248 *in* M. Montellano-Ballesteros and J. Arroyo-Cabrales (coords.), Avances en los Estudios Paleomastozoológicos en México. Seria Arqueología del Instituto Nacional Antropología e Historia (INAH), México D.F.

McDowell, F. W., and S. E. Clabaugh. 1979. Ignimbrites of the Sierra Madre Occidental and their relation to the tectonic history of western Mexico. Pp. 113–124 *in* C. E. Chapin and W. E. Elston (eds.), Ashflow Tuffs. Geological Society of America Special Paper 180.

McDowell, F. W., and R. L. Mauger. 1994. K-Ar and U-Pb zircon chronology of Late Cretaceous and Tertiary magmatism in central Chihuahua State, Mexico. Geological Society of America Bulletin 106: 118–132.

McDowell, F. W., J. Roldán-Quintana, and R. Amaya-Martínez. 1997. Interrelationship of sedimentary and volcanic deposits associated with Tertiary extension in Sonora, Mexico. Geological Society of America Bulletin 109: 1349–1360.

Miller, W. E. 1980. The Late Pliocene Las Tunas local fauna from southernmost Baja California, Mexico. Journal of Paleontology 54: 762–805.

Miller, W. E., and O. Carranza-Castañeda. 1984. Late Cenozoic Mammals from central Mexico. Journal of Vertebrate Paleontology 4: 216–236.

Miller, W. E., and O. Carranza-Castañeda. 1996. *Agriotherium schneideri* from the Hemphillian of Central Mexico. Journal of Mammalogy 77: 568–577.

Miller, W. E., and O. Carranza-Castañeda. 2002. Importance of Mexico's Late Tertiary mammalian faunas. Pp. 83–102 *in* M. Montellano-Ballesteros and J. Arroyo-Cabrales (coords.), Avances en los Estudios Paleomastozoológicos en México. Seria Arqueología del Instituto Nacional Antropología e Historia (INAH), México D.F.

Molnar, R. E. 1974. A distinctive theropod dinosaur from the Upper Cretaceous of Baja California (Mexico). Journal of Paleontology 48: 1009–1017.

Montellano-Ballesteros, M. 2002. Panorama de los mamíferos mesozoicos en México. Pp. 11–24 *in* M. Montellano-Ballesteros and J. Arroyo-Cabrales (coords.), Avances en los Estudios Paleomastozoológicos en México. Seria Arqueología del Instituto Nacional Antropología e Historia (INAH), México D.F.

Moore, G., C. Marone, I. S. E. Carmichael, and P. Renne. 1994. Basaltic volcanism and extension near the intersection of the Sierra Madre Volcanic Belt. Geological Society of America Bulletin 106: 383–394.

Mooser, O., and W. W. Dalquest. 1975. Pleistocene mammals from Aguascalientes, Central Mexico. Journal of Mammalogy 56: 781–820.

Morán-Zénteno, J. D. 1984. Geología de la República Mexicana. Universidad Nacional

Autónoma de México, Facultad de Ingienería and Instituto Nacional de Estadística, Geografía, y Informática (INEGI), México D.F.

Morris, W. J. 1967. Baja California: Late Cretaceous dinosaurs. Science 155: 1539–1541.

Morris, W. J. 1973. A review of Pacific Coast hadrosaurs. Journal of Paleontology 47: 551–561.

Morris, W. J. 1981. A new species of hadrosaurian dinosaur from the Upper Cretaceous of Baja California *Lambeosaurus laticaudus*. Journal of Paleontology 55: 453–462.

Murray, G. E. 1961. Geology of the Atlantic and Gulf Coastal Province of North America. Harper and Row, New York.

Navarro-G., A., and J. Tovar-R. 1974. Stratigraphy and tectonics of the State of Chihuahua, Mexico. Pp. 87–91 *in* Geologic Fieldtrip Guidebook through the States of Chihuahua and Sinaloa, Mexico. West Texas Geological Society Publication 74–63.

Nieto-Obregón, J., L. Delgado-Argote, and P. E. Damon. 1985. Geochronologic, petrologic and structural data related to large morphologic features between the Sierra Madre Occidental and the Mexican Volcanic Belt. Geofísica Internacional 24: 623–665.

Nieto-Samaniego, A. F., L. Ferrari, S. A. Alaníz-Álvarez, G. Labarthe-Hernández, and R. Rosas-Elguera. 1999. Variation of Cenozoic extension and volcanism across the southern Sierra Madre Occidental volcanic province, Mexico. Geological Society of America Bulletin 111: 347–363.

Novacek, J. M., I. Ferrusquía-Villafranca, J. J. Flynn, R. A. Wyss, and A. M. Norell. 1991. Wasatchian (Early Eocene) Mammals and other vertebrates from Baja California, Mexico: The Lomas Las Tetas de Cabra Fauna. American Museum of Natural History Bulletin 208.

Ortega-Gutiérrez, F., L. M. Mitre-Salazar, J. Roldan-Quintana, J. J. Aranda-Gómez, J. D. Moran-Zenteno, S. A. Alanís-Álvarez, and A. F. Nieto-Samaniego. 1992. Carta Geológica de la República Mexicana. Map scale. 1:2,000,000; 5th ed. Consejo de Recursos Minerales and Instituto de Geografía, Universidad Nacional Autónoma de México, México D.F.

Ortlieb, L. 1991a. Quaternary vertical movements along the coasts of Baja California and Sonora. Pp. 447–480 *in* J. P. Dauphin and B. R. T. Simoneit (eds.), The Gulf and Peninsular Province of the Californias. American Association of Petroleum Geologists Memoir 47.

Ortlieb, L. 1991b. Quaternary shorelines along the northeastern Gulf of California; geochronological data and neotectonic implications.

Pp. 95–120 *in* E. Perez-Segura and C. Jaques-Ayala (eds.), Studies of Sonoran Geology. Geological Society of America, Special Paper 254.

Oskin, M., and A. Martín-Barajas. 2003. Continental edge tectonics of Isla Tiburón, Sonora, Mexico. Pp. 53–70 *in* Geologic Transects across Cordilleran Mexico. Guidebook of the Fieldtrips for the 99th Annual Meeting of the Geological Society of America, Cordilleran Section, Puerto Vallarta, Jalisco, Mexico. Publicación Especial 1. Instituto de Geología, Universidad Nacional Autónoma de México, México D. F.

Padilla-Sanchez, R. J., 1985. Las estructuras de la Curvatura de Monterrey, Estados de Coahuila, Nuevo León, Zacatecas y San Luis Potosí. Universidad Nacional Autónoma de México, Instituto de Geología Revista 6: 1–20.

Pasquare, G., L. Ferrari, V. Perazzoli, M. Tiberi, and F. Turchetti. 1987a. Morphological and structural analysis of the central sector of the Trans-Mexican Volcanic Belt. Geofísica Internacional (Mexico) 26: 177–194.

Pasquare, G., F. Forcella, A. Tibaldi, L. Vezzoli, and A. Zanchi. 1986. Structural behavior of a continental volcanic arc: the Mexican Volcanic Belt. Pp. 509–527 *in* F. C. Wezel (ed.), The Origin of Arcs. Developments in Geotectonics. Elsevier, Amsterdam.

Pasquare, G., L. Vezzoli, and A. Zanchi. 1987b. Morphological and structural model of the Mexican Volcanic Belt. Geofísica Internacional (Mexico) 26: 159–176.

Perez-Villegas, G. 1990. Viento Dominante, Map IV.4.2, scale 1: 4,000,000. Atlas Nacional de México. Instituto de Geografía, Universidad Nacional Autónoma de México, México D.F.

Raisz, E. 1964. Landforms of Mexico. Map scale 1:3,000,000; 2nd ed. U.S. Office of Naval Research, Cambridge, Massachusetts.

Remane, J. (comp.), A. Faure-Muret, G. S. Odin, M. B. Cita, J. Dercourt, P. Bouysse and F. L. Reppetto. 2001. International Stratigraphic Chart. Earth Science Division, UNESCO, Paris.

Reynoso-Rosales, V. H. 1996. Middle Jurassic Sphenodon-like sphenodontian (Diapsida: Lepidosauria) from Huizachal Canyon, Tamaulipas. Journal of Vertebrate Paleontology 16: 210–221.

Rodríguez-de la Rosa, R. A., and S. R. S. Cevallos-Ferriz. 1998. Vertebrates of the El Pelillal locality (Campanian, Cerro del Pueblo Formation), southeastern Coahuila, Mexico. Journal of Vertebrate Paleontology 18: 751–764.

Roldan-Quintana, J. 1984. Evolución tectónica del Estado de Sonora. Universidad Nacional Autónoma de México, Instituto de Geología Revista 5: 178–185.

Roldan-Quintana, J., T. Calmus, F. W. McDowell, R. Amaya-Martínez, and H. Bellon. 2003. Geology of northern Sierra Madre Occidental, eastern Sonora and western Chihuahua, México. Pp. 33–52 *in* Geologic Transects across Cordilleran Mexico. Guidebook of the Fieldtrips for the 99th Annual Meeting of the Geological Society of America, Cordilleran Section, Puerto Vallarta, Jalisco, Mexico. Publicación Especial 1. Instituto de Geología, Universidad Nacional Autónoma de México, México D.F.

Ross, M. I., and C. R. Scotese. 1988. A hierarchical tectonic model of the Gulf of Mexico and Caribbean region. Tectonophysics 155: 139–168.

Sedlock, R. L., F. Ortega-Gutiérrez, and R. C. Speed. 1993. Tectonostratigraphic terranes and tectonic evolution of Mexico. Geological Society of America, Special Paper 278.

Shaw, C. A. 1981. The Middle Pleistocene El Golfo local fauna from northwestern Sonora. Master's thesis, California State University, Long Beach.

Smith, J. T. 1991. Cenozoic marine mollusks and paleogeography of the Gulf of California. Pp. 637–666 *in* J. P. Dauphin and B. R. T. Simoneit (eds.), The Gulf and Peninsular Province of the Californias. American Association of Petroleum Geologists Memoir 47.

Stock, C. 1943. The Cave of San Josecito, Mexico: new discoveries of the vertebrate life of the Ice Age. California Institute of Technology, Pasadena, Contribution 361.

Sutter, M. 1980. Tectonics of the external part of the Sierra Madre Oriental foreland thrust-and-fold belt between Xilitla and the Moctezuma River (Hidalgo and San Luis Potosí States). Universidad Nacional Autónoma de México, Instituto de Geología Revista 4: 19–31.

Sutter, M. 1984. Cordilleran deformation along the eastern edge of the Valles-San Luis Potosí carbonate platform, Sierra Madre Oriental fold-thrust belt, East Central Mexico. Geological Society of America Bulletin 95: 1387–1397.

Sutter, M. 1987. Structural traverse across the Sierra Madre Oriental fold-thrust belt, East Central Mexico. Geological Society of America Bulletin 98: 249–264.

Trewartha, G. T. 1968. An Introduction to Weather and Climate. McGraw-Hill, New York.

Vanderhoof, V. L. 1942. An occurrence of the Tertiary marine mammal *Cornwallius* in Lower California. American Journal of Science 240: 298–301.

Van Devender, T. R. 1990. Packrat middens. Pp. 134–165 *in* J. L. Betancourt, T. R. Van Devender, and P. S. Martin (eds.), Late Quaternary Vegetation and Climate of the Sonoran Desert, United States and Mexico. University of Arizona Press, Tucson.

Vidal-Zepeda, R., 1990a. Temperatura Media, Map IV.4.6, scale 1: 4,000,000. Atlas Nacional de México. Instituto de Geografía, Universidad Nacional Autónoma de México, México D.F.

Vidal-Zepeda, R. 1990b. Precipitación, Map IV.4.10, scale 1: 4,000,000. Atlas Nacional de México. Instituto de Geografía, Universidad Nacional Autónoma de México, México D.F.

Viniegra-Osorio, F. 1965. Geología del Macizo de Teziutlán y la Cuenca Cenozoica de Veracruz. Asociación Mexicana de Geólogos Petroleros Boletín 17: 101–163.

Weber, R., and S. R. S. Cevallos-Ferriz, 1994. Perfil actual y perspectivas de la Paleobotánica en Mexico. Sociedad Botanica de México, Boletín V. 55: 141–148.

Wilson, H. H. 1987. The structural evolution of the Golden Lane, Tampico Embayment. Journal of Petroleum Geology 10: 5–40.

2

Northern Mexico's Landscape,
Part II: The Biotic Setting across Time

ISMAEL FERRUSQUÍA-VILLAFRANCA

LAURA I. GONZÁLEZ-GUZMÁN

An understanding of the terrestrial biodiversity and ecosystems of northern Mexico (i.e., the Baja California peninsula and mainland Mexico north of the Trans-Mexican Volcanic Belt; see fig. 1.1) must begin with an appreciation of past and current physical forces. Natural communities and ecosystems are dynamic over a wide range of temporal and spatial scales, and as demonstrated in chapter 1, they have been modified by geological events that occurred over tens of millions of years. Plate tectonic activity, volcanism, marine transgression, and the associated shifts in climate and soil types all have played a large role in establishing the heterogeneity of current habitats and the distribution of contemporary species and communities.

Although northern Mexico's abiotic environments are spatially diverse, they have been relatively constant during the Cenozoic. The present latitudinal position of northern Mexico has essentially not changed since the Late Cretaceous. This geographic position is responsible for constant solar radiation both annually and over the course of a day. It has also resulted in the dominance of arid climates observed in the region. In contrast, the convoluted topography and wide range of elevations cause drastic changes in climate and soils across space (see chapter 1). The environmental diversity within which communities and ecosystems are embedded is evident in the presence of both temperate and tropical biomes.

In this chapter we explore what is known of northern Mexico's paleoenvironments and biotic history, both of which were shaped under the influence of physical forces. We begin with an in-depth discussion of the model of biotic evolution drafted in chapter 1 and incorporate an attempt to trace the origins of the biotas of present-day northern Mexico.

Temporal Dynamics and Origins of Regional Biotas

The evolution of regional biotas over large geographic areas and geological time is strongly influenced by major shifts in relief, rock framework, climate, and soil that have occurred throughout history. Given that such environmental components are driven by physical factors dependent on geological processes, it follows that the evolution of biotas is intimately interwoven with the geologic history of biotic regions.

Reconstructing changes in regional communities over geologic time poses a dual challenge due to the requisite integration of many different fields (geology, geography, ecology, and evolution) and the current incompleteness of the lithic and fossil records. For northern Mexico, the available information is very limited, particularly that pertaining to the fossil record; therefore, few studies have attempted to reconstruct changes in regional communities over geologic time. Most of these analyses have a wide regional focus on general patterns across the whole country. Furthermore, most of them only address the origins of certain lineages within the biota (Ayala et al. 1993; Fa and Morales 1993; Flores-Villela 1993; Graham 1993; Ramamoorthy and Elliot 1993; Rzedowski 1993; Sousa and Delgado 1993; Valdés and Cabral 1993). By comparison, few studies incorporate geologic information to understand the dynamics of regional communities (Maldonado-Koerdell 1964; Wendt 1993). Few works focus specifically on select areas of northern Mexico (e.g., Axelrod 1979, the Sonoran Desert vegetation; Grismer 1994, the Baja California herpetofauna; and Van Devender 2001, La Frontera [U.S.–Mexico border strip] flora and vegetation). Not a single study, then, deals with northern Mexico as a whole, nor systematically incorporates geologic and paleontologic information to describe its biotic evolution.

We attempt to fill this gap by combining the available geographic, geologic, and paleontologic information to understand the dynamics of biotas in northern Mexico. Our interpretations are largely based on the most current and validated information available for the region. We examine 10 major geologic and climatic events (both regional and global) that shaped the evolutionary history of the biota. Some of these events occurred sequentially, but most overlapped in time. They are geochronologically arranged in Fig. 2.1 (see also chart 1.1). We discuss the major environmental and biotic consequences of each event.

Late Cretaceous

The modern evolutionary history of terrestrial organisms in northern Mexico most likely began with geologic and climatic events that occurred during the Late Cretaceous. By this time, the major lithospheric plate arrangement was approaching that of the present, although the distribution of lands and seas was quite different (Smith et al. 1994).

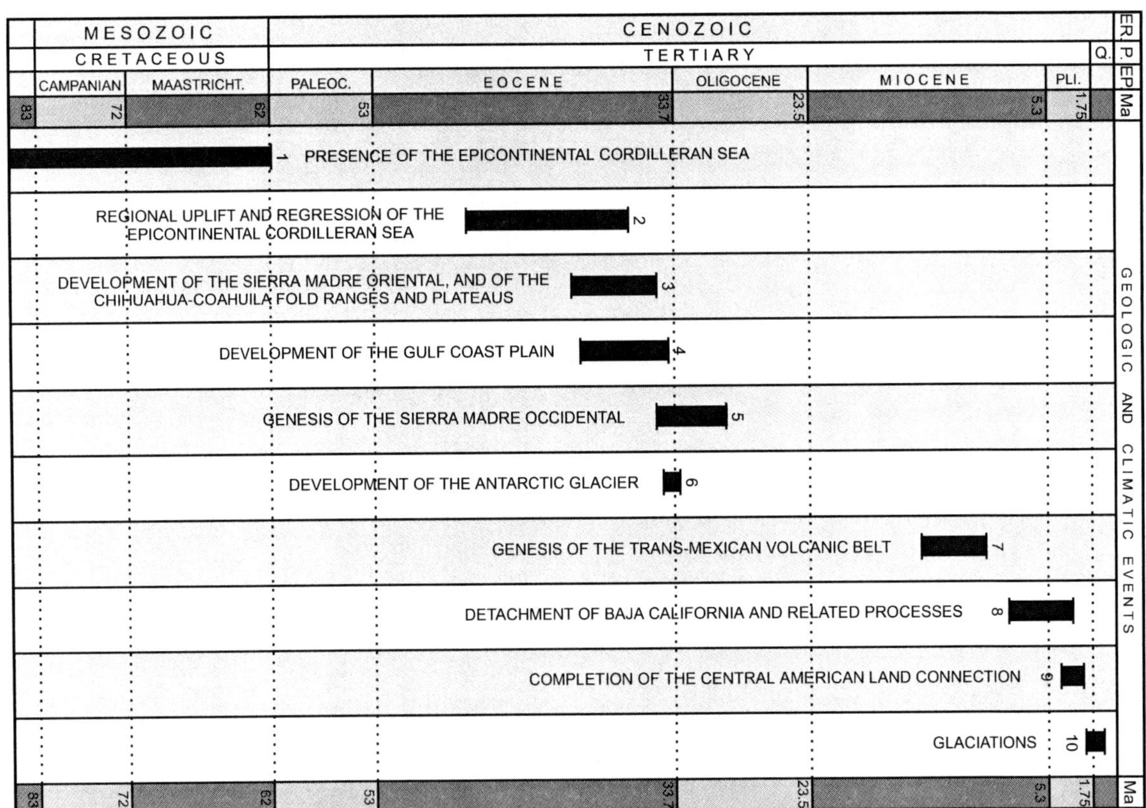

Figure 2.1. Main geologic and climatic events that shaped the biotic history of northern Mexico.

Across the world, the Late Cretaceous landscape consisted of many isolated land masses that had formed as a result of a dramatic rise in sea level. In North America, the Mid-continent Sea or Late Cretaceous Epicontinental Cordilleran Sea divided the region in eastern (connected to "Europe") and western (circumpolarly connected to Asia) zones. Outcrop and subsurface information indicates that the epicontinental sea was either interrupted or very shallow along an east–west belt roughly corresponding to present-day 32°–34° N (i.e., southern Texas-New Mexico-Arizona-California, in the United States, and northeastern Chihuahua-northern Sonora, in Mexico; Cook and Bally 1975). The belt connected the terrestrial regions of northern Mexico to eastern and western North American nonmarine megaregions. The connections facilitated intercontinental biotic exchange, so that the northern Mexican biota included taxa biogeographically related to others in Europe, Asia, or elsewhere, and vice versa. However, the polarity of dispersal for most taxa remains unknown, due to the paucity of the fossil record. In short, northern Mexico's biota could parsimoniously be considered as the southern portion of the biota developed and/or established in North America during the Late Cretaceous.

The available published geologic information does not allow a detailed reconstruction of the paleogeography of northern Mexico (i.e., at a scale 1:250,000 or greater) during the Late Cretaceous. In fact, information is not even adequate to delineate the region's littoral. However, the existence of large land features such as the Coahuila peninsula (see Murray 1961) and of local terrestrial areas, such as those bearing dinosaurs in the states of Baja California, Sonora, Chihuahua, and Coahuila (see chapter 1, fig. 1.7, localities 2–9) is common knowledge. The presence of Late Cretaceous andesitic bodies in the northern section of the Sierra Madre Occidental (McDowell and Clabaugh 1979) and in California, and the implied subduction zone (that of the Kula Plate [in the Pacific] under the North American Plate), has led some geologists to postulate an episode of regional uplift and the emergence of western Mexico (i.e., Baja California, Baja California Sur, Sonora, and Sinaloa) during the Late Cretaceous (Morán-Zenteno 1994). The lack of evidence for this uplift episode makes it hypothetical. If it happened, the evidence either has been eroded away or it lies buried under the Sierra Madre Occidental's Upper Volcanic Complex; in either case it is not available. Under these circumstances, it

would be unjustified to relate this episode as the cause of the presence of Late Cretaceous terrestrial areas in western Mexico.

The climate during the Late Cretaceous was globally warm and moist. In northern Mexico, it is quite possible that the Mid-Continent Sea provided moisture and produced warm and equable climatic conditions in the region. Nonetheless, evidence of these conditions is only found in a few Late Cretaceous records. For example, fossil floras from the early Campanian, currently found in southern California (Holz Shale) but then situated near the present latitude of Vizcaíno Bay, Baja California, suggest riparian forests along water courses and dry tropical forests along well-drained drier slopes. Moreover, fossil records from the Campanian El Rosario Formation (located near the namesake village), Baja California, then placed somewhere between Comondú and Loreto (Baja California Sur), suggest the presence of a tropical savanna (Axelrod 1979). In northern Chihuahua, just south of Big Bend National Park, Texas, a southern extension of the Campanian Aguja Formation (Maxwell and Hazzard 1967; Andrade et al. 2002), bears remains of gymnosperms (Podocarpaceae and Pinaceae), monocots (Arecaceae [Palmae]), other angiosperms (two kinds of pollen grains), and dinosaurs (*Chasmosaurus mariscalensis*, a ceratopsian). This assemblage is suggestive of a transitional subtropical mixed forest/savanna community.

In southern Coahuila, the Campanian, continental clastic Cerro del Pueblo Formation has yielded at El Pellillal monocots (Araceae and Musaceae; Estrada-Ruíz and Cevallos-Ferriz 2002; Pérez-Hernández and Cevallos-Ferriz 2002), dicots (Rhamnaceae, Phytolaccaceae and Hamamelidales; Estrada-Ruíz and Cevallos-Ferriz 2002), dinosaurs, and aquatic reptiles (a soft-shell turtle [Trionychidae], a snapping turtle [*Protochelydra* sp.], and 2 crocodiles [Family Goniopholididae]; Rodríguez de la Rosa and Cevallos-Ferriz 1998]. The biota indicates a tropical to subtropical savanna adjacent to a pond or shallow lake.

Somewhat later, the fossil record discloses the presence of an Early Maastrichtian flora, recovered from the clastic (sandy to muddy), coal-bearing Olmos Formation of northern Coahuila (Weber and Cevallos-Ferriz 1994). The flora includes aquatic ferns (*Dorfiella* and *Salvinia* [extant]), monocots referred to Arecaceae (form-genera *Sabalites* and *Phoenicites*) and to an aquatic Araceae, as well as several dicot taxa, of which those identified are members of the Magnoliaceae and Lauraceae. The Olmos flora records a tropical forest adjacent to a stagnant coastal lagoon, traversed by fluvial currents, set in a rapidly subsiding basin; as water level rose, it eventually asphyxiated the forest. This paleoecological condition is mentioned by Weber (1972).

The Late Cretaceous plant and vertebrate fossil assemblages of North America (northern Mexico included) largely consist of widely distributed taxa and lack significant endemism. The Late Cretaceous floras from the southwestern United States (central California, southern Arizona, southern Nevada, and northern New Mexico) and northern Mexico (Chihuahua and Coahuila) indicate widespread distribution of tropical savannas and of dry tropical and subtropical forests across the region (Axelrod 1979). Likewise, the mammals and dinosaurs from the Campanian El Gallo Formation (located near El Rosario, Baja California) show striking similarities to those of the western United States and the Canadian Foothills (Morris 1972, 1973, 1981; Molnar 1974; Ferrusquía-Villafranca 1978, Helenes and Téllez-Duarte 2002). The same applies to the dinosaurs from the Campanian Cerro del Pueblo Formation in southern Coahuila (Rodríguez de la Rosa and Cevallos-Ferriz 1998).

Paleocene

During the Paleocene, extensive sea regression took place, significantly increasing the land surface toward the eastern section of northern Mexico. The increase in land surface was associated with major changes in marine and atmospheric circulation which caused worldwide climate changes, including an important increase in seasonality (both thermal and pluvial, linked to the latitudinal gradient), and a big decrease in moisture, so that a trend toward aridity became established. These changes must have had a profound influence on the terrestrial biota of the world, which became adapted to these new conditions.

The lack of any terrestrial Paleocene lithic and fossil record for northern Mexico prevents us from knowing the location, size, and configuration of the land and attributes of its biota. The record in the adjacent United States is rather meager; that of Big Bend, Texas (Wilson 1967), discloses the presence of a subtropical community where archaic mammals, such as condylarth and taeniodont mammals thrived.

Eocene and Oligocene

The almost 20-Ma–long Eocene Epoch and the nearly half as long Oligocene Epoch had a complex geologic, climatic, and biotic history. In North America the major geologic event was the genesis and development of the Rocky Mountain Cordillera. In northern Mexico this event corresponded to the genesis and development of the Sierra Madre Oriental, the fold-ranges and high plateaus of Chihuahua-Coahuila, and the development of the Gulf Coast Plain. During the Early Oligocene, the Sierra Madre Occidental, an enormous volcanic field, quickly developed. All these events and resulting major geographic features added much physiographic diversity to the landscape inherited from the Paleocene, deeply influenced the climate (wind circulation, pluvial discharge, and temperature distribution, among other factors), generated barriers, provided corridors, and thus promoted biotic differentiation (regionalization).

The worldwide climate changes involved an Early Eocene warm interval, followed by a stepwise and then abrupt cooling phase at the Middle Eocene end, and a more abrupt one at the Eocene/Oligocene boundary (Berggren and Prothero 1992). The latter is known as the Eocene Terminal Climatic Event (Wolfe 1978), which corresponds to a major land mammal extinction long known as the Grande Coupure, the "big break" (Stehlin 1909). The driving mechanism appears to have been the development of the Antarctic glacier complex (Bartek et al. 1992), which continued operating throughout the Tertiary. The temperature and humidity episodically rose and fell throughout the remaining Tertiary but never reached Early Eocene conditions. In fact, a trend toward aridity took place again, becoming increasingly stronger throughout the Neogene.

In North America, as in the rest of the world, the climatic changes across the Eocene and Oligocene caused major responses in the biota. For example, in the South Dakotan Badlands, where the sedimentary and fossil record is best known, it appears that "Late Eocene forests gave way to latest Eocene dry woodland, then Early Oligocene wooded grassland and finally to Middle Oligocene open grassland" (Berggren and Prothero 1992: 16; see also Retallack 1992; Wolfe 1992). Land mammals followed suit: archaic Eocene forest browsers and tree dwellers were succeeded by Oligocene and younger scrub-grassland dwellers that belonged to extant families.

The paucity of the continental sedimentary and fossil record for northern Mexico allows only glimpses of the biotic responses in the region. The legume cf. *Inga* (Fabaceae) from La Carroza Formation of La Popa Basin in south-central western Nuevo León (Calvillo-Canadell and Cevallos-Ferriz 2002) is the only Eocene angiosperm known in this vast region. The only Early Eocene (Wasatchian) terrestrial vertebrate fauna known is from Punta Prieta, Baja California (Novacek et al. 1991); it consists of taxa well represented in western North America and probably belonged to a subtropical forest or woodland community. The only Middle Eocene land mammal fauna is from Marfil, Guanajuato (Ferrusquía-Villafranca et al. 2002); it consists of archaic mammals, mostly endemic rodents, condylarths, and xenarthranlike forms, which seemingly were subtropical woodland dwellers. The only Late Eocene (Chadronian) fauna known is from Ojinaga, Chihuahua (Ferrusquía-Villafranca et al. 1997); it includes mostly large, browsing herbivores, indicative of a subtropical woodland or forest. Also present were some large snails, characteristic of humid areas (see Evanoff et al. 1992).

Strictly speaking, there are no Oligocene biotas known from northern Mexico. However, drawing on geologic-paleontologic information from the southern United States (see Axelrod 1979; Carroll 1988; Bally and Palmer 1988, for paleobotanical, vertebrate paleontological, and geologic summaries, respectively), and applying it to the known Oligocene physiography of northern Mexico, several inferences can be made. We postulate that the Sierra Madre Occidental and Oriental, as well as the fold-ranges and high plateaus of Chihuahua-Coahuila supported an evergreen conifer forest that might have extended farther south through the eastern highland corridor. Elevationally, such forest was replaced downward by a deciduous forest, and in the moister parts by tropical forest and woodland. The Gulf Coast Plain provided a corridor that allowed tropical communities to reach as far north as present-day Laredo, Texas, in the Late Eocene (see chapter 1; Westgate 1994). In the drier parts and in the internal lowlands (of Chihuahua-Coahuila, and the Central Plateau), a combination of woodland and savanna probably was established, becoming increasingly tropical southward.

It should be noted that in Tepexi de Rodríguez, Puebla, just south of the Trans-Mexican Volcanic Belt (TMVB), the Coatzingo Formation of supposed Oligocene age has yielded a flora indicative of co-existing dry tropical forest and chaparral-like communities that seemingly inhabited an area close to the southern limit of northern Mexico before the development of the TMVB. The flora includes *Eucommia* (a gymnosperm; Weber and Cevallos-Ferriz 1994), *Cedrelospermum* (Ulmaceae; Magallón-Puebla and Cevallos-Ferriz 1994), *Lysiloma*, cf. *Inga*, and cf. *Pithecellobium* (Fabaceae; Weber and Cevallos-Ferriz 1994; Calvillo-Canadell and Cevallos-Ferriz 2002), *Salix* and *Populus* (Salicaceae; Ramírez and Cevallos-Ferriz 2000), and *Pseudosmodingium* (Anacardiaceae; Ramírez et al. 2000). The Oligocene age assignment given to the formation and the flora is not compelling because angiosperms are less precise geochronological indicators than land mammals, and the fossiliferous strata lack independent corroborative dating evidence.

Prior to the TMVB, a broad transition zone probably connected the temperate communities of northern Mexico and the tropical ones of southern Mexico (see chapter 1). Finally, because the communication between mainland Mexico's biota and that of the Baja California peninsula became hampered by the Sierra Madre Occidental, and by the Protogulf, we postulate that the latter biota became progressively isolated. Moreover, we believe that the marine circulation pattern (particularly the Subarctic-California Current) responsible for the peninsular arid weather started to develop by Late Oligocene time and became fully established by Middle Miocene time in mid-latitudes such as that of Baja California (see Barrera et al. 1985; Savin et al. 1985). The Late Oligocene-Early Miocene El Cien Formation, Baja California Sur, has yielded abundant silicified wood referred to *Mimosoxylon* (also known from the Tertiary of Oaxaca) and aff. *Copaifera* (Fabaceae, Weber and Cevallos-Ferriz 1994), and aff. *Tapiria* (an Anacardiaceae genus extant in Oaxaca; Martínez-Cabrera and Cevallos-Ferriz 2002). This xyloflora is suggestive of a scrub community, seemingly already adapted to a semi-arid to arid environment. The phytogeographic distribution of these taxa is concordant with the hypothesis of increasing isolation proposed above.

Miocene and Pliocene

During the Miocene and Pliocene, several geologic and climatic events occurred that influenced biotic development in northern Mexico (and elsewhere). On the mainland, the most important event was the genesis and development of the TMVB by Middle

Miocene time, at last bounding northern Mexico, thus separating it from southern Mexico (largely the Sierra Madre del Sur Morphotectonic Province) and providing a material limit to the temperate and tropical communities inhabiting southern North America. In other words, the Nearctic and Neotropical realms became tangibly separated. This is a biogeographic event of the utmost importance because it set the beginning of the biogeographic and biotic regionalization discernible at present.

In northwestern Mexico several events fostered the current climatic conditions and biogeography observed in the area. On the Baja California peninsula, extensive Middle Miocene volcanic activity generated the Sierra de La Giganta (the southern half of the Peninsular Ranges). During the Pliocene the detachment of Baja California and the development of the Gulf (northwardly enlarging the Protogulf) occurred. These events increased the trend toward aridity established by the Subarctic-California current and drastically reduced moisture in western Sonora. They also further hindered biotic exchange between the peninsula and the mainland, thus intensifying the former's near isolation. The net result was severe regional aridity in Baja California and Sonora, as well as the development of xeric communities, which include a relatively high proportion of endemic taxa, particularly so in Baja California (see Axelrod 1975, 1979; Hall 1981; Grismer 1994; Riddle et al. 2000; Van Devender 2001).

Quaternary

During the Quaternary, 2 major events influenced biotic development in northern Mexico (and elsewhere): the Great American Biotic Interchange (GABI), which was continentwide; and the glaciations, which were of worldwide reach. GABI is the biotic response to the last developmental phase of the Central American Land Connection that eventually joined Central to South America. This interchange largely took place during the Pleistocene, but it had a longer history that goes back to the Pliocene (and probably to the Late Miocene; see Campbell et al. 2000), as the marine barrier that separated Central and South America gradually disappeared.

The Central American Land Connection put an end to the nearly full Cenozoic-long isolation of South America and deeply influenced the whole American continent's biotic makeup, geographic distribution, and evolution (see Patterson and

Pascual 1963, 1972; Ferrusquía-Villafranca 1978; Stehli and Webb 1985). The impact was greater in South America than in North America. In northern Mexico, direct evidence of GABI is largely restricted to the Central Plateau, where Late Pliocene and Pleistocene South American mammals (hydrochoerid rodents and pilosan xenarthrans belonging at least to 3 families; chapter 1) are known. In fact, the second earliest North American record of such mammals (3.9–4.1 Ma), occurs in Guanajuato (Kowallis et al. 1998). Direct evidence from plants and other groups has yet to be produced.

Quaternary glaciations in North America were generated by advances and retreats of the Laurentide Continental Glacier. Advances (glacial periods) produced sea-level drops (sometimes more than 100 m) and decreases in temperature and moisture, so that cold, dry weather conditions prevailed. Sea-level drops caused marine regressions and modest land surface increases, which locally provided land bridges to otherwise isolated areas, such as the Gulf of California islands, facilitating their colonization. Retreats (interglacial periods) produced opposite effects: sea-level rises caused marine transgressions, loss of land, and flooding of land bridges, thus fostering isolation and endemism (see Case and Cody 1983; Case et al. 2002). The climatic changes driven by glaciation (particularly in rainfall) also affected lake levels. Epochs with scarce rainfall ultimately caused severe drops in lake levels, potentially eliminating swamps and shallow lakes; epochs with high rainfall had opposite effects.

The general biotic response to the advances and retreats of the Laurentide Continental Glacier can be summarized as follows: at a given site, communities changed from humid prone-warm prone during interglacial periods to dry prone-cold prone during the glacial periods. In the cordilleras, altitude-related changes also occurred. For instance, the timberline ascended during the interglacial periods and descended during the glacial ones.

Local physiography and latitudinal position altered the trends in vegetation change. For example, the Bolsón de Mapimí in northeastern Durango (around 25°30'–26°00' N), allowed the survival of some desert insects and plants during the Wisconsin Glacial Period; they became extinct elsewhere (Elias 1992). The wealth of endemic species, and to a large extent the distribution of northern Mexico's present-day biota, probably are the net result of the contractions, retractions, and displacements of the geographic ranges of the taxa involved, imposed by the glaciation-driven climatic changes within a physiographically diverse landscape. Changes in the distribution of certain organisms and the climate fluctuations that caused these changes are known only for the few species that left a fossil record. The available information for the Quaternary in northern Mexico chiefly pertains to Late Pleistocene mammals, palynomorphs, and lake sediments of a few scattered sites (e.g., Lindsay 1984; Van Devender et al. 1985; Shaw and McDonald 1987). To better understand this complex biotic evolution, a much greater database is needed.

The Sonoran Desert Region is a notable exception, with much insight gained on its evolutionary history and the influence of climate changes. The tropical affinity of this desert has been examined through the biogeography and phylogenetics of extant species, paleoclimatic reconstructions, the pollen record, and analysis of plant and animal remains in ancient packrat middens from the latest Pleistocene–Early Holocene (e.g., Axelrod 1979; Van Devender 1990; Cornejo 1994). The Sonoran Desert vegetation probably originated by Late Miocene time (Axelrod 1979) as a result of the drying trend mentioned above; it was likely derived from thornscrub, which may have formed on the dry edges of deciduous forests during the Middle Miocene (see Van Devender 2002, for a more detailed account). During most of the Pleistocene, glaciation-driven climate changes pushed the boundaries of the Sonoran Desert southward and down toward lower elevations (Van Devender 1990). Species assemblages seen today appeared during the present interglacial (Holocene), with many of the extant Sonoran Desert plants, especially columnar cacti (Cornejo 1994), having their closest relatives in tropical deciduous forests much farther south along the Pacific slopes of central and southern Mexico.

Sources of Biota

The paucity of northern Mexico's fossil record makes it a highly speculative exercise to attempt tracing the origins of its present-day biota. What we offer here is a parsimonious interpretation of the available pertinent geologic, paleontologic, and neontologic information. We begin by considering the following facts:

- Undisputed angiosperms are known from Aptian (late Early Cretaceous) strata of Europe, Central Asia, and North America (Stewart 1985). In Campanian-Maastrichtian time (Late Cretaceous) they became the dominant component of terrestrial floras throughout the world.

- The Cretaceous world's plate tectonic arrangement resembled that found today, becoming even more similar by the Late Cretaceous (Ross and Scotese 1988). This implies that the major continental plates had approximately the same position as today and that they were largely separated by big oceanic expanses. In contrast, the paleogeography (land, sea, and ocean distribution) was very different (Smith et al. 1994), as epicontinental seas extensively flooded the continents, and the climate tended to be equable, more humid, and less seasonal. Yet, the latitudinal gradient controlled solar energy reception worldwide then, as it does now.

  It should be noted that the presence of Late Cretaceous broadleaf angiosperms and dinosaurs in localities occurring around 50° N (e.g., England and southwestern Canada; see Stewart 1985; Weishampel 1990) has led to the generalized idea that at the time, the whole earth had a tropical-like climate. This is certainly not so. What these paleobiologic climate indicators (as well as some sedimentary and stable isotope evidence) disclose is the greater efficiency during the Late Cretaceous of the mechanism that transfers heat from low latitudes to high latitudes. Such increased efficiency offset, but did not eliminate, the latitudinal gradient effect, causing a wider extension of the tropical-subtropical climate belt than in recent times. Although the mechanism is not well understood, it is supposed to be closely related to the Late Cretaceous land:sea ratio distribution across the earth.

- Given the above conditions and applying the actualistic principle (i.e., past events of the earth's history can be explained in terms of processes still at work), it follows that by Late Cretaceous time, the angiosperms were adapted to diverse local and regional environments, and that each continent had developed its own (indigenous) flora. Each continental flora kept closer affinities with those to which it remained connected biogeographically—for instance, western North America with Asia (through a circumpolar corridor) or eastern North America with "Europe" (due to spatial proximity). The same applies to the Paleogene (particularly the Paleocene-Eocene interval) and is extended to other terrestrial groups, such as invertebrates and vertebrates.

- Late Cretaceous North America was no exception. It was a large continent spreading then from 16° to 70° N, thus having different climatic belts: cold at high latitudes, temperate at intermediate latitudes, and warm at low latitudes, such as the Mexican territory, which was then sort of a peninsula tapering southward. The extensive Cordilleran or Mid-Continent Sea divided North America lengthwise into an eastern mega-region and a western (narrower) one, connected in the south by a narrow stretch of land, located in the present-day southwestern United States–northern Mexico territory (see Cook and Bally 1975). This added geographic diversity and probably promoted some differentiation of the indigenous (in the sense discussed above) Late Cretaceous North American biota.

- Northern Mexico's Late Cretaceous biota is the southern part of the indigenous North American biota. Angiosperm taxa that seem to have inhabited southwestern North America (northern Mexico included) since the Late Cretaceous-Early Paleogene include *Acacia, Atriplex* (and other Chenopodiaceae), *Bursera, Caesalpinia, Ephedra, Frankenia, Menodora, Prosopis, Sophora,* and *Persea* (Axelrod 1979; Rzedowski 1993; Sousa and Delgado 1993; Velasco de León et al. 1998; Burnham and Graham 1999). These taxa typically have very widespread distributions today. Their presence in the Late Cretaceous-Early Paleogene fossil record of this subcontinent makes them the oldest known extant genera for northern Mexico. Their known fossil record does not allow us to objectively identify their exact geographic origin, nor the geologic time of this origin. Whether these genera evolved outside the region at an unspecified pre-Late Cretaceous time, as advocated by some, is of no concern here. They were already established in North America at the time when the differentiation of northern Mexico's modern biotas began, as shown above.

Based on evidence from outside northern Mexico, Axelrod (1979), Rzedowski (1981), and Burnham and Graham (1999) consider that *Bauhinia, Ery-*

thrina, and Cactaceae, to name a few, dispersed from South America sometime during the Tertiary, but before GABI. Related to this event is the arrival of the Bignoniaceae, Malpighiaceae, and Sapindaceae (Burnham and Graham 1999). Finally, origination in situ has produced many endemic taxa in Megamexico (see Rzedowski 1993). Probably among such taxa are *Agave, Fouquieria,* and *Simmondsia* (see Axelrod 1979; Rzedowski 1993; Challenger 1998), but the timing and place of such originations remain unknown, due to the absence of fossil records.

The faunistic component of northern Mexico's biota has already been discussed in the preceding section (see also chapter 1) and is briefly summarized here. Late Cretaceous, Paleogene, and Early Neogene land mammal taxa showed North American affinities, were archaic, and had no direct relationship with extant mammals (see Ferrusquía-Villafranca 1978, 2003; Ferrusquía-Villafranca et al. 2002). In contrast, Late Neogene and Quaternary taxa are truly modern: all belong to extant families, and in most cases also to extant genera (see Carranza-Castañeda and Ferrusquía-Villafranca 1978, 1979; Miller and Carranza-Castañeda 1984; Arroyo-Cabrales et al. 2002). In North America some of these families and genera are now extinct. South American immigrants appeared in the Late Pliocene and Quaternary, but of these the only surviving genera in northern Mexico (and in temperate North America) today are the marsupial *Didelphis* and the edentate *Dasypus.* The Late Neogene mammals of northern Mexico's southern part appear to have been dwellers of a subtropical woodland, associated with an open country subtropical savanna, as the climate was more humid than it is today; their abundant and diverse fossil record (including many species of large animals) indicates that the community to which they belonged had much greater productivity than the one inhabiting this region at present. High productivity requires relatively humid conditions.

Other vertebrate groups are much less known in the fossil record. Amphibians such as Cryptobranchidae, Salamandridae (Urodela) and Ranidae (Anura), as well as reptiles such as Trionychidae (Chelonia), and Iguanidae (Lacertilia) have inhabited southwestern North America (northern Mexico included) at least since the Paleogene (see Fries et al. 1955, for the Middle Eocene iguanid record of Marfil, Guanajuato; also Hutchinson 1992). The avian fossil record in northern Mexico is scarce; it goes back to the latest Miocene (Hemphillian) of Yepómera, Chihuahua where a flamingo and other aquatic birds are known. The record shows a big gap, so that bird remains appear again only during the Late Pleistocene at several localities. The record is skewed toward cave dwellers such as owls (see Ferrusquía-Villafranca 1978) that lived in open areas and chiefly preyed on small mammals.

Origination in situ produced endemic taxa, such as the Middle Eocene rodents of Marfil, Guanajuato (see Fries et al. 1955; Black and Stephens 1973; Ferrusquía-Villafranca 1989), the highly specialized Early Miocene camels of Sonora and Aguascalientes (see Ferrusquía-Villafranca 1990), or the bird *Callipepla* (Escalante et al. 1993). Finally, terrestrial invertebrates have such a meager fossil record in northern Mexico that is not appropriate even to speculate about their possible biotic evolution. Some examples are Late Eocene land snails from Nuevo León (Gardner 1945) and from northeastern Chihuahua (Ferrusquía-Villafranca et al. 1997) and adjacent Texas (Roth 1984) and Early Holocene arthropods from northwestern Sonora and adjacent Arizona (Hall et al. 1988; Van Devender and Hall 1993).

Conclusions

Through the Late Cretaceous-Cenozoic interval, at least 10 major geologic and climatic events contributed to shape northern Mexico's paleoenvironments, thereby influencing biotic evolution. Among all those events, the origin and development of the TMVB was of utmost importance because it provided a tangible separation of the Nearctic and Neotropical biotas. With greater geographic heterogeneity, a trend toward increased provincialism became established. The increasing aridity overprinted most of the territory, so that arid and semiarid biomes now predominate in the region. The spatial distribution of northern Mexico's contemporary taxa is the result of the complex interplay between the changing geographic scene and the evolving biota, the latter responding by becoming differentiated throughout the Cenozoic and forming several broad assemblages occupying particular regions.

Acknowledgments

B. Simpson, T. Wendt, R. S. Felger, and J. Henrickson helped locate relevant information. Enrique Martínez-Hernández, Z. de Cserna, J.-L. Cartron, R. S. Felger, B. Simpson, and T. Wendt provided many helpful comments on previous versions of the chapter. José Ruiz ably drafted the figure.

Literature Cited

Andrade-Ramos, P., M. Montellano-Ballesteros, S. R. S. Cevallos-Ferriz, and S. Lozano-García, 2002. Nueva localidad del Cretácico Superior, Altares, Chihuahua, México. Pp. 27 *in* Ponencias, Sociedad Mexicana de Paleontología, VIII Congreso Nacional de Paleontología, Guadalajara, Jalisco.

Arroyo-Cabrales, J., O. J. Polaco, and E. Johnson. 2002. La mastofauna del Cuaternario Tardío en México. Pp. 103–123 *in* M. Montellano-Ballesteros and J. Arroyo-Cabrales (eds.), Avances en los Estudios Paleomasto-zoológicos en México. Instituto Nacional de Antropología e Historia, Serie Arqueológica, México D.F.

Axelrod, D. I. 1975. Evolution and biogeography of Madrean-Tethyan sclerophyll vegetation. Annals of the Missouri Botanical Gardens 62: 280–334.

Axelrod, D. I. 1979. Age and origin of Sonoran Desert vegetation. Occasional Papers of the California Academy of Sciences no. 132.

Ayala, R., T. L. Griswold, and S. H. Bullock. 1993. The native bees of Mexico. Pp. 179–227 *in* T. P. Ramamoorthy, R. Bye, A. Lot, and J. Fa (eds.), Biological diversity of Mexico: Origins and Distribution. Oxford University Press, New York.

Bally, A. W., and A. R. Palmer (eds.). 1988. The Geology of North America, vol. A. Geological Society of America, Boulder, Colorado.

Barrera, E., G. Keller, and S. M. Savin. 1985. Evolution of the Miocene Ocean in the eastern North Pacific as inferred from oxygen and carbon isotopic ratios of foraminifera. Pp. 83–102 *in* J. P. Kennett (ed.), The Miocene Ocean: Paleoceanography and Biogeography. Geological Society of America Memoirs 163.

Bartek, R. L., L. C. Sloan, J. B. Anderson, and M. I. Ross. 1992. Evidence from the Antarctic continental margin of Late Paleogene Ice Sheets: a manifestation of plate reorganization and synchronous changes in atmospheric circulation over the emerging southern ocean? Pp. 131–159 *in* D. R. Prothero and W. A. Berggren (eds.), Eocene-Oligocene Climatic and Biotic Evolution. Princeton University Press, Princeton, New Jersey.

Berggren, W. A., and D. R. Prothero. 1992. Eocene-Oligocene climatic and biotic evolution: an overview. Pp. 1–28 *in* D. R. Prothero and W. A. Berggren (eds.), Eocene-Oligocene Climatic and Biotic Evolution. Princeton University Press, Princeton, New Jersey.

Black, C. C., and J. J. Stephens III. 1973. Rodents from the Paleogene of Guanajuato, Mexico. Occasional Papers, Museum of Texas Tech University 14: 1–10.

Burnham, R. J., and A. Graham. 1999. The history of Neotropical vegetation: new developments and status. Annals of the Missouri Botanical Gardens 86: 546–589.

Calvillo-Canadell, L., and S. R. S. Cevallos-Ferriz. 2002. *Inga* y *Pithecellobium* en el Terciario de México. Pp. 37 *in* Ponencias, Sociedad Mexicana de Paleontología, VIII Congreso Nacional de Paleontología, Guadalajara, Jalisco.

Campbell, K. C., M. Frailey, L. Romero-Pittman, S. B. Ingement, and D. R. Prothero. 2000. Late Miocene dynamics of the Great American Faunal Interchange: waifs are out. Journal of Vertebrate Paleontology 20 (suppl.): 33A.

Carranza-Castañeda, O., and I. Ferrusquía-Villafranca. 1978. Nuevas investigaciones sobre la fauna del Rancho El Ocote, Plioceno Medio de Guanajuato, México. Informe Preliminar. Universidad Nacional Autónoma de México, Instituto de Geología Revista 2: 163–168.

Carranza-Castañeda, O., and I. Ferrusquía-Villafranca. 1979. El Género *Neohipparion* (Mammalia-Perissodactyla) de la fauna Local Rancho El Ocote, Plioceno Medio, de Guanajuato, México. Universidad Nacional Autónoma de México, Instituto de Geología Revista 3: 29–38.

Carroll, R. L. 1988. Vertebrate Paleontology and Evolution. W. H. Freeman, New York.

Case, T. J., and M. L. Cody (eds.). 1983. Island Biogeography in the Sea of Cortéz. University of California Press, Berkeley.

Case, T. J., M. L. Cody, and E. Ezcurra (eds.). 2002. A New Island Biogeography of the Sea of Cortés. Oxford University Press, New York.

Challenger, A. 1998. Utilización y Conservación de los Ecosistemas Terrestres de México: Pasado, Presente y Futuro. Comisión Nacional para el Conocimiento y Uso de la Biodiversidad, Universidad Nacional Autónoma de México, Instituto de Biología and Agrupación Sierra Madre, México D.F.

Cook, T. D., and A. W. Bally (eds.). 1975. Stratigraphic Atlas of North and Central America. Princeton University Press, Princeton, New Jersey.

Cornejo, D. O. 1994. Morphological evolution and biogeography of Mexican columnar cacti, Tribe Pachycereeae, Cactaceae. Ph.D. dissertation, University of Texas, Austin.

Elias, S. A. 1992. Late Quaternary zoogeography of the Chihuahuan Desert insect fauna, based on fossil records from packrat middens. Journal of Biogeography 19: 285–297.

Escalante Pliego, P., A. G. Navarro Sigüenza, and A. T. Peterson. 1993. A geographic, ecological, and historical analysis of land bird diversity in Mexico. Pp. 281–317 *in* T. P. Ramamoorthy, R. Bye, A. Lot, and J. Fa (eds.), Biological Diversity of Mexico: Origins and Distribution. Oxford University Press, New York.

Estrada-Ruíz, E., and S. R. S. Cevallos-Ferriz. 2002. Estructuras reproductivas del Orden Arales en la formación Cerro del Pueblo (Cretácico, Campaniano) de Coahuila, México. Pp. 45 *in* Ponencias, Sociedad Mexicana de Paleontología, VIII Congreso Nacional de Paleontología, Guadalajara, Jalisco.

Evanoff, E., D. R. Prothero, and R. H. Lander. 1992. Eocene-Oligocene climatic change in North America: the White River Formation near Douglas, East-Central Wyoming. Pp. 116–130 *in* D. R. Prothero and W. A. Berggren (eds.), Eocene-Oligocene Climatic and Biotic Evolution. Princeton University Press, Princeton, New Jersey.

Fa, J. E., and L. M. Morales. 1993. Patterns of mammalian diversity in Mexico. Pp. 319–361 *in* T. P. Ramamoorthy, R. Bye, A. Lot, and J. Fa (eds.), Biological Diversity of Mexico: Origins and Distribution. Oxford University Press, New York.

Ferrusquía-Villafranca, I. 1978. Distribution of Cenozoic vertebrate faunas in Middle America and problems in migration between North and South America. Pp. 193–321 *in* I. Ferrusquía-Villafranca (ed.), Conexiones Terrestres entre Norte y Sudamérica. Boletín 101, Universidad Nacional Autónoma de México, México D.F.

Ferrusquía-Villafranca, I. 1989. A new rodent genus from Central Mexico and its bearing on the origin of the Caviomorpha. Pp. 91–117 *in* C. C. Black and M. R. Dawson (eds.), Papers on Fossil Rodents in Honor of Albert Elmer Wood. Natural History Museum of Los Angeles County Science Series 33.

Ferrusquía-Villafranca, I. 1990. Biostratigraphy of the Mexican Continental Miocene. Paleontología Mexicana 56: 1–149.

Ferrusquía-Villafranca, I. 2003. Mexico's Middle Miocene mammalian assemblages: an overview. Vertebrate Fossils and their Context: Contributions in Honor of Richard H. Tedford. American Museum of Natural History Bulletin 279: 321–347.

Ferrusquía-Villafranca, I., C. Galindo-Hernández, and H. Barrios-Rivera. 1997. Los mamíferos oligocénicos de México: Revisión y adición a la fauna local Rancho Gaitán, Formación Prietos, Chadroniano de Chihuahua Nororiental. Pp. 97–134 *in* J. Arroyo-Cabrales and O. J. Polaco (eds.), Homenaje al Profesor Ticul Alvarez. Instituto Nacional de Antropología e Historia, Colecciones Científicas, México D.F.

Ferrusquía-Villafranca, I., E. Jiménez-Hidalgo, J. A. Ortiz-Mendieta, and V. M. Bravo-Cuevas. 2002. El registro paleogénico de mamíferos de México y su significación geológico-paleontológica. Pp. 25–45 *in* M. Montellano-Ballesteros and J. Arroyo-Cabrales (eds.), Avances en los Estudios Paleomastozoológicos en México. Instituto Nacional de Antropología e Historia, Serie Arqueologica, México D.F.

Flores-Villela, O. 1993. Herpetofauna of Mexico: distribution and endemism. Pp. 253–280 *in* T. P. Ramamoorthy, R. Bye, A. Lot, and J. Fa (eds.), Biological Diversity of Mexico: Origins and Distribution. Oxford University Press, New York.

Fries, C., Jr., W. C. Hibbard, and D. H. Dunkle. 1955. Early Cenozoic vertebrates in the red conglomerate at Guanajuato, Mexico. Smithsonian Miscellaneous Collection 123: 1–25.

Gardner, J. 1945. Mollusca of the Tertiary formations of Northeastern Mexico. Geological Society of America Memoirs 11.

Graham, A. 1993. Historical factors and biological diversity in Mexico. Pp. 109–127 *in* T. P. Ramamoorthy, R. Bye, A. Lot, and J. Fa (eds.), Biological Diversity of Mexico: Origins and Distribution. Oxford University Press, New York.

Grismer, L. L. 1994. The origin and evolution of the peninsular herpetofauna of Baja California, Mexico. Herpetological Natural History 2: 51–106.

Hall, E. R. 1981. The Mammals of North America, vols. I and II. Ronald Press, New York.

Hall, W. E., T. R. Van Devender, and C. A. Olson. 1988. Late Quaternary arthropod remains from Sonoran Desert packrat middens, Southwestern Arizona and Northwestern Sonora. Quaternary Research 29: 277–293.

Helenes, J., and M. A. Téllez-Duarte. 2002. Paleontological evidence of the Campanian

to early Paleocene paleogeography of Baja California. Palaeogeography, Palaeocimatology, Palaeoecology 186: 61–80.

Hutchinson, J. H. 1992. Western North American reptile and amphibian record across the Eocene/Oligocene boundary and its climatic implications. Pp. 451–463 *in* D. R. Prothero and W. A. Berggren (eds.), Eocene-Oligocene Climatic and Biotic Evolution. Princeton University Press, Princeton, New Jersey.

Kowallis, B. J., C. C. Swisher, O. Carranza-Castañeda, W. E. Miller, and G. D. Tingey. 1998. Preliminary radiometric dates in selected Late Tertiary vertebrate faunas from Mexico. Instituto de Investigaciones de Ciencias de la Tierra, Universidad Autónoma del Estado de Hidalgo, Avances en Investigación, Publicación Especial 1. Paleontología de Vetebrados 1: 103–108.

Lindsay, E. H. 1984. Late Cenozoic mammals from northwestern Mexico. Journal of Vertebrate Paleontology 4: 208–215.

Magallón-Puebla, S., and S. R. S. Cevallos-Ferriz. 1994. Fossil legume fruits from Tertiary strata of Puebla, Mexico. Canadian Journal of Botany 72: 1027–1038.

Maldonado-Koerdell, M. 1964. Geohistory and paleogeography of Middle America. Pp. 3–32 *in* R. Wauchope and R. C. West (eds.), Handbook of Middle American Indians, vol. I. Natural Environments and Early Cultures. University of Texas Press, Austin.

Martínez-Cabrera, I., and S. R. S. Cevallos-Ferriz. 2002. *Tapirira* (Anarcadiaceae) de la Formación el Cien (Oligoceno-Mioceno), Baja California Sur: Consideraciones biogeográficas y evolutivas. Pp. 63 *in* Ponencias, Sociedad Mexicana de Paleontología, VIII Congreso Nacional de Paleontología, Guadalajara, Jalisco.

Maxwell, R. A., and R. T. Hazzard. 1967. Stratigraphy. Pp. 23–156 *in* R. A. Maxwell, J. T. Lonsdale, R. T. Hazzard, and J. A. Wilson (eds.), Geology of Big Bend National Park, Brewster County, Texas. University of Texas Publication no. 6711. Bureau of Economic Geology, Austin, Texas.

McDowell, F. W., and S. E. Clabaugh. 1979. Ignimbrites of the Sierra Madre Occidental and their relation to the tectonic history of western Mexico. Pp. 113–124 *in* C. E. Chapin and W. E. Elston (eds.), Ashflow Tuffs. Geological Society of America Special Paper 180.

Miller, W. E., and O. Carranza-Castañeda. 1984. Late Cenozoic Mammals from central Mexico. Journal of Vertebrate Paleontology 4: 216–236.

Molnar, R. E. 1974. A distinctive theropod

dinosaur from the Upper Cretaceous of Baja California (Mexico). Journal of Paleontology 48: 1009–1017.

Morán-Zenteno, D. 1994. The geology of the Mexican Republic. American Association of Petroleum Geologists, Studies in Geology no. 39. American Association of Petroleum Geologists, Tulsa, Oklahoma.

Morris, W. J. 1972. A giant hadrosaurian dinosaur from Baja California. Journal of Paleontology 46: 777–779.

Morris, W. J. 1973. A review of Pacific Coast hadrosaurs. Journal of Paleontology 47: 551–561.

Morris, W. J. 1981. A new species of hadrosaurian dinosaur from the Upper Cretaceous of Baja California: *Lambeosaurus laticaudus*. Journal of Paleontology 55: 453–462.

Murray, G. E. 1961. Geology of the Atlantic and Gulf Coastal Province of North America. Harper and Row, New York.

Novacek, J. M., I. Ferrusquía-Villafranca, J. J. Flynn, R. A. Wyss, and A. M. Norell. 1991. Wasatchian (Early Eocene) Mammals and other vertebrales from Baja California, Mexico: the Lomas Las Tetas de Cabra Fauna. American Museum of Natural History Bulletin 208: 1–88.

Patterson, B., and R. Pascual. 1963. The extinct land mammals of South America. Pp. 138–148 *in* 16th International Congress of Zoology, Program Volume, Washington, D.C.

Patterson, B., and R. Pascual. 1972. The fossil mammal fauna of South America. Pp. 247–309 *in* A. Keast, F. C. Erk, and B. Glass (eds.), Evolution, Mammals and Southern Continents. State University of New York Press, Albany, New York.

Pérez-Hernández, B. R., and S. R. S. Cevallos-Ferriz. 2002. Estructura de Phytolaccaceae (Caryophyllales) del Cretácico Superior (Campaniano) de Coahuila, México. Pp. 69 *in* Ponencias, Sociedad Mexicana de Paleontología, VIII Congreso Nacional de Paleontología, Guadalajara, Jalisco.

Ramamoorthy, T. P., and M. Elliott. 1993. Mexican Lamiaceae: diversity, distribution, endemism, and evolution. Pp. 513–539 *in* T. P. Ramamoorthy, R. Bye, A. Lot, and J. Fa (eds.), Biological Diversity of Mexico: Origins and Distribution. Oxford University Press, New York.

Ramírez, J. L., and S. R. S. Cevallos-Ferriz. 2000. Leaves of Salicaceae (*Salix* and *Populus*) from Oligocene sediments near Tepexi de Rodríguez, Puebla, Mexico. International Journal of Plant Sciences 161: 521–534.

Ramírez, J. L., S. R. S. Cevallos-Ferriz, and A. Silva-Pineda. 2000. Reconstruction of the leaves of two new species of *Pseudosmodingium* (Anacardiaceae) from Oligocene strata of Puebla, Mexico. International Journal of Plant Sciences 161: 509–519.

Retallack, G. J. 1992. Paleosols and changes in climate and vegetation across the Eocene/Oligocene Boundary. Pp. 382–398 *in* D. R. Prothero, and W. A. Berggren (eds.), Eocene-Oligocene Climatic and Biotic Evolution. Princeton University Press, Princeton, New Jersey.

Riddle, B. R., D. J. Hafner, L. F. Alexander, and J. R. Jaeger. 2000. Cryptic vicariance in the historical assembly of a Baja California peninsular desert biota. Proceedings of the National Academy of Sciences 97:V14438–14443.

Rodríguez-de la Rosa, R. A., and S. R. S. Cevallos-Ferriz. 1998. Vertebrates of the El Pelillal locality (Campanian, Cerro del Pueblo Formation), southeastern Coahuila, Mexico. Journal of Vertebrate Paleontology 18: 751–764.

Ross, M. I., and C. R. Scotese. 1988. A hierarchical tectonic model of the Gulf of Mexico and Caribbean region. Tectonophysics 155: 139–168.

Roth, B. 1984. *Lysinoe* (Gastropoda: Pulmonata) and other land snails from Eocene-Oligocene of Trans-Pecos Texas, and their paleoclimatic significance. The Veliger 27: 200–218.

Rzedowski, J. 1981. Vegetación de México. Limusa-Noriega Editores, México D.F.

Rzedowski, J. 1993. Diversity and origins of the phanerogamic flora of Mexico. Pp. 129–144 *in* T. P. Ramamoorthy, R. Bye, A. Lot, and J. Fa (eds.), Biological Diversity of Mexico: Origins and Distribution. Oxford University Press, New York.

Savin, S. M., L. Abel, E. Barrera, D. Hodell, G. Séller, J. P. Kennett, J. Killingley, M. Murphy, and E. Vincent. 1985. The evolution of Miocene surface and near-surface marine temperatures: oxygen isotopic evidence. Pp. 49–82 *in* J. P. Kennett (ed.), The Miocene Ocean: Paleoceanography and Biogeography. Geological Society of America Memoirs 163.

Shaw, C. A., and H. G. McDonald. 1987. First record of giant anteater (Xenartha, Myrmecophagidae) in North America. Science 26: 186–188.

Smith, G. A., D. G. Smith, and B. F. Funnell. 1994. Atlas of Mesozoic and Cenozoic Coastlines. Cambridge University Press, Cambridge.

Sousa, M., and A. Delgado. 1993. Mexican Leguminosae: Phytogeography, endemism, and origins. Pp. 459–511 *in* T. P. Ramamoorthy, R. Bye, A. Lot, and J. Fa (eds.), Biological Diversity of Mexico: Origins and Distribution. Oxford University Press, New York.

Stehli, F. G., and S. D. Webb (eds.). 1985. The Great American Biotic Interchange. Plenum Press, NewYork.

Stehlin, H. G. 1909. Remarques sur les faunules de mammifères des couches eocènes et oligocènes du Basin de Paris. Bulletin de la Société Geologique de France 9: 488–520.

Stewart, W. N. 1985. Paleobotany and the Evolution of Plants. Cambridge University Press, Cambridge.

Valdés, J., and I. Cabral. 1993. Chorology of Mexican grasses. Pp. 439–446 *in* T. P. Ramamoorthy, R. Bye, A. Lot, and J. Fa (eds.), Biological Diversity of Mexico: Origins and Distribution. Oxford University Press, New York.

Van Devender, T. R. 1990. Late Quaternary vegetation and climate of the Sonoran Desert, United States and Mexico. Pp. 134–165 *in* J. L. Betancourt, T. R. Van Devender, and P. S. Martin (eds.), Packrat Middens: The Last 40,000 Years of Biotic Change. University of Arizona Press, Tucson.

Van Devender, T. R. 2001. Deep history and biogeography of La Frontera. Pp. 56–66 *in* G. L. Webster and C. J. Bahre (eds.), Changing Plant Life of La Frontera. University of New Mexico Press, Albuquerque.

Van Devender, T. R. 2002. Environmental history of the Sonoran Desert. Pp. 3–24 *in* T. H. Fleming and A. Valiente-Banuet (eds.), Columnar Cacti and Their Mutualists: Evolution, Ecology, and Conservation. University of Arizona Press, Tucson.

Van Devender, T. R., and W. E. Hall. 1993. Fossil arthropods from the Sierra Bacha, Sonora, Mexico. Ecológica (Centro Ecológico de Sonora, Hermosillo, Sonora) 3: 1–12.

Van Devender, T. R., A. M. Rea, and M. L. Smith. 1985. The Sangamon interglacial vertebrate fauna from Rancho La Brisca, Sonora. Transactions of the San Diego Society of Natural History 21: 23–55.

Velasco de León, P., S. R. S. Cevallos-Ferriz, and A. Silva-Pineda. 1998. Leaves of *Karwinskia axamilpense* sp.nov. (Rhamnaceae) from Oligocene sediments, near Tepexi de Rodríguez, Puebla, Mexico. Canadian Journal of Botany 76: 410–419.

Weber, R. 1972. La vegetación maestrichtiana de la Formación Olmos de Coahuila, Mexico. Boletín de la Sociedad de Geológica Mexicana 33: 5–19.

Weber, R., and S. R. S. Cevallos-Ferriz. 1994. Perfil actual y perspectivas de la Paleobotánica en México. Boletín de la Sociedad Botánica de México 55: 141–148.

Weishampel, D. B. 1990. Dinosaurian distribution. Pp. 63–139 *in* D. B. Weishampel, P. Dodson, and H. Osmolska (eds.), The Dinosauria. University of California Press, Berkeley.

Wendt, T. 1993. Composition, floristic affinities, and origins of the canopy tree flora of the Mexican Atlantic slope rain forests. Pp. 595–680 *in* T. P. Ramamoorthy, R. Bye, A. Lot, and J. Fa (eds.), Biological Diversity of Mexico: Origins and Distribution. Oxford University Press, New York.

Westgate, J. W. 1994. A new leptochoerid from Middle Eocene (Uintan) deposits of the Texas Coastal Plain. Journal of Vertebrate Paleontology 14: 296–299.

Wilson, J. A. 1967. Early Tertiary mammals. Pp. 157–169 *in* R. A. Maxwell, J. T. Lonsdale, R. T. Hazzard, and J. A. Wilson (eds.), Geology of Big Bend National Park, Brewster County, Texas. University of Texas Publication no. 6711, Austin.

Wolfe, J. A. 1978. A paleobotanical interpretation of Tertiary climates in the Northern Hemisphere. American Scientist 66: 694–703.

Wolfe J. A. 1992. Climatic, floristic and vegetational changes near the Eocene/Oligocene Boundary in North America. Pp. 437–450 *in* D. R. Prothero and W. A. Berggren (eds.), Eocene-Oligocene Climatic and Biotic Evolution. Princeton University Press, Princeton, New Jersey.

3

Recent History of Natural Resource Use and Population Growth in Northern Mexico

SCOTT H. STOLESON

RICHARD S. FELGER

GERARDO CEBALLOS

CAROL RAISH

MICHAEL F. WILSON

ALBERTO BÚRQUEZ

Long a remote frontier region, northern Mexico (fig. 3.1)—especially the border states, Tamaulipas, Nuevo León, Coahuila, Chihuahua, Sonora, and Baja California, from east to west—now supports booming human populations and plays a key role in the industrial and agricultural sectors of the country's economy (INEGI 1993). Patterns of demographic and economic growth and their impacts in northern Mexico have been shaped and constrained by a variety of factors. These include the country's land tenure history, with the allocation of vast tracts of lands (including most forested lands) to peasants under communal landholding regimes. Also important are the influence of the United States and growing interdependence of the 2 countries, which make the northern border states a destination in a pilgrimage toward perceived improvements in the economic milieu of millions of Mexicans from the central and southern parts of the country. In this chapter we present a brief and necessarily selective overview of anthropogenic effects on natural environments of northern Mexico and the animals and plants that inhabit the region. Agriculture, fishing, mining, and forestry are all important primary sector activities in the region. Other regional activities with important effects on natural areas include unregulated hunting, sports fishing, tourism, and the trade in wild plants and birds.

Contemporary Demographic and Economic Growth

Human population growth is the leading force underlying environmental problems in Mexico (Ehrlich and Ceballos 1997; Kemper and Alvarado 2001). The population quadrupled from 25 million in 1950 to 100 million in 2000, and nearly doubled in the 25 years between 1975 and 2000 (INEGI 2001). Although for much of its history northern Mexico remained sparsely populated, the human population grew exponentially in the late twentieth century (table 3.1). During the early decades of the twenty-first century, the northern border region could grow 2.3–2.5 times faster than the national average (Kemper and Alvarado 2001).

Except for El Paso/Ciudad Juarez, Tijuana, and Sonoyta, all settlements on the U.S.–Mexico border west of the Rio Grande (Río Bravo) were established in the late nineteenth and twentieth centuries (Bahre and Hutchinson 2001). The Mexican population along the U.S. border grew 830% from 1930 to 1990 (Kaye 1995). Municipal areas on the U.S.–Mexico border had a population growth rate of 57.1% in the years from 1980 to 1995, while Mexico as a whole grew by 36.4% (INEGI 2001; table 3.1). Of 79 border municipalities, 20 surpassed the national growth rate, and these were responsible

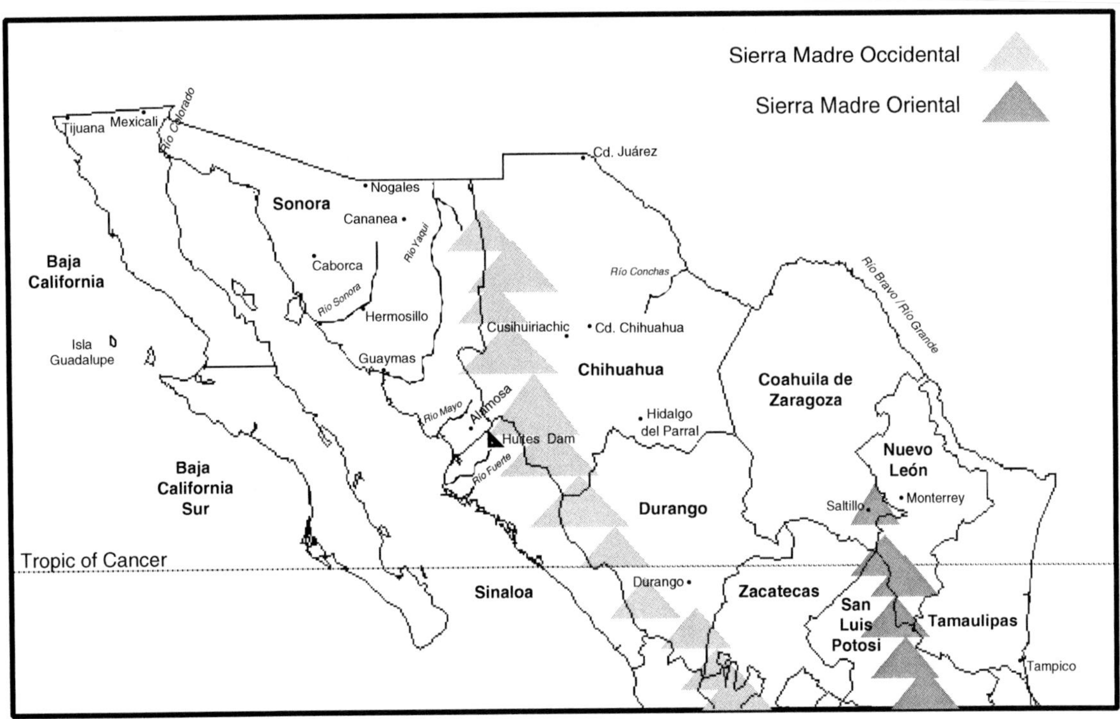

Figure 3.1. General map of northern Mexico indicating states, cities, and major geographical features mentioned in this chapter.

for 83.3% of the growth during 1980–1995. Seventy-two percent of these border municipalities had average annual growth rates below the national average, and 25 of them had negative growth rates during this 15-year period (SEMARNAT 2003). Of the border states, only Coahuila grew at a rate below the national average in the 1990s (INEGI 2001). The particularly high growth rate in the border states is due mostly to in-country migration and is concentrated in urban areas, where there is an ever-increasing demand for labor near the U.S. border (Contreras 1987; INEGI 1993; Anguiano-Téllez 1998; CONABIO 1998). In recent decades border towns have become among the fastest growing in all of Mexico, and all but Tijuana, a sprawling city of more than 1.2 million people, are larger than their U.S. counterparts. Between 1980 and 1995, populations of these urban areas grew by an average of 41% (Ingram 1998). Ciudad Juárez has tripled in size since 1980, creating with El Paso a cross-border urban area of 4 million people. Northern Mexico is now the most urbanized region of the country after the Distrito Federal, with more than 75% of the population in urban areas (table 3.1; Cavazos-Doria 1997; Simon 1997).

The geographic distribution of the population within the northern states is highly heterogeneous. Rapid growth in irrigated agriculture and fisheries has concentrated much of the recent growth in Sonora, Sinaloa, and Tamaulipas along the arid coasts. For example, the population of Sonora has doubled since 1970, but most of that growth (85%) has been in the arid coastal lowlands (West 1993). In contrast, in Baja California and Chihuahua the human population is concentrated in the border regions. Despite these considerable increases in population the states of Baja California Sur, Chihuahua, Coahuila, Durango, and Sonora remain among the least densely populated in the country (CONABIO 1998).

Cross-border economic disparity is the primary driving force behind demographic and economic growth in northern Mexico. U.S. goods exported to Mexico were valued at US$101.5 billion in 2001, while imports to the United States from Mexico were $131.4 billion (USTR 2003). The United States is Mexico's largest trading partner and provides a major source of employment for Mexican workers. Since World War II, three agreements have

Table 3.1. Population trends and annual growth rates, 1970–2020, in Mexico's northern states compared to the nation as a whole, in thousands.

State[a]	1970	1990	% Annual Growth, 1970–1990	2000	% Annual Growth, 1990–2000	Estimated, 2020	Estimated % Annual Growth	% Urban in 2000
BC	870.2	1,660.9	3.21	2,487.7	4.98	3,440.1	3.83	91.7
BCS	128.0	317.8	4.52	423.5	3.33	498.1	1.76	81.3
CHI	1,612.5	2,441.9	2.06	3,047.9	2.48	3,940.8	2.93	82.5
COA	1,115.0	1,972.3	2.83	2,295.8	1.64	2,800.8	2.20	89.4
DGO	939.2	1,349.4	1.80	1,445.9	0.72	1,709.4	1.82	63.8
NL	1,694.7	3,098.7	3.00	3,826.2	2.35	4,850.3	2.68	93.4
SIN	1,266.5	2,204.1	2.75	2,534.8	1.50	2,631.3	0.38	67.3
SON	1,098.7	1,823.6	2.52	2,213.4	2.14	2,739.8	2.38	83.1
TAM	1,456.9	2,249.6	2.16	2,747.1	2.21	3,398.9	2.37	85.4
Nation	48,225.2	81,249.6	2.59	97,361.7	1.98	122,602.7	2.59	74.7

Data and projections from INEGI (2001).

[a]States: BC = Baja California, BCS = Baja California Sur, CHI = Chihuahua, COA = Coahuila, DGO = Durango, NL = Nuevo León, SIN = Sinaloa, SON = Sonora, TAM = Tamaulipas.

especially strengthened economic ties between the United States and Mexico, reflecting the interdependence of the 2 countries: the Bracero Program, the Border Industrialization Program, and the North American Free Trade Agreement (NAFTA).

The Bracero Program of the 1940s–1960s was instituted to bring Mexican farm laborers to the United States to alleviate labor shortages caused by World War II (Kaye 1995), and the program continued until 1964 (Barry et al. 1994). At its peak almost one-half million braceros worked legally in the United States. After its termination by the U.S. government, many workers remained near the border. Largely as a response to the demise of the bracero program, in 1965, Mexico and the United States initiated the Border Industrialization Program to promote the establishment of assembly plants in border towns. These mostly foreign-owned factories, or *maquiladoras*, take advantage of the relatively lax regulation and lower wages in Mexico to provide economically attractive labor for the global manufacturing market. They began booming in the 1980s with the devaluation of the peso (Barry et al. 1994). Between 1983 and 1993, the number of maquilador workers increased 14.4% per year, and the number of factories increased by 11% per year (Ingram 1998). By 1999, more than 3000 factories were established along the border, employing more than 500,000 workers (Mumme 1999). Fourteen municipalities adjacent to the United States

contained 65.1% of all maquiladora workers and just 2 of the cities, Ciudad Juarez and Tijuana, were home to 35% of all these employees (SEMARNAT 2003). By the early 1990s the maquiladoras contributed one-third of the trade between the United States and Mexico (Barry et al. 1994). Besides providing employment opportunities to Mexican workers, the maquiladoras provide much-needed hard currency and are vitally important to Mexico's economy (Ingram 1998). Maquiladoras are a major source of industrial waste and other pollution and a primary cause of degradation of borderland waterways (Barry et al. 1994). In the early 2000s, significant numbers of border maquiladoras began closing due to cheaper labor and bigger markets in Asia, creating economic hardship (INEGI 2003).

In December 1993, Mexico, Canada, and the United States signed the NAFTA treaty, which went into effect in January 1994 (Foster 1997). Under NAFTA, price supports and tariffs have been dropped in Mexico, prompting an acceleration of industrial growth. The United States continues to subsidize its agriculture while Mexican price supports for agriculture are unilaterally phased out, putting many Mexican farmers at a serious disadvantage. It is predicted that 5–15 million Mexican farmers will abandon their fields, resulting in a leap in the rate of rural migration to the northern border to seek employment (Simon 1997; Bracamonte 2000). The full environmental repercussions of the treaty remain to be seen but are a major concern (Mumme 1999; Búrquez and Martínez Yrízar 2000).

Land Tenure: The *Ejido* Regime

Land tenure is a major factor in many of Mexico's environmental problems. The Mexican Revolution of 1910 originated in the inequitable distribution of arable lands, most of which were controlled by huge haciendas. The revolution led to the creation of new political structures and the constitution of 1917 (Ewing 1966; Foster 1997). Article 27 of the new constitution dealt with land reform. Its purpose was to eliminate monopolies on water and mineral resources and give the Mexican nation rights to such resources. Land reform came to the fore especially during the presidency of Lázaro Cárdenas (1934–1940) with the redistribution of nearly 18 million hectares of expropriated land to peasants under the *ejido* form of land tenure.

Government policies after 1940 alternated between promoting a private sector and more land expropriation and redistribution, with the last important landholding redistribution occurring during the presidency of Luis Echeverría (1970–1976). Nationwide, there were about 30,000 *ejidos* at the end of the twentieth century (Warman 1994). Many of the *ejidos* were established in the most marginal of Mexico's arable land and were barely sufficient to produce a subsistence crop (Simon 1997). In the arid north, most *ejidos* were geared to cattle production with no arable land use whatsoever. Some, like Ejido Sierrita del Rosario in extreme northwestern Sonora, were vast extensions of sandy desert that served a political purpose but did not allow a subsistence economy, much less development (Búrquez and Castillo 1994). In the *ejidos* that managed to persist, the problem was later compounded by population growth, often leading to increasing land parceling, with many *ejidatarios* holding individual lots too small to support a family. Increased production on those marginal lands also came at the expense of traditional practices and was often heavily dependent on government-supplied fertilizers and pesticides. In some cases, this dependence has led to declining land productivity, environmental damage, and untenable agricultural practices when government support was withdrawn (Simon 1997).

Under Article 27 of the Mexican Constitution, *ejidatarios* were granted both the right to farm and otherwise use and pass their usufruct rights through inheritance, but they were not granted ownership rights (Cornelius and Myhre 1998). The *ejido* has been described as a "form of common property with private appropriation" (Córdova 1973; De Janvry et al. 1997). Until recently, *ejidatarios* managed individual plots of land as well as land held in common (Lewis 2002). A far-reaching amendment to Article 27 in 1992 created programs and institutions to officially certify and record *ejidatarios* rights to their land (i.e., PROCEDE; Programa de Certificación de Derechos Ejidales y Titulación de Solares Urbanos) and set the stage for an accelerated shift toward private land ownership (Lewis 2002). This amendment authorized *ejidatarios* to sell their lands or to pledge them as collateral for loans (Foley 1995), and has resulted in substantial and previously "off limits" land being bought by developers and corporations (Yetman and Búrquez 1998).

Another form of collective land use is through *comunidades*, a form of land tenure unique to indigenous communities. These lands can be more difficult to transfer or sell than ordinary *ejido* lands. These represent about 20% of the collectively owned land in the country but vary widely among and within states (e.g., 5.6% of the state of Sonora, 7.4% of Chihuahua, 0.5% of Coahuila, but 48% of Oaxaca; De Gortari 2000). As with the *ejidos*, *comunidades* have an internal governing body where all sorts of decisions are made, from purely pragmatic and secular to sacred and ritual.

The largest areas of forests and wildlands are usually communal lands. Seventy percent or more of forest area in Mexico is held by *ejidos* (Segura 2000; Molnar and White 2001). For example, in Chihuahua, 37% of the total area is held by *ejidos*, with another 2% occupied by *comunidades* (Challenger 1998). A similar situation is present throughout Mexico's arid northern states. However, this scenario is rapidly changing, and SEMARNAT (2000a) reports that by 2002 only 16% of the *ejido* land of Chihuahua (about 7.4 million ha) was still devoted to forestry. Most likely the remaining land has been transferred to private hands.

In summary, it is important to emphasize the earlier connection between the absorption of peasants (*campesinos*) into the proletariat and the more recent privatization of the countryside (Toledo 1996). It is ironic that ultimately the *campesinos*'s land is being wrested away from them by giving them legal title to their lands. The resulting cultural erosion and loss of traditional knowledge goes hand in hand with the loss of biodiversity (Yetman and Búrquez 1998; Búrquez and Yetman 2001; Laird 2002).

Industries and Resource Uses

Mining

Mexico's tremendous mineral wealth has played a vital role in the history and development of the country. The Mexican mining industry is among the top 13 for world production of 18 minerals and worldwide leader in production of silver, celestite, and bismuth (Salas 1991; Roldán and Clark 1992; Secretaría de Economía 2000). The majority of mining in Mexico, both in volume and economic importance, is located in the northern tier of states (fig. 3.2). Although production volumes vary from year to year, the main mining states are Sonora, generally the leader in the production of gold, copper, graphite, molybdenum, and wollastonite; Coahuila, first in the production of celestite, coal, dolomite, and coke; Chihuahua, main producer of lead, zinc, and cadmium; and Zacatecas, the most important silver producer. In Mexico, silver deposits are found mainly along the Sierra Madre Occidental from Sonora to the Isthmus of Tehuantepec. Although coal deposits have been identified throughout the country, most of the known and utilized deposits are in the north, particularly in Coahuila (Salas 1991). As of 1985, there were an estimated 4.4 billion metric tons of coal reserves in the country, of which 2.8 billion (63%) were in the northern tier of states: Chihuahua, Coahuila, Nuevo León, Sonora, and Tamaulipas (Salas 1991). With respect to the distribution of the surface concessioned to each state for mining, Sonora represents the largest mining surface in Mexico (Secretaría de Economía 2001).

During the 1960s, the government progressively "Mexicanized" the mining industry, most of which had previously been foreign-owned. Beginning in 1961 foreigners were restricted to owning no more than 49% of the shares of Mexican mining companies, making Mexico unattractive for foreign investment. The 1975 mining law restricted foreign companies to 34% participation in mining concessions

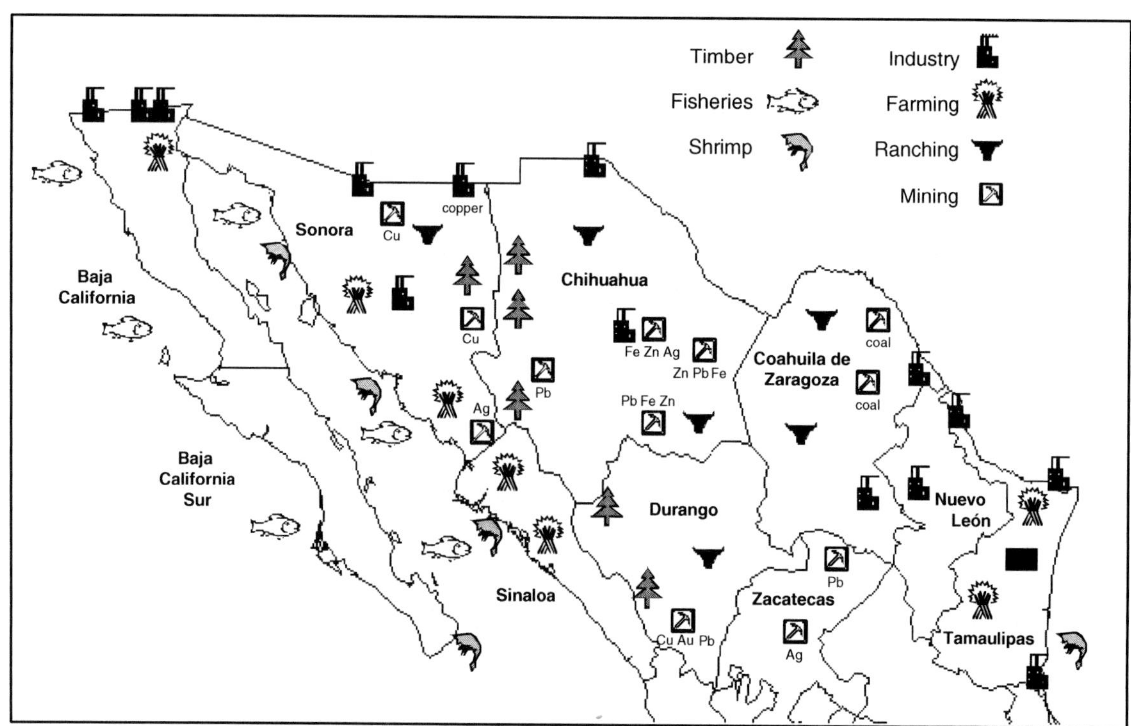

Figure 3.2. Major areas of natural resource uses and industry in northern Mexico.

on national reserves and in exploitation of certain minerals such as coal and iron ore. Mexico radically overhauled the nationalistic mining law structure in 1990–1992 by modifying Article 27 of the Mexican constitution, privatizing virtually all government mining interests and actively retreating from competition with the private sector in mining exploration and development. These modifications allowed foreigners to indirectly hold the remaining 51% of the shares through the establishment of trusts, spurring a boom in foreign exploration investment (Sanchez-Mejorada 2000). Superceding previous legislation, the mining law of 1992 included no restrictions for ownership other than complying with the Foreign Investment Law, which in 1993 allowed full foreign investment in mining but always through Mexican companies. Within the 3 years following the passage of the new mining law, more than 70 foreign companies, dominated by U.S. and Canadian concerns, established offices in Hermosillo, Sonora (Border Ecology Project 1995). Although amendments to the Foreign Investment Law published on December 24, 1996 made it legally possible for foreigners to hold mining concessions regardless of the provisions of the law denying them that right, the executive's constitutionally allowed discretion continues to bar all foreigners from directly obtaining mining concessions in Mexico (Sanchez-Mejorada 2000).

From initial exploration until long after closure, mining can affect the environment in ways that vary from the obvious to the insidious. Mineral extraction can affect wildlife, vegetation, groundwater, surface water, soils, air, and cultural resources (Environmental Law Institute 2000). Obviously, mines vary markedly in their size, proximity to sensitive areas, materials extracted, composition and nature of waste, timing and duration of operation, and other factors. Therefore, specific environmental issues vary from mine to mine and can involve the contamination and drawdown of groundwater aquifers, erosion, acid mine drainage, dust, and air pollution from tailings and road surfaces, smelter emissions, and landscape clearing and other degradation (e.g., due to the cutting and burning of wood to power smelters or mining equipment), among other problems (National Research Council 1999). Mined areas are generally not only heavily polluted, but also effectively sterilized, as rehabilitation of old pit mines is virtually nonexistent and generally not feasible (Búrquez and Martínez-Yrízar 1997, in press). The bulk of Mexico's copper output comes from Sonora, where the country's 3 largest mines— La Caridad, Cananea, and Minera María—are located. North America's oldest copper mine, one of the world's largest and the second largest in Mexico, Cananea, exemplifies not only many of the pollution problems associated with mining but also many of the social and economic issues related to the industry. A century's worth of mining and smelter wastes has accumulated at the headwaters of the San Pedro River and Río Sonora near the mine, where it threatens both watersheds (Arias Rojo 2000).

The major impacts from commercial mining come from direct destruction of land and indirectly from the resources expended to support the mines. Near mines and smelters, the demand for building materials and fuels decimates local woodlands and forests. The current composition of woodlands reflects the selective nature of usage related to mining. For example, there are few old trees in the woodlands and forests near smelters at Cananea, Sonora, although they remain abundant in similar nearby areas (McPherson and Villanueva-Díaz 2001). Valuable tropical dry forest hardwood trees such as *amapa* (*Tabebuia* spp.) seem to be recolonizing the hills around Alamos, Sonora, where the trees were cut out mostly in the nineteenth century for supports for tunnels in the silver mines as well as for roof beams (Felger et al. 2001). The same trend can be seen around San Javier, a small mining town in central Sonora, where *amapa* now covers extensive areas where once it was virtually extirpated (Varela 2004).

Besides direct damage to the environment, some of the more insidious problems associated with mining include the contamination of water, soil, or air with dangerous materials. Some sources of pollution remain long after mine closure; for instance, lead mines that were used during the Roman Empire still produce acid drainage today, 2000 years after ceasing operations. In Mexico, mercury illustrates well the historical burden of mining when no oversight or controls are available (National Research Council 1999). For almost 4 centuries, mercury was widely used in Mexico to extract silver and gold from ore-bearing rock. Although Mexico has mercury deposits, the production of mercury in Mexico was prohibited by Spain from 1680 to 1811, and during the colonial period Spain sent some 50,000 tons of mercury to Mexico. In the technique known as the amalgam process, mercury is combined with metallic ore, and an amalgamation of mercury and other metallic minerals is formed.

Mercury is then usually burned off, leaving the other metals. It is always lost to the environment during this procedure (Acosta-Ruíz 2001). Mercury pollution has had serious health and environmental consequences in many places in the world. Although natural sources of mercury also cause pollution, as, for instance, contamination of the Río Hardy from the Cerro Prieto geothermal field in northwestern Sonora (Gutiérrez-Galindo et al. 1988), the most serious cases of mercury pollution are anthropogenic. The San Juan River basin, covering portions of Coahuila, Nuevo León and Tamaulipas and merging with the Rio Grande, is polluted with mercury, with the highest concentrations at 11 µg/l (CINVESTAV 1994). One of the most infamous instances of mercury pollution in Mexico is near Zacatecas City, where more than 20 million tons of mercury-containing mining waste have been redeposited by stream flow into the El Pedernalillo basin and affect an area exceeding 12 square miles (INE 1996). Corn from the basin has exhibited mercury concentrations as high as 1 part per 10,000 (Stevenson 2002).

Mercury is still used in many developing countries to isolate precious metals. However, after the late 1800s, the amalgam process was largely replaced by processes using cyanide. Wet processing of precious metal ores uses a sodium cyanide solution that is allowed to percolate through the crushed stone dissolving gold and silver. Cyanide rapidly degrades into carbon and nitrogen in sunlight or in contact with the atmosphere and can also be rapidly neutralized by hydrogen peroxide and other chemicals. Although cyanide itself may be a problem in certain areas, particularly when discharged directly into aquatic environments where it is extremely damaging to aquatic organisms, cyanide frees other metals such as mercury, cadmium, and lead, which in turn can contaminate groundwater (Marr et al. 1996).

Sources of chemical pollution of water, air, and soil that remain scarcely documented are the many small-scale mining enterprises and individual prospectors (*gambusinos*) who operate throughout Mexico, though primarily in the northern Sierra Madre Occidental (Sánchez-Críspin 1994). Mexican gold production from artisanal miners is not inconsequential, and during 1994 almost 5 tons of gold were produced by rudimentary processes. The official number of these miners in Mexico is believed to be around 2000, but the actual number is unknown, and this figure is considered to be an un-

derestimate (Veiga 1997). The use of cyanide is the preferred technique for small gold processors, and though mercury may not be commonly used by placer miners, data are unavailable. Obviously, the presence of so many point sources of pollution at the headwaters of most of the important western rivers may have serious consequences.

Though there are legal safeguards in Mexico to protect the environment from mining activities, mining associations and organizations constitute a powerful political group. The 1992 mining law directed the National Institute of Ecology to review applications for claims within only 90 days, negated requirements for public participation, and eliminated restrictions on the size of explorations. The establishment by decree through the federal government of biosphere or ecological reserves prohibits or severely restricts mining activities within the boundaries of these reserves, but these restrictions, too, can be modified when mining associations influence legislation. For instance, certain proposed protected areas have been reduced in size to protect mining claims from interference, such as the Ajos-Bavispe Protected Natural Area (PNA) in the state of Sonora (Búrquez and Martínez Yrízar in press). The surface initially contemplated to be part of this protected area was reduced by 48%, and, as a result, the expansion plan for the reserve was reduced to one-third of the original area. Only 104 mining sites were affected instead of 291 (Secretaría de Economía 2001), and even with these concessions, the mining lobby still contests the designation of this protected area (Búrquez and Martínez Yrízar in press). A 1998 regulation (SEMARNAP 1998) establishes environmental protection requirements for direct mining exploration activities in dry and hot regions, where vegetation consists of xerophilous, deciduous tropical, coniferous, or oak forests. Especially for northern Mexico, this legislation would be extremely important. However, it is unclear at this time to what degree regulations will be enforced and if there will be adequate compliance by industry and individuals to make these laws meaningful. In the meantime, mining exploration projects continue to submit mandatory preventive reports on their activities to comply with the environmental permits.

Agriculture

In northwestern Mexico, the first large-scale commercial agriculture began around 1890–1910, with the development of large irrigation systems along major rivers by a variety of private, U.S.-backed businesses (West 1993). After the 1910 revolution, the federal government assumed control of these irrigation systems, and following World War II began extensive dam and reservoir construction (West 1993). From the 1940s to the 1970s, agriculture in northern Mexico was modernized and intensified through the Green Revolution, which had its beginnings at the International Maize and Wheat Improvement Center (CIMMYT) at Ciudad Obregón, Sonora. The Green Revolution used new high-yield varieties of wheat, coupled with irrigation and intensive use of fertilizers and pesticides, to greatly increase crop production (Glaeser 1987; Conway 1998). In the 1940s, Mexico imported half of its wheat from the United States and Canada, but by the mid-1960s began to export wheat, due primarily to vastly increased yields (Brown 1970).

In 1980, the Sistema Alimentario Méxicano (Mexican Food System, or SAM) was initiated in part to achieve further self-sufficiency in the production of basic food crops (maize, wheat, beans; Fernández-González 1980). The area under cultivation in northern Mexico increased significantly under SAM. For example, total cropland (irrigated and dry farming) in Tamaulipas increased from 3% to 16% of the state's land area between the 1950s and the 1980s (Purdy and Tomlinson 1991); by 1992 this area had exceeded 28% (Flores-Villela and Geréz 1994). Most of this increase has been in irrigated cropland, which was around 98,000 ha in 1960 and 371,000 ha in 1982 (Contreras 1987).

In addition, the scale of cultivation has changed dramatically. Agriculture in the northern states of Mexico has increasingly shifted from traditional small farms geared toward local markets to large-scale commercial agro-industry. Conversion to large-scale agriculture has been concentrated in the lowland areas of the northwest and northeast, particularly in Sinaloa, Sonora, and Tamaulipas (Simon 1997; Anguiano-Téllez 1998; fig. 3.2). Much of the production is exported to the United States, including increasing quantities of nontraditional crops (e.g., fruits, nuts, and especially winter vegetables including tomatoes), especially since NAFTA. In addition, agro-industry in the northern states supplies much of Mexico's domestic food production. For example, Sonora alone grows 40% of the nation's wheat crop (West 1993). By the 1990s large commercial farms controlled about half of Mexico's farmland (Simon 1994).

The agribusiness model developed in the United States and Canada appears ill-suited for much of Mexico (Klein-Robbenhaar 1995). This model relies on large-scale production of single crops, with heavy use of irrigation, pesticides, and fertilizers. In arid northern Mexico, irrigation is essential for agricultural production (Ezcurra and Montaña 1990; Klein-Robbenhaar 1995; Búrquez and Martínez Yrízar 2000). Therefore intensive agriculture is concentrated in the lowlands where arable land and irrigation water are available—for example, along the Río Colorado delta in Baja California Norte and Sonora, the Mayo, Yaqui, Sonora, and Magdalena rivers in western Sonora, and the Fuerte delta in Sinaloa (Bojorquez-Tapia et al. 1985; West 1993). Long-term effects of irrigation include desiccation of wetlands and riparian areas, soil erosion, water pollution, salinization, and sedimentation of waterways (Lemly et al. 2000). There is great turnover as new lands are cleared and planted to make up for lands abandoned due to desertification, water loss, and salinization. Estimates of areas under cultivation can be misleading because simple statistics may not consider these abandoned lands (CONABIO 1998).

Increasing use of diversion irrigation and groundwater pumping have drastically lowered water tables and increased soil salinity in many areas (West 1993; Contreras-Balderas and Lozano-Vilano 1994; Cavazos-Doria 1997). For example, in arid western Sonora the pumping of water from deep aquifers during the second half of the twentieth century greatly exceeded rates of replenishment, and in some places water tables have dropped as much as 1–2 ms per year (West 1993; Simon 1997; Moreno 2000). As fresh water became unavailable, tens of thousands of hectares of irrigated cropland were abandoned (Ezcurra and Montaña 1990), and between 1973 and 1992 the area under cultivation in central Sonora declined more than 24% (Valdez-Zamudio et al. 2000). Of the original 150,000 ha cleared for irrigated agriculture along the Costa de Hermosillo, only 70,000 ha remain under cultivation (Búrquez and Martínez-Yrízar 1997, in press; Moreno 2000). In this area the water table has dropped as much as 50 m below sea level, and in places has been contaminated by the intrusion of salt water from the Gulf of California up to 30 km inland (Steinich et al. 1998).

Large-scale commercial agriculture relies on heavy use of pesticides and fertilizers (Simonian 1988; Barry et al. 1994). Numerous pesticides outlawed in the United States were still used in Mexico at the end of the twentieth century, including DDT, chlordane, aldicarb, lindane, paraquat, and pentachlorophenol (Barry et al. 1994; Simon 1997; CICOPLAFEST 2002). At least until the late 1980s, large industrial farms in northern Mexico regularly used 36 chemicals considered to be extremely hazardous by the United Nations (Simonian 1988). Most of the DDT used in Mexico was produced domestically by Fertimex, which was the last commercial producer in the Western Hemisphere (Barry et al. 1994). However, as part of the North American Regional Action Plan, in 1997 Mexico began to phase out domestic use of DDT and chlordane, to be completed by 2006, and Fertimex closed its DDT facilities in 1999 (Anonymous 1997). Since the late twentieth century there has been more emphasis on alternative pesticides including more toxic but short-lived organophosphates (Barry et al. 1994). Intense use of agrochemicals has long-term consequences for productivity, the environment, and human health (e.g., Restrepo 1988; Albert 1990; Guillette et al. 1998). In response to changing U.S. market demands, there is a small but increasing organic farming industry in northern Mexico.

Large areas of tropical deciduous forest, thornscrub, arid scrub, desert, and riparian woodlands have been converted to agricultural production in the northern states. This destruction and fragmentation of natural areas is the major cause of extirpation and extinction of populations and species (Ceballos 1993; Challenger 1998). In many regions, ecologically important species have been extirpated, and since many have marginal populations in Mexico, local extinctions often mean countrywide extirpations (Ceballos et al. 1998). For example, agriculture in the San Quintin Valley of Baja California has led to the apparent extinction of the endemic kangaroo rat *Dipodomys gravipes* (Ceballos and Navarro 1991). Similarly, thousands of hectares of prairie dog towns have been lost in central and southern Coahuila and Nuevo León in the last 3 decades (Ceballos 1993; Treviño-Villarreal and Grant 1998). The Llanos de San Juan Bautista in the delta/plains of the Río Sonora once supported extensive mesquite (*Prosopis glandulosa* var. *torreyana*) forests and a rich associated fauna and flora, most of which disappeared with the development of the Costa de Hermosillo Irrigation District (Felger and Lowe 1976; Búrquez and Martínez-Yrízar 1997). The loss of Tamaulipan thornscrub to commercial agriculture has profoundly affected wildlife populations, including commercially important game birds such as the white-winged dove (*Zenaida asiatica*; Purdy and Tomlinson 1991), and endemic species such as the green-cheeked Amazon (*Amazona viridigenalis*), Altamira yellowthroat (*Geothlypis flavovelata*), and Tamaulipas pocket gopher (*Geomys tropicalis*; Ceballos and Navarro 1991; Ceballos and Márquez 2000).

Livestock

Cattle ranching occupies the greatest land base of any industry in northern Mexico, where it has flourished for centuries. Despite agrarian reform and communal ownership of some rangelands, most ranches are privately owned and vary greatly in size (Perramond 1999). In general, the ranches of the arid north are larger than those of the more southern regions. Cattle ranching in the north is extractive in nature, with the animals far less integrated into field agriculture than in the south (Perramond 1999). Historically, there was little direct cattle herd management, with no institutions responsible for cattle ranching until cattlemen's associations were founded in the 1920s and 1930s. Although modernization of ranching practices has occurred, serious environmental and social concerns remain surrounding the industry in the region.

Drought in Mexico's northernmost 6 states caused the number of cattle to drop by 75% between 1902 and 1923, from 1.25 million to 312,000 head (Ezcurra and Montaña 1990). Since World War II, however, the livestock industry in northern Mexico has expanded considerably. Nationwide, the number of cattle grew from 10 million head in 1940 to 37.5 million in 1983, a population growth rate greatly exceeding that of Mexico's human population (Toledo 1990). Although the livestock industry remains concentrated in the north (CONABIO 1998; fig. 3.2), there has been a recent shift in focus of cattle grazing to the dwindling forests and rainforests of the southern states (Klein-Robbenhaar 1995). Currently, an estimated 38% of the cattle, 50% of the goats, and 25% of the sheep in Mexico are grazed in the northern states (Ezcurra and Montaña 1990; Cavazos-Doria 1997). The percentage of combined agricultural and rangelands devoted to livestock is highest in the north: 92% in Chihuahua, 79% in Sonora, 73% in Tamaulipas, and 70% in Coahuila (Toledo 1990). This northern concentration of livestock is largely due to ready access to U.S. markets, the major importer of Mexican cattle and beef

(Toledo 1990) and the availability of relatively inexpensive semiarid land. Grazing is less widespread in the Baja California peninsula due to general isolation, aridity, and lack of water (chapter 18).

Unlike the southwestern United States, where most grasslands and woodlands used for grazing are managed by government agencies, most grazing in northern Mexico is on private ranches or *ejidos* (CONABIO 1998). Stocking rates and grazing intensity in the United States are usually based on agency policies, whereas in Mexico, ranchers often tend to maximize stocking levels to obtain the greatest income, which inexorably leads to overgrazing (Bahre and Hutchinson 2001). For example, the density of livestock on northern ranges increased from 8.14 ha per head in 1960 to 6.99 ha per head in 1980 (Toledo 1990). These trends are exacerbated by economic problems, the small size of most Mexican holdings, and environmental fluctuations (Bahre and Hutchinson 2001). For example, 70% of Sonora's Yaqui River drainage is grazed, much of it heavily; one estimate suggests that grazing intensities range from 2 to 10 times the proper carrying capacities for the region (Bojorquez-Tapia et al. 1985). Despite changes in the state's economy, there has not been significant overall improvement in rangeland quality in Sonora since the late 1800s, in part because grazing intensity remains high (McPherson and Villanueva-Díaz 2001). In Tamaulipas, overgrazing is considered the major problem affecting the livestock industry, prompting government efforts to limit livestock numbers (Fulbright 2001).

Economic issues are changing the nature of both agricultural and ranching operations in northern Mexico, and both industries have grown increasingly interdependent. As with agriculture, ranching in northern Mexico is shifting toward very large holdings, often owned by absentee owners or corporations. Some of these large operations are reputedly owned by drug traffickers, who use legitimate cattle operations as a means of laundering illicit income (e.g., Martin and Yetman 2000). Much of the decline during the last decades of the twentieth century in the production of human food staples has been due to the transformation of many farms to the production of commodities for agro-industry, especially for cattle feedlots, as well as for poultry and farm animals in industrial feeding systems (Becket and Oltjen 1993; Goodland and Pimentel 2000). Although more and more land has been devoted for forages, the consumption of beef is declining in the United States and even more so in Europe, and Mexico may be following the trend.

Until about the 1970s most beef was produced on open ranges that primarily supplied cattle for internal consumption. By the end of the 1990s open-range ranches primarily supplied young calves and heifers ("feeder cattle," ca. 150 and 200 kg in weight, respectively), both for the feedlot industry, and also for export to the United States (UGRS 2001). The number of feeder cattle exported from Mexico to the United States was 1.13 million in 2001 (Hawkes 2004). The explosive growth in the number of feed lots for industrial beef and pork production has caused serious pollution problems. For example, the Abelardo L. Rodríguez Dam above Hermosillo, Sonora, used largely for domestic water, was polluted during the 1990s by effluents from feedlots upstream of the reservoir. The actual impact has never been assessed, but pollution created a public outcry that was widely publicized in newspapers (A. Martínez-Yrízar, pers. comm.).

The impacts of livestock grazing on native ecosystems can vary considerably, depending on type of habitat, grazing intensity and management, and site history. In general, overgrazing can induce changes in plant species composition and vegetation cover and structure and cause soil compaction and erosion (Hobbs 1996; Manzano and Návar 2000). In fire-prone areas, overgrazing can indirectly affect fire regimes by altering the quantity and quality of fuels (Hobbs 1996). These impacts tend to be most severe under heavy grazing (e.g., Kauffman and Krueger 1984; Bock et al. 1993; Jemison and Raish 2000). Large areas of shrublands and chaparral are burned and seeded each year to promote conversion to grasslands for grazing; an estimated 108,000 ha are converted each year (Cavazos-Doria 1997).

Larger and larger areas of scrubland, desert, and tropical deciduous forest in northern Mexico are being converted to pastures of exotic grasses for grazing (Ceballos and García 1995; Martin and Yetman 2000). A particularly insidious problem is buffelgrass (*Pennisetum ciliare*), a highly variable Old World grass, widely introduced in hot, semiarid regions of the world for forage and fodder. Plantings in northern Mexico began in the 1950s, especially in Sonora and Tamaulipas, to boost rangeland production (Perramond 2000; Búrquez et al. 2002). In the state of Sonora at least 600,000 ha that once supported Sonoran Desert vegetation have been legally cleared and seeded with buffelgrass (Búrquez et al. 1998), and as much as 1.6 million ha may have been illegally planted in Sonora alone; these figures do not include the extensive areas where buffelgrass has naturalized (Búrquez et al. 2002). The federal agency Comisión Técnico Consultiva de Coeficientes de Agostadero (COTECOCA) has determined that about 4.8 million ha, or one-third of the state's area, is suitable for conversion to buffelgrass (Búrquez et al. 2002). Much of the unique and increasingly rare tropical dry forest habitat in southern Sonora and northern Sinaloa is threatened by conversion to large ranches planted in buffelgrass monoculture (Ceballos and García 1995; Martin and Yetman 2000). In many areas, buffelgrass has escaped from planted pastures to become dominant over large areas, and its range continues to expand rapidly. This grass increases litter levels substantially, is highly flammable, and resprouts rapidly after burning (Búrquez et al. 1998). Consequently, it has created a new pattern of frequent ground fires in desert habitats, killing the native vegetation that is not adapted to fire (Perramond 2000). Over time, desert invaded by buffelgrass shifts toward single-species stands of exotic grassland with greatly reduced species diversity, productivity, and structural complexity (Búrquez and Martínez-Yrízar 1997, in press; Búrquez et al. 2002).

Heavily overgrazed native grasslands tend to shift toward shrub-dominated habitats, perhaps ironically given that native shrublands and deserts are being actively converted to exotic grasslands (e.g., Humphrey 1987; Turner et al. 2003). Along portions of both sides of the U.S.–Mexico border, a century of heavy grazing has resulted in replacement of perennial grasses by woody shrubs such as mesquite, snakeweed (*Gutierrezia* spp.), and cholla (*Cylindropuntia* spp.). Heavy grazing also has promoted the spread of numerous invasive exotics, such as the purposely introduced Lehmann lovegrass (*Eragrostis lehmanniana*), among others (Tellman 2002). Heavy grazing, primarily by sheep, in the mountain meadows of Baja California has shifted the composition of those habitats from mostly perennial grasses to mostly weedy annuals (e.g., *Bromus tectorum, Erodium cicutarium*; Sosa-Ramírez and Franco-Vizcaíno 2001). The overgrazing of those meadows has caused the extirpation of species such as the California meadow mouse (*Microtus californicus*) from Mexico and is putting at risk endemic species such as the Baja California mole (*Scapanus anthonyi*; Ceballos and Navarro 1991; Yates and Salazar in press).

Excessive livestock grazing in Tamaulipan thornscrub has altered the original landscape from a mosaic of mixed woodlands and grassland to almost pure thornscrub (Reid et al. 1990; Fulbright 2001). Heavy browsing in woodlands creates a browse line and inhibits seedling recruitment (Bahre and Minnich 2001). In piñon–juniper woodlands, grazing reduces the amount of fine fuels; a reduction in ground fires promotes a shift from open woodland to closed-canopy woodland (Gottfried and Pieper 2000). In many of the few relatively intact riparian areas remaining in northern Mexico, grazing has reduced or eliminated recruitment of cottonwoods (*Populus* spp.) and willows (*Salix* spp.; Bojorquez-Tapia et al. 1985; De la Torre 2003). Riparian grazing also accelerates bank erosion, contaminates water, and can reduce fish, mammal, and bird populations (Kauffmann and Krueger 1984; Cartron et al. 2000).

It is difficult to generalize how grazing impacts wildlife, in part because of the broad range of habitat needs of different wildlife species, and because most impacts are indirect (Bock et al. 1984; Hayward et al. 1997). The conversion of grasslands to shrublands does not necessarily produce a decline in bird abundance, but it can cause a nearly complete turnover in species from grassland specialists to birds that are generalist shrub-associated species (Pidgeon et al. 2001). The composition and diversity of mammal communities in arid grasslands also changes substantially with overgrazing (see chapter 21). Most grassland species are able to survive in other habitats, but some habitat specialists, such as the white-sided jackrabbit (*Lepus callotis*), have disappeared over large areas of their geographic range (Ceballos and Navarro 1991). Similarly, the role of dominant seed predators in Chihuahuan Desert grasslands shifts from ants to rodents with shrub encroachment (Kerley and Whitford 2000). Non-native ungulates brought to northern Mexico by the Spanish arrived with a whole suite of diseases to which the native species were not resistant. The spread of cattle, sheep, and goats reduces the numbers of native animals such as the bighorn sheep (*Ovis canadensis*; chapter 19).

Private ranchers and some government officials accuse the *ejidatarios* of overgrazing their ranges, cluttering them with nonmarketable animals such as horses and burros, and not practicing selective breeding. *Ejidatarios* counter that private ranchers overgraze their ranges; both U.S. and Mexican environmentalists claim that private ranchers are destroying immense tracts of desert, thornscrub, and tropical deciduous forest to plant buffelgrass (Sheridan 2001). Desmond et al. (chapter 22) present a contrasting view that grasslands on private ranches are generally in better condition than those owned by *ejidos* and Mennonite colonies because rotational grazing systems are generally used on the private ranches. The *ejidos*, on the other hand, tend to use continuous grazing regimes with intensive summer grazing. Ranch size and numbers of participating members in the operations may be conditioning these choices. In contrast, a recent quantitative study in Sonora found no evidence that range condition was related to land tenure system (Coronado-Quintana and McClaran 2001). Clearly, conditions and cases vary across the board. Finally, it is important to recognize that some ranchers are moving toward more responsible practices including strong conservation measures (e.g., ANGADI 2003).

Forestry

Most of northern Mexico is arid, and as a consequence, woodlands and forests occur primarily at higher elevations. Wood for timber and other products is an important commercial resource in these highlands. Pines (*Pinus* spp.) are the most important timber trees. Mexico's montane coniferous forests are dominated by pines and are home to more than 45% of the world's pine species (Styles 1993). Middle elevations support a high diversity of oaks (*Quercus* spp.), which are often favored for fuelwood and pulp (Spellenberg 2001). Many of the diverse woody trees and shrubs of lower elevations are heavily exploited as well. Besides saw timber and fuelwood, other wood products extracted from northern Mexico's forests include charcoal, paper pulp, and pine resin.

Commercial timber extraction from the Sierra Madre forests began in the latter half of the nineteenth century. The first sawmill in the Sierra Madre Occidental was opened in Madera in northern Chihuahua, in 1880 (Lammertink et al. 1996). Many of the early harvests provided raw materials for industry in the United States as well as fuel for the railroads in Mexico. Large timber concessions were granted to U.S. businesses and were expropriated after the 1910 revolution, first to large Mexican concessions, and then parceled out to *ejidos*. Large concessions and state-controlled logging led to overharvesting and poor management (Guerrero et al. 2000a). Since the 1880s, most of the old-growth forests have been logged, and the remaining extensive primary forests were heavily exploited between 1950 and 1975 (Lammertink et al. 1996). In 1986, Mexico passed the Forestry Law to regulate the timber industry and diminish environmental impacts (Guerrero et al. 2000b). A 1992 law, however, reduced controls on logging activities and regulation, resulting in increased rates of logging (PROFEPA 1998; Guerrero et al. 2000b).

Chihuahua and Durango are Mexico's largest producers of wood products and together account for almost 50% of pine products in the nation (fig. 3.2), while Sonora has a smaller, but growing, timber industry (Bojorquez-Tapia et al. 1985). Chihuahua has 7.6 million ha of forested lands—more than any other Mexican state (SEMARNAP 1999). Most of these forested lands are owned by *ejidos*, which account for 90% of Chihuahua's timber production. Most of the timber cut (60%) goes to lumber for both internal use and export. An additional 29% is used for pulp and paper production, and the remainder is for plywood, chipped products, and other uses (Fisher et al. 1995).

Before NAFTA, most timber and wood products were intended for domestic use (SEMARNAP 1999). With NAFTA came an elimination of tariffs on wood products imported into Mexico and a subsequent increase in both imports and exports. The United States is the largest importer of Mexico's wood products (85–96%), with most of these originating in Chihuahua, northern Baja California, and Tamaulipas (Guerrero et al. 2000a). The increased demand has induced rapid growth in the forestry industry in northern Mexico. In the period 1993–1999, the number of sawmills in the region increased by 250%, and the value of wood-product exports increased by 73% (Guerrero et al. 2000a). Large multinational corporations have bought and consolidated many of the small pulp and paper operations of the region. This rapid growth in the industry also resulted in increases in the number of cutting permits issued and has sparked competition among the *ejidos* and private cutters for the best wood, putting increasing stress on forest ecosystems. Illegal cutting has increased tremendously, and it is estimated that the volume approaches 50% of the permitted cut (Guerrero et al. 2000a,b).

Since the 1960s, drug producers have cleared more and more areas of the Sierra Madre Occidental forests for marijuana and opium production. The region is considered the second-largest illicit drug-producing region in the world (Simon 1997). Poppy and marijuana cultivation in Mexico uses small, widely dispersed fields in remote and inaccessible regions of the western sierras. Since land used for illicit cultivation is subject to seizure, drug crops are usually cultivated on public or communal lands. Approximately 60% of the marijuana production in Mexico is concentrated in the northern states. Government eradication efforts use manual labor by the army or employ the use of helicopters to spray herbicides. The government eradicated nearly 15,000 ha of *Cannabis* during 2002 (DEA 2003). It is not known what effects these actions by both the cultivators and government cause, but the initial clearing, increased access, and eradication efforts cannot be beneficial for forests. As they have with cattle ranches, drug traffickers have purchased legitimate timber operations as a means to launder money (Guerrero et al. 2000b).

In the Sierra Madre Occidental, rotation cycles for logging are usually less than 20 years, and non-target species (i.e., non-pines) are often eliminated. These practices tend to produce young, even-aged stands of pure pine. Cutting is especially concentrated in the mesa forests on high plateaus and moderate slopes in the highest part of the sierra (2400–3000 m) because of their large, valuable trees and relatively easy access (Lammertink et al. 1996). Snags are increasingly cut for the production of paper pulp. Where cutting is less intense or more selective, mature oaks, madroño (*Arbutus* spp.), and noncommercial conifers like *Abies* may remain to provide an older component to the forest (Lammertink et al. 1996). At lower elevations in oak and oak–pine forests, relatively little large-scale logging occurs (Lammertink et al. 1996). However, recent increases in the demand for paper pulp have resulted in the commercial exploitation of large oaks in the Sierra Madre Occidental (Spellenberg 2001). Between 1980 and 1990, the annual rate of deforestation in northern Mexico's coniferous forests was 1.4%, compared to 6.6% in the tropical forests along the Gulf of Mexico (Deininger and Minten 1999). Perhaps surprisingly, timbering operations account for only a portion of the annual loss of coniferous forests and oak woodlands in northern Mexico. Substantial areas of forest are lost to high-intensity fires (up to 50% of the forested land total annual loss) or are cleared for grazing (28% of the forested land total annual loss; INE 1993).

As of 1997, Mexico produced 420,000 metric tons of fuelwood and charcoal annually, of which 53% comes from the northern states, especially Sonora and Tamaulipas (Cavazos-Doria 1997). In part because of the geographical and topographical distribution of people in northern Mexico, scrub and desert ecosystems and species tend to be overutilized as sources of fuelwood and charcoal (Bojorquez-Tapia et al. 1985). For example, many areas of Tamaulipan thornscrub have shifted from species-rich shrub communities to low-diversity mesquite or cactus scrub primarily due to fuelwood cutting of preferred woody species. Trees preferred for fuelwood include *Acacia, Celtis, Condalia, Haematoxylum, Pithecellobium* (sensu lato), *Prosopis*, and *Quercus* (Cavazos-Doria 1997; Spellenberg 2001).

Mesquite, the preferred cooking fuel from ancient to present times, accounts for the majority of charcoal production in arid regions of northern Mexico, especially the Sonoran Desert (Nabhan and Carr 1994; Felger et al. 2001). Mesquites, among the most important structural components of the arid and semiarid regions of North American (Simpson 1977), are harvested both legally and illegally from northern Mexico. For example, many of the extensive riparian and floodplain forests of mesquites in Sonora have disappeared due to long-term local cutting for heating and cooking fuel (Felger 1977), and remaining stands are rapidly disappearing due to demand for charcoal in northern Mexico and U.S. markets (Nabhan and Carr 1994; Búrquez and Martínez-Yrízar 1997, in press). Major sources of charcoal include mesquites and desert ironwood trees (*Olneya tesota*) bulldozed to clear desertscrub and thornscrub for buffelgrass pasture (Búrquez and Martínez-Yrízar in press). In addition, desert ironwood has been increasingly exploited for carvings by the Seri Indians (Felger and Moser 1985) and by Seri imitators for the tourist trade (Suzán et al. 1997). Desert ironwoods, among the oldest individual plants in the Sonoran Desert, as well as mesquites, have experienced shifts in population structure toward smaller sizes (Búrquez et al. 2002; Búrquez and Martínez-Yrízar in press). Old-growth desert ironwood is a major community structuring element across much of the Sonoran Desert, allowing the persistence of many species and forming islands of diversity under its canopy (Búrquez and Quintana 1994; Búrquez and Martínez-Yrízar 1997, in press; Suzán et al. 1997; Búrquez et al. 2002).

Since the early 1970s *vara blanca* (*Croton fantzianus* in Sonora and Sinaloa and related species in Sinaloa) has become an important resource from the tropical deciduous forest of southeastern Sonora

(Muro Garcia 1987; Bye 1995; Lopez Urquidez 1997; Lindquist 2000; Felger et al. 2001). The slender, straight stems are harvested in great quantity for support stakes for chilies, cucumbers, eggplants, grape, and tomatoes in northwestern Mexico including Sonora, Sinaloa, and both states of Baja California. *Vara blanca* is a structurally important small tree in the tropical deciduous forest of the region and the current extraction rates cannot be sustained (Lindquist 2000).

Wood-harvesting practices in northern Mexico have altered the structure and composition of forests and woodlands. Certain tree species and sizes are selectively exploited, and as a consequence they are decreasing in abundance (Reid et al. 1990; Bahre and Hutchinson 2001). By 1995, the original old-growth pine-oak forests of the Sierra Madre Occidental had been reduced to 571 km^2, or about 0.6% of the original forest area (Lammertink et al. 1996). Clandestine cutting threatens the 3 rare, endemic spruce (*Picea*) species of the Sierra Madre Occidental and Oriental (Ledig et al. 2000). Logged forests support extensive networks of logging roads, promoting erosion and homesteading. Forest regeneration lags well behind the rate of loss due to cutting, fire, and conversion (Fisher et al. 1995). Oak and oak–pine woodlands have essentially disappeared from many areas due to intensive cutting for fuelwood, such as near smelters and urban areas. In particular, these woodlands were gone from the vicinity of Ciudad Chihuahua by 1945 (Spellenberg 2001). Even relatively low-intensity cutting of oaks for fuelwood changes the structure from large individual trees to many-stemmed, shrubby thickets (Bahre and Minnich 2001).

Changes in forests and woodlands also affect species that depend on these environments. In many areas, large snags have disappeared through direct harvesting and a lack of regeneration due to short rotations. The imperial woodpecker (*Campephilus imperialis*), the world's largest woodpecker, once ranged throughout the high elevation Sierran forests (Leopold 1949; Tanner 1964; Fuller 2001). The loss of tree snags on which it foraged, loss and modification of the old-growth forests, and hunting all seem to have contributed to its extinction by the mid-twentieth century (Lammertink et al. 1996; Ceballos and Márquez 2000). The thick-billed parrot (*Rhynchopsitta pachyrhyncha*) has declined drastically as well, the most serious threat being loss of nesting sites by cutting of snags and mature pines (Lanning and Shiflett 1983). Impacts on smaller, less

prominent species of old-growth forests remain largely unknown.

Nontimber Products

Although timber and other wood products dominate international commerce, the harvesting of nontimber products is important, especially in domestic markets. Nontimber products may constitute the primary income in certain economically marginal regions. These products include resins, gums, oils, waxes, fibers, food, pharmaceuticals, and other materials. They are primarily produced in Michoacán, Tamaulipas, Zacatecas, Baja California (Norte), Coahuila, Veracruz, San Luis Potosí, and Nayarit. This commerce declined from almost 200,000 tons in 1990 to about 100,000 tons in 1995. Much of the reported nontimber production in Mexico is pine resin, and the decrease is mainly due to the replacement of these products by petroleum derivatives (SEMARNAP 1999).

Candelilla (*Euphorbia antisyphilitica*) for wax and guayule (*Parthenium argentatum*) as a source of natural rubber have been important wild harvests in the Chihuahuan Desert (Sheldon 1980). Candelilla has been extirpated from sizeable areas due to overharvesting for expanding export markets (Cavazos-Doria 1997). Guayule is no longer a wild-harvested crop, although there is interest in the plant as an agronomic crop in part because the latex is nonallergenic (Wood 2002). A few other examples of wild harvests in northern Mexico include fiber from yuccas (*Yucca* spp.) and ixtle fibers from lechuguilla (*Agave lechuguilla*; Gentry 1982). The harvest of mescal primarily for alcoholic beverages is prevalent in the mountains of mainland northern Mexico, for which Gentry (1982) lists 11 species commonly harvested. Most famous is *bacanora*, named after the town of the same name, made from wild populations of *A. vivipera* (*A. angustifolia*) in eastern Sonora. The plants are harvested as they begin to develop their inflorescences, thus preventing seed production. Nabhan (1985) estimated one-half million plants are harvested each year. Although there is concern over potential overexploitation, Nabhan (cited in Hodgson 2001) argues that at least some Sonorans know how to manage the plants in favor of regeneration (see also Valenzuela-Zapata and Nabhan 2004).

Other natural products include herbs and condiments such as Mexican *orégano* (*Lippia berlandieri*) and Chihuahuan *orégano* (*Aloysia wrightii*). The fiery little chilies, known in Mexico and the southwestern United States as chiltepines (*Capsicum annuum* var. *aviculare*) continue to be wild harvested in the mountains of northwestern Mexico (Felger et al. 1997). They are an indispensable part of local and many international cuisines. During a dry year, roughly 20 tons of dry fruits of wild chiltepines have been harvested in Sonora, and as much as 50 tons might be harvested during a wet year (De Witt 1991). The 1990 export to the United States was about 6 tons, with a retail price of US$72/pound (Nabhan 1990). Techniques for commercial wild harvest include tearing away large parts of the plant in a rather destructive harvest method, although smaller branches are more often harvested with less impact. Currently there is a movement to supplement or replace wild harvest with cultivated plants.

The fruits or pods of several species of mesquites (*Prosopis*, section *Algarobia*) in northern Mexico have been harvested and processed primarily for cattle feed as well as high-quality mesquite flour for human consumption (Silbert 1988; Figueiredo 1990). Processing of the pods has declined in the Chihuahuan Desert Region, but in northwestern Sonora, for example, Ivan Aguirre (pers. comm.) of Rancho la Inmaculada milled about 4 tons of mesquite flour in 2003 and is supplying the non-profit organization "Native Seed/SEARCH" in Tucson and others. Harvest of mesquite pods is likely to expand significantly and represents a significant nontimber, renewable resource (e.g., Felger 1990a).

Acorns, or *bellotas*, from Emory oak (*Quercus emoryi*) in northeastern Sonora and northwestern Chihuahua and adjacent Arizona and New Mexico are wild harvested each summer (Felger et al. 1997). A favorite of Sonorans and many Arizonans, the seeds are eaten fresh. They are sold in Sonoran markets and sometimes can be purchased in southern Arizona. The retail price in 2002 was US$5 per pound. The shelled acorns are edible with no preparation due to a relatively low tannin content (in contrast to those of many other oak species). These acorns are dietetically significant because they have an extremely high glycemic index value (Brand et al. 1990).

A number of species of columnar cacti in Mexico yield delicious fruits and are seasonably available in local markets. The demand is high. Desirable species in the northern part of the country include *garambullo* (*Myrtillocactus geometrizans*) in the Chihuahuan Desert (e.g., Anderson 2001) and organ pipe or *pitaya dulce* (*Stenocereus thurberi*) and *pitaya agria* (*S. gummosus*) in the Sonoran Desert (Felger and Moser 1985; Nabhan 1985; Felger et al. 1997, 2001; Paredes et al. 2000). Many other species likewise yield delicious fruits, but these seldom are seen in markets (Felger et al. 1997; Bustamante and Búrquez in press). The *pitaya agria* is one of the most delicious fruits in the world (Felger et al. 2001). *Pitaya agria* ice cream, sometimes available in Baja California Sur, is an extraordinary treat (Reina-Guerrero et al. 2000).

Several species of barrel cacti (e.g., *Ferocactus haematacanthus*, *F. histrix*, and *F. pilosus*) in the Chihuahuan Desert Region are widely exploited for their edible flower buds and fruits, which are sometimes canned (Anderson 2001; del Castillo and Trujillo 1991). The fruits are also used to make alcoholic and nonalcoholic drinks. Certain large barrel cacti such as *F. histrix* and *Echinocactus platyacanthus* (a giant barrel cactus with a massive stem 0.5–2.5 m tall, called *biznaga de dulce* or *biznaga gigante*) are destroyed for cattle fodder and to make a candy called *dulce de acitrón* (Bravo-Hollis and Sánchez-Mejorada 1991). Their extensive exploitation is a conservation concern.

Medicinal Plants

Of some 22,000 species of plants in Mexico, it is estimated that up to 5000 species may have medicinal uses (Lozoya 1994; Aguilar 1999). Of 3000 Mexican species with documented medicinal uses, only about 1% have been studied in any detail for their pharmacologic properties (Argueta 1994). An estimated 10% of the flora in arid regions of northwestern Mexico has been used by local people for medicinal purposes, and many continue to be used regularly (Felger and Moser 1985; Bañuelos and Búrquez 1996). The market for medicinal plants from Mexico is significant for local economies (Hersch-Martínez 1995), and like other medicinal markets elsewhere, exports will only increase. For a variety of reasons, however, among them the generality of import categories (in the United States medicinal plants may be classified in data and analysis as NESOI—an acronym which means "not elsewhere specified or indicated"), real figures for exports of medicinal plants from Mexico are difficult to obtain. During 2002, the declared value of Mexican imports into the United States of plant parts used primarily for "perfumery, pharmacy or for insecticides" approached US$7 million (ITA 2003), but this seems an underestimate.

The world market for herbal remedies has increased 10% per year for the past decade and now approaches US$20 billion. In a survey conducted between 1998 and 1999, about 49% of American adults, some 100 million people, tried remedies from plants (Castleman 2001). The use of alternative medicine, of which herbal remedies are a part, is growing rapidly, and the use of "nutritional supplements," many of which have herbal components, is likewise growing rapidly (Foster and Tyler 1999; Karch 1999).

As important as the harvest and sale of medicinal plants is or will become, however, the collection of medicinal plants is not without effects. Two-thirds of the estimated 50,000 medicinal plants used around the world are harvested from the wild. Collection is usually carried out by low-income and socially disadvantaged groups. In Mexico, the number of people collecting medicinal plants is increasing. New collectors pushed into trade by unfavorable economic circumstances do not observe traditional harvest practices and thus imperil medicinal plant populations (Hersch-Martínez 1995). Up to a fifth of wild medicinal plants worldwide may be in varying states of jeopardy because of unsustainable collection (Reuters News Service 2004). In northern Mexico, for example, osha, or *chuchupate* (*Ligusticum porteri*), is a robust perennial herb with a large, thickened, and aromatic root. It grows in the mountains of the southwestern United States and eastern Sonora and Chihuahua and is one of the best known medicinal herbs in the region (Yetman and Felger 2002). *Chuchupate* is used to treat a wide range of ailments (Bye 1986; Rea 1997) and is harvested for domestic use and exported to such countries as the United States, Japan, and Germany (Felger et al. 1997). Overcollection has seriously diminished many populations, and this moisture- and shade-requiring plant also is declining as forests are cleared and arroyo heads erode. TRAFFIC North America classified populations of *chuchupate* as declining (Robbins 1999), and there have been efforts to list the plant under CITES Appendix II (University of Maryland Program in Sustainable Development and Conservation Biology 1999).

Bioprospection in the Tarahumara region, with support from the International Cooperative Biodiversity Group via University of Arizona and the Jardín Botánico del Instituto de Biología, Universidad Autónoma de México, in Mexico City, has focused on various plants of potential pharmaceutical and agrochemical potential (Bye 2003). This project is establishing the parameters for the sustainable use of medicinal plants including *Ligusticum*. Mexican organizations working with indigenous people desire to carry the commercialization to a higher level, but an American corporation has patented the active compounds of *Ligusticum*, posing a serious bioethical problem (R. Bye, pers. comm.).

Fisheries

Mexico's largest commercial fisheries are in the northern coastal states of Baja California, Baja California Sur, Sinaloa, Sonora, and Tamaulipas (SEMARNAT 2000b; fig. 3.2). The Gulf of California in particular abounds with fish and shellfish (chapter 9). It was not until the mid-twentieth century, however, that fisheries became a major commercial activity in Sonora and Sinaloa (West 1993; Doode and Bañuelos 1995). Shrimp and sardines make up the bulk of the catch in these 2 states, and shrimp predominates in Tamaulipas (SEMARNAT 2000b). Three-quarters of the shrimp caught in these 3 states is exported to the United States. Pelagic fisheries are also important. Mexico takes a significant portion of the tunas and sharks caught in the Americas (FAO 1997) and ranks fourth in the world in shark harvests (CONABIO 1998). The intensive and largely unregulated overfishing of large sharks and rays is alarming, with the take in most areas being unsustainable and in urgent need of control (e.g., Castillo-Geniz et al. 1998; Márquez-Farías 2000; SAGARPA 2003). Continued harvests of these top predators at current levels will certainly have impacts on entire marine ecosystems (e.g., Applegate et al. 1993).

Functionally unregulated commercial and sports fisheries have had a very pronounced impact on marine life in the Gulf of California. Most commercially important species have been harvested well above the level of sustainability; these include shrimp, groupers, snappers, corvinas (sea trouts), yellowtail, billfishes, and sharks (Thomson et al. 2000). Shrimp trawling results in large volumes of bycatch of nontarget species, most of which is discarded, and has destroyed large areas of the seafloor ecosystems in northern Mexico (Brusca in press; chapter 9). As a result of all these factors, fish and shellfish stocks have declined, and harvests have plummeted (Barry et al. 1994; Búrquez and Martínez-Yrízar 1997, in press; chapter 9).

The vaquita (*Phocena sinus*), the smallest species of porpoise and the most critically endangered marine cetacean, is endemic to the northernmost Gulf of California, but is rapidly disappearing largely because of bycatch in gillnets (D'Agrosa et al. 2000; chapter 14). Attempts to outlaw gillnets or regulate the fishing industry have met with little success. The major target species of the gillnetters in the upper Gulf of California was the totoaba (*Totoaba macdonaldi*), which is also endangered by overfishing, bycatch of young by shrimpers, and especially habitat degradation of its spawning grounds (Cisneros-Mata et al. 1995; Hastings and Findley in press; chapter 9). Despite legislation to protect the species in Mexico, in the United States, and internationally, some poaching of totoaba continues (e.g., Hastings and Findley in press). In the Gulf of Mexico, shrimp trawlers constitute the greatest anthropogenic cause of mortality for sea turtles (McDaniel et al. 2000; see also chapter 20).

In recent years, the wholesale destruction of scarce coastal wetlands, estuaries, lagoons, mangroves, and other coastal habitats for saltwater aquaculture has grown exponentially, primarily for the production of shrimp and to a lesser extent for shellfish (Búrquez and Martínez-Yrízar 1997, in press; CONABIO 1998; see chapter 15). These ecosystems provide critical breeding, spawning, and nursery grounds for numerous marine species (chapter 9), and continuing losses of these habitats are likely to have severe impacts. The development of shrimp aquaculture contributes to locally intense pollution of coastal waters from the discharge of nutrients and organic matter (Paez-Osuna et al. 1998). Freshwater aquaculture is also increasing in Mexico, where the major harvests are carp (*Cyprinus carpio* and *Ctenopharyngodon idella*), tilapia (*Oreochromis* spp.), and various catfishes destined for the domestic market (Hernández-Rodríquez et al. 2001).

Hunting

Unregulated subsistence, predator control, and trophy hunting are widespread in northern Mexico. The effects of such activities have strong impacts on wildlife. Subsistence hunting has decimated many species of medium- and large-size birds and mammals, and its negative effects have increased with increasing human population (Leopold 1959; Aranda 1991; Ceballos and Navarro 1991). Predator control and unregulated sport hunting also have had severe impacts on a variety of large species of mammals, including the Mexican wolf (*Canis lupus baileyi*), black bear and grizzly bear (*Ursus americanus* and *U. arctos*), jaguar (*Panthera onca*), desert pronghorn antelope (*Antilocapra americana*), and bighorn sheep (Brown 1985; Cavazos-Doria 1997; Kourous 1998; chapter 19).

Regulated hunting is an economic activity that can create benefits for both local communities and wildlife, especially if hunting quotas are properly implemented. Several species such as bighorn sheep have recovered from the brink of extinction with incentives of regulated hunting and solid scientific research (chapter 19). Each year more than 10,000 sport hunters come from the United States to hunt in Mexico (SEMARNAT 2001), with unknown effects on wildlife. The long-term effects of trophy hunting may have undesirable consequences. For example, a decline in horn size for bighorn rams in Alberta, Canada, due to trophy hunting (Coltman et al. 2003) has been called a "decline of the fittest" (Whitfield 2003).

International Trade in Wildlife and Plants

The U.S.–Mexico border is a hot spot of trade in exotic wildlife species (Barry et al. 1994). This trade is second only to drug smuggling in terms of volume and money involved, and it has been predicted to increase with NAFTA (Barry et al. 1994). Traffic in pelts of endangered felines, such as jaguar and ocelot, was contributing to the disappearance of these animals in Mexico (Redford and Robinson 1991), but such traffic has decreased since the Convention on International Trade of Endangered Species (CITES) entered into force (Ceballos and Eccardi 2003). Products made from crocodile and sea turtles pose threats to those species throughout the country (chapter 20). Trade in these species is prohibited by CITES (Hemley 1994), but regulatory enforcement remains inadequate (Iñigo-Elias and Ramos 1991).

Wild birds make up the bulk of international traffic in live wildlife, and of these, parrots are the most economically important (Beissinger 2001; Wright et al. 2001). Mexico was one of the largest legal exporters of Neotropical parrots until 1982, when it imposed a ban on the export of indigenous species (Thomsen and Mulliken 1992). Meanwhile, the rate of importation of birds by the United States declined in 1992 after the passage of the Wild Bird Conservation Act (Jorgenson and Thomsen 1987; Hemley 1994). Illegal trade continues, however, and

Mexico is not only the source of many wild-caught parrots smuggled into the United States, but also serves as a conduit for species from other countries (Iñigo-Elias and Ramos 1991). An estimated 150,000 exotic birds are smuggled into the United States each year (Barry et al. 1994; CONABIO 1998). Smuggling of birds has the added risk of transporting such potentially devastating pathogens as Newcastle disease (Cooper 1989).

Although all 21 Mexican parrot species are exploited by the pet trade to varying degrees, not all species are equally affected. Those hardest hit tend to be the largest and least abundant species that command the highest prices in the United States and Europe, particularly macaws (*Ara* spp.), the thick-billed parrot, and Amazon (*Amazona* spp.) parrots (Wright et al. 2001). In the north, Amazon parrots are extensively harvested in Sinaloa, Nuevo León, and Tamaulipas (Iñigo-Elias and Ramos 1991). Up to 60% of birds taken from the wild die of suffocation, starvation, or inhumane treatment in attempts to bring the birds into the United States (Iñigo-Elias and Ramos 1991). The level of harvest for domestic markets is unknown. Thus, U.S. import statistics grossly underestimate the true impact of the trade on wild populations. For example, the 3 *Amazona* species found in Tamaulipas are now rare because of excessive trapping and deforestation (Thomsen and Brautigam 1991; Wright et al. 2001).

The exploitation of Mexico's indigenous biota is not limited to birds. Many reptile species have been commercially important in northern Mexico. An alarming number of endangered, vulnerable, and insular endemic reptiles of the Baja California peninsula and Gulf of California islands are threatened by illegal trade for the exotic pet market (Mellink 1995a; Grismer 2002). Certain areas have been heavily degraded by collectors breaking apart rock piles searching for high-priced species such as the California mountain kingsnake (*Lampropeltis zonata*), rosy boa (*Lichanura trivirgata*), and banded rock lizards (*Petrosaurus* spp.; Mellink 1995a). The exotic pet trade is thought to negatively affect approximately 32% of reptile species in Mexico (CONABIO 1998). The use of crocodilians for luxury leather has declined markedly since the first half of the twentieth century due to overexploitation (Flores-Villela 1993). The subsequent disastrous illicit and legal trade in sea turtles was due in part to Mexican government promotion because of the serious decline in crocodile resources, resulting in great demand for sea turtle hides (chapter 20).

There is an extensive trade, much of it illegal, in plants such as cycads, orchids, and cacti collected from the wild (Hemley 1994; Floyd 1998; chapter 13). There is a great demand for cacti among collectors in the United States, Europe, and Japan (Floyd 1998). Although most cacti in trade are now artificially propagated, an estimated 15% of the 7.5 million cacti traded internationally each year come from the wild (Hemley 1994). Since Mexico became signatory to CITES in 1992, legal exports of most kinds of cacti have been halted; exceptions include prickly pears (*Opuntia* spp.) and limited exports for scientific research under a highly restrictive permit system. Smuggling, however, is rampant (Floyd 1998). Partly because of well-enforced regulations within the United States, various plants destined for the landscape trade in the southwestern United States or for collectors in Europe are often removed from the deserts of Mexican border states, especially in the Chihuahuan Desert. There seems to have been a reduction in plants smuggled into the United States, especially cactus and succulents, since the 1990s (Robbins 2003), but this is difficult to gauge. Most of the numerous cactus and succulent societies and sponsored exhibits discourage or no longer admit wild-collected plants, but the interest in wild-collected plants elsewhere remains strong.

There is considerable illicit trafficking in decorative architectural products such as giant cactus "ribs" from saguaro (*Carnegiea gigantea*) and cardón (*Pachycereus pringlei*) in northwestern Mexico. In addition to collecting the dry, woody ribs from dead plants, there is an alarming practice of girdling living cacti and returning later to harvest the ribs (Felger unpubl. field notes; J. Floyd, pers. comm.). These are collected without proper permits and are covered by CITES regulations in the same way as living plants.

Ecosystem-Level Consequences of Land Use Impacts

A direct consequence of northern Mexico's expanding human population and burgeoning economy is the loss and conversion of native habitats (Ingram 1998). Nearly all natural areas have been affected, but the magnitudes vary with the degree and intensity of human use as well as with resilience of particular habitats. For example, much desert habitat is being converted into non-native grassland, and native grasslands are being converted to shrublands due to grazing pressure and changes in fire regimes. Still, northern Mexico retains some of the continent's largest and healthiest expanses of desert grassland (chapter 21). Wooded habitats are especially hard hit due to wood harvesting, and Mexico has one of the highest rates of deforestation in Latin America (CONABIO 1998; Gómez-Pompa and Kaus 1999). The net rate of deforestation accelerated markedly after 1980, largely because the rate of cutting increased (Fisher et al. 1995). In central Sonora the area covered by Madrean evergreen woodland and Sonoran Desert scrub decreased 28% and 31%, respectively, between 1973 and 1992 (Valdez-Zamudio et al. 2000). During this same period, Mexico lost 65% of its mangrove communities to direct exploitation, agricultural, and urban development (Herrera-Silveira and Ceballos-Cambranis 2000), and in northern Mexico major losses have also been due to development for tourism and shrimp aquaculture (chapter 15).

Riparian woodlands have suffered severe impacts due to the concentration of agriculture and urbanization along watercourses, construction of dams, and woodcutting. The condition of northern Mexico's riparian habitats is poorly documented compared to those in the southwestern United States, but it has been suggested that they are perhaps in even worse shape (Kourous 1998). In Sonora, for example, only vestiges of once extensive cottonwood-willow forests remain along the major rivers, and the forests of the lower Río Colorado (Colorado River) died during the first half of the twentieth century (Felger 2000; Felger et al. 2001). In general, riparian areas in the southwestern United States and northwestern Mexico are considered to be among the most threatened habitats in North America (e.g., Ceballos 1985; Manci 1989). Although they make up only a small fraction of the land area, riparian habitats support a disproportionate amount of the region's biodiversity, and their destruction has an especially profound impact on wildlife and plant populations (Ezcurra et al. 1988; Cartron et al. 2000).

Erosion and Desertification

As the suite of human land uses continues, the impacts of those uses will increase in severity and extent. Since the 1970s, the rate of soil erosion in Mexico has accelerated dramatically and is now recognized as one of the most critical problems facing the country (CONABIO 1998). A countrywide study categorized 15% of Mexico's land as "totally eroded," another 26.2% as suffering from "advanced erosion," and 24% as showing "moderate erosion" (Klein-Robbenhaar 1995). More than 20 million ha of land in the northern tier of states is considered to be at high risk of erosion (SEMARNAP 1999).

Northern Mexico suffers one of the world's highest rates of desertification, a process defined as land degradation in arid and semiarid regions, with loss or reduction of biological or economic productivity and complexity of natural ecosystems, croplands, and pastures (Landa et al. 1997). Within Mexico, the process of desertification is most extreme in the northwestern states of Sonora and northern Sinaloa (CONABIO 1998). The many causes of desertification include land clearing for buffelgrass; clearing for agriculture, which is often abandoned in a few decades; woodcutting; and overgrazing (e.g., Felger et al. 2001; Búrquez et al. 2002). As a result of desertification (including soil salinization), approximately 2250 km^2 of potentially productive farmlands in Mexico are taken out of production or abandoned each year. Balling (1988) compared soil and surface air temperatures between adjacent areas on the 2 sides of the Sonora-Arizona border. The higher temperatures detected in the overgrazed and devegetated areas on the Sonora side are likely caused by the lack of shading vegetation, which leads to higher soil evaporation, a factor contributing to desertification. One consequence is the local acceleration of global warming trends: a study in Sonora and Arizona documented significantly higher rates of warming in Sonora (Klopatek et al. 1996). In Sonora, the combined effects of invasion by exotic buffelgrass and overgrazing give the perception of worsening drought conditions, a perception that is not substantiated by climate data (Búrquez and Martínez-Yrízar 1997, in press).

Fresh Water Resources: Rivers, Fish, and People

The exploding human population in northern Mexico, coupled with pollution generated by urban areas, industry, and agriculture, is straining the already inadequate water resources in this arid region. Population growth has occurred nationwide but has been greater in the semiarid and arid northern and central regions, which are precisely the regions where major water shortages occur (Kemper and Alvarado 2001). As in the southwestern United

States, the few rivers are grossly overexploited. There are severe water deficits across all of northern Mexico (CONABIO 1998), and continued uncontrolled growth will only exacerbate existing problems. The great aquifers along the Mexico–U.S. border are being tapped at unsustainable levels. For example, the Hueco Bolson under Ciudad Juárez/El Paso is currently pumped at a rate predicted to exhaust the aquifer before 2035 (Kaye 1995; Ingram 1998).

As in the southwestern United States, all major waterways in northern Mexico are dammed (see chapter 7). Mexico ranks seventh globally in the number of water projects constructed for irrigation, with more than 1330 large- and medium-sized dams, and in general the environmental impacts are poorly documented (Tortajada 2002). Perhaps typical is the large Huites Dam on the Río Fuerte in Sinaloa, constructed between 1992 and 1995, to provide irrigation water, hydroelectric power, and flood control. The dam created a 16,000-ha lake that destroyed extensive areas of tropical deciduous and riparian forest habitats, as well as indigenous communities. The lake has been stocked with a variety of non-native game fish (bass, carp, catfish, and tilapia) to attract foreign sportsmen (El Pescador Tour Services 2002), with unknown effects on the local fish fauna. Tragically, the rich biota of the Huites canyon system remained undocumented at the time of its annihilation.

The drastic changes of the Colorado River and its delta have parallels with other large rivers in arid regions, not only in northern Mexico but worldwide (Sykes 1937; Felger 2000; Bergman 2002). The Colorado delta once supported vast riparian forests and wetlands (Sykes 1937; Felger 2000), and during the nineteenth century steamships regularly plied the river from its mouth to Yuma (Lingenfelter 1978). Since construction of twentieth-century dams and diversions, virtually all the water has been apportioned for upstream use, and during most years the Mexican portion of the river is dry. The great cottonwood (*Populus fremontii*) and willow (*Salix gooddingii*) forests and their associated biota disappeared, while areas not under irrigation came to be dominated by invasives such as tamarisk (*Tamarix ramosissima*; Felger 2000). Famous places like Leopold's (1949) "Green Lagoons" in the Ciénega de Santa Clara all but disappeared. Since the 1980s, however, due to bizarre political and economic developments involving the Yuma desalination plant, the 20,000-ha Ciénega de Santa Clara

wetland has been resuscitated (Glenn et al. 1992, 1996; Felger 2000). Unfortunately, the U.S. Bureau of Reclamation has been directed by the Energy and Water Development Appropriations Act of 2004 (H. R. 2754) to expedite its modifications of the Yuma desalination plant necessary to allow operation of the facility (Pitt et al. 2002). Without mitigation, this development will most probably bring the brief resurrection of the Ciénega de Santa Clara to an end. The Ciénega, one of the largest and most important wetlands in northwestern Mexico, supports rich and varied wildlife including the endangered desert pupfish (*Cyprinodon macularis*) and an enormous and diverse avifauna including the endangered Yuma clapper rail (*Rallus longirostris yumanensis*; Glenn et al. 1992; Felger 2000; Hinojosa-Huerta et al. 2001, in press; Felger and Broyles in press).

Since about 1981, some intermittent flows have returned to the Mexican portion of the Colorado River, including periodic overbank flooding, promoting the return of a small fraction of the native-tree vegetation (Glenn et al. 2001; Zamora-Arroyo 2001). Some animals such as muskrat (*Ondatra zibethicus*) and beaver (*Castor canadensis*), considered "At Risk" by the Mexican federal government continue to survive in the remaining but rather artificially supported wetlands (Mellink 1995*b*; Mellink and Luévano 1998).

Irrigation and its attendant problems of water loss, salinization, and pollution, tend to drastically reduce or alter native fish communities (Williams et al. 1989; Lemly et al. 2000). Indeed, most recent extinctions in Mexico have been freshwater fishes, with at least 25 recorded such events (Ceballos 1993; Contreras-Balderas et al. 2003). For example, *Cyprinodon inmemoriam*, from a tiny desert spring in the isolated Bolson de la Sandía, Nuevo León, was described after it became extinct (Lozano-Vilano and Contreras-Balderas 1993). Native freshwater fishes are now absent or generally extremely rare or threatened in the lower Colorado River, with the exception of the amazingly adaptable desert pupfish in artificially maintained wetlands (Minckley 2002; Glenn and Nagler in press; Hastings and Findley in press). Species such as Colorado pikeminnow (*Ptychocheilus lucius*), the largest native fish in the system, known to attain 1.5 m in length, have vanished from the Mexican part of the river. Instead of the native freshwater fish fauna is a bewildering array of 38 introduced species (Minckley 2002; Mueller and Marsh 2002). "The future is grim for native fish in the Lower Colorado River" (Mueller and Marsh 2002: 65).

Most towns in northern Mexico, including large cities like Ciudad Juárez, lack adequate wastewater treatment (Barry et al. 1994). Uncontrolled population growth, agriculture, and industry have outpaced the capacity of local governments to manage the associated sewage and pollution. The result has been the devastation of most freshwater systems. The level of untreated sewage released into the Tijuana River in northwestern Baja California grew fourfold between 1983 and 1992 to reach 12+ million gallons/day (Castillo and Perry 1992). The New River, in the Colorado River Valley, drains sewerless parts of Mexicali and has become one of the continent's most polluted waterways (Barry et al. 1994). Sewage and industrial effluents in the Río San Juan below Monterrey have created an especially toxic environment (Contreras-Balderas and Lozano-Vilano 1994).

The pollution problem is exacerbated by illegal dumping of industrial waste, much of it created by maquiladoras (Lenderking 1996). In 1990, 1035 of the 1850 maquiladora factories in northern Mexico were identified as major generators of hazardous waste (Castillo and Perry 1992). For example, more than 120 volatile organic compounds were detected in the New River, where it flows into the United States (Castillo and Perry 1992).

Losses of Native Species

Obviously, as habitats are destroyed, converted, or polluted, the flora and fauna dependent on those habitats have declined. Salinization of the lower Rio Grande has become so severe that 32 native freshwater fish species have been replaced by 54 marine or salt-tolerant species as far as 400 km from the ocean (Contreras-Balderas and Lozano-Vilano 1994; Kaye 1995; Kourous 1998). As of 1989, at least 11 native freshwater fish became extinct in northern Mexico, and at least another 120 are threatened (Miller et al. 1989; Contreras-Balderas and Lozano-Vilano 1994).

It is often difficult to determine extinction statistics exclusively for the northern states for other taxa. For Mexico as a whole, the World Conservation Union (IUCN) identifies 22 plant and animal species that have become extinct in historical times; another 7 persist only in captivity, 68 are Critically Endangered, 151 Endangered, and 259 are classified as Vulnerable (Hilton-Taylor and Mittermeier

2000). A more precise estimate indicates that between 47 and 52 vertebrate species have become extinct or extirpated from Mexico since 1600; 19 are freshwater fishes (Ehrlich and Ceballos 1997; CONABIO 1998; Ceballos and Eccardi 2003). Unfortunately, these numbers underestimate the actual extinction problem because they do not take into consideration species such as the grizzly bear that have disappeared in Mexico but still survive in other countries. And these numbers do not consider the enormous number of species with populations having compromised geographic or ecological ranges that also face untimely extinction. Overall, 28% of the vertebrates are threatened to some degree (SEMARNAT 2001). Freshwater fish and endemic cacti make up a disproportionate number of threatened taxa. Perhaps what is remarkable is that the loss of biodiversity has remained as low as it has, given the extreme diversity and very high levels of endemism in Mexico (Gómez-Pompa and Kaus 1999).

Exotic Species

Invasive exotic plant species are a major ecological and economic problem worldwide (Pimentel et al. 2000). Northern Mexico is no exception, although some areas are less hard hit than similar habitats in adjacent parts of the United States (Felger 2000; Pimentel et al. 2000; Tellman 2002). For example, comparisons between montane meadows in California and Baja California show that exotics dominate the U.S. habitats but tend to be less common on the Mexican side (Sosa-Ramírez and Franco-Vizcaíno 2001; chapter 18). Some of the more arid, remote, and undisturbed parts of northern Mexico indeed have surprisingly low percentages of non-native plants (Wilson et al. 2002). For example, no non-native plants were found among the flora of 111 species in the Sierra del Rosario, a small range in the extremely arid Pinacate Biosphere Reserve in northwestern Sonora, although several non-natives occur on nearby dunes (Felger 2000). In undisturbed, desert areas of northwestern Sonora, only 20 non-native plant species, or 3.8% of the total flora, are present as well-established, reproducing populations, and including "natural" and disturbed but nonirrigated habitats, 58 (11%) of the species are non-native (Felger 2000). In the Municipio de Yécora (3300 km^2) in eastern Sonora, 14% of the grass species are non-native, and of those only buffelgrass and Natal grass (*Melinis repens*) represent ecological threats (chapter 10). The mid-twentieth century flora of the Sonoran Desert (Wiggins 1964) included approximately 146 non-native species, or 5.7% of the flora (Felger 1990*b*). By the end of the twentieth century there were approximately 233 non-native species, or 11.6% of the flora (Wilson et al. 2002). Similar trends can be expected across northern Mexico with a steady increase in the numbers of non-natives.

Besides buffelgrass, invasive Old World species in northern Mexico capable of carrying fire into non-fire–adapted ecosystems include red brome (*Bromus rubens* [*B. madritensis* subsp. *rubens*]; e.g., Solis-Garza and Jenkins 1998; Esque and Schwalbe 2002; Salo 2004). Tamarisk has become established in many of the lower-elevation riparian and semi-riparian habitats to the exclusion of other plants (Felger 2000; Felger et al. 2001). In many arid grasslands in northern Mexico the Old World exotic grass *Eragrostis lehmanniana* and others have become a serious problem (Solis-Garza and Jenkins 1998; Tellman 2002). On Isla Guadalupe (see below), 61 non-native plants have been recorded, many of them aggressive weeds (León de la Luz et al. 2003). One of the more chilling aspects of invasive plant ecology is the phenomenon known as the period of latency—non-native plants may spend years as small and seemingly innocuous populations and undergo explosive population expansion as much as one-half century later (Burgess et al. 1991; Búrquez et al. 1998). Continued disturbance in northern Mexico will increase the severity of biological invasions (Hobbs and Huenneke 1992).

Non-native animals are widespread but have had especially severe impacts on fragile island ecosystems. On some islands of the Gulf of California the introduction of Old World rats (*Rattus rattus*) by guano miners has had drastic effects on nesting seabirds (chapter 23). One of the most severe cases is on Isla Guadalupe, 260 km west of the Baja California peninsula. Goats and cats, introduced during the nineteenth century, have devastated the endemic fauna and flora (e.g., Anthony 1925; Jehl and Everett 1985; Moran 1996; Ceballos and Márquez 2000; León de la Luz et al. 2003). The Guadalupe storm petrel (*Oceanodroma macrodactyla*), which suffered heavy mortality caused by cat predation, is believed to be extinct (Anthony 1925; Jehl and Everett 1985). In 2000, there were an estimated 4000 feral goats on the island (León de la Luz et al. 2003), and the goat population may have reached 100,000 around 1870 (Moran 1996). The remaining endemic trees (such as the pine *Pinus radiata* var. *binata*, palm *Brahea edulis*, and cypress *Cupressus guadalupensis* var. *guadalupensis*), as well as a major portion of the remaining native flora, face imminent extinction if the goats, which have destroyed most of the native vegetation, are not removed (Moran 1996; León de la Luz et al. 2003).

River basin ecosystems in northern Mexico have been severely affected by exotics. Many native fishes and amphibians have been extirpated from most areas where game fish, mosquito fish (*Gambusia affinis*), sunfish (*Lepomis macrochirus*), and non-native bullfrogs (*Rana catesbeiana*) and crayfish have been introduced (Contreras-Balderas and Escalante 1984; Edwards and Contreras-Balderas 1991; Flores-Villela 1993; Rosen and Schwalbe 2002). Largemouth bass (*Micropterus salmoides*), one of the most coveted and widely introduced fishes in reservoirs of northern Mexico, prey on small native fishes, causing population declines or extirpations and extinctions (Minckley 1973). Other invasive, commonly introduced game fish of concern include smallmouth bass (*Micropterus dolomieui*) and common tilapia. Introduced fish are a threat to the native freshwater fish fauna of Tamaulipas in particular (chapter 7).

Prospects for the Future

What will happen to the diverse ecosystems of northern Mexico in the future? Mexico's gross domestic product increased at a rate of 7% per year in the last decade (FAO 2001). The population growth rate is slowing, but it is still expected to exceed a 2%/year increase through 2020 (table 3.1; INEGI 2001). Continued and perhaps accelerated growth and development seem likely, especially in the Gulf of California-Baja California peninsula region, where the *Escalera Nautica* project (described in chapter 9 and 15; see also Búrquez and Martínez Yrízar 2000), now being implemented, calls for the development of a network of marinas, new highways, and resorts and associated developments such as golf courses. Thus, it is also likely that the negative impacts on ecosystems and biodiversity briefly described in this chapter will continue to increase in extent and severity. As a result, the rate of extinction and endangerment of northern Mexico's diverse flora and fauna will increase as well. There may be additional impacts that are not predictable; seemingly benign actions or changes may prove to be damaging to some component of an ecosystem. For example, the recent electrification of rural areas in northern Chihuahua has had unforeseen consequences for local and migrant raptors (chapter 17).

But the prospects are not entirely bleak. Growing environmental problems and constraints are causing officials on both sides of the Mexico–U.S. border to address common issues of limited water resources, uncontrolled population growth, and ecosystem health. Already, concerns over the environmental impacts of NAFTA prompted the creation of two binational organizations: the Border Environment Cooperation Commission (BECC) and the Commission for Environmental Cooperation (CEC; Liverman et al. 1999). The Mexican government has increasingly enacted and enforced environmental legislation both internally and internationally. For example, the country became signatory to CITES in 1992. Mexico has taken the lead internationally in the development of biosphere reserves (Gómez-Pompa and Kaus 1999). There is a growing environmental movement in Mexico, and nongovernmental organizations play an increasingly important role in land use policy (Liverman et al. 1999). The country has a large cadre of dedicated scientists working in conservation biology whose research documents the distribution and status of species, as well as conservation threats.

Protected natural areas in Mexico include Mexican biosphere reserves, national parks and monuments, sanctuaries, areas for the protection of natural resources, areas for the protection of flora and fauna, and state parks and reserves (Gomez-Pompa and Kaus 1999; see also Introduction and chapter 4). The largest protected areas are, or once were, the most remote, least populated, and most arid areas of least economic value. Total area and percentage of state lands are greatest for the two Baja California states and are drastically reduced eastward and southward (table 3.2). Apart from Chihuahua and Sonora, the areas of protection range from small to truly insignificant with Sinaloa and Durango having less than 1% of lands set aside as natural areas. Many additional areas are under consideration, such as the long-proposed Laguna Madre, Tamaulipas, with 613,052 ha (Cantú et al. 2003). Additionally, there are substantial areas of "paper preserves"—reserves in name but not reality (Búrquez and Martínez-Yrízar 1997, in press). The challenges are to add much needed reserves as recommended by Mexican scientists and conserva-

Table 3.2. Area in mainland federal and state protected areas for the northern states of Baja California, Baja California Sur, Sonora, Sinaloa, Chihuahua, Durango, Coahuila, Nuevo León, and Tamaulipas, by aerial extent and as a percentage of total state area.

State	Total Protected Area (km²)	Percentage of State Area
Baja California	30,602	42.10
Baja California Sur	30,046	40.25
Sonora	18,463	9.98
Sinaloa	512	0.90
Chihuahua	28,228	11.52
Durango	880	0.71
Coahuila	6,500	4.30
Nuevo León	2,545	3.93
Tamaulipas	2,154	2.67
Islas del Golfo de California[a]	3,580	na

Data from Vargas-Márquez et al. (2001), Cantú et al. (2003), CONANP (2003), and UNEP-WCMC (2003).

[a]The Islas del Golfo de California is a federally designated Area for the Protection of Flora and Fauna not included in estimates of mainland area of states (see also table I.2).

tionists and to maintain the integrity of current reserves, and provide sufficient funding for management (Cantú et al. 2003). "The allocation of funds is still precarious and some reserves have no protection, except for that given by the edicts and their natural isolation" (Búrquez and Martínez-Yrízar in press).

Environmental activism and public opinion are prompting the federal and local governments to better protect the country's natural environment. A turning point was reached in 2000, when President Ernesto Zedillo announced the government's decision to cancel Mitsubishi's planned construction of the world's largest salt evaporation works at Laguna San Ignacio, part of the Vizcaino Biosphere Reserve, a UNESCO World Heritage Site in Baja California Sur (Reuters News Service 2000). Environmentalists and scientists argued that it posed a threat to one of Latin America's biggest wildlife sanctuaries, an important breeding and nursery ground for the gray whale (*Eschrichtius robustus*). The decision was the result of a 5-year legal and public campaign by local fishermen and residents, the general public, and Mexican and international NGOs (Russell 2001). This was the first large, successful conservation effort to stop a government initiative.

The problems associated with rising human population and overuse of natural resources are not unique to Mexico; rather, they are shared by the rest of the world. Megadiversity and high endemism in Mexico and our own special sense of responsibility heightens the call to rally against untimely extinction, loss of resources, and human dignity.

Acknowledgments

We thank the library staff of the Rocky Mountain Research Station for their excellent literature reviews and document retrieval. We also thank James Bauml, Jean-Luc Cartron, Diana Hadley, Exequiel Ezcurra, Pesach Lubinsky, Wallace J. Nichols, Thomas E. Sheridan, Pamela Stoleson, and an anonymous reviewer for providing significant assistance, comments, and suggestions. Richard Felger and Michael Wilson thank the Wallace Research Foundation for support.

Literature Cited

Acosta-Ruíz, G. 2001. Inventory of sites in Mexico with elevated concentrations of mercury. Report to the North American Commission for Environmental Cooperation, Montreal. Acosta y Asociados, Agua Prieta, Sonora, Mexico.

Aguilar, A. 1999. Plantas Medicinales del Sur de México. Guías prácticas, México Desconocido, México D.F.

Albert, L. (ed.). 1990. Los Plaguicidas, El Ambiente y La Salud. Centro de Ecodesarrollo, México D.F.

Anderson, E. F. 2001. The Cactus Family. Timber Press, Portland, Oregon.

ANGADI. 2003. Asociación Nacional de Ganaderos Diversificados Criadores de Fauna, Nuevo Laredo, Tamaulipas, Mexico. [Available: http://www.angadi.com.mx/whoare.htm]

Anguiano-Téllez, M. E. 1998. Migración a la frontera norte de México y su relación con el mercado de trabajo regional. Papeles de Población 4: 63–79.

Anonymous. 1997. Mexico moves to phase out DDT and chlordane. Environmental Health Perspectives 105: 788.

Anthony, A. W. 1925. Expedition to Guadalupe Island, Mexico in 1922: The birds and mammals. Proceedings of the California Academy of Sciences 14: 277–320.

Applegate S. P., F. Sotelo-Macías, and L. Espinosa-Arrubarena. 1993. An overview of Mexican shark fisheries, with suggestions for shark conservation in Mexico. Pp. 31–37 *in*

S. Branstetter (ed.), Biology of Elasmobranchs. NOAA Technical Report NMFS-115. National Oceanic and Atomspheric Administration, Washington, D.C.

Aranda, M. 1991. Wild mammal skin trade in Chiapas, Mexico. Pp. 175–177 *in* J. G. Robinson and K. H. Redford (eds.), Neotropical Wildlife Use and Conservation. University of Chicago Press, Chicago.

Argueta A. 1994. Atlas de las Plantas de la Medicina Tradicional Mexicana, 3 vols. Instituto Nacional Indigenista, México D.F.

Arias Rojo, H. 2000. International groundwaters: the upper San Pedro River Basin case. Natural Resources Journal 40: 199–222.

Bahre, C. J., and C. F. Hutchinson. 2001. Historic vegetation change in *La Frontera* west of the Rio Grande. Pp. 67–83 *in* G. L. Webster and C. J. Bahre (eds.), Changing Plant Life of La Frontera: Observations on Vegetation in the United States/Mexico Borderlands. University of New Mexico Press, Albuquerque.

Bahre, C. J., and R. A. Minnich. 2001. Madrean oak woodlands along the Arizona/Sonora boundary. Desert Plants 17: 3–14.

Balling, R. C. 1988. The climatic impact of a Sonoran vegetation discontinuity. Climatic Change 13: 99–109.

Bañuelos, N., and A. Búrquez. 1996. Las plantas: una estrategia de salud en la medicina doméstica Mayo. Estudios Sociales 6(12): 163–189.

Barry, T., H. Browne, and B. Sims. 1994. The Great Divide: The Challenge of U.S.–Mexico Relations in the 1990s. Grove Press, New York.

Beckett J. L., and J. W. Oltjen. 1993. Estimation of the water requirement for beef production in the United States. Journal of Animal Science 71: 818–826.

Beissinger, S. R. 2001. Trade of live wild birds: potentials, principles and practices of sustainable use. Pp. 182–202 *in* J. D. Reynolds, G. M. Mace, K. H. Redford, and J. G. Robinson (eds.), Conservation of Exploited Species. Cambridge University Press, Cambridge.

Bergman, C. 2002. Red Delta: Fighting for Life at the End of the Colorado River. Fulcrum Publishing, Golden, Colorado.

Bock, C. E., J. H. Bock, W. R. Kenney, and V. M. Hawthorne. 1984. Response of birds, rodents, and vegetation to livestock exclosure in a semidesert grassland site. Journal of Range Management 37: 239–242.

Bock, C. E., V. A. Saab, T. D. Rich, and D. S. Dobkin. 1993. Effects of livestock grazing on neotropical migratory landbirds in western North America. Pp. 296–309 *in* D. M. Finch and P. W. Stangel (eds.), Status and Management of Neotropical Migratory Birds. General Technical Report RM-229, U.S. Department of Agriculture, Forest Service, Rocky Mountain Research Station, Fort Collins, Colorado.

Bojorquez-Tapia, L. A., R. Aguirre, and A. Ortega. 1985. Rio Yaqui watershed, northwestern Mexico: use and management. Pp. 475–478 *in* R. R. Johnson, C. D. Ziebel, D. R. Patton, P. F. Ffolliott and R. H. Hamre (eds.), Riparian Ecosystems and Their Management: Reconciling Conflicting Uses. General Technical Report RM-120, U.S. Department of Agriculture, Forest Service, Rocky Mountain Forest and Range Experiment Station, Fort Collins, Colorado.

Border Ecology Project. 1995. Environmental/social impacts of multi-national mining investments in Sonora, Mexico. Border Ecology Project, Bisbee, Arizona.

Bracamonte Sierra, A. 2000. Crisis y modernización del campo. Pp. 207–224 *in* I. Almada Bay (ed.), Sonora 2000 a Debate: Problemas y Soluciones, Riesgos y Oportunidades. Editorial Cal y Arena, México D.F.

Brand, J. C., B. J. Snow, G. P. Nabhan, and A. S. Truswell. 1990. Plasma glucose and insulin responses to traditional Pima Indian meals. American Journal of Clinical Nutrition 5: 416–420.

Bravo-Hollis, H., and H. Sánchez-Mejorada R. 1991. Las Cactaceas de México, vol. 2. Universidad Nacional Autónoma de México, México D.F.

Brown, D. E. 1985. The Grizzly in the Southwest. University of Oklahoma Press, Norman.

Brown, L. R. 1970. Seeds of Change; the Green Revolution and Development in the 1970's. Overseas Development Council, New York.

Brusca, R. C. In press. Invertebrate Biodiversity in the Northern Sea of Cortez. *In* R. S. Felger and B. Broyles (eds.), Dry Borders: Great Natural Areas of the Gran Desierto and Upper Gulf of California. University of Utah Press, Salt Lake City.

Burgess, T. L., J. E. Bowers, and R. M. Turner. 1991. Exotic plants at the Desert Laboratory, Tucson. Madroño 38: 96–114.

Búrquez, A., and C. Castillo. 1994. Reserva de la biosfera El Pinacate y Gran Desierto de Altar: entorno biológico y social. Estudios Sociales 5(9): 9–64.

Búrquez, A., and A. Martínez-Yrízar. 1997. Conservation and landscape transformation in Sonora, Mexico. Journal of the Southwest 39: 371–398.

Búrquez, A., and A. Martínez-Yrízar. 2000. El Desarrollo Económico y la Conservación de los Recursos Naturales. Pp. 267–333 *in* I.

Almada Bay (ed.), Sonora 2000 a Debate: Problemas y Soluciones, Riesgos y Oportunidades. Editorial Cal y Arena, Mexico D.F.

Búrquez, A., and A. Martínez-Yrízar. In press. Conservation and landscape transformation in northwestern Mexico. *In* R. S. Felger and B. Broyles (eds.), Dry Borders: Great Natural Areas of the Gran Desierto and Upper Gulf of California. University of Utah Press, Salt Lake City.

Búrquez, A., A. Martínez-Yrízar, M. E. Miller, K. Rojas, M. A. Quintana, and D. Yetman. 1998. Mexican grasslands and the changing aridlands of Mexico: an overview and a case study in northwestern Mexico. Pp. 21–44 *in* B. Tellman, D. M. Finch, C. Edminster, and R. Hamre (eds.), The Future of Arid Grasslands: Identifying Issues, Seeking Solutions. Proceedings RMRS-P-3, U.S. Department of Agriculture, Forest Service, Rocky Mountain Research Station, Fort Collins, Colorado.

Búrquez, A., M. E. Miller, and A. Martínez-Yrízar. 2002. Mexican grasslands, thornscrub, and the transformation of the Sonoran Desert by invasive exotic buffelgrass (*Pennisetum ciliare*). Pp. 126–146 *in* B. Tellman (ed.), Invasive Exotic Species in the Sonoran Region. University of Arizona Press and Arizona-Sonora Desert Museum, Tucson.

Búrquez, A., and M. A. Quintana. 1994. Islands of diversity: ironwood ecology and the richness of perennials in a Sonoran Desert biological reserve. Pp. 9–27 *in* G. P. Nabhan and J. L. Carr (eds.), Ironwood: An Ecological and Cultural Keystone of the Sonoran Desert. Occasional Papers in Conservation Biology no. 1. Conservation International, Washington D.C.

Búrquez, A., and D. Yetman. 2001. Veintisiete hectáreas: consecuencias ecológicas de las modificaciones a la Ley Agraria: un estudio de caso en Sonora. Cuadernos de Trabajo 1: 58–70. (Lo que el 27 se llevó . . . Cambios urbanos, mineros, pesqueros, ganaderos y ecológicos en los ejidos Sonorenses). Sonoran Institute, Tucson, and Red Fronteriza de Salud y Ambiente, Colegio de Sonora, Hermosillo, Mexico.

Bustamante, E., and A. Búrquez. In press. Fenología y biología reproductiva de las cactáceas columnares. Cactaceas y Suculentas Mexicanas.

Bye, R. 1986. Medicinal plants of the Sierra Madre: comparative study of Tarahumara and Mexican market plants. Economic Botany 40: 103–124.

Bye, R. 1995. Ethnobotany of the Mexican tropical dry forests. Pp. 423–438 *in* S.H. Bullock, H. A. Mooney, and E. Medina (eds.), Seasonally Dry Tropical Forests. Cambridge University Press, Cambridge.

Bye, R. 2003. Beyond the desert: ethnobotanical studies in the Sierra Tarahumara, Chihuahua (abstract). Society for Economic Botany, Program 44th Annual Meeting, Tucson.

Cantú, C., R. G. Wright, J. M. Scott, and E. Strand. 2003. Conservation assessment of current and proposed nature reserves of Tamaulipas State, Mexico. Natural Areas Journal 23: 220–228.

Cartron, J.-L. E., S. H. Stoleson, P. L. L. Stoleson, and D. W. Shaw. 2000. Riparian areas. Pp. 281–327 in R. Jemison and C. Raish (eds.), Livestock Management in the American Southwest: Ecology, Society, and Economics. Elsevier, Amsterdam, Netherlands.

Castillo, V. M., and D. Perry. 1992. Environmental implications of the free trade agreement in the maquiladora industry. Transboundary Resources Report 6: 3–5.

Castillo-Geniz, J. L., J. F. Márquez-Farías, M. C. Rodríguez de la Cruz, E. Cortés, and A. Cid del Prado V. 1998. The Mexican artisanal shark fishery in the Gulf of México: towards a regulated fishery. Australian Journal of Marine and Freshwater Research 49: 611–620.

Castleman, M. 2001. The New Healing Herbs. Rodale Press, Emmaus, Pennsylvania.

Cavazos-Doria, J. R. 1997. Uso multiple de los agostaderos en el norte de México. Revista de Ciencia Forestal en México 22: 3–26.

Ceballos, G. 1985. The importance of riparian habitats for the conservation of endangered mammals in Mexico. Pp. 96–100 in R. R. Johnson, C. D. Ziebell, D. R. Patton, P. R. Ffolliott, and R. H. Hamre (eds.), Riparian Ecosystems and Their Management: Reconciling Conflicting Uses. Proceedings of the First North American Riparian Conference. General Technical Report RM-120, U.S. Department of Agriculture, Forest Service, Rocky Mountain Forest and Range Experiment Station, Fort Collins, Colorado.

Ceballos, G. 1993. Especies en peligro de extinción. Revista Ciencias (no. especial) 7: 5–10.

Ceballos, G., and F. Eccardi. 2003. Los Animales de México en Peligro de Extinción. Fundación IUSA, México D.F.

Ceballos, G., and A. García. 1995. Conserving neotropical diversity: the role of dry forests in western Mexico. Conservation Biology 9: 1349–1356.

Ceballos, G., and L. Márquez. 2000. Las Aves de México en Peligro de Extinción. CONABIO, Universidad Autónoma de México, Fondo de Cultura Economica, México D.F.

Ceballos, G., and D. Navarro. 1991. Diversity and conservation of Mexican mammals. Pp. 167–198 in M. A. Mares and D. J. Schmidly (eds.), Latin American Mammalogy: History, Biodiversity, and Conservation. University of Oklahoma Press, Norman.

Ceballos, G., P. Rodríguez, and R. Medellín. 1998. Assessing conservation priorities in megadiverse Mexico: mammalian diversity, endemicity, and endangerment. Ecological Applications 8: 8–17.

Challenger, A. 1998. Utilización y Conservación de los Ecosistemas Terrestres de México. CONABIO, México D.F.

CICOPLAFEST (Comisión Intersecretarial para el control del procesoyuso de Plaguicidas, Fertilizantes y Sustancias Tóxicas). 2002. Catálogo de Plaguicidas. Secretaría de Salud, México D.F.

CINVESTAV (Centro de Investigación y de Estudios Avanzados del IPN). 1994. Evaluación de los posibles efectos sobre de salud. SEDESOL, Instituto Nacional de Ecología, México D.F.

Cisneros-Mata, M. A., G. Montemayor-López, and M. J. Román-Rodríguez. 1995. Life history and conservation of Totoaba macdonaldi. Conservation Biology 9: 806–814.

Coltman, D. W., P. O'Donoghue, J. T. Jorgenson, J. T. Hogg, C. Strobeck, and M. Festa-Bianchet. 2003. Undesirable evolutionary consequences of trophy hunting. Nature 426: 655–658.

CONABIO. 1998. La Diversidad Biológica de México: Estudio de País, 1998. Comisión Nacional para el Conocimiento y Uso de la Biodiversidad, México D.F.

CONANP (Comisión Nacional de Áreas Naturales Protegidas). 2003. Áreas Naturales Protegidas. [Available: www.conanp.gob.mx/]

Contreras, J. L. 1987. Mecanización Agrícola, Empleo y Migración en El Norte de Tamaulipas. El Colegio de la Frontera Norte (COLEF), Tijuana, Mexico.

Contreras-Balderas, S., P. Almada-Villela, M. L. Lozano-Vilano, and M. E. García-Ramírez. 2003. Freshwater fish at risk or extinct in Mexico. Reviews in Fish Biology and Fisheries 12: 241–251.

Contreras-Balderas S., and M. A. Escalante. 1984. Distribution and known impacts of exotic fishes in Mexico. Pp. 102–130 in W. R. Courtenay, Jr. and J. R. Stauffer (eds.), Distribution, Biology, and Management of Exotic Fishes. Johns Hopkins University Press, Baltimore, Maryland.

Contreras-Balderas, S., and M. L. Lozano-Vilano. 1994. Water, endangered fishes, and development perspectives in arid lands of Mexico. Conservation Biology 8: 79–387.

Conway, G. 1998. The Doubly Green Revolution: Food for all in the 21st Century. Cornell University Press, Ithaca, New York.

Cooper, J. E. 1989. Birds and zoonoses. Ibis 132: 181–191.

Córdova, A. 1973. La ideología de la revolución Mexicana, 2nd ed. Ediciones Era, México D.F.

Cornelius, W. A., and D. Myhre. 1998. Introduction. Pp. 1–24 in W. A. Cornelius and D. Myhre (eds.), The Transformation of Rural Mexico: Reforming the Ejido Sector. Center for U.S.–Mexican Studies, University of California, San Diego.

Coronado-Quintana, J. A., and M. P. McClaran. 2001. Range condition, tenure, management, and bio-physical relationships in Sonora, Mexico. Journal of Range Management 54: 31–38.

D'Agrosa, C., C. E. Lennert-Cody, and O. Vidal. 2000. Vaquita bycatch in Mexico's artisanal gillnet fisheries: driving a small population to extinction. Conservation Biology 14: 1110–1119.

DEA. 2003. Mexico: country profile for 2003. Drug Intelligence Report DEA-03047. U.S. Drug Enforcement Administration. Washington, D.C.

De Gortari, L. 2000. Comunidad como forma de tenencia de la tierra. [Available: http://www.pa.gob.mx/publica/pa070806.htm]

Deininger, K. W., and B. Minten. 1999. Poverty, policies, and deforestation: the case of Mexico. Economic Development and Cultural Change 47: 313–344.

De Janvry, A., G. Gordillo, and E. Sadoulet. 1997. Mexico's second agrarian reform: household and community responses, 1990–1994. Ejido Reform Research Project, Center for U.S.–Mexican Studies, University of California, San Diego.

De la Torre Sánchez, N. 2003. Efecto de la restauración en la dinámica de comunidades riparias del Valle de San Bernardino, Sonora, México/Arizona, Estados Unidos. B.Sc. thesis, Centro de Estudios Superiores del Estado de Sonora, Hermosillo, Mexico.

Del Castillo, R. F., and S. Trujillo. 1991. Ethnobotany of Ferocactus histrix and Echinocactus platyacanthus (Cactaceae) in the semiarid central Mexico: past, present, and future. Economic Botany 45: 495–502.

De Witt, D. 1991. Yo soy un chiltepinero. Chile Pepper 5(3): 22–30.

Doode, S., and N. Bañuelos. 1995. Reflexiones sobre la conceptualización y manejo de la pesquería de sardin en Sonora. Pp. 218–233 in M. Camberos, V. Salazar, P. L. Salido,

and S. Sandoval (eds.), Las Consecuencias de la Modernización y el Desarrollo Sustentable. Programa Universitario de Alimentos, UNAM/CIAD, Hermosillo, Mexico.

Edwards, R. J., and S. Contreras-Balderas. 1991. Historical changes in the ichthyofauna of the lower Rio Grande (Rio Bravo del Norte), Texas and Mexico. Southwestern Naturalist 36: 201–212.

Ehrlich, P., and G. Ceballos. 1997. Población y medio ambiente: ¿que nos espera? Revista Ciencia 48 (4): 19–30.

El Pescador Tour Services. 2002. El Fuerte, Sinaloa. [Available: http://thepescador.com/lake_huites.htm]

Environmental Law Institute. 2000. Towards a Regional Framework for Pollution Prevention in the Mining Sector. Environmental Law Institute, Washington, D.C.

Esque, T. C., and C. R. Schwalbe. 2002. Alien annual grasses and their relationships to fire and biotic change in Sonoran desertscrub. Pp. 164–194 in B. Tellman (ed.), Invasive Exotic Species in the Sonoran Region. Arizona-Sonora Desert Museum and University of Arizona Press, Tucson.

Ewing, R. C. 1966. Major historical themes. Pp. 1–63 in R. C. Ewing (ed.), Six Faces of Mexico. University of Arizona Press, Tucson.

Ezcurra, E., R. S. Felger, A. Russell, and M. Equihua. 1988. Freshwater islands in a desert sand sea: The hydrology, flora, and phytogeography of the Gran Desierto oases of northwestern Mexico. Desert Plants 9: 35–44, 55–63.

Ezcurra, E., and C. Montaña. 1990. Los recursos naturales renovables en el norte árido de México. Pp. 297–327 in E. Leff (ed.), Medio Ambiente y Desarrollo en México. Centro de Investigaciones Interdisciplinarias en Humanidades, México D.F.

FAO. 1997. Review of the state of world fishery resources: marine fisheries. FAO Fisheries Circular no. 920. Food and Agriculture Organization of the United Nations, Rome.

FAO. 2001. State of the World's Forests 2001. Food and Agriculture Organization of the United Nations, Rome.

Felger, R. S. 1977. Mesquite in Indian cultures of Southwestern North America. Pp. 150–176 in B. B. Simpson (ed.), Mesquite: Its Biology in Two Desert Scrub Ecosystems. Dowden, Hutchinson, and Ross, Stroudsburg, Pennsylvania.

Felger, R. S. 1990a. Mesquite, a world food crop. Aridus 2(1): 1–3.

Felger, R. S. 1990b. Non-native plants of Organ Pipe Cactus National Monument. Technical Report 31, Cooperative National Park

Resource Studies Unit, University of Arizona, Tucson.

Felger, R. S. 2000. Flora of the Gran Desierto and Río Colorado Delta of Northwestern Mexico. University of Arizona Press, Tucson.

Felger, R. S., and B. Broyles (eds.). In press. Dry Borders: Great Natural Areas of the Gran Desierto and Upper Gulf of California. University of Utah Press, Salt Lake City.

Felger, R. S., M. B. Johnson, and M. F. Wilson. 2001. Trees of Sonora, Mexico. Oxford University Press, New York.

Felger, R. S., and C. H. Lowe. 1976. The island and coastal vegetation and flora of the Gulf of California, Mexico. Contributions in Science 285. Natural History Museum of Los Angeles County, California.

Felger, R. S., and M. B. Moser. 1985. People of the Desert and Sea: Ethnobotany of the Seri Indians. University of Arizona Press, Tucson.

Felger, R. S., G. P. Nabhan, and R. Bye. 1997. The Apachian/Madrean Region of south-western North America as a Center of Plant Diversity. Pp. 172–180 *in* S. D. Davis et al. (eds.), Centres of Plant Diversity: A Guide and Strategy for their Conservation, vol. 3. The Americas. World Wide Fund for Nature and International Union for Conservation and Nature. IUCN Publications Unit, Cambridge.

Fernández-González, R. 1980. La metodología del sistema alimentario Méxicano. Ingenieria Agronomica 5: 4–21.

Figueiredo, A. A. 1990. Mesquite: history, composition and food uses. Food Technology 44: 118–128.

Fisher, J. T., P. A. Glass, and J. T. Harrington. 1995. Temperate pines in northern Mexico: their use, abuse, and regeneration. Pp. 165–173 *in* L. F. DeBano, P. F. Ffolliott, A. Ortega-Rubio, G. J. Gottfried, R. H. Hamre, and C. B. Edminster (eds.), Biodiversity and Management of the Madrean Archipelago: The Sky Islands of Southwestern United States and Northwestern Mexico. General Technical Report RM-264, U.S. Department of Agriculture, Forest Service, Rocky Mountain Forest and Range Experiment Station, Fort Collins, Colorado.

Flores-Villela, O. 1993. Herpetofauna of Mexico: distribution and endemism. Pp. 253–280 *in* T. P. Ramamoorthy, R. Bye, A. Lot, and J. Fa (eds.), Biological Diversity of Mexico: Origins and Distribution. Oxford University Press, New York.

Flores-Villela, O., and P. Geréz, 1994. Biodiversidad y Conservación en México: Vertebrados, Vegetación y Uso del Suelo. CONABIO/Universidad Aútónoma de México, México D.F.

Floyd, J. P. 1998. Legal and illegal trade between Mexico and the U.S. in plants and plant products listed under CITES. Pp. 129–138 *in* G. J. Gottfried, C. B. Edminster, and M. C. Dillon (eds.), Cross Border Waters: Fragile Treasures for the 21st Century: Ninth U.S./Mexico Border States Conference on Recreation, Parks and Wildlife: Tucson, Arizona, June 3–6, 1998. Proceedings RMRS-P-5, U.S. Department of Agriculture, Forest Service, Rocky Mountain Research Station, Fort Collins, Colorado.

Foley, M. W. 1995. Privatizing the countryside: the Mexican peasant movement and neo-liberal reform. Latin American Perspectives (issue 84) 22: 59–76.

Foster, L. V. 1997. A Brief History of Mexico. Facts on File, New York.

Foster, S., and V. Tyler. 1999. Tyler's Honest Herbal. Haworth Herbal Press, New York.

Fulbright, T. E. 2001. Human-induced vegetation changes in the Tamaulipan scrub of *La Frontera*. Pp. 166–175 *in* G. L. Webster and C. J. Bahre (eds.), Changing Plant Life of La Frontera: Observations on Vegetation in the United States/Mexico Borderlands. University of New Mexico Press, Albuquerque.

Fuller, E. 2001. Extinct Birds, revised ed. Cornell University Press, Ithaca, New York.

Gentry, H. S. 1982. The Agaves of Continental North America. University of Arizona Press, Tucson.

Glaeser, B. (ed.). 1987. The Green Revolution Revisited: Critique and Alternatives. Allen and Unwin, Boston.

Glenn, E. P., R. S. Felger, A. Búrquez, and D. S. Turner. 1992. Ciénega de Santa Clara: Endangered wetland in the Colorado River Delta, Sonora, Mexico. Natural Resources Journal 32(4): 817–824.

Glenn, E. P., C. Lee, R. S. Felger, and S. Zengel. 1996. Effects of water management on the wetlands of the Colorado River Delta, Mexico. Conservation Biology 10: 1175–1186.

Glenn, E. P., and P. L. Nagler. In press. New Life for the Colorado River Delta. *In* R. S. Felger and B. Broyles (eds.), Dry Borders: Great Natural Areas of the Gran Desierto and Upper Gulf of California. University of Utah Press, Salt Lake City.

Glenn, E. P., F. Zamora-Arroyo, P. L. Nagler, M. Briggs, W. Shaw, and K. Flessa. 2001. Ecology and conservation biology of the Colorado River Delta, Mexico. Journal of Arid Environments 49: 5–15.

Gómez-Pompa, A., and A. Kaus. 1999. From pre-Hispanic to future conservation alternatives: lessons from Mexico. Proceedings of the National Academy of Science USA 96: 5982–5986.

Goodland, R., and D. Pimentel. 2000. Sustainability and integrity in the agriculture sector. Pp. 121–137 *in* D. Pimentel, L. Westra, and R. F. Noss (eds.), Ecological Integrity: Integrating Environment, Conservation and Health. Island Press, Washington D.C.

Gottfried, G. J., and R. D. Pieper. 2000. Pinyon-juniper rangelands. Pp. 153–212 *in* R. Jemison and C. Raish (eds.), Livestock Management in the American Southwest: Ecology, Society, and Economics. Elsevier, Amsterdam, Netherlands.

Grismer, L. L. 2002. Amphibians and Reptiles of Baja California including Its Pacific Islands and the Islands in the Sea of Cortez. University of California Press, Berkeley.

Guerrero, M. T., F. de Villa, M. Kelly, C. Reed, and B. Vegter. 2000a. The Forestry Industry in the State of Chihuahua: Economic, Ecological, and Social Impacts post-NAFTA. Commission for Environmental Cooperation, Washington, D.C.

Guerrero, M. T., C. Reed, B. Vegter, and G. Kourous. 2000b. The timber industry in northern Mexico: social, economic, and environmental impacts. Borderlines 64: 1–4, 15–16.

Guillette, E.A., M. M. Meza, M. G. Aguilar, A. D. Soto, and I. E. Garcia. 1998. An anthropological approach to the evaluation of preschool children exposed to pesticides in Mexico. Environmental Health Perspectives 106: 347–353.

Gutiérrez-Galindo, E. A., G. Flores Muñoz, and A. Aguilar Flores. 1988. Mercury in freshwater fish and clams from Cerro Prieto geothermal field of Baja California, Mexico. Bulletin of Environmental Contamination and Toxicology 41: 201–207.

Hastings, P. A., and L. T. Findley. In press. Marine Fishes of the Upper Gulf Biosphere Reserve, Northern Gulf of California. *In* R. S. Felger and B. Broyles (eds.), Dry Borders: Great Natural Areas of the Gran Desierto and Upper Gulf of California. University of Utah Press, Salt Lake City.

Hawkes, L. 2004. Cattle imports into U.S. up by 2%: Influx of imports trouble U.S. producers. [Available: http://agriculture.about.com]

Hayward, B., E. J. Heske, and C. W. Painter. 1997. Effects of livestock grazing on small mammals at a desert cienega. Journal of Wildlife Management 61: 123–129.

Hemley, G. (ed.). 1994. International Wildlife Trade: a CITES Sourcebook. Island Press, Washington, D.C.

Hernández-Rodríguez, A., C. Alceste-Oliviero, R. Sánchez, D. Jory, L. Vidal, and L. F. Constain-Franco. 2001. Aquaculture development trends in Latin America and the Caribbean. Pp. 317–340 *in* R. P. Subasinghe, P. Bueno, M. J. Phillips, C. Hough, S. E. McGladdery, and J. R. Arthur (eds.), Aquaculture in the Third Millennium: Technical Proceedings of the Conference on Aquaculture in the Third Millennium. Food and Agricultural Organization, Rome.

Herrera-Silveira, J., and E. Ceballos-Cambranis. 2000. Manglares: ecosistemas valiosos. Biodiversitas 19: 1–10.

Hersch-Martínez, P. 1995. Commercialization of wild medicinal plants from southwest Puebla, Mexico. Economic Botany 49: 197–206.

Hilton-Taylor, C., and R. Mittermeier (eds.). 2000. 2000 IUCN Red List of Threatened Species. IUCN-World Conservation Union, Gland, Switzerland.

Hinojosa-Huerta, O., S. DeStefano, and W. W. Shaw. 2001. Distribution and abundance of the Yuma clapper rail (*Rallus longirostris yumanensis*) in the Colorado River delta, Mexico. Journal of Arid Environments 49: 171–182.

Hinojosa-Huerta, O., J. García-Hernández, Y. Carrillo-Guerrero, and E. Zamora-Hernández. In press. Hovering over the Alto Golfo: status and conservation of birds from the Río Colorado to the Gran Desierto. *In* R. Felger and B. Broyles (eds.), Dry Borders: Great Natural Areas of the Gran Desierto and Upper Gulf of California. University of Utah Press, Salt Lake City.

Hobbs, N. T. 1996. Modification of ecosystems by ungulates. Journal of Wildlife Management 60: 695–713.

Hobbs, R. J., and L. F. Huenneke. 1992. Disturbance, diversity, and invasion: implications for conservation. Conservation Biology 6: 324–337.

Hodgson, W. 2001. Food Plants of the Sonoran Desert. University of Arizona Press, Tucson.

Humphrey, R. R. 1987. 90 Years and 535 Miles. University of New Mexico Press, Albuquerque.

INE (Instituto Nacional de Ecología). 1993. Informe de la Situación General en Material del Equilibrio Ecológico y la Protección al Ambiente 1991–1992. Instituto Nacional de Ecología, México D.F.

INE. 1996. Mexican Group on Mercury—Mercury in Zacatecas. Instituto Nacional de Ecología, México D.F.

INEGI. 1993. Niveles de Bienestar en México. Instituto Nacional de Estadística Geografía e Informática, Aguascalientes, Mexico.

INEGI. 2001. XI Censo general de población y vivienda. Instituto Nacional de Estadística Geografía e Informática, Aguascalientes, Mexico.

INEGI. 2003. Balanza comercial de México. Noviembre 2003. Instituto Nacional de Estadística Geografía e Informática, Aguascalientes, Mexico.

Ingram, H. 1998. Planning for "natural" disaster: unsustainable development of the U.S./ Mexico border. Pp. 3–13 in G. J. Gottfried, C. B. Edminster, M. C. Dillon (eds.), Cross Border Waters: Fragile Treasures for the 21st Century: Ninth U.S./Mexico Border States Conference on Recreation, Parks and Wildlife: Tucson, Arizona, June 3–6, 1998. Proceedings RMRS-P-5, U.S.D.A. Forest Service, Rocky Mountain Research Station, Fort Collins, Colorado.

Iñigo-Elias, E. E., and M. A. Ramos. 1991. The psittacine trade in Mexico. Pp. 380–392 in J. G. Robinson and K. H. Redford (eds.), Neotropical Wildlife Use and Conservation. University of Chicago Press, Chicago.

ITA (International Trade Administration). 2003. U.S. Department of Commerce, U.S. Trade Quick-Reference Tables: December 2002 Imports. [Available: http://www.ita.doc.gov/ td/industry/otea/Trade-Detail/Latest-December/Imports/12/121190.html]

Jehl, J. R., Jr., and W. T. Everett. 1985. History and status of the avifauna of Isla Guadalupe, Mexico. Transactions of the San Diego Society of Natural History 20: 313–336.

Jemison, R., and C. Raish. 2000. Livestock Management in the American Southwest: Ecology, Society, and Economics. Elsevier, Amsterdam, Netherlands.

Jorgenson, A., and J. B. Thomsen. 1987. Neotropical parrots imported into the United States, 1981 to 1985. TRAFFIC 7: 3–8.

Karch, S. 1999. Consumer's Guide to Herbal Medicine. Advanced Research Press, New York.

Kauffman, J. B., and W. C. Krueger. 1984. Livestock impacts on riparian ecosystems and streamside management implications—a review. Journal of Range Management 37: 430–438.

Kaye, G. 1995. The Mexico-U.S.A. border region: the filling of an empty land. Transborder Resources Report 9: 4–7.

Kemper, K. E., and O. E. Alvarado. 2001. Water. Pp. 619–644 in M. M. Giugale, O. Lafourcade, and V. H. Nguyen (eds.), Mexico: A Comprehensive Development Agenda for the New Era. World Bank, Washington, D.C.

Kerley, G. I. H., and W. G. Whitford. 2000. Impact of grazing and desertification in the Chihuahuan Desert: plant communities, granivores and granivory. American Midland Naturalist 144: 78–91.

Klein-Robbenhaar, J. F. I. 1995. Agro-industry and the environment: the case of Mexico in the 1990s. Agricultural History 69: 395–412.

Klopatek, J. M., R. C. Balling, A. W. Bravel, J. Franklin, and C. J. Watts. 1996. Land-use change along the United States-Mexican border: ecosystem effects and climate feedbacks. Project ES PP96IV-2, Southwest Center for Environmental Research and Policy (SCERP), Phoenix, Arizona.

Kourous, G. 1998. Borderlands biodiversity: walking a thin line. Borderlines 43: 1–4.

Laird, S. A. (ed.). 2002. Biodiversity and Traditional Knowledge: Equitable Partnerships in Practice. Earthscan Publications, London.

Lammertink, J. M., J. A. Rojas, F. M. O. Casillas, and R. L. Otto. 1996. Status and conservation of old-growth forests and endemic birds in the pine-oak zone of the Sierra Madre Occidental, Mexico. Verslagen en Technische Gegevens 69: 1–89.

Landa, R., Meave, J., and J. Carabias. 1997. Environmental deterioration in rural Mexico: an examination of the concept. Ecological Applications 7: 316–329.

Lanning, D. V., and J. T. Shiflett. 1983. Nesting ecology of thick-billed parrots. Condor 85: 66–73.

Ledig, F. T., M. Mápula-Larreta, B. Bermejo-Velázquez, V. Reyes-Hernández, C. Flores-López, and M. Capó-Arteaga. 2000. Locations of endangered spruce populations in Mexico and the demography of *Picea chihuahuana*. Madroño 47: 71–88.

Lemly, A. D., R. T. Kingsford, and J. R. Thompson. 2000. Irrigated agriculture and wildlife conservation: conflict on a global scale. Environmental Management 25: 485–512.

Lenderking, B. 1996. The U.S.-Mexican border and NAFTA: problem or paradigm? Pp. 190–202 in O. J. Martínez (ed.), U.S.-Mexico Borderlands: Historical and Contemporary Perspectives. Scholarly Resources, Wilmington, Delaware.

León de la Luz, J. L., J. P. Rebman, and T. Oberbauer. 2003. On the urgency of conservation on Guadalupe Island, Mexico: is it a lost paradise? Biodiversity and Conservation 12: 1073–1082.

Leopold, A. 1949. A Sand County Almanac, and Sketches Here and There. Oxford University Press, New York.

Leopold, A. S. 1959. Wildlife of Mexico. The Game Birds and Mammals. University of California Press, Berkeley.

Lewis, J. 2002. Agrarian change and privatization of ejido land in northern Mexico. Journal of Agrarian Change 2: 401–419.

Lindquist, C. A. 2000. Dimensions of sustainability: use of vara blanca as a natural resource in the tropical deciduous forest of

Sonora, Mexico. Ph.D. dissertation, University of Arizona, Tucson.

Lingenfelter, R. E. 1978. Steamboats on the Colorado River. University of Arizona Press, Tucson.

Liverman, D. M., R. G. Varady, O. Chávez, and R. Sánchez. 1999. Environmental issues along the United States-Mexico border: drivers of change and responses of citizens and institutions. Annual Review of Energy and Environment 24: 607–643.

Lopez Urquidez, G. A. 1997. Distribución y abundancia de la *vara blanca* (*Croton* sp.) en el Estado de Sinaloa. Master's thesis, Universidad Autónoma de México, México D.F.

Lozano-Vilano, M. L., and S. Contreras-Balderas. 1993. Two new species of *Cyprinodon* from southern Nuevo León, Mexico, with a key to *C. eximius* complex (Teleostei: Cyprinodontidae). Ichthyological Exploration of Freshwaters 4: 294–308.

Lozoya, X. 1994. Two decades of Mexican ethnobotany and research on plant derived drugs. Pp. 130–140 *in* G. T. Prance, D. Chadwick, and J. Marsh (eds.), Ethnobotany and the Search for New Drugs. Ciba Foundation Symposium 185. Wiley, New York.

Manci, K. M. 1989. Riparian ecosystem creation and restoration: a literature summary. U.S. Fish and Wildlife Service Biological Report 89 (20): 1–59.

Manzano, M. G., and J. Návar. 2000. Processes of desertification by goats overgrazing in the Tamaulipan thornscrub (*matorral*) in northeastern Mexico. Journal of Arid Environments 44: 1–17.

Márquez-Farías, J. F. 2000. Tiburones del Golfo de California. Pp. 238–257 *in* Sustentabilidad y Pesca Responsable en México: Evaluación y Manejo, 1999–2000. Instituto Nacional de Pesca, Secretaría de Medio Ambiente y Recursos Naturales. [Available: http://inp.semarnat.gob.mx/Publicaciones/sustentabilidad/Pacifico/tibucal.pdf]

Marr, J. C. A., J. Lipton, D. Cacela, J. A. Hansen, H. L. Bergman, J. S. Meyer, and C. Hogstrand. 1996. Relationship between copper exposure duration, tissue copper concentration, and rainbow trout growth. Aquatic Toxicology 36: 17–30.

Martin, P. S., and D. A. Yetman. 2000. Introduction and prospect: secrets of a tropical deciduous forest. Pp. 3–18 *in* R. H. Robichaux and D. A. Yetman (eds.), The Tropical Deciduous Forests of Alamos: Biodiversity of a Threatened Ecosystem in Mexico. University of Arizona Press, Tucson.

McDaniel, C. J., L. B. Crowder, and J. A. Priddy. 2000. Spatial dynamics of sea turtle abundance and shrimping intensity in the U.S. Gulf of Mexico. Conservation Ecology 4: 15.

McPherson, G. R., and J. Villanueva-Díaz. 2001. Land use, climate, soils, and forest structure in the Animas Mountains and the Sierra Los Ajos. Pp. 143–155 *in* G. L. Webster and C. J. Bahre (eds.), Changing Plant Life of La Frontera: Observations on Vegetation in the United States/Mexico Borderlands. University of New Mexico Press, Albuquerque.

Mellink, E. 1995a. The potential effect of commercialization of reptiles from Mexico's Baja California Peninsula and its associated islands. Herpetological Natural History 3: 95–99.

Mellink, E. 1995b. Status of the muskrat in the Valle de Mexicali and Delta del Río Colorado, Mexico. California Fish and Game 81: 33–38.

Mellink, E., and J. Luévano. 1998. Status of beaver (*Castor canadensis*) in Valle Mexicali, Mexico. Bulletin of the Southern California Academy of Sciences 97: 115–120.

Miller, R. R., J. D. Williams, and J. E. Williams. 1989. Extinctions of North American fishes during the past century. Fisheries 14: 22–38.

Minckley, W. L. 1973. Fishes of Arizona. Arizona Game and Fish Department, Phoenix.

Minckley, W. L. 2002. Fishes of the Lowermost Colorado River, its delta and estuary: a commentary on biotic change. Pp. 63–78 *in* M. L. Lozano-Vilano (ed.), Libro Jubilar en Honor al Dr. Salvador Contreras Balderas. Ediciones Universidad Autónoma de Nuevo León, Monterrey, Mexico.

Molnar, A., and T. A. White. 2001. Forestry and land management. Pp. 669–690 *in* M. M. Giugale, O. Lafourcade, and V. H. Nguyen (eds.), Mexico: A Comprehensive Development Agenda for the New Era. World Bank, Washington, D.C.

Moran, R. 1996. The flora of Guadalupe Island, Mexico. Memoirs of the California Academy of Sciences no. 19.

Moreno Vázquez, J. L. 2000. Apropiación y sobreexplotación del agua subterránea en la Costa de Hermosillo: 1945–2000. Ph.D. dissertation, CIESAS Occidente/Universidad de Guadalajara, Guadalajara.

Mueller, G. A., and P. C. Marsh. 2002. Lost, a desert river and its native fishes: a historical perspective of the Lower Colorado River. U.S. Geological Survey, Biological Resources Division, Information and Technology Report USGS/BRD/ITR2002–0010: 1–69. U.S. Geological Survey, Fort Collins,

Colorado. [Available: http://www.fort. usgs.gov/products/Publications/10026/10026.asp.

Mumme, S. 1999. NAFTA's environmental side agreement: almost green? Borderlines 7: 1–4.

Muro Garcia, B. 1987. El proceso de producción de la "Vara Blanca" (*Croton* sp.) en el Estado de Sinaloa. Memoria de Seminario de Titulación. Universidad Autónoma de Chapingo, Division de Ciencias Forestales, Chapingo, Mexico.

Nabhan, G. P. 1985. Gathering the Desert. University of Arizona Press, Tucson.

Nabhan, G. P. 1990. Conservationists and Forest Service join forces to save wild chilies. Diversity 6: 47–48.

Nabhan, G. P., and J. L. Carr (eds.). 1994. Ironwood: an ecological and cultural keystone of the Sonoran Desert. Conservation International, Occasional Papers in Conservation Biology No. 1.

National Research Council. 1999. Hardrock Mining on Federal Lands Committee. National Research Council, Washington, D.C.

Paez-Osuna, F., S. R. Guerrero-Galván, and A. C. Ruiz-Fernández. 1998. The environmental impact of shrimp aquaculture and the coastal pollution in Mexico. Marine Pollution Bulletin 36: 65–75.

Paredes Aguilar, R., T. R. Van Devender, and R. S. Felger. 2000. Cactaceas de Sonora, México: Su Diversidad, Uso y Conservación. Arizona-Sonora Desert Museum Press, Tucson.

Perramond, E. P. 1999. Desert meadows: the cultural, political, and ecological dynamics of private cattle ranching in Sonora, Mexico. Ph.D. dissertation, University of Texas, Austin.

Perramond, E. P. 2000. A preliminary analysis of soil erosion and buffelgrass in Sonora, Mexico. Yearbook, Conference of Latin Americanist Geographers 26: 131–138.

Pidgeon, A. M., N. E. Mathews, R. Benoit, and E. V. Nordheim. 2001. Response of avian communities to historic habitat change in the northern Chihuahuan Desert. Conservation Biology 15: 1772–1788.

Pimentel, D., L. Lach, R. Zuniga, and D. Morrison. 2000. Environmental and economic costs of nonindigenous species in the United States. BioScience 50: 53–65.

Pitt, J., C. W. Fitzer, and L. Force. 2002. New water for the Colorado River: economic and environmental considerations for replacing the bypass flow. Water Law Review 6: 68–86.

PROFEPA (Procuraduría Federal de Protección al Ambiente). 1998. Informe trianual: 1995–1997. SEMARNAP, México D.F.

Purdy, P. C., and R. E. Tomlinson. 1991.The eastern white-winged dove: factors influencing use and continuity of the resource. Pp. 255–265 *in* J. G. Robinson and K. H. Redford (eds.), Neotropical Wildlife Use and Conservation. University of Chicago Press, Chicago.

Rea, A. 1997. At the Desert's Green Edge: An Ethnobotany of the Gila River Pima. University of Arizona Press, Tucson.

Redford, K. H., and J. G. Robinson. 1991. Subsistence and commercial uses of wildlife in Latin America. Pp. 6–23 *in* J. G. Robinson and K. H. Redford (eds.), Neotropical Wildlife Use and Conservation. University of Chicago Press, Chicago.

Reid, N., J. Marroquín, and P. Beyer-Münzel. 1990. Utilization of shrubs and trees for browse, fuelwood and timber in the Tamaulipan thornscrub, northeastern Mexico. Forest Ecology and Management 36: 61–79.

Reina-Guerrero, A. L., T. R. Van Devender, D. Yetman, and R. S. Felger. 2000. Usos etnobótanicos de las cactáceas en Sonora, México. Pp. 191–202 *in* D Vásquez del Castillo, M. Ortegas N., R. A. Corella B., and R. A. Castillo G. (eds.), II Simposium Internacional sobre la Utilización y Aprovechamiento de la Flora Silvestre de Zonas Aridas. Departamento de Investigaciones Científicas y Tecnólogicas de la Universidad de Sonora, Hermosillo, Mexico.

Restrepo, I. 1988. Naturaleza Muerta, los Plaguicidas en México. Editorial Planeta, México D.F.

Reuters News Service. 2000. Mexico scraps salt works near whale sanctuary. Reuters News Service, March 3.

Reuters News Service. 2004. Report: Herbal remedy boom threatens plants. Reuters News Service, January 9. [Available: http://www .cnn.com/2004/TECH/science/01/09/herbal.remedies.reut/]

Robbins, C. S. (ed). 1999. Medicine from U.S. Wildlands: An assessment of native plant species harvested in the United States for medicinal uses and trade and evaluation of the conservation and management implications. TRAFFIC North America, prepared for The Nature Conservancy. [Available: http://www.nps.gov/plants/medicinal/pubs/traffic.htm]

Robbins, C. S. (ed.). 2003. Prickly trade: trade and conservation of Chihuahuan Desert cacti. TRAFFIC North America. World Wildlife Fund, Washington D.C. [Available:

http://www.traffic.org/news/press-releases/
prickly_trade.html]

Roldán Quintana, J., and K. F. Clark. 1992. An overview of the geology and mineral deposits of northern Sierra Madre Occidental and adjacent areas. Pp. 39–65 *in* K. F. Clark, J. Roldán Quintana, and R. H. Schmidt (eds.), Geology and Mineral Resources of Northern Sierra Madre Occidental, Mexico, 1992 Field Conference Guidebook, El Paso Geological Society, El Paso, Texas.

Rosen, P. C., and C. R. Schwalbe. 2002. Widespread effects of introduced species on reptiles and amphibians in the Sonoran Desert Region. Pp. 220–240 *in* B. Tellman (ed.), Invasive Exotic Species in the Sonoran Region. Arizona-Sonora Desert Museum and University of Arizona Press, Tucson.

Russell, D. 2001. Eye of the Whale. Simon and Schuster, New York.

SAGARPA. 2003. [Abstracts] I Foro de Intercambio Científico sobre Tiburones y Rayas, 29–30 de Mayo de 2003, Guaymas, Sonora. Secretaría de Agricultura, Ganadería, Desarrollo Rural, Alimentación y Pesca/ Centro Regional Pesquera del Instituto Nacional de Pesca. [Available: http:// inp.semarnat.gob.mx/Publicaciones/ Publicaciones_Especiales/Foros/foros%20gy/ tiburones/Portada.htm]

Salas, G. P. (ed.). 1991. Economic Geology, Mexico. The Geology of North America, vol. P-3. Geologic Society of America, Boulder, Colorado.

Salo, L. F. 2004. Population dynamics of red brome (*Bromus madritensis* subsp. *rubens*): times for concern, opportunities for management. Journal of Arid Environments 57: 291–296.

Sánchez-Críspin, A. 1994. Mining municipios of Mexico: accessibility and socio-economic marginality. Randol at Vancouver, 1994: Latin American Mining Opportunities. Randol International, Golden, Colorado.

Sanchez-Mejorada V., R. 2000. Mining law in Mexico. Mineral Resources Engineering 9: 129–139.

Secretaría de Economía. 2000. Mexican mining industry report 2000. Secretaría de Economía, México D.F. [Available: http:// www.economia.gob.mx/?P=1033>http:// www.economia.gob.mx/?P=1033]

Secretaría de Economía. 2001. Mexican mining industry report 2001. Secretaría de Economía, México D.F.

Segura, G. 2000. Mexico's forest sector and policies: a general perspective. Presented at Constituting the Commons: Crafting Sustainable Commons in the New Millennium, the Eighth Conference of the International Association for the Study of Common Property, Bloomington, Indiana, May 31– June 4. [Available: http://dlc.dlib.indiana. edu/documents/dir0/00/00/03/41/index.html]

SEMARNAP. 1998. NORMA Oficial Mexicana NOM-120-ECOL-1997. Diario Oficial de la Federación, November 19, 1998. Secretaría del Medio Ambiente, Recursos Naturales y Pesca. [Available: http://www.economia. gob.mx/?P=1066]

SEMARNAP. 1999. El sector forestal de México: situación actual y perspectivas. Secretaría del Medio Ambiente, Recursos Naturales y Pesca, México D.F. [Available: http:// www.ine.gob.mx/ueajei/publicaciones/ consultaPublicacion.html?id_pub=109]

SEMARNAT. 2000a. Compendio de Estadísticas Ambientales. Secretaría de Medio Ambiente y Recursos Naturales, México D.F. [Available: http://carpetas.semarnat.gob.mx/ estadisticas_2000/Presentacion/index.shtml]

SEMARNAT. 2000b. Anuario estadístico de Pesca 1999. Secretaría de Medio Ambiente y Recursos Naturales, México D.F. [Available: http://www.sernapesca.cl/paginas/ publicaciones/anuarios/ index_anuario.php?&control=2]

SEMARNAT. 2001. Programa de Conservación de La Vida Silvestre y Diversificación Productiva en El Sector Rural 1997–2000. Secretaría de Medio Ambiente y Recursos Naturales, México D.F. [Available: http:// www.ine.gob.mx/ueajei/publicaciones/ consultaPublicacion.html?id_pub=279]

SEMARNAT. 2003. Reporte del Estado Ambiental y de Los Recursos Naturales en La Frontera Norte de México. Produced by the Instituto Tecnológico y de Estudios Superiores de Monterrey and the Instituto de Información Fronteriza México-Estados Unidos. Instituto Nacional de Ecología, México D.F. [Available: http://www. ine.gob.mx/ueajei/publicaciones/ consultaPublicacion.html?id_pub=109]

Sheldon, S. 1980. Ethnobotany of *Agave lechuguilla* and *Yucca carnerosana* in Mexico's Zona Ixtlera. Economic Botany 34: 376–390.

Sheridan, T. E. 2001. Cows, condos, and the contested commons: the political ecology of ranching on the Arizona-Sonora borderlands. Human Organization 60(2): 141–152.

Silbert, M. S. 1988. Mesquite pod utilization for livestock feed: an economic development alternative in central Mexico. Master's thesis, University of Arizona, Tucson.

Simon, J. 1997. Endangered Mexico: An Environment on the Edge. Sierra Club Books, San Francisco, California.

Simonian, L. 1988. Pesticide use in Mexico: decades of abuse. Ecologist 18: 82–87.

Simpson, B. B. (ed.). 1977. Mesquite: its biology in two desert scrub ecosystems. Dowden, Hutchinson, and Ross, Stroudsburg, Pennsylvania.

Solis-Garza, G., and P. Jenkins. 1998. Riparian vegetation on the Río Santa Cruz, Sonora. Pp. 100–118 in G. J. Gottfried, C. B. Edminster, M. C. Dillon (eds.), Cross Border Waters: Fragile Treasures for the 21st Century: Ninth U.S./Mexico Border States Conference on Recreation, Parks and Wildlife: Tucson, Arizona, June 3–6, 1998. Proceedings RMRS-P-5, U.S. Department of Agriculture, Forest Service, Rocky Mountain Research Station, Fort Collins, Colorado.

Sosa-Ramírez, J., and E. Franco-Vizcaíno. 2001. Grazing impacts on mountain meadows of the peninsular ranges in La Frontera. Pp. 156–165 in G. L. Webster and C. J. Bahre (eds.), Changing Plant Life of La Frontera: Observations on Vegetation in the United States/Mexico Borderlands. University of New Mexico Press, Albuquerque.

Spellenberg, R. 2001. Oaks of La Frontera. Pp. 176–186 in G. L. Webster and C. J. Bahre (eds.), Changing Plant Life of La Frontera: Observations on Vegetation in the United States/Mexico Borderlands. University of New Mexico Press, Albuquerque.

Steinich, B., O. Escolero, and L. E. Marin. 1998. Salt-water intrusion and nitrate contamination in the Valley of Hermosillo and El Sahuaral coastal aquifers, Sonora, Mexico. Hydrogeology Journal 4: 518–526.

Stevenson, M. 2002. Mercury from colonial Spanish silver mines poses threat in Mexico. Associated Press News, November 14.

Styles, B. T. 1993. Genus Pinus: a Mexican purview. Pp. 397–420 in T. P. Ramamoorthy, R. Bye, A. Lot, and J. Fa (eds.), Biological Diversity of Mexico: Origins and Distribution. Oxford University Press, New York.

Suzán, H., D. T. Patten, and G. P. Nabhan. 1997. Exploitation and conservation of ironwood (Olneya tesota) in the Sonoran Desert. Ecological Applications 7: 948–957.

Sykes, G. 1937. The Colorado Delta. American Geographical Society Special Publication 19, Carnegie Institution of Washington, Washington, D.C., and the American Geographical Society of New York.

Tanner, J. T. 1964. The decline and present status of the imperial woodpecker of Mexico. Auk 81: 74–81.

Tellman, B. 2002. Human introduction of exotic species in the Sonoran region. Pp. 25–46 in B. Tellman (ed.), Invasive Exotic Species in the Sonoran Region. Arizona-Sonora Desert Museum and University of Arizona Press, Tucson.

Thomsen, J. B., and A. Brautigam. 1991. Sustainable use of neotropical parrots. Pp. 359–379 in J. G. Robinson and K. H. Redford (eds.), Neotropical Wildlife Use and Conservation. University of Chicago Press, Chicago.

Thomsen, J. B., and T. A. Mulliken. 1992. Trade in neotropical psittacines and its conservation implications. Pp. 221–239 in S. R. Beissinger and N. F. R. Snyder (eds.), New World Parrots in Crisis: Solutions from Conservation Biology. Smithsonian Institution Press, Washington, D.C.

Thomson, D. A., L. T. Findley, and A. N. Kerstitch. 2000. Reef Fishes of the Sea of Cortez. University of Texas Press, Austin.

Toledo, V. M. 1990. El proceso de ganaderización y la destrucción biológica y ecológica de México. Pp. 191–225 in E. Leff (ed.), Medio Ambiente y Desarrollo en México. Centro de Investigaciones Interdisciplinarias en Humanidades, México D.F.

Toledo, V. 1996. Las consecuencias ecológicas de la Ley Agraria de 1992. Estudios Agrarios 4. [Available: http://www.pa.gob.mx/]

Tortajada, C. 2002. Environmental Impact Assessment of Water Projects in Mexico. Centro del Tercer Mundo para el Manejo de Agua, México D.F.

Treviño-Villarreal, J., and W. E. Grant. 1998. Geographic range of the endangered Mexican prairie dog (Cynomys mexicanus). Journal of Mammalogy 79: 1273–1287.

Turner, R. M., R. H. Webb, J. E. Bowers, and J. R. Hastings. 2003. The Changing Mile Revisited: an Ecological Study of Vegetation Change with Time in the Lower Mile of an Arid and Semiarid Region. University of Arizona Press, Tucson.

UGRS. 2001. Se cierra el ciclo 2000–2001 con buenos resultados en exportación. Unión Ganadera Regional de Sonora, Hermosillo, Sonora. [Available: http://www.unionganadera.com/paginas/nota01.html]

UNEP-WCMC. 2003. United Nations Environment Program, World Conservation Monitoring Centre, Protected Areas Database. [Available: http://www.unep-wcmc.org/protected_areas/data/nat2.htm]

University of Maryland Program in Sustainable Development and Conservation Biology. 1999. Draft Proposal to list Ligusticum porteri in Appendix II of the Convention on International Trade in Endangered Species of Wild Fauna and Flora. University of Maryland, College Park.

USTR. 2003. National trade estimate report on foreign trade barriers (NTE). Office of the U.S. Trade Representative, Washington, D.C. [Available: http://www.ustr.gov/reports/nte/2003/mexico.pdf]

Valdez-Zamudio, D., A. Castellanos-Villegas, and S. E. Mash. 2000. Land cover changes in central Sonora, Mexico. Pp. 349–351 *in* P. F. Ffolliott, M. B. Baker, C. B. Edminster, M. C. Dillon, and C. L. Mora (eds.), Land Stewardship in the 21st Century: The Contributions of Watershed Management. Proceedings, RMRS-P-13, U.S. Department of Agriculture, Forest Service, Rocky Mountain Research Station, Fort Collins, Colorado.

Valenzuela-Zapata, A. G., and G. P. Nabhan. 2004. ¡Tequila! A Natural and Cultural History. University of Arizona Press, Tucson.

Varela Espinosa, L. 2004. Flora y vegetación de la región de San Javier, Sonora, México. B.Sc. Thesis, Facultad de Ciencias, Universidad Nacional Autónoma de México, México D.F.

Vargas Márquez, F., R. de la Maza Elvira, and R. Marcó del Pont Lalli. 2001. Áreas Naturales Protegidas de México con Decretos Estatales. Instituto Nacional de Ecología-SEMARNAP, México D.F.

Veiga, M. M. 1997. Introducing new technologies for the abatement of global mercury pollution in Latin America. United Nations Industrial Development Organization/UBC/CETEM/CNP, CETME, Rio de Janeiro.

Warman Gryj, A. 1994. La reforma al Artículo 27 constitucional. La Jornada, April 8. (also included in Estudios Agrarios, no. 2, January-March 1996; http://www.pa.gob.mx]

West, R. C. 1993. Sonora: Its Geographic Personality. University of Texas Press, Austin.

Whitfield, J. 2003. Sheep horns downsized by hunter's taste for trophies. Nature 426: 595.

Wiggins, I. L. 1964. Flora of the Sonoran Desert. Pp. 189–1740 *in* F. Shreve and I. L. Wiggins, Flora and Vegetation of the Sonoran Desert, 2 vols. Stanford University Press, Stanford, California.

Williams, J. E., J. E. Johnson, D. A. Hendrickson, S. Contreras-Balderas, J. D. Williams, M. Navarro Mendoza, D. E. McAllister, and J. E. Deacon. 1989. Fishes of North America endangered, threatened, or of special concern. Fisheries 14: 2–20.

Wilson, M. F., L. Leigh, and R. S. Felger. 2002. Invasive exotic plants in the Sonoran Desert. Pp. 81–90 *in* B. Tellman (ed.), Invasive Exotic Species in the Sonoran Region. University of Arizona Press and Arizona-Sonora Desert Museum, Tucson.

Wood, M. 2002. Desert shrub may help preserve wood. Agricultural Research 50(4): 10–11.

Wright, T. F., C. A. Toft, E. C. Enkerlin-Hoeflich, J. González-Elizando, M. Albornoz, A. Rodríguez-Ferraro, et al. 2001. Nest poaching in neotropical parrots. Conservation Biology 15: 710–720.

Yates, T., and J. Salazar. In press. A revision of *Scapanus latimanus*, with the revalidation of a Mexican species. *In* R. Medellín and V. Sánchez Cordero (eds.), Homenaje al Dr. Bernardo Villa. Universidad Nacional Autónoma de México, México D.F.

Yetman, D., and A. Búrquez. 1998. Twenty-seven: a case study in ejido privatization in Mexico. Journal of Anthropological Research 54: 73–95.

Yetman, D., and R. S. Felger. 2002. Ethnoflora of the Guarijíos. Pp. 174–230 *in* D. Yetman, Guarijíos of the Sierra Madre: The hidden people of Northwestern Mexico. University of New Mexico, Albuquerque.

Zamora-Arroyo, F., P. L. Nagler, M. Briggs, D. Radtke, H. Rodriquez, J. Garcia, C. Valdes, A. Huete, and E. P. Glenn. 2001. Regeneration of native trees in response to flood release from the United States into the delta of the Colorado River, Mexico. Journal of Arid Environments 49: 49–64.

4

Mexico's Legal and Institutional Framework for the Conservation of Biodiversity and Ecosystems

ALBERTO SZÉKELY

LUIS OCTAVIO MARTÍNEZ MORALES

MARK J. SPALDING

DOMINIQUE CARTRON

Mexico's environmental legislation has grown rapidly in the last 15 years to progressively develop constitutional provisions regarding the conservation of natural resources that had been in force since 1917. In particular, 2 important amendments to the federal constitution were adopted. The amendment to Article 27, published in the *Federal Official Gazette (Diario Oficial de la Federación)* on August 10, 1987 (DOF 1987), establishes the power to dictate the necessary measures to preserve and restore ecological balance. At the same time, the new Section XXIX-G of Article 73 grants the federal congress the power to enact laws to ensure concurrence among the federal, state, and municipal governments, when acting under their respective jurisdictions, in the areas of environmental protection and preservation and restoration of ecological balance. The second amendment, published in the *Federal Official Gazette* on June 28, 1999 (DOF 1999f), creates a new paragraph five in Article 4 that establishes the right of every person to an environment appropriate for his or her development and well-being. These constitutional provisions are the basis of all environmental legislation enacted in recent years, particularly in the field of protection of biodiversity and ecosystems.

This chapter provides an overview of the legal framework for the protection of biodiversity and ecosystems in Mexico and details some of the problems that arise in the implementation of those provisions and the institutions responsible for their enforcement. The true challenge for Mexico is not the continuous enactment of laws, administrative rules, norms, or provisions, but the strengthening of the country's rule of law to ensure that effective enforcement and implementation of environmental legislation becomes the rule rather than the exception, as is the case now.

Legal Framework

General Act for Ecological Balance and the Protection of the Environment

Published in the *Federal Official Gazette* on January 28, 1988 (DOF 1988), the General Act for Ecological Balance and the Protection of the Environment (LGEEPA) is the most important piece of environmental protection legislation ever enacted in Mexico. According to its Article 1, the LGEEPA develops the constitutional provisions pertaining to the preservation and restoration of ecological balance and the protection of the environment. The goal of the law is to promote sustainable development through several mechanisms that include (1) guaranteeing the right of every person to live in an environment appropriate to his or her development, health, and well-being; (2) defining the principles of environmental policy and the instruments for its application to ensure the preservation, restoration, and improvement of the environment; (3) providing for the preservation and protection of biodiversity; (4) regulating the establishment and management of natural protected areas; (5) seeking sustainable utilization, preservation, and restoration of soil, water, and other natural resources in a manner that conciliates economic profit and social activity with the preservation of ecosystems; and (6) ensuring the prevention and control of air, water, and soil pollution.

Since 1988, several modifications have been made to the LGEEPA. The most significant modification, which was published in the *Federal Official Gazette* on December 13, 1996 (DOF 1996k), significantly amends the LGEEPA.

The LGEEPA Act establishes in Articles 15 to 41 (DOF 1988) the principles and instruments of environmental policy, which the authorities of the different levels of the government (federal, state and municipal) must apply. Among those instruments, the following are relevant to the protection and conservation of biodiversity and ecosystems: ecological land zoning (*ordenamiento ecológico*); environmental impact assessment, and Mexican Official Norms (*Normas Oficiales Mexicanas*).

Ecological Land Zoning

Ecological land zoning is defined by Article 3/ XXIII (DOF 1988) as the "instrument of environmental policy whose objective is to regulate or induce the use of land and productive activities towards the protection of the environment and the preservation and sustainable use of natural resources."

According to Article 19 Bis, environmental land zoning is made through the formulation and enactment of the following programs:

- General Ecological Land Zoning Program for the National Territory (national level)
- Regional Ecological Land Zoning Program (state level, it may include part of the territory of a state or part of the territory of several states)
- Local Ecological Land Zoning Program (municipal level)
- Marine Ecological Land Zoning Program.

The general and marine programs are formulated, enacted, and published by the Secretaría de Medio Ambiente y Recursos Naturales (SEMARNAT), in conformity with the LGEEPA and the Planning Law, with the participation of public and private organizations, academic and research institutions, and other interested stakeholders. The regional and local programs are prepared by the state governments and the municipal authorities, respectively, and according to the applicable state and municipal ecological laws and regulations.

Article 19 of the LGEEPA states that in the formulation of the ecological land zoning (ELA), the following criteria shall be taken into account:

1. The nature and characteristics of ecosystems in the national territory and the zones over which Mexico has jurisdiction;
2. The vocation of each zone or region, based on its natural resources, population distribution, and prevailing economic activities;
3. Imbalances in the ecosystems due to human settlement, economic activities, or other human activities or natural phenomena;
4. The balance that should exist between human settlements and the environment; and
5. The environmental impact of new human settlements, communication infrastructure, and other public works or activities.

Ecological land zoning has the potential to become the most effective instrument of environmental policy to protect the ecosystems existing in the country because, at the municipal program level (which must be consistent with the regional and national programs), it may establish permissible land uses in a determined area (forestry, agriculture, cattle raising, mining, human settlements, conservation, protection, restoration, etc.) and the limits to the implementation of such activities (both in terms of density and intensity). Once the ELA establishes what uses are permissible, compatible, or forbidden for a certain area, every permit authorizing different types of activities, including infrastructure projects, should be consistent with the limitation imposed by the ELA program.

Despite the potential and the importance of ELA as an instrument to protect biodiversity and ecosystems, only a few programs have been developed so far (table 4.1). In the 15 years since the LGEEPA was issued, the federal government has failed to formulate and enact a General Ecological Land

Table 4.1. Ecological land zoning programs.

Program Type	State and Year
Regional ecological land zoning programs (entire states)	Baja California (1995), Colima (1993), Guanajuato (1999), Hidalgo (2001), Jalisco (2001), México (1999) and Tlaxcala (2002)
Regional ecological land zoning programs (parts of states)	Corredor Tijuana-Ensenada (1995), Corredor Costero San Felipe Puertecitos, B. C. (1997), Costa de Jalisco (1999), Lázaro Cárdenas, Mich. (1998), Corredor Cancún-Tulúm, Q. Roo (1994/2001), Sian Ka'an, Q. Roo (2002) and Tula-Tepeji, Hgo (2002)
Local ecological land zoning programs	Distrito Federal (2000), Corredor Los Cabos, B.C.S (1994), Municipal de Los Cabos (1995), Cuatrociénegas, Coah (1997), Municipal de Huasca, Hgo (2000), Bahía de Banderas, Nay (1992), Costa Maya, Q. Roo (2000), Isla Mujeres, Q. Roo (2002), Sistema Lagunar de Nichupté, Q. Roo (1994).

Zoning Program, which should be the starting point for all the other ecological land zoning programs, as well as for other important instruments such as the Forestry Zoning Program.

Environmental Impact Assessment

The environmental impact assessment procedure is regulated by Articles 28–35 Bis 3 of the LGEEPA and the Administrative Rules on Environmental Impact Assessment published in the *Federal Official Gazette* on May 30, 2000 (DOF 2000a).

According to Article 28 of the LGEEPA, the

environmental impact assessment is the procedure through which SEMARNAT shall establish the conditions to which public works and activities that might cause ecological imbalance or exceed the limits and conditions established in the provisions applicable to the protection of the environment and the preservation and restoration of the ecosystems, shall be subject for the purpose of avoiding or minimizing their negative effects on the environment.

Every work, activity or infrastructure project listed in Article 28 (table 4.2) requires an authorization of environmental impact granted by SEMARNAT prior to its undertaking. Public works, activities, and infrastructure projects that require authorization are further specified in Article 5 of the Administrative Rules on Environmental Impact Assessment.

To obtain this authorization it is necessary to submit an environmental impact statement (*manifestación de impacto ambiental*) to SEMARNAT.

During the evaluation procedure, any person from the community may request a public inquiry and has the right to formulate opinions and to propose the imposition of specific mitigation measures.

SEMARNAT is to issue a resolution within 60 days. This period could be extended for another 60 days if, given the complexity of the project, the reviewing authority requires more time. In its resolution, SEMARNAT may (1) approve the public work, activity, or infrastructure project; (2) authorize the work, activity, or infrastructure project with conditions; or (3) deny the authorization. The state and municipal governments are responsible for the evaluation and approval of the public works and activities not listed in Article 28 of LGEEPA, in accordance with the respective ecology law in force in every state.

In practice, the environmental impact assessment process has several limitations. First, the list included in Article 28 of the LGEEPA excludes from the environmental impact assessment process government programs, some of which may have a great impact on the environment (such as the National Development Program). Second, few professionals are capable of preparing an environmental impact statement (and even fewer have sufficient moral integrity to withstand any pressure from the owner of the project, who pays for the study). The government has even fewer experts capable of evaluating environment impact statements, which means only a very limited number of proposed projects are effectively evaluated. Further, many projects proceed without authorization of any kind. In fact, it is a common practice (and more convenient in economic terms) to apologize and pay a fine, rather than to submit a project proposal for an environmental impact assessment.

Table 4.2. Type of public work, activity, or infrastructure project listed in Article 28 and requiring an environmental impact authorization granted by SEMARNAT.

I	Hydraulic works, general means of communication, pipelines, gas lines, coal chutes, and multipurpose lines
II	Petroleum, petrochemical, chemical, steel, paper, sugar, cement, and electrical industries
III	Exploration, extraction, and mining of minerals and substances reserved to the federation in the terms of the Mining Laws and Regulation of Constitutional Article 27 on Nuclear Material
IV	Installations of treatment, storage, or elimination of hazardous wastes, as well as radioactive wastes
V	Use of forests in tropical jungles and of species of difficult regeneration
VI	(Repealed)
VII	Changes of use of the land in forest areas, as well as in jungles and arid zones
VIII	Industrial parks where there are provisions for highly risky activities to be carried out
IX	Building developments that affect coastal ecosystems
X	Works and activities in marshes, mangrove swamps, lagoons, rivers, lakes, and estuaries connected with the sea, as well as along their littorals or in associated federal zones
XI	Works in protected natural areas under the jurisdiction of the Federation
XII	Fishing, fish farming, or land and cattle-rising activities that may put the preservation of one or more species in danger or cause damages to the ecosystems
XIII	Works or activities that correspond to matters under federal jurisdiction that may cause serious or irreparable ecological imbalances, damages to the public health or to the ecosystems, or exceed the limits and conditions established in the legal provisions relative to the preservation of ecological balance and environmental protection

At the state level the situation is even worse. The state and municipal governments do not have the ability to evaluate the environmental impact of proposed projects, not to mention the capacity to enforce their own impact assessment resolutions. Many states face a serious problem of corruption and collusion between environmental authorities and the promoters of projects.

Nevertheless, the environmental impact assessment is the instrument of environmental policy which, in relative terms, is more effectively used, particularly in the case of important projects that are subject to public scrutiny.

Mexican Official Norms

In accordance with the LGEEPA and the Federal Law on Metrology and Normalization (*Ley Federal sobre Metrología y Normalización*), the executive branch can issue technical standards known as "Mexican Official Norms" (*Normas Oficiales Mexicanas*) on different subjects, including the environment. Technical standards are documents that specify criteria for a practice or system of practices. To date, 6 norms (plus 1 proposed additional norm) have been issued to protect ecosystems and biodiversity (table 4.3). Seven norms have been published concerning environmental impact assessments (table 4.4; see table 4.5 for norms on forestry issues).

Natural Protected Areas

One of the most important instruments established in the LGEEPA to protect biodiversity and ecosystems is the establishment of natural protected areas (Articles 44–77).

Article 3/II defines natural protected areas as "zones in the national territory and those areas where the nation exercises its sovereignty and jurisdiction, in which the original environment has not been significantly altered by man or which have been preserved and restored and are subject to the system provided in this law."

The constitutional basis for the establishment of natural protected areas is in the third paragraph of Article 27 of the federal constitution, which allows the nation to impose, over private property, the restrictions and limits (modalities) required by the public interest. In that regard, Article 44 states:

Those who own, possess or hold other rights on lands, waters and forests included within protected natural areas, shall be subject to modalities. In accordance with [the LGEEPA], the modalities are established by the decrees through which those areas are established, as well as by other provisions contained in the management program of the area and in the corresponding ecological land zoning programs.

Table 4.3. Mexican Official Norms on biodiversity and ecosystems.

Mexican Official Norm	Date issued	Description	Reference
NOM-059-ECOL-2001	March 6, 2002	Establishes environmental protection for native plant and wildlife species from Mexico; includes risk categories and specifications for their inclusion, exclusion, or change in status and lists species at risk	DOF 2002b
NOM-060-ECOL-1994	May 13, 1994	Establishes the specifications for mitigating the adverse effects of forestry exploitation on soils and water bodies	DOF 1994a
NOM-061-ECOL-1994	May 13, 1994	Establishes the specifications for mitigating the adverse effects of forestry exploitation on wild flora and fauna	DOF 1994b
NOM-062-ECOL-1994	May 13, 1994	Establishes the specifications for mitigating the adverse effects on biodiversity of land-use changes from forestry to agriculture	DOF 1994c
NOM-126-ECOL-2000	March 20, 2001	Establishes the specifications for scientific collection of biological material of plant and wildlife species and other biological resources on the national territory	DOF 2001a
NOM-056-FITO-1995	July 11, 1996	Establishes the phyto-sanitary requirements for the domestic shipment, importation, and establishment of field trials with organisms that have been manipulated through genetic engineering	DOF 1996j
NOM-70-FITO-1995 (proposed norm)	April 25, 1996	Establishes the requirements and specifications for the importation, introduction, transport, and release of biological control agents	DOF 1996a

Mexican Official Norms are published in the *Diario Oficial de la Federación* (DOF).

In accordance with Article 45, the purpose of establishing natural protected areas is to

1. Preserve natural environments that represent the different biogeographical and ecological regions and the most fragile ecosystems to ensure equilibrium and continuity of evolutionary and ecological processes;
2. Safeguard the genetic diversity of wildlife species and ensure the preservation and sustainable use of nature on the national territory, especially with regard to species that are in danger of extinction or that are threatened, endemic, or rare;
3. Ensure the sustainable use of ecosystems and their components.
4. Provide areas for scientific research on ecosystem health;
5. Discover and publicize knowledge, practices, and technology that contribute to rational sustained use of biodiversity;
6. Protect towns, communication routes, industrial facilities, and agricultural activities from the risk of floods, through protection of montane forests; and
7. Protect natural environments in areas that have archeological and historical significance, as well as areas of importance to tourism, recreation, culture, and national identity and areas important to indigenous populations.

At present, Article 46 of the LGEEPA establishes 8 types of natural protected areas, 6 under federal jurisdiction, 1 under state jurisdiction, and 1 under municipal jurisdiction (state and municipal natural protected areas are regulated by the ecological legislation in each state):

- Biosphere Reserves
- National Parks
- Natural Monuments
- Areas for the Protection of Natural Resources
- Areas for the Protection of Flora and Fauna
- Sanctuaries

Table 4.4. Mexican Official Norms on environmental impact assessment.

Mexican Official Norm	Date of Publication	Description	Reference
NOM-113-ECOL-1998	October 26, 1998	Establishes the environmental protection specifications for the planning, design, construction, operation, and maintenance of electric generating and transmission substations in urban, suburban, rural, farming, industrial, urban-equipment, service, and tourism areas	DOF 1998d
NOM-114-ECOL-1998	November 23, 1998	Establishes the environmental protection specifications for the planning, design, construction, operation, and maintenance of transmission and subtransmission electric lines in urban, suburban, rural, farming, industrial, urban-equipment, service, and tourism areas	DOF 1998f
NOM-115-ECOL-1998	November 25, 1998	Establishes the environmental protection specifications for drilling land oil wells for exploration and production in agricultural, cattle-raising, and uncultivated areas	DOF 1998i
NOM-116-ECOL-1998	November 24, 1998	Establishes the environmental protection specifications for land seismic surveys in agricultural, cattle-raising, and uncultivated areas	DOF 1998g
NOM-117-ECOL-1998	November 24, 1998	Establishes the environmental protection specifications for the installation and major maintenance of systems for the transport and distribution of liquid and gaseous petrochemicals and hydrocarbons performed on existing land rights-of-way in agricultural, cattle-raising, and uncultivated areas	DOF 1998h
NOM-120-ECOL-1997	November 19, 1998[a]	Establishes the environmental protection specifications for mining exploration activities in areas with dry and temperate climates and with xerophyllous scrub vegetation, tropical deciduous woodland, and coniferous or oak woodlands	DOF 1998e
NOM-130-ECOL-2000	March 23, 2001	Establishes the environmental protection specification for the planning, design, site preparation, construction, operation, and maintenance of telecommunication fiberoptic network systems	DOF 2001b

Mexican Official Norms are published in the *Diario Oficial de la Federación* (DOF).

[a]A note explaining this norm was published on January 6, 1999 (DOF 1999).

- State Parks and Reserves
- Ecological Preservation Zones in Population Centers.

Originally, the LGEEPA provided for 2 additional types of natural protected areas, but both designations have now been repealed. For the establishment of natural protected areas under federal jurisdiction, Article 58 requires the preparation of a "previous justifying study" that provides the technical and sci-entific information to justify the establishment of the area. Once this study is finished, SEMARNAT, through the National Commission for Natural Protected Areas (CONANP), publishes a notice in the official gazette that provides interested parties with an opportunity to submit their comments. Opinions from the state governments involved, agencies of the federal government, public and private organizations, indigenous people, and academic institutions are all expected.

Subsequently, a presidential decree is issued establishing the natural protected area. That decree must be published in the *Federal Official Gazette* and owners and landholders whose property is affected must be notified. In case their address is unknown, a second notification must be published in the official gazette.

In accordance with Article 60, the decree must include the following:

- The delineation of the area, indicating its size and boundaries, and, as applicable, the corresponding zoning
- The modalities to which the use and exploitation of natural resources are subjected
- The description of the activities that can be carried out in the area, as well as the modalities and limitations to which they are subject
- In the case that the establishment of the area involves land expropriation, the public utility cause of such measure
- The guidelines and general principles for the attainment of the preservation, restoration, and sustainable use of the natural resources inside the boundaries of the area, the management and monitoring of the area, and the preparation of the area's management program and management rules.

For the modification of an established natural protected area, the same procedure and requirements that the LGEEPA requires for establishing such an area must be satisfied.

Once a natural protected area has been established, a management program and management rules are formulated by SEMARNAT, with the participation of the owners and landholders of properties within the area, state and municipal governments, and agencies of the federal government, as well as public and private organizations interested in the area.

In accordance with Article 67 of the LGEEPA, SEMARNAT can transfer the management of a natural protected area to state and municipal government and also to private individuals or entities (e.g., *ejidos*, agrarian communities, indigenous people, private organizations), through specific agreements.

In terms of Article 76, SEMARNAT has created a National System of Natural Protected Areas with the purpose of including those areas which, because of their biodiversity and ecological characteristics, are considered of special importance to the country.

Restoration Programs and Restoration Zone Orders

Article 78 of the LGEEPA states that SEMARNAT shall formulate and implement ecological restoration programs in areas that show signs of degradation or desertification or serious ecological imbalances, in order to recover and reestablish the conditions that are propitious for the evolution and the continuity of natural processes. According to Article 78, participation of land owners, users, public or private social organizations, indigenous people, local governments, and any other interested parties should be promoted by SEMARNAT during the formulation, implementation, and continuation of ecological restoration programs.

Likewise, Article 78 Bis states that, in cases where processes of desertification or degradation are accelerated, which imply the loss of resources that are difficult to regenerate, recover, or reestablish, or irreversible consequences to ecosystems or their components, an executive presidential order may be issued establishing an "ecological restoration zone." According to Article 78 Bis,

> Prior to issuing the order, [SEMARNAT] is required to conduct the necessary studies that justify such an order. The orders must be published in the *Federal Official Gazette*, and recorded in the corresponding Public Registry of Property. The orders may totally or partially include lands subject to any system of ownership and must include:
> I. The delimitation of the zone subject to ecological restoration, indicating its size, location, and boundaries;
> II. The necessary actions to regenerate, recover, or reestablish the natural conditions of the zone;
> III. The conditions to which uses of the land and exploitation of natural resources and of the flora and fauna shall be subject, as well as the realization of any type of work or activity within the zone;
> IV. The guidelines for the elaboration and implementation of the corresponding ecological restoration program, as well as for the participation in those activities by owners, users, public or private social organizations, indigenous people, local governments and other interested parties, and
> V. The period of implementation of the respective ecological restoration program.

According to Article 78 Bis 1, through the executive order the president may impose modalities and restrictions to the use of the land involved.

To date, only 1 Restoration Program has been implemented, while only 1 Ecological Zone Order has been issued, in both cases mostly in response to forest fires in 1998 (DOF 1998a,b, 1999b,c,d).

General Wildlife Act

The General Wildlife Act (*Ley General de Vida Silvestre* or LGVS), published in the *Federal Official Gazette* on July 3, 2000 (DOF 2000b), is governed by Articles 27 and 73/XXIX-G of the federal constitution. Its objective is to establish the concurrence of federal, state, and municipal governments, under their own jurisdiction, when taking actions related to the conservation and sustainable exploitation of wildlife and its habitat in the Mexican Republic territory, and in the areas where the nation exercises jurisdiction.

The LGVS abrogates the Federal Hunting Act, whose publication dated back to January 5, 1952 (DOF 1952), and which was much more of a conservation law than the LGVS. In fact, the Federal Hunting Act allowed the exploitation of wildlife only through sport hunting. The LGVS, in contrast, allows not only sport hunting, but also the exploitation of wildlife for commercial purposes (including hunting, capture, and collection of specimens of wildlife). It establishes a market for wildlife, and, in general terms, tries to promote the exploitation of wildlife, which represents a significant departure from the traditional environmental policy of the Mexican government. The LGVS was amended by a decree published in the *Federal Official Gazette* on January 10, 2002 (DOF 2002a). Everything related to the exploitation of forests (regulated by the General Law on Sustainable Forestry Development) and to aquatic species (regulated by the Fisheries Law) is excluded from the application of this law (with the exception of species at risk).

Sustainable Exploitation of Wildlife

Articles 18–55 of the LGVS's Title 5 establish general provisions for the conservation and sustainable exploitation of wildlife. Articles 82–103 in Title 7 establish specific provisions for the exploitation of wildlife.

Among the provisions of Title 5, those related to the "Management Units for the Conservation of Wildlife" (UMAS) are especially relevant. UMAS are defined by Article 3 of the LGVS as the lands and facilities registered to operate with an approved management plan, in which there is continuous monitoring by the state of the habitat as well as of wildlife and wildlife populations.

Article 39 states that the landowners and landholders of places where activities of conservation and/or exploitation of wildlife are to be conducted must register them with SEMARNAT as UMAS. These UMAS might have specific purposes of restoration, protection, maintenance, recovery, reproduction, repopulation, reintroduction, research, rescue, guarding, rehabilitation, exhibition, recreation, environmental education, or sustainable exploitation of wildlife.

Once the petition for registration, accompanied by a management plan, has been submitted, SEMARNAT determines whether to grant the registration. Registration may be conditioned by changes in the management plan. SEMARNAT may also deny the registration because a legal provision has been violated in executing the management plan.

Specific provisions to regulate the exploitation of wildlife are established in Title 7. The LGVS establishes 2 types of exploitation: "extractive," which includes the collection, capture, and hunting of animals, and "nonextractive." Likewise, there are special provisions regarding exploitation by local communities and indigenous people for survival purposes and for the scientific collection of specimens.

Even though the UMAS system seems to be a strong instrument to conciliate the exploitation of wildlife (in an alleged sustainable manner) with habitat conservation, in practice, SEMARNAT lacks the capacity to effectively evaluate all management programs submitted. Additionally, almost any facility may be registered as UMAS, simulating sustainable exploitation. Finally, the problem of illegal exploitation of wildlife is increasing, and the Procuraduría Federal de Protección al Ambiente (PROFEPA) appears to be overwhelmed.

Conservation of Wildlife

The provisions regarding the conservation of wildlife are established in Articles 56–81 of the LGVS. These provisions include rules for declaring species and populations as "At Risk" (Endangered,

Threatened, or Subject to Special Protection). According to Article 56, SEMARNAT is responsible for identifying those species. This identification process has been completed and the resulting list issued through Mexican Official Norm NOM-059-ECOL-2001 in the *Federal Official Gazette* on March 6, 2002 (DOF 2002b).

Species At Risk are to be the focus of conservation and recovery projects, establishment of special management measures, conservation of critical habitats, establishment of refuge areas (for the protection of aquatic species), permanent monitoring programs, and certification of sustainable exploitation. Article 63 states that SEMARNAT has the discretion to declare that certain areas are "critical habitats for the conservation of wildlife." In accordance with Article 64 of the LGVS, within these areas SEMARNAT may agree with the landowners to special management and conservation measures. Alternatively, it may file, before the president of the republic, the imposition of limits to domain rights in accordance with the Expropriation Law. Furthermore, Article 65 states that SEMARNAT might establish "refuge areas" with the purpose to protect aquatic wildlife species that live in federal jurisdictional waters. With regard to other conservation measures, such as restoration programs or the establishment of prohibitions (*vedas*), the LGVS refers to the provisions of the LGEEPA.

General Act for Sustainable Forestry Development

Published in the *Federal Official Gazette* on February 25, 2003 (DOF 2003), the General Act for Sustainable Forestry Development (*Ley General de Desarrollo Forestal Sustentable*; LGDFS) replaces the 1992 Forestry Act. It is one of the most significant laws for the conservation of biodiversity.

The LGDFS is governed by Article 27 of the federal constitution, and its purpose is to regulate and promote the conservation, protection, restoration, production, arrangement, cultivation, management, and exploitation of Mexico's forest ecosystems and associated resources, as well as to distribute jurisdiction among the federal, state, and municipal governments, under the principle of concurrence established by Section XXIX of constitutional Article 73, with the goal of promoting sustainable forestry development.

Like the LGEEPA, the LGDFS also establishes a series of principles, criteria, and the following instruments of forestry policy:

- Forestry development planning
- National system of forests information
- National Inventory of Forests and Soil
- Forest zoning
- National forestry record
- National system of forestry promotion
- Mexican Official Norms on forestry matters.

Among these instruments of policy, only the National Inventory of Forests and Soil, forest zoning, and Mexican Official Norms, if used in a proper way, can contribute to the protection of forested habitats and their biodiversity. However, the LGDFS also has other instruments more relevant to the protection of forests, such as authorizations on forestry matters, forest bans, and economic instruments for the promotion of forestry industry.

National Inventory of Forests and Soil

Article 44 of the LGDFS states that SEMARNAT must regulate the procedures and methodology that the National Forestry Commission uses to integrate the National Inventory of Forests and Soil. This inventory incorporates information regarding

- The surface area and location of all forest lands of the country, with the purpose of integrating statistical data and elaborating maps at different levels of management
- Temporary land forests, including surface area and location
- The types of forest vegetation and soils, including location, conformations, and classes, with projections and tendencies that allow the classification and delimitation of status of degradation, as well as zones for conservation, protection, restoration, and forest production, zone identified on the basis of hydrological-forest basins, ecological regions, permanent forestry areas, and natural protected areas
- Monitoring data on the modification of forest vegetation, with the purpose of knowing and evaluating deforestation and degradation rates and disturbances, and recording their principal causes

- The quantification of forest resources, including the value of goods and environmental services generated by forest ecosystems, as well as the impacts that affect them
- The criteria and the index of sustainability and degradation of forest ecosystems
- The inventory of the existing forestry infrastructure
- Other information to be determined in administrative regulations.

According to Article 46, the information listed above is the basis for (1) formulating, executing, controlling, and supervising programs and actions related to forestry issues, (2) calculating the volume of wood or standing forest biomass, its increase, or the potential volume of exploitation, (3) elaborating forest zoning, forest ordinance, and ecological land zoning, and (4) evaluating and monitoring long-term, mid-term, and short-term plans.

Forest Zoning

Article 48 of the LGDFS states that forest zoning is to be used to identify, group, and order forest lands and "preferably forest lands" within the hydrological-forest basins, sub-basins and micro-basins, on the basis of their functions and biological, environmental, socioeconomic, recreational, protective and restoration subfunctions, with management purposes and with the object to promote better management and to contribute to sustainable forestry practices. The National Forestry Commission is in charge of carrying out forest zoning, which must be approved by SEMARNAT.

Mexican Official Norms in Forestry Matters

To date, as many as 18 Mexican Official Norms (plus 1 proposed additional norm) related to forestry have been issued (table 4.5).

Authorizations on Forestry Matters

The granting of forestry authorizations is the main tool that the federal government has to regulate forestry. The LGDFS empowers SEMARNAT to establish the specific conditions to be followed by any person or legal entity interested in the exploitation of forests, the performance of forestation and refor-estation activities, or the transportation, storage, and transformation of forestry raw materials.

In accordance with Article 73 of the LGDFS, an authorization is required to exploit forest resources. The process includes approval of a forest management program and the authorization of an environmental impact study.

To obtain this authorization, the applicant must specifically submit

- A simplified forestry management program, in the case of exploiting an area < 20 ha
- An intermediate forestry management program; in the case of exploiting an area > 20 ha and < 250 ha
- An advanced forestry management program; in the case of exploiting an area > 250 ha.

Once the authorization is requested and the corresponding management program is submitted, SEMARNAT is required to issue a resolution within 30 days. However, if an environmental impact authorization is required, there is a 60-day time period, which could be extended for another 60 days depending on the complexity of the project. In the resolution SEMARNAT may (1) authorize the forestry management program, (2) authorize the forestry management program with the imposition of additional conditions or mitigation measures, or (3) deny the authorization.

For commercial forest plantations with an area < 800 ha, the submission of a written notification is the only requirement listed in the LGDFS. In the case of an area > 800 ha, the LGDFS requires the submission of a forestry management program for authorization.

Forestation projects with conservation and restoration purposes, as well as forestation activities and agro-silviculture practices in degraded forest lands, do not require an authorization and are only subject to the provisions of applicable Mexican Official Norms (Article 131).

The change in the use of soil in forest land (which implies the removal of vegetation of forest land to perform nonforestry activities), is allowed only exceptionally. Article 117 states that a previous technical opinion by the Forest State Council and justifying studies are required. Those studies must prove that there is no danger to biodiversity, no risk of soil erosion, and no loss in quantity or quality of water and that the proposed activities are more productive in the long term.

Table 4.5. Mexican Official Norms on forestry issues.

Mexican Official Norm	Date Issued	Description	Reference
NOM-001-RECNAT-1995	December 1, 1995	Establishes the types of markings that wood veneer in rolls should have, such as guidelines for its use and control	DOF 1995
NOM-002-RECNAT-1996	May 30, 1996	Establishes procedures, criteria, and specifications for the use, extraction, transport, and storage of pine resin	DOF 1996c
NOM-003-RECNAT-1996	June 5, 1996	Establishes procedures, criteria, and specifications for the use, extraction, transport, and storage of soil	DOF 1996d
NOM-004-RECNAT-1996	June 24, 1996	Establishes procedures, criteria, and specifications for the use, transport, and storage of roots and rhizomes of forest vegetation	DOF 1996e
NOM-005-RECNAT-1997	May 20, 1997	Establishes procedures, criteria, and specifications for the [sustainable] use, transportation, and storage of bark, stems, and whole forest plants	DOF 1997d
NOM-006-RECNAT-1997	May 28, 1997	Establishes procedures, criteria, and specifications for the use, transportation, and storage of palm leaves	DOF 1997e
NOM-007-RECNAT-1997	May 30, 1997	Establishes the procedures, criteria, and specifications for the use, transportation, and storage of branches, leaves, flowers, fruits, and seeds	DOF 1997f
NOM-008-RECNAT-1996	June 24, 1996	Establishes procedures, criteria, and specifications for the use, transportation, and storage of plant buds or shoots	DOF 1996f
NOM-009-RECNAT-1996	June 26, 1996	Establishes procedures, criteria, and specifications for the use, transport, and storage of latex derived from trees	DOF 1996g
NOM-010-RECNAT-1996	May 28, 1996[a]	Establishes procedures, criteria, and specifications for the use, transportation, and storage of fungi	DOF 1996b
NOM-011-RECNAT-1996	June 26, 1996	Establishes procedures, criteria, and specifications for the use, transportation, and storage of moss, hay, and *doradilla*	DOF 1996h
NOM-012-RECNAT-1996	June 26, 1996[b]	Establishes procedures, criteria, and specifications for the domestic use of firewood	DOF 1996i
NOM-013-RECNAT-1997	September 28, 1998	Sanitary importation regulation of natural Christmas trees of the species *Pinus sylvestris* and *Pseudotsuga menziesii* and the genus *Abies*	DOF 1998c
NOM-015-SEMARNAP/SAGAR-1997	March 2, 1999	Regulates the use of fire in forests and agricultural areas and establishes the procedures, criteria, and specification for organizing public and governmental participation in detecting and fighting forest fires	DOF 1999e
NOM-018-RECNAT-1999	October 27, 1999	Establishes procedures, criteria, and technical and administrative specifications for the sustainable use of *candelilla* (*Euphorbia antisyphilitica*) and for the storage and transport of *cerote*	DOF 1999h
NOM-019-RECNAT-1999	October 25, 2000	Establishes technical guidelines for the control of and fighting against insects eating the bark of coniferous trees	DOF 2000c
NOM-020-RECNAT-2001	December 10, 2001	Establishes procedures and guidelines for the restoration, improvement, and conservation of forest pastures	DOF 2001d
NOM-023-RECNAT-2001	December 10, 2001	Establishes technical specifications for the inventory, classification, and mapping of soils	DOF 2001e
PROY-NOM-025-RECNAT-2001 (proposed norm)	December 7, 2001	Establishes procedures and specifications for the collection and distribution of commercial and/or research forest germplasm intended for reforestation	PROY 2001c

Mexican Official Norms are published in the *Diario Oficial de la Federación* (DOF).

[a] A clarification of this norm was published on May 6, 1997 (DOF 1997b).
[b] A clarification of this norm was published on May 13, 1997 (DOF 1997c).

Forestry Bans

In accordance with Article 128 of the LGDFS, once a justifying study has been conducted and the opinion of the Forest Council has been requested, a presidential decree establishing a forestry ban (*veda forestal*) may be issued. A draft of the decree is first published in the *Federal Official Gazette* and notice sent to the potentially affected parties (owners, land holders, authorization holders, etc.) so that they can exercise their constitutional right to a hearing. If the address of these individuals is unknown, a second publication in the *Federal Official Gazette* must be issued. Subsequently, the final presidential decree is published twice in the *Federal Official Gazette* and once in a newspaper with wide distribution in the state where the affected lands are located.

Fisheries Act

The Fisheries Act (*Ley de Pesca* or LP) was published in the *Federal Official Gazette* on June 25, 1992 (DOF 1992a). Article 27 of the federal constitution also governs this law, whose purpose is to guarantee the conservation, preservation, and reasonable exploitation of fishery resources and to establish the principles for their proper promotion and management. The provisions of the LP apply to waters under federal jurisdiction and to Mexican ships that perform fishing activities in the high seas or in waters under the jurisdiction of a foreign country under an authorization, permit, or concession granted by that foreign country to Mexico or to the Mexican people.

Due to an amendment to the Federal Public Administration Organic Act and to the Fisheries Act, published in the *Federal Official Gazette* on November 30, 2000 (DOF 2000d), the application of the provisions of the LP is now under the jurisdiction of the Secretaría de Agricultura, Ganadería, Desarrollo Rural, Pesca y Alimentación (SAGARPA). In accordance with Article 4 of the LP, a permit, concession, or authorization granted by SAGARPA is required to exploit fisheries resources. An exception to this rule is established in Article 4, which states that fishing activities for domestic consumption, sport fishing performed from the land, and aquaculture activities performed in water bodies outside federal jurisdiction do not need an authorization.

Concessions may be granted for a period between 5 and 20 years and, in the case of aquaculture, for a maximum of 50 years. The term may be extended for the same period for which the concession was granted. Permits are valid only for four years.

Concession and permit holders are under obligation to report to SAGARPA their methods and fishing gears, findings, investigations, studies, and new projects related to fishing activities. They must also submit a record of the information previously determined by SAGARPA. According to Article 7 of the LP, the grant of concessions and permits is subject to the availability and conservation of the respective resources and the modalities dictated by public interest.

Additional requirements and obligations for concession and permit holders, as well as for aquaculture and fisheries activities, are established and regulated in the Fisheries Act Administrative Rules (*Reglamento de la Ley de Pesca*; RLP), published in the *Federal Official Gazette* on September 29, 1999 (DOF 1999g). The administrative rules provide that fisheries activities are classified in

- Commercial fisheries (*Pesca comercial*)
- Promotional fisheries (*Pesca de fomento*)
- Didactic fisheries (*Pesca didáctica*)
- Sport fisheries (*Pesca deportivo-recreativa*)
- Domestic consumption fisheries (*Pesca de consumo doméstico*).

Aquaculture activities are classified in

- Farming or aquaculture with commercial purposes (*Cultivo o acuacultura con fines comerciales*)
- Farming or aquaculture with promotional purposes (*Cultivo o acuacultura con fines de fomento*)
- Farming or aquaculture with teaching and demonstration purposes (*Cultivo o acuacultura con fines didácticos*).

According to the LP, concessions and permits to conduct commercial fishing may be granted to physical or juridical Mexican persons. Exceptionally, in conformity with international treaties in force, the Mexican government may declare the existence of a surplus of a species and allow foreign ships to partake of that surplus in the "exclusive economic zone" under the requirements and conditions established by SAGARPA and in conformity with the 1986 Federal Marine Act.

Authorizations, concessions, and permits may be revoked by SAGARPA when the holders (Article 17 of the LP):

- Damage the ecosystem or put it at imminent risk
- Fail to provide necessary information within the terms and deadlines required by SAGARPA, or if false information is provided
- Fail to obey the general technical conditions imposed by SAGARPA
- Transfer their authorizations
- Transfer the rights of a concession or permit without the permission of SAGARPA
- Fall into bankruptcy, disappearance, or dissolution in the case of legal entities.

The main purpose of the Fisheries Act is to regulate the exploitation of fisheries resources, which explains why there are just a few provisions regarding the conservation of marine biodiversity. In fact, the LP is very weak with regard to conservation and biodiversity. The LP only empowers SEMARNAT to dictate measures to protect turtles, sea mammals, and aquatic species requiring special protection or at risk of extinction and to establish bans on the exploitation of those species.

SAGARPA has responsibility for

- Establishing methods for the conservation of fisheries resources and for the repopulation of fishing areas
- Creating refuge areas to protect aquatic species
- Determining the maximum allowable catch
- Regulating instruments, equipment, fishing gear, personnel, and fishing techniques
- Determining the number of ships and their characteristics involved in the exploitation of a given species or group of species
- Establishing the catch season, and the size, or minimum weight of specimens, and proposing management, conservation, and transportation measures.

Finally, the LP requires that SAGARPA formulate, publish, and update the National Fisheries Charter (*Carta Nacional Pesquera*), which must include an inventory of fisheries resources in waters under federal jurisdiction susceptible to exploitation. The charter also contains the guidelines, strategies, and other provisions related to the conservation, protection, restoration, and exploitation of

aquatic resources. It contains guidelines for the performance of activities that might damage ecosystems and lists of acceptable fishing methods and equipment (see Article 18 of the LP's Regulations).

As mentioned previously, at the beginning of the current administration, the jurisdiction over fisheries matters passed from SEMARNAP to SAGARPA. This change proved to be a crucial mistake because SAGARPA has demonstrated that it lacks the capacity or perhaps the willingness to protect and conserve marine biodiversity. Proof of this was the publication in the *Federal Official Gazette* on July 12, 2002 of Official Norm NOM-029-PESC-2000 (specifications for the exploitation of sharks; DOF 2002c), which allowed fishing equipment forbidden almost everywhere else. This norm provoked such controversy and negative reactions from the public that SAGARPA had to revoke it immediately (in the official gazette on October 11, 2002; DOF 2002d). Additionally, SAGARPA appears to be granting fishing permits and concessions without any restrictions and supervision.

National Waters Act

Published in the *Federal Official Gazette* on December 1, 1992 (DOF 1992b), the National Waters Act (*Ley de Aguas Nacionales*; LAN) is governed by the provisions of Article 27 of the federal constitution. The objective of the LAN is to regulate the exploitation, use, distribution, and control of national waters, as well as the preservation of the quantity and quality of water necessary to achieve sustainable development.

In accordance with Article 27 of the federal constitution, the water ownership regime in Mexico is structured as described below.

National water is water from watercourses that flow to the sea or that cross state lines or international borders or that constitute the international border or the limits between two Mexican states. It includes waters from international watercourses; water bodies declared national property; waters within the territorial sea; water from lagoons and wetlands that are permanently or intermittently linked to the sea and that form lakes that are linked to constant watercourses; and water of lakes, lagoons, and wetlands that are national property.

Underground water is considered as national, but the landowner can exploit it freely unless there is a ban, a declared reserve, or an administrative rule for the use of the water in a specific area. These limitations must be established in a presidential decree. In the event that these limitations exist, the landowner must obtain a concession or permit to exploit the water (Article 18 of LAN) from the National Water Commission (CNA).

Private waters are found within the boundaries of a single piece of property. If the water crosses through more than one property within the same state, then its exploitation may be subject to state control.

Use and Exploitation of National Waters

To use or exploit national waters, it is necessary to obtain, from CNA, a concession or permit for a specific use, as provided by Article 20 of the LAN. An exception to Article 20 is that when the water is used for domestic purposes or for cattle raising, a permit from CNA is not needed (Article 17).

To obtain a permit, a petition must be submitted to the CNA containing the following information (Article 21):

- Name and address of the petitioner
- Basin, region, and locality
- The point of diversion of national waters for which a permit is requested
- The consumption volume requested (the CNA will establish the volume approved in the concession title)
- The initial intented use of the water
- The point of discharge, with the conditions of quantity and quality
- A description of the proposed or existing diversion and discharge structures
- The term for which the concession is requested.

The CNA, in accordance with Article 22, must respond to the petition within 90 working days from the submission date. Granting of the permit is subject to the provisions of the LAN and its regulations and takes into account the availability of water in accordance with the hydrologic regime of the water source, existing uses, the Public Registry of Water Rights, and existing legal bans and declared reserves. The permit can be granted, in accordance with Article 24, for a period of 5–50 years. The term can be extended for the same period for which the permit was granted. The permit title must be registered in the Public Registry of Water Rights, as well as every act or agreement related to the total or partial transfer of the permit title (Article 30).

Taxes must be paid by users for the exploitation and use of national waters, including underground water, as well as for the discharge of those waters, in accordance with tax legislation (Article 112).

Reserves of National Waters

Article 38 of the LAN provides that the executive branch may declare a reserve or a ban or issue administrative rules to regulate the use and exploitation of national waters to

- Prevent or compensate for the overexploitation of underground water
- Protect or restore ecosystems
- Preserve drinkable water sources or protect them from pollution
- Preserve and control the quality of water
- Address extraordinary drought.

The "reserve" of national waters is of special relevance for the protection of ecosystems because, according to Article 78/IV of the administrative rules of the LAN, a reserve could be established to protect, conserve, or restore an aquatic ecosystem, including wetlands, lakes, lagoons, and estuaries, as well as any other aquatic ecosystems that have historical, tourism, or recreational values.

To declare this type of reserve, technical studies that can justify its establishment are required. The development of these studies should involve the participation through "watershed councils" of the water users whether individuals or organizations. Afterward, a presidential decree must be published in the *Federal Official Gazette* and recorded in the Public Registry of Waters Rights.

Prevention and Control of Pollution of National Waters

According to Article 88 of the LAN, a permit from the CNA is required to discharge wastewater in national water bodies, including the sea. Discharges must comply with the Mexican Official Norm NOM-001-ECOL-1996 (published in the *Federal Official Gazette* on January 6, 1997; DOF 1997a), which establishes the discharge limitations for specific pollutants. This norm forces the users to treat any contaminated water before it is discharged into a national water body.

At the present time, water represents a national security problem due to existing pollution. In the future it is going to be a question of survival. Aside from problems due to the inequitable distribution of water (drought in the developed north and abundance in the poor south), and the fact that almost every basin in the country is polluted, it represents a serious water management crisis.

Water issues have resulted mainly from the legal framework, with all national water being managed by the federal government through the CNA. The CNA has become a powerful bureaucratic authority that manages the water discretionally and without the intervention of any other part of the federal government (SEMARNAT included), or the state and municipal governments. The CNA grants permits at its discretion and without any restriction. This often encourages corruption.

Moreover, not only does the CNA lack the capability to enforce the provisions regarding prevention and control of water pollution, but in many cases it simply decides not to enforce the law and not to sanction those responsible for illegal acts (including municipal and state governments, farmers organizations, etc.), just to avoid political problems. This comes at a high price for the conservation of biodiversity and ecosystems.

Concluding Remarks

On paper, Mexico has successfully internalized the various international treaties and conventions it has signed in the area of wildlife conservation. From this overview of ecological laws in Mexico, it should be apparent that the country has the necessary legal instruments to ensure the conservation of biodiversity and ecosystems. However, because of the poor situation of the rule of law in the country, most of those laws in practice remain dead letters.

Literature Cited

DOF. 1952. Ley Federal de Caza. Diario Oficial de la Federación 5 Enero.

DOF. 1987. Decreto por el que se reforma el párrafo tercero del artículo 27; y se adiciona una fracción XXIX-G al artículo 73 de la Constitución Política de los Estados Unidos Mexicanos. Diario Oficial de la Federación 10 Agosto.

DOF. 1988. Ley General del Equilibrio Ecológico y la Protección al Ambiente. Diario Oficial de la Federación 28 Enero.

DOF. 1992a. Ley de Pesca. Diario Oficial de la Federación 25 Junio.

DOF. 1992b. Ley de Aguas Nacionales. Diario Oficial de la Federación 1 Diciembre.

DOF. 1994a. Norma Oficial Mexicana NOM-060-ECOL-1994, que establece las especificaciones para mitigar los efectos adversos ocasionados en los suelos y cuerpos de agua por el aprovechamiento forestal. Diario Oficial de la Federación 13 Mayo.

DOF. 1994b. Norma Oficial Mexicana NOM-061-ECOL-1994, que establece las especificaciones para mitigar los efectos adversos ocasionados en la flora y fauna silvestres por el aprovechamiento forestal. Diario Oficial de la Federación 13 Mayo.

DOF. 1994c. Norma Oficial Mexicana NOM-062-ECOL-1994, que establece las especificaciones para mitigar los efectos adversos sobre la biodiversidad que se ocasionen por el cambio de uso del suelo de terrenos forestales a agropecuarios. Diario Oficial de la Federación 13 Mayo.

DOF 1995. Norma Oficial Mexicana NOM-001-RECNAT-1995, Que establece las características que deben de tener los medios de marqueo de la madera en rollo, así como los lineamientos para su uso y control. Diario Oficial de la Federación 1 Diciembre.

DOF 1996a. Proyecto de Norma Oficial Mexicana NOM-70-FITO-1995, Que establece los requisitos y especificaciones fitosanitarios para la importación, introducción, movilización y liberación de agentes de control biológico. Diario Oficial de la Federación 25 Abril.

DOF. 1996b. Norma Oficial Mexicana NOM-010-RECNAT-1996, Que establece los procedimientos, criterios y especificaciones para realizar el aprovechamiento, transporte y almacenamiento de hongos. Diario Oficial de la Federación 28 Mayo.

DOF. 1996c. Norma Oficial Mexicana NOM-002-RECNAT-1996, Que establece los procedimientos, criterios y especificaciones para realizar el aprovechamiento, transporte y almacenamiento de resina de pino. Diario Oficial de la Federación 30 Mayo.

DOF. 1996d. Norma Oficial Mexicana NOM-003-RECNAT-1996, Que establece los procedimientos, criterios y especificaciones para realizar el aprovechamiento, transporte y almacenamiento de tierra de monte. Diario Oficial de la Federación 5 Junio.

DOF. 1996e. Norma Oficial Mexicana NOM-004-RECNAT-1996, Que establece los procedimientos, criterios y especificaciones para realizar el aprovechamiento, transporte y almacenamiento de raíces y rizomas de vegetación forestal. Diario Oficial de la Federación 24 Junio.

DOF. 1996f. Norma Oficial Mexicana NOM-

008-RECNAT-1996, Que establece los procedimientos, criterios y especificaciones para realizar el aprovechamiento, transporte y almacenamiento de cogollos. Diario Oficial de la Federación 24 Junio.

DOF. 1996g. Norma Oficial Mexicana NOM-009-RECNAT-1996, Que establece los procedimientos, criterios y especificaciones para realizar el aprovechamiento, transporte y almacenamiento de látex y otros exudados de vegetación forestal. Diario Oficial de la Federación 26 Junio.

DOF. 1996h. Norma Oficial Mexicana NOM-011-RECNAT-1996, Que establece los procedimientos, criterios y especificaciones para realizar el aprovechamiento, transporte y almacenamiento de musgo, heno y doradilla. Diario Oficial de la Federación 26 Junio.

DOF. 1996i. Norma Oficial Mexicana NOM-012-RECNAT-1996, Que establece los procedimientos, criterios y especificaciones para realizar el aprovechamiento de leña para uso doméstico. Diario Oficial de la Federación 26 Junio.

DOF. 1996j. Norma Oficial Mexicana NOM-056-FITO-1995, Por la que se establecen los requisitos fitosanitarios para la movilización nacional, importación y establecimiento de pruebas de campo de organismos manipulados mediante la aplicación de ingeniería genética. Diario Oficial de la Federación 11 Julio.

DOF. 1996k. Decreto que reforma, adiciona y deroga diversas disposiciones de la Ley General del Equilibrio Ecológico y la Protección al Ambiente. Diario Oficial de la Federación 13 Diciembre.

DOF. 1997a. Norma Oficial Mexicana NOM-001-ECOL-1996, Que establece los límites máximos permisibles de contaminantes en las descargas de aguas residuales en aguas y bienes nacionales. Diario Oficial de la Federación 6 Enero.

DOF. 1997b. Aclaración a la Norma Oficial Mexicana NOM-010-RECNAT-1996, que establece los procedimientos, criterios y especificaciones para realizar el aprovechamiento, transporte y almacenamiento de hongos, publicada el 28 de mayo de 1996. Diario Oficial de la Federación 6 Mayo.

DOF. 1997c. Aclaración a la Norma Oficial Mexicana NOM-012-RECNAT-1996, Que establece los procedimientos, criterios y especificaciones para realizar el aprovechamiento de leña para uso doméstico, publicada el 26 de junio de 1996. Diario Oficial de la Federación 13 Mayo.

DOF. 1997d. Norma Oficial Mexicana NOM-005-RECNAT-1997, Que establece los procedimientos, criterios y especificaciones para realizar el aprovechamiento, transporte y almacenamiento de corteza, tallos y plantas completas de vegetación forestal. Diario Oficial de la Federación 20 Mayo.

DOF. 1997e. Norma Oficial Mexicana NOM-006-RECNAT-1997, Que establece los procedimientos, criterios y especificaciones para realizar el aprovechamiento, transporte y almacenamiento de hojas de palma. Diario Oficial de la Federación 28 Mayo.

DOF. 1997f. Norma Oficial Mexicana NOM-007-RECNAT-1997, Que establece los procedimientos, criterios y especificaciones para realizar el aprovechamiento, transporte y almacenamiento de ramas, hojas o pencas, flores, frutos y semillas. Diario Oficial de la Federación 30 Mayo.

DOF. 1998a. Acuerdo que tiene por objeto fijar los lineamientos a que deberá sujetarse el programa de restauración ecológica denominado campaña para evitar el cambio de uso de suelo por los incendios forestales. Diario Oficial de la Federación 22 Junio.

DOF. 1998b. Decreto por el que se declara zona de restauración ecológica diversas superficies afectadas por los incendios forestales de 1998. Diario Oficial de la Federación 23 Septiembre.

DOF. 1998c. Norma Oficial Mexicana NOM-013-RECNAT-1997, Que regula sanitariamente la importación de árboles de navidad naturales de las especies *Pinus sylvestris, Pseudotsuga menziesii* y del género *Abies.* Diario Oficial de la Federación 28 Septiembre.

DOF. 1998d. Norma Oficial Mexicana NOM-113-ECOL-1998, Que establece las especificaciones de protección ambiental para la planeación, diseño, construcción, operación y mantenimiento de subestaciones eléctricas de potencia o de distribución que se pretendan ubicar en áreas urbanas, suburbanas, rurales, agropecuarias, industriales, de equipamiento urbano o de servicios y turísticas. Diario Oficial de la Federación 26 Octubre.

DOF. 1998e. Norma Oficial Mexicana NOM-120-ECOL-1997, Que establece las especificaciones de protección ambiental para las actividades de exploración minera directa, en zonas con climas secos y templados en donde se desarrolle vegetación de matorral xerófilo, bosque tropical caducifolio, bosques de coníferas o encinos. Diario Oficial de la Federación 19 Noviembre.

DOF. 1998f. Norma Oficial Mexicana NOM-114-ECOL-1998, Que establece las especificaciones de protección ambiental para la planeación, diseño, construcción, operación y mantenimiento de líneas de transmisión y de subtransmisión eléctrica que se pretendan ubicar en áreas urbanas, suburbanas, rurales, agropecuarias, industriales, de equipamiento urbano o de servicios y turísticas. Diario Oficial de la Federación 23 Noviembre.

DOF. 1998g. Norma Oficial Mexicana NOM-116-ECOL-1998, Que establece las especificaciones de protección ambiental para prospecciones sismológicas terrestres que se realicen en zonas agrícolas, ganaderas y eriales. Diario Oficial de la Federación 24 Noviembre.

DOF. 1998h. Norma Oficial Mexicana NOM-117-ECOL-1998, Que establece las especificaciones de protección ambiental para la instalación y mantenimiento mayor de los sistemas para el transporte y distribución de hidrocarburos y petroquímicos en estado líquido y gaseoso, que se realicen en derechos de vía terrestres existentes, ubicados en zonas agrícolas, ganaderas y eriales. Diario Oficial de la Federación 24 Noviembre.

DOF. 1998i. Norma Oficial Mexicana NOM-115-ECOL-1998, Que establece las especificaciones de protección ambiental que deben observarse en las actividades de perforación de pozos petroleros terrestres para exploración y producción en zonas agrícolas, ganaderas y eriales. Diario Oficial de la Federación 25 Noviembre.

DOF. 1999a. Aclaración a la Norma Oficial Mexicana NOM-120 ECOL-1997, Que establece las especificaciones de protección ambiental para las actividades de exploración minera directa en zonas con climas secos y templados en donde se desarrolle vegetación de matorral xerófilo, bosque tropical caducifolio, bosque de coníferas y encinos, publicada el 19 de noviembre de 1998. Diario Oficial de la Federación 6 Enero.

DOF. 1999b. Aviso por el que se informa al público en general, que la Secretaría de Medio Ambiente, Recursos Naturales y Pesca ha concluido los avances de los programas de restauración de las áreas siniestradas por los incendios forestales por la temporada de 1998. Diario Oficial de la Federación 19 Enero.

DOF 1999c. Aviso mediante el cual se informa al público en general que la Secretaría de Medio Ambiente, Recursos Naturales y Pesca ha concluido los avances de los programas de restauración de las áreas siniestradas por los incendios forestales de la temporada de 1998, con el objeto de que los interesados puedan vertir opciones que enriquezcan el contenido de los avances citados. Diario Oficial de la Federación 1 Febrero.

DOF. 1999d. Aclaración al Aviso mediante el cual se informa al público en general que la Secretaría de Medio Ambiente, Recursos Naturales y Pesca ha concluido los programas de restauración definitivos de las áreas siniestradas por los incendios forestales de la temporada de 1998, con objeto de que los interesados puedan consultarlos y, en su caso, motiven los comentarios y opiniones a que haya lugar, publicado el 1 de febrero de 1999. Diario Oficial de la Federación 8 Febrero.

DOF. 1999e. Norma Oficial Mexicana NOM-015–SEMARNAP/SAGAR-1997, Que regula el uso del fuego en terrenos forestales y agropecuarios, y que establece las especificaciones, criterios y procedimientos para ordenar la participación social y de gobierno en la detección y el combate de los incendios forestales. Diario Oficial de la Federación 2 Marzo.

DOF. 1999f. Decreto por el que se declara la adición de un párrafo quinto al artículo 4o. Constitucional y se reforma el párrafo primero del artículo 25 de la Constitución Política de los Estados Unidos Mexicanos. Diario Oficial de la Federación 28 Junio.

DOF. 1999g. Reglamento de la Ley de Pesca. Diario Oficial de la Federación 29 Septiembre.

DOF. 1999h. Norma Oficial Mexicana NOM-018-RECNAT-1999, Que establece los procedimientos, criterios y especificaciones técnicas y administrativas para realizar el aprovechamiento sostenible de la hierba de candelilla, transporte y almacenamiento del cerote. Diario Oficial de la Federación 27 Octubre.

DOF. 2000a. Reglamento de la Ley General del Equilibrio Ecológico y la Protección al Ambiente en Materia de Evaluación del Impacto Ambiental. Diario Oficial de la Federación 30 Mayo.

DOF. 2000b. Ley General de Vida Silvestre. Diario Oficial de la Federación 3 Julio.

DOF. 2000c. Norma Oficial Mexicana NOM-019-RECNAT-1999, Que establece los lineamientos técnicos para el combate y control de los insectos descortezadores de las coníferas. Diario Oficial de la Federación 25 Octubre.

DOF. 2000d. Decreto por el que se reforman, adicionan y derogan diversas disposiciones de la Ley Orgánica de la Administración Pública Federal, de la Ley Federal de Radio y Televisión, de la Ley General que establece las Bases de Coordinación del Sistema Nacional de Seguridad Pública, de la Ley de la Policía Federal Preventiva y de la Ley de Pesca. Diario Oficial de la Federación 30 Noviembre.

DOF. 2001a. Norma Oficial Mexicana NOM-126-ECOL-2000, Por la que se establecen las especificaciones para la realización de actividades de colecta científica de material biológico de especies de flora y fauna silvestres y otros recursos biológicos en el territorio nacional. Diario Oficial de la Federación 20 Marzo.

DOF. 2001b. Norma Oficial Mexicana NOM-130-ECOL-2000, Protección ambiental-Sistemas de telecomunicaciones por red de fibra óptica-Especificaciones para la planeación, diseño, preparación del sitio, construcción, operación y mantenimiento. Diario Oficial de la Federación 23 Marzo.

DOF. 2001c. Proyecto de Norma Oficial Mexicana PROY-NOM-025-RECNAT-2001, Que establece los procedimientos y especificaciones para la recolección y distribución del germoplasma forestal con fines comerciales o de investigación que tenga como destino la forestación o re-forestación. Diario Oficial de la Federación 7 Diciembre.

DOF. 2001d. Norma Oficial Mexicana NOM-020-RECNAT-2001, Que establece los procedimientos y lineamientos que se deberán observar para la rehabilitación, mejoramiento y conservación de los terrenos forestales de pastoreos. Diario Oficial de la Federación 10 Diciembre.

DOF. 2001e. Norma Oficial Mexicana NOM-023-RECNAT-2001, Que establece las especificaciones técnicas que deberá contener la cartografía y la clasificación para la elaboración de los inventarios de suelos. Diario Oficial de la Federación 10 Diciembre.

DOF. 2002a. Decreto por el que se reforman diversas disposiciones de la Ley General de Vida Silvestre. Diario Oficial de la Federación 10 Enero.

DOF. 2002b. Norma Oficial Mexicana NOM-059-ECOL-2001, Protección ambiental-Especies nativas de México de flora y fauna silvestres-Categorías de riesgo y especificaciones para su inclusión, exclusión o cambio-Lista de especies en riesgo. Diario Oficial de la Federación 6 Marzo.

DOF. 2002c. Norma Oficial Mexicana NOM-029-PESC-2000, Pesca responsable de tiburón y especies afines. Especificaciones para su aprovechamiento. Diario Oficial de la Federación 12 Julio.

DOF. 2002d. Cancelación de la Norma Oficial Mexicana NOM-029-PESC-2000, Pesca responsable de tiburón y especies afines. Especificaciones para su aprovechamiento, publicada el 12 de julio de 2002. Diario Oficial de la Federación 11 Octubre.

DOF. 2003. Ley General de Desarrollo Forestal Sustentable. Diario Oficial de la Federación 25 Febrero.

II

PATTERNS OF SPECIES DIVERSITY AND ECOLOGICAL IMPORTANCE OF NATURAL ECOSYSTEMS

5

Distribution and Diversity of Grasses in the Yécora Region of the Sierra Madre Occidental of Eastern Sonora, Mexico

THOMAS R. VAN DEVENDER

JOHN R. REEDER

CHARLOTTE G. REEDER

ANA LILIA REINA G.

Howard Gentry's *Río Mayo Plants* (1942) was a major contribution to the knowledge of the flora of northwestern Mexico. His Río Mayo Region included a broad area in southern Sonora and adjacent Chihuahua in the Río Fuerte, Río Mayo, and Río Yaqui drainages. Beginning in the 1970s, a new network of roads provided access to many new areas (Búrquez et al. 1992). A revision of the Río Mayo flora by Martin et al. (1998) recorded 2835 taxa and incorporated the Chihuahuan treatments for the plants of the Cascada de Basaseachi (Spellenberg et al. 1996) and Nabogame (Laferriére 1994). The flora of the Río Mayo included 71 genera and 246 taxa of Gramineae, 8.7% of the total. In 1995, we began an intensive survey of the Municipio (= county) de Yécora (Reina et al. 1999). This area in the Sierra Madre Occidental of eastern Sonora became more accessible with the completion of Mexico Federal Highway (MEX) 16 in 1992. The diversity of habitats and plants had great potential to augment the floras of the Río Mayo and the state of Sonora. In this chapter, we present an analysis of diversity of the grasses in the Municipio de Yécora and compare the results with other floras in northwestern Mexico and the southwestern United States.

Study Area and Methods

Study Area

The Municipio de Yécora is an area of 3300 km² along MEX 16 in eastern Sonora (fig. 5.1). One of the few highways crossing the extensive north–south-oriented Sierra Madre Occidental, MEX 16 connects Hermosillo, Sonora, and La Junta, Chihuahua (Búrquez et al. 1992). Most of the Yécora region is in the drainage of the Río Yaqui; only the southeastern edge of Mesa del Campanero is in the Río Mayo basin. Elevation in the Municipio ranges from 480 m along Arroyo Tepoca near Curea to 2140 m on Mesa del Campanero, an increase of 1660 m in 27 km (by air).

Vegetation

There are important changes in vegetation with elevation that are most easily seen along MEX 16 from northwest of Tepoca to the Chihuahua border (fig. 5.2). Foothills thornscrub (*matorral espinoso*) occurs at 460–550 m elevation in the rain-shadow valley at Curea (fig. 5.3). Tropical deciduous forest (*selva baja caducifolia*) is found in a broad band

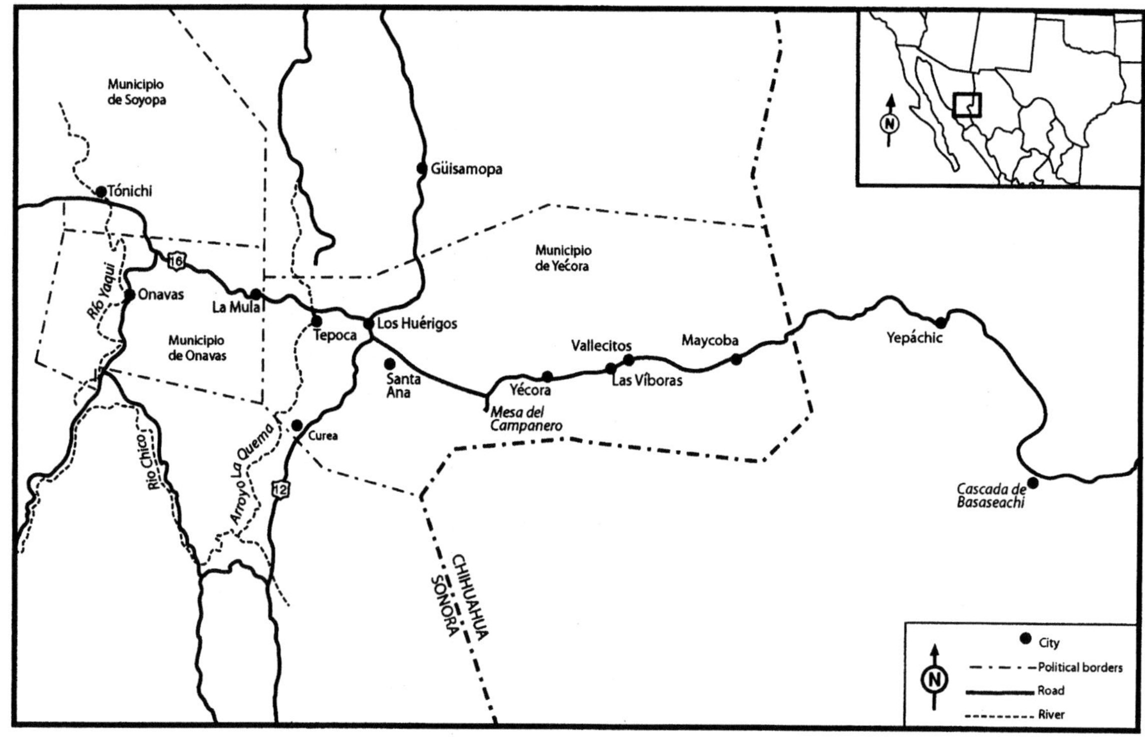

Figure 5.1. Map of the Municipio de Yécora, Sonora, and adjacent Chihuahua.

at 500–1160 m. Oak woodland (*bosque de encino*) is present at 1050–1700 m, often in a mosaic with pine–oak forest (*bosque de pino-encino*) at 1220–2240 m (fig. 5.4). Grassland (*pastizal*) occurs in high valleys at 1200–1700 m surrounded by oak woodland or pine-oak forest. Mixed-conifer forest (*bosque de coníferas mixtas*) with Durango fir (*pinabete duranguense, Abies durangenis*) at 1900–2100 m in Barranca El Salto on the west side of Mesa del Campanero is the only eastern Sonora example of the vegetation type.

Special habitats in the Municipio include riparian areas, *ciénegas*, and gossans at various elevations. The Ciénega de Camilo at 1550 m elevation east of Maycoba is a spring mound formed by peat (*musgo, Sphagnum palustre*), a new habitat for Sonora and the Sierra Madre Occidental (Van Devender et al. 2003). Gossans are relatively bare areas of reddish, highly acidic (pH as low as 4.0) soils derived from hydrothermally-altered volcanic rocks (Goldberg 1982). Gossans at 820–920 m elevation support oak woodland or pine–oak forest

surrounded by tropical deciduous forest on unaltered soils.

Methods

During an intensive floristic survey of the Municipio, plants were collected on 29 field trips between May 1995 and May 2004 (see Reina et al. 1999; Van Devender et al. 2003). Father Bill Trauba, residing in Yécora in 1996–1998, collected additional material. Specimens were deposited into herbaria at the University of Arizona (ARIZ), the Universidad de Sonora (USON), and 14 other institutions in the United States and Mexico. Additional records of grasses from the Municipio were from Martin et al. (1998), mostly with vouchers in ARIZ. Specimens of important records cited in Beetle and Johnson (1991) were borrowed for verification from the Comisión Técnico Consultiva de Coeficientes de Agostadero (COTECOCA, the COCA herbarium) in Hermosillo, Sonora. A total of 702 records provide the basis for the analyses

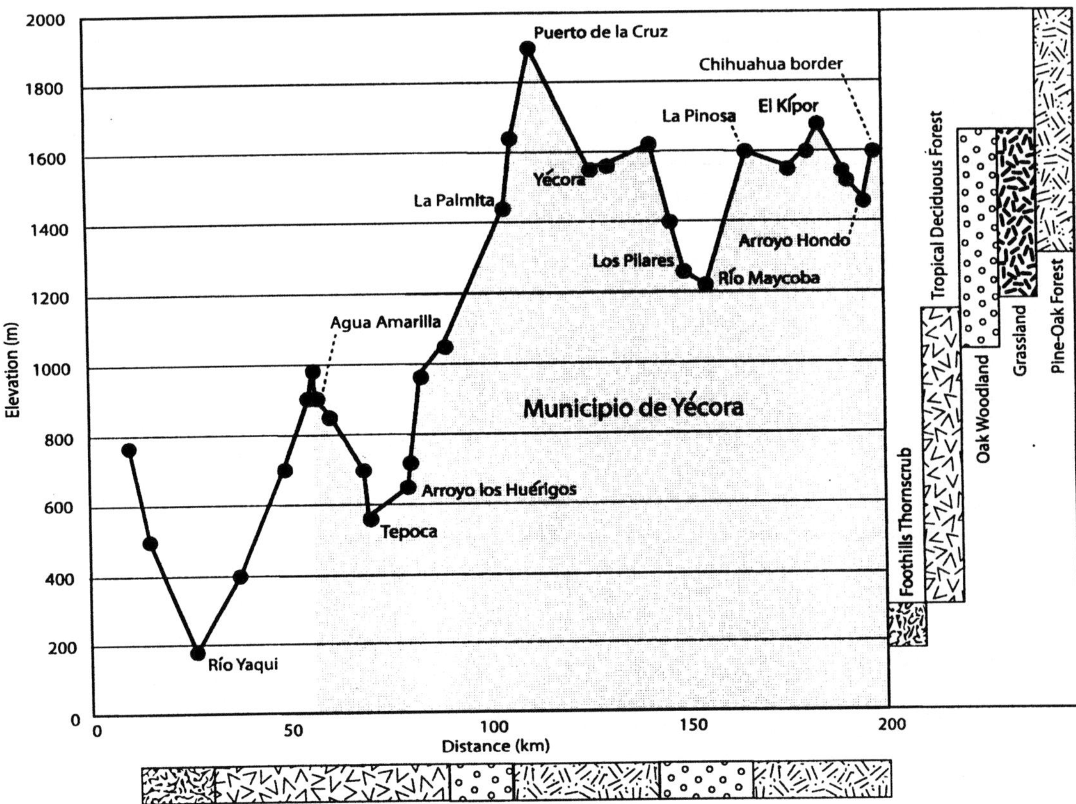

Figure 5.2. Elevation and vegetation profile along Mexico Highway 16 in eastern Sonora, Mexico. Wide elevational overlap of vegetation zones reflects strong slope aspect differences and broad ecotones between grassland and oak woodland. Only the dominant vegetation types are represented along the x-axis. See text for elevational range of vegetation types in the Municipio de Yécora as a whole.

presented in this chapter; 86.3% of them are our collections.

Results and Discussion

Flora

Martin et al. (1998) reported 2835 plant taxa for the Río Mayo region, which includes the entire Municipio de Yécora. With our work, the current total for the flora of the Municipio de Yécora is 1696, raising the total for the Río Mayo Region to more than 3000. The additional new taxa reflect the intensive collecting in a relatively small area and suggest that many more species are to be found in the region. The largest families in the Municipio de Yécora flora are Compositae (252 taxa, 14.9%),

Gramineae (186 taxa, 11%), Leguminosae (184 taxa, 10.8%), Cyperaceae (60 taxa, 3.5%), and Euphorbiaceae (56 taxa, 3.3%).

Grass Flora

The grass flora of the Municipio de Yécora is diverse with at least 186 taxa (183 species plus 3 varieties and subspecies) in 60 genera (table 5.1). Of these taxa, 85.5% had been reported previously for the Río Mayo region (Martin et al. 1998). With our new records, however, the total known for that region is now 273 taxa. The 27 additions to the Río Mayo flora include 1 genus (*Imperata*), 25 species, and 2 varieties; 6 of them are introduced species. Species added to the flora of Sonora include *Brachiaria plantaginea*, *Echinochloa holciformis*, *Eragrostis pringlei*, *Muhlenbergia diver-*

Figure 5.3. Foothills thornscrub is only found in the Municipio de Yécora at 460–550 m elevation near Curea. Note that the organpipe cacti (*Stenocereus thurberi*) are taller than the other trees and shrubs. Only 19 grass taxa were found in this habitat.

siglumis, M. laxiflora, M. lucida, and *M. schmitzii. Paspalum palmeri* from Agua Amarilla was the first Sonoran collection since the 1890 type collection by Edward Palmer from Alamos 165 km to the south (Chase 1942). The only other known locality for this species is near Autlán, Jalisco (McVaugh 1983).

The taxonomy of the Gramineae appears to be well known, and in the Municipio de Yécora the genera with the most taxa are *Muhlenbergia* (32), *Paspalum* (13), *Aristida* (10), *Eragrostis* (10), *Digitaria* (9), and *Bouteloua* (8). These 6 genera contain 44.1% of the Yécora grasses. *Muhlenbergia* with 32 taxa is the most diverse genus in the entire angiosperm flora. Annual grasses are less common in Yécora than perennials (58.4%), although some such as crinkleawn (*Trachypogon montufari*), tanglehead (*Heteropogon contortus*), *zacate duro* (*Trisetum deyeuxioides*), and others may flower vigorously the first year.

An additional 38 taxa of grasses are to be expected in the Municipio de Yécora (table 5.2) based on their presence in nearby areas within the Río Mayo region with similar tropical or montane habitats. These areas include Basaseachi (Spellenberg et al. 1996), Nabogame (Laferriére 1994), and Talayotes (Martin et al. 1998) in Chihuahua, and Cañón Estrella, Guajaráy, and Tesopaco (Martin et al. 1998) in Sonora. Other potential Municipio de Yécora species are based on our collections from La Barranca (Municipio de San Javier) and near Tónichi along the Río Yaqui (Municipio de Soyopa). The potential total of at least 224 taxa (186 documented, 38 expected) suggests that the diversity of grasses in the Municipio de Yécora is as high as anywhere in Mexico.

The numbers of *Muhlenbergia* (muhly) and *Paspalum* (camalote) in this portion of the Sierra Madre Occidental indicate that it is a major center of diversity for these grasses. The total of 43 recorded (table 5.1) and expected (table 5.2) taxa of *Muhlenbergia* is higher than for any of the various floras in northwestern Mexico and southern Arizona presented in table 5.3. Other floras with relatively many

Figure 5.4. Pine–oak forest at 1220–2240 m elevation in the Municipio de Yécora is very diverse, with 9 species of pines and 13 species of oaks. The grass flora, with 94 taxa, is also very diverse in this habitat.

Muhlenbergia are the Cascada de Basaseachi, Chihuahua (23), the Sierra de Manantlán, Jalisco (21), and the Chiricahua (21) and Huachuca (20) mountains, Arizona, reflecting the importance of muhly grasses in montane woodland and forest habitats. Tropical and desert floras have relatively few *Muhlenbergia*.

The total of 13 taxa of *Paspalum* in the Municipio de Yécora is also noteworthy. In the floras surveyed (table 5.3), only the Sierra de Manantlán, with 25 taxa, had more. The relatively low numbers in most of the other areas including the sky-island mountains of southeastern Arizona suggest that many *Paspalum* species are more tropical grasses.

Several species reported for the Municipio de Yécora by Beetle and Johnson (1991) were found to be misidentified and excluded from the flora. Specimens in the COCA herbarium were annotated to the species in parentheses: *Brachiaria arizonica* (= *B. fasciculata*), *Leptochloa scabra* Nees (= *L. panicoides*), *Muhlenbergia quadridentata* (H.B.K.) Trinius (= *M. trifida*), *M. robusta* (= *M. emersleyi*), and *Setaria longipila* E. Fourn. (= *S. grisebachii*).

Diversity of Grasses Along an Elevation–Rainfall Gradient

Although grasses were present in all habitats within the Municipio, there was an obvious increase in diversity with elevation and rainfall. The numbers of grasses collected in habitats from low to high elevation were 19 (thornscrub), 65 (grassland), 69 (tropical deciduous forest), 94 (oak woodland) and 94 (pine–oak forest). Many grasses (41.9%) were only recorded in a single vegetation type: thornscrub (3), tropical deciduous forest (27), oak woodland (14), pine–oak forest (25), and grassland (9).

Characteristic Grasses

Spider grass (*Aristida ternipes* var. *ternipes*) is the only common perennial grass in foothills thornscrub, while fluffgrass (*Erioneuron pulchellum*) and sprucetop grama (*Bouteloua chondrosioides*) are often locally common. Panicgrass (*Panicum hirticaule* var. *hirticaule*) is a common annual, while Parry grama (*Bouteloua parryi*) is occasional.

Table 5.1. Grasses collected in the Municipio de Yécora, Sonora, Mexico.

Species [Common Name]	Species [Common Name]
Aegopogon cenchroides Humb. & Bonpl. ex Willd.	*Brachiaria plantaginea* (Link) Hitchc.
Aegopogon tenellus (DC.) Trinius	*Bromus anomalus* Rupr. ex E. Fourn. [nodding brome]
Agrostis exarata Trinius	*Bromus carinatus* Hooker & Arn. [California brome]
Agrostis scabra Willd.	**Bromus catharticus* Vahl; *Bromus uniloides* H.B.K., *B. willdenovii* Kunth [rescue grass]; native to South America
Andropogon glomeratus (Walt.) B.S.P. [bushy beardgrass]	*Bromus ciliatus* L. [fringed brome] (COCA)
Aristida adscensionis L. [six-weeks threeawn]	*Cathestecum brevifolium* Swallen [false grama; *zacate liebrero, zacate, borreguero*]
Aristida appressa Vasey	*Cenchrus echinatus* L. [sandbur; *guachapor*]
Aristida arizonica Vasey	*Cenchrus spinifex* Cav.; *C. pauciflora* Bentham, *C. incertus* M.A. Curtis [sandbur; *guachapor*]
Aristida divaricata Humb. & Bonpl. ex Willd. [poverty threeawn]	*Chloris chloridea* (J. Presl) Hitchc.
Aristida jorullensis Kunth	*Chloris submutica* H.B.K. [*papalote verdillo*]
Aristida laxa Cav. [threeawn]	*Chloris virgata* Swartz [feather fingergrass]
Aristida marginalis Ekman [threeawn]	**Cynodon dactylon* (L.) Pers. [Bermuda grass]; native to the Old World
Aristida schiedeana Trinius & Rupr.; *A. orcuttiana* Vasey	**Dactyloctenium aegyptium* (L.) P. Beauv. [crowfoot grass]; native to the Old World
Aristida ternipes Cav. var. **gentilis** (Henrard) Allred; *A. hamulosa* Henrard	*Diectomis fastigiata* (Swartz) P. Beauv.
Aristida ternipes Cav. var. **ternipes** [spider grass]	*Digitaria argillacea* (Hitchc. & Chase) Fernald
Arundinella palmeri Vasey ex Beal	*Digitaria bicornis* (Lam.) Roemer & Schultes
**Arundo donax* L. [giant cane; *carrizo*]; native to the Old World	*Digitaria ciliaris* (Retz.) Koeler
**Avena fatua* L. [wild oats; *avena*]; native to Europe	*Digitaria horizontalis* Willd.
**Avena sativa* L. [oats; *avena*]; escaped cultivar from Europe	*Digitaria insularis* (L.) Mez ex Ekman; *Trichachne insularis* (L.) Nees
Axonopus compressus (Swartz) P. Beauv.	**Digitaria ischaemum* (Schreber) Schreber [smooth crabgrass]; native to Eurasia
Blepharoneuron tricholepis (Torrey) Nash [pine dropseed] (COCA)	*Digitaria panicea* (Swartz) Urban
Bothriochloa barbinodis (Lag.) Herter; *Andropogon barbinodis* Lag. [cane beardgrass]	**Digitaria sanguinalis* (L.) Scop. [common crabgrass]; native to Europe
**Bothriochloa ischaemum* (L.) Keng; *Andropogon ischaemum* L.; native to the Old World	**Digitaria ternata* (A. Rich.) Stapf; native to Africa
Bouteloua aristidoides (H.B.K.) Griseb. [six-weeks needle grama; *zacate saitilla*]	**Echinochloa colonum* (L.) Link [jungle rice]; native to the Old World
Bouteloua chondrosioides (H.B.K.) Bentham ex S. Watson [sprucetop grama]	**Echinochloa crusgalli* (L.) P. Beauv. [barnyard grass]; native to the Old World
Bouteloua curtipendula (Michx.) Torrey [sideoats grama; *zacate de navaja, navajitas*]	*Echinochloa holciformis* (H.B.K.) Chase
Bouteloua gracilis (H.B.K.) Lag. ex Griffiths [blue grama]	**Eleusine indica* (L.) Gaertn. [goose grass; *zacate escoba*]; native to Eurasia
Bouteloua hirsuta Lag. [hairy grama]	*Elymus arizonicus* (Scribner & J.G. Smith) Gould; *Agropyron arizonicum* Scribner & J.G. Smith [wheat grass]
Bouteloua parryi (E. Fourn.) Griffiths [Parry grama; *zacate gusano*]	**Eragrostis barrelieri* Daveau; native to southern Europe
Bouteloua radicosa (E. Fourn.) Griffiths [*zacate sabanilla, navajitas*]	**Eragrostis cilianensis* (All.) Vign.-Lutati ex Janchen [stink grass]; native to the Old World
Bouteloua repens (H.B.K.) Scribner & Merrill. [slender grama; *zacate sabanilla, navajitas*]	*Eragrostis intermedia* Hitchc. [plains lovegrass]
Brachiaria fasciculata (Swartz) Parodi; *Panicum fasciculatum* Swartz, *Urochloa fasciculata* (Swartz) R.D. Webster [browntop panicgrass]	

Eragrostis lugens Nees

Eragrostis maypurensis (H.B.K.) Steudel (COCA)

Eragrostis mexicana (Hornem.) Link subsp. *mexicana*. [Mexican lovegrass]

Eragrostis pectinacea (Michx.) Nees var. *miserrima* (E. Fourn.) J. Reeder; *Eragrostis arida* Hitchc., *E. tephrosanthos* Schultes

Eragrostis pectinacea (Michx.) Nees var. *pectinacea*

Eragrostis pringlei Mattei; *Eragrostis scribneriana* Hitchc.

Eragrostis secundiflora J. Presl subsp. *oxylepis* (Torrey) S.D. Koch; *Eragrostis oxylepis* (Torrey) Torrey [red lovegrass] (COCA)

Eriochloa acuminata (J. Presl) Kunth var. *minor* (Vasey) R.B. Shaw; *Eriochloa gracilis* (E. Fourn.) Hitchc. var. *minor* (Vasey) Hitchc. [cupgrass]

Eriochloa aristata Vasey [bearded cupgrass]

Eriochloa lemmonii Vasey & Scribner; small southwestern cupgrass

Erioneuron pulchellum (H.B.K.) Tateoka; *Dasyochloa pulchella* (H.B.K.) Steudel, *Tridens pulchellus* H.B.K.) Hitchc.[fluffgrass; *zacate borreguero*] (COCA)

Gouinia virgata (J. Presl) Scribner [*sorguillo*]

Hackelochloa granularis (L.) Kuntze [pitscale grass]; native to the Old World

Heteropogon contortus (L.) P. Beauv. ex Roemer & Schultes [tanglehead]; tropical and warm regions worldwide

Heteropogon melanocarpus (Ell.) Bentham [sweet tanglehead]; tropical and warm regions worldwide

Hilaria belangeri (Steudel) Nash [curly mesquite]

Hilaria cenchroides H.B.K. [Madrean mesquite-grass]

Hordeum murinum L. subsp. *leporinum* (Link) Arcangeli; *Hordeum leporinum* Link [wild barley; *cebada*]; native to Europe

Imperata brevifolia Vasey [satintail]

Lasiacis ruscifolia (H.B.K.) Hitchc. [*negrito*]

Leptochloa dubia (H.B.K.) Nees [green sprangletop]

Leptochloa fusca (L.) Kunth subsp. *uninervia* (J. Presl) N. Snow; *L. uninervia* (J. Presl) Hitchc. & Chase [Mexican sprangletop]

Leptochloa panicea (Retz.) Ohwi subsp. *brachiata* (Steudel) N. Snow; *L. filiformis* (Lam.) P. Beauv., *L. mucronata* (Michx.) Kunth (misapplied) [red sprangletop]

Leptochloa panicoides (J. Presl) Hitchc. [sprangletop]

Luziola gracillima Prodoehl [*pastito de agua*]

Lycurus phalaroides H.B.K. [wolftail]

Lycurus phleoides H.B.K. [wolftail]

Lycurus setosus (Nutt.) C. Reeder [wolftail]

Melinis repens (Willd.) Zizka subsp. *repens*; *Rhynchelytrum repens* (Willd.) C.E. Hubbard [Natal grass; *zacate rosado, espiga colorada, alucema*]; introduced from South Africa

Microchloa kunthii Desv. [*zacate encorvado*]

Muhlenbergia alamosae Vasey

Muhlenbergia annua (Vasey) Swallen [*zacate liebrero*]

Muhlenbergia argentea Vasey

Muhlenbergia arizonica Scribner

Muhlenbergia diversiglumis Trinius

Muhlenbergia dumosa Scribner ex Vasey [bamboo muhly; totchkam (Pima; C. J. Pennington unpubl. data); *carricillo, otatillo*]

Muhlenbergia elongata Scribner ex Beal [tropical cliff muhly]

Muhlenbergia eludens C. Reeder

Muhlenbergia emersleyi Vasey [bullgrass; *cola de zorra*]

Muhlenbergia flavida Vasey

Muhlenbergia fragilis Swallen [*zacate liebrero*]

Muhlenbergia gooddingii Soderstrom

Muhlenbergia grandis Vasey [*cola de zorra*] (ASU)

Muhlenbergia implicata (H.B.K.) Trinius

Muhlenbergia laxiflora Scribner ex Beal

Muhlenbergia longiligula Hitchc.

Muhlenbergia lucida Swallen

Muhlenbergia microsperma (DC.) Trinius [littleseed muhly]

Muhlenbergia minutissima (Steud.) Swallen [*zacate liebrero*]

Muhlenbergia montana (Nutt.) Hitchc. [mountain muhly]

Muhlenbergia palmeri Vasey (COCA)

Muhlenbergia pectinata C. Goodding

Muhlenbergia polycaulis Scribner [cliff muhly]

Muhlenbergia rigens (Bentham) Hitchc. [deer grass; *escobón, hierba del paisano; norr/nol* (Pima; Reina 1993)]

Muhlenbergia rigida (H.B.K.) Trinius [purple muhly; norr/nol (Pima)]

Muhlenbergia schmitzii Hackel

Muhlenbergia scoparia Vasey

Muhlenbergia shepherdii (Vasey) Swallen

Muhlenbergia straminea Hitchc. [screwleaf muhly]

Muhlenbergia tenella (H.B.K.) Trinius

Muhlenbergia texana Buckley [Texas muhly]

Muhlenbergia trifida Hackel

Oplismenus burmannii (Retz.) P. Beauv. [basket grass; *zacate salado*]

Oplismenus hirtellus (L.) P. Beauv. [basket grass; *zacate salado*]

Otatea acuminata (Munro) Calderón & Soderstrom subsp. *aztecorum* Guzmán, Anaya C. & Santana M. [bamboo; *otate, bambu*]

Panicum acuminatum Swartz

(continued)

Table 5.1. Continued

Species [Common Name]	Species [Common Name]
Panicum bulbosum H.B.K. [bulb panicgrass; *nor sha'i* (Pima; Laferriére 1994)]	*Schizachyrium mexicanum* (Hitchc.) A. Camus; *Andropogon mexicanus* Hitchc.
Panicum hallii Vasey [Hall's panicgrass]	*Schizachyrium sanguineum* (Retz.) Alston var. *hirtiflorum* (Nees) Hatch;
Panicum hians Elliott	*Andropogon hirtiflorus* (Nees) Kunth
Panicum hirticaule J. Presl var. *hirticaule* [panicgrass]	*Schizachyrium semitectum* (Swallen) J. Reeder; *Andropogon semitectus* Swallen
Panicum trichoides Swartz [*zacate salado*]	*Schizachyrium tenerum* Nees; *Andropogon tener* (Nees) Kunth
Paspalum botterii (E. Fourn.) Chase	*Setaria grisebachii* E. Fourn. [Grisebach bristlegrass; *zacate cola de zorra*] (COCA)
Paspalum convexum Humb. & Bonpl. ex Willd. [*camalote anual*]	*Setaria liebmannii* E. Fourn. [bristlegrass; *cola de zorra*]
Paspalum distichum L. [knotgrass]	*Setaria macrostachya* H.B.K. [plains bristlegrass]
Paspalum humboldtianum Flügge	*Setaria parviflora* (Poiret) Kerguélen; *S. geniculata* (Lam.) P. Beauv. [knotroot
Paspalum langei (E. Fourn.) Nash	bristlegrass]
Paspalum lentiginosum J. Presl vel aff.	**Setaria pumila* (Poiret) Roemer & Schultes; *S. glauca* L., *S. lutescens* (Weigel) F.T.
Paspalum palmeri Chase	Hubbard [yellow bristlegrass]]; native to Europe
Paspalum paniculatum L.	*Setariopsis auriculata* (E. Fourn.) Scribner [*plumerillo*]
Paspalum pubiflorum Rupr. ex E. Fourn. var. *pubiflorum*	*Setariopsis latiglumis* (Vasey) Scribner
Paspalum setaceum Michx. var. *muhlenbergii* (Nash) D.J. Banks	*Sorghastrum nudipes* Nash (COCA)
Paspalum squamulatum E. Fourn.	*Sorghastrum nutans* (L.) Nash [Indian grass; *zacate dorado*]
Paspalum tenellum Willd. (COCA)	**Sorghum bicolor* (L.) Moench [milo, sorghum; *malo maíz, sorgo*]; escaped
Paspalum virletii E. Fourn.	cultivar, originally from the Old World
**Pennisetum ciliare* (L.) Link; *Cenchrus ciliaris* L. [buffelgrass]; *zacate buffel*];	**Sorghum halepense* (L.) Pers. [Johnson grass; zacate Johnson]; native to the
introduced from eastern Africa	Old World
Pennisetum durum Beal. Cited in Martin et al. (1998). (ASU)	*Sporobolus indicus* (L.) R. Br. [dropseed, smutgrass]
Pereilema crinitum J. Presl	*Trachypogon montufari* (H.B.K.) Nees
**Phalaris minor* Retz. [littleseed canary grass]; native to the Mediterranean region	*Trachypogon secundus* (J. Presl) Scribner [crinkleawn]
Piptochaetium fimbriatum (H.B.K.) Hitchc. [pinyon ricegrass; *hierba de la vejiga*]	*Tripsacum dactyloides* (L.) L. [gamagrass; *maizillo, zacate maíz*]
Piptochaetium virescens (H.B.K.) Parodi [ricegrass]	*Tripsacum lanceolatum* Rupr. ex E. Fourn. [Mexican gamagrass; *maizillo, zacate maíz*]
**Poa annua* L. [annual bluegrass]; native to Europe	*Tripsacum zopilotense* Hernández X. & Randolph [gamagrass; *maizillo, zacate maíz*]
Poa bigelovii Vasey & Scribner [Bigelow bluegrass]	*Trisetum deyeuxioides* (H.B.K.) Kunth [*pajiza del bosque*]
Polypogon elongatus H.B.K. (COCA)	**Triticum aestivum* L. [wheat; *trigo*]. Escaped cultivar, originally from the Middle East
**Polypogon monspeliensis* (L.) Desf. [rabbitfoot grass; *zacate cola de zorra, hierba*	*Vulpia octoflora* (Walter) Rydb. var. *octoflora*; *Festuca octoflora* Walter [six-weeks
del caso]; native to Europe	fescue]
Schizachyrium brevifolium (Swartz) Nees ex Büse; *Andropogon brevifolius* Swartz	**Zea mays* L. subsp. *mays* [corn; *hu'un* (Pima; Reina 1993); *maíz*]; escaped
Schizachyrium cirratum (Hackel) Wooton & Standley; *Andropogon cirratum*	cultivar, from Mexico
Hackel [Texas beardgrass]	**Zea mays* L. subsp. *mexicana* (Schrad.) H.H. Iltis; *Euchlaena mexicana* Schrad.
Schizachyrium condensatum (H.B.K.) Nees; *Andropogon condensatus* H.B.K., *A.*	[*teosinte, maíz cócono*]; formerly cultivated by Maycoba Pima (C. W.
microstachyus Desv. ex Ham.	Pennington pers. comm. 1996)

*Definite and probable non-native species. Selected synonyms are given. The Spanish common names are mostly from our interviews with residents in the Río Mayo region. A few are from Beetle and Johnson (1991). Voucher specimens are at the herbarium of the University of Arizona (ARIZ) unless otherwise indicated. Other herbaria: ASU = Arizona State University; COCA = Comisión Técnico Consultiva de Coeficientes de Agostadero (COTECOCA).

Table 5.2. Additional species to be expected in the Municipio de Yécora based on their presence in other areas in the Río Mayo region.

Species [Common Name]	Area in Río Mayo Region
Agrostis hyemalis (Walter) B.S.P. [bentgrass]	Basaseachi
Aristida pansa Wooton & Standley [Wooton threeawn]	Talayotes
Bouteloua barbata Lag. var. *barbata* [six-weeks grama]	Tesopaco
Bouteloua barbata Lag. var. *sonorae* (Griffiths) Gould [Sonoran grama]	Tónichi
Bouteloua rothrockii Vasey; *B. barbata* Lag. var. *rothrockii* (Vasey) Gould [Rothrock grama]	La Barranca
Brachiaria arizonica (Scribner & Merrill) S. T. Blake; *Panicum arizonicum* Scribner & Merrill [Arizona panicgrass]; cited for the Municipio de Yécora in Beetle and Johnson (1991) but misidentified: voucher specimen is *B. fasciculata* (COCA)	Tónichi
Deschampsia elongata (Hooker) Munro ex Bentham	Basaseachi
Dichanthium annulatum (Forssk.) Stapf; Native to tropical Africa and southeastern Asia	Guajaráy
Echinochloa muricata (P. Beauv.) Fernald; native to the Old World	Basaseachi, Nabogame
Eleusine multiflora Hochst. ex A. Rich; *E. tristachya* (Lam.) Lam.; native to the Old World	Basaseachi, Nabogame
Elymus trachycaulus (Link) Gould ex Shinners; *Agropyron trachycaulum* (Link) Malte, *A. pauciflorum* (Keating) Hitchc. [slender wheatgrass]	Basaseachi
Elyonurus barbiculmis Hackel	Nabogame
Eragrostis erosa Scribner [lovegrass]	Basaseachi
Eriochloa acuminata (J. Presl) Kunth var. *acuminata*; *E. gracilis* (E. Fourn.) Hitchc. [cupgrass]	Nabogame
Festuca arundinacea Schreber; *F. elatior* L. [reed, tall fescue]; Native to Europe	Basaseachi
Festuca breviglumis Swallen [fescue]	Basaseachi
Festuca rubra L. [red fescue]	Basaseachi
Leptochloa viscida (Scribner) Beal [sticky sprangletop]	Tónichi
Muhlenbergia ciliata (H.B.K.) Trinius; cited for the Municipio de Yécora in Beetle and Johnson (1991) but misidentified: voucher specimen is *M. pectinata* (COCA)	Basaseachi
Muhlenbergia crispiseta Hitchc.	Basaseachi
Muhlenbergia durangensis Herrera	Nabogame
Muhlenbergia leptoura (Piper) Hitch; cited for the Municipio de Yécora in Beetle and Johnson (1991) but COCA specimen not seen	Basaseachi Nabogame
Muhlenbergia flaviseta Scribner	Basaseachi
Muhlenbergia pauciflora Buckley [New Mexico muhly]; cited for the Municipio de Yécora in Beetle and Johnson (1991) but COCA specimen not seen	Basaseachi
Muhlenbergia pubescens (H.B.K.) Hitchc.	Basaseachi
Muhlenbergia strictior Scribner ex Beal	Basaseachi
Muhlenbergia tenuifolia (H.B.K.) Trinius [mesa muhly]	Basaseachi
Muhlenbergia vaginata Swallen	Basaseachi
Muhlenbergia wolfii (Vasey) Rydb.	Basaseachi
Panicum sphaerocarpon Elliott	Basaseachi
Paspalum setaceum Michx. var. *stramineum* (Nash) D.J. Banks; *P. stramineum* Nash	Nabogame
Pennisetum karwinskyi Schrad.	Cañón Estrella
Poa fendleriana (Steudel) Vasey var. *fendleriana* [Fendler muttongrass]	Basaseachi
Polypogon viridis (Gouan) Breistr.; *Agrostis semiverticillata* (Forssk.) C. Christ., *A. verticillata* Vill., *Polypogon semiverticillata* (Forssk.) Hylander [water polypogon]; native to the Old World	Nabogame
Schizachyrium scoparium (Michx.) Nash; *Andropogon scoparius* Michx. [little bluestem]	Basaseachi
Trisetum filifolium Scribner ex Beal	Basaseachi
Trisetum viride (H.B.K.) Kunth; *T. palmeri* Hitchc.; cited for the Municipio de Yécora in Beetle and Johnson (1991) but COCA specimen not seen	Basaseachi
Vulpia octoflora (Walt.) Rydb. var. *hirtella* (Piper) Henrard; *Festuca octoflora* (Walter) Rydb. subsp. *hirtella* Piper [six-weeks fescue]	Tónichi

*Non-native. COCA = Comisión Técnico Consultiva de Coeficientes de Agostadero (COTECOCA) herbarium

Typical perennials in tropical deciduous forest are *negrito* (*Lasiacis ruscifolia*) and *sorguillo* (*Gouinia virgata*). Slender grama (*Bouteloua repens*) and spider grass are also common in grassland and Sonoran desertscrub. The northernmost known stands of bamboo (*otate, Otatea acuminata*) are near Curea (fig. 5.1). *Camalotes* (*Paspalum*), with 7 species, and especially *P. langei*, are common in moist areas. Muhlies, with only 4 species, are less common in tropical deciduous forest than in oak woodland and pine–oak forest, but bamboo muhly (*M. dumosa*) and tropical cliff muhly (*M. elongata*) are large, easily seen species. Common annuals include bristlegrass/*cola de zorra* (*Setaria liebmannii*), browntop panicgrass (*Brachiaria fasciculata*), *plumerillo* (*Setariopsis auriculata*), and *zacate salado* (*Panicum trichoides*). Littleseed muhly (*Muhlenbergia microsperma*), red sprangletop (*Leptochloa panicea* subsp. *brachiata*), and six-weeks needle grama (*Bouteloua aristidoides*) are annuals also common in the Sonoran Desert.

In oak woodland, the perennial grasses are diverse, but few species stand out as dominant. Bullgrass (*Muhlenbergia emersleyi*) and Mexican gamagrass (*zacate maíz* or *maizillo, Tripsacum lanceolatum*) are large and locally common. Muhlies (12 species), camalotes (7 species), and beardgrasses (*Schizachyrium*, 4 species) are important. Annuals are locally common, especially in open, disturbed areas. Annual muhlies (9 species), and especially Texas muhly (*M. texana*), are common. *Eragrostis maypurensis* and *M. eludens* were only found in oak woodland.

In pine–oak forest, perennials are diverse, but only crinkleawn (*Trachypogon secundus*), Indian grass (*zacate dorado, Sorghastrum nutans*), and 2 threeawns (*Aristida arizonica, A. schiedeana*) stand out as typical. Common annuals include *Aegopogon cenchroides, A. tenellus, camalote anual* (*Paspalum convexum*), and sweet tanglehead (*Heteropogon melanocarpus*). Muhlies with 15 perennial and 9 annual species are diverse. The perennials *M. alamosae, M. laxiflora, M. rigida,* and *M. trifida* and the annuals *M. diversiglumis, M. minutissima, M. shepherdii,* and *M. texana* are especially common. The annual *Eragrostis pringlei* was only found in pine–oak forest.

Important perennials in grassland include crinkleawn (*Trachypogon montufari*), curly mesquites (*Hilaria belangeri, H. cenchroides*), hairy grama (*Bouteloua hirsuta*), *pajiza del bosque* (*Trisetum deyeuxioides*), plains lovegrass (*Eragrostis inter-*

media), *papalote verdillo* (*Chloris submutica*), and *zacate encorvado* (*Microchloa kunthii*). *Echinochloa holciformis* and *pastito de agua* (*Luziola gracillima*) are only known from moist soils in grassland. Annuals are present in bare areas but are more common in other vegetation types.

Comparisons with Other Floras

We compared the Yécora grass flora with 13 floras from Jalisco north to southern Arizona and southern California (table 5.3). Although comparisons are difficult because of enormous differences in area (total richness increases, but species/areas ratios inherently decrease with larger areas), topography, geology, and climate among the sites, the numbers of species shared with other floras provide insight into biogeographic affinities between these floras. The Municipio de Yécora differs from most areas in that both tropical and montane habitats are present. The greatest number of shared species of grasses was with the rich tropical flora of the Sierra de Manantlán in Jalisco (77; Vázquez G. et al. 1995), followed in no particular order by those along the Río Cuchujaqui, a tropical river in southern Sonora (66; Van Devender et al. 2000); the Cascada de Basaseachi (63; Spellenberg et al. 1996) and Nabogame (56; Laferriére 1994) in the Sierra Madre Occidental in Chihuahua; the Cape Region (73; Lenz 1992; León de la Luz et al. 1999) of tropical and desert Baja California Sur; and the Huachuca (67; Bowers and McLaughlin 1996) and Chiricahua (66; Bennett et al. 1996) mountains in southeastern Arizona. Relatively few Yécora grasses were shared with the tropical deciduous forests of Chamela, Jalisco (28; Lott 1993; E. J. Lott pers. comm. 1998); coastal thornscrub in southern Sonora (25; Friedman 1996), and the Gran Desierto portion of the Sonoran Desert in northwestern Sonora (37; Felger 2000). Moderate numbers of Yécora grasses were shared with the Mediterranean flora of San Diego, California (41; Beauchamp 1986); Sonoran desertscrub in the Tucson Mountains (45; Rondeau et al. 1996); and the desert grassland/oak woodland of Sycamore Canyon, Arizona (47; Toolin et al. 1979; Van Devender unpubl. data).

Considering that none of these floras shared more than 41.2% of the Yécora grasses, the flora of the Municipio is very diverse, with elements shared with many biotic regions: the New World tropics to the west and south; the woodlands and forests of the main Sierra Madre Occidental to the east and south,

Table 5.3. Grasses in the Municipio de Yécora shared with other floras.

FLORA (Reference)[a]	Area (km²)	No. of Taxa	No. of Grass Taxa	Grass Taxa/Area	% of Flora	Muhlenbergia	Paspalum	No. of Taxa Shared with Yécora	% Yécora Grasses Shared with Other Flora
Municipio de Yécora, Sonora (15)	3,300	1,696	186	0.06	11	32	13	11	
Tropical Floras									
Sierra de Manantlán, Jalisco (14)	1,400	2,774	225	0.16	8.1	21	25	77	41.2
Chamela, Jalisco (9)	350	1,146	57	0.16	5.0	0	4	28	15.1
Río Cuchujaqui, Sonora (13)	46	744	84	1.83	11.2	7	7	66	35.5
Coastal thornscrub, Sonora (5)	350	1,146	45	0.06	5.6	0	0	25	13.5
Sonoran Desert Floras									
Cape Region, Baja California Sur (7, 8)	8,500	1,124	141	0.02	12.5	13	6	73	39.2
Gran Desierto, Sonora (4)	15,000	589	75	0.01	12.7	2	0	37	19.9
Tucson Mountains, Arizona (10)	404	610	95	0.24	15.6	3	0	45	24.2
Mediterranean Flora									
San Diego County, California (1)	120	2,210	191	1.59	8.6	5	2	41	22
Sierra Madre Occidental Floras									
Cascada de Basaseachi, Chihuahua (11)	65	823	83	1.28	10.1	23	3	63	33.9
Nabogame, Chihuahua (6)	40	601	65	1.63	10.8	14	3	56	30.1
Sky Island Floras									
Chiricahua Mountains, Arizona (2)	1,540	1,261	148	0.10	11.7	21	2	66	35.5
Huachuca Mountains, Arizona (3)	315	994	123	0.39	12.4	20	1	67	36
Sycamore Canyon, Arizona (12)	9	624	80	8.89	12.5	15	4	47	25.3

[a]References: (1) Beauchamp 1986; (2) Bennett et al. 1996; (3) Bowers and McLaughlin 1996; (4) Felger 2000; (5) Friedman 1996; (6) Laferriére 1994; (7) Lenz 1992; (8) León de la Luz et al. 1999; (9) Lott 1993 and pers. comm. 1998; (10) Rondeau et al. 1996; (11) Spellenberg et al. 1996; (12) Toolin et al. 1979 and Van Devender unpubl. data; (13) Van Devender et al. 2000; (14) Vázquez G. et al. 1995; (15) this report.

and their sky island outliers to the north in Arizona; the temperate grasslands and Chihuahuan Desert to the east; the Sonoran Desert lowlands to the west; and the Mediterranean chaparral to the west in Baja California and southern California.

Non-native Species

Only 28 of the Yécora grasses (15.1%) are not native. A few additional species may or may not be native: they occur throughout the warm parts of the world today (e.g., the tanglehead, *Heteropogon melanocarpus*), and it is difficult to know their original distributions. Most of the introduced grasses are from the Old World (89.3%). Introduced taxa from the New World are corn (*maíz, Zea mays* subsp. *mays*) and teosinte (*maíz cócono, Zea mays* subsp. *mexicana*) from Mexico and rescue grass (*Bromus catharticus*) from South America. Of the Old World grasses, 3 are restricted to Africa, 3 to the Mediterranean region, 7 to Europe, and 2 to Eurasia, while 10 are widespread.

Non-native species arrived in the Municipio de Yécora by various mechanisms. Only buffelgrass (*Pennisetum ciliare*) has been deliberately planted for cattle forage. Many species disperse along riparian arroyos, streams, and so on, or in pseudo-riparian habitats along highways. Six (21.4%) of the introduced grasses are primarily found in moist soil microhabitats in arroyos or *ciénegas*. One grass (*Bothriochloa ischaemum*) was found for the first time in 2001 on a roadside near Restaurant Puerto de la Cruz. This is a popular truck stop at the summit of MEX 16 west of Yécora, suggesting the probable source. Wild barley (*Hordeum murinum*) is only known from the yard of Capuchin Missionaries in Yécora, collected in May 1996 and 1998. The priests appear to have been the unintentional dispersal agent for this Mediterranean weed from their home base in Solvang, California, to Sonora. Most of the other introduced grasses are easily dispersed by wind, water, or animals to disturbed soils, which are so common in grazed landscapes. Some grass florets are likely contaminants in bales of hay or alfalfa (*Medicago sativa*) brought to the area from other areas for cattle feed. Corn (*Zea mays* subsp. *mays*), sorghum (*milo maíz, Sorghum bicolor*), and oats (*Avena sativa*) are cultivars that were found growing on roadsides away from cultivation.

The only introduced grasses that appear to represent ecological problems are Natal grass (*Melinis repens*) and buffelgrass, both from Africa. A dramatic invasion by Natal grass is in progress. In 1998, this species was widespread in the Municipio but was especially invasive in the grasslands in the Los Pilares area west of Maycoba and near Santa Ana. By September 2003, Natal grass had expanded along roadsides and into natural habitats in essentially all areas below pine–oak forest. Buffelgrass is a stout perennial that has been planted in cleared areas in tropical deciduous forest near Tepoca. As elsewhere, it is spreading into other areas along roads and trails. However, it has not yet caused the extensive conversion of native vegetation into Africanized savannah, as seen in other parts of Sonora (Búrquez et al. 1998), most likely because of the shady understory conditions in intact tropical deciduous forest. Additional ecological impacts are to be expected as both species increase in abundance and expand their ranges.

Some exotic grasses not found in the Yécora area deserve to be mentioned here. In the southwestern United States, red brome (*Bromus rubens*), cheatgrass (*B. tectorum*), wild barley, Mediterranean grass (*Schismus barbatus*), fountain grass (*Pennisetum setaceum*), and Lehmann's lovegrass (*Eragrostis lehmanniana*) are serious threats to the native ecosystems, often causing large-scale ecological changes. In years with good winter-spring rainfall in the Sonoran Desert in Arizona, red brome, wild barley, and wild oats, annual grasses native to southern Europe, form dense ground covers in paloverde (*Parkinsonia microphylla*)–sahuaro (*Carnegiea gigantea*) desertscrub. In the heat and aridity of the foresummer in May and June, wild fires fueled by these dried grasses have been increasingly frequent, with devastating impacts (Van Devender et al. 1997). Cheatgrass causes similar impacts in the Joshua tree (*Yucca brevifolia*)–creosotebush (*Larrea divaricata*) desertscrub in the Mohave Desert in Nevada and California. In these areas, fire-sensitive desertscrub communities are converted to annual grasslands after fire. Mediterranean grass is a common spring annual in drier desertscrub habitats in the Lower Colorado River Valley of southwestern Arizona and southeastern California. Fountain grass, a larger relative of the African buffelgrass, is invasive in rocky habitats in the Sonoran Desert near Phoenix and Tucson, Arizona (Van Devender and Dimmitt 2000). Lehmann's lovegrass is a perennial grass native to South Africa that has been introduced as forage in the southwestern United States. It is an aggressive competitor of native bunch grasses and often becomes dominant in desert grassland (McClaran and Anable 1992). The success of these various non-native species in other areas suggests that introduced grasses could have greater impacts in southeastern Sonora in the future.

Ethnobotany

The Mountain Pima Indians live in small communities in the Sierra Madre Occidental from near Yécora eastward to Madera and Yepáchic, Chihuahua. Reina (1993) recorded medicinal uses of several grasses in the Yécora–Maycoba area. As with most Pima herbal remedies, infusions were prepared from the grasses to be drunk as tea. *Mal de orín* (urinary tract infection) is treated with teas brewed from the stems and blades of deer grass (*escobón, Muhlenbergia rigens*), pinyon ricegrass (*hierba de la vejiga, Piptochaetium fimbriatum*), and rabbit-foot grass (*zacate cola de zorra, Polypogon monspeliensis*), and from corn silk.

Teas from the stems and blades of deer grass (*Muhlenbergia rigens*) and Natal grass or *espiga colorada* are used to treat stomachache (gastritis). Corn kernels are roasted and ground with salt to prepare a drink that serves as a remedy for *empacho*, an impaction of food in the intestines. Grass teas are also used to treat back pains (Natal grass), heart-related pains (bamboo muhly [*Muhlenbergia dumosa*] and corn silk), pneumonia (bamboo muhly and corn silk), and premenstrual cramps (bamboo muhly). For the pneumonia remedy, the corn silk is mixed with the bark of guasaraco (*Parthenium tomentosum* DC. var. *stramonium* [E. Greene] Rollins). Interestingly, the guasaraco (a tree composite) bark was transported to the Maycoba area from tropical deciduous forest near Moris, Chihuahua, about 30 km to the south.

In 1968 and 1970, Campbell W. Pennington (pers. comm. 1996) interviewed the Pima Indians at Maycoba about their cultivated crops, including corn, sorghum, and wheat (*Triticum aestivum*). Wheat had been more commonly cultivated earlier when more Pimas had oxen for plowing. At least four pre-Columbian varieties of corn for flour, pinole, and popcorn were cultivated. Two of these varieties (*maíz reventador* and *maíz blanco*) were thought to have been developed through hybridization involving teosinte. The Pimas said that *maíz cócono*, as they called teosinte (the wild relative of corn), was formerly grown in the midst of the corn, a practice likely learned from the Tepehuanes of neighboring Chihuahua. Pimas in Los Pilares said that the native *zacate maíz* (*Tripsacum lanceolatum*) was crossing with the corn in the fields (Lucio Sierra, pers. comm. 1992). Interviews in September 2003 revealed that the old land-races of corn have been lost in El Kípor and Los Pilares and that Pima farmers do not know teosinte.

Considering the 186 taxa in the Municipio de Yécora, relatively few grasses have ethnobotanical uses. Of these, 3 are native grasses, 2 introduced, and 2 cultivated.

Conclusions

The diversity of grasses in the Municipio de Yécora is very high and is certainly higher than comparable areas in tropical lowland and desertscrub communities in northwestern Mexico. However, putting the Yécora grass flora into a broader context is difficult because we are not aware of similar analyses of grasses along elevation–vegetation gradients. Summaries of the grasslands of Mexico (Shreve 1942; Rzedowski 1978) only present lists of important species in broad areas. The grass floras of Sonora (Beetle and Johnson 1991), Chihuahua (Lebgue and Valerio 1986), and Durango (Herrera A. 2001) are taxonomic accounts and do not summarize the grasses in local areas within the states. The Sierra de Manantlán in Jalisco with 225 grass taxa in 1400 km² in tropical and montane forests (Vázquez G. et al. 1995) appears to have a richer grass flora than the Yécora area.

The 186 grass taxa reported for the Municipio de Yécora represent 57% of the total of 326 grasses species known from Sonora (Beetle and Johnson 1991). The 94 taxa in oak woodland and the 94 taxa in pine–oak forest in the Municipio de Yécora may reflect the general richness of grasses in these communities in the northern Sierra Madre Occidental. Although Yécora likely has the highest grass diversity in Sonora, comparisons with several other areas in the state would be interesting. The grasslands of *la frontera* along the Arizona border have been floristically neglected, especially the Altar and Santa Cruz river valleys in north-central Sonora and the western base of the Sierra San Luis in northeastern Sonora.

Acknowledgments

We appreciate Father Bill Trauba's help and companionship in the field and his providing shelter in Yécora. George Ferguson, Mark Fishbein, and Richard Spellenberg helped in the field and shared their collection data with us. Celia C. Bujdud helped arrange a loan of grass specimens in the COTECOCA herbarium in Hermosillo. Phil Jenkins accompanied John Reeder on a visit to COTECOCA to examine grass specimens. The careful and insightful editing of Richard Spellenberg, Jean-Luc Cartron, and Richard Felger helped polish the chapter. Campbell W. Pennington graciously shared his unpublished notes on the Mountain Pima uses of corn. Gerald Dawavendewa drafted the figures.

Literature Cited

Beauchamp, R. M. 1986. A Flora of San Diego County, California. Sweetwater River Press, National City, California.

Beetle, A. A., and D. J. Johnson. 1991. Gramineas de Sonora. Secretaría de Agrícultura y Recursos Hídraulicos, Hermosillo, Mexico.

Bennett, P. S., R. R. Johnson, and M. R. Kunzmann. 1996. An annotated list of vascular plants of the Chiricahua Mountains. U.S. Geological Survey Biological Resources Division Cooperative Park Studies Unit Special Report No. 12. University of Arizona, Tucson, Arizona.

Bowers, J. E., and S. P. McLaughlin. 1996. Flora of the Huachuca Mountains, a botanically significant sky island in Cochise County, Arizona. Journal of the Arizona-Nevada Academy of Sciences 29: 66–107.

Búrquez, A., A. Martínez Y., and P. S. Martin. 1992. From the high Sierra Madre to the coast: changes in vegetation along highway 16, Maycoba-Hermosillo. Pp. 239–252 *in* K. F. Clark, J. Roldán, and R. H. Schmidt (eds.), Geology and Mineral Resources of the Northern Sierra Madre Occidental. Guidebook, El Paso Geological Survey Publication No. 24, El Paso, Texas.

Búrquez, A., A. Martínez Y., M. E. Miller, K. Rojas, M. A. Quintana, and D. A. Yetman. 1998. Mexican grasslands and changing aridlands of Mexico: an overview and a case study in northwestern Mexico. Pp. 21–32 *in* B. Tellman, D. Finch, R. Hamre, and C. Edminster (eds.), The Future of Arid Grasslands: Identifying Issues, Seeking Solutions. USDA Forest Service Proceedings RMRS-P-3. Rocky Mountain Forest and Experiment Station, Fort Collins, Colorado.

Chase, A. 1942. The North American species of *Paspalum*. Contributions to the United States Herbarium 28: 109–110.

Felger, R. S. 2000. Flora of the Gran Desierto and Río Colorado of Northwestern Mexico. University of Arizona Press, Tucson.

Friedman, S. L. 1996. Vegetation and flora of the coastal plains of the Río Mayo region, southern Sonora, Mexico. M.S. thesis, Arizona State University, Tempe.

Gentry, H. S. 1942. Río Mayo Plants. A study of the Río Mayo, Sonora. Carnegie Institution of Washington Publication 527, Washington, D.C.

Goldberg, D. E. 1982. The distribution of evergreen and deciduous trees relative to soil type: an example from the Sierra Madre, Mexico, and a general model. Ecology 63: 942–951.

Herrera A., Y. 2001. Las Gramineas de Durango. Instituto Politénico Nacional and Comisión Nacional para el Conocimiento y Uso de la Biodiversidad, Durango, Mexico.

Laferriére, J. E. 1994. Vegetation and flora of the Mountain Pima village of Nabogame, Chihuahua, Mexico. Phytologia 77: 102–140.

Lebgue, T., and A. Valerio. 1986. Manual para Indentificar las Gramineas de Chihuahua. Talleres Gráficos del Gobierno del Estado de Chihuahua, Cd. Chihuahua, Mexico.

Lenz, L. W. 1992. An Annotated Catalogue of the Plants of the Cape Region, Baja California Sur, Mexico. Cape Press, Claremont, California.

León de la Luz, J. L., J. L. Pérez N., M. Domínguez L., and R. Domínguez C. 1999. Flora de la región del Cabo de Baja California Sur. Listados Florísticos de México XVIII. Instituto de Biología, Universidad Autónoma de México, D.F.

Lott, E. J. 1993. Annotated checklist of the vascular plants of the Chamela Bay region, Jalisco, Mexico. Occasional Papers of the California Academy of Sciences 148: 1–60.

Martin, P. S., D. Yetman, M. Fishbein, P. Jenkins, T. R. Van Devender, and R. K. Wilson. 1998. Gentry's Río Mayo Plants. The Tropical Deciduous Forest and Environs of Northwest Mexico. University of Arizona Press, Tucson, Arizona.

McClaran, M. P., and M. E. Anable. 1992. Spread of introduced Lehmann lovegrass along a grazing intensity gradient. Journal of Applied Ecology 29: 92–98.

McVaugh, R. 1983. Flora Novo-Galiciana. A

Descriptive Account of the Vascular Plants of Western Mexico, Vol. 14. Gramineae. University of Michigan Press, Ann Arbor.

Reina G., A. L. 1993. Contribución a la introducción de nuevos cultivos en Sonora: las plantas medicinales de los Pimas Bajos del Municipios de Yécora. Undergraduate thesis, Universidad de Sonora, Hermosillo, Mexico.

Reina G., A. L., T. R. Van Devender, W. Trauba, and A. Búrquez M. 1999. Caminos de Yécora. Notes on the vegetation and flora of Yécora, Sonora. Pp. 137–144 *in* D. Vásquez del Castillo, M. Ortega N., and C. A. Yocupicio C. (eds.), Symposium Internacional sobre la Utilización y Aprovechamiento de la Flora Silvestre de Zonas Aridas. Universidad de Sonora, Hermosillo, Mexico.

Rondeau, R., T. R. Van Devender, C. D. Bertelsen, P. Jenkins, R. K. Wilson, and M. A. Dimmitt. 1996. Annotated flora and vegetation of the Tucson Mountains, Pima County, Arizona. Desert Plants 12: 3–46.

Rzedowski, J. 1978. Vegetación de México. Editorial Limusa, México D.F.

Shreve, F. 1942. Grassland and related vegetation in northern Mexico. Madroño 6: 190–198.

Spellenberg, R. S., T. Lebgue, and R. Corral D. 1996. A specimen-based, annotated checklist of the plants of Parque Nacional "Cascada de Basaseachi" and immediately adjacent areas, southwestern Chihuahua, México, XIII. Listadas Florísticas de México, Instituto de Biología, Universidad Nacional Autónoma de México, México D.F.

Toolin, L. J., T. R. Van Devender, and J. M. Kaiser. 1979. The flora of Sycamore Canyon, Pajarito Mountains, Santa Cruz County, Arizona. Journal of the Arizona-Nevada Academy of Sciences 14: 66–74.

Van Devender, T. R., and M. A. Dimmitt. 2000. Desert grasses. Pp. 265–280 *in* S. J. Phillips and P. W. Comus (eds.), A Natural History of the Sonoran Desert. Arizona-Sonora Desert Museum Press, Tucson.

Van Devender, T. R., R. S. Felger, and A. Búrquez M. 1997. Exotic plants in Sonora, Mexico. *In* M. Kelley, E. Wagner, and P. Warner (eds.), Proceedings California Exotic Pest Plant Council Symposium 3: 10–15.

Van Devender, T. R., A. L. Reina G., M. C. Peñalba G., and C. I. Ortega R. 2003. The Ciénega de Camilo: a threatened habitat in the Sierra Madre Occidental of eastern Sonora, Mexico. Madroño 50: 187–195.

Van Devender, T. R., A. C. Sanders, R. K. Wilson, and S. A. Meyer. 2000. Vegetation, flora, and seasons of the Río Cuchujaqui, a tropical deciduous forest near Alamos, Sonora. Pp. 36–101 *in* R. H. Robichaux and D. A. Yetman (eds.), The Tropical Deciduous Forest of Alamos. Biodiversity of a Threatened Ecosystem in Mexico. University of Arizona Press, Tucson.

Vázquez G., J. A., R. Cuevas G., T. S. Cochrane, H. H. Iltis, F. J. Santana M., and L. Guzmán H. 1995. Flora de Manantlán. Botanical Research Institute, Sida, Botanical Miscellany No. 13, Fort Worth, Texas.

6

Scorpion Biodiversity and Patterns of Endemism in Northern Mexico

W. DAVID SISSOM

BRENT E. HENDRIXSON

Although comprising a relatively small group of terrestrial arthropods, scorpions are subjects of considerable interest to both the scientist and layperson. In some areas of the world, including much of Mexico, they are feared by the general public because of their highly toxic venoms. To those uneducated about them, scorpions are all assumed to be lethal; the truth of the matter is, only about two dozen of the 1200 or so species are considered dangerous by medical experts. About 7 species of *Centruroides* in Mexico are known to cause human mortality; Simard and Watt (1990) estimated that about 100,000 stings occur in Mexico each year, and possibly as many as 800 people die. Most mortality involves young children and the elderly.

Scorpions are ancient arthropods, derived from aquatic ancestors that lived in the Silurian Period, more than 400 million years ago. These ancient forms closely resemble modern scorpions in details of basic anatomy. Apparently, the scorpion body plan that developed so long ago is a highly successful one. It is interesting to note that a few of the fossil scorpions were approximately 1 m in length, far greater in size than the largest of today's species, which are just longer than 20 cm.

Scorpions are also interesting animals because of their reproductive biology and ecology. All species are viviparous, with embryos developing in the female's reproductive tract and receiving nourishment from maternal tissues (Francke 1982). Polis and Farley (1980) and Polis (1990) have suggested that scorpions have very low reproductive rates in comparison with most animals. A low r_{max} (maximum rate of population increase) for several species is in part due to long generation times and low survivorship among sexually immature females. Gestation periods may be quite long, ranging from several months in many buthids to more than a year and a half in certain scorpionoids (Polis and Sissom 1990). The young are fairly large at birth, and time to maturity is at the minimum about 2 years (in some species, maturity may not be reached for at least 6–8 years). Average longevity is probably around 4 years, but *Hadrurus* spp. may live 25 years (Polis 1990). In general, then, scorpions resemble long-lived vertebrate species in their life-history traits and should be regarded as "K-selected." Such species are often of great concern to conservation biologists because they cannot replace their populations rapidly.

Scorpions are important components of arid and semiarid ecosystems, but they are not limited to these areas. Other habitats in which they may be found include deciduous and pine forests, grasslands, the floors and trees of tropical jungles, high mountain slopes in the Himalayas and Andes (up to 14,000 feet elevation), deep subtropical and tropical caves, and even certain rocky intertidal habitats at the ocean's edge. Despite this, we still tend to think of scorpions as desert animals, and current evidence suggests that deserts harbor more species than other ecosystems. Scorpions may be very abundant; Levy and Amitai (1980) measured densities of 1.12/m² in the Middle Eastern *Leiurus quinquestriatus*, and Polis (1990) reported densities of 8–12/m² in *Serradigitus littoralis*, an intertidal scorpion from the Baja California peninsula.

The diversity of scorpions in Mexico, as judged by world standards, is exceptional. Currently, more than 200 species and subspecies of scorpions are known from Mexico, more than any other country in the world (Lourenço and Sissom 2000). Mexico is a land of great diversity in landforms, elevation, climate, and vegetation (chapters 1 and 2). Opportunities for prolific speciation of Mexican scorpions have undoubtedly resulted from the country's complex geological history (chapter 1), which must have repeatedly fragmented ancestral scorpion populations.

Our purpose here is to review scorpion biodiversity of the Baja California peninsula and the states of Sonora, Sinaloa, Durango, Chihuahua, Coahuila, Nuevo León, and Tamaulipas on the mainland of Mexico. First, a brief history of taxonomic research and a synopsis of the scorpion fauna are provided, followed by an assessment of the patterns of distribution and endemism (as currently understood) and the role scorpions might play in future conservation strategies. Needs for further studies of scorpion biodiversity in northern Mexico are identified.

Status of Biodiversity Research in Northern Mexico

Brief Historical Review of Efforts to Catalog Diversity

Pocock (1898, 1902) was among the earliest taxonomists to work extensively with scorpions from Mexico. The first comprehensive assessment of scorpion biodiversity in Mexico was published by Hoffmann (1931, 1932); in his monograph, Hoffmann synthesized all of the information known about Mexican scorpions, described a number of new taxa, and listed 15 species from the northern states (several of which were misidentified and 4 for which he relied solely on literature reports). Hoffmann focused primarily on mainland Mexico and had very little material from the Baja California peninsula.

Only a handful of papers pertaining to the Mexican fauna were published in the next few decades after Hoffmann's monograph, but during the 1960s and 1970s, several U.S. taxonomists (O. F. Francke, W. J. Gertsch, M. E. Soleglad, H. L. Stahnke, and S. C. Williams) took an interest in the scorpions of Mexico, including those of the northern states. The publications of Stanley Williams, in particular, were based on extensive fieldwork on the Baja California peninsula using ultraviolet detection techniques. He culminated this research with an important monograph of the Baja California fauna (Williams 1980), representing one of the most thorough studies of a regional scorpion fauna ever conducted. Considering only currently valid species, Williams nearly tripled the number known from the peninsula, describing as new 35 (58.3%) of the species.

The study of the mainland fauna has consisted largely of sporadic papers describing newly discovered species. Williams (1968a) partially described the vaejovid fauna of the Cuatrociénegas area in Coahuila; of 5 species reported, 1 is apparently endemic. Mitchell (1968) described the remarkable troglobytic genus *Typhlochactas*, with 1 of its original species (*T. rhodesi* Mitchell) recorded from La Cueva de la Mina in Tamaulipas. Díaz Najera (1964, 1975) published numerous new distributional records for many species in Mexico. Since the 1980s, a few papers have dealt with the scorpion biodiversity of mainland Mexico. These papers described new species and/or provided partial taxonomic revisions. Two papers provided assessments of the fauna of the state of Sonora (Sissom 1991; Sissom and Stockwell 1991). Relatively detailed distributional information for 2 species in northern Mexico has also been presented (Shelley and Sissom 1995; Yahia and Sissom 1996).

A number of complete studies of the Mexican scorpion fauna have appeared in the last few years: catalogs of the world fauna (Fet et al. 2000) and of the Mexican fauna (Beutelspacher 2000), as well as a synoptic review of the entire Mexican fauna (Lourenço and Sissom 2000). Two additional species of *Vaejovis* have been described from Sonora (Capes 2001; Hendrixson 2001) since publication of the *Catalog of the Scorpions of the World (1758–1998)* (Fet et al. 2000).

Approaches to Biodiversity Analysis

Almost all of the early information on scorpion diversity was obtained by the conventional rock-rolling technique, which entails looking underneath rocks, logs, and other surface cover to find scorpions. In using this method it is important to return these objects to their original positions after rolling them to look for scorpions. Many different kinds of organisms are dependent on surface objects for food and shelter (because of the favorable microclimate offered). Failure to replace such surface objects will cause many of these organisms to die unnecessarily.

The rock-rolling technique appreciably underestimates the total scorpion diversity in an area, but diversity assessment can be greatly improved by using the ultraviolet light detection method (Williams 1968b). The method enables researchers to locate scorpions when they are active at night and, consequently, provides more information on ecology and behavior than do other methods. Use of this tool throughout Mexico will reveal significant numbers of new species, especially in the genera *Diplocentrus* and *Vaejovis*. UV light detection, although the method of choice for most scorpion researchers, is nevertheless dependent on the activity patterns of the scorpions. Owing to the sporadic nature of nightly and seasonal surface occurrence of some species, not all of the species at a given location may be collected unless sampling is conducted periodically. Some species are only sporadically active on the surface throughout the warmer months of the year, whereas others exhibit brief periods of intense activity. In general, if sampling large areas (e.g., one or more geopolitical states) black-lighting gives more return per unit effort than the other methods (both in numbers of species and numbers of individuals). In our UV light-sampling regime for roadside collecting in the United States, for example, we have sampled at least 5 localities per night (double that if we are able to divide into 2 separate collecting parties). Consequently, a 2-week collecting trip can yield large numbers of specimens from many localities, as demonstrated by the work of S. C. Williams on the Baja California peninsula.

Pitfall trapping, a more labor-intensive method, is also quite productive. There are some limitations to this method as well. In montane habitats and other rocky soils, it may be difficult to dig the holes necessary to set up the traps. Lithophiles and burrowing forms, which typically wait near their crevices or burrows for prey without moving around much on the surface, are not as likely to be taken by this method, except during the mating season when males are actively moving in search of females. Pitfall trapping is also affected by seasonal activity patterns of species; trapping must be conducted throughout the warm season to maximize the species catch at a given location. Unlike UV-light sampling, it is not feasible to thoroughly sample large geographical areas with pitfall trapping, each trap transect takes considerable time to set up, and this factor restricts the number of localities that can be sampled on a collecting trip. The traps must be left open long enough to generate a catch, and the sites must be revisited to obtain the specimens. Pitfall trapping seems most useful when the investigator is working at 1 or several study sites over an extended period. Another limitation of the method is that specimens collected in antifreeze/alcohol traps (necessary to prevent cannibalism) are generally not well preserved.

Obtaining a complete accounting of the species present is only one of the problems in assessing scorpion biodiversity in northern Mexico. Accurate estimates of the geographical distributions of most species in mainland Mexico are lacking. This is primarily due to inadequate sampling, but it is also due in part to the rarity or sporadic surface occurrence of some species. Given that few sites have been intensively investigated, mainly isolated records of individual species have accumulated. The problem can be illustrated by fieldwork conducted at Big Bend Ranch State Park in southwestern Texas (Sissom and Henson, unpubl. data), an area similar to many mountainous sites in neighboring states of northern Mexico. At this site, a series of transects were monitored for 4 years. Based on observations made on a 20 m × 100 m rocky slope transect over 63 nights, the probability of finding the uncommon *Vaejovis intermedius* Borelli on a given night was 11.1%, and no more than 3 individuals were seen on any night. On the same transect, the probability of finding another uncommon species, *Pseudouroctonus apacheanus* (Gertsch and Soleglad), was 22.2%. Of the 2 species, at least *V. intermedius* occurs in northeastern Mexico, and both taxa possibly exhibit similar densities over much of their range. It is interesting that, although 8 species were observed on that transect over the 4 years, an average of only 2.7 species (range 0–6) were observed nightly.

As sampling in Mexico continues, some species currently thought to be endemics may be found to exhibit a broader geographical distribution. Therefore, to understand patterns of endemism more fully, filling out the distributional data for the scorpion species in northern Mexico should be given high priority.

Scorpion Diversity in Mexico

Mexico is a medium-sized country, and its scorpion diversity is still impressive when one standardizes species diversity according to area. An arbitrary measure of species density used here is the number of species per 100,000 km². Scorpion diversity of 15 representative countries from 5 continents is tabulated in table 6.1 to facilitate raw and standardized comparisons.

The present state of knowledge suggests that Mexico has more families, genera, and species than any other country in the world (table 6.1). Species density, however, falls below that of several small countries, namely Cuba, Venezuela, and Ecuador.

At the same time, it should be noted that the species density of the Baja California peninsula is approximately twice that of the most scorpion-rich country listed in the table. The scorpion fauna of the peninsula appears to be the most diverse in the world (at least among areas > 10,000 km²). It is likely that other parts of Mexico will exhibit similar species densities once they become better sampled; based on existing museum collections, states such as Oaxaca, Guerrero, and Michoacan show considerable promise. However, it is difficult to say with any degree of certainty what percentage of the fauna remains to be discovered and described.

Several factors should be considered in making these comparisons. First, whereas estimates of species

Table 6.1. Comparison of world scorpion diversities using representative countries.

Region	No. of Families	No. of Genera	No. of Species	No. of Species/10^5 km²
North America and Antilles				
Mexico (total)	7	22	197	10.06
Baja California peninsula	6	14	60	41.64
United States (total)	5	12	84	0.9
Southwestern states (TX, NM, AZ, CA)	5	12	80	4.67
Cuba	2	10	24	21.65
South America				
Argentina	2	9	50	1.81
Brazil	4	16	93	1.09
Colombia	4	9	41	3.60
Ecuador	4	8	36	12.70
Venezuela	4	18	124	13.60
Africa				
Egypt	3	10	20	2.00
Namibia	4	7	58	7.04
South Africa and Lesotho	3	10	101	8.07
Asia				
India	6	21	98	2.98
Saudi Arabia	4	14	19	0.88
Yemen	4	14	32	6.06
Australasia				
Australia	4	6	29	0.38

Numbers of species are based on existing published records and do not include known introduced species. The list also does not include subspecies (see text). Scorpions are very unevenly distributed across the United States, with 80 of 84 species occurring in the southwestern border states. For this reason, separate species densities are given for the country as a whole and for those southwestern states.

density will likely increase considerably in Mexico, the smaller countries that have been reasonably sampled will have a larger percentage of their species already known; with less geographical area, there is less potential to contain many large unexplored areas that can hold a sizable number of endemics. Second, subspecies and synonyms are not considered in this analysis. There seems to be a general trend to reconsider the status of subspecies (e.g., Prendini 2001), and it is expected that many subspecies will eventually be considered valid species, whereas others will be synonymized; in addition, some old synonyms will be revived and considered valid species. Subspecies were especially proliferated over the years by taxonomists who worked with Old World scorpions (e.g., according to Fet et al. [2000], the genus *Scorpio* Linnaeus currently has 1 species divided into 19 subspecies, and *Euscorpius* Thorell has 5 species and 42 subspecies). This trend was not as prevalent in the New World (including Mexico), and only a moderate number of subspecies has been described. Adjustments in subspecies and synonymies will influence species density calculations, but because subspecies typically make up a minor percentage of a country's fauna, this influence will not be great. As an example, South Africa represented a country with a large number of subspecies. Prendini's (2001) reassessment of described *Opistophthalmus* scorpions (which constitute about one-third of the known South African scorpion fauna) raised the total number of recognized species in that country from 92 to 101 and changed the species density from 7.35 to 8.07. There are 20 remaining subspecies from other genera in South Africa; in comparison, Mexico has 18.

Obviously, the scorpion faunas of some parts of the world are better known than those of others. The fauna of the Baja California peninsula is certainly one of the best known (although not completely inventoried), but there are areas outside of Mexico that have also enjoyed a great amount of taxonomic attention (e.g., Cuba, Namibia, South Africa, Venezuela). Generally speaking, however, most of the countries listed in table 6.1 are inadequately sampled, and the mainland of Mexico is certainly comparable to any of those.

Lourenço (1994) suggested that the tropics (particularly in South America) will eventually hold the greatest scorpion diversity, and perhaps this will eventually be proven true as in many other taxonomic groups. In fact, Lourenço proposed that an endemic area in the tropical Andes encompassing all of Ecuador, about 40% of Colombia, and about 25% of Peru has the highest diversity in the world, even compared directly to that of the Baja California peninsula. The tropical Andean area indeed demonstrates high diversity (6 families, 12 genera, and 68 species), but its estimated surface area is approximately 1,060,000 km^2—more than 7 times the area of the Baja California peninsula. Calculating species density as above, the known diversity of the tropical Andes endemic area is 6.4 species/ 100,000 km^2. Even if the cataloged diversity of this area triples, as Lourenço suggests is possible once the fauna is more completely known, it will still reach only half the species density of Baja California (assuming no additional species are found on the peninsula, which is unlikely).

Scorpion Diversity in Northern Mexico

A current checklist of scorpions known from the states of northern Mexico is provided in table 6.2. The fauna in this part of Mexico comprises 6 families, 16 genera, and 109 species. Based on the number of known species, northern Mexico has more than half the total species (109 out of 197) in the country. The better sampled fauna of the Baja California peninsula contributes 60 of the 109 species in the north.

Family Buthidae

The family Buthidae is represented in northern Mexico by a single genus, *Centruroides* Marx (fig. 6.1, *1*). In eastern border states (i.e., Chihuahua, Coahuila, Nuevo León, Tamaulipas) there are only 3 species, *C. vittatus* (Say), *C. rileyi* Sissom, and *C. gracilis* (Latreille). *Centruroides vittatus* is also widely distributed in the central United States, and *C. gracilis* has a very wide range, occurring in Florida, the Caribbean, Central America, and northern South America. In the westernmost areas of northern Mexico (states of Durango, Sonora, and Sinaloa, and the Baja California peninsula), 6 species are recorded: *C. exilicauda* (Wood), *C. infamatus* (Koch), *C. pallidiceps* Pocock, *C. suffusus* Pocock, *C. margaritatus* (Gervais), and *C. vittatus*. Several of these are highly venomous and are known to contribute to human mortality. The taxonomy of the genus is poorly understood, with highly variable species diagnosed primarily on the basis of color and morphometrics. Evaluating the fauna of Durango

Table 6.2. Checklist of the scorpion species of northern Mexico.

Taxon	BC	BCS	SON	SIN	DGO	CHIH	COAH	NL	TAMPS
Family Buthidae									
Centruroides exilicauda	BC	BCS	SON			?			
Centruroides gracilis									TAMPS
Centruroides infamatus					DGO				
Centruroides margaritatus				SIN					
Centruroides pallidiceps			SON	SIN					
Centruroides rileyi									TAMPS
Centruroides suffusus					DGO				
Centruroides vittatus					DGO	CHIH	COAH	NL	TAMPS
Family Chactidae									
Nullibrotheas allenii		BCS							
Family Diplocentridae									
Bioculus caboensis		BCS							
Bioculus cerralvensis		BCS							
Bioculus comondae		BCS							
Bioculus cruzensis		BCS							
Diplocentrus colwelli								NL	
Diplocentrus diablo									TAMPS
Diplocentrus ferrugineus								NL	
Diplocentrus gertschi			SON	SIN					
Diplocentrus lindo							COAH	NL	
Diplocentrus spitzeri			SON						
Diplocentrus whitei						CHIH	COAH	NL	
Diplocentrus williamsi			SON						
Family Iuridae									
Anuroctonus phaiodactylus	BC								
Hadrurus arizonensis	BC		SON						
Hadrurus concolor	BC	BCS							
Hadrurus hirsutus		BCS							
Hadrurus pinteri	BC	BCS							
Family Superstitioniidae									
Superstitionia donensis	BC	BCS	SON						
Typhlochactas cavicola									TAMPS
Typhlochactas rhodesi									TAMPS
Family Vaejovidae									
Paravaejovis pumilis		BCS							
Paruroctonus arnaudi	BC								
Paruroctonus baergi			SON						
Paruroctonus bajae	BC								
Paruroctonus borregoensis	BC		SON						
Paruroctonus coahuilanus							COAH		
Paruroctonus gracilior							COAH		
Paruroctonus luteolus	BC								
Paruroctonus nitidus	BC								
Paruroctonus pseudopumilis		BCS							
Paruroctonus silvestrii	BC								
Paruroctonus stahnkei			SON						
Paruroctonus surensis		BCS							
Paruroctonus utahensis						CHIH			
Paruroctonus ventosus	BC								
Paruroctonus xanthus			SON						
Pseudouroctonus andreas	BC								
Pseudouroctonus cazieri	BC								
Pseudouroctonus chicano						CHIH			
Pseudouroctonus lindsayi		BCS							
Pseudouroctonus rufulus	BC								
Serradigitus bechteli		BCS							
Serradigitus adcocki		BCS							
Serradigitus agilis			SON						

(continued)

Table 6.2. Continued

Taxon	BC	BCS	SON	SIN	DGO	CHIH	COAH	NL	TAMPS
Serradigitus allredi			SON						
Serradigitus armadentis		BCS							
Serradigitus baueri	BC								
Serradigitus calidus							COAH		
Serradigitus dwyeri		BCS							
Serradigitus gertschi	BC	BCS							
Serradigitus gigantaensis		BCS							
Serradigitus haradoni		BCS							
Serradigitus harbisoni	BC								
Serradigitus hearnei	BC	BCS	SON						
Serradigitus littoralis	BC	BCS							
Serradigitus minutus		BCS							
Serradigitus pacificus	BC								
Serradigitus polisi			SON						
Serradigitus subtilimanus			SON						
Serradigitus yaqui			SON						
Smeringurus grandis	BC								
Smeringurus mesaensis	BC		SON						
Syntropis macrura		BCS							
Vaejovis bilineatus							COAH	NL	TAMPS
Vaejovis bruneus		BCS							
Vaejovis cazieri							COAH	NL	
Vaejovis coahuilae					DGO	CHIH	COAH		
Vaejovis confusus	BC		SON						
Vaejovis crassimanus					DGO			NL	
Vaejovis decipiens			SON			CHIH			
Vaejovis diazi		BCS							
Vaejovis eusthenura		BCS							
Vaejovis galbus		BCS							
Vaejovis globosus					DGO		COAH		
Vaejovis gravicaudus	BC	BCS							
Vaejovis hirsuticauda	BC								
Vaejovis hoffmanni	BC	BCS							
Vaejovis insularis		BCS							
Vaejovis intermedius					DGO	CHIH	COAH	NL	
Vaejovis janssi		BCS							
Vaejovis magdalensis		BCS							
Vaejovis mauryi			SON						
Vaejovis minckleyi							COAH		
Vaejovis pattersoni		BCS							
Vaejovis peninsularis		BCS							
Vaejovis pequeno			SON						
Vaejovis platnicki									TAMPS
Vaejovis punctipalpi		BCS							
Vaejovis puritanus	BC	BCS							
Vaejovis rossmani								NL	TAMPS
Vaejovis rubrimanus								NL	
Vaejovis sonorae			SON						
Vaejovis spinigerus	BC		SON						
Vaejovis sprousei								NL	TAMPS
Vaejovis vaquero						CHIH			
Vaejovis viscainensis	BC	BCS							
Vaejovis vittatus		BCS							
Vaejovis waeringi	BC								
Vaejovis waueri					DGO	CHIH	COAH	NL	
Vejovoidus longiunguis		BCS							
Total Species/State	32	41	24	3	8	9	13	13	10

Subspecies are not included. Abbreviations for state names: BC = Baja California (northern state of the Baja California peninsula); BCS = Baja California Sur; SON = Sonora; SIN = Sinaloa; DGO = Durango; CHIH = Chihuahua; COAH = Coahuila; NL = Nuevo León; TAMPS = Tamaulipas.

Figures 6.1. Representative scorpions from northern Mexico. *1*, Buthidae: *Centruroides exilicauda* (Wood) from Arizona, New Mexico, the Baja California peninsula, and Sonora. *2*, Diplocentridae: *Diplocentrus whitei* (Gervais) from Texas, Chihuahua, Coahuila, and Nuevo León. *3*, Iuridae: *Hadrurus concolor* Stahnke from the Baja California peninsula. *4*, Superstitioniidae: *Superstitionia donensis* Stahnke from the southwestern United States, Sonora, and the Baja California peninsula. *5*, Vaejovidae: *Paruroctonus gracilior* (Hoffmann) from Texas and New Mexico southwest to Aguascalientes (in northern Mexico documented only in Coahuila). *6*, Vaejovidae: *Vaejovis crassimanus* Pocock from Texas, New Mexico, Arizona, Durango, and Nuevo León. (All photos by W. D. Sissom.)

should be especially interesting. A map generated by Hoffmann (1938) indicated that the ranges of *C. suffusus* and *C. infamatus* were contiguous, but no precise localities were published to support the distributional boundaries. Most recently (de Armas and Frias 2000), *C. suffusus chiaravigli* Borelli was proposed as a junior synonym of *C. vittatus*, placing a third species in Durango.

Centruroides scorpions, which range in adult body size from about 35 to 110 mm, have slender pedipalps and metasomas ("tails"; fig. 6.1, *1*); the metasoma is especially elongated in the male. Most species have a tooth or tubercle underneath the curvature of the stinger. Many *Centruroides* species are yellowish, with or without dark stripes or mottling on the dorsum; some of the larger species (such as *C. gracilis* and *C. margaritatus*) are dark brown to blackish in coloration. The species tend to be ecologically plastic, occurring in a wide range of habitats, including desert flats and rocky slopes, arid and semiarid grasslands, scrubland, deciduous woods, and montane slopes at least up to 2500 m in elevation. Unlike most other scorpions in North America, they are active foragers and are commonly encountered climbing in vegetation. In daytime, they use surface cover, cracks and crevices, or existing burrows of other animals as retreats. Their high population densities and their climbing tendencies often place them on the walls and roofs of human habitations, and they are among the most common scorpions encountered in households.

Family Chactidae

The family Chactidae is primarily South American, but 1 genus is endemic to Baja California Sur. *Nullibrotheas* Williams is represented there by a single species, *N. allenii* (Wood). This is a small brownish, burrowing scorpion with robust pedipalps. Its closest relatives are in South America.

Family Diplocentridae

In northern Mexico, there are 2 genera of diplocentrid scorpions, *Bioculus* Stahnke and *Diplocentrus* Peters. *Bioculus*, recently revalidated as a monophyletic group (Stockwell 1992; Prendini 2000), consists of 4 species endemic to southern Baja California. *Diplocentrus* (fig. 6.1, *2*) is a heterogeneous assemblage of species widely distributed

from the southern United States (southern parts of Arizona, New Mexico, and Texas) throughout Mexico and into Central America. Eight species occur in northern Mexico (table 6.2).

In diplocentrids, the pedipalp chelae are robust, often bearing strong carinae and reticulations in the male (mostly smooth and lustrous in the female), and the telson bears a pronounced cone-shaped tubercle underneath the curvature of the sting. Adults are commonly reddish brown to blackish in coloration; juveniles are yellowish. Most species are around 45–60 mm in length, but a few reach 80 mm or more in length. Diplocentrid scorpions are obligate burrowers with a more K-selected life history (i.e., low reproductive rates, long gestation periods, long time to reach maturity, longer life spans). The natural history of only a few species has been examined.

Family Iuridae

The family Iuridae consists of 2 genera in northern Mexico: *Anuroctonus* Pocock and *Hadrurus* Thorell. *Anuroctonus* is currently recognized as monotypic. The species *A. phaiodactylus* (Wood) is distributed in Arizona, California, Nevada, and Utah in the United States and into the state of Baja California. *Hadrurus* (fig. 6.1, *3*) is represented by 4 species in northern Mexico: *H. arizonensis* Ewing, *H. concolor* Stahnke, *H. hirsutus* (Wood), and *H. pinteri* Stahnke (the latter 3 endemic to the Baja California peninsula). These "giant desert hairy scorpions" are the largest scorpions in the region, with several species reaching a length of approximately 120 mm; they are also robust in body form. Despite their large size, their venom is not very potent.

Family Superstitioniidae

The Superstitioniidae Stahnke includes the genus *Superstitionia* Stahnke and the remarkable Mexican troglobites of the genera *Typhlochactas* Mitchell, *Sotanochactas* Francke, and *Alacran* Francke. *Superstitionia* (fig. 6.1, *4*), a small scorpion with a median dark stripe on the tergites, is known to occur in Baja California (both states) and Sonora. Two of the species of *Typhlochactas* are known from caves in Tamaulipas. These troglobites are small, whitish, eyeless scorpions with moderate attenuation of the appendages.

Family Vaejovidae

The Vaejovidae is the most diverse family of North American scorpions (Sissom 2000), with 147 species (14 of which are polytypic) distributed from southern Canada to Guatemala. Approximately three-fourths of the scorpion species in the northern states of Mexico are vaejovids.

Most *Paruroctonus* Werner (fig. 6.1, 5) are psammophiles and consequently exhibit a certain degree of endemism in various sand dune systems. The genus is well known on the Baja California peninsula, where 7 species have been recorded in the northern state and 2 in Baja California Sur. One of these species and 3 more are also recorded from Sonora; all 4 species are also known from dunes of southeastern California and southwestern Arizona. The records in Sonora are based on materials collected mainly in the Puerto Peñasco area; there are apparently no published records for the extensive dunes of the Gran Desierto in the extreme northwestern corner of the state. The fauna of the rest of the northern mainland states is less well known; in fact, outside of the Baja California peninsula, only 3 species are known from a small handful of records. A fourth species, *P. boquillas* Sissom and Henson, described from Boquillas Canyon in Big Bend National Park in Texas (Sissom and Henson 1998), would be expected to occur in the sand dunes across the Rio Grande in Mexico.

Vaejovis C. L. Koch (fig. 6.1, 6) is a large genus (71 total species), clearly not monophyletic (Sissom 1985), which is very widely distributed in North America. A little more than half of the species in this genus are recorded in the area under consideration here. The genus includes burrowing forms (e.g., the *eusthenura*, *intrepidus*, and *punctipalpi* groups) and crevice dwellers (the *mexicanus* group [*sensu lato*] and the *nitidulus* group). Preliminary phylogenetic analysis (Sissom 1985) suggests that at least several of these species groups should be elevated to genus level.

Other vaejovid genera in northern Mexico include *Pseudouroctonus* Stahnke (5 of 13 total species in North America), *Serradigitus* Stahnke (19 of 24 species; all lithophiles), *Smeringurus* Haradon (2 of 4 species; large species closely related to *Paruroctonus*, although only 1 is a true psammophile), *Syntropis* Kraepelin (containing a single large lithophilic species from the Baja California peninsula and several islands in the Gulf of California),

Vejovoidus Stahnke (a single psammophilic species endemic to the sand dunes of the Vizcaino Desert on the Baja California peninsula), and *Paravaejovis* Williams (with 1 species endemic to Baja California Sur).

Patterns of Diversity and Endemism

Although it is biologically more proper to regard endemism in relation to true biogeographical units, rather than in relation to geopolitical boundaries, it is a fact that conservation issues (e.g., setting aside reserves for protection of species and/or habitats) are decided by governments. Consequently, the following discussion refers to both natural biogeographical areas and geopolitical states.

General Patterns

Scorpions typically do not exhibit high vagility, although some species are readily transported by humans to new areas (e.g., several *Centruroides* spp., *Isometrus maculatus*). In general, specialized burrowing forms (especially psammophiles) and montane species are more likely to exhibit restricted distributions and high levels of endemism. Psammophilic forms are restricted to loose, sandy soils, and their occurrence in a particular area can usually be predicted if sand dunes are present. As seems typical of the biota of this region, mountains harboring species restricted to higher elevations often represent isolated "islands" surrounded by inhospitable habitats.

Some large-scale geographical trends in the fauna of northern Mexico and the southwestern United States are evident. Common to both geographical areas are the Sonoran and Chihuahuan deserts. The 2 deserts more or less interdigitate in southeastern Arizona and southwestern New Mexico, and there the faunas overlap. Interestingly, in this area there is a general tendency for the Sonoran forms (e.g., *Centruroides exilicauda*, *Vaejovis spinigerus*, *Superstitionia donensis*) to inhabit rocky outcrops, and the Chihuahuan forms (*Paruroctonus gracilior*, *Vaejovis coahuilae*, *V. crassimanus*, *V. russelli*) to inhabit the low, often sandy, scrub habitats. The habitat associations to the south of Arizona and New Mexico in northern Mexico remain unknown at this time. Overall, however, the Sonoran and Chihuahuan deserts are very different in species composition, and they consequently represent 2 distinct, major areas of scorpion endemism in northern Mexico and the southwestern United States. Although there is very little overlap in the species, most of the genera are held in common and possess certain species counterparts in the 2 regions. For example, *Centruroides exilicauda* is the Sonoran counterpart of the Chihuahuan (actually, central U.S.) species *C. vittatus*, and *Vaejovis spinigerus* appears to be the Sonoran counterpart of *V. coahuilae*.

Significant endemism is expected in the mountainous areas within the deserts. In particular, the members of the *Vaejovis vorhiesi* complex seem to have responded evolutionarily to mountaintop isolation in Arizona and New Mexico. These species are generally restricted to elevations above 2000 m, commonly on steep slopes forested with pines and oaks. It is expected that related forms will be found in the northern part of the Sierra Madre Occidental and nearby isolated ranges. A very small amount of unprocessed museum material documents this expectation. The lower arid and rocky slopes over the same area are inhabited by *Serradigitus* spp. and members of the *Vaejovis nitidulus* group. New species of these groups should be found in significant numbers in northwestern Mexico. Most of the known species of *Serradigitus* are found on the Baja California peninsula and in Sonora, with a single isolated species in the Cuatrocienegas area. In the case of the *V. nitidulus* group, there are 3 species from the northeastern states, 2 from southwestern Chihuahua and southeastern Sonora, and 1 from the Baja California peninsula. The area east of the Continental Divide to Coahuila contains a great deal of suitable habitat for species of both groups.

Baja California Peninsula and the Islands in the Gulf of California

Due primarily to the works of Stanley Williams, the scorpion fauna of the Baja California peninsula and its associated islands, with 60 known species, is perhaps the most thoroughly assessed scorpion fauna in the world (Williams 1980). (Note: Williams [1980] counted *Vaejovis janssi* from Isla Socorro in the Islas Revillagigedos as part of the Baja California fauna and its associated islands. This island is technically part of the state of Colima, and *V. janssi* is omitted from the peninsular fauna in this discussion.) Baja California seems quite extraordinary compared to adjacent areas (table 6.3), but this will prove in part to be a sampling artifact. The northern states of mainland Mexico have been the focus of limited sampling with the UV technique (the fauna of the southwestern United States, in contrast, is relatively well sampled with the UV technique, but many new species and new state records await publication).

In any case, the fauna of the Baja California peninsula is very diverse, especially considering the size of its geographical area (60 species in an area consisting of approximately 140,000 km²). Using the previously recognized 8 biogeographic provinces of the peninsula, Williams (1980) pointed out that the highest diversity was observed in the Island Province (32 species, omitting *V. janssi*) and the Volcanic Province (25 species); least diversity was seen in the Vancouveran (6 species) and Magdalena Plain provinces (10 species).

Forty-five (75%) species of scorpions are endemic to the Baja California peninsula and the islands in the Gulf of California (Williams 1980). The highest levels of endemism are observed in the Magdalena Plain (90%), Cape (87%), Volcanic (80%), and Island (76%) provinces. The Californian, Vancouveran, and Colorado Desert provinces exhibit the lowest degree of endemism (collectively, 7 of 30 species; 23%), as they share species with adjoining southern California.

Mainland Mexico

Despite our still-fragmentary knowledge of the scorpion fauna of northern mainland Mexico, some general biogeographical observations can be made. The fauna of Sonora is better known than that of the rest of the northern states, but the data derive mostly from rock-rolling. Some UV searches have been conducted in the Alamos area in the southeastern part of the state and along the coast. In particular, the mountainous interior has been poorly sampled, and this area has the greatest potential for the discovery of additional species. Most of Sonora lies within the Sonoran Desert and, consequently, shares two-thirds of its 24 known species with either Arizona or northern Baja California, or both. Several other species are shared with Sinaloa or Chihuahua, leaving only 5 species that are potential endemics (based on current distributional information).

Little is known of the fauna of Sinaloa and Durango (tables 6.2 and 6.3). For both, but especially the former, there is only a small to moderate amount of material available in museums, and much

Table 6.3. Comparison of scorpion diversities across northern Mexico and the southwestern United States.

Region	No. of Families	No. of Genera	No. of Species	No. of Species/10^5 km^2
Northern Mexico (total)	6	16	109	10.39
Baja California	4	9	32	44.68
Baja California Sur	6	12	41	56.58
Chihuahua	3	5	9	3.66
Coahuila	3	5	13	8.64
Durango	2	2	8	6.48
Nuevo León	3	3	13	19.97
Sinaloa	2	2	3	5.13
Sonora	5	7	24	13.15
Tamaulipas	4	4	10	12.56
Southwestern U.S. border states (total)	5	12	80	4.67
Arizona	4	9	35	11.85
California	4	11	49	11.92
New Mexico[a]	4	7	12	3.81
Texas	3	6	19	2.75

Numbers of species are based on existing published records, and the list does not include subspecies (see text). The data point out the great disparity in the known faunas of the states of northern Mexico.
[a]With misidentifications corrected.

of it is unstudied. For example, there are no published records for *Vaejovis* in Sinaloa, although a few Sinaloan specimens of that genus have been observed in collections (D. Sissom per obs.). As already indicated, the Sierra Madre Occidental should be rich in species of the genus *Serradigitus* and members of the *Vaejovis vorhiesi* complex. The vicinity of Durango City is home to the highly venomous species *Centruroides suffusus*. Another highly venomous species, *C. infamatus*, occurs in the southern part of the state. Eastern Durango is in the southwestern part of the Chihuahuan Desert, and some of the species known from there (*Centruroides vittatus*, *Vaejovis coahuilae* Williams, *V. crassimanus* Pocock, *V. globosus* Borelli, *V. intermedius* Borelli, and *V. waueri* [Gertsch and Soleglad]) are the same as those that occur over much of the Chihuahuan Desert to the north (i.e., southeastern Arizona, southern New Mexico, and Trans-Pecos Texas). Undoubtedly, other species common in the northern Chihuahuan Desert, such as *Paruroctonus gracilior* (Hoffmann) and *V. russelli* Williams, should also occur there. The apparent disjunction of these Durango populations is clearly not real, but rather is the result of our poor knowledge of the scorpion fauna of the states of Chihuahua and Coahuila.

In the central part of northern Mexico, occupied by the expansive Chihuahuan Desert and its associated nondesert sky islands, there is a strong tendency for species of the lowland scrub desert areas to be widely distributed. These species include *C. vittatus*, *V. coahuilae*, *V. crassimanus*, and *P. gracilior* (*C. vittatus*, an ecologically plastic species, is usually more abundant in rocky outcroppings within this area). Another common species of rocky, arid slopes is the small *Vaejovis waueri*. Possible endemics occur in biogeographic islands within the Chihuahuan Desert Region—for example, *Pseudouroctonus chicano* (Gertsch and Soleglad) and *V. vaquero* in Chihuahua and *V. minckleyi* Williams and *S. calidus* Soleglad in the Cuatrociénegas area (Williams 1968a; Gertsch and Soleglad 1972; Soleglad 1974). *Paruroctonus coahuilanus* Haradon is also currently known only from the Cuatrociénegas area (Haradon 1985), an area of outstanding interest that has been investigated for many years. A complete assessment of the scorpiofauna of Cuatrociénegas is unfinished, although it was begun by Williams (1968a); this author made significant collections at the site, but only published on the vaejovids. Once a faunal inventory is completed for Cuatrociénegas, this area will undoubtedly prove to harbor very high local diversity.

The easternmost part of northern Mexico lies in the coastal plain of the Gulf of Mexico. This plain meets the Sierra Madre Oriental in southern Nuevo León and western Tamaulipas. The coastal plain appears to be quite depauperate in terms of species diversity, with only *Centruroides vittatus*, *Diplocentrus diablo* Stockwell and Nillson, and *Vaejovis bilineatus* Pocock (the eastern counterpart of *V. waueri*) currently known (in the adjoining Rio Grande Valley in Texas, *V. crassimanus* and *V. waueri* are also found). From the foothills of the Sierra Madre Oriental westward, however, several endemic forms are encountered: *Vaejovis rossmani* Sissom, *V. sprousei* Sissom, *Diplocentrus colwelli* Sissom, *D. ferrugineus* Fritts and Sissom, and *V. rubrimanus* Sissom. The large, black *Centruroides gracilis*, a common species that ranges into Central America (and introduced to Venezuela) also gets into extreme southern Tamaulipas, as does the small, mottled *C. rileyi*.

Patterns of Local Diversity

Another way to consider diversity is to examine the number of species in small localized areas (i.e., the number of sympatric species that might occupy 1 km^2). Typically, small localized areas in the Sonoran and Chihuahuan deserts harbor between 4 and 8 species, but some localities have fewer or more. For example, on the Baja California peninsula, the vicinity of El Faro in the extreme northeast has 5 species; the Santo Tomas area, 7; San Borja area, 7; Bahía de los Angeles area, 9; La Paz area, 8; Todos Santos area, 6; and the Cabo San Lucas area, 9 (Polis 1990). In the vicinity of Puerto Escondido/ Loreto there is extraordinary local diversity (by scorpion standards), with 13 sympatric species. In contrast, some locations in the Sierra de San Pedro Mártir have as few as 3 species (Polis 1990).

The islands in the Gulf of California commonly exhibit high diversity. One of the most remarkable sites is Isla Danzante just off the coast from Puerto Escondido; this small island, only 4.49 km^2 in size, has been documented to harbor 10 species (Williams 1980; Due 1992). Some of the other (larger) islands also have high diversity: Isla San Jose, 12 species; Isla Partida Sur, 10 species; Isla Carmen, 10 species; and Isla San Marcos, 10 species. Some of the largest islands (e.g., Isla Angel de la Guarda, Isla Cerralvo, Isla Tiburón) have fewer species—this may reflect inadequate sampling, but in the case of Angel de la Guarda, the degree of isolation may also be a contributing factor.

On the mainland, local diversity is poorly understood, with only a few tentative examples worth presenting. In Sonora, there are 3 reasonably well-sampled sites: the areas around Puerto Peñasco in the north and Alamos in the southeast are known to have 6 species, and the vicinity of Guaymas has 7 species (Williams and Hadley 1967; Haradon 1984; Sissom 1991; Sissom and Stockwell 1991; Sissom and Wheeler 1995). The Cuatrociénegas Basin in Coahuila has at least 10 species and, based on knowledge of the surrounding areas, undoubtedly more will be documented (Williams 1968a; Soleglad 1974; Haradon 1985). Comparable regions in southwest Texas may give additional insight: the Sauceda Ranch Headquarters of Big Bend Ranch State Park has 9 species, and in Big Bend National Park, the Chisos Mountains and Basin and Rio Grande Village each have 9.

Some of the sites listed above exhibit the highest known local diversities in the world, with the record apparently being 13 in the Loreto area in Baja California Sur (Polis 1990). Unfortunately, little has been published on local scorpion diversity for the rest of the world, and to generate estimates would require scouring the records sections of numerous taxonomic descriptions, revisions, and regional studies. Polis (1990), however, reported that some tropical localities (e.g., in Trinidad, Venezuela, Brazil, French Guiana, and Costa Rica) have between 3 and 7 sympatric species.

Scorpions as Subjects of Conservation Biology

Scorpions are certainly not the "warm fuzzy" animals that the average person finds so appealing, and it is doubtful that the lay community will ever be much concerned about their conservation. Despite this, some scorpion species may eventually be threatened with extinction due to habitat destruction. Of particular concern would be the troglobites; these species are known only from single caves or cave systems and are typically rare. Cave environments can suffer ill effects from pollution and development, and species might be lost in localized situations. General habitat destruction, such as the clearing of thornscrub and tropical deciduous forest habitats, probably contributes to the demise of certain scorpion species, but may promote others. Lourenço and Cloudsley-Thompson (1996) demonstrated that habitat destruction in eastern Brazil associated with urbanization led to a marked decrease in abundance of the buthid *Tityus stigmurus*, but a dramatic increase in *T. serrulatus* (a dangerous, parthenogenetic species). Given the ecological plasticity of *Centruroides* spp. and the highly venomous nature of some, habitat destruction might warrant concern in parts of Mexico.

A large problem in assessing the need for conservation is the paucity of ecological data available for scorpions. Only a few common species have been studied extensively enough to understand their population dynamics. As mentioned above, an interesting fact of scorpion life-history is that these animals tend toward K-selected characteristics (Polis and Sissom 1990). There is, of course, a gradient in these life-history characteristics, with the members of the Scorpionoidea (Diplocentridae, Scorpionidae, Ischnuridae, etc.) exhibiting the strongest K-selected traits. *Hadrurus* (Iuridae), with most of its species in northern Mexico and the southwestern United States, also appears to fit in this category. A potential concern with these species is their appearance in the pet trade. Almost all, if not all, of the scorpions sold in the pet trade are wild-caught specimens. Currently, the more exotic species (e.g., the African *Pandinus* Thorell and *Hadogenes* Kraepelin; the Asian *Heterometrus* Ehrenberg) are the most popular, but *Hadrurus* is not uncommon in pet stores. At this time, we have no data to indicate whether collecting for the pet trade has any impact on the populations of scorpions, but the volume of trade for some species is significant. Approximately 105,000 *Pandinus imperator* (Koch) were exported in 2 years time (1995–1996) from 3 small countries in western Africa (M. Haywood, pers. comm.), indicating the magnitude of trade in this particular species. Although it seems doubtful that levels of trade in *Hadrurus* will reach that of *Pandinus*, one should remember that trade in the Mexican *Brachypelma* tarantulas was judged significant enough to warrant listing of these spiders on Appendix II of CITES (containing species whose trade is to be rigorously monitored).

Scorpions also have potential use as tools for identifying areas of endemism to be considered for conservation efforts. They can and should be added to the growing number of taxa suitable for this approach to conservation biology. Scorpions exhibit a number of features that would make them useful in this regard. First, although some species are easily and widely transported by humans (e.g., *Isometrus maculatus* [De Geer], certain species of *Centruroides*), many species typically exhibit low vagility and often have restricted distributions. Second, scorpions in most habitats are readily surveyed by the ultraviolet detection technique, which renders sampling and monitoring relatively easy and inexpensive. This would be especially true of the desert areas of northern Mexico.

Based on more than a decade of research by W. R. Lourenço in tropical South America (Lourenço 2001), scorpions exhibit geographical patterns of endemism comparable to woody plants, birds, butterflies, and other organisms. The knowledge of scorpion endemism, combined with similar data for other taxa, has contributed to the recognition of 25 refugia in the Neotropics. In fact, endemism of scorpions and other indicator species has played a role in proposing several conservation areas in Guyana, French Guiana, and the tropical Andes (Lourenço 2001). Whether scorpions realize their potential to contribute to conservation efforts in northern Mexico will most likely depend on the interest and efforts of future researchers.

Conclusions

In conclusion, the scorpion fauna of Mexico is one of the richest in the world, if not the richest. It currently has more families, genera, and species of scorpions than any other country. The fauna of the Baja California peninsula and its associated islands is very well known, but additional species should be collected there in the future. In comparison, the fauna of the mainland is very poorly known, and a more complete assessment of its species diversity should be treated as a priority. In this regard, there should be 3 specific objectives: (1) generating a more accurate species list; (2) mapping out the distributions of all species; and (3) obtaining information on population biology and life-history patterns. As these goals are reached, it will be possible to determine centers of endemism and which, if any, species may potentially be threatened with extinction. Scorpions can play a significant role in shaping conservation strategies, as has already been shown in Amazonia. The low vagility and habitat specialization exhibited by many species may prove useful in helping to identify general areas of endemism.

Acknowledgments

We are grateful to Jean-Luc Cartron, Gerardo Ceballos, and Richard Felger for the opportunity to contribute this chapter to the book, and especially to Dr. Cartron for his encouragement and patience throughout the project. We also thank Dr. Victor Fet of Marshall University for providing a review of the manuscript.

Literature Cited

Beutelspacher, C. R. 2000. Catálogo de los alacranes de México. Universidad Michoacana de San Nicolás de Hidalgo, Morelia, Michoacán, Mexico.

Capes, E. M. 2001. Description of a new species in the *nitidulus* group of the genus *Vaejovis* (Scorpiones, Vaejovidae). Journal of Arachnology 29: 42–46.

de Armas, L. F., and E. M. Frias. 2000 (1999). Presencia del alacran *Centruroides vittatus* (Say) (Scorpiones, Buthidae) en el estado de Durango, México. Revista Nicaraguense de Entomología 47: 11–13.

Díaz Najera, A. 1964. Alacranes de la Republica Mexicana. Identificación de ejemplares capturados en 235 localidades. Revista del Instituto de Salubridad y Enfermedades Tropicales 24: 15–30.

Díaz Najera, A. 1975. Listas y datos de distribución geográfica de los alacranes de México. Revista del Instituto de Salubridad y Enfermedades Tropicales 35: 1–36.

Due, A. D. 1992. Biogeography of scorpions in Baja California, Mexico and an analysis of the insular scorpion fauna in the Gulf of California. Ph.D. dissertation, Vanderbilt University, Nashville, Tennessee.

Fet, V., W. D. Sissom, G. Lowe, and M. Braunwalder. 2000. Catalog of the Scorpions of the World (1758–1998). New York Entomological Society, New York.

Francke, O. F. 1982. Parturition in scorpions (Arachnida, Scorpiones): a review of the ideas. Revue Arachnologique 4: 27–37.

Gertsch, W. J., and M. E. Soleglad. 1972. Studies of North American scorpions of the genera *Uroctonus* and *Vejovis*. Bulletin of the American Museum of Natural History 148(4): 549–608.

Haradon, R. M. 1984. New and redefined species belonging to the *Paruroctonus borregoensis* group (Scorpiones, Vaejovidae). Journal of Arachnology 12: 317–339.

Haradon, R. M. 1985. New groups and species belonging to the nominate subgenus *Paruroctonus* (Scorpiones, Vaejovidae). Journal of Arachnology 13: 19–42.

Hendrixson, B. E. 2001. A new species of *Vaejovis* (Scorpiones, Vaejovidae) from Sonora, Mexico. Journal of Arachnology 29: 47–55.

Hoffmann, C. C. 1931. Monografias para la entomología médica de México. Monografía no. 2, Los escorpiones de México. Primera parte: Diplocentridae, Chactidae, Vejovidae. Annales del Instituto de Biología, Universidad Nacional Autónoma de México 2(4): 291–408.

Hoffmann, C. C. 1932. Monografias para la entomología médica de México. Monografía no. 2, Los escorpiones de México. Segunda parte: Buthidae. Annales del Instituto de Biología, Universidad Nacional Autónoma de México 3(3): 243–282; (4): 283–361.

Hoffmann, C. C. 1938. Nuevas consideraciones acerca de los alacranes de México. Annales del Instituto de Biología, Universidad Nacional Autónoma de México 9: 318–337.

Levy, G., and P. Amitai. 1980. Fauna Palaestina. Arachnida I. Scorpiones. Israel Academy of Sciences and Humanities, Jerusalem.

Lourenço, W. R. 1994. Diversity and endemism in tropical versus temperate scorpion communities. Biogeographica 70(3): 155–160.

Lourenço, W. R. 2001. Scorpion diversity in tropical South America: implication for conservation programs. Pp. 406–415 *in* P. Brownell and G. Polis (eds.), Scorpion Biology and Research. Oxford University Press, Oxford.

Lourenço, W. R., and J. L. Cloudsley-Thompson. 1996. Effects of human activities on the environment and on the distribution of dangerous species of scorpions. Pp. 49–60 *in* C. Bon and M. Goyffon (eds.), Envenomings and Their Treatments. Edition Fondation Marcel Merieux, Lyon, France.

Lourenço, W. R., and W. D. Sissom. 2000. Scorpiones. Pp. 115–135 *in* J. L. Bousquets, E. González Soriano, and N. Papavero (eds.), Biodiversidad, Taxonomía y Biogeographía de Artrópodos de México: Hacia una Síntesis de su Conocimiento. Vol. II. Universidad Nacional Autónoma de México, México D.F.

Mitchell, R. W. 1968. *Typhlochactas*, a new genus of eyeless cave scorpion from Mexico (Scorpionida, Chactidae). Annales de Spéléologie 23(4): 753–777.

Pocock, R. I. 1898. The scorpions of the genus *Vaejovis* contained in the collection of the British Museum. Annual Magazine of Natural History (ser. 7) 1: 394–400.

Pocock, R. I. 1902. Arachnida, Scorpiones, Pedipalpi and Solifugae. Biología Centrali-Americana. Taylor and Francis, London.

Polis, G. A. 1990. Ecology. Pp. 247–293 *in* G. A. Polis (ed.), The Biology of Scorpions. Stanford University Press, Palo Alto, California.

Polis, G. A., and R. D. Farley. 1980. Population biology of a desert scorpion: survivorship, microhabitat, and the evolution of life history strategy. Ecology 61: 620–629.

Polis, G. A., and W. D. Sissom. 1990. Life history. Pp. 161–223 *in* G. A. Polis (ed.), The Biology of Scorpions. Stanford University Press, Stanford, California.

Prendini, L. 2000. Phylogeny and classification of the superfamily Scorpionoidea Latreille 1802 (Chelicerata, Scorpiones): an exemplar approach. Cladistics 16: 1–78.

Prendini, L. 2001. A review of synonyms and subspecies in the genus *Opistophthalmus* C. L. Koch (Scorpiones: Scorpionidae). African Entomology 9: 1–48.

Shelley, R. M., and W. D. Sissom. 1995. Distributions of the scorpions *Centruroides vittatus* (Say) and *C. hentzi* (Banks) in the United States and Mexico. Journal of Arachnology 23: 100–110.

Simard, J. M., and D. D. Watt. 1990. Venoms and Toxins. Pp. 414–444 *in* G. A. Polis (ed.), The Biology of Scorpions. Stanford University Press, Stanford, California.

Sissom, W. D. 1985. Systematics of the *nitidulus* group of the genus *Vaejovis*, with comments on phylogenetic relationships within the family Vaejovidae (Arachnida: Scorpiones). Ph. D. dissertation, Vanderbilt University, Nashville.

Sissom, W. D. 1991. The genus *Vaejovis* in Sonora, Mexico (Scorpiones, Vaejovidae). Insecta Mundi 5: 215–225.

Sissom, W. D. 2000. Family Vaejovidae. Pp. 503–553 *in* V. Fet, W. D. Sissom, G. Lowe, and M. E. Braunwalder (eds.), Catalog of the Scorpions of the World (1758–1998). New York Entomological Society, New York.

Sissom, W. D., and R. N. Henson. 1998. A new species of *Paruroctonus* (Scorpiones, Vaejovidae) from Big Bend National Park, Texas. Entomological News 109: 240–246.

Sissom, W. D., and S. A. Stockwell. 1991. The genus *Serradigitus* in Sonora Mexico (Scorpiones, Vaejovidae), with descriptions of four new species. Insecta Mundi 5: 197–214.

Sissom, W. D., and A. L. Wheeler. 1995. Scorpions of the genus *Diplocentrus* (Diplocentridae) from Sonora, Mexico, with description of a new species. Insecta Mundi 9(3/4): 309–316.

Soleglad, M. E. 1974. *Vejovis calidus*, a new species of scorpion from Coahuila, Mexico (Scorpionida, Vejovidae). Entomological News 85: 108–115.

Stockwell, S. A. 1992. Systematic observations on North American Scorpionida with a key and checklist of the families and genera. Journal of Medical Entomology 29: 407–422.

Williams, S. C. 1968a. Scorpions from Northern Mexico: Five new species of *Vejovis* from Coahuila, Mexico. Occasional Papers of the California Academy of Sciences 68: 1–24.

Williams, S. C. 1968b. Methods of sampling scorpion populations. Proceedings of the California Academy of Sciences (ser. 4) 36: 221–230.

Williams, S. C. 1980. Scorpions of Baja California, Mexico and adjacent islands. Occasional Papers of the California Academy of Sciences 135: 1–127.

Williams, S. C., and N. F. Hadley. 1967. Scorpions from the Puerto Peñasco area (Cholla Bay), Sonora, Mexico with a description of *Vaejovis baergi* n.sp. Proceedings of the California Academy of Sciences 35(4): 103–116.

Yahia, N., and W. D. Sissom. 1996. Studies on the systematics and distribution of the scorpion *Vaejovis bilineatus* Pocock (Vaejovidae). Journal of Arachnology 24: 81–88.

7

Fishes of the Continental Waters of Tamaulipas: Diversity and Conservation Status

FRANCISCO J. GARCÍA DE LEÓN

DELLADIRA GUTIÉRREZ TIRADO

DEAN A. HENDRICKSON

HÉCTOR ESPINOSA PÉREZ

With an origin dating back 400 million years, fishes represent the most ancient group of vertebrates (Helfman et al. 1997). They are also the most diverse, with more than 25,000 species. Of the more than 2200 species known from Mexico, about 500 live in freshwater. The Mexican Official Norm NOM-059-ECOL-2001 lists only 186 fish species among the 1515 vertebrates "At Risk" in Mexico (SEMARNAT 2002; see chapter 4). Fishes thus account for only 12.3% of all listed species in Mexico, compared to 30.8% for reptiles, 24.8% birds, and 19.5% mammals.

Why are so few fishes listed in Mexico? The answer probably has little to do with actual conservation status and more to do with other factors. First, the great taxonomic diversity of fishes renders any comprehensive evaluation of their conservation status quite daunting. Not only are fishes more than half of all vertebrate species, but new species continue to be described every year (Helfman et al. 1997). Because fishes live only in water, they are more difficult to observe than are most other vertebrates. Finally, fishes show a high degree of intraspecific phenotypic variation that makes them highly sensitive to environmental factors and often difficult to identify (Allendorf et al. 1987; Allendorf 1988).

Fishes are important to humans because they represent an important source of food. Their commercial and recreational value has led to fish farming on an industrial scale, both for easy exploitation and as a means to recover overharvested natural populations. Scientific interest in fishes is also considerable. Those species easy to manage in captivity can be used in laboratory experiments. Additionally, freshwater fishes in particular can be used as biogeographic indicators, contributing important information to our understanding of the history of river basins and serving as indicators of aquatic ecosystem health. Though their aquatic habitats perhaps make wild fish populations more difficult to study than terrestrial organisms, they clearly deserve greater emphasis in the field of biological conservation.

The northern part of Mexico harbors 3 aquatic ecoregions known as the Sonoran, Chihuahuan-Potosian, and Tamaulipan regions (Contreras-Balderas 1969). The Tamaulipan ecoregion is located between the Sierra Madre Oriental and the Gulf of Mexico, within the Mexican states of Coahuila, Nuevo León, and Tamaulipas (CONABIO 2000); the last of these states is the focus of this chapter. To the north, Tamaulipas is bounded by the Rio Grande (Río Bravo), which marks the border with Texas. To the west, Tamaulipas is bounded by the Mexican states of Nuevo León and San Luis Potosí, to the east by the Gulf of Mexico, and to the south by the states of Veracruz and San Luis Potosí (fig. 7.1).

The geomorphology of watersheds influences species richness (Eadie et al. 1986). For example, river discharge is a direct measure of availability of habitat for freshwater fishes (Livingstone et al.

"

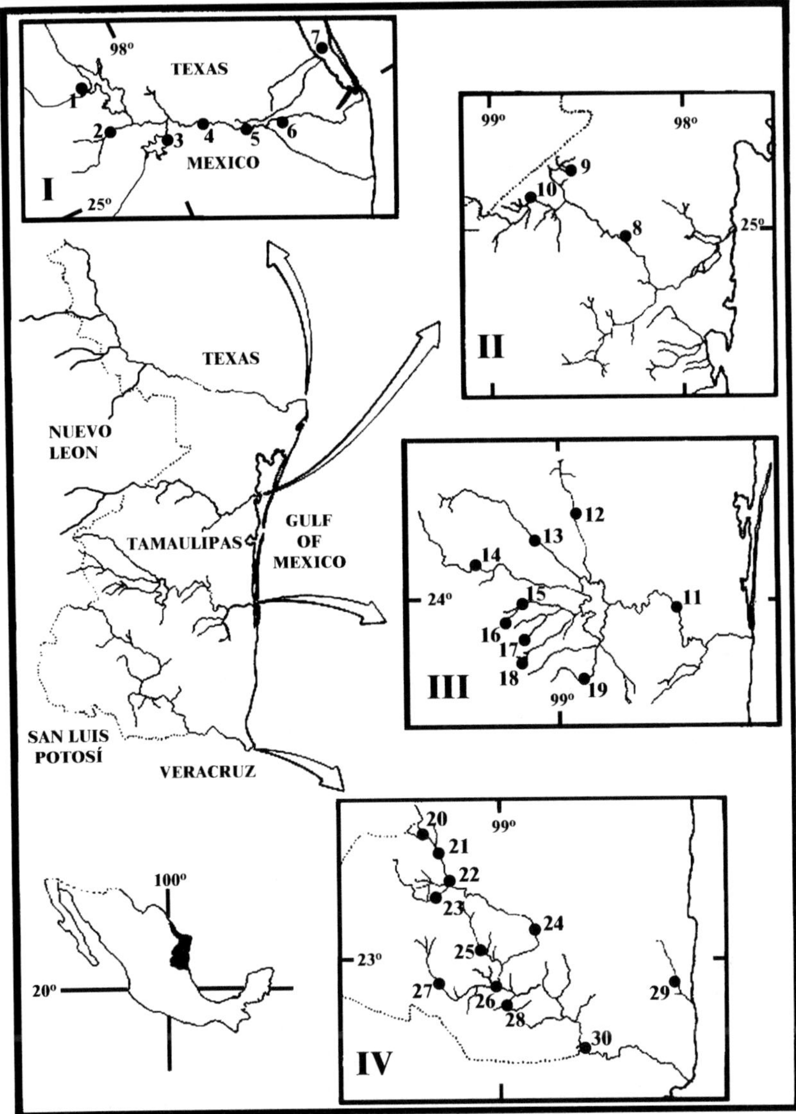

Figure 7.1. Hydrologic systems of Tamaulipas: *I*, Lower Río Bravo (Rio Grande); *II*, Río San Fernando; *III*, Río Soto la Marina; *IV*, Río Guayalejo-Tamesí. Major hydrologic features: *1*, mouth of Río Salado at Falcón Reservoir; *2*, Río Álamo; *3*, Marte R. Gómez Reservoir on Río San Juan; *4*, Río Bravo (Rio Grande); *5*, Anzalduas Dam; *6*, Retamal Dam; *7*, Laguna Madre; *8*, Río San Fernando; *9*, Río Potosí; *10*, Río Linares; *11*, Río Soto la Marina; *12*, Río San Carlos; *13*, Río Pilón; *14*, Río Purificación; *15*, Río Corona; *16*, Río San Felipe; *17*, Río Santa Ana; *18*, Río San Marcos; *19*, Arroyo Grande; *20*, Río Chiue; *21*, Arroyo los Ángeles; *22*, Río Nogales; *23*, Río San Vicente; *24*, Río Jaumave; *25*, Río Guayalejo; *26*, Río Sabinas; *27*, Río Frío; *28*, Río Las Flores; *29*, Río Mante; *30*, Río Tigre; *31*, Río Tamesí.

1982), and there is a positive correlation between species richness and surface area of a river basin (Horwitz 1978). Thus, anthropogenic alterations of a watershed can drastically reduce its associated biological diversity (Sheldon 1987). Decreases in total habitat area and habitat fragmentation (typically a result of dams) occur conjunctively, reducing not only the size of many populations but also the potential for dispersal and genetic flow (Frankham et al. 2001). Additionally, diversion canals linking once separate waterways, and the introduction of exotic species, are both leading to homogenization of aquatic faunas (Sheldon 1988).

Due to the rapid increase in human populations, northern Mexico has experienced major, human-induced alterations and fragmentation of its watersheds and associated changes in the distribution of taxa and loss of biodiversity (Contreras-Balderas 1978). Clearly, any conservation effort requires an inventory of the ichthyofauna using a taxonomic and biogeographic approach, focusing on documenting and maintaining overall biodiversity, but also including the rare and endangered species.

The specific objectives of the study described in this chapter were to evaluate the diversity of freshwater fishes in Tamaulipas, to characterize each watershed and analyze the status of its ichthyofauna, and to determine the level of anthropogenic impact on freshwater fish communities statewide. We begin with a description of the watersheds of Tamaulipas, then present a synthesis of the state of knowledge of taxonomy, biology, genetics, evolution, exploitation, and conservation of fishes in Tamaulipas and provide the first list of freshwater fishes assembled for the state.

Tamaulipas River Basins

Tamaulipas is drained by 4 major river systems (fig. 7.1). From north to south, these are: (1) the lower Rio Grande and its tributaries, the Ríos Álamo and San Juan; (2) the Río San Fernando and its major tributary the Río Conchos, (3) the Río Soto la Marina, and (4) the Río Pánuco, with its major tributaries the Ríos Guayalejo and Tamesí (INEGI 1983). Each of these receive water from outside of the state, but at least in the case of the Ríos San Fernando and Soto la Marina, by far the majority of the drainage is contained within Tamaulipas. All drainages have been modified by humans to varying degrees. Besides the 3 major dams of the state,

Falcon (Falcón), Marte R. Gómez, and Vicente Guerrero, there are 13 reservoirs of lesser importance distributed among the state's river systems.

Rio Grande (Río Bravo)

On its way to the Gulf of Mexico, the Rio Grande crosses the largest desert of North America, the Chihuahuan Desert (Miller 1978). Based on fish distributions, Smith and Miller (1986) divided the Rio Grande watershed into the upper Rio Grande (mainly in New Mexico and Colorado), the Pecos River, the Río Conchos (in Chihuahua), Mexican interior basins, and the lower Rio Grande. The lower reach of the Rio Grande stretches from about Del Rio, Texas, and Ciudad Acuña, Coahuila, to the river's delta, crossing a slightly undulating plain where local relief seldom exceeds 100 m above sea level (Smith and Miller 1986). Four main tributaries discharge into the lower Rio Grande, 3 from Mexico (the Ríos Salado, Álamo, and San Juan), and 1 from the United States (Arroyo Los Olmos) (Edwards and Contreras-Balderas 1991). Four dams have been built along this reach: Falcon Dam, fed by the Rio Grande and Río Salado, and situated above the mouth of the Río Álamo; Marte R. Gómez Dam on the Río San Juan; and, farther downstream, Anzalduas and Retamal Dams, both fed by the Rio Grande (fig. 7.1). In this chapter, only the lower Rio Grande and the Ríos Álamo and San Juan are considered.

Río San Fernando

The headwaters of the Río San Fernando are in the state of Nuevo León where countless canyons and small valleys collect atmospheric moisture intercepted by the Sierra Madre Oriental. The uppermost parts of the basin drain the area around Galeana, Nuevo León, through the Ríos Potosí (Cabezones), Linares, Conchos, San Lorenzo, and Cruillas. As the Río Conchos enters the state of Tamaulipas it becomes known as the Río San Fernando, which continues into the Gulf of Mexico through the Laguna Madre (fig. 7.1). In Tamaulipas the Río San Fernando Basin falls within an area framed by the coordinates 24°39'–25°39' N and 97°48'–99°00' W (INEGI 1983). The whole watershed has an area of 15,600 km².

Soto La Marina River System

The Río Soto La Marina constitutes the state's largest watershed, lying within an area delineated by the coordinates 23°13'50"–24°41'59" N and 97°45'00"–98°39'50" W, with a total area of 22,600 km². It originates in the state of Nuevo León, where it is called Río Blanco; downstream it becomes known as the Río de la Cruz, and upon entering Tamaulipas becomes the Río Purificación (INEGI 1983). The complex Río Soto la Marina system is formed by 8 tributaries, which are from north to south: Ríos San Carlos, Pilón, Purificación, Corona, Santa Ana, San Felipe, San Marcos, and Arroyo Grande (INEGI 1998). These streams all end at Vicente Guerrero Dam, below which the name Río Soto La Marina is applied to the river, which then discharges into the Gulf of Mexico in the southern part of the Laguna Madre (INEGI 1997, 1998). Vicente Guerrero Dam, the system's main impoundment, has greatly modified the flow of water and fragmented the watershed, thereby changing the distribution of freshwater fishes (García de León et al. 2002).

Guayalejo-Tamesí River System

The port city of Tampico receives water from the south via the Río Pánuco and from the north via the Río Tamesí, the 2 rivers joining shortly before entering the Gulf of Mexico (22°12'34" N; 97°52'00" W; INEGI 1983). Several tributaries of the Río Pánuco have headwaters in the highlands of the Central Valley of Mexico, crossing through impressive canyons in the Sierra Madre Oriental, often with spectacular waterfalls. Lying outside of Tamaulipas, these are not discussed here; however, we do discuss both the Río Tamesí and its northernmost major tributary, the Río Guayalejo (Darnell 1962), which together cover 15,200 km² of the state of Tamaulipas. The Río Guayalejo has headwaters north of Miquihuana at an elevation of 3400 m, where it is known as the Río Alamar, then the Río Chihue after receiving the waters of the Río Maravillas. Farther downstream its name changes to Río Los Angeles, Río Nogales, and Río San Vicente, until it joins the Río Jaumave (23°44'03" N; 99°16'06" W), after which the name Río Guayalejo is finally applied. During its southward journey, the river receives the waters of other tributaries, which from north to south are the Ríos Sabinas, Frio, and Mante. The Río Sabinas is fed by the Río Forlón and the Río El Encino; the Río Frio by the Río Boquillas and the Río Mante. These tributaries all flow from large springs that discharge waters originating in the highlands of the adjacent Sierra Guatemala. They are intermittent streams at first, but have a more regular flow of water below the springs and across the plains. Since the early 1950s, the Río Mante has been extensively modified for irrigation.

At the southern end of the watershed, the Río Tamesí serves as a political boundary between the states of Tamaulipas and Veracruz. As it heads toward the coast, with low velocity but substantial discharge, it passes through swamps and lagoons. After crossing Laguna de Chairel it joins the Río Pánuco, and the 2 rivers flow as 1 into the Gulf of Mexico (Darnell 1962). This system has 2 very important artificial reservoirs, Xicotencatl and Ramiro Caballero Dams.

History of Research on the Freshwater Fishes of Tamaulipas

There is a substantial body of literature focusing directly or indirectly on the ichthyofauna of Tamaulipas. Much of this literature, however, consists of simple mentions of Tamaulipan species in more broadly focused works, such as North American fish catalogues and monographs or isolated species descriptions. Another large portion of this literature, unfortunately, corresponds to unpublished documents and dissertations.

The first scientific expeditions along the U.S.–Mexico border date back to the 1800s (Girard 1859), and one of the researchers associated with that effort was the first to write about fishes of Tamaulipas (Jordan 1885). By the end of the nineteenth century and beginning of the twentieth century, much research was being conducted on the fishes of North and Central America by Jordan and Evermann (1896–1900), Jordan and Snyder (1899), Meek (1904), Regan (1906–1908), and Jordan and Dickerson (1908). Several decades later, Darnell (1962) initiated the first research effort focusing on the Río Tamesí, and Miller (1966) focused on the geographical distribution of fishes from Central America, including species and groups whose range extended to Tamaulipas, providing useful data regarding zoogeography and distribution at the family level. Miller (1976) provided a taxonomic update and a valuable analysis of Meek's (1904) book on the fishes of Mexico. Miller and Smith (1986) described the origin and geographic distribution of Mexican fishes. Papers by Hubbs et al. (1991), Lee et al. (1980), and McEachran and Fechhelm (1998) cover an area broader than Tamaulipas, but are nonetheless useful references for the state.

Other studies have focused on select taxa or groups represented in Tamaulipas. Research on Cyprinidae includes systematic studies on *Notropis* and *Cyprinella* (Hubbs and Hubbs 1958; Hubbs and Miller 1975, 1978; Chernoff et al. 1982; Mayden 1989) and *Dionda* (Hubbs and Miller 1974, 1977; Mayden et al. 1992). Hubbs (1926) conducted a thorough review of the order Cyprinodontiformes. Within this order, the biogeography and systematics of the Poeciliidae have been studied by Rosen and Bailey (1963); their genetics by Hubbs and Gordon (1943), Gordon (1947), Gordon and Gordon (1957), and Schartl (1995); the genus *Gambusia* (Family Poeciliidae) by Rivas (1963), Miller and Minckley (1970), and Rauchenberger (1989); *Poecilia formosa* and *P. mexicana* by Darnell and Abramoff (1968), Menzel and Darnell (1973), and Miller (1983); and the genus *Xiphophorus* by Rauchenberger et al. (1990). Systematics of the cichlids was researched by Regan (1905) and Taylor and Miller (1983).

There are also excellent contributions from Mexican authors who focused on the study of Mexican fishes in general, providing important information on the distribution of Tamaulipan species. Álvarez del Villar (1950) produced a bibliography on freshwater ichthyology in Mexico. De Buen (1946) conducted an interesting analysis of fish biogeography, and in 1947 published a catalog of fishes from the Nearctic region of Mexico. Contreras-Balderas (1962) began research on the fishes of northeastern Mexico, and later on, with the help of his students over more than 20 years, explored the rivers of Coahuila, Nuevo León, and Tamaulipas. At least for Tamaulipas, results of this research are available mostly in the form of unpublished theses (Rivera-Teillery 1971; Verduzco-Martínez 1972; Rodríguez-Olmo 1975; Ruíz-Campos 1982; Villarreal-Treviño 1983; Gómez-Soto 1988). Finally, Espinosa-Pérez et al. (1993) published a list of freshwater fishes from Mexico, and Castro-Aguirre (1978) and Castro-Aguirre et al. (1999) published taxonomic keys for estuarine species with distribution notes.

Some papers deal specifically with the fishes of Tamaulipas. Treviño-Robinson (1959), Smith and Miller (1986), and Edwards and Contreras-Balderas (1991) reported on the lower Rio Grande. Rivera-Teillery (1971), and more recently Smith and Miller (1986), published on fishes of the San Fernando and Soto la Marina watersheds; and Darnell (1962) and Verduzco-Martínez (1972) reported on fishes of the Guayalejo-Tamesí system. All of these papers represent basic references on the diversity and biogeography of Tamaulipan fishes. It must be noted that studies of fishes of the Ríos San Fernando and Soto la Marina are scarce, and those that exist have focused almost exclusively on the lowermost reaches of these rivers. Almost all of the information available on the fishes of Tamaulipas so far (except for the lower Rio Grande) has been on distribution and taxonomy, with very little work on ecology.

For this chapter we obtained fish distribution information from the literature described above, from scientific collections that include fishes from Tamaulipas, from personal communications, and from 111 sampling efforts carried out by staff of the Integrative Biology Laboratory of the Technological Institute of Ciudad Victoria (ITCV). These samples were taken between February 1999 and July 2001 at 86 locations along the Ríos San Fernando, Soto La Marina, and Guayalejo-Tamesí, including previously unsampled upper reaches of some of these rivers. We did not sample the lower Rio Grande, and all data presented here for that system are compiled from Smith and Miller (1986) and Edwards and Contreras-Balderas (1991).

Fish Biodiversity

Table 7.1 provides the first published list of the freshwater fishes of Tamaulipas and forms the basis for the following discussion of the state's ichthyofauna. A total of 198 freshwater fishes representing 51 families and 23 orders are known from Tamaulipas. Using Myers' (1938) classification, 56 of these are primary freshwater fishes: they have a long evolutionary history in freshwater and physiological inability to survive in the sea. Of the others, 31 are secondary (generally restricted to freshwater but occasionally entering seawater), and the remaining 111 are euryhaline (with wide salinity tolerance). Some euryhaline fishes are freshwater species capable of living in coastal waters, but most are marine species that can penetrate epicontinental waters (Castro-Aguirre et al. 1999).

There are 5 species endemic to Tamaulipas. Three are found in the Guayalejo-Tamesí system: *Prietella lundbergi*, "*Cichlasoma*" (*Herichthys*) *pantostictus* (we agree with Miller [in press] that until revisionary studies produce a stable classification, *Cichlasoma* should be used in quotes to indicate its uncertain, but probably non-monophyletic status, followed by the probable future genus name in parentheses), and *Poecilia latipunctata*. One is found in the Soto la Marina river system (*Xiphophorus xiphidium*), and 1 is previously known only from the Soto la Marina system, but we also collected it in the Río San Fernando (*Notropis aguirrepequenoi*). Three Tamaulipas species, all of them from the lower Rio Grande, have been extirpated from the state or are now extinct: *Acipenser oxyrinchus* (an anadromous species), *Hybognathus amarus*, and *Notropis orca*.

The 18 species that have been introduced in Tamaulipas are mostly natives of Africa, Asia, and the United States. There are 10 introduced species in the lower Rio Grande, 4 in the Río San Fernando watershed, 8 in the Soto la Marina watershed, and 6 in the Guayalejo-Tamesí river system. In addition, 5 species native to the state (*Micropterus salmoides*, *Ameiurus melas*, *Ictalurus punctatus*, *I. furcatus*, and *Atractosteus spatula*) have been introduced into river systems where they were previously not documented to occur. In some cases, the introduction involved transplantations of populations native to the state, but in other cases a non-native species or stocks from non-native population(s) were introduced. In the Soto La Marina system, both *M. salmoides* (a native taxon, but stocks from outside of Tamaulipas have been introduced) and *M. floridanus* (native to Florida; note that we follow Kassler et al. [2002] in referring to this taxon, often before referred to *M. salmoides floridianus*, as a full species) were introduced, as were non-native U.S. (or domesticated aquaculture varieties of) populations of the wide-ranging *Ictalurus furcatus*.

All rivers of Tamaulipas comply with global patterns of progressively changing species assemblages as one moves downstream from headwaters to the mouth, where marine and euryhaline taxa become dominant, and in having species diversity increase generally downstream. Tamaulipan rivers, however, differ among one another, with each tending to have its own, fairly distinctive ichthyofauna, with the exception of the San Fernando and Soto la Marina systems, which are quite similar. Excluding introduced species, Jaccard's similarity index values (de la Cruz Agüero 1994) among the fish faunas of the state's 4 major basins are no greater than 0.45 (index values range from 0 to 1). Similarity of the fish fauna of the Río Soto La Marina to faunas of both the Guayalejo-Tamesí system and the lower Río Bravo is comparable (0.35 and 0.41, respectively), but the Río San Fernando fauna is clearly quite different from the faunas of both the Guayalejo-Tamesí (0.19) and Río Bravo (0.20). When only primary and secondary species are analyzed, the Soto la Marina and San Fernando systems are quite similar (0.69). Although these similarity index values must be interpreted with caution (estuarine fishes may be important in determining similarity but are not well known in Tamaulipas), this analysis does run counter to Smith and Miller's (1986) conclusion that the freshwater icththyofaunas of the Ríos Soto la Marina and San Fernando are similar to the continental ichthyofauna of the lower Rio Grande. Clearly, additional sampling is required to fully understand interbasin faunal resemblance patterns throughout the region.

Lower Rio Grande

The ichthyofauna of the lower Rio Grande consists of 136 species from 40 families. Forty-two of these are primary, 14 secondary, 12 estuarine, 66 essentially marine, and 2 vicariant (of marine origin but now confined to freshwater) (table 7.1). Of the 136 species, 10 are introduced. Smith and Miller (1986) mention that 7 species are endemic to the lower Rio Grande; *Dionda diaboli*, *Gila conspersa*, *Cyprinella rutila*, *N. saladonis*, *Gambusia gaigei*, *G. krumholzi*, and *Xiphophorus couchianus*, of which only *X. couchianus* might be found in Tamaulipas (not yet documented). Thus, the reach referred to as the lower Rio Grande has relatively high endemism, but since these endemic species occur above Tamaulipas, none of them can be considered Tamaulipas endemics. According to Edwards and Contreras-Balderas (1991), the upper reach of the lower Rio Grande (mostly just above Tamaulipas) is dominated (in terms of abundance) first by *Dorosoma cepedianum*, then *D. petenense*, *Cyprinella lutrensis*, *Pimephales vigilax*, *Strongylura marina*, *Fundulus grandis*, *Poecilia latipinna*, *P. formosa*, *Gambusia affinis*, *Menidia beryllina*, *Lepomis macrochirus* (introduced), "*Cichlasoma*" (*Herichthys*) *cyanoguttatum* and *Oreochromis aureus* (introduced). During the last 100 years an important reduction has been observed in numbers of *Macrhybopsis aestivalis* and *Hybognathus amarus*, the latter species extirpated from the entire river basin except near Albuquerque, New Mexico (Bestgen and Propst 1996). Population declines have been observed also for *Notropis jemezanus*, *N. orca* (now treated as *N. simus orca* by Chernoff et al. [1982] and Gilbert [1998]), and possibly *Ictalurus furcatus*, *Notropis buchanani*, *N. amabilis*, *Moxostoma congestum*, *Aplodinotus grunniens*, *Mugil cephalus*, *Atractosteus spatula*, *Lepisosteus oculatus*, and *L. osseus*. In the lower reaches of the lower Rio Grande

Table 7.1. Fishes of Tamanlipas.

	Ecology[a]	Introduced	Euryhaline	Lowland	Upland	River	Stream
				Habitat[b]			
Carcharhinidae							
Carcharhinus leucas	2A			3			
Dasyatidae							
Dasyatis americana	2B		T	T		T	
Dasyatis sabina	1B		T	T		T	
Acipenseridae							
Acipenser oxyrinchus	3B			1		1	
Lepisosteidae							
Atractosteus spatula	S		1,2,4	1,2,3,4		1,2,4	2
Lepisosteus oculatus	S		1	1,2		1,2	1,2
Lepisosteus osseus	S		1,2,4	1,2,3,4		1,2,4	1,2
Elopidae							
Elops saurus	2A		1,2,4,T	1,2,4,T		2,4,T	
Megalopidae							
Megalops atlanticus	2A		1,T	1,T		1,T	
Albulidae							
Albula vulpes	2B		T	T		T	
Anguillidae							
Anguilla rostrata	3A		1,2,4,T	1,2,4,T	1	1,2,4,T	1
Ophichthidae							
Myrophis punctatus	2A		2,T	2,T		2,T	
Engraulidae							
Anchoa mitchilli	2A		1,2,4,T	1,2,3,4,T		1,2,T	1,4
Anchoa hepsetus	2A		2,T	2,3,T		2,T	
Clupeidae							
Alosa chrysochloris	2A		2	2		2	
Brevoortia gunteri	2A		2,T	2,T		2,T	
Brevoortia patronus	2A		2,T	2,T		2,T	
Dorosoma cepedianum	1A		1,3,4,T	1,2,3,4,T		1,2,3,4,T	1,2,4
Dorosoma petenense	1A		1,3,4,T	1,2,3,4,T	4	1,2,4,T	1,2,4
Etrumeus teres	2B		2	2		2	
Harengula jaguana	2A		2,T	2,T		2,T	
Opisthonema oglinum	2B		T	T		T	
Cyprinidae							
Campostoma anomalum	P			1,T			1,T
Campostoma ornatum	P				1		1
Carassius auratus	P	I		2		2	2
Ctenopharyngodon idella	P	I		T		T	
Cyprinella lutrensis	P			T	T	T	T
Cyprinella venusta	P	I		1			1
Cyprinus carpio	P	I		2,T		2,T	2,T
Dionda episcopa	P				1,T		1,T
Dionda erimyzonops	P			4			4
Dionda ipni	P			4	T		4,T
Hybognathus amarus	P			1,2	1	1,2	1,2
Hypophthalmichthys molitrix	P	I		T		T	
Macrhybopsis aestivalis	P			1,2,T	1,T	1,2	1,2,T
Notemigonus crysoleucas	P	I		2			

Reservoir	Spring	Subterranean	LRB 1	LRB 2	LRB Consensus	RSF 1	RSF This Study	RSF Consensus	SLM 1	SLM This Study	SLM Consensus	SGT 3	SGT 4	SGT This Study	SGT Consensus
3												*			*
														*	*
										*	*				
			Ex		Ex										
3,T			*	*	*					*	*	*	*		*
			*	*	*	*		*							
3			*	*	*							*	*		*
			*	*	*					*	*		*	*	*
									*	*	*			*	*
										*	*				
			*	*	*				*	*	*			*	*
				*	*					*	*				
3			*	*	*				*	*	*	*	*		*
3				*	*				*	*	*	*			
				*	*										
				*	*					*	*			*	*
				*	*					*	*				
2,3,T			*	*	*	*	*	*	*	*	*	*	*	*	*
2,3,4,T			*	*	*	*	*			*	*	*	*	*	*
				*	*					*	*				
				*	*					*	*				
										*	*				
			*		*					*	*				
			*		*										
				I	I										
T										I	I			I	I
							*	*		*	*				
			I		I										
2,T				I	I					I	I			I	I
			*		*					*	*				
													*		*
													*	*	*
				*	*										
														I	I
2			*	*	*	*	*	*							
2				I	I										

(continued)

Table 7.1. Continued

	Ecology[a]	Introduced	Euryhaline	Lowland	Upland	River	Stream
Notropis aguirrepequenoi	P			1	1,T	1	1,T
Notropis amabilis	P			2	1	1,2	1,2
Notropis braytoni	P			1,2	1	1,2	1,2
Notropis buchanani	P			1,2		2	1,2
Notropis chihuahua	P				1		1
Notropis jemezanus	P			1,2	1	1,2	1,2
Notropis orca	P			1,2	1	1,2	1,2
Notropis stramineus	P			1		1	1
Notropis tropicus	P			4	T		4,T
Pimephales vigilax	P			1,2		2	1,2
Rhinichthys cataractae	P				1	1	1
Catostomidae							
Carpiodes carpio	P			1,2	1	1,2	1,2
Cycleptus elongatus	P			1	1	1	1
Ictiobus bubalus	P			1,3,4,T	1	1,4,T	1,3,T
Ictiobus labiosus	P			4		4	
Ictiobus niger	P			1	1	1	1
Moxostoma austrinum	P				1	1	1
Moxostoma congestum	P			1,2	T	1,2	1,2,T
Characidae							
Astyanax mexicanus	P		3	1,2,3,T	1,3,4,T	2,3,4,T	1,2,3,4,T
Astyanax jordani	P				4,T		
Ictaluridae							
Ameiurus melas	P	I		1			1
Ictalurus australis	P			3	3		3
Ictalurus furcatus	P			1,2,3,4	3	1,2,3,4	2,3
Ictalurus lupus	P				1,T		1,T
Ictalurus cf. lupus	P				1		
Ictalurus mexicanus	P			3,4	3,4,T	3,4,T	3,4,T
Ictalurus punctatus	P			1,2,4		1,2,4	1,2
Ictalurus sp.	P				T		T
Prietella lundbergi	P				T		
Pylodictis olivaris	P			1,4,T	1,T	1,4,T	1
Ariidae							
Ariopsis felis	2A		1,2,4,T	1,2,3,4,T		1,2,4,T	4
Synodontidae							
Synodus foetens	2A		2,T	2,T		2,T	
Phycidae							
Urophycis floridana	2B		2	2		2	
Batrachoididae							
Opsanus beta	2A		T	3,T		T	
Porichthys plectrodon	2A		T	T		T	
Antennariidae							
Histrio histrio	2B		2	2		2	
Mugilidae							
Agonostomus monticola	V		1,2,4,T	1,2,4	1,3,4,T	2,4,T	1,3,4,T
Mugil cephalus	2A		1,2,4	1,2,3,4		1,2,4	1,4
Mugil curema	2A		2,4,T	2,3,4,T		2,4,T	4

Drainage Occurrence[c]

Reservoir	Spring	Subterranean	LRB 1	LRB 2	LRB Consensus	RSF 1	RSF This Study	RSF Consensus	SLM 1	SLM This Study	SLM Consensus	SGT 3	SGT 4	SGT This Study	SGT Consensus
							*	*	E	*	*				
			*	*	*										
2			*	*	*										
			*	*	*										
			*		*										
			*	*	*										
			Ex	*	Ex										
			*		*										
												*		*	*
2			*	*	*										
			*		*										
2			*	*	*	*		*	*		*				
			*		*										
			*		*	*	*	*	*		*	*		*	*
														*	*
			*		*				*		*				
			*		*										
			*	*	*	*	*	*	*	*	*				
2,3,T	T		*	*	*	*	*	*	*	*	*	*	*	*	*
		4,T											*	*	*
			*		*								I	I	
												*			*
2,3,T			*	*	*		*	*	*		*	*	*		*
			*		*	*		*	*	*	*			*	*
			*		*										
												*	*	*	*
2,T			*	*	*	*	*	*	*	*	*			*	*
										*	*				
		T												E	*
T			*		*		*	*		*	*			*	*
3,T			*		*		*	*	*		*			*	*
				*	*					*	*				
				*	*										
3										*	*			*	*
										*	*				
				*	*										
4			*	*	*				*	*	*	*	*	*	*
3,4			*	*	*				*		*	*	*		*
3,4				*	*					*	*	*	*		*

(*continued*)

147

Table 7.1. Continued

	Ecology[a]	Introduced	Euryhaline	Lowland	Upland	River	Stream
Atherinidae							
Menidia beryllina	2A		1,2,4	1,2,4,T		1,2,T	
Menidia peninsulae	1B		2,4	2,4		2	
Membras martinica	2A		1,2	1,2		2	1
Membras vagrans	2A		T	T		T	
Belonidae							
Platybelone argalus	2B		2	2		2	
Strongylura marina	2A		1,2	1,2		1,2	
Strongylura timucu	2A		T	T		T	
Hemiramphidae							
Hemiramphus brasiliensis	2B		T	T		T	
Fundulidae							
Fundulus grandis	1B		1,2,3,4,T	1,2,3,4	1,T	2	1,4,T
Fundulus similis	1B		1,2,3,4	1,2,3,4		2	
Lucania parva	1B		1,2,3,4,T	1,2,3,4	1,T	1,2	1,T
Poeciliidae							
Gambusia affinis	S		4,T	1,2,3,4,T	T	2,3,T	1,2,4,T
Gambusia atrora	S			4			
Gambusia aurata	S			4	T		4,T
Gambusia panuco	S			4	3,T		3,4,T
Gambusia regani	S			4	1,3,4,T		3,4,T
Gambusia senilis	S				1		1
Gambusia speciosa	S				1,T		1,T
Gambusia vittata	S			4,T	3,T	3	3,4,T
Heterandria jonesii	S			4	4		4
Heterophallus marshi	S				1		
Poecilia formosa	S		1,2,T	1,2,3,4,T	T	2	1,3,4,T
Poecilia latipinna	S		1,2,4,T	1,2,3,4,T	T	2	T
Poecilia latipunctata	S				3,4,T		3,4,T
Poecilia mexicana	S		4,T	3,4,T	1,3,T	3,4,T	1,3,4,T
Xiphophorus couchianus	S				1		1
Xiphophorus nezahualcoyotl	S				3,T		3,T
Xiphophorus pygmaeus	S			4	T		4,T
Xiphophorus variatus	S			4	1,3,T	4	3,4,T
Xiphophorus xiphidium	S				1.T		1,T
Cyprinodontidae							
Cyprinodon eximius	S				1		1
Cyprinodon variegatus	1B		1,2,3,4,T	1,2,3,4,T		1,2,T	1
Syngnathidae							
Microphis brachyurus lineatus	2A		2,4	2,3,4		2	4
Syngnathus louisianae	2A		2	2		2	
Syngnathus scovelli	2A		2	2		2	
Triglidae							
Prionotus tribulus	2A		T	T		T	
Centropomidae							
Centropomus parallelus	2A		2,4	2,4		2,4	
Centropomus undecimalis	2A		2,4,T	2,4,T		2,4,T	

| | | | Drainage Ocurrence[c] | | | | | | | | | | | | |
Reservoir	Spring	Subterranean	LRB 1	LRB 2	LRB Consensus	RSF 1	RSF This Study	RSF Consensus	SLM 1	SLM This Study	SLM Consensus	SGT 3	SGT 4	SGT This Study	SGT Consensus
2,4,T			*	*	*	*		*	*	*	*		*		*
2,4				*	*								*		*
2			*	*	*				*		*				
T										*	*				
				*	*										
			*	*	*				*		*				
										*	*				
										*	*				
2,3,4			*	*	*	*	*	*	*	*	*	*	*		*
3			*	*	*							*			*
2,3,4			*	*	*	*		*	*	*	*	*	*		*
2,3,4			*	*	*	*	*	*	*	*	*	*	*	*	*
													*	*	*
							*	*		*	*	*	*	*	*
							*	*	*	*	*	*	*	*	*
	1		*		*										
			*		*	*	*	*	*		*				
												*	*	*	*
												*			*
			*		*										
1,2,4,T			*	*	*	*	*	*	*	*	*	*	*	*	*
1,2,3,4			*	*	*	*	*	*	*	*	*	*	E	*	*
3,4,T			*		*	*	*	*	*	*	*	*	*	*	*
			*		*										
												*		*	*
												*		*	*
										*	*	*		*	*
									E	*	*				
			*		*										
2,3,4			*	*	*	*	*	*					*		*
				*	*										
3				*	*										
				*	*										
				*	*										
										*	*				
				*	*								*		*
				*	*					*	*		*	*	*

(continued)

Table 7.1. Continued

| | Ecology[a] | Introduced | Habitat[b] | | | | |
			Euryhaline	Lowland	Upland	River	Stream
Moronidae							
Morone chrysops	P	I		2		2	2
Serranidae							
Epinephelus nigritus	2A		2	2		2	
Mycteroperca bonaci	2B		T	T		T	
Serranus atrobranchus	2B		T	T		T	
Centrarchidae							
Lepomis auritus	P	I		2			
Lepomis cyanellus	P	I		1,2,T	1	1,2	1,2,T
Lepomis gulosus	P	I		1,2	1	1,2	1,2
Lepomis macrochirus	P	I		1,2,T	1	1,2,T	1,2,T
Lepomis megalotis	P			1	1	1	1
Lepomis microlophus	P	I		2		2	2
Micropterus floridanus	P	I		1,2,T	1,T	1,2,T	1,2,T
Micropterus salmoides	P	I		1,2,T	1,T	1,2,T	1,2,T
Pomoxis annularis	P	I		2		2	2
Percidae							
Etheostoma grahami	P				1		1
Percina macrolepida	P				1		1
Rachycentridae							
Rachycentron canadum	2B		2	2		2	
Carangidae							
Caranx crysos	2B		T	T		T	
Caranx hippos	2A		2,T	2,3,T		2,T	
Caranx latus	2A		4,T	3,4,T		4,T	4
Chloroscombrus chrysurus	2A		2	2		2	
Hemicaranx amblyrhynchus	2B		2	2		2	
Oligoplites saurus	2A		2,T	2,T		2,T	
Selene vomer	2A		2	2		2	
Trachinotus carolinus	2B		2,T	2,T		2,T	
Trachinotus falcatus	2B		2,T	2,T		2,T	
Lutjanidae							
Lutjanus analis	2A		2	2		2	
Lutjanus apodus	2A		2	2		2	
Lutjanus campechanus	2B		2,T	2,T		2,T	
Lutjanus cyanopterus	2A		T	3,T		T	
Lutjanus griseus	2A		1,2,T	1,2,3,T		2,T	1
Lutjanus synagris	2B		2	2		2	
Gerreidae							
Diapterus auratus	2A		1,2,T	1,2,T		1,2,T	1
Diapterus rhombeus	2A		2,T	2,T		2,T	
Eucinostomus argenteus	2A		1,2	1,2		1,2	1
Eucinostomus melanopterus	2A		1,2,T	1,2,3,T		2,T	1
Eucinostomus gula	2A		2	2		2	
Eugerres brasilianus	2A			3			
Eugerres plumieri	2A			3			
Gerres cinereus	2A		T	T		T	

Drainage Occurrence[c]

			LRB			RSF			SLM			SGT			
Reservoir	Spring	Subterranean	1	2	Consensus	1	This Study	Consensus	1	This Study	Consensus	3	4	This Study	Consensus
2				I	I										
				*	*										
										*	*				
										*	*				
2				I	I										
2			*	*	*		I	I		I	I				
2			*	I	I										
2,T			*	*	*		I	I		I	I				
2			*	*	*										
2				I	I										
2,T	T						I	I		I	I			I	I
2,T	T		*	*	*	*	*	*	*	*	*			I	I
2				I	I										
			*		*										
			*		*										
			*		*										
										*	*			*	*
2,3				*	*					*	*	*			*
3				*	*					*	*	*	*		*
			*		*										
			*		*										
			*		*					*	*				
			*		*										
			*		*					*	*				
			*		*					*	*				
			*		*										
			*		*										
			*		*					*	*				
										*	*	*			*
3			*		*				*	*	*	*			*
3			*		*										
			*	*	*					*	*				
										*	*				
			*	*	*				*						
3				*	*				*	*	*	*			*
				*	*										
3										*	*				*
3										*	*			*	*
									*		*				

(continued)

Table 7.1. Continued

	Ecology[a]	Introduced	Habitat[b]				
			Euryhaline	Lowland	Upland	River	Stream
Haemulidae							
Anisotremus surinamensis	2B		T	T		T	
Conodon nobilis	2B		2	2		2	
Orthopristis chrysoptera	2A		2	2		2	
Pomadasys crocro	2A		2,T	2,T		2,T	
Sparidae							
Archosargus probatocephalus	2A		T	3,T		T	
Lagodon rhomboides	2A		2,T	2,T		2,T	
Polynemidae							
Polydactylus octonemus	2A		1,2	1,2		1,2	
Sciaenidae							
Aplodinotus grunniens	V		T	1,2,4,T	1	1,2,4,T	1,2,T
Bairdiella chrysoura	2A		T	T		T	
Bairdiella ronchus	2A		2,T	2,3,T		2,T	
Cynoscion arenarius	2A		2,T	2,T		2,T	
Cynoscion nebulosus	2A		2,T	2,3,T		2,T	
Leiostomus xanthurus	2A		2,T	2,T		2,T	
Micropogonias undulatus	2A		1,2,T	1,2,3,T		1,2,T	
Pogonias cromis	2A		1,2,T	1,2,3,T		2,T	1
Sciaenops ocellatus	2A		2	2		2	
Cichlidae							
"Cichlasoma" (Nandopsis) bartoni	S				4		4
"Cichlasoma" (Herichthys) labridens	S			4,T	4,T		4,T
"Cichlasoma" (Herichthys) pantostictus	S			4,T	4		4
"Cichlasoma" (Herichthys) cyanoguttatum	S			1,2,3,4,T	1,3,T	1,2,3,T	1,2,3,4,T
"Cichlasoma" (Herichthys) carpintis	S			T		T	
"Cichlasoma" (Theraps) steindachneri	S			3	3,4		3,4
Oreochromis aureus	S	I		2		2	2
Oreochromis mossambicus	S	I		T		T	T
Eleotridae							
Eleotris amblyopsis	1B		4	4			4
Eleotris pisonis	1B		4	3,4			4
Erotelis smaragdus	1B		2	2		2	
Dormitator maculatus	1B		1,2,4,T	1,2,3,4,T		1,2,T	1,4
Gobiomorus dormitor	1B		1,2,4,T	1,2,3,4,T	1,3,T	1,2,3,T	1,3,4,T
Gobiidae							
Awaous banana	1B		1,2,4,T	1,2,4,T		1,2,4,T	4
Bathygobius soporator	2A		2	2		2	
Ctenogobius boleosoma	2A		1,2	1,2,3		1,2	
Ctenogobius claytoni	1B			3			
Evorthodus lyricus	1B		2,4	2,3,4		2,4	4
Gobionellus oceanicus	2A		2,T	2,T		2,T	
Gobiosoma bosc	2A		2	2,3		2	
Ephippidae							
Chaetodipterus faber	2A		T	3,T		T	

Drainage Ocurrence[c]

Reservoir	Spring	Subterranean	LRB			RSF			SLM			SGT			
			1	2	Consensus	1	This Study	Consensus	1	This Study	Consensus	3	4	This Study	Consensus
										*	*				
				*	*										
				*	*										
				*	*									*	*
3										*	*	*			*
				*	*					*	*				
			*	*	*										
2			*	*	*		*	*	*		*			*	*
										*	*			*	*
3				*	*					*	*	*			*
				*	*					*	*				
3				*	*					*	*	*			*
				*	*					*	*				
3			*	*	*					*	*	*			*
3				*	*				*	*	*	*			*
				*	*										
4														*	*
4										*	*			*	*
4,T										*	*		E	*	*
2,3,T			*	*	*	*	*	*	*	*	*	*	*	*	*
T										*	*			*	*
3												*	*		*
2				I	I		I	I		I	I			I	I
T										I	I				
														*	*
3												*	*		*
				*	*										
3			*	*	*				*	*	*	*	*	*	*
3			*	*	*				*	*	*	*	*	*	*
			*	*	*									*	*
				*	*										
3			*	*	*							*			*
3												*			*
3,4				*	*							*	*		*
				*	*					*	*				
3				*	*							*			*
3												*		*	*

(continued)

Table 7.1. Continued

	Ecology[a]	Introduced	Habitat[b]				
			Euryhaline	Lowland	Upland	River	Stream
Trichiuridae							
Trichiurus lepturus	2B		T	T		T	
Scombridae							
Scomberomorus regalis	2B		T	T		T	
Paralichthydae							
Citharichthys spilopterus	2A		1,2,4	1,2,3,4		1,2,4	4
Cyclopsetta fimbriata	2B		2	2		2	
Etropus crossotus	2B		2,T	2,T		2,T	
Paralichthys lethostigma	2B		2	2		2	
Achiridae							
Achirus lineatus	2A		2	2		2	
Cynoglossidae							
Symphurus civitatus	2B		T	T		T	
Symphurus plagiusa	2A		2	2		2	

Taxonomy follows Eschmeyer's *Catalog of Fishes* (Eschmeyer 1998), except for the families Cichlidae, where we apply the nomenclature of Miller (in press), and Gobiidae, where we follow Pezold (1984). Assignment of species to habitats was based on frequency of occurrence in our own collections or on the basis of literature.

[a]Ecology: P = primary freshwater species; S = secondary freshwater species; 1A = seasonal inhabitant of estuarine/laguna systems; 1B = permanent inhabitant of estuarine/laguna systems; 2A = euryhaline marine species; 2B = stenohaline marine species; 3A = catadromous; 3B = anadromous; V = vicariant species.

(including Tamaulipas), the original dominant fauna included several cyprinids (*Macrhybopsis aestivalis*, *Hybognathus amarus*, and *Notropis jemezanus*), the characid *Astyanax mexicanus*, the catostomid *Carpiodes carpio*, various secondary taxa such as the clupeid *Dorosoma*, poeciliids, and cyprinodontids; and marine forms, including members of the families Sciaenidae, Gobiidae, and Gerreidae. This fauna has now been replaced by marine-estuarine and coastal species as a result of the river's increased salinity, due to both a decrease in the volume of fresh-water discharge and increasing upstream salt-water penetration.

San Fernando

Compared to the lower Rio Grande, the Río San Fernando has much lower fish diversity, with an ichthyofauna composed of 33 species in 13 families. Among them, 15 species are primary, 10 secondary, 5 estuarine, 2 marine, and 1 vicariant (table 7.1). Of the 33 species, 4 are introduced. The dominant species (predominantly of the order Cyprinodontiformes) are typical of lower reaches but have a wide distribution. Species from the mouth of the river have not been well collected; thus information regarding fishes of estuarine and marine origin is still incomplete.

Upstream, the ichthyofauna of the Río San Fernando is dominated by an abundance of *Fundulus grandis* and *Poecilia mexicana*. Downstream, the dominant fishes are *Astyanax mexicanus*, *Cyprinella lutrensis*, and *Micropterus*, represented mainly by introduced populations of *M. salmoides* and *M. floridanus* and hybrids of these two species (possibly still undescribed) (Rodríguez-Martínez 2001). Less common species, in decreasing order of abundance, are *Poecilia mexicana*, *Macrhybopsis aestivalis*, *Gambusia sp.* (probably *G. speciosa*), "*Cichlasoma*" (*Herichthys*) *cyanoguttatum*, and *Moxostoma congestum*, this last species being uncommon. Near the mouth of the river, the follow-

Drainage Ocurrence[c]

Reservoir	Spring	Subterranean	LRB 1	LRB 2	LRB Consensus	RSF 1	RSF This Study	RSF Consensus	SLM 1	SLM This Study	SLM Consensus	SGT 3	SGT 4	SGT This Study	SGT Consensus
														*	*
										*	*				
3,4			*	*	*							*	*	*	*
				*	*										
				*	*					*	*				
				*	*										
				*	*										
										*	*				
				*	*										

[b]Habitat and occurrence are as reported in the following publications: 1 = Smith and Miller (1986); 2 = Edwards and Contreras-Balderas (1991) and citations therein; 3 = Darnell (1962); 4 = Miller and Smith (1986). T = new collections from this study.

[c]Drainages: LRB = Lower Río Bravo (Rio Grande); RSF = Río San Fernando; SLM = Soto la Marina System; SGT = Guayelejo-Tamesí System. Consensus = basin-by-basin occurrence determination based on all evidence considered in this study; * = reported/present; Ex = extinct or extirpated; E = endemic; I = introduced.

ing species appear to be the most abundant: *A. mexicanus*, *M. aestivalis*, *C. lutrensis*, and "*Cichlasoma*" (*Herichthys*) *cyanoguttatum*. As mentioned before, *Notropis aguirrepequenoi* is a new record for this river. Typical estuarine and coastal faunal elements are rather scarce in this basin, as are *Ictiobus bubalus*, *Lepomis macrochirus*, *Poecilia latipinna*, *Gambusia speciosa*, and *Cyprinodon variegatus*. No alligator gars (*Atractosteus spatula*) or long-nosed gars (*Lepisosteus osseus*) were captured in this basin during our collecting efforts, nor were any *Ictalurus lupus*, *Lucania parva*, or *Menidia beryllina*, despite the fact that all 5 species have been documented from this river in the literature.

Soto la Marina

The Soto la Marina river system harbors a very diverse fauna of 98 species in 38 families. Twenty-one are primary freshwater species; 15 are secondary, 7 estuarine, and 55 of marine origin (including 2 vicariant, i.e., isolated from the ocean). Of the total 98 species, 8 are introduced. *Notropis aguirrepequenoi* was previously considered endemic to this system, but our collections demonstrate the distribution of this species to be broader. *Xiphophorus xiphidium* is the only species endemic to this river (Smith and Miller 1986; table 7.1).

There are 6 dominant species in the upper Soto la Marina watershed. In order of decreasing abundance they are *Poecilia mexicana*, *Astyanax mexicanus*, "*Cichlasoma*" (*Herichthys*) *cyanoguttatum*, *Gambusia regani*, *Poecilia formosa*, and *Notropis aguirrepequenoi*. Other taxa occurring in lesser abundance, but still important, are *Xiphophorus xiphidium*, *Micropterus salmoides* (natural genetic populations of this species have been found in the upper reaches of this watershed; Rodríguez-Martínez 2001), *Gambusia affinis*, *Moxostoma congestum*, *Oreochromis mossambicus* (introduced), *Gambusia aurata*, and *Dionda* sp. Our samples of an unidentified member of the genus *Ictalurus* from this basin may represent a new species, or hybrids of a native form and *I. punctatus*. This last species is a native fish harvested extensively in the lower reach of the river. Further taxonomic and genetic studies are required to determine the status of *Ictalurus* populations throughout the Soto la Marina system.

The most common species in the lower reach of Soto la Marina are *Dorosoma petenense*, *D. cepedianum*, *Astyanax mexicanus*, *Ariopsis felis*, and *Gambusia affinis*. Species introduced in San Vicente Reservoir for commercial purposes are carps (*Cyprinus carpio* and *Ctenopharyngodon idella*) and tilapia (*Oreochromis mossambicus*), and the native species also exploited are catfishes *Ictalurus punctatus* and *I. furcatus*. There are no data from this basin regarding present existence or status of native populations of either largemouth bass (*Micropterus salmoides*), a species important for sport fishing, or alligator gar (*Atractosteus spatula*). Both are believed to be native, but conspecific stocks have also been introduced from elsewhere. We did not collect any *Lepisosteus* species in this river basin, nor *Ictiobus bubalus*, *Fundulus grandis*, *Mugil cephalus*, *Eucinostomus* sp., or *Gobiomorus dormitory*, all of which have been reported in previous studies.

Species of estuarine and marine origin are abundant at the mouth of the river, although some freshwater taxa with the ability to withstand high salinity levels also live there. Taxa that dominated our collections were, in decreasing order of abundance, *Gambusia affinis*, *Eucinostomus melanopterus*, *Membras vagrans*, *Gobiomorus dormitor*, *Etropus crossotus*, *Poecilia formosa*, *Harengula jaguana*, *Strongylura timucu*, "*Cichlasoma*" (*Herichthys*) *labridens*, *Trachinotus carolinus*, and *Ariopsis felis*.

Guayalejo-Tamesí

The ichthyofauna of the Guayalejo-Tamesí watershed consists of 93 species in 33 families. There are 19 primary freshwater species, 23 secondary, 14 estuarine, and 37 of marine origin (2 vicariant). Of the 93 species, 6 are introduced. As in most watersheds of the state, species assemblages here are transitional between the Nearctic and the Neotropics. According to Darnell (1962), southern (tropical) freshwater species slightly outnumber northern (temperate) species by a 13:10 ratio.

In the upper reaches of the Guayalejo-Tamesí watershed (Ríos Chihue, Sabinas, Frío, Mante, and Guayalejo) dominant taxa in our collections were, in decreasing order of abundance, *Astyanax mexi-*

canus, *Poecilia mexicana*, "*Cichlasoma*" (*Herichthys*) *labridens*, *Poecilia formosa*, *Dionda ipni*, and "*Cichlasoma*" (*Herichthys*) *cyanoguttatum*. Our finding in this watershed of *Xiphophorus nezahualcoyotl* represents a new record, not only for Tamaulipas, but also San Luis Potosí (see Rauchenberger et al. 1990). Two cave-adapted species are also found in the upper watershed. *Prietella lundbergi* is endemic to Tamaulipas, but the other, *Astyanax mexicanus* (= *A. jordani*), is found also outside of the state. We consider the blind form of the Mexican "tetra" (or blind tetra) as a separate species (see below).

In the lower reaches of the watershed (Ríos Tigre and Tamesí) the dominant fishes, in order of decreasing abundance, are *Gambusia vittata*, *Astyanax mexicanus*, *Poecilia mexicana*, *Gambusia aurata*, *Poecilia formosa*, *Xiphophorus variatus*, *Oreochromis aureus* (introduced) and "*Cichlasoma*" (*Herichthys*) *labridens*. The following species reported for this system were not collected during our research: *Atractosteus spatula*, *Ictiobus bubalus*, *Ictalurus punctatus*, *Lucania parva*, *Cyprinodon variegatus*, *Gambusia speciosa*, "*Cichlasoma*" (*Theraps*) *steindachneri*, "*Cichlasoma*" (*Herichthys*) *labridens*, and *Eleotris pisonis*. Near the mouth of the river, species communities are dominated by euryhaline and marine fishes, as well as some secondary freshwater taxa. These, in order of decreasing abundance, include *Mugil cephalus*, *Ariopsis felis*, the introduced *Hypophthalmichthys molitrix*, *Eugerres plumieri*, *Elops saurus*, *Centropomus undecimalis*, and *Bairdiella chrysoura*.

Origin and Biogeography

Freshwater assemblages of fishes through much of Tamaulipas are composed of a mix of relict and recent species with Nearctic and Neotropic affinities. There are also many species of restricted distribution in this transitional area, as well as many marine fishes that penetrate continental waters. The relict fishes of pre-Tertiary origin are represented by sturgeons (currently extirpated from Tamaulipan waters), and the gars (Lepisosteidae). There are also taxa whose origins date back to Eocene or early Miocene, such as the primitive catostomids (*Carpiodes*, *Ictiobus*) (Smith 1992), catfishes (*Ictaluridae*) (Lundberg 1992), and some Centrarchidae (*Micropterus*) (Briggs 1986), though these dates are from fossils found farther north in North America, and existence of these groups in Tamaulipas at those times is not documented by fossils. Taxa with a probable origin during the Oligocene, Miocene, Pliocene, and Pleistocene (Miller 1986) include cyprinids and modern catostomids (*Moxostoma*). It is possible that the Tamaulipan (or phantom) blindcat (*Prietella lundbergi*; Walsh and Gilbert 1995) and Neotropical taxa such as the characid *Astyanax*, cichlids (*"Cichlasoma"* [*Herichthys*]), and poeciliids (*Poecilia, Xiphophorus*) also originated at that time.

During the Pleistocene, sea levels were repeatedly up to 120 m lower than they are at present, causing the coastal line of the Gulf of Mexico to move much farther down the continental shelf (Flint 1971). During these times, rainfall was abundant, leading to increased water drainage across coastal floodplains (Bailey et al. 1954; Conner and Suttkus 1986), and connections (now broken) may have existed among the watersheds of Tamaulipas. It is also likely that increased flow of freshwater into the gulf lowered salinities of estuarine and coastal lagoon habitats, permitting more extensive dispersal of freshwater fishes from one watershed to another. Such hydrographic and ecological histories could explain the current wide distributions of many fishes associated with these areas.

Freshwater fish distributions are controlled not only by physiography, but also by physical, chemical, and ecological conditions within a river basin. Most floodplains represent a natural barrier greatly reducing the potential for species of the upper part of the watershed to disperse downstream (and vice versa). At higher elevations (typically about 200–2000 m in Tamaulipas exclusive of the Rio Grande basin), watercourses have a greater proportion of rapids compared to pools, and the most common habitats are mid-sized springs and streams. Waters in these areas are typically clear, with relatively constant temperatures. At mid-elevations (typically 50–199 m in Tamaulipas), rivers have left mountainous areas, the elevation gradient is not as steep, and there are no rapids. Flow velocities decrease and turbidities increase. In Tamaulipas, dams, built for flow regulation and flood control, are typically in these areas. Finally, the coastal area or river mouth (< 50 m above sea level) is under the strong influence of adjacent marine waters and includes a variety of channels (natural and artificial) and lagoons. The rivers of the higher areas have an average width of less than 10 m, and widths of coastal and valley rivers are either narrow (< 10 m) or moderate (10–70 m). In Tamaulipas, the effect of elevational gradient is correlated with that of latitude, as the higher

elevations are found mostly in the central and southern parts of the state.

The presence of reservoirs like Falcon, Marte R. Gómez, and Anzalduas along the lower Rio Grande, Vicente Guerrero on the Río Soto la Marina, and Xicotencatl and Ramiro Caballero Dams in the Guayalejo-Tamesí system, is important. All of these reservoirs act as artificial barriers effectively preventing dispersal of both freshwater and marine fishes.

Description of Taxonomic Groups Present in Tamaulipas

Several groups of continental fishes represented in Tamaulipas are in need of taxonomic revision. Other groups have benefited from recent genetic studies that have often proven important to conservation issues, and/or offer insight into evolutionary processes. A few examples are provided here.

The North American catfishes of Mexico are generally poorly known. During our research, various specimens from the upper reaches of the Soto la Marina and Guayalejo-Tamesí watersheds could not be identified with existing keys (Miller in press). It appears to us that our inability to determine these to species may relate to the fact that they represent undescribed species, that variation in described species has not been adequately studied and described, or that these specimens represent hybrids with introduced species. Catfishes are commercially important in the Río Soto la Marina watershed, where both blue catfish (*Ictalurus furcatus*) and channel catfish (*I. punctatus*) are harvested for export to the United States. There was no initial baseline evaluation of the native populations of these 2 species in Tamaulipas, and now domesticated stocks of both species have been introduced and have surely escaped. Impacts of this on native ictalurid populations requires further attention. Another catfish, the recently discovered Tamaulipan blindcat (*Prietella lundbergi*), is endemic to Tamaulipas, living in caves of the Sierra Madre Oriental near the city of Mante. It remains very poorly known, but recent expeditions and collections have provided new morphological, ecological, and genetic information (Hendrickson et al. 2001; Wilcox et al. 2004). The Tamaulipan blindcat and other troglodytic organisms can be used as potential indicators of human impacts on aquifers, and studies of such species can shed light on difficult questions about aquifers and groundwater interconnections (Hendrickson and Krejca 2000).

The genus *Dionda* (Cyprinidae) of streams in Tamaulipas and the southwestern United States represents a biological model for the study of speciation and historical biogeography. Its characteristic distribution pattern, with species occurring in pairs, either sympatrically or allopatrically, begged the question of whether speciation has occurred allopatrically or sympatrically in this group. Mayden et al. (1992) conducted a phylogenetic analysis of all *Dionda* species and found that allopatric speciation with subsequent dispersal resulting in sympatry is the most likely history. According to these authors, speciation in *Dionda* may be related to geological events that took place in the Rio Grande and Río Pánuco regions. By repeating this analysis for other species that show similar distributions, a general explanation regarding the biogeographic history of fishes could be formulated for all of northeastern Mexico.

Cichlids are another poorly known group requiring further study in our region. We collected in the upper reaches of the Guayalejo-Tamesí and Soto la Marina watersheds, mainly upstream along the San Marcos River, where we found forms that appear to be hybrids of *"Cichlasoma"* (*Herichthys*) *cyanoguttatum* and *"Cichlasoma"* (*Herichthys*) *labridens*. If this finding is confirmed, it is possible that hybrid zones could exist in the region. Hybrid zones are considered natural laboratories for the study of microevolution, and hybridization is also often an indicator of environmental change.

The family Poeciliidae is particularly interesting with regard to hybridization. Members of the genus *Poecilia* have also been studied, with recent contributions toward documenting and describing geographic variation in the genus in Mexico by Daza-Zepeda (1999). Within this genus, *P. formosa* is asexual, with only female individuals. This is a rare phenomenon among vertebrates (Vrijenhoek et al. 1989), and this was the first vertebrate species for which gynogenetic reproduction was reported (Hubbs and Hubbs 1932). This species evolved as a result of hybridization between *P. mexicana* and *P. latipinna* (Schlupp et al. 1998). *Poecilia formosa* is abundant in northeastern Mexico, where it is sympatric with *P. mexicana* and *P. latipinna* (Darnell and Abramoff 1968; Schlupp et al. 2002). A great diversity of *P. formosa* genotypes has been reported in the Soto la Marina watershed, specifically in the Río Purificación, where diploid and triploid forms and clones with microchromosomes have been observed, in addition to another triploid with microchromosomes (Balsano et al. 1989). Another interesting phenomenon within the genus *Poecilia* is the rare occurrence of females with male phenotypes, called pseudo-males. These are rare in the wild but can be produced in a laboratory (Turner 1984; Schartl et al. 1991; Schlupp et al. 1992). The diversity of *P. formosa* genotypes in the Soto La Marina watershed represents a natural laboratory with a model system of biological evolution (Vrijenhoek 1994).

Other poeciliids that have caught the attention of the scientific community are the hybrids of different species within the genus *Xiphophorus*. These hybrids are generated in laboratories to study human melanomas (Schartl et al. 1995; Kazianis et al. 1996). There are various *Xiphophorus* species in Tamaulipas, but their geographic distributions are not well known. One of us (García de León) has worked on the genetic, morphologic, and behavioral characterization of a recently discovered hybrid of *X. birchmanni* and *X. malinche* in the state of Hidalgo (Rosenthal et al. 2003). As previously reported by Rauchenberger et al. (1990), southern Tamaulipas has a population of *X. variatus* and *X. nezahualcoyotl* hybrids that remains unstudied. Finally, *Gambusia* is yet another problematic Poeciliid genus in need of thorough taxonomic revision, despite the extensive research by Rauchenberger (1989).

The Mexican tetra, *Astyanax mexicanus*, is widely distributed in Mexico and is found in every major drainage in Tamaulipas. Despite some efforts to study its genotypic and phenotypic variation in Mexico (Álvarez del Villar 1970; Paulo Maya 1994), many questions remain unanswered. A blind morph of the Mexican tetra is known worldwide as a popular aquarium fish, and as an evolutionary enigma through the publications of Mitchell et al. (1977), Wilkens (1988), and many others. It is one of the most studied hypogean (subterranean) taxa, with much research having been done on the genetics of ocular regression and loss of body pigmentation (Langecker et al. 1993, 1995; Behrens et al. 1997, 1998) in particular. The origin of Mexican tetra populations and their genetic relationships were recently analyzed by Espinasa and Borowsky (2001) and by Dowling et al. (2002). Although the surface and hypogeal forms do indeed correspond to 1 single species, it is important to conserve blind lineages and, though recognizing that it is not monophyletic, we retain use of the single name for the blind populations to call attention to these important components of biocomplexity and to facilitate and promote their conservation within the largely species-oriented political/bureaucratic systems currently operating in our region. Overexploitation of groundwater can lead quickly to extinction of subterranean forms, and inaccessibility of their habitats greatly impedes study and monitoring. In general, a thorough systematic and taxonomic reevaluation of the genus *Astyanax* throughout its wide range in Mexico remains necessary to clarify many still-unknown aspects of its evolutionary history and to facilitate science-based conservation efforts.

The suckers ("matalotes") of the family Catostomidae have been the focus of many phylogenetic studies (e.g., Hubbs 1930; Jenkins 1970; Smith 1992; Harris and Mayden 2001). Taking into consideration morphological and biochemical characters and historical information, Smith (1992) argued that the genus *Moxostoma* was not monophyletic, dividing it instead into *Moxostoma* and *Scartomyzon*. However, Harris et al. (2002) sequenced cytochrome b of the mitochondrial DNA and concluded that this split is not valid. Evidently, more work is needed on the systematics of the family. There are populations of *Moxostoma congestum*, *Ictiobus bubalus*, and *Carpiodes carpio* in Tamaulipas, but as with many other species of the region, there are few data regarding biology and behavior of these taxa in natural environments.

The largemouth or black bass has been the focus of numerous systematic studies over many years (e.g., Hubbs and Bailey 1940; De Buen 1942; Bailey and Hubbs 1949; Álvarez del Villar 1970). In 1949, Bailey and Hubbs described 2 subspecies, *Micropterus salmoides salmoides* (with a more northern distribution), and *M. s. floridanus* (from Florida, and, as noted before, now considered a valid species). Genetics of the species has also been thoroughly investigated (Philipp et al. 1981, 1983; Carmichael et al. 1986; Williamson et al. 1986; Morizot et al. 1991; Philipp and Whitt 1991; Nedbal and Philipp 1994; Neff et al. 1999); however, all of this research was conducted on U.S. populations, and populations of northeastern Mexico remain little studied. The distribution of *Micropterus salmoides* extends naturally to the Rio Grande and the Soto la Marina and San Fernando watersheds. In Vicente Guerrero and other reservoirs *M. salmoides* is an important sport-fishing species. When native populations began to decline, *M. floridanus* was introduced. Recently, García de León et al. (2001b) used biochemical markers to analyze the impact of

introduced populations in the Soto la Marina watershed on isolated remnant native populations. They mention that the native populations are being strongly introgressed by alleles of *M. floridanus* and urgently recommend creation of a sanctuary to protect the genetic integrity of this valuable native resource.

Fish of the family Lepisosteidae, known as gars, have important characteristics that are unique or rare among living fishes—for example, ganoid scales and a vascularized gas bladder. Also, because they have experienced a low rate of evolutionary change since the Cretaceous (120 mya), gars are considered to be "living fossils" (Wiley 1976; Wiley and Hans-Peter 1984). The basic biology of a majority of gar species has not been adequately studied, and absence of gars from our recent collections in Tamaulipas seems to indicate they are now threatened in the region. Research focusing on biology of the alligator gar has begun in Tamaulipas (García de León et al. 2001a). Another form reported from southern Texas by Suttkus (1963) remains undescribed, but it may also extend into Tamaulipas.

Conservation

There are efforts to regulate commercial and sport fishing of black bass and alligator gar populations introduced in Vicente Guerrero Dam. With that exception, however, fish conservation efforts in Tamaulipas are practically nonexistent. Yet 30 freshwater fish species from Tamaulipas are found in the Mexican Official Norm NOM-059-ECOL 2001 (SEMARNAT 2002); 3 are listed as subject to Special Protection, 17 as Threatened, 7 as Endangered, and 3 are reported as probably extinct. Of the 7 Endangered species, 3 are from the lower Rio Grande bordering Tamaulipas: *Dionda episcopa* (also found in the Río Soto la Marina; see table 7.1), *Gambusia speciosa* (the distribution of which includes the Ríos San Fernando and Soto la Marina; table 7.1), and *Xiphophorus couchianus*. The other 4 are found in the Guayalejo-Tamesí system: "*Cichlasoma*" (*Nandopsis*) *bartoni*, *H. labridens* (also found in the Río Soto la Marina watershed; table 7.1), "*Cichlasoma*" (*Theraps*) *steindachneri*, and *Poecilia latipunctata*.

In Tamaulipas, human settlements, unsustainable tourism, and agricultural and industrial activities are the most important threats to freshwater fish communities. As in many other regions, the rivers of Tamaulipas have been altered profoundly by diversion channels and dams. Human-induced flow regulation along the lower Rio Grande and resulting changes in fish communities were the focus of an analysis by Edwards and Contreras-Balderas (1991). Even upstream of Tamaulipas, the Rio Grande watershed's fish fauna has lost many of its original elements; indigenous species have been largely replaced by estuarine species, and the fauna of the lower river along Tamaulipas has few freshwater taxa remaining, with most now replaced by estuarine and marine species. This pattern of change in fish fauna composition is correlated with decreasing river flow associated with an increase in pollution by chemical substances of agricultural and petrochemical origin and increased salinities. In spite of all these well-documented heavy impacts, we are not aware of any restoration plan at the ecosystem level for the lower Rio Grande. A recovery plan exists, but only for the silvery minnow (*Hybognathus amarus*), a species presumably extirpated from the lower Rio Grande (U.S. Fish and Wildlife Service 2002).

The Río San Fernando watershed has not been altered extensively in Tamaulipas, but in Nuevo León 2 dams (José López Portillo and Porvenir) have been recently built near Linares. Although we collected along this watershed at only a few sites, the ichthyofauna of this watershed appears to remain healthy, but impacts of the recent upstream reservoirs may have not yet been expressed. We suggest that conservation measures and monitoring should be implemented soon, before fish faunas are heavily impacted.

The upper reaches of the Río Soto la Marina watershed are threatened by tourism, mining, and forestry. Fortunately, these upper reaches are still relatively pristine, but a conservation strategy is necessary, as is greater government control of mining and forestry activities. The Vicente Guerrero Reservoir is a major hydraulic impact that has modified fish distribution and abundance in the floodplain. Human settlement and unsustainable fishing and tourism are some of the threats to fish communities near the river's mouth. Despite the dam and other threats, there is no major industrial activity in this watershed, and it is still in relatively good condition. Nevertheless, industrial activities are projected to grow, with the potential of considerable watershed impact.

The upper reaches of the Guayalejo-Tamesí watershed are in relatively good condition, but as with the Río Soto la Marina, irresponsible tourism could seriously impact this area. The presence of diversions and dams (Xicotencatl, Ramiro Caballero, and Española), the use of fertilizers and pesticides in sugarcane and orange crops, cattle ranching, and the mostly non–point-source contaminants such as agricultural pesticides and nutrients all represent threats to the ichthyofauna of the floodplain area. Near the river's mouth (Tampico and Ciudad Madero), the most important impacts are probably increased sediment loads and contaminants associated with intensive petrochemical industry activities. Overall, the Guayalejo-Tamesí watershed is more heavily and widely impacted by humans than are either the Soto la Marina or San Fernando watersheds, and consequently its fish fauna conservation status is less satisfactory.

Final Comments

With the exception of the lower Rio Grande and the coastal region of the Río Pánuco, the conservation status of the ichthyofauna of Tamaulipas remains perhaps somewhat better than the status of related fish faunas of rivers of the southern and southwestern United States. Nonetheless, this evaluation must be tempered by the fact that we still lack considerable, important information on freshwater fishes of the state. Despite collection of much information on the state's ichthyofauna in general, and on the diversity and biology of some taxonomic groups in particular during the second half of the twentieth century, species previously unknown to science are still being discovered in the area, and surely new taxa will be described from Tamaulipas in the near future. Even after such discoveries, we will still lack much knowledge about the basic biology and ecology, as well as aspects of genetics and population structure, of not only the new species, but also those long known from the region.

Along with introduced species, threats to the state's native continental ichthyofauna include water diversion or use for agriculture, as well as impacts associated with livestock, recreational activities, and urban areas, and oil and other industrial contamination of coastal areas. For the immediate future, research should continue to focus on cataloging and monitoring the state's species and communities to identify conservation needs. Investigative efforts in population biology, coupled with a clear understanding of ecosystems, will be of utmost importance for the conservation of the ichthyofauna.

Only sustained financial support can permit the type of more complex, interdisciplinary research necessary to help guide the conservation of species, habitats, and ecosystems. An important challenge for the federal and state governments will be the creation of environmental career opportunities for trained personnel to monitor and manage rivers, lakes, lagoons, and springs, and their associated faunas.

Acknowledgments

We thank the members of the Biología Integrativa lab group at ITCV for their assistance in the field and lab: Verónica Martínez Vázquez, Mireya Olazarán Sánchez, Fernando Vanoye Eligio, Andrés I. Hernández Sandoval, Omar Alcocer Torres, Pedro Castillo Castillo, Rocío I. Rodríguez Martínez, Xochitl F. De la Rosa Reyna, José M. Herrera Castillo, Aldo Guevara Carrizales, Carmen Lesso García, Lizeth Martínez García, Yara Sánchez, Gilberto Herrera Patiño, Ana María Hernández Mendoza, and Rafael Valencia. We are also grateful to Leticia Huidobro, Angélica Daza, and Felipe Villegas from IBUNAM for help with IBUNAM collections data and production of a draft of figure 7.1. Research published here was supported by CONACyT and SIREYES (contract no. SIRREYES19980606034). We thank SEMARNAP for issuing the fish-collecting permit (no. 041099-213-03); the Smith-Root Company for donating electric fishing equipment and accessories; Edmundo Díaz Pardo for comments that helped improve the text; and Lorie McCracken and Jean-Luc Cartron for translating the manuscript into English.

Literature Cited

Allendorf, F. W. 1988. Conservation biology of fishes. Conservation Biology 2: 145–148.

Allendorf, F. W., N. Ryman, and F. M. Utter. 1987. Genetics and fishery management. Past, present, and future. Pp. 1–19 *in* N. Ryman and F. Utter (eds.), Population Genetics and Fishery Management. University of Washington Press, Seattle.

Álvarez del Villar, J. 1950. Ictiología dulceacuícola mexicana. II. Bibliografía taxonómica. Revista de la Sociedad de Historia Natural 11: 201–216.

Álvarez del Villar, J. 1970. Claves para peces Mexicanos. Instituto Nacional de Investigaciones Biológico Pesqueras, Servicio de Investigación Pesquera.

Bailey, R. M., and C. L. Hubbs. 1949. The black basses (*Micropterus*) of Florida, with description of a new species. Occasional Papers of the Museum of Zoology, University of Michigan, 516.

Bailey, R. M., H. E. Winn, and C. L. Smith. 1954. Fishes from Escambia River, Alabama and Florida, with ecology and taxonomic notes. Proceedings of the Academy of Natural Sciences of Philadelphia 56: 109–164.

Balsano, J. S., E. M. Rasch, and P. J. Monaco. 1989. The evolutionary ecology of *Poecilia formosa* and its triploid associate. Pp. 277–298 *in* G. K. Meffe and F. F. Snelson (eds.), Ecology and Evolution of Livebearing Fishes (Poeciliidae). Prentice Hall, Englewood Cliffs, New Jersey.

Behrens, M., T. G. Langecker, H. Wilkens, and H. Schmale. 1997. Comparative analysis of Pax-6 sequence and expression in the eye development of the blind cave fish *Astyanax fasciatus* and its epigean conspecific. Molecular Biology and Evolution 14: 299–308.

Behrens, M., H. Wilkens, and H. Schmale. 1998. Cloning of the alpha A-crystallin genes of a blind cave form and the epigean form of *Astyanax fasciatus*: a comparative analysis of structure, expression and evolutionary conservation. Gene 216: 319–326.

Briggs, J. C. 1986. Introduction to the zoogeography of North America fishes. Pp. 1–16 *in* C. H. Hocutt and E. O. Wiley (eds.), The Zoogeography of North American Freshwater Fishes. John Wiley and Sons, New York.

Bestgen, K. R., and D. L. Propst. 1996. Redescription, geographic variation, and taxonomic status of Rio Grande silvery minnow, *Hybognathus amarus* (Girard, 1856). Copeia 1996: 41–55.

Carmichael, G. J., H. Williamson, M. E. Schmidt, and D. C. Morizot. 1986. Genetic marker identification in largemouth bass with electrophoresis of low-risk tissues. Transactions of the American Fisheries Society 113: 455–459.

Castro-Aguirre, J. 1978. Catálogo Sistemático de los Peces Marinos que Penetran a las Aguas Continentales de México, con Aspectos Zoogeográficos y Ecológicos. Departamento de Pesca, Serie Científica 19. México D.F.

Castro-Aguirre, J., H. S. Espinosa Pérez, and J. J. Schmitter Soto. 1999. Ictiofauna Estuarino-Lagunar y Vicaria de México. Ed. Limusa, México D.F.

Chernoff, B., R. R. Miller, and G. R. Gilbert. 1982. *Notropis orca* and *Notropis simus*, cyprinid fishes from the American Southwest, with a description of a new subspecies. Occasional Papers of the Museum of Zoology, University of Michigan, 698: 1–49.

CONABIO. 2000. Estrategia Nacional Sobre la Biodiversidad de México. Comisión Nacional para el Conocimiento y Uso de la Biodiversidad, México D.F.

Conner, J. V., and R. D. Suttkus. 1986. Zoogeography of freshwater fishes of the western Gulf slope. Pp. 413–456 *in* C. H. Hocutt and E. O. Wiley (eds.), Zoogeography of North American Freshwater Fishes. John Wiley and Sons, New York.

Contreras-Balderas, S. 1962. Contribución al Conocimiento de la Ictiofauna del Río San Juan, Provincia del Río Bravo, México. Universidad de Nuevo León, Facultad de Ciencias Biológicas.

Contreras Balderas, S. 1969. Perspectivas de la ictiofauna en las zonas áridas de México. Pp. 293–304 *in* T. W. Box and P. Rojas-Mendoza (eds.), Proceedings: International Symposium on Increasing Food Production in Arid Lands, 22–25 April 1968, Monterrey, Mexico. Texas Technological College, Lubbock, Texas.

Contreras-Balderas, S. 1978. Speciation aspects and man-made community composition changes in Chihuahuan Desert fishes. Pp. 405–431 *in* R. H. Wauer and D. H. Riskind (eds.), Symposium on the Biological Resources of the Chihuahuan Desert Region, United States and Mexico. National Park Service Transactions and Proceedings Series 3.

Darnell, R. M. 1962. Fishes of the Rio Tamesí and related coastal lagoons in east-central Mexico. Publications of the Institute of Marine Science, University of Texas 8: 300–361.

Darnell, R. M., and P. Abramoff. 1968. Distribution of the gynogenetic fish, *Poecilia formosa*, with remarks on the evolution of the species. Copeia 1968: 354–361.

Daza-Zepeda, D. A. 1999. Variación morfológica de *Poecilia mexicana* (Pisces: Poeciliidae) en la vertiente Atlántica de México. Licienciatura thesis, Universidad Nacional Autónoma de México, México D.F.

De Buen, F. 1942. Segunda contribución al estudio de la ictiofauna mexicana. Investigaciones de la Estación Limnológica de Pátzcuaro II (3): 25–55.

De Buen, F. 1946. Ictiogeografía continental mexicana. Revista de la Sociedad Mexicana de Historia Natural VII (1–4): 87–137.

De Buen, F. 1947. Investigación sobre ictiología mexicana. I. Catalogo de los peces de la región neartica en suelo mexicano. Anales del Instituto de Biología, Universidad Nacional Autónoma de México, XVIII (1): 257–348.

De la Cruz-Agüero, G. 1994. Sistema para el análisis de comunidades (ANACOM).

Departamento de Pesquerías y Biología Marina, CICIMAR-IPN.

Dowling, T. E., D. P. Martasian, and W. R. Jeffery. 2002. Evidence for multiple genetic forms with similar eyeless phenotypes in the blind cavefish, *Astyanax mexicanus*. Molecular Biology and Evolution 19: 446–455.

Eadie, J. M., T. A. Hurley, R. D. Montgomerie, and L. Teather. 1986. Lakes and rivers as islands: species–area relationships in the fish faunas of Ontario. Environmental Biology of Fishes 15: 81–89.

Edwards, R. J., and S. Contreras-Balderas. 1991. Historical changes in the ichthyofauna of the lower Rio Grande (Río Bravo del Norte), Texas and Mexico. Southwestern Naturalist 36: 201–212.

Eschmeyer, W. N. (ed.). 1998. Catalog of Fishes. California Academy of Sciences, San Francisco.

Espinasa, L., and R. B. Borowsky. 2001. Origins and relationship of cave populations of the blind Mexican tetra, *Astyanax faciatus*, in the Sierra de El Abra. Environmental Biology of Fishes 62: 233–237.

Espinosa-Pérez, H., M. T. Gaspar, and P. Fuentes. 1993. Peces Dulceacuícolas Mexicanos. Listados Faunísticos de México, Universidad Nacional Autónoma de México, 3: 1–99. [Available: http://www.ibiologia.unam.mx/publicaciones/lf3.html]

Flint, R. F. 1971. Glacial and Quaternary Geology. John Wiley and Sons, New York.

Frankham, J. D., J. D. Ballou, and D. A. Briscoe. 2001. Introduction to Conservation Genetics. Cambridge University Press, London.

García de León, F. J., L. González-García, J. M. Herrera-Castillo, K. O. Winemiller, and A. Banda-Valdez. 2001a. Ecology of the alligator gar, *Atractosteus spatula*, in the Vicente Guerrero Reservoir, Tamaulipas, Mexico. Southwestern Naturalist 46: 151–157.

García de León, F. J., J. M. Herrera Castillo, and A. Banda Valdez. 2002. La Presa Vicente Guerreo Tamaulipas. Pp 504–543 *in* G. De la Lanza Espino and J. L. García C (eds.), Lagos y Presas de México, 2nd ed. AGT Editor, S.A., México D.F.

García de León, F. J., R. I. Rodríguez Martínez, and D. C. Morizot. 2001b. Do natural populations of largemouth bass (*Micropterus salmoides*) still exist in northeastern Mexico? Abstracts of the 131st American Fisheries Society Annual Meeting, 19–23 August 2001, Phoenix, Arizona. American Fisheries Society, Bethesda, Maryland.

Gilbert, C. R. 1998. Type catalog of recent and fossil North American freshwater fishes: families Cyprinidae, Catostomidae, Ictaluridae, Centrarchidae and Elassomatidae. Florida Museum of Natural History, Special Publication 1.

Girard, C. F. 1859. Ichthyology. Pp. 1–85 *in* United States and Mexican boundary survey, under the order of Lieut. Col. W. H. Emory, Major First Cavalry and United States commissioner. Fishes Mexican Boundary Survey, vol. 2 (pt 2). Fishes Pls. 1–41. U.S. Senate, 34th Cong. 1st sess., Ex. Doc. 108. A.O.P. Nicholson (printer), Washington, D.C.

Gómez-Soto, A. 1988. Ictiofauna y Recursos Ictiofaunísticos Pesqueros Actuales en la Laguna Madre, Tamaulipas, México. Licenciatura thesis, Universidad Autónoma de Nuevo León, Monterrey, Mexico.

Gordon, M. 1947. Genetics of *Platypoecilius maculatus*. IV. The sex determining mechanism in two wild populations of the Mexican platyfish. Genetics 32: 8–17.

Gordon, H., and M. Gordon. 1957. Maintenance of polymorphism by potentially injurious genes in eight natural populations of the platyfish, *Xiphophorus maculatus*. Journal of Genetics 55: 1–44.

Harris, P. M., and R. L. Mayden. 2001. Phylogenetic relationships of major clades of Catostomidae (Teleostes: Cypriniformes) as inferred from mitochondrial SSU and LSU rDNA sequences. Molecular and Phylogenetic Evolution 20: 225–237.

Harris, P. M., R. L. Mayden, H. Espinosa Pérez, and F. J. García de León. 2002. Phylogenetic relationships of redhorse (*Moxostoma*) and jumprock (*Scartomyzon*) suckers (Cypriniformes: Catostomidae) based on mitochondrial cytochrome b sequence data. Journal of Fish Biology 61: 1433–1452.

Helfman, G. S., B. B. Collette, and D. E. Facey. 1997. The Diversity of Fishes. Blackwell Science, Malden, Massachusetts.

Hendrickson, D. A., and J. K. Krejca. 2000. Cavefish and subterranean freshwater biodiversity in northeastern Mexico and Texas. Pp. 41–43 *in* R. A. Abell et al. (eds.), Freshwater Ecoregions of North America: A Conservation Assessment. Island Press, Washington, D.C.

Hendrickson, D. A., J. K. Krejca, and J. M Rodríguez. 2001. Mexican blindcats, genus *Prietella* (Ictaluridae): review and status based on recent explorations. Environmental Biology of Fishes 62: 315–337.

Horwitz, R. J. 1978. Temporal variability patterns and the distributional patterns of stream fishes. Ecological Monographs 48: 307–321.

Hubbs, C., R. J. Edwards, and G. P. Garrett. 1991. An annotated checklist of the freshwa-ter fishes of Texas. Texas Journal of Science (suppl.) 43 (4): 1–56.

Hubbs, C. L. 1926. Studies of the fishes of the order Cyprinodontes. VI. Miscellaneous Publications of the Museum of Zoology, University of Michigan 16: 1–86.

Hubbs, C. L. 1930. Materials for a revision of the Catostomid fishes of eastern North America. Miscellaneous Publications of the Museum of Zoology, University of Michigan 20: 1–47.

Hubbs, C. L., and R. M. Bailey. 1940. A revision of the black basses (*Micropterus* and *Huro*) with description of four new forms. Occasional Papers of the Museum of Zoology, University of Michigan 48: 1–51.

Hubbs, C. L., and M. Gordon. 1943. Studies of cyprinodont fishes. XIX. *Xiphophorus pygmaeus*, new species from Mexico. Copeia 1943: 31–33.

Hubbs, C. L., and L. C. Hubbs. 1932. Apparent parthenogenesis in nature in a form of fish of hybrid origin. Science 76: 628–630.

Hubbs, C. L., and C. Hubbs. 1958. *Notropis saladonis*, a new cyprinid fish endemic to the Rio Salado of northeastern Mexico. Copeia 1958: 297–307.

Hubbs, C. L., and R. R. Miller. 1974. *Dionda erimyzonops*, a new dwarf cyprinid inhabiting the gulf coastal plain of Mexico. Occasional Papers of the Museum of Zoology, University of Michigan 671: 1–17.

Hubbs, C. L., and R. R. Miller. 1975. *Notropis tropicus*, a new cyprinid fish from eastern Mexico. Southwestern Naturalist 20: 121–131.

Hubbs, C. L., and R. R. Miller. 1977. Six distinctive cyprinid fish species referred to *Dionda* inhabiting segments of the Tampico embayment drainage of Mexico. Transactions of the San Diego Society of Natural History 18: 267–336.

Hubbs, C. L., and R. R. Miller. 1978. *Notropis panareys*, n. sp., and *N. prosperinus*, cyprinid fishes of subgenus *Cyprinella*, each inhabiting a discrete section of the Rio Grande complex. Copeia 1978: 582–592.

INEGI. 1983. Síntesis Geográfica del Estado de Tamaulipas. Secretaría de Programación y Presupuesto, México D.F.

INEGI. 1997. Anuario Estadístico del Estado de Tamaulipas. Secretaría de Programación y Presupuesto, México D.F.

INEGI. 1998. Anuario Estadístico del Estado de Tamaulipas. Secretaría de Programación y Presupuesto, México D.F.

Jenkins, R. E. 1970. Systematic studies of the Catostomid fish tribe Moxostomatini. Ph.D. dissertation, Cornell University, Ithaca, New York.

Jordan, D. S. 1885. A catalog of the fishes known to inhabit the waters of North America north of the Tropic of Cancer. Report of the Commission of Fish and Fisheries for 1885.

Jordan, D. S., and M. C. Dickerson. 1908. Notes on a collection of fishes from the Gulf of Mexico at Vera Cruz and Tampico. Proceedings of the U.S. National Museum 34: 11–22.

Jordan, D. S., and B. W. Evermann. 1896–1900. The fishes of North and Middle America. Bulletin of the U.S. National Museum 47: 1–3313. Pt. 1, pp. 1–1230, 1896; Pt. 2, pp. 1231–2183, 1898; Pt. 3, pp. 2l83a-3136, 1899; Pt. 4., pp. 3137–3313, 1900.

Jordan D. S., and J. O. Snyder. 1899. Notes on a collection of fishes from rivers of Mexico, with description of twenty new species. Bulletin of the U.S. Fish Commission 19: 115–147.

Kassler, T. W., J. B. Koppelman, T. J. Near, C. B. Dillman, J. M. Levengood, D. L. Swofford, J. L. VanOrmann, J. E. Claussen, and D. P. Philipp. 2002. Molecular and morphological analysis of the black basses (Micropterus): implications for taxonomy and conservation. American Fisheries Society Symposium 31: 291–322.

Kazianis, S., D. C. Morizot, B. B. McEntire, R. S. Nairn, and R. L. Borowsky. 1996. Genetic mapping in Xiphophorus hybrid fish: assignment of 43 AP-PCR/RAPD and isozyme markers to multipoint linkage groups. Genome Research 6: 280–289.

Langecker, T. G., H. Schmale, and H. Wilkens. 1993. Transcription of the opsin gene in degenerate eyes of cave-dwelling Astyanax fasciatus (Teleostei, Characidae) and of its conspecific epigean ancestor during early ontogeny. Cell Tissue Research 273: 183–192.

Langecker, T. G., H. Wilkens, and H. Schmale. 1995. Developmental constraints in regressive evolution: studies of the expression of the gamma s-crystallin gene in the developing lens of cave-dwelling Astyanax fasciatus (Cuvier, 1819) (Teleostei, Characidae) by in situ hybridization. Journal of Zoology, Systematics, and Evolutionary Research 33: 123–128.

Lee, D. S., C. R. Gilbert, C. H. Hocutt, R. E. Jenkins, D. E. McAllister, and J. R. Stauffer, Jr. 1980. Atlas of North American freshwater fishes. North Carolina Biological Survey, Raleigh.

Livingstone, D. A., M. Rowland, and P. E. Bailey. 1982. On the size of African riverine fish faunas. American Zoologist 22: 361–369.

Lundberg, J. G. 1992. The phylogeny of ictalurid catfishes: a synthesis of recent work. Pp. 392–420 in R. L. Mayden (ed.), Systematics, Historical Ecology, and North American Freshwater Fishes. Stanford University Press, Palo Alto, California.

Mayden, R. L. 1989. Phylogenetic studies of North American minnows, with emphasis on the genus Cyprinella (Teleostei: Cypriniformes). Miscellaneous Publications of the Museum of Natural History, University of Kansas 80: 1–189.

Mayden, L. M., R. H. Maston, and D. M. Hillis. 1992. Speciation in the North American genus Dionda (Teleostei: Cypriniformes). Pp 710–746 in R. L. Mayden (ed.), Systematics, Historical Ecology, and North American Freshwater Fishes. Stanford University Press, Palo Alto, California.

McEachran, J. D., and J. D. Fechhelm. 1998. Fishes of the Gulf of Mexico, vol. 1. Myxiniformes to Gasterosteiformes. University of Texas Press, Austin.

Meek, S. E. 1904. The freshwater fishes of Mexico north of the Isthmus of Tehuantepec. Field Columbian Museum of Natural History Publications (Zoological Series) 5.

Menzel, B. W., and R. M. Darnell. 1973. Systematics of Poecilia mexicana (Pisces: Poeciliidae) in northeastern Mexico. Copeia 1973: 225–237.

Miller, R. R. 1966. Geographical distribution of Central American freshwater fishes. Copeia 1966: 773–802.

Miller, R. R. 1976. An evaluation of Seth E. Meek's contributions to Mexican ichthyology. Fieldiana Zoology 69: 1–31.

Miller, R. R. 1978. Composition and derivation of native fish fauna of the Chihuahuan Desert region, U.S. and Mexico. Pp 365–381 in R.H. Wauer and D.H. Riskind (eds.), Symposium on the Biological Resources of the Chihuahuan Desert Region, United States and Mexico. U.S. National Park Service Transactions and Proceedings Series 3.

Miller, R. R. 1983. Checklist and key to the mollies of Mexico (Pisces: Poeciliidae: Poecilia, subgenus mollienesia). Copeia 1983: 817–822.

Miller, R. R. 1986. Composition and derivation of the freshwater fish fauna of Mexico. Anuario de la Escuela Nacional de Ciencias Biológicas, Instituto Politécnico Nacional de México 30: 121–153.

Miller, R. R. In press. Fishes of Mexico. University of Chicago Press.

Miller, R. R., and W. L. Minckley. 1970. Gambusia aurata, a new species of poeciliid fish from northeastern Mexico. Southwestern Naturalist 15: 249–259.

Miller, R. R., and M. L. Smith. 1986. Origin and geography of the fishes of Central Mexico. Pp. 487–517 in C. H. Hocutt and E. O.

Wiley (eds.), The Zoogeography of North American Freshwater Fishes. John Wiley and Sons, New York.

Mitchell, R. W., W. H. Russell, and W. R. Eliott. 1977. Mexican eyeless characin fishes, genus *Astyanax*: environment, distribution, and evolution. Special Publications of the Museum of Texas Tech University 12.

Morizot, D. C., S. W. Calhoun, L. L. Cepper, M. E. Schmidt, J. H. Williamson, and G. J. Carmichael. 1991. Mutispecies hybridization among native and introduced centrarchid basses in central Texas. Transactions of the American Fisheries Society 120: 283–289.

Myers, G. S. 1938. Fresh-water fishes and West Indian zoogeography. Annual Report of the Smithsonian Institution 1937: 339–364.

Nedbal, M. A., and D. P. Philipp. 1994. Differentiation of mitochondrial DNA in largemouth bass. Transactions of the American Fisheries Society 123: 460–468.

Neff, B. D., P. Fu, and M. R. Gross. 1999. Microsatellite evolution in sunfish (Centrarchidae). Canadian Journal of Fisheries and Aquatic Sciences 56: 1198–1205.

Paulo-Maya, J. 1994. Análisis morfométrico del género *Astyanax* (Pisces: Characidae) en México. M.S. thesis, Escuela Nacional de Ciencias Biológicas, Instituto Politécnico Nacional, México D.F. 59 pp.

Pezold, F. L. III. 1984. A revision of the gobioid fish genus Gobionellus. Ph.D. dissertation, University of Texas at Austin.

Philipp, D. P., W. F. Childers, and G. S. Whitt. 1981. Management implications for different genetic stocks of largemouth bass (*Micropterus salmoides*) in the United States. Canadian Journal of Fisheries and Aquatic Sciences 38: 1715–1723.

Philipp, D. P., W. F. Childers, and G. S. Whitt. 1983. A biochemical genetic evaluation of the Northern and Florida subspecies of largemouth bass. Transactions of the American Fisheries Society 112: 1–20.

Philipp, D. P., and G. S. Whitt. 1991. Survival and growth of Northern, Florida and reciprocal F1 hybrid largemouth bass in central Illinois. Transactions of the American Fisheries Society 1: 59–64.

Rauchenberger, M. 1989. Systematics and biogeography of the genus *Gambusia* (Cyprinodontiformes: Poeciliidae). American Museum Novitates 2951: 1–74.

Rauchenberger, M., K. D. Kallman, and D. C. Morizot. 1990. Monophyly and geography of the Río Panuco basin swordtails (Genus *Xiphophorus*) with descriptions of four new species. American Museum Novitates 2975: 1–44.

Regan, C. T. 1905. A revision of the fishes of the American cichlid genus *Cichlasoma* and of the allied genera. Annals and Magazine of Natural History (ser. 7) 16: 60–77, 225–243, 432–445.

Regan, C. T. 1906–1908. Pisces. Biologia Centrali-Americana 8: 1–203.

Rivas, L. R. 1963. Subgenera and species groups in the poeciliid fish genus *Gambusia* Poey. Copeia 1963: 331–347.

Rivera-Teillery, R. 1971. Ictiofauna de los Ríos San Fernando y Soto la Marina, Estados Unidos de Nuevo León y Tamaulipas. Licenciatura thesis, Universidad Autónoma de Nuevo León, Monterrey, Mexico.

Rodríguez-Martínez, R. I. 2001. Estudio morfológico y genético de las poblaciones de Lobina Negra (*Micropterus salmoides*) en el Noreste de México. Licienciatura thesis, Instituto Tecnológico de Ciudad Victoria, Tamaulipas, Mexico.

Rodríguez-Olmo, G. 1975. Campos en la composición de especies de peces en comunidades del bajo Río Bravo, México-Estados Unidos. M.S. thesis, Universidad Autónoma de Nuevo León, Monterrey, Mexico.

Rosen, D. E., and R. M. Bailey. 1963. The Poeciliid fishes (Cyprinodontiformes), their structure, zoogeography and systematic. Bulletin of the American Museum of Natural History 126: 1–176.

Rosenthal, G. G., X. F. de la Rosa Reyna, D. C. Morizot, S. Kazianis, M. J. Stephens, M. J. Ryan, and F. J. García de León. 2003. Dissolution of sexual signal complexes in a hybrid zone between the swordtails *Xiphophorus birchmanni* and *X. malinche* (Poeciliidae). Copeia 2003: 299–307.

Ruiz-Campos, G. 1982. Datos ictiobiológicos del Río Álamo, Subcuenca del Río Bravo, Noreste de México. Licienciatura thesis, Universidad Autónoma de Nuevo León, Monterrey, Mexico.

Schartl, M. 1995. Platyfish and swordtails: a genetic system for the analysis of molecular mechanisms in tumor formation. Trends in Genetics 11: 185–189.

Schartl, M., I. Schlupp, A. Schartl, M. Meyer, I. Nanda, M. Schmid, J. T. Epplen and J. Parzefall. 1991. On the stability of dispensable constituents of the eukaryotic genome: stability of coding sequences versus truly hypervariable sequences in a clonal vertebrate, the Amazon Molly, *Poecilia formosa*. Proceedings of the National Academy of Sciences USA 88: 8759–8763.

Schartl, M., B. Wilde, I. Schlupp, and J. Parzefall. 1995. Evolutionary origin of a parthenoform, the amazon molly *Poecilia formosa* on the basis of a molecular genealogy. Evolution 49: 827–835.

Schlupp, I., I. Nanda, M. Döbler, D. K. Lamatsch, J. Epplen, J. Parzefall, M. Schmid, and M. Schartl. 1998. Dispensable and indispensable genes in an ameiotic fish, the Amazon molly *Poecilia formosa*. Cytogenetics and Cell Genetics 80: 193–198.

Schlupp, I., J. Parzefall, J. T. Epplen, I. Nanda, M. Schmid, and M. Schartl. 1992. Pseudo-male behaviour and spontaneous masculin-ization in the all-female teleost *Poecilia formosa* (Teleostei: Poeciliidae). Behaviour 122: 88–104.

Schlupp, I., J. Parzefall, and M. Schartl. 2002. Biogeography of the unisexual Amazon molly, *Poecilia formosa*. Journal of Biogeography 29: 1–6.

SEMARNAT. 2002. Norma Oficial Mexicana NOM-059-ECOL-2001, Protección ambiental—Especies nativas de México de flora y fauna silvestre – Categorías de riesgo y especificaciones para su inclusión, exclusión o cambio – Lista de especies en riesgo. Diario Oficial de la Federación, México (Miercoles 6 de marzo 2002).

Sheldon, A. L. 1987. Rarity: patterns and consequences for stream fishes. Pp 203–209 *in* W. J. Matthews and D. C. Heins (eds.), Community and Evolutionary Ecology of North American Stream Fishes. University of Oklahoma Press, Norman.

Sheldon, A. L. 1988. Conservation of stream fishes: patterns of diversity, rarity and risk. Conservation Biology 2: 149–156.

Smith, G. R. 1992. Phylogeny and biogeography of the Catostomidae, freshwater fishes of North America and Asia. Pp. 778–826 *in* R. L. Mayden (ed.), Systematics, Historical Ecology, and North American Freshwater Fishes. Stanford University Press, Palo Alto, California.

Smith, L. M., and R. R. Miller. 1986. The evolution of the Rio Grande basin as inferred from its fish fauna. Pp 457–485 *in* C. H. Hocutt and E. O. Wiley (eds.), The Zoogeography of North American Freshwater Fishes. John Wiley and Sons, New York.

Suttkus, R. D. 1963. Order Lepisostei. Pp. 61–88 *in* Fishes of the Western North Atlantic. Memoir 1, Part 3. Sears Foundation for Marine Research, Yale University, New Haven, Connecticut.

Taylor, J. N., and R. R. Miller. 1983. Cichlid fishes (Genus *Cichlasoma*) of the Rio Panuco basin, eastern Mexico, with description of a new species. Occasional Papers of the Museum of Natural History, University of Kansas 104: 1–24.

Treviño-Robinson, D. T. 1959. The ichthyofauna of the lower Rio Grande, Texas and Mexico. Copeia 1959: 253–256.

Turner, B. J. (ed.). 1984. Evolutionary Genetics of Fishes. Plenum Press, New York.

U.S. Fish and Wildlife Service. 2002. 50 CFR Part 17: Endangered and Threatened Wildlife and Plants; Designation of Critical Habitat for the Rio Grande Silvery Minnow; Proposed Rule. Federal Register 67 (109): 39205–39235.

Verduzco-Martínez, J. A. 1972. Ictiofauna del Río Panuco, Noreste de México. Licen-ciatura thesis, Universidad Autónoma de Nuevo León, Monterrey, Mexico.

Villarreal-Treviño, C. M. 1983. Cambios en las comunidades de peces por factores físico-químicos en el Río San Juan, provincia del Río Bravo, Noreste de México. Licienciatura thesis, Universidad Autónoma de Nuevo León, Monterrey, Mexico.

Vrijenhoek, R. C. 1994. Unisexual fish: model systems for studying ecology and evolution. Annual Review of Ecology and Systematics 25: 71–96.

Vrijenhoek, R. C., R. M. Dawley, C. J. Cole, and J. P. Bogart. 1989. A list of known unisexual vertebrates. Pp 19–23 *in* R. M. Dawley and J. P. Bogart (eds.), Evolution and Ecology of Unisexual Vertebrates. New York State Museum, Albany.

Walsh, J., and C. R. Gilbert. 1995. New species of troglobitic catfish of the genus *Prietella* (Siluriformes: Ictaluridae) from Northeastern Mexico. Copeia 1995: 850–861.

Wilcox, T. P., F. J. García de León, D. A. Hendrickson, and D. M. Hillis. 2004. Convergence among cave catfishes: long-branch attraction and a Bayesian relative rates test. Molecular Phylogenetic Evolution 31: 1101–1113.

Wiley, E. O. 1976. The phylogeny and biogeog-raphy of fossil and recent gars (Actinop-terigii: Lepisosteidae). Miscellaneous Publications of the Museum of Natural History, University of Kansas 64: 1–11.

Wiley, E. O., and S. Hans-Peter. 1984. Family Lepisosteidae (gars) as living fossils. Pp. 160–165 *in* N. Eldredge and S. M. Stanley (eds.), Living Fossils. Springer Verlag, New York.

Wilkens, H. 1988. Evolution and genetics of epigean and cave *Astyanax fasciatus* (Characidae, Pisces). Evolutionary Biology 23: 271–367.

Williamson, J. H., G. J. Carmichael, M. E. Scmidt, and D. C. Morizot. 1986. New biochemical genetic markers for largemouth bass. Transactions of the American Fisheries Society 115: 460–465.

8

Historical and Ecological Biogeography of the Terrestrial Herpetofauna of Northern Baja California

L. LEE GRISMER

ERIC MELLINK

After traveling along the entire length of the Baja California peninsula in 1837, the German naturalist F. Deppe described it as having very little wildlife, especially reptiles and insects (Lichtenstein 1839). As it turned out, Deppe's impression of the regional herpetofauna was not borne out by subsequent research. A large body of studies during the twentieth century has revealed relatively few amphibian species but a fairly high diversity of reptiles, especially among lizards and snakes (for a comprehensive overview see Grismer 2002). This is hardly a surprise, given that the diversity of habitats occurring in Baja California is among the highest in North America. With its southern tip reaching below the tropic of cancer, the peninsula spans more than 10° in latitude. The high variety of local climates is the result of both a complex topography and the influence of two very dissimilar bodies of water nearly surrounding it, the Pacific Ocean and the Gulf of California. All of these factors have created and maintain very different habitats within close proximity of each other.

The area from El Rosario in the west and Bahía de San Luis Gonzaga in the east, northward to the U.S.–Mexico border, has amazing herpetofaunal diversity (fig. 8.1; Islas Coronado and Islas Todos Santos, offshore from Tijuana and Ensenada, respectively, are also included within our focus area). In addition to the bullfrog (*Rana catesbeiana*), excluded from this account, the area supports 86 (84 extant) species, or >90% of the peninsula's amphib-

ians and reptiles. The regional herpetofauna consists specifically of 18 (16 extant) amphibians, 2 turtles, 35 lizards, and 31 snakes. The diversity of amphibians in particular contrasts with that of Baja California Sur, which has only 3 native species.

In this chapter, we examine the historical and ecological biogeography of northern Baja California reptiles and amphibians. A history of vicariant and dispersal events bears testimony to geologic events and associated climate changes. Patterns of distribution observed today are the results of those events. At the same time, they reflect the current, sharp delineation of Pacific coastal, montane, and desert habitats.

Although our focus here is unrelated to the conservation of Baja California's herpetofauna, some of the species we mention are threatened by humans, primarily through illegal collecting and associated habitat degradation (Mellink 1995; Grismer 2002). Some areas of northern Baja California (Sierra Juárez, Sierra San Pedro Mártir, along Mexican Highway 1) bear the mark of collectors breaking apart rock piles in search of California mountain kingsnakes (*Lampropeltis zonata*), rosy boas (*Lichanura trivirgata*), and banded rock lizards (*Petrosaurus mearnsi*). The bullfrog has expanded its range through introductions and natural dispersal and is present in the Baja California peninsula including the area north of El Rosario and Bahía San Luis Gonzaga (Grismer 2002). It is negatively affecting some native species such as the two-striped garter *snake* (*Thamnophis*

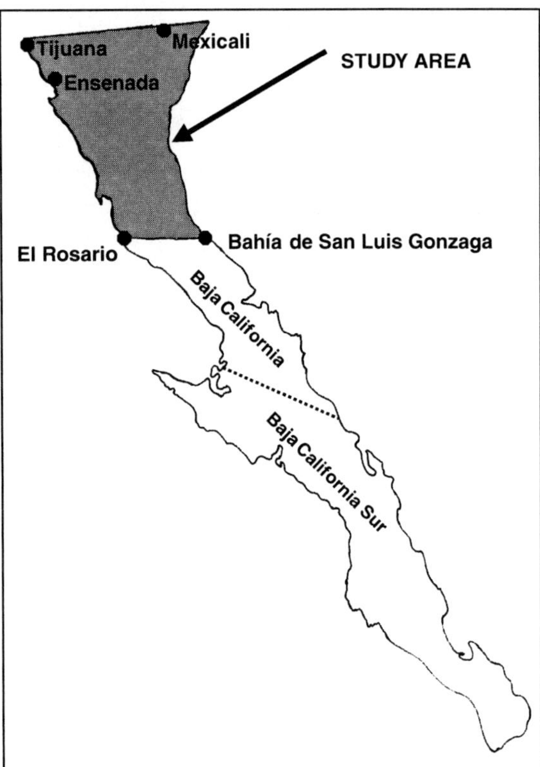

Figure 8.1. Baja California peninsula. The study area spans from the U.S.–Mexico border south to Santa Rosalia in the west and Bahía de San Luis Gonzaga in the east. Islas Coronado and Islas Todos Santos (not on this map) are also included in the study area. These 2 groups of small islands are located offshore from Tijuana and Ensenada, respectively.

hammondii). For more information on amphibian and reptile conservation in Baja California, see Grismer (2002) and chapter 20.

Historical Biogeography

The main goal of historical biogeography is to link the regional occurrence of a species with past environmental events. It requires knowledge of the region's environmental history (paleogeography and paleoclimate) and of species' phylogenies.

Two geologic events have greatly influenced the herpetofauna of Baja California as a whole: the northward-trending separation of the peninsula from mainland Mexico and the formation of the Gulf of California. Before regressing to its current position, the Gulf of California extended farther north, into northern California. Species once continuously distributed became divided into allopatric eastern and western populations. The southward regression of the Gulf of California brought many of these populations back in contact, but in some cases not before speciation had occurred. With the regression of the Gulf of California and the simultaneous formation of mid-peninsular deserts, xerophilic species that had evolved with deserts of the southwestern United States apparently invaded the Baja California peninsula.

Grismer (1994a, 2002) divided the peninsular herpetofauna into various biogeographical complexes based on hypotheses concerning their evolutionary origin. Five extant species are endemic or nearly endemic to northern Baja California (including Islas Coronado and Islas Todos Santos): the Islas Coronado alligator lizard (*Elgaria nana*), Sierra los Cucapás collared lizard (*Crotaphytus grismeri*; fig. 8.2), banded rock lizard, Islas Todos Santos mountain kingsnake (*Lampropeltis herrerae*), and Islas Coronado rattlesnake (*Crotalus caliginis*). Two other species are not considered here. The Rio Grande leopard frog (*Rana berlandieri*) is a recent probable addition to the Baja California herpetofauna (Grismer 2002). The spiny soft-shell turtle (*Apalone spinifera*) is common in northeastern Baja California but was introduced in the mid-1940s (Grismer 2002). The rest of the northern peninsular herpetofauna (n = 77 extant and 2 extirpated species; 1 species is rep-

Figure 8.2. The Sierra los Cucapás collared lizard (*Crotaphytus grismeri*). This reptile is endemic to northern Baja California. (Photo by L. Grismer.)

resented in 2 of the 6 complexes) are classified as below.

Southern Miocene Vicariant Complex

Members of the southern Miocene vicariant complex are endemic to Baja California and apparently evolved in southern Baja California as the peninsula was separating from mainland Mexico during the Miocene. All species (or monophyletic groups of endemic species) of this complex have close relatives in southwestern Mexico. This group is composed of the orange-throated whiptail (*Cnemidophorus hyperythrus*), western skink (*Eumeces skiltonianus*), peninsular leaf-toed gecko (*Phyllodactylus xanti*), granite spiny lizard (*Sceloporus orcutti*), black-tailed brush lizard (*Urosaurus nigricaudus*), Baja California rat snake (*Bogertophis rosaliae*), western black-headed snake (*Tantilla planiceps*), and Baja California rattlesnake (*Crotalus enyo*).

Northern Pliocene Vicariant Complex

Amphibians and reptiles of the northern Pliocene vicariant complex were widespread in northern Baja California, the southwestern United States, and northwestern Mexico when northern Baja California existed as a mere western extension of mainland Mexico. The distribution of these forms became bi-sected with the northern extension of the newly forming Protogulf of California in the late Miocene to early Pliocene when it extended at least as far north as the Banning Pass in Riverside County, California. This separation prompted divergence or even speciation on either side of the northern portion of the Gulf of California. When the Gulf receded to its current position in the Pleistocene, the new sister taxa came into contact with one another and are now parapatric near the head of the Gulf of California or occur on opposite sides of the Coachella Valley in California. This group is composed of western banded gecko (*Coleonyx variegatus*) sister populations, Baja California collared lizard (*Crotaphytus vestigium*) group/Great Basin collared lizard (*C. bicinctores*; not found on the peninsula), Baja California leopard lizard (*Gambelia copeii*)/long-nosed leopard lizard (*G. wislizenii*), desert spiny lizard (*Sceloporus magister*)/Baja California spiny lizard (*S. zosteromus*), banded sand snake (*Chilomeniscus cinctus*) sister populations, coachwhip (*Masticophis flagellum*)/Baja California coachwhip (*M. fuliginosus*), ground snake (*Sonora semiannulata*) sister populations, lyre snake (*Trimorphodon biscutatus*) sister populations, and western diamondback rattlesnake (*Crotalus atrox*)/red diamond rattlesnake (*C. ruber*).

Western Desert Complex

Western desert complex species are currently distributed continuously around the head of the Gulf of

California in close association with the Sonoran, Mojave, and Great Basin deserts. Presumably they evolved outside of Baja California in association with the formation of these deserts and entered Baja California from the northeast subsequent to the regression of the Gulf of California. As such, their geographic variation on the peninsula is generally weak and clinal, and their closest relatives occur in northern mainland Mexico. A portion of this group includes species that are situated at the head of the Gulf of California and only marginally enter northeastern Baja California. These species are closely associated with the low-lying, sandy regions left behind after the regression of the Gulf or the flora therein and have not dispersed farther south into Baja California. This group is composed of the zebra-tailed lizard (*Callisaurus draconoides*), desert iguana (*Dipsosaurus dorsalis*), flat-tailed horned lizard (*Phrynosoma mcallii*), desert horned lizard (*Phrynosoma platyrhinos*), northern chuckwalla (*Sauromalus obesus*), Colorado Desert fringe-toed lizard (*Uma notata*), long-tailed brush lizard (*Urosaurus graciosus*), western shovel-nosed snake (*Chionactis occipitalis*), spotted leaf-nosed snake (*Phyllorhynchus decurtatus*), and western patch-nosed snake (*Salvadora hexalepis*).

Río Colorado Complex

In the Río Colorado complex are species restricted to the northeast of the peninsula but with diverse origins outside of Baja California. On the peninsula, these forms are strictly associated with the Colorado River, its drainages, and/or its floodplain vegetational associates, and they probably entered Baja California after the regression of the Gulf. The Río Colorado complex is composed of the Colorado River toad (*Bufo alvarius*), Great Plains toad (*Bufo cognatus*), Woodhouse's toad (*Bufo woodhousii*), Yavapai leopard frog (*Rana yavapaiensis*, extirpated), tree lizard (*Urosaurus ornatus*), Sonoran gopher snake (*Pituophis catenifer affinis*), and checkered garter snake (*Thamnophis marcianus*).

Northwestern Complex

The northwestern complex consists of mesophilic forms found in northwestern Baja California. The member species appear to have entered from California, their dispersal facilitated by the uplift of coastal mountains in southern California during the Pleistocene. This uplift may have forced the Gulf to recede to its present position and, at the same time, provided a mesic corridor through which these species could disperse southward. The fact that they generally dispersed south only to El Rosario was likely the result of the simultaneous formation of the arid Vizcaíno Region, which would have served as an effective xeric barrier to further dispersal, just as it does today. The closest relatives of these forms are found in the Pacific Northwest and/or central United States. This group is the most diverse of the 6 complexes, with 17 extant and 1 extirpated species. It is composed of the arboreal salamander (*Aneides lugubris*), Monterey ensatina (*Ensatina eschscholtzii*), large-blotched ensatina salamander (*Ensatina klauberi*), California treefrog (*Hyla cadaverina*), Pacific treefrog (*Hyla regilla*), red-legged frog (*Rana aurora*), foothill yellow-legged frog (*Rana boylii*, extirpated), western spadefoot (*Spea hammondii*), western pond turtle (*Clemmys marmorata*), southern alligator lizard (*Elgaria multicarinata*), coast horned lizard (*Phrynosoma coronatum*), western fence lizard (*Sceloporus occidentalis*), southern sagebrush lizard (*Sceloporus vandenburgianus*), ringneck snake (*Diadophis punctatus*), western terrestrial garter snake (*Thamnophis elegans*), two-striped garter snake, and western rattlesnake (*Crotalus viridis*). The final member of this group is the gopher snake (*Pituophis catenifer*), except for its Sonoran race, which belongs to the Río Colorado complex (see above).

Chaparral-Madrean Woodland Complex

In the chaparral-madrean woodland complex are mesophilic species restricted to cismontane and/or montane woodland mesic environments of northern Baja California. These species were presumably widespread through North America during cooler and wetter periods, but with the onset of Pleistocene drying trends became fragmented and restricted to higher elevations where these cool and moist climates still prevail. The closest relatives of these species occur in the northern portions of the Sierra Madre Occidental and the mountains of central Arizona. This group is composed of the California toad (*Bufo californicus*), Baja California whiptail (*Cnemidophorus labialis*), Gilbert's skink (*Eumeces gilberti*), and California mountain kingsnake.

Species of Unknown Evolutionary Origin in the Region

There is also a group of ubiquitous species that are widespread throughout Baja California and that are found also in the United States and mainland Mexico, along with their closest living relatives. Their widespread distribution makes it difficult to form viable hypotheses concerning their historical origin in Baja California. This group consists of the red-spotted toad (*Bufo punctatus*), Couch's spadefoot (*Scaphiopus couchii*), side-blotched lizard (*Uta stansburiana*), glossy snake (*Arizona elegans*), night snake (*Hypsiglena torquata*), and common kingsnake (*Lampropeltis getula*).

Finally, the phylogenetic placement and geographic distribution of several species remains poorly understood, and they cannot yet be classified into historical biogeographic units: garden slender salamander (*Batrachoseps major*), western toad (*Bufo boreas*), Baja California legless lizard (*Anniella geronimensis*), California legless lizard (*Anniella pulchra*), western whiptail (*Cnemidophorus tigris*), barefoot banded gecko (*Coleonyx switaki*), granite night lizard (*Xantusia henshawi*), desert night lizard (*Xantusia vigilis*), sidewinder (*Crotalus cerastes*), speckled rattlesnake (*Crotalus mitchellii*), western blind snake (*Leptotyphlops humilis*), rosy boa, California striped racer (*Masticophis lateralis*), and long-nosed snake (*Rhinocheilus lecontei*).

Ecological Biogeography

As already mentioned, the high diversity of the herpetofauna of northern Baja California is also due to the wide variety of current habitats, which is in turn the result of physical factors such as climate and topography. The distribution and genetic structuring of regional species reflects the sharp boundaries between Pacific coastal, montane, and desert habitats.

Climate

Northern Baja California receives the majority of its precipitation from winter storms that usually originate in the western Pacific and sweep southeastward over the peninsula (Hastings and Turner 1965). These storm fronts lose their influence as they stretch south (Humphrey 1974) and usually do not extend past Laguna San Ignacio. The gradual rise in elevation of the western slopes of the Sierra Juárez and the Sierra San Pedro Mártir from the Pacific coast to their crests induces much precipitation as rain and snow, relieving these passing storm fronts of the majority of their moisture. Consequently, little rain falls in areas to the east of these mountains (Hastings and Humphrey 1969; Markham 1972), although their associated low-pressure systems cause strong winds. This rainshadow effect has resulted in the adjacent desert regions being the driest area in North America, receiving about one-half as much annual rainfall as Death Valley, California (Markham 1972). They also represent the hottest area in North America (Meigs 1953).

As northern desert regions of North America heat up during the summer months, the rising air creates a low-pressure cell, which draws in warm, moist air from tropical Pacific areas farther south. As this air mass gains momentum and surges across the Gulf of California, it picks up additional moisture. When it moves onto land, it slams into the precipitous eastern face of the Peninsular Ranges, forcing this moisture-laden air to ascend as it crosses the mountains. As this air rapidly rises, it cools and causes its moisture to precipitate in the form of heavy, and often violent, rain showers with spectacular displays of thunder and lightning. When tropical moisture is present, such storms (*aguaceros*) can occur almost daily in the higher elevations of the Sierra San Pedro Mártir.

Hurricanes and hurricane-fringe storms also impact the region during the summer and fall. From July through November these storms may sweep up the Gulf of California, hitting the east coast of the peninsula. Hurricanes that do not come onto shore but continue up the Gulf of California can increase in intensity as they move northward, potentially generating winds of more than 200 kph. Hurricanes are usually associated with large quantities of rain (Hastings and Humphrey 1969; Markham 1972) and can cause extensive damage when they move onto land.

Chubascos or *toritos* are smaller anticyclonic systems that usually develop more locally within the Gulf of California. They, too, are characterized by strong winds and locally heavy rains and can cause extensive localized damage when they move onto land. Although they generally do not last for more than a few hours, they create some of the most dangerous conditions in the Gulf of California.

Differences in ambient temperature across Baja California result from the dissimilarities of its surrounding bodies of water. The Pacific Ocean dominates the temperature regime of western Baja California with its southerly flowing, cold California Current. This current is responsible for relatively high amounts of coastal advection fog (Shreve 1951) and cloud cover, which, with steady onshore breezes, keep the coastal temperatures relatively low.

Freezing temperatures are rare in the western coastal areas but quite common in the northern mountains. Snow begins to blanket these ranges in late November but usually does not persist for more than a few days, except at the higher elevations in the Sierra San Pedro Mártir, where snow may last considerably longer and fall as late as March. From March to November, the northwest portion of the peninsula generally has windy days and cloudy nights. Although little rain falls (Humphrey 1974), early morning low cloud cover usually extends far enough inland to blanket the foothills of the Peninsular Ranges and adds a significant amount of moisture to this region's precipitation (Markham 1972). Much of this cloud cover comes in the form of radiational fog, which usually burns off by mid-morning.

The temperature of the eastern portions of Baja California are primarily controlled by the Gulf of California (Sea of Cortez). This is a much warmer body of water than the nearby Pacific Ocean (Robinson 1973) and offers little precipitation or cooling to the peninsula. The areas lying east of the Peninsular Ranges receive no cooling from onshore Pacific breezes, and, consequently, the Gulf coast becomes extremely hot during summer months (Markham 1972).

Physiography and Phytogeography of Baja California North of 30° Latitude

Baja California's wide range of climatic conditions and well-sculpted topography support a clearly defined array of phytogeographic regions (Shreve and Wiggins 1964; Wiggins 1980; Grismer 1994b; fig. 8.3). These regions are good indicators of natural biotic provinces because they are ecological reflections of the interactions of climate, topography, and soil. Phytogeographic regions are set apart from one another with the assumption that many of the plants therein share the same ecological and environmental history. A comparison of the distribution and geographic variation of the herpetofauna with the phytogeographic regions of Baja California reveals that the two coincide very closely (Grismer 1994b) as a result of the ecological restrictions imposed on the biota. For example, *Lampropeltis zonata*, the California mountain kingsnake, occurs only in the Sierras Juárez and San Pedro Mártir; *Aneides lugubris*, the arboreal salamander, only in the California region; and *Callisaurus draconoides*, the zebra-tailed lizard, only in the deserts.

Northern Baja California has 3 phytogeographic regions (Cody et al. 1983; Grismer 1994b, 2002; adapted from Shreve and Wiggins 1964; Wiggins 1980; Turner and Brown 1982). Their delineation in turn reflects that of 3 physiographic regions. Two of the 3 physiographic and phytographic regions are also discussed by Minnich and Franco-Vizcaíno in chapter 18.

The Cismontane and California Regions

The cismontane region of Baja California extends from the foothills of the Sierra Juárez and the Sierra San Pedro Mártir westward to the Pacific coast. This region consists of low foothills, plains, and mesas of varying sizes, which are occasionally interrupted by wide canyons and deep arroyos. This region descends gently westward toward the Pacific coast from the crest of the Peninsular Ranges, in some places terminating in precipitous, coastal bluffs. The most dominant geographical feature of this region is a small range of 3 narrowly connected mountains, which lie inland from Punta Santo Tomás, approximately 110 km below the U.S.–Mexico border. From north to south, these mountains are Sierra Peralta, Sierra Warner, and Sierra San Miguel. The foothills of these ranges stop short of the Pacific and give way to a series of narrow, coastal plains known collectively as the San Quintín Plain. This plain extends as far south as Rancho Socorro, where it develops into an even narrower plain extending nearly to El Rosario. The Sierra Warner and the Sierra Peralta form a portion of the northwestern border of Valle de Trinidad (see below). Between the Sierra San Miguel in the west and Sierra San Pedro Mártir in the east lie a series of narrow, arid valleys collectively known as Valle de San José. These valleys are buffered from cool Pacific breezes by the western ranges and, consequently, heat up quickly during the summer months (Shreve 1951).

Phytogeographically, the California Region occupies the cismontane areas of northern Baja Cali-

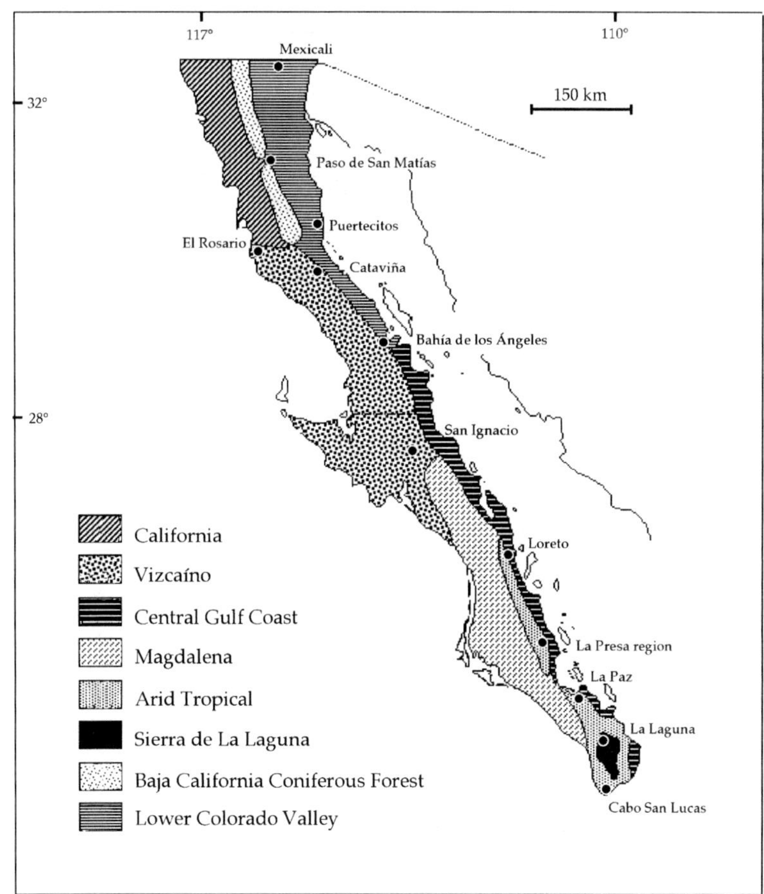

Figure 8.3. Phytographic regions of the Baja California peninsula.

fornia from the U.S.–Mexico border southward to the vicinity of Rancho Socorro (fig. 8.3). Its eastern border is situated on the lower western slopes of the Sierra Juárez and the Sierra San Pedro Mártir at the edge of the Jeffrey Pine Belt. This region is a southern extension of the coastal scrub and chaparral communities of southern California. The Pacific coastal scrub occurs in the sporadic, flat, low-lying western portions of this region from sea level to the edge of the chaparral belt at approximately 300–600 m in elevation. Among the dominant plant species here are drought-resistant deciduous forms such as California sagebrush (*Artemisia californica*), white and black sage (*Salvia apiana* and *S. mellifera*), California buckwheat (*Eriogonum fasciculatum*), and coastal agave (*Agave shawii*).

The chaparral portion of the California Region occurs primarily in canyons in the coastal regions and at the higher elevations, in the foothills of the Peninsular Ranges to the east. Chaparral assemblages are characterized by larger (1–3 m) evergreen shrubs such as chamise (*Adenostoma fasciculatum*), hoary-leaf lilac (*Ceanothus crassifolia*), chaparral ash (*Fraxinus trifoliata*), toyon (*Heteromeles arbutifolia*), scrub oak (*Quercus berberidifolia*), laurel sumac (*Malosma laurina* [= *Rhus l.*]), lemonade berry (*Rhus integrifolia*), and red shank (*Adenostoma sparsifolium*). Although the California Region reaches to El Rosario in the south (Mellink 2002), inland from Cabo Colonet it begins a gradual transition into the more arid, southerly Vizcaíno Region.

The Montane and Baja California Coniferous Forest Regions

The Montane Region of northern Baja California comprises 2 major mountain ranges, the Sierra

Juárez and the Sierra San Pedro Mártir, which are an extension of a series of mountains continuing southward out of southern California. They gently grade into coastal cismontane areas to the west, but abruptly transform into desert transmontane areas to the east. The Sierra Juárez reaches an elevation of nearly 1410 m, and at its upper elevations is flat and has an ephemeral body of water, Laguna Hanson. The Sierra San Pedro Mártir is a much more dominating feature. Its highest peak, Picacho del Diablo, reaches 3096 m, and from this elevation one is able to see both coastlines of Baja California as well as the coastline of Sonora. Between the southern end of the Sierra Juárez and northern end of the Sierra San Pedro Mártir lies Paso de San Matías, at the eastern end of Valle de Trinidad. Northwest of Valle de Trinidad is Valle de San Rafael (currently Ojos Negros), and north of Ensenada, Valle de Guadalupe.

The Baja California Coniferous Forest Region is composed of 2 disjunct sections that represent the southernmost portion of the broader and more inclusive Sierran Montane Conifer Forest (Pase 1982). This is a cool, mesic area at the upper elevations of the Sierra Juárez and the Sierra San Pedro Mártir, above the chaparral belt of the California Region in the west and the creosote bush scrub of the Lower Colorado Valley Region in the east. This region extends southward from the U.S.–Mexico border about 300 km to approximately Cerro Matomí. This region is bisected between the Sierra Juárez and Sierra San Pedro Mártir by Paso de San Matías (San Matías Pass).

The Baja California Coniferous Forest Region receives more precipitation than any other area in northern Baja California. The majority of this comes from cold, northern winter storms, but a significant portion also results from convectional, southern summer storms. Consequently, the floristic composition is relatively diverse and composed of many large shrubs and trees, but with a conspicuous lack of understory vegetation. The dominant species of this region are piñon pine (*Pinus quadrifolia*) and Jeffrey pine (*P. jeffreyi*), the former being more prevalent in the Sierra Juárez and the latter being more prevalent in the Sierra San Pedro Mártir in the higher elevations. Other trees in the area include lodgepole pine (*P. contorta* ssp. *murayana*), sugar pine (*P. lambertiana*), white fir (*Abies concolor*), quaking aspen (*Populus tremuloides*), and incense cedar (*Calocedrus decurrens*).

Transmontane and Lower Colorado Valley Regions

The Transmontane or desert portion of Baja California lies to the east of the northern Peninsular Ranges. It is a low-elevation area with small, arid, rocky mountains. East of the Sierra Juárez are the Sierra de Los Cucapás, Sierra El Mayor, and Sierra Las Pintas. The Sierra de Los Cucapás and Sierra El Mayor, the northernmost of these mountain ranges, form a contiguous, isolated mass extending approximately 80 km below the U.S.–Mexico border. They are separated from the Sierra Juárez by the Laguna Salada (or Macuata) basin. These ranges support only scant vegetation and a depauperate herpetofauna. Their highest peak reaches an elevation of 1000 m, and the entire range is surrounded by the ancient Río Colorado floodplain.

Sixty kilometers south of the Sierra El Mayor and across the southwestern margin of the Laguna Salada basin lies the northern end of the Sierra Las Pintas. The Sierra Las Pintas is a contorted, jagged, colorful, volcanic mountain range whose peaks protrude through the southern end of the dry lake bed. The Sierra Las Pintas and Sierra Juárez are linked at their southern bases by a wide set of rocky, volcanic hills that come together in the north to form the Sierra Las Tinajas. This latter range then continues northward into the Laguna Salada basin as an isolated finger of land between the Sierra Juárez and Sierra Las Pintas. The southern connection of the Sierra Juárez, Sierra Las Pintas, and Sierra Las Tinajas may account, in part, for the greater floral and faunal diversity of the latter two as compared to that of the more insular Sierra de los Cucapás and Sierra El Mayor.

Immediately south of the Sierra Las Pintas and east of the Sierra de San Pedro Mártir lies the Sierra de San Felipe. This range reaches 1332 m in elevation and is separated from the Sierra San Pedro Mártir to the west by the Laguna del Diablo basin. This is an ephemeral dry lake, which is filled primarily by runoff from the eastern slopes of the Sierra San Pedro Mártir during bouts of heavy summer precipitation. The Sierra San Felipe angles away from the Sierra San Pedro Mártir to the southeast and generally runs uninterrupted past Bahía de San Felipe. At this point it grades into a small series of extremely rugged and broken volcanic mesas and canyons known as the Sierra Santa Isabel and Sierra Santa Rosa. These ranges reach 1200 m in elevation and slope gently toward the Gulf of California.

The Lower Colorado Valley Region is the largest subdivision of the Sonoran Desert, but only a thin, southern extension of its western portion enters Baja California. Here, it occupies the areas between the Río Colorado and the western coastline of the Gulf of California and the eastern foothills of the Peninsular Ranges, from the U.S.–Mexico border southward to just below Bahía de los Ángeles. In its northern portion, the region is a low-lying, flat, and sandy area in the rainshadow of the Sierras Juárez and San Pedro Mártir. South of here from Puertecitos to Bahía de los Ángeles, it is very rugged and composed of many small, eroded volcanic terraces and mountain ranges.

The Lower Colorado Valley Region of Baja California is extremely hot and arid and dominated by various small-leafed, drought-resistant plants. The most common of these in the flat, sandy, northern areas are creosote bush (*Larrea divaricata*), white bursage (*Ambrosia dumosa*), ocotillo (*Fouquieria splendens*), brittlebush (*Encelia farinosa*), and desert agave (*Agave deserti*). In the arroyos and rugged foothill areas, mesquite (*Prosopis glandulosa*), smoke tree (*Psorothamnus spinosus*), little-leaf palo verde (*Cercidium microphyllum*), ironwood (*Olneya tesota*), and chollas (*Cylindropuntia* spp.) are common.

Ecological Groups

Although some species comprising the Baja California herpetofauna are ubiquitous in distribution, many others can be placed into well-defined ecological groups in accordance with their general natural history and distribution (Grismer 1994b). Thus, unlike historical biogeography, which groups taxa according to their geographic or phylogenetic history, ecological biogeography (or ecogeography) groups species in agreement with contemporary environmental factors. Looking at the herpetofauna in terms of its ecological biogeography is useful because it underscores the diversity of this region and the wide array of adaptive types that inhabit it. Grouping taxa in this way produces 8 ecogeographic groups represented in northern Baja California (Grismer 1994b, 2002). Among these ecological groups, 5 are composed of reptiles and amphibians whose distributions reflect strongly the delineation of northern Baja California's 3 physiographic and 3 phytographic regions (fig. 8.4), as described below.

Northwestern Mesophilic Group

The Northwestern Mesophilic group contains cismontane species whose distributions closely follow or are contained within the California Region. These species avoid the more arid portions of the peninsula and, thus, generally range no farther south than the vicinity of El Rosario. This group is composed of the arboreal salamander, garden slender salamander, Monterey ensatina, western toad, California toad, red-legged frog, western spadefoot, California treefrog, western pond turtle, California legless lizard, Baja California legless lizard, southern alligator lizard, Islas Coronado alligator lizard, Gilbert's skink, western skink, western fence lizard, granite night lizard, Baja California whiptail, ringneck snake, Islas Todos Santos mountain kingsnake, gopher snake (with the exception of the Sonoran subspecies), Islas Coronado rattlesnake, and western rattlesnake.

Northern Montane Group

The Northern Montane species assemblage occurs in the Baja California Coniferous Forest Region of the Sierra Juárez and Sierra San Pedro Mártir. Some of the members of this group are endemic to these mountains and all represent the fragmented, southern distribution of more widely ranging northerly species or closely related forms. This group is composed of the large-blotched ensatina, foothill yellow-legged frog (extirpated), southern sagebrush lizard, California mountain kingsnake, and western terrestrial garter snake.

Northeastern Xerophilic Group

The Northeastern Xerophilic group occupies northeastern Baja California in the Lower Colorado Valley Region east of the Peninsular Ranges and generally extends no farther south than Bahía de San Luis Gonzaga. Most of the members of this group have various specializations (e.g., fringed toes, sidewinding locomotion, counter-sunk lower jaws) for living in the open, sandy, arid extremes of this area or on or within the vegetation of this region. This group includes the Sierra los Cucapás collared lizard, the only Baja California endemic reptile truly restricted to the northern part of the peninsula (with the exception of insular species). The other species of the northeastern xerophilic group are the flat-tailed horned lizard, desert horned lizard, desert

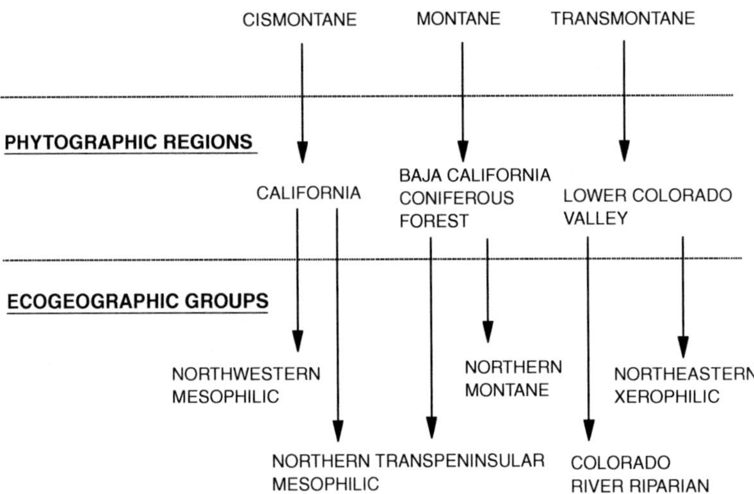

Figure 8.4. Northern Baja California's 3 physiographic and phytographic regions, with corresponding ecological groups of amphibians and reptiles. Other groups are represented in the region, but they are affiliated with physiographic/phytographic regions farther south, or they have no obvious affiliation.

spiny lizard, Colorado Desert fringe-toed lizard, long-tailed brush lizard, western shovel-nosed snake, coachwhip, Sonoran gopher snake, sidewinder, and western diamondback rattlesnake.

Colorado River Riparian Group

The distribution of the Colorado River Riparian group in Baja California is contained entirely within the Lower Colorado Valley Region. Species of this group extend into Baja California along the Río Colorado and its tributaries. Some are strictly aquatic, and others find shelter and food in the vegetation along the watercourses or irrigation canals. None of the members of this group would occur in Baja California if it were not for these water sources. This group is composed of the Colorado River toad, Great Plains toad, Woodhouse's toad, Yavapai leopard frog (extirpated), Rio Grande leopard frog, spiny soft-shell turtle, tree lizard, and checkered garter snake.

Northern Transpeninsular Mesophilic Group

The Northern Transpeninsular Mesophilic group contains species that are widespread in the mesic California and/or Baja California Coniferous Forest Regions of the north. As they extend southward into the more arid regions of the peninsula, their distributions become highly fragmented and restricted to mesic refugia such as springs, oases, and mountaintops. This group is composed of the Pacific treefrog, California striped racer, western black-headed snake, and two-striped garter snake.

Transpeninsular Xerophilic Group

The Transpeninsular Xerophilic group is similar to the Transpeninsular Mesophilic group but is more widely distributed, ranging throughout the arid Lower Colorado Valley, Vizcaíno, Central Gulf Coast, and Magdalena Regions of the peninsula. Members of this group are conspicuously absent from the mesic California and Baja California Coniferous Forest regions of the northwest and, to a large extent, the Sierra de La Laguna Region of the Cape. This group is composed of the red-spotted toad, Couch's spadefoot, zebra-tailed lizard, desert iguana, long-nosed leopard lizard, Baja California spiny lizard, desert night lizard, banded sand snake, spotted leaf-nosed snake, and ground snake.

Transpeninsular Saxicolus Group

The Transpeninsular Saxicolus group is generally restricted to the more arid, rocky areas of the Peninsular Ranges. Members of this group range throughout many phytogeographic regions and show little in the way of geographic variation. The species of this group are usually endemic to Baja California. The limiting factor of their distribution seems to be the presence of rock. This group is composed of the Baja California collared lizard, banded rock lizard, peninsular leaf-toed gecko, northern chuckwalla, granite spiny lizard, black-tailed brush lizard, speckled rattlesnake, and lyre snake.

Southern Xerophilic Group

The species included in the Southern Xerophilic group are endemic to Baja California and range through the xeric regions of the central and southern peninsula. One species, the Baja California rattlesnake, just reaches northern areas of the western portion of the peninsula.

In addition to these groups, there are a number of ubiquitous species: the western whiptail, western banded gecko, Baja California leopard lizard, side-blotched lizard, glossy snake, red diamond rattlesnake, night snake, common kingsnake, western blind snake, rosy boa, and western patch-nosed snake. Finally, knowledge of some other species is incomplete, and they cannot be properly classified: the orange-throated whiptail, barefoot banded gecko, coast horned lizard, Baja California rat snake, Baja California coachwhip, and long-nosed snake.

Valle de la Trinidad and Paseo de San Matías: An Invitation to Herpetological Research

Paso de San Matías is the largest low-elevation mountain pass in the Peninsular Ranges. It is situated at the eastern end of Valle de la Trinidad and delimits the southern end of the Sierra Juárez and the northern end of the Sierra San Pedro Mártir. Valle de la Trinidad is a 100-km southeast–northwest tending valley between 800 m in elevation at its northwestern end just east of Ensenada and 975 m in elevation at its southeastern terminus at Paseo de San Matías. This valley is constricted at the town of Valle de la Trinidad by a small, northeast–southwest tending range, the Sierra Warner.

This area provides unique opportunities for studying the comparative ecology and physiology of closely related reptiles. The narrow, low-elevation Paseo de San Matías cuts through the northern Peninsular Ranges, allowing cismontane and transmontane species to meet in a narrow contact zone. There are no other places where Pacific coastal and desert species occur sympatrically to any extent, and this intermixing allows us to test several aspects of the biology of these species. For example, to what extent do competitive exclusion and physiological constraints limit the incursion of coastal species into desert habitats? The same can be asked of desert species extending into coastal areas. To date, however, no studies have been conducted to address these questions.

What have been done are studies focusing on the evolutionary issues of species identity. This is one of few places where taxonomic issues of subspecies-versus-species can be directly addressed. Grismer (1994a) found that the whipsnakes *Masticophis flagellum piceus* and *M. f. fuliginosus* met in Valle de la Trinidad and exhibited no signs of intergradation and recognized the Baja California subspecies as the full species *M. fuliginosus*, the Baja California coachwhip. Grismer and McGuire (1996) demonstrated that the desert spiny lizards *Sceloporus magister uniformis* and *S. m. rufidorsum* were actually syntopic in Paseo de San Matías, and they elevated the latter to *S. zosteromus*. Grismer (1997, 2001) used the distribution and morphology of gopher snakes (*Pituophis*) in Valle de Trinidad as evidence to recognize the Baja California endemic *P. vertebralis* as a full species. McGuire (1996) showed that the leopard lizards *Gambelia wislizenii wislizenii* and *G. w. copei* were syntopic in Paseo de San Matías and consequently elevated the Baja California form *C. w. copei* to full-species status.

These are the only taxa that have been studied, but many more are awaiting close scrutiny of their relations (cismontane taxon of the pair listed first): western banded geckos *Coleonyx variegatus abbotti* and *C. v. variegatus*; western whiptail lizards *Cnemidophorus tigris stejnegeri* and *C. t. tigris*; western blind snakes *Leptotyphlops humilis humilis* and *L. t. cahuliae*; glossy snakes *Arizona elegans occidentalis* and *A. e. eburnata*; night snakes *Hypsiglena torquata klauberi* and *H. t. deserticola*; and the entire western *Pituophis* complex of gopher snakes, which has *P. catenifer annectens*, *P. c. affinis* (which may represent a full species) and, perhaps *P. vertebralis*, although it has not be found in the area yet, contacting one another in Valle de Trinidad.

The uniqueness of the area should not be overlooked. There are no other regions in northern Mexico or in the southwestern United States where we can study the evolutionary history and comparative ecology of sister taxa (i.e., each other's closest relatives) from coastal and desert regions occurring on opposite sides of such a formidable barrier as the northern Peninsular Ranges. Resolving these issues will require additional field and museum work as well as molecular data sets.

Acknowledgments

We thank J.-L. Cartron and R. Whitmore for their assistance with some of the figures. E. Gergus, G. Ceballos, J.-L. Cartron, and R. Felger provided helpful comments on an earlier draft of the chapter.

Literature Cited

Cody, M. L., R. Moran, and H. Thompson. 1983. The plants. Pp. 49–97 in T. J. Case and M. L. Cody (eds.), Island Biogeography in the Sea of Cortéz. University of California Press, Berkeley.

Grismer, L. L. 1994a. The origin and evolution of the peninsular herpetofauna of Baja California, Mexico. Herpetological Natural History 2: 51–106.

Grismer, L. L. 1994b. Ecogeography of the peninsular herpetofauna of Baja California, Mexico, and its utility in historical biogeography. Pp. 89–125 in Herpetology of the North American Deserts. Special Publication 5. Southwestern Herpetologists Society. Van Nuys, California.

Grismer, L. L. 1997. The distribution of *Pituophis melanoleucus* and *P. vertebralis* in northern Baja California, Mexico. Herpetological Review 28: 68–70.

Grismer, L. L. 2001. Comments on the taxonomy of gopher snakes from Baja California, Mexico: A reply to Rodriguez-Robles and de Jesús-Escobar. Herpetological Review 32: 81–83.

Grismer, L. L. 2002. Amphibians and Reptiles of Baja California, Its Associated Pacific Islands, and Islands in the Sea of Cortez. University of California Press, Berkeley.

Grismer, L. L., and J. A. McGuire. 1996. Taxonomy and biogeography of the *Sceloporus magister* complex (Squamata: Phrynosomatidae) in Baja California, Mexico. Herpetologica 52: 416–427.

Hastings, J. R., and R. R. Humphrey. 1969. Climatological data and statistics for Baja California. University of Arizona Institute of Atmospheric Physics, Technical Reports on the Meteorology and Climatology of Arid Regions 18: 1–96.

Hastings, J. R., and R. M. Turner. 1965. Seasonal precipitation regimes in Baja California, Mexico. Geografiska Annaler 47 (ser. A): 204–223.

Humphrey, R. R. 1974. The Boojum and Its Home. University of Arizona Press, Tucson, Arizona.

Lichtenstein, H. 1839. Beitrag zur Ornithologischen Fauna von Californien Nebst Bemerkungen über die Artkennzeichen der Pelicane und Über Einige Vögel von den Sandwitch-Inseln. Abhandlungen der Königlichen Akademie der Wissenschaften, Berlin, Germany, pp. 417–451.

Markham, C. G. 1972. Baja California's climate. Weatherwise 25: 66–101.

McGuire, J. A. 1996. Phylogenetic systematics of crotaphytid lizards (Reptilia: Iguania: Crotaphytidae). Bulletin of the Carnegie Museum of Natural History 32: 1–143.

Meigs, P. 1953. World distribution of arid and semi-arid homoclimates. Pp. 202–210 in Reviews of Research on Arid Zone Hydrology. Arid Zone Programme 1. UNESCO, Paris, France.

Mellink, E. 1995. The potential effect of commercialization of reptiles from Mexico's Baja California Peninsula and its associated islands. Herpetological Natural History 3: 95–99.

Mellink, E. 2002. El límite sur de la región mediterránea de Baja California, con base en sus tetrápodos endémicos. Acta Zoológica Mexicana 85: 11–23.

Pase, C. P. 1982. Sierran montane conifer forest. Pp. 49–51 in D. E. Brown (ed.), Biotic Communities of the American Southwest—United States and Mexico. Desert Plants 4: 1–342.

Robinson, M. D. 1973. Chromosomes and systematics of the Baja California whiptail lizards *Cnemidophorus hyperythrus* and *C. ceralbensis* (Reptilia: Teiidae). Systematic Zoology 22: 30–35.

Shreve, F. 1951. Vegetation and Flora of the Sonoran Desert, vol. 1. Vegetation. Carnegie Institute of Washington Publications 591: 1–192.

Shreve, F., and I. R. Wiggins. 1964. Vegetation and Flora of the Sonoran Desert, vols. I and II. Stanford University Press, Stanford, California.

Turner, R. M., and D. E. Brown. 1982. Tropical-subtropical desertlands. Pp. 180–221 in D. E. Brown (ed.), Biotic Communities of the American Southwest—United States and Mexico. Desert Plants 4: 1–342.

Wiggins, I. L. 1980. Flora of Baja California. Stanford University Press, Stanford, California.

9

Macrofaunal Diversity in the Gulf of California

RICHARD C. BRUSCA

LLOYD T. FINDLEY

PHILIP A. HASTINGS

MICHEL E. HENDRICKX

JORGE TORRE COSIO

ALBERT M. van der HEIDEN

History of Research on the Fauna of the Gulf of California

The Gulf of California (Sea of Cortez) has held a growing fascination for naturalists over the past 150 years (Lindsay 1983). The first serious collectors of marine life in the region were not professional biologists. One was John Xantus (de Vesey), a U.S. government tidal observer stationed at the tip of the Baja California peninsula (April 1859 to mid-1861), most of whose collections are now at the Smithsonian Institution (e.g., Gill 1862–1863; Jordan and Gilbert 1882). Another was Frederick Reigen, a Belgian citizen who lived in Mazatlán from 1848 to 1850 and amassed one of the largest collections of marine molluscs of all time—14 tons of specimens! The Reigen collection found its way to Liverpool, and from there it was partly dispersed. Much of it was published on by Philip Carpenter (Carpenter 1857; see also Hendrickx and Toledano-Granados 1994). During the next 4 decades, a few ichthyologists made collections of fishes at some readily accessible sites, notably Guaymas and Mazatlán, that were reported on mainly by David Starr Jordan and colleagues (e.g., Streets 1877; Jenkins and Evermann 1889; Evermann and Jenkins 1891; Jordan 1895). Oceanographic data were recorded and marine organisms trawled by the U.S. Fish Commission steamer *Albatross* in the late 1880s/early 1890s, and again in 1911, and most of

these specimens are also at the Smithsonian Institution (e.g., Gilbert 1892).

Modern oceanography in the Gulf of California began with the 1939 *E.W. Scripps* cruise to the region, which made 53 detailed hydrographic stations throughout the Gulf and sampled both phytoplankton and zooplankton (Sverdrup 1941; Roden and Groves 1959). In 1940, modern marine biology in the Gulf of California had its birth with the remarkable pioneering expedition of Edward F. Ricketts and John Steinbeck aboard the *Western Flyer*, a purse seiner out of Monterey, California. The biology (and philosophy) of that amazing voyage is chronicled in *Sea of Cortez. A Leisurely Journal of Travel and Research* (Steinbeck and Ricketts 1941; see also Astro and Hayashi 1971; Hedgpeth 1978a,b; Brusca 1993). Ricketts pioneered the concept of "community ecology" on the Pacific coast of America, bringing it west from his experience with W. C. Allee at the University of Chicago (Allee 1923; Hedgpeth 1978a,b). Ricketts was perhaps also the first person to codify the concept of intertidal zonation, based in part on his research in the Gulf of California, and many of his ideas were liberally borrowed and published upon by scientists with university degrees, such as M. Doty and T. A. A. Stephenson.

Expeditions from Scripps Institution of Oceanography, the University of California at Los Angeles, Stanford University, the California Academy of Sciences, and the University of Southern California's

Allan Hancock Foundation in the 1940s and 1950s ushered in an era of organized research effort in the Gulf. The fieldwork and taxonomic publications of the former Allan Hancock Foundation stand above all others in documenting the invertebrate biodiversity of the Gulf (Brusca 1980a). Between 1942 and 1983, the Hancock publications on Pacific marine life produced an astonishing 23,000 pages of primarily taxonomic text that was a watershed in marine biodiversity research (U.S.C. Press 1985). Since the late 1960s, our knowledge of the Gulf of California and its biodiversity has increased substantially through research by scientists at a number of U.S. and Mexican institutions (appendix 9.1). This body of work has resulted in many publications describing the flora, fauna, and environment of the region, much of it cataloged in Schwartzlose et al. (1992). However, compared to many of the world's seas and coastlines, exploration and documentation of the biodiversity of the Gulf of California are still in their early stages, and we estimate that more than half of its macrofauna is yet to be described, and the natural history of almost all species remains unknown.

The information in this chapter was derived largely from 2 projects. First, the Macrofauna Golfo Project has been a 10-year effort, which we have led, and it has produced a comprehensive database of the macrofauna of the Gulf of California containing taxonomic, distributional, and ecological information. "Macrofauna" is defined as those animals visible to the naked eye and generally larger than a few millimeters in size (but excluding copepods and ostracods). This database is planned for publication by Conservation International's Center for Applied Biodiversity Science as a CD-ROM and website component (Findley et al. in press). Data for the Macrofauna Golfo Project were derived from a variety of sources, including published literature, museum collections, and the personal field notes and records of the 6 principal investigators (having >150 years of collective research experience in the Gulf). The second source of data is a case study on the Upper Gulf of California/Colorado River Delta Biosphere Reserve, prepared for the UNESCO 2000 Conference on Biodiversity and Society (Brusca et al. 2001; Brusca and Bryner 2003).

Geography of the Region

Baja California encloses the Gulf of California and is one of the most remote peninsular areas in the world, exceeded in length only by the Malay and Kamchatka peninsulas. The Gulf is a large, semi-enclosed sea exceeding 1100 km in length, 100–200 km in width, with 258,593 km^2 (99,843 mi^2) of surface area (calculation by F. Zamora and S. Carroll), spanning more than 9° of latitude to traverse the Tropic of Cancer in its southern reaches, which extend to Cabo San Lucas (Baja California Sur) and Cabo Corrientes (Jalisco). The Gulf is home to more than 900 islands and islets, creating a region rich in habitat diversity and ripe for the forces of evolution to shape its flora and fauna.

In this chapter, we divide the Gulf into 3 faunal regions following the Macrofauna Golfo Project and based on the principal faunal regions established by Walker (1960) as modified by Thomson et al. (1979; see also Castro-Aguirre et al. 1995; fig. 9.1). The Northern Gulf extends from (and includes) the marine-influenced Colorado River Delta, southward to (and including) the Midriff Islands or las Islas del Cinturón or las Grandes Islas del Golfo (the largest being Islas Tiburón and Ángel de la Guarda), and to Bahía San Francisquito (Baja California) and Bahía Kino (Sonora). The Central Gulf ranges from the southern limit of the Northern Gulf to Guaymas (Sonora) and to Punta Coyote (Baja California Sur, north-northeast of La Paz). The Southern Gulf extends from the southern limit of the Central Gulf southward to Cabo Corrientes, Jalisco (the southern limit of the large Bahía Banderas) on the mainland, and on the Baja California peninsula to Cabo San Lucas. A number of species in the Southern Gulf have distributions extending around the Cape Region and up the southwestern coast of Baja California Sur. Many workers consider this area on the outer coast of the Baja California peninsula from Cabo San Lucas to Punta Eugenia (just below 28° N) to be a region of overlap (mixed tropical and temperate species), whereas others regard the Bahías Magdalena–Almejas lagoon complex as the northernmost boundary of the Tropical Eastern Pacific Fauna (reviewed in Brusca and Wallerstein 1979; Brusca 1980b; see also Hubbs 1960; Castro-Aguirre et al. 1992, 1993; Castro-Aguirre and Torres-Orozco 1993; Hastings 2000). Bahía Tortuga (Bahía Tortola) is the northernmost location on the west coast of Baja California Sur where the number of tropical species outnumbers temperate species (and where the giant North Pacific kelp, *Macrocystis*, makes its last southern stand), and Bahía San Ignacio (to the south) is home

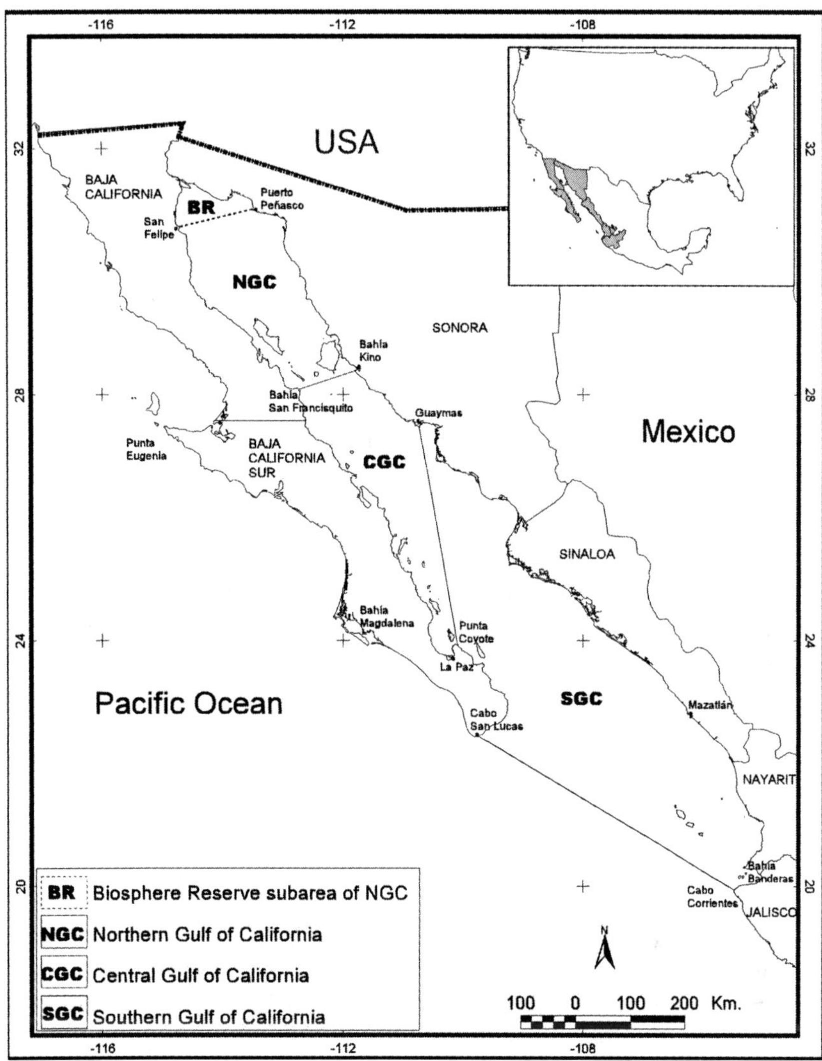

Figure 9.1. Faunal regions of the Gulf of California.

to the northernmost stand of mangroves along this coast (Brusca 1975).

We recorded biodiversity in the Upper Gulf of California/Colorado River Delta Biosphere Reserve (hereafter the "Biosphere Reserve"), at the head of the Gulf, as a subregion of the Northern Gulf. The southern boundary of the Biosphere Reserve forms a line extending from Punta Pelícano (Roca del Toro; the southern margin of Bahía Cholla and the larger Bahía de Adair), Sonora, across the Gulf to Punta El Machorro (Punta San Felipe), at San Felipe, Baja California.

Oceanography of the Gulf of California

The Northern Gulf covers about 60,000 km^2 (24,000 mi^2) of ocean surface, spans 3° latitude, and is a unique body of water in many ways. The climate is very dry, with an annual rainfall of less than 100 mm. The range of monthly mean air temperatures in the Northern Gulf is 18°C. The Northern Gulf has some of the greatest tides in the world. The annual tidal range (amplitude) at San Felipe and Puerto Peñasco is about 7 m, and on the Colorado River Delta at the head of the Gulf it is nearly 10 m.

The low delta islands of Montague and Pelícano (Isla Gore) are largely under water during high spring tides. Most of the Northern Gulf (north of the Midriff Islands) is shallow, less than 50 m deep, with the deepest areas reaching about 200 m in the small Wagner Basin, and in the larger Delfín Basin above Isla Ángel de la Guarda and extending into the deeper Salsipuedes Basin that separates the island from the peninsula (Alvarez-Borrego 1983; Maluf 1983; Lavín et al. 1998). Coastal seawater temperatures throughout the Northern Gulf are very low in the winter, dropping to 8°–12°C (equivalent to southern California shores), but rise to 30°C or more in the summer. Circulation patterns in the Northern Gulf are still not well understood, although there is evidence of a net counterclockwise rotation (Alvarez-Borrego et al. 1975; Lepley et al. 1975; Lavín et al. 1998; Carriquiry and Sánchez 1999).

Nutrient levels are high year-round and show little seasonality, although in recent years the primary sources of nutrients have probably been from agricultural drainage and the release of nutrients trapped in Colorado River sediments that are now eroding. Bray and Robles (1991) argue that influx of cold, deep water into the Southern Gulf brings nutrients into the Sea of Cortez and elevates productivity, but it is not clear to what extent this effect might reach the Northern Gulf. Although lower in biodiversity compared to the Central and Southern Gulf, high nutrient levels, shallow waters, and strong tidal mixing combine to make the Northern Gulf one of the most productive marine regions in the world. Standing crops of both phytoplankton and zooplankton are high year-round. Large fishes, sea turtles, and at least 12 species of whales and dolphins, including the critically endangered vaquita porpoise (Vidal et al. 1999; chapter 14), exploit these productive waters. Suspension-feeding clams, crustaceans, and polychaete worms also occur in great abundance throughout this region.

The Central and Southern Gulf maintain more oceanic conditions. South of the Midriff Islands depth increases quickly. The basins of the Central and Southern Gulf reach considerable depths, such as the Guaymas Basin (2000 m), Carmen Basin (2700 m), Farallón Basin (3150 m), and Pescadero Basin (3700 m) (Maluf 1983; Bray and Robles 1991). With a length of 220 km, the Guaymas Basin is the largest depression in the Gulf; it is an active, spreading center with hydrothermal vents and their unique community of benthic organisms. The

Carmen, Farallón, and Pescadero Basins together form a trough along the long axis of the Gulf, and these also are probably spreading centers. At the mouth of the Gulf, the trough approaches Cabo Corrientes, the southwestern extremity of the large Bahía Banderas and the probable original attachment site of the Cape Region of Baja California (Gastil et al. 1983). The famous sublittoral "sand falls" of Baja California's Cape Region lead to deep submarine canyons (to 2400 m) that extend off the tip of the peninsula. Rainfall is fairly high in the Southern Gulf, averaging 1000 mm per year along the coast of Sinaloa/Nayarit. The mean annual air temperature range at Cabo Corrientes is just 6°C. Tides in the Southern Gulf have much lower amplitudes than in the Northern Gulf; in Sinaloa they have a maximum annual range of 2.7 m, and at La Paz (on the Baja California peninsula) only 2.3 m. The mouth of the Gulf is a complex mix of waters from several sources, including cold California Current water, warm eastern tropical Pacific water, and warm saline Gulf of California water, and below these are Equatorial (Tropical) Subsurface Water, Antarctic Intermediate Water, and Pacific Bottom Water. The formation of fronts at the entrance to the Gulf is a major feature of the region and one that has attracted the attention of numerous commercial and sport fisheries. Because evaporation exceeds precipitation in the Gulf, there is a net influx of cold Pacific water into the basin, estimated at 1.7×10^4 m³/s (Roden 1958).

Oxygen concentrations tend to be high throughout the Northern Gulf, decreasing from about saturation values at the surface (5–6 ml/l) to about 1 ml/l at 300–500 m depth in the Delfín Basin (north of Isla Ángel de la Guarda). However, in the Central and Southern Gulf low oxygen concentrations are typical at intermediate depths (Alvarez-Borrego 1983). Concentrations in these regions often fall to less than 0.5 ml/l below 150 m. Deep waters in the Central and Southern Gulf have an intrusion of the Equatorial Subsurface Water Mass (from the south) characterized by an oxygen minimum layer, and oxygen concentrations of less than 0.2 ml/l have been found between 450 and 750 m (and occasionally deeper) in many areas (Robison 1972). Below about 750 m, oxygen begins increasing, reaching about 2.4 ml/l at 3500 m.

Few data exist for bottom conditions in the Gulf. Although strong tidal currents in the Northern Gulf keep the water column well mixed (Alvarez-Borrego 1983), it seems almost certain that bottom areas chronically disturbed by the numerous shrimp trawl-ers (and accumulation of their discarded bycatch on the sea floor) experience hypoxia (less than 0.2 ml/l dissolved oxygen), at least periodically, or even anoxia, but few data are available, and no reports of anoxia have yet been published for this region. In the southeastern Gulf, near anoxic values have occasionally been recorded at bottom depths of 60 m (Hendrickx 2001), but it is not known whether this is a natural phenomenon or directly related to bottom disturbance by trawling activities. In deeper waters, a wide fringe of oxygen-depleted bottom water occurs along the east coast of the Central and Southern Gulf (Parker 1964; Díaz and Rosenberg 1995). In this region, hypoxia or anoxia generally occurs from 100 m to about 700 m on the bottom (and in the water column), and benthic macrofaunal biomass in this region is very low. A highly toler-ant (surviving in oxygen concentrations of 0.3 to 0.8 ml/l oxygen) and distinct benthic invertebrate and fish community is found on the outer edge of this oxygen minimum zone in water deeper than 800 m (Hendrickx 2001).

Because evaporation exceeds freshwater input, the entire Gulf is regarded as an evaporation basin, particularly its northern part (Bray and Robles 1991). The estimated mean evaporation rate for the Northern Gulf is 1.1 m/year, while precipitation is only 4–8 cm/year (Alvarez-Borrego 1983; Lavín et al. 1998). Salinities have increased here in response to a dramatic reduction of freshwater discharge over the past 70 years (loss of river input), the increase of saline agricultural drainage, and probably global warming (enhancing evaporation). Summer surface salinities may reach 39 parts per thousand (ppt) in the shallow coastal areas and inner channels of the Río Colorado Delta region, and in Northern Gulf *esteros* (hypersaline coastal lagoons). Over deeper water in the Northern Gulf surface salinities are 35.3–37.2 ppt (Lavin et al. 1998), whereas in the Central and Southern Gulf they are closer to typi-cal oceanic waters (35.0–35.8 ppt).

Biogeography

The Gulf of California is a subtropical system with exceptionally high rates of primary productivity. Alvarez-Borrego (1983) concluded that the rates of primary productivity in the Gulf are comparable to those of the Bay of Bengal and to the upwelling areas off the west coast of Baja California or North Africa. Productivity rates are 2 or 3 times greater than that of the open Atlantic or open Pacific at similar latitudes (Zeitzschel 1969).

The flora and fauna that inhabit the Gulf arrived there from diverse sources: tropical Central and South America, the Caribbean Sea (before the up-lift of the Panama Isthmus), the temperate shores of California (during past glacial periods), and even across the vast stretch of the Pacific Ocean from the tropical Indo-West Pacific (Walker 1960; Rosen-blatt 1967; Briggs 1974; Thomson et al. 1979; Brusca 1980b; Rosenblatt and Waples 1986; Castro-Aguirre et al. 1995; Hastings 2000; Bernardi et al. 2003). Community composition at any given locality in the Gulf comprises a reasonably predictable mix of species, combined with a much larger suite of "un-predictable" species, the unpredictability being driven by complex networks of interacting physi-cal and biological factors. However, relative species richness (diversity) is predictable and largely a func-tion of habitat and substrate type. Benthic species richness is highest on reefs, relatively stable shores, and intertidal or shallow bottoms composed of softer sedimentary rocks such as sandstones or eroded vol-canic tuffs and rhyolites. Benthic diversity is low-est on beaches composed of smooth, hard rocks such as granites and basalts and on unstable beaches of sand or cobble, the latter perhaps having the low-est (benthic) diversity of any coastal habitat. Areas that have a variety of substrate types harbor more species than do more homogeneous ones (Walker 1960; Parker 1964; Brusca 1980b, 1989). Man-grove estuaries (including true positive estuaries) and *esteros* (moderately hypersaline coastal la-goons, or "negative estuaries") are notably diverse areas, and these habitats provide important nurs-ery and feeding grounds for the young of many coastal fishes and shellfish, including most commer-cial finfish and shrimps (Findley 1976; Cervantes et al. 1992; Flores-Verdugo et al. 1993; Güereca-Hernández 1994). Whitmore et al. (chapter 15) re-port 160 fishes and 214 invertebrates occurring in the mangrove lagoons of Baja California Sur. The islands of the Gulf also harbor an extraordinarily high species diversity (Thomson and Gilligan 1983), and these areas serve as important refugia for spe-cies that have lost habitat or been extirpated on the mainland coast, as well as being important seabird breeding sites (chapter 23). The Gulf is-lands also tend to harbor a fauna more typical of mainland coastal communities hundreds of kilo-meters to the south.

Species diversity is also influenced by seasonal oceanographic conditions, especially in the Northern Gulf where marked seasonal changes occur. This shallow region is strongly influenced by the climate of the surrounding Sonoran Desert and, as noted above, it experiences extreme annual variations in seawater temperatures. As a result, it is essentially a warm-temperate marine environment during the winter, but a tropical marine environment during the summer. Here, distinct seasonal changes occur with respect to many invertebrates, algae, and some vertebrates as certain tropical species disappear or emigrate during the cold winters and other, more temperate species, vanish during the warm summers (Thomson and Lehner 1976; Brusca 1980b). The Southern Gulf, with its greater area, deeper basins, and proximity to the open Pacific, is strongly influenced by the open ocean and is largely a stable tropical environment year round. As one moves northward from the Midriff Islands, benthic species diversity gradually declines, reaching its minimum in the homogenous mud bottoms of the Colorado River Delta. Based on the above, Brusca (1989) speculated that the Northern Gulf is a more physically controlled environment, whereas the Southern Gulf is a more biologically accommodated environment.

The rich pelagic waters of the Gulf are famous for supporting large numbers of fishes, marine mammals, sea turtles, and marine birds. Nearly 50% of Mexico's fisheries production comes from the Gulf, and 15% comes from the Northern Gulf alone (Cudney-Bueno and Turk Boyer 1998; Cudney-Bueno 2000). In general, benthic communities throughout the Gulf are species rich, although in subtidal areas that are susceptible to bottom trawling much diversity has been lost over the past 50 years due to high disturbance (Nava-Romo 1994; Nava and Findley 1994). However, we have almost no knowledge regarding community composition and food web structure for the Gulf's offshore benthic or pelagic communities. Notably high biodiversity occurs on the very limited intertidal beachrock ("coquina") formations that occur at just 4 sites in the Northern Gulf: Puerto Peñasco and Punta Borrascoso (Sonora), and San Felipe and Coloradito (Baja California). These small, rare, eroding beachrock habitats harbor a disproportionately high species diversity, giving them high priority for protection.

Macrofaunal Biodiversity in the Gulf of California

The following information, as noted above, derives primarily from the Macrofauna Golfo Project and from a Case Study on the Upper Gulf of California/Colorado Delta Biosphere Reserve. The marine macrofauna of the Gulf is diverse, comprising at least 5969 named species and subspecies: 4854 invertebrates and 1115 vertebrates (891 fishes; 224 nonfish vertebrates) (tables 9.1–9.5). Due to the presence of many undescribed invertebrate species, including many members of the planktonic community, this total is estimated to be less than half of the actual animal diversity of the Gulf. Overall faunal diversity decreases from the south to the north (table 9.1), the highest diversity being in the Southern Gulf, with 4095 taxa (69% of the total Gulf diversity). From the Central Gulf, 4025 taxa have been recorded (67% of the Gulf's diversity). The Northern Gulf houses 2802 taxa (47% of the Gulf's diversity), 1457 of which are within the Biosphere Reserve (24% of the Gulf's total diversity). One of

Table 9.1. Summary of macrofaunal diversity in the Gulf of California by region.

Region[a]	Fishes	Nonfish Vertebrates	Invertebrates	Totals
SGC	778	204	3113	4095
CGC	562	170	3293	4025
NGC	367	177	2258	2802
BR	258	149	1050	1457

[a]SGC = Southern Gulf of California; CGC = Central Gulf of California; NGC = Northern Gulf of California; BR = Upper Gulf/Delta Biosphere Reserve subregion of northern Gulf.

the richest areas in the entire Gulf is the Cabo Pulmo Reef (between La Paz and Cabo San Lucas, Baja California Sur), and this is the only true coral reef in the Sea of Cortez. We are aware of only 1 published faunal inventory of this reef, which reported 121 species of invertebrates and 108 species of fishes from cursory surveys (Brusca and Thomson 1975; also see Squires 1959; Glynn and Wellington 1985; Brusca 1985). Recent coastal and offshore (shelf) surveys in Sinaloa have revealed an unexpectedly high invertebrate diversity in the southeastern Gulf. For example, 300 species of decapod crustaceans have now been collected from coastal Sinaloa (Hendrickx 1996, 2001; Hendrickx and Brusca 2002), and the Sinaloan fish fauna comprises at least 600 species (van der Heiden and Findley 1990). Other notably high-diversity regions in the Gulf include Cabo San Lucas, Bahía Banderas, several stretches of rocky shoreline and coastal lagoons in Sonora and Sinaloa, and most of the Gulf islands. Puerto Refugio, at the northern end of Isla Ángel de la Guarda, and the isolated Rocas Consag, have long been recognized as biodiversity hot spots (Thomson et al. 1979; Thomson and Gilligan 1983).

Of the 891 fish taxa recorded from the Gulf, 801 are bony fishes (ray-finned fishes, or Actinopterygii) belonging to the Teleostei or "higher bony fishes" (table 9.2). Of these 801 taxa, 719 are neritic (continental shelf) and 82 are deep sea (strictly oceanic, mostly mesopelagic). In addition, there are 3 hagfishes (Myxini) and 87 other cartilaginous fishes (Chondrichthyes) in 3 groups: 3 chimaeras (rattails, Holocephali), 46 sharks, and 38 rays (Findley et al. 1996). About 10% of the fish fauna is endemic to the Gulf (87 taxa), comprising 80 species and 4 subspecies of bony fishes, 2 chondrichthians, and 1 hagfish (*Eptatretus sinus*, known only from deep

waters of the Northern and Central Gulf). Although 13 fish species are endemic to the Northern Gulf, only 1 of them is strictly endemic to the Biosphere Reserve (the delta silverside, or *pejerrey delta*, *Colpichthys hubbsi*), thought to be a euryhaline relict from truly estuarine conditions before the Colorado River was dammed (Crabtree 1989; Hastings and Findley, in press). However, the fish fauna of the Reserve also includes the endangered totoaba (*Totoaba macdonaldi*) and the currently overfished gulf corvina (*corvina golfina*, *Cynoscion othonopterus*) (Cisneros et al. 1995; Findley et al. 1996; Hastings and Findley in press). An estimated 10,000 tons of this corvina were taken by fishers in the Reserve between 1996 and 2000 (José Campoy-Favela and Martha Román-Rodriguez, pers. comm.; see also Román-Rodríguez et al. 1998).

Although 8 fish species and 1 subspecies are recorded as endemic to the Southern Gulf, they are poorly known, reported from only a few specimens, and often from deep water. The same can be said for the Central Gulf, where 5 endemic fish species are recorded, most of them poorly known, including 3 species of rockfishes (*rocotes*, *Sebastes*). In contrast, the more isolated Northern Gulf contains 13 endemic fish species, including a number of soft-bottom–associated forms with very restricted distributions, and another 2 species of *Sebastes* restricted to deeper (cold) waters around the Midriff Islands (Chen 1975; Rócha-Olivares et al. 1999). A majority of endemic fishes in the Gulf occur in more than 1 region, and their diversity is especially notable in rocky-shore, small-bodied species of the families Gobiidae, Chaenopsidae, Labrisomidae, and Gobiesocidae (Walker 1960; Findley et al. 1997). Although not endemic, several fishes in the Northern Gulf are members of the Gulf's "disjunct fauna."

Table 9.2. Fish diversity in the Gulf of California.

	Total for Gulf	Present in SGC	Present in CGC	Present in NGC	Present in BR
Myxini (hagfishes)	3 (1)	1 (0)	2 (0)	1 (0)	0
Chondrichthyes (sharks, rays, chimaeras)	87 (2)	80 (0)	72 (0)	58 (0)	39 (0)
Actinopterygii: Teleostei (bony fishes)	801 (84)[a]	697 (9)[b]	488 (5)	308 (13)	219 (1)[c]
Totals	891 (87)[a]	778 (9)[b]	562 (5)	367 (13)	258 (1)

Numbers of endemic species/subspecies per region given in parentheses. SGC = Southern Gulf of California; CGC = Central Gulf of California; NGC = Northern Gulf of California; BR = Upper Gulf/Delta Biosphere Reserve subregion of Northern Gulf.
[a]Includes 4 subspecies-rank taxa.
[b]Includes 1 subspecies-rank taxon.
[c]*Colpichthys hubbsi* (Atherinopsidae), also included in count for NGC.

These are eurythermal or cold-water adapted species found in the Northern and (sometimes Central) Gulf, but not to the south, reappearing again in the colder waters of the California Current on the outer coast of the Baja California peninsula and northward to California (Walker 1960; Thomson and Gilligan 1983; Present 1987; Castro-Aguirre et al. 1995; Terry et al. 2000; Huang and Bernardi 2001; Bernardi et al. 2003).

Of the 224 nonfish marine vertebrates known from the Gulf, 181 are birds, 36 are mammals, and 7 are reptiles (table 9.3). The aquatic bird fauna includes species in 10 orders, the most diverse being the Charadriiformes (including plovers, gulls, terns and sandpipers) with 76 species, and the Anseriformes (ducks and geese) with 35 species. The Southern Gulf houses the most diverse bird fauna, with 165 resident and seasonal species, whereas 131 species have been reported from the Biosphere Reserve. Although no aquatic birds are wholly restricted in their range to the Gulf, 1 species is essentially endemic, with only a few records outside the Gulf: the yellow-footed gull (*Larus livens*). This bird, plus 4 other species (least storm petrel, *Oceanodroma microsoma*; Heermann's gull, *Larus heermanni*; elegant tern, *Sterna elegans*; Craveri's murrelet, *Synthliboramphus craveri*) rely almost wholly on the Gulf for their breeding sites, and 90% (or more) of their breeding populations are found in the Gulf, mostly on only a few small islands (Anderson 1983; chapter 23).

The marine mammal fauna of the gulf is surprisingly diverse (Vidal et al. 1993; chapter 14), with 36 species representing 31 cetaceans (whales, dolphins, porpoises), 4 pinnipeds (sea lions, seals), and 1 bat (the coastal fishing bat, *Myotis vivesi*; Patten and Findley 1970; Bogan 1999). Among the cetaceans, the Odontoceti (toothed whales, most in the family Delphinidae) are represented by 23 species, 8 of which have been recorded from the Biosphere Reserve. The Mysticeti (baleen whales) are represented by 8 species, 5 of which enter the Biosphere Reserve. Four species of pinnipeds have been recorded from the Gulf, with the California sea lion, *Zalophus californianus*, being the only true resident and by far the most abundant and ubiquitous, occurring in all regions and seasonally occupying several important breeding sites on Gulf islands and coastal headlands (Le Boeuf et al. 1983; Vidal et al. 1993; Aurioles-Gamboa and Zavala 1999). The presence of 36% (13 species) of the Gulf's marine mammal fauna, either permanently or seasonally, in the Biosphere Reserve demonstrates the importance of this area to conservation efforts in the eastern Pacific. One of the cetaceans of the Biosphere Reserve is the vaquita (*Phocoena sinus*), the world's smallest and most endangered marine cetacean (Vidal et al. 1999; Rojas-Bracho and Taylor 1999; chapter 14). This rare porpoise is endemic to the uppermost part of the Northern Gulf, where its critical habitat appears to straddle the southern boundary of the Biosphere Reserve in the small area west of Rocas Consag (Gallo-Reynoso 1998). Only 2 species of marine mammals are endemic to the Gulf, the vaquita and the fishing bat. Although the latter occurs in scattered colonies throughout the Gulf, especially in the Midriff Islands area, it is rare north of there and has not been recorded from the Biosphere Reserve.

The 7 marine reptiles of the Gulf comprise now-small populations of 5 sea turtles, 1 sea snake (*Pelamis platurus*), and 1 crocodile (*Crocodylus acutus*). Four of the turtles (all of which are threatened or endangered) are recorded from the Biosphere Reserve, although their numbers are few throughout the Northern Gulf due to historical

Table 9.3. Nonfish vertebrate diversity (aquatic birds, reptiles, mammals) in the Gulf of California.

	Total for Gulf	Present in SGC	Present in CGC	Present in NGC	Present in BR
Reptiles (sea turtles, sea snake, crocodile)	7 (0)	7	6	7	5
Aves (birds)	181 (1)[a]	165	135	146	131
Mammals (cetaceans, pinnipeds, fishing bat)	36 (2)	32	29	24 (1)	13
Totals	224 (3)	204	170	177 (1)	149

Numbers of endemic species per region in parentheses. NGC = Northern Gulf of California; CGC = Central Gulf of California; SGC = Southern Gulf of California; BR = Upper Gulf/Delta Biosphere Reserve subregion of Northern Gulf.
[a]Essentially endemic.

fishing pressure and modern incidental take in gill-nets and shrimp trawls. The crocodile is now present only in a few estuaries of the mainland side of the Southern Gulf (Navarro-Serment 2002). The yellowbelly sea snake (a tropical Indo-Pacific species) occurs infrequently in the Central and Northern Gulf, but is increasingly common southward all the way to Ecuador.

Invertebrate diversity is highest in the Central Gulf (3293 taxa), and lowest in the Northern Gulf (2258 taxa; table 9.4). The Biosphere Reserve is home to 1050 invertebrate taxa, or 22% of all invertebrates known from the Gulf. For benthic invertebrates, highest species diversity occurs in shallow coastal regions, particularly in the Northern Gulf and along the coasts of Sinaloa and Nayarit (fig. 9.2). For pelagic invertebrates, highest diversity occurs along the Gulf coast of Baja California Sur, from Bahía Concepción southward to La Paz; lowest species diversity occurs in the Northern Gulf, along the coast of Sonora, and in the extreme Southern Gulf (fig. 9.3).

Among the invertebrates, the highest diversity occurs with the Mollusca (2193 taxa) and Arthropoda (1051 taxa; table 9.4). Within the Mollusca, the gastropods and bivalves stand out with 1530 and 565 taxa, respectively. More than 20% (460 taxa) of the molluscs known from the Gulf are endemic to that region, including 396 gastropods (of which 39 are opisthobranchs [sea slugs] and 3 are marine pulmonates), 44 bivalves, 15 chitons, 4 cephalopods (3 octopuses; 1 squid, *Loliolopsis chiroctes*), and a scaphopod (tables 9.4 and 9.5). Among the Arthropoda, the brachyuran crabs (Decapoda) and amphipods (Peracarida) are most diverse, with 301 and 232 taxa, respectively (table 9.4).

Overall invertebrate endemicity in the Gulf is 16% (766 taxa). At the phylum level, the highest levels of endemism occur in the Brachiopoda (80%), Ctenophora (50%), Platyhelminthes (41%), Echiura (25%), and Mollusca (21%). At lower taxonomic levels, highest endemism occurs among Anthozoa (34%), Polyplacophora (26%), Gastropoda (26%), and Cumacea (25%). However, several of these figures should be viewed with caution because some taxa are very poorly studied in the Gulf and tropical eastern Pacific in general (e.g., Brachiopoda, Cnidaria, Ctenophora, Platyhelminthes, Echiura, Cumacea, Tanaidacea, micromolluscs, Urochordata, Hemichordata). Only 1 hemichordate (*Ptychodera flava*) and 1 cephalochordate (the lancelet, *Branchiostoma californiense*) are recorded from the Gulf.

Among the 128 invertebrates endemic to the Northern Gulf are the unique carpet anemone (*Palythoa ignota*) and the giant aphroditid polychaetes (*Aphrodita mexicana, A. sonorae*), all of which appear to be greatly reduced in numbers and threatened due to excessive bottom trawling. In addition, 7 species of pea crabs (Pinnotheridae) are endemic to the northern Gulf, as are 2 goneplacid crabs (*Glyptoplax consagae, Speocarcinus spinicarpus*), 11 species of sea slugs, the cone snail *Conus angulatus* (previously considered a synonym of *C. regularis*), and the scallop *Leptopecten palmeri*.

Threats to Biodiversity in the Gulf of California

There are many threats to biodiversity in the Gulf. Not all are being driven from within Mexico, and most are influenced by economic or environmental pressure from the United States. One of the greatest threats comes from the disruption of rivers that once flowed into the Gulf. Although virtually all of the rivers that once reached the Gulf have been altered or destroyed by overdraft and diversion (e.g., Ríos Fuerte, Mayo, Yaqui, Sonora, and Concepción), the most significant is the Colorado River which, before construction of Hoover (Boulder) Dam, provided most of the fresh water supply to the Northern Gulf (Brusca et al. 2001; Brusca and Bryner 2003; Brusca 2004). Historically, an average of 16.7 million acre-feet (maf) of water reached the Colorado Delta annually from the river. A hundred years ago, riverboats steamed from the Gulf of California up the Lower Colorado/Gila River system into Arizona. Until completion of Hoover Dam in 1935 (creating Lake Mead), fresh water from the Colorado River flowed into the Northern Gulf throughout the year, with great seasonal floods resulting from spring snowpack melt in the Rocky Mountains. By the time Glen Canyon Dam was completed in 1962, input of Colorado River water to the Delta and Northern Gulf had completely ceased. For 20 years after completion of Glen Canyon Dam, as Lake Powell filled, virtually no water from the river reached the sea. In 1968, flow readings at the southernmost measuring station on the river were discontinued, since there was nothing left to measure. Today, 20 dams (58 dams if tributaries are included) and thousands of kilometers of canals, levies, and dikes have converted the Colorado River

Table 9.4. Species diversity of major invertebrate taxa in the Gulf of California.

Major Taxa	Totals for Gulf	SGC	CGC	NGC	BR	END	PEL	BEN
Porifera	86	36	42	46	19	16	0	86
Cnidaria	253	112	146	114	33	47	20	209
Hydrozoa	146	55	74	69	11	12	15	113
Anthozoa	102	53	72	44	21	35	0	96
Scyphozoa	5	4	0	1	1	0	5	0
Ctenophora	4	2	3	1	1	2	4	0
Platyhelminthes	22	5	16	12	10	9	0	22
Nemertea	17	5	8	6	2	2	0	17
Annelida	717	442	436	287	117	79	21	675
Oligochaeta	1	0	0	1	1	0	0	0
Polychaeta	716	442	436	286	116	79	21	675
Sipuncula	11	8	10	5	5	0	0	11
Echiura	4	2	2	2	0	1	0	4
Pogonophora	1	0	1	0	0	0	0	1
Arthropoda	1051	785	713	508	248	118	154	861
Pycnogonida	15	4	9	10	9	0	0	15
Cirripedia	43	30	22	14	5	9	4	37
Stomatopoda	28	22	17	8	2	3	0	28
Peracarida	328	223	224	174	59	30	116	189
Mysida	6	5	3	0	0	1	5	0
Amphipoda	232	163	155	126	30	17	109	113
Isopoda	80	53	62	41	27	10	2	67
Tanaidacea	2	2	2	1	0	0	0	2
Cumacea	8	0	2	6	2	2	0	7
Euphausiacea	14	14	8	4	1	0	14	0
Decapoda	623	492	433	298	172	76	20	592
Dendrobranchiata	32	31	24	16	10	1	8	24
Stenopodidea	2	2	2	0	0	0	0	2
Caridea	132	80	95	40	17	14	10	120
Astacidea	1	1	0	0	0	0	0	1
Thalassinidea	19	14	8	8	5	3	0	17
Palinura	8	8	4	2	1	0	0	8
Anomura	128	106	87	65	41	18	0	127
Brachyura	301	250	213	167	98	40	2	293
Mollusca	2193	1386	1560	1000	542	460	11	1965
Monoplacophora	1	1	0	0	0	0	0	1
Polyplacophora	57	25	44	38	20	15	0	55
Gastropoda	1530	938	1073	656	360	396	3	1317
Bivalvia	565	392	415	285	150	44	0	561
Scaphopoda	20	14	14	15	8	1	0	19
Cephalopoda	20	16	14	6	4	4	8	12
Bryozoa (Ectoprocta)	169	96	147	119	10	10	0	165
Brachiopoda	5	3	0	2	2	4	0	5
Echinodermata	262	207	180	138	56	16	0	262
Chaetognatha	20	17	14	7	1	0	20	0
Hemichordata	1	0	1	1	1	0	0	1
Chordata	38	6	13	10	3	3	21	14
Ascidia	16	5	12	9	2	3	0	13
Appendicularia	21	0	0	0	0	0	21	0
Cephalochordata	1	1	1	1	1	0	0	1
Totals	4854	3113	3293	2258	1050	766	251	4299

NGC = Northern Gulf of California; CGC = Central Gulf of California; SGC = Southern Gulf of California; BR = Upper Gulf/ Delta Biosphere Reserve subregion of Northern Gulf; END = endemic to the Gulf; PEL = pelagic species only; BEN = benthic species only.

into a highly controlled plumbing system in which every drop of water is carefully managed, and only about 4 maf/year reaches the Delta (and then, only during wet years). In addition, most of the Delta's wetlands have been converted into farmland or urban sprawl. What was once 2 million acres of wetlands has been reduced to less than 60,000 acres of freshwater wetlands (much of this in the "recreated" 30,000 acre Ciénega de Santa Clara) and 130,000 acres of coastal salt marsh. Due to the greatly reduced freshwater flow, the powerful tides of this region now overwhelm the river channel. During spring tides, seawater creates an estuarine basin for 50–60 km up-river, averaging 2–8 km in width and 16 km wide at the mouth. This marine intrusion has killed most of the freshwater flora and fauna that used to live along the lowermost river corridor (Brusca et al. 2001; Brusca and Bryner 2003; Brusca 2004).

Before construction of Hoover Dam, the annual sediment discharge from the Colorado River into the Gulf was enormous, estimated to range from 45 to 455 million metric tons/year. Indeed, the entire Northern Gulf is considered the "Colorado River Sedimentary Province." The name of the river itself, Colorado, is Spanish for a red or ruddy color, and the first name given to the Gulf of California was the Vermilion Sea (by Francisco de Ulloa, the first Spanish navigator to sail there under orders from the conquistador Hernán Cortés). The reduction of freshwater input and sediment discharge since 1935 has modified the hydrography of the Colorado River Delta/Northern Gulf system, initiating a regime of deltaic erosion. Deltaic deposition no longer takes place, and the entire Delta is now exposed to destructive hydrodynamic forces of tides and storms, promoting resuspension, erosion of ancient river sediments, and the gradual export of sediments to the southwest and eventually out of the Northern Gulf. These changes are altering the littoral wetlands and biological equilibrium of the region (Brusca et al. 2001; Brusca and Bryner 2003; Brusca 2004).

The single most serious threat to the integrity of the Delta's natural communities is from Colorado River water management decisions made in the United States. A 1944 water treaty guarantees Mexico 1.5 maf/year from the river (plus an additional 200,000 acre-feet when surpluses are declared), and a 1973 amendment to the treaty guarantees Mexico relatively pure water. However, virtually all of the Colorado River water crossing the border is diverted for urban and agriculture use in the Mexicali Valley, where a half-million acres are under irrigation.

During most years, no Colorado River water reaches the Sea of Cortez, nor does the Delta have any explicit ecological water entitlement. Water that has reached its riparian corridor in recent years has done so solely because infrequent U.S. flood releases have exceeded the use and diversion capacity of upstream users (Brusca et al. 2001; Glenn et al. 2001).

It is likely that the reduction of freshwater input into the Northern Gulf, in combination with other anthropogenic factors, has driven some species to (or nearly to) extinction. However, we have so few historical or baseline data for marine organisms of this region that extinctions (or local extirpations) would go unnoticed for commercially unimportant or otherwise little-known species. There has never been a comprehensive, dedicated survey of the marine fauna of the Northern Gulf and Colorado River Delta ecosystem.

The delta clam, *Mulinia coloradoensis*, used to be one of the most abundant animals of the northernmost Gulf. Windrows of its shells line the beaches of the Delta and northwestern shores of the Northern Gulf. This species was thought to be extinct until its recent rediscovery in small numbers near the mouth of the river (Kowalewski et al. 2000; Rodríguez et al. 2001). The near demise of this species has been suggested to be the result of decreased benthic productivity resulting from upstream diversion of the Colorado River's flow (Kowalewski et al. 2000; Rodríguez et al. 2001). However, there is no indication that nutrient levels (and hence productivity) have decreased significantly in the Northern Gulf, and nutrients that have been lost by depletion of riverine sediment input may have been regained in the form of agricultural runoff and deltaic erosion (releasing ancient trapped nutrients). Hence, the near extinction of this clam may be linked to another, as yet unknown factor related to reduction of freshwater input to the Delta.

The recent return in large numbers of the once "commercially extinct" gulf corvina (*corvina golfina, Cynoscion othonopterus*) to the northernmost Gulf may be tied to increased Colorado River flows during wet years since the 1980s (Román-Rodríguez et al. 1998; Rowell et al. in press). Freshwater input from the Colorado River is also important to the life history of commercial shrimps of the region. Commercial shrimp catches have been falling since the 1960s, due to a combination of overfishing and loss of habitat for young (reviewed in Brusca et al. 2001). It has been estimated that an influx of just 250,000 acre-feet/year of Colorado River water

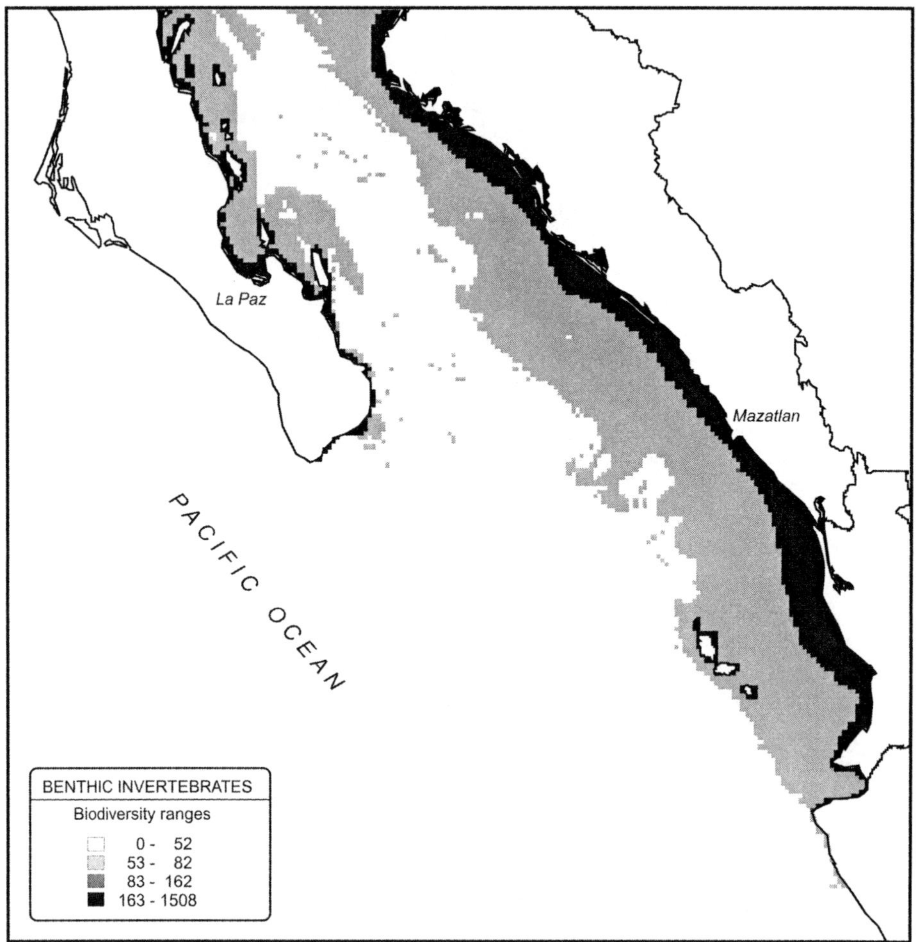

Figure 9.2a. Species diversity of benthic invertebrates in the Gulf of California.

would double shrimp production in the Northern Gulf (Galindo-Bect et al. 2000). The young of these shrimps use the shallow wetlands and *esteros* of the region, including the tidelands of the delta, as a nursery, migrating into these areas subsequent to their offshore planktonic larval phase. When the shrimp reach juvenile or subadult stage, they migrate offshore once again.

Reduction of the brackish estuarine habitat likely also has, in combination with historical overfishing and continuing capture of juveniles in shrimp nets, driven the large, corvinalike totoaba to near extinction (Hastings and Findley, in press). Continued absence of freshwater input could drive the endemic Palmer's saltgrass (*Distichlis palmeri*), which apparently needs periodic freshwater flooding to germinate, to extinction (see Felger 2000, for

a summary of the biology of this species). The same may be true of the endemic delta silverside fish (*pejerrey delta, Colpichthys hubbsi*). Aquatic birds also rely heavily on the Gulf's coastal lagoons and wetlands, all of which are on the western flyway for migratory waterfowl.

Today, every fishery in the Gulf is probably overfished (Greenberg and Vélez-Ibáñez 1993; Musick et al 2000; Brusca et al. 2001; Greenberg, in press). The American Fisheries Society (AFS) official list of North American Marine Fishes at Risk of Extinction reports (an underestimated) 11 at-risk species in the Gulf of California. Five of these are large serranids (groupers, *meros, cabrillas*) and sciaenids (corvinas, *berrugatas, chanos*), some of which are endemic or nearly endemic to the Gulf. These species are sensitive to overharvesting because of their

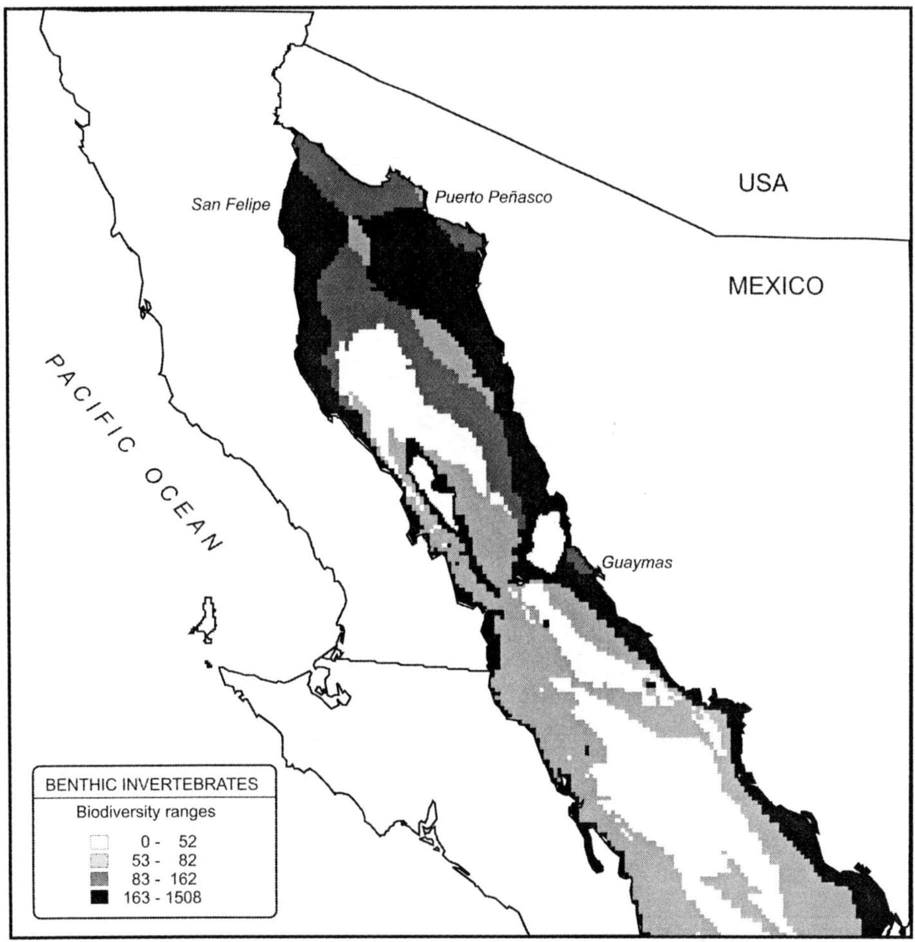

Figure 9.2b. Species diversity of benthic invertebrates in the Gulf of California.

late maturity and formation of localized spawning aggregations. In addition, most (if not all) of the serranids are protogynous, and the sciaenids require estuarine habitats once provided by the rapidly diminishing Colorado River Delta for spawning and nursery grounds. The AFS also lists the Gulf, especially its northern part, as 1 of 5 geographic hotspots in North America where numerous fish species are at risk; certainly, the same could be said for the invertebrates of this region.

The vaquita is a small, endemic porpoise that lives only in the northernmost Gulf. With the most recent estimate of vaquita abundance at only 567 individuals, and bycatch mortality at an estimated 39–84 deaths per year, this porpoise is the most endangered marine cetacean in the world (Vidal 1995; Jaramillo-Legorreta et al. 1999; Rojas-Bracho

and Taylor 1999; Vidal et al. 1999; D'Agrosa et al. 2000; chapter 14). The primary cause of vaquita mortality is incidental capture in gillnets, and unless this type of fishing gear is banned in the Biosphere Reserve and in its critical habitat to the south (Gallo-Reynoso 1998), the vaquita will almost certainly be extinct in a few years.

Many once abundant but less visible species, such as the threatened giant brown sea cucumber (*Parastichopus fuscus*), are now practically gone from the Gulf. Sea cucumbers have vanished at the hands of Mexican and Japanese fishers who collect them for Asian food markets. Even though this sea cucumber is now protected by Mexican law, it continues to be harvested from the offshore islands in the Gulf, the last remaining refugium for this and many other species that were once abundant along

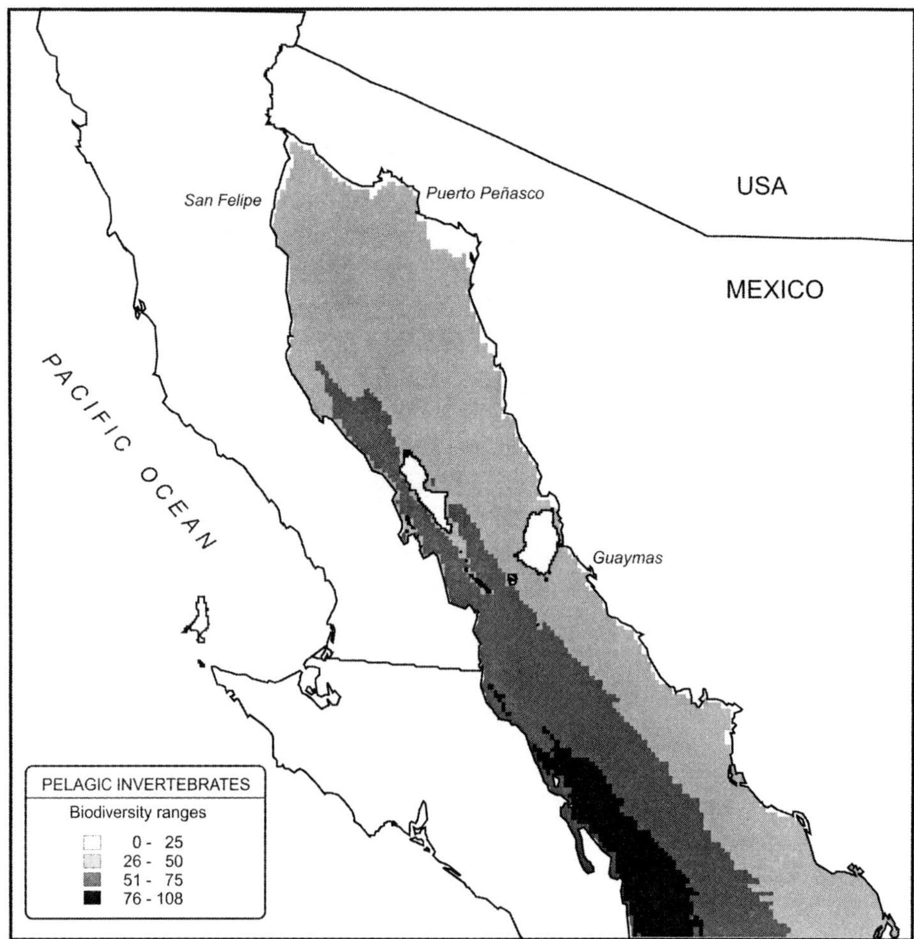

Figure 9.3a. Species diversity of pelagic invertebrates in the Gulf of California.

the mainland and peninsular coasts. In areas of heavy and increasing tourism, such as Puerto Peñasco, San Felipe, Guaymas/San Carlos, Mazatlán, Loreto, La Paz, and Los Cabos, littoral biodiversity is but a shadow of what it was just 20 years ago. Part of the tourism-driven loss is hand collecting of animals by visitors (and perhaps the trampling underfoot of fragile habitats exposed at low tide), but equally important is the collection of large molluscs and echinoderms by residents for sale to tourists as curios or to local restaurants where they are served in seafood cocktails (e.g., large bivalve and gastropod molluscs, octopuses). In the Southern Gulf (and to the south) many large molluscs are disappearing due to subsistence (artisanal) fishing. For example, the large fasciolarid snail *Pleuroploca princeps*, the large chiton *Chiton articulatus,* and giant limpet

Patella mexicana (the largest living limpet; to 15 cm) have disappeared from accessible shores, and today they are found almost exclusively on island refugia or highly inaccessible stretches of the mainland coast.

Industrial and artisanal shrimp fishing also exacts a harsh toll on the Gulf's marine environment, as more than a 1000 large shrimp trawlers annually rake an area of sea floor equivalent to twice the total size of the Gulf (Brusca et al. 2001). This high rate of bottom trawling damages fragile benthic habitats, and during the mid-1960s to late 1970s trawlers generally captured an average of 10 kg of fish and invertebrate bycatch for every kilogram of commercial shrimp, with a range of about 1.2–35 kg of bycatch (per kilogram shrimp) in 95% of trawl hauls analyzed (Pérez-Mellado and Findley 1985; van der Heiden 1985). Today, bottom trawlers aver-

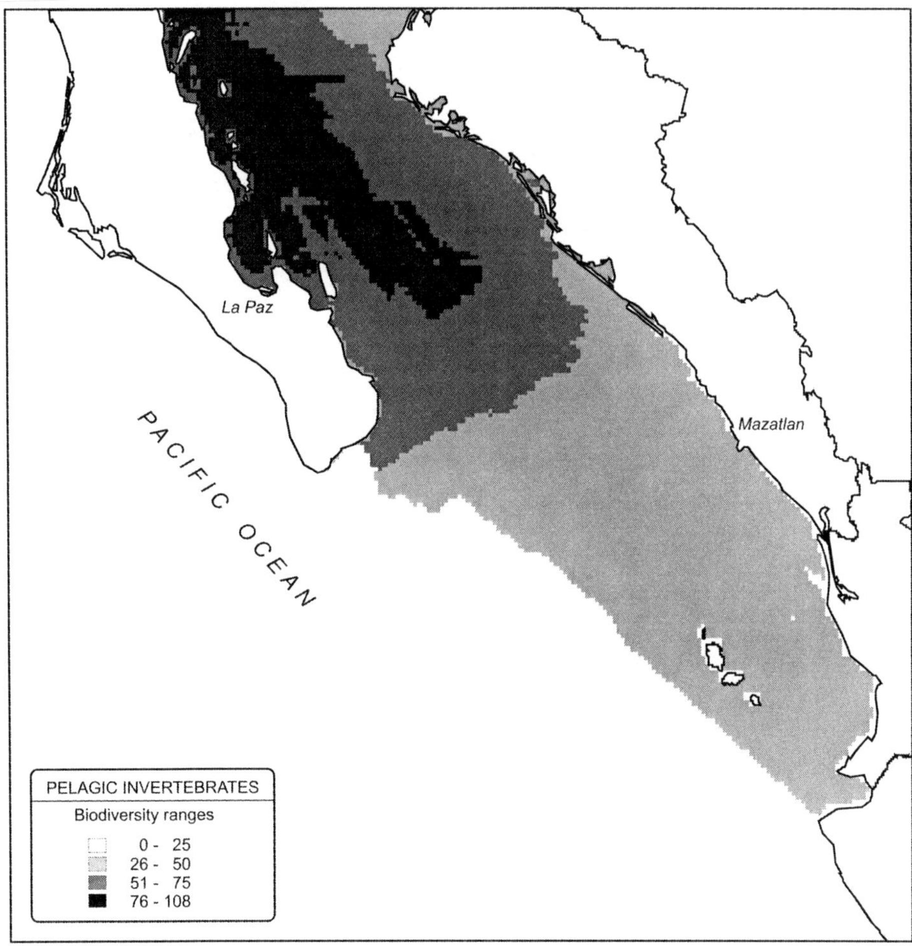

Figure 9.3b. Species diversity of pelagic invertebrates in the Gulf of California.

age 10–30 kg bycatch per kilogram shrimp (depending on the time of year) in the Northern Gulf (R. Brusca, interviews with Puerto Peñasco shrimp fishers). The number of commercial shrimp trawlers in the Gulf grew from 700 in 1970 to a high of 1400 in 1983, then decreased to 1144 in 1997, then increased again to 1470 in 2000 (García-Caudillo and Gómez-Palafox, in press), despite warnings as early as the 1970s of a possible crisis resulting from overexploitation (e.g., Snyder-Conn and Brusca 1977). Catch-per-unit-effort has been declining for decades, while government subsidies artificially sustain the overcapacity of the industrial fishing fleet. Without these government subsidies, the current level of commercial trawling would not be economically feasible. Limited scientific and anecdotal information suggests that sweeping changes in

benthic/demersal community structure have taken place over the past 50 years as a result of this disturbance (van der Heiden 1985; Nava and Findley 1994; Nava-Romo 1994), including an apparent accelerating decrease in the diversity and biomass of the bycatch, possibly heralding an early stage in regional benthic/demersal ecosystem collapse.

Increasing loss of coastal habitats due to development, including poorly designed marinas and aquaculture installations lacking environmental controls, has reduced the rich *esteros*, estuaries, and mangrove communities of the Gulf that serve as critical spawning and nursery grounds for shrimp and other invertebrate and fish species. Ninety percent of Mexico's cultivated shrimp production is in southern Sonora, Sinaloa, and Nayarit, all of which have rapidly growing, semi-intensive, and spatially

Table 9.5. Endemicity of major invertebrate taxa in the Gulf of California.

Major taxa	END	SGC	CGC	NGC	ND
Porifera	16	5	3	6	0
Cnidaria	47	4	16	12	0
Hydrozoa	12	1	3	5	0
Anthozoa	35	3	13	7	0
Scyphozoa	0	0	0	0	0
Ctenophora	2	0	1	0	0
Platyhelminthes	9	1	1	4	0
Nemertea	2	0	2	0	0
Annelida	79	9	34	20	2
Oligochaeta	0	0	0	0	0
Polychaeta	79	9	34	20	2
Sipuncula	0	0	0	0	0
Echiura	1	0	0	0	0
Pogonophora	0	0	0	0	0
Arthropoda	118	11	15	31	3
Pycnogonida	0	0	0	0	0
Cirripedia	9	2	1	0	1
Stomatopoda	3	0	2	0	0
Peracarida	30	1	4	14	0
Mysida	1	0	0	0	0
Amphipoda	17	0	0	10	0
Isopoda	10	1	3	3	0
Tanaidacea	0	0	0	0	0
Cumacea	2	0	1	1	0
Euphasiacea	0	0	0	0	0
Decapoda	76	8	8	17	2
Dendrobranchiata	1	0	0	0	0
Stenopodidea	0	0	0	0	0
Caridea	14	3	6	1	0
Astacidea	0	0	0	0	0
Thalassinidea	3	1	0	1	0
Palinura	0	0	0	0	0
Anomura	18	2	0	1	2
Brachyura	40	2	2	14	0
Mollusca	460	111	70	51	22
Monoplacophora	0	0	0	0	0
Polyplacophora	15	0	0	4	0
Gastropoda	396	101	62	42	21
Bivalvia	44	10	7	5	1
Scaphopoda	1	0	0	0	0
Cephalopoda	4	0	1	0	0
Bryozoa (Ectoprocta)	10	0	3	1	0
Brachiopoda	4	2	0	2	0
Echinodermata	16	12	2	0	0
Chaetognatha	0	0	0	0	0
Hemichordata	0	0	0	0	0
Chordata	3	0	0	1	0
Ascidiacea	3	0	0	1	0
Appendicularia	0	0	0	0	0
Cephalochordata	0	0	0	0	0
Totals	766	155	147	128	27

NGC = Northern Gulf of California; CGC = Central Gulf of California; SGC = Southern Gulf of California; END = total number of Gulf endemics; ND = specific locality data lacking.

expansive coastal pond infrastructure. Loss of these wetlands also reduces important stopover sites for migratory birds (see chapter 15). Mexico's Ministry of Tourism's planned "Nautical Ladder" (*Escalera Náutica*) proposes 23 marinas to be in place by 2006 around both sides of the Baja California peninsula and southward on the mainland all the way to Tea-capán (Sinaloa). The marinas likely will cause permanent loss of wetlands, and constructing the infrastructure required to connect them with roads and services could also be damaging. Areas that have experienced rapid growth of tourism and "vacation home" development (mainly for visitors from the United States) such as Puerto Peñasco and San Carlos in Sonora, and the coastal strip from San Felipe to Puertecitos (Baja California), have been hard hit ecologically by human perturbation. The intertidal zones of these areas have lost almost all of their larger-bodied fauna (especially echinoderms and molluscs) and now harbor only small remnants of their past biological diversity.

The coral reef at Cabo Pulmo is an area of extreme concern. Although the reef is included in the 7111-ha Cabo Pulmo National Marine Park, created in 1995, it has never had proper protection. The reefs are in shallow water, which means they are quickly affected by increased sea-surface temperatures associated with global warming and from sediment and wastewater runoff from coastal development. Charter sport fishing boats and small-scale Mexican fishers exploit the reef, taking fish and causing anchor damage. Tropical fishes are also removed from the reef habitat by the commercial aquarium trade. The maritime boundaries of the park are not marked. The reef is also easily (and heavily) accessed by sport divers, and as commercial sport diving operations in the area grow, damage to the reefs is inevitable unless strict regulations and educational programs are put in place. The growth of vacation homes along the shoreline of Cabo Pulmo could also result in nutrient enrichment by way of sewage runoff (above ground or below ground), resulting in overgrowth of the reef by algae and eventual eutrophication. Despite regional concern by universities, nongovernmental organizations, and local residents, local politics and lack of protection present major threats to the survival of this rare ecosystem—the Gulf of California's only coral reef.

Commercial fishing boats using large gillnets and long-lines with many hooks overexploit offshore waters, and small boat (*panga*) fishers often take shrimp and finfish from shallower coastal waters, estuaries, and other coastal lagoons before they have reached reproductive maturity. Foster et al. (1997) described the overexploitation of the scallop population (*Argopectin circularis*) around Isla El Requé-son (in Bahía Concepcón, Baja California Sur), where the commercial take fell from 1.5 million kg in 1991 to just 1080 kg in 1994.

Narcotraffickers using the Gulf of California as a corridor to transport illegal drugs from Mexico to the United States present a new and growing threat to biodiversity. They often camp on islands, damaging coastal environments, and after a run abandon (or trade) their *pangas* (skiffs) in the upper Gulf in such high numbers that the local fishers have greatly increased their boat presence, and ecological impact, in the region. Many islands of the Gulf are also threatened by introduction of exotic/domestic animals (e.g., rats, cats, ants, cockroaches) and plants (e.g., buffelgrass) that impact native plant, reptile and bird populations (Bahre 1983).

Living rhodolith beds occur at several shallow-water localities along the southeastern coast of Baja California Sur and infrequently from Puerto Lobos (Cabo Tepoca, Sonora) to at least Bahía Banderas (Jalisco). These are unique living habitats composed of unattached spherical nodules (to 10 cm) of free-living calcareous red algae (Corallinales, Rhodophyta) that can cover large areas of the shallow sea bed (Riosmena-Rodriguez et al. 1999; Foster 2001). Rhodolith beds in the Gulf harbor a diverse associated fauna, although a full survey of their associated animal communities has yet to be undertaken (see Cintra-Buenrostro et al. 2002, for molluscs; Reyes-Bonilla et al. 1997, for corals; and Clark 2000, for chitons). Because the beds are sites of high scallop abundance, disturbance and damage associated with scallop fishing (hookah hoses and anchors dragging the bottom) are a threat to these unique communities, and Foster et al. (1997) recorded considerable damage from scallop fishers in rhodolith beds around Isla El Requéson. And, of course, offshore rhodolith beds can be greatly damaged by shrimp trawlers (video footage of this is seen in the Howard Hall production "Shadows in a Desert Sea"; Hall 1992). Because of their strong three-dimensionality and conversion of soft-bottom habitats to hard-bottom ecosystems, rhodolith beds are probably important sites for larval recruitment, including commercial species. The extensive nearshore Pliocene and Pleistocene carbonate deposits of the southeastern Baja California peninsula indicate that rhodolith communities have been an important part of the shallow marine environment of this region for millions of years.

In summary, biodiversity in the Gulf of California is threatened by reduction of freshwater inflow, chemical pollution from agriculture, runoff, and sewage from urban areas, coastal habitat destruction, inadequate fisheries regulation and historical overfishing, lack of reliable scientific data upon which to base management decisions, uncontrolled tourism, narcotrafficking, and introduction of exotic species. In combination, these factors have resulted in the near extinction of such highly visible or charismatic endemic species as the totoaba and vaquita, near extirpation of 5 species of sea turtles, destruction of biodiversity in littoral and offshore benthic habitats, and decimation of benthic/demersal ecosystems due to commercial bottom trawling for shrimp and finfish (including substantial reductions in the Gulf's important commercial shrimp populations). However, despite these threats, in the Gulf of California one still can find a number of coastal refugia, areas not easily accessible by road or large fishing boats, which serve as important shelters for species extirpated elsewhere in the Gulf. Current discussions on a biodiversity action and sustainable management plan for the Gulf, spearheaded by regional nongovernmental conservation organizations as well as several government agencies, are focusing on protection of these refugia, as well as the islands, estuaries, and other coastal lagoons of this tremendously diverse region. Two recently enacted laws in Mexico prohibit the use of gillnets with mesh sizes greater than 6 inches and prohibit the use of bottom trawling in federally protected areas, such as the Upper Gulf/Delta Biosphere Reserve. These new laws (*normas*) are major steps toward a meaningful conservation effort in the Gulf of California.

Acknowledgments

Many people have been involved in the Macrofauna Golfo Project, which has run from 1994 to the present. We thank the technicians and specialists who aided in data compilation and review of data sets, including Juan Antonio Barragán, Dawn Breese, Bruce Collette, Mercedes Cordero, S. Yvonne Delgado, J. Alonso Esparza, David Catania, Héctor Espinosa-Pérez, María del Carmen Espinosa-Pérez, Phil Heemstra, Jaqueline García-Hernández, Rocio Güereca, Cynthia Klepadlo, Linda Yvonne Maluf, José Manuel Nava, Carlos Navarro-Serment, Mauricia Pérez-Tello, Héctor G. Plascencia, Richard H. Rosenblatt, Jeff Seigel, Bill Smith-Vaniz, Bernie Tershy, Marisol Tordesillas, Sandra Trautwein, H. J. Walker, and Regina Wetzer. For database support and advice, we thank Francisco Zamora-Arroyo, Germán Ramírez, Ernesto Bolado, Roberto González, Rocio Brambila, Monica Guzmán, and especially Alvaro Espinel and Silvio Olivieri of Conservation International's (CI) Center for Applied Biodiversity Science. Special thanks are due to Alejandro Robles of CI, who conceived the Macrofauna Golfo Project along with L. T. Findley and Juan Carlos Barrera (formerly of CIDESON). Very special thanks go to María de los Ángeles (Machangeles) Carvajal Rascón, director of CI's Gulf of California Program, who has been the project's staunchest supporter over the years, along with A. Robles and (more recently) Inocencio Higuera and Alfonso Gardea of CIAD. We thank Roy Houston, Tom Van Devender, and Wendy Moore for their careful review of the manuscript. The initial project was funded by grants from CONABIO (Comisión Nacional para el Conocimiento y Uso de la Biodiversidad), the Instituto Nacional de la Pesca-SEPESCA (via Pronatura), and from USAID and the Homeland Foundation (via CI's Región Golfo de California). Continuing support came from CEMEX and Ford Motor Company (via CI), CI, CIAD, CIDESON, and the Arizona-Sonora Desert Museum (Tucson).

Literature Cited

Allee, W. C. 1923. Studies in marine ecology: 1. The distribution of common littoral invertebrates of the Woods Hole region. Biological Bulletin 18: 167–191.

Alvarez-Borrego, S. 1983. Gulf of California. Pp. 427–449 *in* B. H. Ketchum (ed.), Ecosystems of the World 26. Estuaries and Enclosed Seas. Elsevier Scientific, New York.

Alvarez-Borrego, S., B. P. Flores-Báez, and L. A. Galindo-Bect. 1975. Hidrológia del alto Golfo de California II. Condiciones durante invierno, primavera y verano. Ciencias Marinas 2(1): 21–36.

Anderson, D. W. 1983. The seabirds. Pp. 246–264 *in* T. J. Case and M. L. Cody (eds.), Island Biogeography in the Sea of Cortéz. University of California Press, Berkeley.

Astro, R., and T. Hayashi (eds.). 1971. Steinbeck: The Man and His Work. Oregon State University Press, Corvallis.

Aurioles-Gamboa, D., and G. A. Zavala. 1999. Algunos factores ecológicos que determinan la distribución y abundancia del lobo marino *Zalophus californianus*, en el Golfo de California. Ciencias Marinas 20(4): 535–553.

Bahre, C. 1983. Human impact: the Midriff Islands. Pp. 290–306 *in* T. J. Case and M. L. Cody (eds.), Island Biogeography in the Sea of Cortéz. University of California Press, Berkeley.

Bernardi, G., L. Findley, and A. Rocha-Olivares. 2003. Vicariance and dispersal across Baja California in disjunct marine fish populations. Evolution 57: 1599–1609.

Bogan, M. A. 1999. Family Vespertilionidae. Pp. 139–181 *in* S. T. Alvarez-Castañeda and J. L. Patton (eds.), Mamíferos del Noroeste de México, vol. 1. Centro de Investigaciones Biológicas del Noroeste, S. C., La Paz, Mexico.

Bray, N. A., and J. M. Robles. 1991. Physical oceanography of the Gulf of California. Pp. 511–553 *in* J. P. Dauphin and B. R. T. Simoneit (eds.), The Gulf and Peninsular Province of the Californias. American Association of Petroleum Geologists, Memoir 47.

Briggs, J. C. 1974. Marine Zoogeography. McGraw-Hill, New York.

Brusca, R. C. 1975. Zoological classification of the Scripps Alpha Helix Baja California Expedition. Pp. 72–73 *in* Alpha Helix Research Program 1972–1974. Scripps Institution of Oceanography, University of California, San Diego.

Brusca, R. C. 1980a. The Allan Hancock Foundation of the University Southern California. Association of Systematics Collections Newsletter 8(1): 1–7.

Brusca, R. C. 1980b. Common Intertidal Invertebrates of the Gulf of California. University of Arizona Press, Tucson.

Brusca R. C. 1985. Corals and Coral Reefs of the Galapagos Islands, by P.W. Glynn and G. M. Wellington [book review]. Environmental Conservation (Geneva) 12(2): 191–192.

Brusca, R. C. 1989. Foreword. Pp. i–iii *in* A. Kerstitch, Sea of Cortez Marine Invertebrates. A Guide for the Pacific Coast, Mexico to Ecuador. Sea Challengers Press, Monterey, California.

Brusca, R. C. 1993. The Arizona/Sea of Cortez years of J. Laurens Barnard. Journal of Natural History 27: 727–730.

Brusca, R. C. 2004. The Gulf of California: an overview. Pp. 1–8 *in* R. C. Brusca, E. Kimrey, and W. Moore, Seashore Guide to the Northern Gulf of California. Arizona-Sonora Desert Museum, Tucson, Arizona.

Brusca, R. C., and G. C. Bryner. 2003. A case study of two Mexican biosphere reserves: the Upper Gulf of California/Colorado River Delta and Pinacate/Gran Desierto de Altar Biosphere Reserves. Pp. 28–64 *in* N. E. Harrison and G. C. Bryner (eds.), Science and Politics in the International Environment. Rowman and Littlefield, Lanham, Maryland.

Brusca, R. C., J. Campoy Favela, C. Castillo Sánchez, R. Cudney-Bueno, L. T. Findley, J. Garcia-Hernández, E. Glenn, I. Granillo, M. E. Hendrickx, J. Murrieta, C. Nagel, M. Román, and P. Turk-Boyer. 2001. A Case Study of Two Mexican Biosphere Reserves. The Upper Gulf of California/Colorado River Delta and Pinacate/Gran Desierto de Altar Biosphere Reserves. 2000 UNESCO Conference on Biodiversity and Society, Columbia University Earthscape (an electronic journal). [Available: www.earthscape.org/rr1/cbs01.html]

Brusca, R. C., and D. A. Thomson. 1975. Pulmo Reef: the only "coral reef" in the Gulf of California. Ciencias Marinas 2(2): 37–53.

Brusca, R. C., and B. Wallerstein. 1979. Zoogeographic patterns of idoteid isopods in the northeast Pacific, with a review of shallow water zoogeography for the region. Bulletin of the Biological Society of Washington 3: 67–105.

Carpenter, P. P. 1857. Report on the state of our knowledge with regard to the Mollusca of the west coast of North America. Report, British Association for the Advancement of Science, for 1856: 159–368.

Carriquiry, J. D., and A. Sánchez. 1999. Sedimentation in the Colorado River Delta and Upper Gulf of California after nearly a century of discharge loss. Marine Geology 158: 125–145.

Castro-Aguirre, J. L., E. F. Balart, and J. Arvizu-Martínez. 1995. Contribución al conocimiento del origen y distribución de la ictiofauna del Golfo de California, México. Hidrobiológica 5 (1–2): 57–78.

Castro-Aguirre, J. L., J. C. Ramírez-Cruz, and M. A. Martínez-Muñoz. 1992. Nuevos datos sobre la distribución de lenguados (Pisces: Pleuronectiformes) en la costa oeste de Baja California, México, con aspectos biológicos y zoogeográficos. Anales de la Escuela Nacional de Ciencias Biológicas de México 37: 97–119.

Castro-Aguirre, J. L., J. J. Schmitter-Soto, E. F. Balart, and R. Torres-Orozco. 1993. Sobre la distribución geográfica de algunos peces bentónicos de la costa oeste de Baja California Sur, México, con consideraciones ecológicas y evolutivas. Anales de la Escuela Nacional de Ciencias Biológicas de México 38: 75–102.

Castro-Aguirre, J. L., and R. Torres-Orozco. 1993. Consideraciones acerca del origen de la ictiofauna de Bahía Magdalena-Almejas, un sistema lagunar de la costa occidental de Baja California Sur, México. Anales de la Escuela Nacional de Ciencias Biológicas de México 38: 67–73.

Cervantes, M., L. T. Findley, K. H. Holtschmit, F. Manrique, C. Pantoja, A. Robles, G. Soberón-Chávez, and O. Vidal. 1992. Importancia ecológica del Estero del Soldado. Pp. 73–89 in J. L. Moreno (ed.), Ecología, Recursos Naturales y Medio Ambiente en Sonora. Secretaría de Infraestructura Urbana y Ecología, y El Colegio de Sonora, Hermosillo, Mexico.

Chen, L.-C. 1975. The rockfishes, genus *Sebastes* (Scorpaenidae), of the Gulf of California, including three new species with a discussion of their origin. Proceedings of the California Academy of Sciences 40: 109–141.

Cintra-Buenrostro, C. E., M. S. Foster, and K. H. Meldahl. 2002. Response of nearshore marine assemblages to global change: a comparison of molluscan assemblages in Pleistocene and modern rhodolith beds in the southwestern Gulf of California, Mexico. Paleo 183: 299–320.

Cisneros-Mata, M. A., G. Montemayor-López, and M. J. Román-Rodríguez. 1995. Life history and conservation of *Totoaba macdonaldi*. Conservation Biology 94: 806–814.

Clark, R. N. 2000. The chiton fauna of the Gulf of California rhodolith beds (with the descriptions of four new species). Nemouria 43: 1–18.

Crabtree, C. B. 1989. A new silverside of the genus *Colpichthys* (Atheriniformes: Atherinidae) from the Gulf of California, México. Copeia 1989(3): 558–568.

Cudney-Bueno, R. 2000. Management and conservation of benthic resources harvested by small-scale hookah divers in the northern Gulf of California, Mexico: the black murex snail fishery. M.Sc. thesis, University of Arizona, Tucson.

Cudney-Bueno, R., and P. J. Turk Boyer. 1998. Pescando entre mareas del alto Golfo de California. Una guía sobre la pesca artesanal, su gente y sus propuestas de manejo. CEDO Technical Series (Puerto Peñasco, Sonora), no. 1.

D'Agrosa, C., C. Lennert-Cody, and O. Vidal. 2000. Vaquita bycatch in Mexico's artisanal gillnet fisheries: driving a small population to extinction. Conservation Biology 14: 1110–1119.

Díaz, R. J., and R. Rosenberg. 1995. Marine benthic hypoxia: a review of its ecological effects and the behavior responses of benthic macrofauna. Oceanography and Marine Biology. Annual Review 33: 245–303.

Evermann, B. W., and O. P. Jenkins. 1891. Report upon a collection of fishes made at Guaymas, Sonora, Mexico, with descriptions of new species. Proceedings of the U.S. National Museum 14(846): 121–165.

Felger, R. S. 2000. Flora of the Gran Desierto and Río Colorado of Northwestern Mexico. University of Arizona Press, Tucson.

Findley, L. T. 1976. Aspectos ecológicos de los esteros con manglares en Sonora y su relación con la explotación humana. Pp. 95–105 in B. Braniff and R.S. Felger (eds.), Sonora: Antropología del Desierto. Instituto Nacional de Antropología e Historia, México D.F., Serie Científica no. 27.

Findley, L. T., P. A. Hastings, A. M. van der Heiden, R. Güereca, J. Torre, and D. A. Thomson. 1997. Distribution of endemic fishes of the Gulf of California, Mexico. Pp. 130 in Abstracts, 76th Annual Meeting, American Society of Ichthyologists and Herpetologists, 26 June–2 July 1997, Seattle, Washington.

Findley, L. T., M. E. Hendrickx, R. C. Brusca, A. M. van der Heiden, P. A. Hastings, and J. Torre. In press. Macrofauna del Golfo de California [Macrofauna of the Gulf of California]. CD-ROM version 1.0. Macrofauna Golfo Project. Conservation International, Center for Applied Biodiversity Science, Washington, D.C. [in Spanish and English]

Findley, L. T., J. Torre, J. M. Nava, A. M. van der Heiden, and P. A. Hastings. 1996. Preliminary ichthyofaunal analysis from a macrofaunal database on the Gulf of California, Mexico. Pp. 138 in Abstracts, 75th Annual Meeting, American Society of Ichthyologists and Herpetologists, 13–19 June 1996, New Orleans, Louisiana.

Flores-Verdugo, F., F. González-Farias, and U. Zaragoza-Araujo. 1993. Ecological parameters of the mangroves of semi-arid regions of Mexico. Importance for ecosystem management. Pp. 123–132 in H. Lieth and A. Masoom (eds.), Towards the Rational Use of High Salinity Tolerant Plants, vol. 1. Kluwer Academic Publishers, Dordrecht, the Netherlands.

Foster, M. S. 2001. Rhodoliths: between rocks and soft places. Journal of Phycology 37: 659–667.

Foster, M. S., R. Riosmena-Rodríguez, D. L. Steller and W. J. Woelkerling. 1997. Living rhodolith beds in the Gulf of California and their implications for paleoenvironmental interpretations. Pp. 127–139 in M. E.

Johnson and J. Ledesma-Vázquez (eds.), Pliocene Carbonates and Related Facies Flanking the Gulf of California, Baja California, México. Geological Society of America, Special Papers no. 318, Boulder, Colorado.

Galindo-Bect, M. S., E. P. Glenn, H. M. Page, K. Fitzsimmons, L. A. Galindo-Bect, J. M. Hernández-Ayon, R. L. Petty, J. García-Hernández, and D. Moore. 2000. Penaeid shrimp landings in the upper Gulf of California in relation to Colorado River freshwater discharge. Fisheries Bulletin 98: 222–225.

Gallo-Reynoso, J. P. 1998. La vaquita marina y su hábitat crítico en el alto Golfo de Califor-nia. Gaceta Ecológica (Instituto Nacional de Ecología, SEMARNAP, México, D.F.) 47: 29–44.

García-Caudillo, J. M., and J. V. Gómez-Palafox. In press. La Valoración Económica Ambiental de la Captura Incidental en la Pesquería de Camarón en el Golfo de California. Instituto Nacional de Ecología, SEMARNAT, México D.F.

Gastil, G., J. Minch, and R. P. Phillips. 1983. The geology and ages of the islands. Pp. 13–25 *in* T. J. Case and M. L. Cody (eds.), Island Biogeography in the Sea of Cortéz. University of California Press, Berkeley.

Gilbert, C. H. 1892. Scientific results of explora-tions by the U. S. Fish Commission steamer *Albatross*. XXII. Descriptions of thirty-four new species of fishes collected in 1888 and 1889, principally among the Santa Barbara Islands and in the Gulf of California. Proceedings of the U.S. National Museum 14 [for 1891](880): 539–566.

Gill, T. H. 1862–1863. Catalogue of the fishes of Lower California in the Smithsonian Institution, collected by Mr. J. Xantus, Parts 1–4. Proceedings of the Academy of Natural Sciences of Philadelphia 1862: 140–151, 242–246, 249–262; 1863: 80–88.

Glenn, E. P., F. Zamora-Arroyo, P. L. Nagler, M. Briggs, W. Shaw, and K. Flessa. 2001. Ecology and conservation biology of the Colorado River Delta, Mexico. Journal of Arid Environments 49: 5–15.

Glynn, P. W., and G. M. Wellington. 1985. Corals and Coral Reefs of the Galapagos Islands (with an Annotated List of the Scleractinian Corals of the Galapagos, by J. W. Wells). University of California Press, Berkeley.

Greenberg, J. B. In press. Territorialization, globalization, and dependent capitalism in the political ecology of fisheries in the upper Gulf of California. *In* A. Biersack and J. B. Greenberg (eds.), Culture, History, Power, Nature: Ecologies for the New Millennium. Bureau of Applied Research in Anthropol-ogy, University of Arizona, Tucson.

Greenberg, J. B., and C. Vélez-Ibáñez. 1993. Community dynamics in a time of crisis: an ethnographic overview of the upper Gulf. *In* T. R. McGuire and J.B. Greenberg (eds.), Maritime Community and Biosphere Reserve: Crisis and Response in the Upper Gulf of California. Occasional Paper no. 2, Bureau of Applied Research in Anthropol-ogy. University of Arizona, Tucson.

Güereca-Hernández, L.P. 1994. Contribuciones para la caracterización ecológica del Estero del Soldado, Guaymas, Sonora, Mexico. M.Sc. thesis, Instituto Tecnológico y de Estudios Superiores de Monterrey-Campus Guaymas.

Hall, H. 1992. Shadows in a Desert Sea (video). Howard Hall Productions, Del Mar, California.

Hastings, P. A. 2000. Biogeography of the tropical eastern Pacific: distribution and phylogeny of chaenopsid fishes. Zoological Journal of the Linnean Society 128: 319–335.

Hastings, P. A., and L. T. Findley. In press. Marine fishes of the Biosphere Reserve, northern Gulf of California. *In* R. S. Felger and B. Broyles (eds.), Dry Borders: Great Natural Areas of the Gran Desierto and Upper Gulf of California. University of Utah Press, Salt Lake City.

Hedgpeth, J. W. 1978a. The Outer Shores. Part 1. Ed Ricketts and John Steinbeck Explore the Pacific Coast. Mad River Press, Eureka, California.

Hedgpeth, J. W. 1978b. The Outer Shores. Part 2. Breaking Through. Mad River Press, Eureka, California.

Hendrickx, M. E. 1996. Habitats and bio-diversity of decapod crustaceans in the southeastern Gulf of California. Revista de Biología Tropical 44(2): 603–617.

Hendrickx, M. E. 2001. Occurrence of a conti-nental slope decapod crustacean community along the edge of the minimum oxygen zone in the southeastern Gulf of California, Mexico. Belgian Journal of Zoology 131 (Supplement 2): 95–110.

Hendrickx, M. E., and R. C. Brusca. 2002. Biodiversidad de los invertebrados marinos de Sinaloa. Pp. 141–163 *in* J. L. Cifuentes Lemus and J. Gaxiola López (eds.), Atlas de los Ecosistemas y la Biodiversidad de Sinaloa. Centro de Ciencias de Sinaloa, Culiacán, Mexico.

Hendrickx, M. E., and A. Toledano-Granados. 1994. Catálogo de moluscos pelecípodos, gasterópodos y poliplacoforos. Colección de referencia, Estación Mazatlán, ICML, UNAM. ICML-UNAM and CONABIO.

Huang, D., and G. Bernardi. 2001. Disjunct Sea of Cortez-Pacific Ocean *Gillichthys mirabilis* populations and the evolutionary origin of their Sea of Cortez endemic relative, *Gillichthys seta*. Marine Biology 138: 421–428.

Hubbs, C. L. 1960. The marine vertebrates of the outer coast. Systematic Zoology 9(3–4): 134–147.

Jaramillo-Legorreta, A.M., L. Rojas-Bracho, and T. Gerrodette. 1999. A new abundance estimate for vaquitas: first step for recovery. Marine Mammal Science 15(4): 957–973.

Jenkins, O. P., and B. W. Evermann. 1889. Description of eighteen new species of fishes from the Gulf of California. Proceedings of the U.S. National Museum 11 [for 1888]: 137–158.

Jordan, D. S. 1895. The fishes of Sinaloa. Proceedings of the California Academy of Sciences 5: 377–514. [Reprinted *in* Contributions to Biology, Hopkins Laboratory of Biology, Stanford University Publications 1: 1–71.]

Jordan, D. S., and C. H. Gilbert. 1882. Catalogue of the fishes collected by Mr. John Xantus at Cape San Lucas, which are now in the United States National Museum, with descriptions of eight new species. Proceedings of the U.S. National Museum 5(290): 353–371.

Kowalewski, M., G. E. Avila Serrano, K. W. Flessa, and G. A. Goodfriend. 2000. Dead delta's former productivity: two trillion shells at the mouth of the Colorado River. Geology 28: 1059–1062.

Lavín, M. F., V. M. Godínez, and L. G. Alvarez. 1998. Inverse-estuarine features of the upper Gulf of California. Estuarine, Coastal, and Shelf Science 47: 769–795.

Le Boeuf, B. J., D. Aurioles, R. Condit, C. Fox, R. Gisiner, R. Romero, and F. Sinsel. 1983. Size and distribution of the California sea lion population in Mexico. Proceedings of the California Academy of Sciences 43: 77–85.

Lepley, L. K., S. P. von der Haar, J. R. Hendrickson, and G. Calderon-Riverol. 1975. Circulation in the northern Gulf of California from orbital photographs and ship investigations. Ciencias Marinas 2(2): 86–93.

Lindsay, G. E. 1983. History of scientific exploration in the Sea of Cortez. Pp. 3–12 *in* T. J. Case and M. L. Cody (eds.), Island Biogeography in the Sea of Cortez. University of California Press, Berkeley.

Maluf, L. Y. 1983. Physical oceanography. Pp. 26–45 *in* T. J. Case and M. L. Cody (eds.), Island Biogeography in the Sea of Cortez. University of California Press, Berkeley.

Musick, J. A., M. M. Harbin, S. A. Berkeley, G. H. Burgess, A. M. Eklund, L. Findley, R. G. Gilmore, J. T. Golden, D. S. Ha, G. R. Huntsman, J. C. McGovern, S. J. Parker, S. G. Poss, E. Sala, T. W. Schmidt, G. R. Sedberry, H. Weeks, and S. G. Wright. 2000. Marine, estuarine, and diadromous fish stocks at risk of extinction in North America (exclusive of Pacific salmonids). Fisheries 25 (11): 6–30.

Nava, J. M., and L. T. Findley. 1994. Impact of the shrimp fishery on faunal diversity and stability in the upper Gulf of California, with special emphasis on the vaquita and totoaba. Project final report to Conservation International-Mexico, Gulf of California Program, Guaymas, Mexico.

Nava-Romo, J. M. 1994. Impactos a corto y largo plazo en la diversidad y otras características ecológicas de la comunidad béntico-demersal capturada por la pesquería de camarón en el norte del alto Golfo de California, México. M.Sc. thesis, Instituto Tecnológico y de Estudios Superiores de Monterrey-Campus Guaymas, Mexico.

Navarro-Serment, C. J. 2002. Abundancia, uso de hábitat y conservación del cocodrilo de río, *Crocodylus acutus* Cuvier, 1807 (Reptilia: Crocodylia) en el Estero El Verde, Sinaloa, México. M.Sc. thesis, Centro de Investigación en Alimentación y Desarrollo-Unidad Mazatlán, Sinaloa, Mexico.

Parker, R. H. 1964. Zoogeography and ecology of some macroinvertebrates, particularly mollusks, in the Gulf of California and the continental slope off Mexico. Videnskabelige Meddelelser fra den naturhistoriske Forening i Kjobenhavn 126: 1–178.

Patten, D. R., and L. T. Findley. 1970. Observations and records of *Myotis* (*Pizonyx*) *vivesi* Menegeaux (Chiroptera: Vespertiliontidae). Contributions in Science (Los Angeles County Museum of Natural History) 183: 1–9.

Pérez-Mellado, J., and L. T. Findley. 1985. Evaluación de la ictiofauna acompañante del camarón capturado en las costas de Sonora y norte de Sinaloa, México. Pp. 201–254 *in* A. Yañéz-Arancibia (ed.), Recursos Potenciales de México: La Pesca Acompañante del Camarón. Programa Universitario de Alimentos, Instituto de Ciencias del Mar y Limnología, e Instituto Nacional de la Pesca. Universidad Nacional Autónoma de México, México D.F.

Present, T. M. 1987. Genetic differentiation of disjunct Gulf of California and Pacific coast populations of *Hypsoblennius jenkinsi*. Copeia 1987(4): 1010–1024.

Reyes-Bonilla, H., R. Riosmena-Rodríguez and M. S. Foster. 1997. Hermatypic corals associated with rhodolith beds in the Gulf of California, México. Pacific Science 5(3): 328–337.

Riosmena-Rodríguez, R., W. J. Woelkerling and M. S. Foster. 1999. Taxonomic reassessment of rhodolith-forming species of *Lithophyllum* (Corallinales, Rhodophyta) in the Gulf of California, Mexico. Phycologia 38(5): 401–417.

Robison, B. H. 1972. Distribution of the midwater fishes of the Gulf of California. Copeia 1972(3): 448–461.

Rocha-Olivares, A., R. H. Rosenblatt, and R. D. Vetter. 1999. Molecular evolution, systematics, and zoogeography of the rockfish subgenus *Sebastomus* (*Sebastes*, Scorpaenidae) based on mitochondrial cytochrome *b* and control region sequences. Molecular Phylogenetics and Evolution 11: 441–458.

Roden, G. I. 1958. Oceanographic and meteorological aspects of the Gulf of California. Pacific Science 12(1): 21–45.

Roden, G. I., and G. W. Groves. 1959. Recent oceanographic investigations in the Gulf of California. Marine Research Journal 18(1): 10–35.

Rodríguez, C. A., K. W. Flessa, and D. L. Dettman. 2001. Effects of upstream diversion of Colorado River water on the estuarine bivalve mollusc *Mulinia coloradoensis*. Conservation Biology 15: 249–258.

Rojas-Bracho, L., and B. L. Taylor. 1999. Risk factors affecting the vaquita (*Phocoena sinus*). Marine Mammal Science 15(4): 974–989.

Román-Rodríguez, M., J. C. Barrera-G., and J. Campoy-F. 1998. La curvina golfina: ¿Volvió para quedarse? Voces del Mar (CEDO, Puerto Peñasco, Sonora) 1: 1–2.

Rosenblatt, R. H. 1967. The zoogeographic relationships of the marine shore fishes of tropical America. Studies in Tropical Oceanography 5: 570–592.

Rosenblatt, R. H., and R. S. Waples. 1986. A genetic comparison of allopatric populations of shore fish species from the eastern and central Pacific Ocean: dispersal or vicariance? Copeia 1986(2): 275–284.

Rowell, K., K. W. Flessa, and D. Dettman. In press. Oxygen isotopes in otoliths document that Gulf corvina use Colorado River habitat [abstract in English and Spanish] *in* D. A. Hendrickson and L. T. Findley (eds.), Proceedings of the Desert Fishes Council, vol. 34, 2002 Annual Symposium, 14–17 November, San Luis Potosí, Mexico. Desert Fishes Council, Bishop, California.

Schwartzlose, R. A., D. Alvarez-Millán, and P. Brueggeman. 1992. Golfo de California: Bibliografía de las Ciencias Marinas. Instituto de Investigaciones Oceanológicas, Universidad Autónoma de Baja California, Ensenada, Mexico.

Snyder-Conn, E., and R. C. Brusca. 1977. Shrimp population dynamics and fishery impact in the northern Gulf of California. Ciencias Marinas 1(3): 54–67.

Squires, D. 1959. Results of the Puritan-American Museum Natural History Expedition to Western Mexico. 7. Corals and coral reefs in the Gulf of California. Bulletin of the American Museum of Natural History 118(7): 367–432.

Steinbeck, J., and E. F. Ricketts. 1941. Sea of Cortez. A Leisurely Journal of Travel and Research. Viking Press, New York.

Streets, T.H. 1877. Contributions to the natural history of the Hawaiian and Fanning Islands and Lower California, made in connection with the U.S. North Pacific surveying expedition 1873–1875. Bulletin of the U.S. National Museum no. 7.

Sverdrup, H. U. 1941. The Gulf of California; preliminary discussion on the cruise of the *E.W. Scripps* in February and March 1939. Proceedings of the 6th Pacific Science Congress 3: 161–166.

Terry, A., G. Bucciarelli, and G. Bernardi. 2000. Restricted gene flow and incipient speciation in disjunct Pacific Ocean and Sea of Cortez populations of a reef fish species, *Girella nigricans*. Evolution 54: 652–659.

Thomson, D. A., L. T. Findley, and A. N. Kerstitch. 1979. Reef Fishes of the Sea of Cortez: The Rocky-Shore Fishes of the Gulf of California. John Wiley & Sons, New York. [Revised edition, 2000, University of Texas Press, Austin.]

Thomson, D. A., and M. R. Gilligan. 1983. The rocky-shore fishes. Pp. 98–129 *in* T. J. Case and M. L. Cody (eds.), Island Biogeography in the Sea of Cortez. University California Press, Berkeley.

Thomson, D. A., and C. E. Lehner. 1976. Resilience of a rocky-intertidal fish community in physically unstable environment. Journal of Experimental Marine Biology and Ecology 22: 1–29.

U.S.C. (University of Southern California) Press. 1985. Catalog of Allan Hancock Foundation Publications. University of Southern California.

van der Heiden, A. M. 1985. Taxonomía, biología y evaluación de la ictiofauna demersal del Golfo de California. Pp. 149–200 *in* A. Yañéz-Arancibia (ed.), Recursos Pesqueros Potenciales de México: La Pesca Acompañante del Camarón. Programa Universitario de Alimentos, Instituto de

Ciencias del Mar y Limnología, e Instituto Nacional de la Pesca. Universidad Nacional Autónoma de México, México D.F.

van der Heiden, A. M., and L. T. Findley. 1990. Lista de los peces marinos del sur de Sinaloa, México. Anales del Instituto de Ciencias del Mar y Limnología, Universidad Nacional Autónoma de México 15 [for 1988] (2): 209–223.

Vidal, O. 1995. Population biology and incidental mortality of the vaquita, *Phocoena sinus*. Reports of the International Whaling Commission, special issue 16: 247–272.

Vidal, O., R. L. Brownell Jr., and L. T. Findley. 1999. Vaquita, *Phocoena sinus* Norris and McFarland, 1958. Pp. 357–378 *in* S. H. Ridgway and R. Harrison (eds.), Handbook of Marine Mammals, vol. 6: The Second Book of Dolphins and the Porpoises. Academic Press, San Diego, California.

Vidal, O., L. T. Findley, and S. Leatherwood. 1993. Annotated checklist of the marine mammals of the Gulf of California. Proceedings of the San Diego Society of Natural History 28: 1–16.

Walker, B. W. 1960. The distribution and affinities of the marine fish fauna of the Gulf of California. Systematic Zoology 9(3–4): 123–133.

Zeitzschel, B. 1969. Primary productivity in the Gulf of California. Marine Biology 3(3): 201–207.

Appendix 9.1: Primary Institutions Conducting Zoological Research in the Gulf of California Since 1965

1. Arizona-Sonora Desert Museum (ASDM), Tucson, Arizona
2. California State University at Long Beach (CSULB), Long Beach, California
3. Centro de Investigación Científica y de Educación Superior de Ensenada (CICESE), Baja California
4. Centro de Investigación en Alimentación y Desarrollo (CIAD), Guaymas (Sonora) and Mazatlán (Sinaloa)
5. Centro de Investigaciones Biológicas del Noroeste (CIBNOR), La Paz (Baja California Sur) and Guaymas (Sonora)
6. Centro Interdisciplinario de Ciencias Marinas del Instituto Politécnico Nacional (CICIMAR-IPN), La Paz, Baja California Sur
7. Instituto del Medio Ambiente y Desarrollo Sustentable del Estado de Sonora (IMADES; formerly CIDESON), Hermosillo and El Golfo de Santa Clara, Sonora
8. Instituto Tecnológico y de Estudios Superiores de Monterrey (ITESM)-Campus Guaymas, Sonora
9. Scripps Institution of Oceanography (SIO), University of California at San Diego, La Jolla, California
10. Universidad Autónoma de Baja California (UABC), Ensenada, Baja California
11. Universidad Autónoma de Baja California Sur (UABCS), La Paz, Baja California Sur
12. Universidad de Sonora (UNISON), Hermosillo, Sonora
13. Universidad Nacional Autónoma de México (UNAM): Facultad de Ciencias, Instituto de Ciencias del Mar y Limnología's (ICML-UNAM) Mazatlán field station, and Instituto de Biología (IB-UNAM)
14. University of Arizona (UAZ), Tucson
15. University of California at Los Angeles (UCLA)
16. University of California at Santa Barbara (USB), Marine Science Institute
17. University of California at Santa Cruz (UCSC)

Hummingbird Communities along an Elevational Gradient in the Sierra Madre Occidental of Eastern Sonora, Mexico

THOMAS R. VAN DEVENDER

KAREN KREBBS

JEAN-LUC E. CARTRON

ANA LILIA REINA G.

WILLIAM A. CALDER

With an area of about 185,430 km², Sonora is the second largest state in Mexico. It has a convoluted topography, and in particular its eastern portion is dominated by multiple north–south-trending mountain ranges belonging to the Sierra Madre Occidental. Four main river systems (from north to south, Río Colorado, Río Yaqui, Río Mayo, and Río Fuerte) drain the state to the Gulf of California, which lies to the west. Sonora's rich flora includes about 5000 vascular plant species of mixed temperate and tropical affinity or origin (Felger and Wilson 1995; Felger et al. 2001). Besides vascular plants, the overall high biological diversity of Sonora includes more than 500 species of (year-round) resident and migratory birds, including 16 hummingbirds (Russell and Monson 1998).

As part of a Migratory Pollinator Program at the Arizona-Sonora Desert Museum, we studied hummingbird communities in eastern Sonora from October 1999 to September 2001. Much remains to be learned about the seasonal movements and biology of hummingbirds in Sonora (Russell and Monson 1998). Hummingbird food plants are also important to document in the state, especially given the threat of habitat loss, which worldwide is disrupting plant–pollinator interactions (Kearns et al. 1998; Lennartsson 2002). For hummingbirds, whose life cycles often appear to be strongly determined by nectar availability, the loss of important patches of nectar flowers has the potential to seriously limit seasonal movements, survivorship, and/or reproductive success. With few roads cutting across the Sierra Madre Occidental, many mountainous areas in eastern Sonora remain difficult to access. However, the completion in 1992 of Mexican Federal Highway 16 (MEX 16) connecting Hermosillo, Sonora, to La Junta, Chihuahua, provides a new opportunity for more hummingbird research in the region.

In this chapter, we describe changes in hummingbird species assemblages along an elevation gradient from thornscrub through tropical deciduous forest and oak woodland to pine–oak forest. We provide a few noteworthy natural history observations, including first nesting records for the state of Sonora. For each vegetation type, we list nectar plants visited by hummingbirds, with special emphasis on those used by multiple species. We frame the discussion of our results within the context of migration patterns, as currently understood. As part of the discussion, a few important observations from the Río Magdalena and Río Sonora valleys in north-central Sonora are presented as well.

Study Area and Methods

Study Area

From the Río Yaqui near Tónichi east to Yécora, MEX 16 provides an ideal transect to study biological diversity along an elevational–vegetational gradient (Búrquez et al. 1992; fig. 10.1). Most of the area crossed by MEX 16 is in the Municipio de Yécora, a 3300 km² county located within the Río Yaqui drainage portion of the broad Río Mayo Region (Martin et al. 1998; Reina et al. 1999; see also chapter 5). The western end of the transect from Tónichi to northwest of Tepoca is in the Municipios de Onavas and Soyopa. The vegetation changes from foothills thornscrub (FTS; 180–550 m elevation) and tropical deciduous forest (TDF; 550–1160 m) in the tropical zone to oak woodland (OW; 1050–1700 m) or pine–oak forest (POF; 1220–2240 m) in the Sierra Madre Occidental.

In Sonora, legumes dominate both FTS and TDF, although the 2 vegetation types appear different, with trees generally standing taller than the columnar cacti in TDF, but not in FTS (Gentry 1942). Along the Tónichi–Yécora elevation gradient, dominant plant species in FTS include tree ocotillo (*Fouquieria macdougalii*), organpipe cactus (*Stenocereus thurberi*), brea (*Parkinsonia praecox*), gatuño (*Mimosa distachya*), sámota (*Coursetia glandulosa*), papelío (*Jatropha cordata*), and brasil (*Haematoxylum brasiletto*). In TDF, the vegetation often consists of mauto (*Lysiloma divaricatum*), feather tree/tepeguaje (*Lysiloma watsonii*), tree morning glory/palo santo (*Ipomoea arborescens*), boatthorn acacia/güinolo (*Acacia cochliacantha*), palo zorillo (*Senna atomaria*), kapok/pochote (*Ceiba acuminata*), and torote (*Bursera fagaroides*). In OW, Arizona white oak/encino (*Quercus arizonica*), Chihuahua oak/encino peludo (*Q. chihuahuensis*), Mexican blue oak/encino azul (*Q. oblongifolia*), alligator bark juniper/táscate (*Juniperus deppeana*), piojilla (*Mandevilla foliosa*), Chihuahua pine/pino chino (*Pinus chihuahuana*), gatuño (*Mimosa dysocarpa*), and jehuite (*Montanoa leucantha*) are all well represented. Dominant plants in POF include Yécora pine/pino colorado (*Pinus yecorensis*), Apache pine/pino blanco (*Pinus engelmannii*), pino chino (*Pinus maximinoi*), madrone/madroño (*Arbutus xalapensis*), manzanita/manzanilla

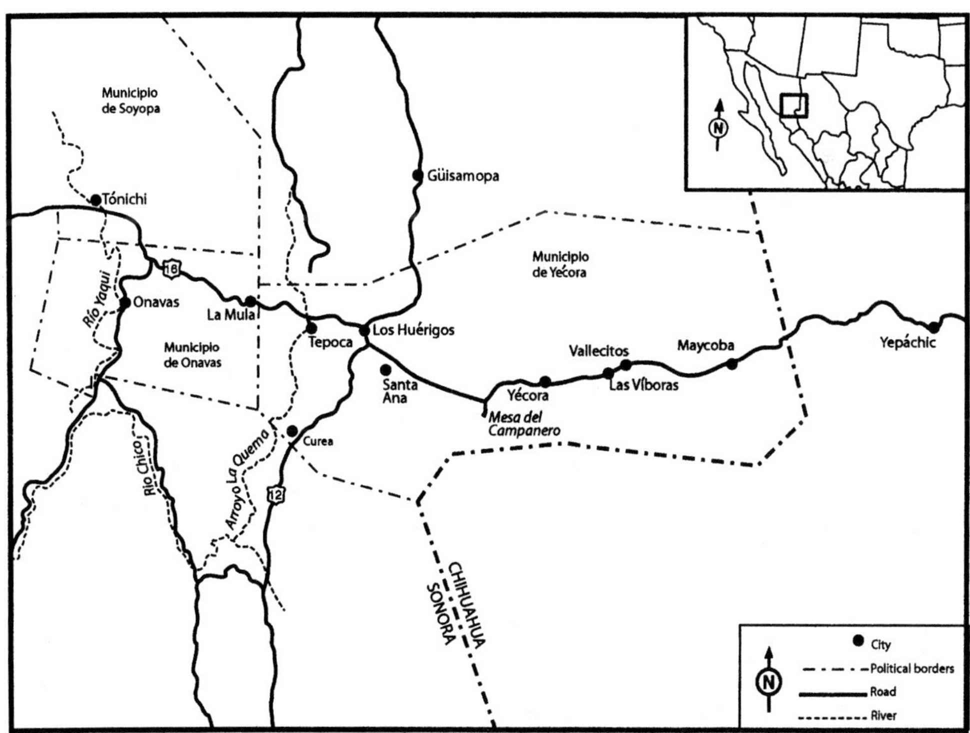

Figure 10.1. Map of the Tónichi-Yécora area in eastern Sonora.

(*Arctostaphylos pungens*), hueja (*Quercus coccolobifolia* and *Q. mcvaughii*), silverleaf oak/*cusi prieto* (*Q. hypoleucoides*), and deer brush (*Ceanothus coeruleus*).

Several of our sites were along drainages (e.g., Río Yaqui, Río Maycoba, Arroyo Pilladito). In all 4 general vegetation types, riparian habitats are linear and more mesic than adjacent slope vegetation. Often these habitats are less distinctive than the typical cottonwood (*Populus fremontii*)-willow (*Salix* spp.) gallery forests of the southwestern United States and northern Sonora. Riparian trees including *amolillo* (*Sapindus saponaria*), Goodding willow/*sauce* (*S. gooddingii*), guamúchil (*Pithecellobium dulce*), and *uvalama* (*Vitex mollis*) are occasional along the Río Yaqui near Tónichi in the FTS zone. In TDF, riparian trees include *bebelama* (*Sideroxylon persimile*), Bonpland willow (*Salix bonplandiana*), figs/*higuera* (*Ficus pertusa, F. petiolaris*, and *F. trigonata*), *guásima* (*Guazuma ulmifolia*), huérigo (*Populus monticola*), palms (*Brahea aculeata*), Mexican bald cypress/*sabino* (*Taxodium distichum* var. *mexicanum*), and *uvalama*. Riparian trees in OW include *madroño* (*Clethra mexicana*), Maycoba juniper/*sabino* (*Juniperus mucronata*), and sycamore/*aliso* (*Platanus racemosa*). Additional riparian trees in POF are alder/*alamillo* (*Alnus oblongifolia*), cherries (*Prunus gentryi, P. serotina*), Durango fir/*pinabete* (*Abies durangensis*), holly (*Ilex tolucana*), and Arizona cypress/*sabino* (*Cupressus arizonica*).

Methods

Hummingbirds were observed along the Tónichi-Yécora transect (loosely delineated based in part on accessibility; see below) during 15 separate field trips between October 1999 and September 2001. A total of 8 trips conducted in February–early May ("spring") presumably captured hummingbird spring migration through the area and encompassed the nesting season of low-elevation species. Five trips in August–early October ("summer-fall") coincided with the nesting season of high-elevation hummingbirds and the late summer migration of species such as the rufous hummingbird (*Selasphorus rufus*). Two field trips in December and January ("winter") documented wintering of some species in the area.

During each trip, which typically lasted about a week, we surveyed hummingbird communities in all 4 vegetation types. Some of the areas visited were near roads accessed from MEX 16, while others required hiking. The number of observers per trip ranged from 2–10. Data were collected using a variety of methods, including mist netting and point-count surveys. We observed hummingbird communities at a series of regular stations (feeders, patches of nectar plants, etc.) and camps, with a special effort made to include patches of flowers of all potential food plants (Van Devender et al. 2004). At each site (station or camp), we recorded the hummingbird species present, their activities (feeding, courtship, fighting, etc.), and the food plants they visited. The time spent at any site ranged from 30 minutes to several hours depending on the abundance and diversity of hummingbirds and nectar flowers. Negative data (absence of hummingbirds) were not recorded.

Our methodology was suited primarily for a study of hummingbird species assemblages and their associated food plants in each vegetation type. Abundance data were often recorded but not used in any statistical analysis within and among vegetation types, due to many potential confounding variables. Because hummingbirds are nectar feeders, their spatial and seasonal distributions can be strongly correlated with patches of food plants and thus may be nonrandom. Feeders were more effective in attracting hummingbirds at some sites than at others. Not all hummingbirds are equally attracted to feeders, and some species aggressively defend them, inhibiting other species from visiting. We tried to have as many skilled observers as possible on each trip. This maximized the chances of observing rare species, but different numbers of observers, of course, resulted in different numbers of observer hours per sampling site. Certain sites generally yielding higher numbers of hummingbirds were visited more frequently during the study period. Sampling at banding sites was longer than at other field sites, typically 3 hours per session, and often involved work early in the morning and again in late afternoon. There were also interannual differences in hummingbird abundance: the Maycoba sage (*Salvia betulaefolia*) patch at Maycoba and the Indian paintbrush/*periquito* (*Castilleja patriotica*) patch on Mesa de los Coronados had very different numbers of species and individuals on sampling days one year apart in 2000 and 2001.

For each vegetation type, we report the number of species detected during the study period. Very coarse estimates of abundance (see above) are provided using 2 different methods. We use site-specific abundance data, with information provided in a table on date and survey effort (number of observers and time spent at the site). We also report the number of sites × days ("number of observations") any given species was observed during a season or the entire study period. The occurrence of a hummingbird species at a patch of flowers, whether 1 or 20 individuals were present that day, counted as 1 observation of that species. The occurrence of a species at 1 site on 2 different days amounted to 2 observations of that species. The number of observations for any given species was tabulated separately for winter, spring, and summer-fall and for each of the 4 vegetation types. Due to the potential biases mentioned above (and others), all hummingbird numbers reported here must be interpreted with caution.

Hummingbird Communities

A total of 418 observations documented the presence of 12 species along the Tónichi–Yécora transect. A 13th species, the plain-capped starthroat (*Heliomaster constantii*), was also documented locally during the study period (see further on), but not by us, and thus is not tabulated here. Observed species assemblages varied across vegetation types and according to season (table 10.1), with an obvious difference between tropical lowland (FTS, TDF) and Madrean highlands (OW, POF). Below is a description of hummingbird species communities in each of the 4 main vegetation types.

Foothills Thornscrub

We observed 6 species of hummingbirds in FTS (table 10.1; fig. 10.2; see also fig. 5.3). All of these 6 species were recorded at least once during the spring, with most (71%) observations during that season being of the broad-billed hummingbird (*Cynanthus latirostris*) and Costa's hummingbird (*Calypte costae*; fig. 10.3). At some locations these 2 species seemed quite abundant, based on our counts of individuals (table 10.2). Although apparently less abundant than the broad-billed hummingbird and Costa's hummingbird, the violet-crowned hummingbird (*Amazilia violiceps*) was also frequently recorded during the spring (table 10.1). The rufous hummingbird (11 observations, including 1 observation of 33 individuals along the Río Yaqui at Loma Maderista; table 10.2) seemed common during the spring, but only between mid-March and

early April. The white-eared hummingbird (*Hylocharis leucotis*) was recorded only once in this vegetation type, in mid-March 2001. Similarly, the black-chinned hummingbird (*Archilochus alexandri*) was observed only once.

In the summer-fall, we had far fewer observations in FTS (table 10.1). Of the 5 species recorded during that time, only the broad-billed hummingbird appeared to be present in fairly high numbers (we counted 37 individuals during a 5-hour period along the Río Yaqui, on August 29, 2000; table 10.2). Observations of this species accounted for 45% of the total in FTS during the summer-fall. There were only 4 observations of the violet-crowned hummingbird, and also of the Costa's hummingbird, and where recorded these 2 species seemingly occurred in small numbers (≤ 2–3 individuals counted). Black-chinned and rufous hummingbirds were observed only once and twice during the summer-fall, respectively. Eight rufous hummingbirds were counted along the Río Yaqui on August 29, 2000. There was no observation of the white-eared hummingbird in FTS in the summer-fall.

Three species were observed during the winter (table 10.1). Eleven (92%) of the 12 observations during that season were of 2 species, the broad-billed and Costa's hummingbirds.

A broad-billed hummingbird nest was located in FTS on March 11, 2000, in a shrub dangling down a steep shady bank of a side canyon of the Río Yaqui near Tónichi. The female was seen feeding 2 nestlings about 14 days old several times. On April 1, the nest was empty, but the female was nearby feeding a fledgling. At Tónichi (28°35'55" N, 109°33'50" W, 200 m, FTS) a broad-billed hummingbird female was observed building a nest on March 12, 2000.

A Costa's hummingbird nest with 2 nestlings was observed on February 22 and March 12, 2000, in a torote (*Bursera laxiflora*) in FTS along the Río Yaqui. On February 22, a male was flying in courting loops. On March 12, the female fed the young several times at 20- to 25-minute intervals. Rogelio Martínez B., a school teacher at Curea (FTS, 28°18'42" N, 109°16'42" W, 490 m), observed a Costa's hummingbird nest with 2 chicks on April 4, 2001.

Tropical Deciduous Forest

Nine species of hummingbirds were observed in TDF (table 10.1; fig. 10.4). The broad-billed

Table 10.1. Numbers of observations of hummingbirds in 1999–2001 along the Tónichi-Yécora transect in eastern Sonora, Mexico.

	Foothills Thornscrub				Tropical Deciduous Forest				Oak Woodland				Pine-Oak Forest			
Common Names	Spring	Summer-Fall	Winter	Total	Spring	Summer-Fall	Winter	Total	Spring	Summer-Fall	Winter	Total	Spring	Summer-Fall	Winter	Total
Broad-billed hummingbird	37	9	5	51	30	8	10	48	1	7	0	8	0	8	0	8
White-eared hummingbird	1	0	0	1	1	0	1	2	5	4	0	9	10	22	2	34
Berylline hummingbird	0	0	0	0	4	0	0	4	0	13	0	13	1	17	0	18
Violet-crowned hummingbird	12	4	1	17	10	1	1	12	0	13	0	13	0	9	0	9
Blue-throated hummingbird	0	0	0	0	0	0	0	0	3	2	0	5	5	13	0	18
Magnificent hummingbird	0	0	0	0	0	0	0	0	5	0	0	5	0	10	0	10
*Black-chinned hummingbird	1	1	0	2	0	1	0	1	1	12	0	13	0	12	0	12
Anna's hummingbird	0	0	0	0	0	0	1	1	0	0	0	0	0	2	0	2
Costa's hummingbird	25	4	6	35	1	0	7	8	0	0	0	0	0	3	0	3
*Calliope hummingbird	0	0	0	0	0	0	0	0	0	1	0	1	0	1	0	1
*Broad-tailed hummingbird	0	0	0	0	2	0	0	2	2	2	0	4	2	5	0	7
*Rufous hummingbird	11	2	0	13	7	1	0	8	2	3	0	5	2	13	0	15
Totals = 418	87	20	12	119	55	11	20	86	19	57	0	76	20	115	2	137

Spring = February–May; summer-fall = August–October; winter = December–January.* = long-distance migrants. Hummingbird species listed by taxonomic order (AOU 1998).

Figure 10.2. Foothills thornscrub at 200 m elevation on the Río Yaqui near Tónichi. Hummingbirds including long-distance migrants are common visitors to tree ocotillos (*Fouquieria macdougalii*) in spring.

hummingbird was again the species most often recorded, with 30 spring, 8 summer-fall, and 10 winter observations.

Seven species were detected during the spring (table 10.1). Thirty (55%) of the 55 observations during that season were of the broad-billed hummingbird. At Rancho La Mula on February 4, 2000 we counted 15 individuals of this species during a 3-hour period; 10 broad-billed hummingbirds were detected along Arroyo Pilladito on April 1, 2000 (table 10.3). Other species frequently detected were the violet-crowned and rufous hummingbirds. The violet-crowned hummingbird seemed common at Rancho La Mula and Arroyo los Huérigos on February 4 and February 26, 2000, respectively (table 10.3). The earliest spring observation of a rufous hummingbird in TDF was in late February. The other species found in TDF during the spring were the berylline (*Amazilia beryllina*), broad-tailed (*Selasphorus platycercus*), Costa's, and white-eared hummingbirds (table 10.1). Where detected, these species seemingly occurred in relatively low numbers (table 10.3).

Our number of hummingbird observations was notably lower during the summer-fall than during the spring. Besides the broad-billed hummingbird, which during this season seemed the most common species, 3 other hummingbirds were recorded, all of them only once: the black-chinned, rufous, and violet-crowned hummingbirds.

Five species were detected during the winter (table 10.1). Two species, the broad-billed and Costa's hummingbird, accounted for 85% of all observations. The Anna's (*Calypte anna*), violet-crowned, and white-eared hummingbirds were detected only once in the winter in TDF. The Anna's hummingbird was the most common hummingbird at Rancho la Palma Agujerada on December 10, 1999 (table 10.3).

A broad-billed hummingbird nest with 2 eggs was found in TDF west of Tepoca on January 9, 2001 (fig. 10.5). The nest had been built on a dead *trompillo* (*Schizocarpum palmeri*) vine that was hanging on a shrub (*Brickellia coulteri*) about 3 m above the highway (MEX 16). The weather was cold with heavy rain and hail, but the nest and the eggs

Figure 10.3. A female Costa's hummingbird (*Calypte costae*) visiting a chuparrosa (*Justicia candicans*). This species is common or abundant for most of the year in foothills thornscrub. In tropical deciduous forest it is found chiefly during the winter.

were found undamaged on January 13. On February 12, the nest was empty, but the female was caring for 2 recently-fledged birds. Broad-billed hummingbird females were also observed building nests in the spring of 2000 in TDF: on February 24 at Rancho La Mula; March 11 at Rancho Panzacola near San Javier (28°32'32" N 109°44'32" W, 490 m); and April 1 at Arroyo La Uvalamita (28°29'10" N 109°22'46" W, 685 m). On September 1, 2001, we saw a male broad-billed hummingbird along Arroyo Los Huérigos mobbing a gray hawk (*Asturina nitida*) in a tree and as this raptor flew away.

Oak Woodland

We observed 10 species of hummingbirds in OW (table 10.1). Most (75%) of our observations were in the summer-fall; all others were during the spring. The white-eared and magnificent (*Eugenes fulgens*) hummingbirds were the 2 species most frequently (5 observations) observed in the spring. The other 5 species detected in the spring were the black-chinned, blue-throated (*Lampornis clemenciae*), broad-billed, broad-tailed, and rufous hummingbirds. A single broad-billed hummingbird was observed in the spring in OW (table 10.4).

Nine species were observed in OW in the summer-fall (table 10.1). Like most other species, the broad-billed hummingbird was seen in OW primarily during summer-fall (7 observations). At Maycoba on September 5, 2001, 17 broad-billed hummingbirds were counted during a 90-minute observation period (table 10.4), indicating that the species could even be locally abundant in OW in the summer-fall. However, we observed 3 species more frequently than the broad-billed hummingbird in the summer-fall: the berylline (13 observations), violet-crowned (13 observations), and black-chinned (12 observations) hummingbirds. The violet-crowned hummingbird occurred in high numbers at

Table 10.2. Some observed hummingbird numbers in foothills thornscrub along the Tónichi-Yécora transect, October 1999–September 2001.

Localities and Dates	No. of Observers	Observation Time (h)	Hummingbird Species[a]											
			ANHU	BBLH	BCHU	BEHU	BLUH	BTLH	CAHU	COHU	MAHU	RUHU	VCHU	WEHU
Spring (February–May)														
Río Yaqui, Loma Maderista, 2/11/01	3	1400–1700		9						10			6	
Río Yaqui, Arroyo Garambullo, 2/12/01	4	0600–1000		27						17			4	
Río Yaqui, Loma Maderista, 2/17/01	5	0600–0900, 17–1900		20						27				
Río Yaqui, Arroyo Garambullo, 2/22/00	6	0700–1000								cm[b]				
Río Yaqui, Arroyo Garambullo, 2/28/00	4	0700–0800								1		5	2	
Río Yaqui, Loma Maderista, 3/10/01	5	0600–0930		21						17		33	11	1
Río Yaqui, Arroyo Garambullo, 3/12/00	5	0700–0900		5						7		4	3	
Río Yaqui, Arroyo Garambullo, 3/16/00	5	0700–0800								2		1		
Río Yaqui, Arroyo Garambullo, 4/1/00	4	0700–0800		9						7			2	
Curea (thornscrub), 4/3/00	4	0600–1100		6								4	1	
Río Yaqui, Arroyo Garambullo, 4/29/00	2	0600–1000		4									1	
Río Yaqui, Loma Maderista, 5/6/01	2	0700–0900		7						9			4	
Summer-fall (August–October)														
Río Yaqui, Arroyo Garambullo, 8/4/00	3	0600–1000		2										
Río Yaqui, Arroyo Garambullo, 8/29/00	4	0600–1100		37	2					2		8	1	
Río Yaqui, Loma Maderista, 9/1/01	5	0700–0900		4						3				
Río Yaqui, Arroyo Garambullo, 9/29/00	2	0630–0830, 1700–1730		6									1	
Río Yaqui, Arroyo Garambullo, 9/30/99	3	0730–0800		1										
Río Yaqui, Arroyo Garambullo, 10/4/99	3	1800–1830											1	
Winter (December–January)														
Río Yaqui, Arroyo Garambullo, 12/9/99	3	0700–1000								occ[b]				
Río Yaqui, Arroyo Garambullo, 1/8/01	3	0800–0830		3						5			2	

[a]Hummingbird species: ANHU = Anna's hummingbird (*Calypte anna*); BBLH = broad-billed hummingbird (*Cynanthus latirostris*); BCHU = black-chinned hummingbird (*Archilochus alexandri*); BEHU = berylline hummingbird (*Amazilia beryllina*); BLUH = blue-throated hummingbird (*Lampornis clemenciae*); BTLH = broad-tailed hummingbird (*Selasphorus platycercus*); CAHU = Calliope hummingbird (*Stellula calliope*); COHU = Costa's hummingbird (*Calypte costae*); MAHU = magnificent hummingbird (*Eugenes fulgens*); RUHU = rufous hummingbird (*Selasphorus rufus*); VCHU = violet-crowned hummingbird (*Amazilia violiceps*); WEHU = white-eared hummingbird (*Hylocharis leucotis*).
[b]Abundance status: cm = common; occ = occasional.

Figure 10.4. A palm (*Brahea aculeata*) stand at 900 m elevation in tropical deciduous forest at Rancho La Mula. Violet-crowned hummingbirds (*Amazilia violiceps*) are common in tropical deciduous forest in spring.

Rancho La Palmita on September 1, 2000 (table 10.4). At Maycoba on September 5, 2001, the other abundant species besides the broad-billed hummingbird was the black-chinned hummingbird, with 20 individuals counted during the 90-minute observation period (table 10.4). Six other species were observed in OW in the summer-fall, including the blue-throated (2 observations), broad-tailed (2 observations), and calliope (*Stellula calliope*; 1 observation, early September) hummingbirds (table 10.1).

Calliope hummingbirds observed were a female and a young male with incomplete gorgets. Rufous hummingbirds were also seen. There was no observation of a magnificent hummingbird in OW in the summer-fall.

Pine–Oak Forest

The largest numbers of hummingbird species and observations recorded were in POF (table 10.1; see

Table 10.3. Some observed hummingbird numbers in tropical deciduous forest along the Tónichi-Yécora transect, October 1999–September 2001.

Localities and Dates	No. of Observers	Observation Time (h)	ANHU	BBLH	BCHU	BEHU	BLUH	BTLH	CAHU	COHU	MAHU	RUHU	VCHU	WEHU
Spring (February–May)														
Rancho La Mula, 2/4/00	6	0700–1000		15									10	
Arroyo los Huérigos, 2/12/01	5	1600–1630		4										
Arroyo los Huérigos, 2/26/00	4	0600–1900		9		1						6	10	
Rancho la Palma Agujerada, 2/29/00	3	0900–0930		occ[b]										
Arroyo los Huérigos, 3/10/01	3	1530–1630		5		1							2	1
Arroyo los Huérigos, 3/13/00	5	0600–1900		5									5	
Rancho La Mula, 3/13/00	3	0700–0900		5									8	
Arroyo Pilladito, 4/1/00	4	1500–1530		10		1		2				5		
Arroyo Uvalamita, 4/1/00	4	1200–1400		9						2				
Rancho La Mula, 4/1/00	4	0600–0800		8								2	3	
Curea (tropical deciduous forest), 4/3/00	4	0730–0900		2								2	5	
Rancho la Palma Agujerada, 4/29/00	3	0700–0730		2									1	
Summer-fall (August–October)														
Arroyo Uvalamita, 8/4/00	3	1300–1400										2		
Rancho La Palma Agujerado, 8/5/00	3	0600–0800		2										
Arroyo Uvalamita, 8/29/00	4	1630–1700		1										
Rancho La Mula, 8/30/00	4	0800–1000		3									4	
Arroyo Uvalamita, 9/1/01	2	1400–1430		1										
Arroyo los Huérigos, 9/1/01	5	1400–1430		1										
Winter (December–January)														
Rancho la Palma Agujerada, 12/10/99	3	0700–1000	cm[b]											
Arroyo los Huérigos, 1/10/01	3	1400–1600		2						1			2	1

[a]Hummingbird species: ANHU = Anna's hummingbird (*Calypte anna*); BBLH = broad-billed hummingbird (*Cynanthus latirostris*); BCHU = black-chinned hummingbird (*Archilochus alexandri*); BEHU = berylline hummingbird (*Amazilia beryllina*); BLUH = blue-throated hummingbird (*Lampornis clemenciae*); BTLH = broad-tailed hummingbird (*Selasphorus platycercus*); CAHU = Calliope hummingbird (*Stellula calliope*); COHU = Costa's hummingbird (*Calypte costae*); MAHU = magnificent hummingbird (*Eugenes fulgens*); RUHU = rufous hummingbird (*Selasphorus rufus*); VCHU = violet-crowned hummingbird (*Amazilia violiceps*); WEHU = white-eared hummingbird (*Hylocharis leucotis*).

[b]Abundance status: cm = common; occ = occasional.

Figure 10.5. A female broad-billed hummingbird (*Cynanthus latirostris*) sitting on her nest. Broad-billed hummingbird nests are found in foothills thornscrub and tropical deciduous forest along the Tónichi-Yécora transect from January through April.

fig. 5.4). In fact, all 12 hummingbird species we detected along the Tónichi-Yécora transect were observed at least once in POF. The white-eared hummingbird was the most frequently observed species, with 34 observations (25% of the total in POF). Berylline and blue-throated hummingbirds were the next most frequently observed species in POF.

In the spring, 5 species were detected for a total of 20 observations (table 10.1). Fifty percent and 25% of all observations were of the white-eared and blue-throated hummingbirds, respectively (table 10.1). Of these 2 species, the white-eared hummingbird typically was detected in higher numbers. One example is our count of 15 white-eared and 4 blue-throated hummingbirds at El Aguajito on April 5, 2000, during a 2-hour observation period (table 10.5). The other 3 species we recorded in POF in the spring were the berylline, broad-tailed, and rufous hummingbirds.

Most (84%) of our observations were in the summer-fall (table 10.1). All 12 species were observed, and many of them appeared to be common. The species most often recorded was, again, the white-eared hummingbird (22 observations). However, the berylline, blue-throated, rufous, black-chinned, and magnificent hummingbirds were all also frequently observed. A 5-hour observation period in Barranca El Salto (fig. 10.6) on September 4, 2001 yielded 21 individuals of both blue-throated and rufous hummingbirds (table 10.5). Fairly high numbers of rufous hummingbirds were observed at other locations (El Aguajito and Mesa de los Coronados), also in early September (fig. 10.7). A female blue-throated hummingbird observed on September 1,

2000 had a loud wing buzz, probably due to molting feathers. The white-eared hummingbird was the only species observed in the winter in POF (table 10.1).

A white-eared hummingbird nest was found in an apple orchard on Mesa del Campanero on September 1, 2000. The nest contained 2 nestlings estimated to be at least 2 weeks old. The area was originally pine-oak forest. On August 8, 2000, a berylline hummingbird nest was found in an oak in POF. Later in the month, the nest was found complete but empty. A magnificent hummingbird banded on September 1, 2000 by Steve and Ruth Russell at Oscar Coronado's home on Mesa del Campanero was recaptured at the same feeder on September 3, 2001 by Susan Wethington.

Hummingbird Food Plants

A total of 98 plant species (including those cultivated and introduced) have been documented as sources of nectar for hummingbirds in Sonora (Van Devender et al. 2004). Along the Tónichi-Yécora transect, hummingbirds visited 46 plant species, 8 of which attracted 5 or more hummingbird species (table 10.6). The plants most visited by hummingbirds were Texas betony (*Stachys coccinea*, flowers in summer-fall; 12 hummingbird species) and tree tobacco (*Nicotiana glauca*, all year, non-native; 9 hummingbird species). *Piojilla* (*Mandevilla foliosa*, early summer) and pineapple sage (*Salvia elegans*, all year) were both visited by 6 hummingbird species. Limita (*Anisacanthus andersonii*, spring), Maycoba

Table 10.4. Some observed hummingbird numbers in oak woodland along the Tónichi-Yécora transect, October 1999–September 2001.

Localities and Dates	No. of Observers	Observation Time (h)	Hummingbird Species[a]											
			ANHU	BBLH	BCHU	BEHU	BLUH	BTLH	CAHU	COHU	MAHU	RUHU	VCHU	WEHU
Spring (February–May)														
Rancho la Palmita, 2/13/01	1	0900–0930					1							
Rancho la Palmita, 2/25/00	6	1400–1430												1
Rancho la Palmita, 3/13/00	4	1300–1330									1			2
Rancho la Palmita, 3/19/00	1	ND[b]										2		
Rancho la Palmita, 4/3/00	4	1700–1800					2	1			1			1
Rancho la Palmita, 4/5/00	4	0600–0900		1	1		1	1			3	1		1
Rancho la Palmita, 5/8/01	2	1400–1430									1			1
Summer-fall (August-October)														
Río Maycoba, 8/6/00	1	1200–1300			1									
Rancho la Palmita, 8/6/00	3	0700–1000		1	3	2	1	1					2	5
Rancho la Palmita, 8/8/00	3	1230–1530		4	3	8		1				2	2	1
Rancho la Palmita, 8/31/00	4	0600–1400		2	2	8							12	
Rancho la Palmita, 9/1/00	4	0600–0900, 1700–1830			3	6			2			1	ab[c]	
Maycoba, 9/2/00	2	1245–1330			2								2	1
Rancho la Palmita, 9/2/00	4	1000–1200, 1700–1830		1	4	3						1	8	
Rancho la Palmita, 9/3/01	5	0600–0900		1	2	4							10	1
Maycoba, 9/5/01	5	1400–1530		17	20	5								
Río Maycoba, 9/5/01	1	1500–1600					1							
Rancho la Palmita, 9/7/01	5	0600–0900		1		1							1	
Rancho la Palmita, 9/30/00	2	1900–2000				2							3	

[a]Hummingbird species: ANHU = Anna's hummingbird (*Calypte anna*); BBLH = broad-billed hummingbird (*Cynanthus latirostris*); BCHU = black-chinned hummingbird (*Archilochus alexandri*); BEHU = berylline hummingbird (*Amazilia beryllina*); BLUH = blue-throated hummingbird (*Lampornis clemenciae*); BTLH = broad-tailed hummingbird (*Selasphorus platycercus*); CAHU = Calliope hummingbird (*Stellula calliope*); COHU = Costa's hummingbird (*Calypte costae*); MAHU = magnificent hummingbird (*Eugenes fulgens*); RUHU = rufous hummingbird (*Selasphorus rufus*); VCHU = violet-crowned hummingbird (*Amazilia violiceps*); WEHU = white-eared hummingbird (*Hylocharis leucotis*).
[b]No data.
[c]Abundance status: a[b] = abundant.

Table 10.5. Some observed hummingbird numbers in pine-oak forest along the Tónichi-Yécora transect, October 1999–September 2001.

Localities and Dates	No. of Observers	Observation Time (h)	Hummingbird Species[a]											
			ANHU	BBLH	BCHU	BEHU	BLUH	BTLH	CAHU	COHU	MAHU	RUHU	VCHU	WEHU
Spring (February–May)														
El Aguajito, 2/13/01	3	1200–1230												3
El Aguajito, 2/25/00	5	1500–1530					2							
El Aguajito, 3/11/01	3	1000–1200												6
El Aguajito, 3/13/00	3	1530–1600					2							3
Mesa del Campanero, 3/14/00	5	0600–1100						3						
Mesa del Campanero, 4/2/00	4	1500–1700		1										2
El Aguajito, 4/5/00	2	1300–1500					4					1		15
El Aguajito, 4/29/00	2	1400–1430					2							
El Aguajito, 5/8/01	2	1100–1130				2	2							1
Summer-fall (August–October)														
El Aguajito, 8/5/00	3	1500–1700			2	4	4							2
El Aguajito, 8/8/00	3	0930–1200				5	3							3
El Aguajito, 8/31/00	5	1500–1530				5	1					3	1	2
El Aguajito, 9/1/00	2	0830–0900				10	4						13	12
Mesa de los Coronados, 9/1/00	2	1000–1500		2	1		1	2			2	10		4
El Aguajito, 9/2/01	5	0800–1000, 1700–1900		3		10	2					15	5	7
Mesa de los Coronados, 9/3/01	3	1000–1400			4							1	1	1
Barranca el Salto, 9/4/01	5	1000–1500	2	10	5	2	21	2	1	10	2	21	4	1
El Aguajito, 9/7/01	5	1600–1800		10	3	12							1	10
Mesa de los Coronados, 9/7/01	5	1130–1330			1							1		1
El Aguajito, 9/30/00	2	1100–1200				6	2							6
Winter (December–January)														
El Aguajito, 12/12/99	3	0830–0900												2
El Aguajito, 1/10/01	3	1000–1030												3

[a]Hummingbird species: ANHU = Anna's hummingbird (*Calypte anna*); BBLH = broad-billed hummingbird (*Cynanthus latirostris*); BCHU = black-chinned hummingbird (*Archilochus alexandri*); BEHU = berylline hummingbird (*Amazilia beryllina*); BLUH = blue-throated hummingbird (*Lampornis clemenciae*); BTLH = broad-tailed hummingbird (*Selasphorus platycercus*); CAHU = Calliope hummingbird (*Stellula calliope*); COHU = Costa's hummingbird (*Calypte costae*); MAHU = magnificent hummingbird (*Eugenes fulgens*); RUHU = rufous hummingbird (*Selasphorus rufus*); VCHU = violet-crowned hummingbird (*Amazilia violiceps*); WEHU = white-eared hummingbird (*Hylocharis leucotis*).

Figure 10.6. A local area of mixed-conifer forest within pine-oak forest at 1900–2100 m elevation in Barranca El Salto on the west edge of Mesa del Campanero. On 3 days in September 2001, 11 or 12 species of hummingbirds were observed visiting Texas betony (*Stachys coccinea*).

sage (summer-fall), tree morning glory (winter and spring), and wild *jícama* (*Ipomoea bracteata*, spring) all attracted 5 hummingbird species. Among all nectar plants, the only species found to attract hummingbirds in all 4 vegetation types was the tree tobacco (table 10.6).

The most important hummingbird localities along the Tónichi–Yécora transect were located in major patches of plants with nectar-rich flowers. Tree ocotillo was an important food plant in spring on the Río Yaqui near Tónichi (FTS, 180–200 m elevation, 28°34'15–40" N, 109°33'09" W). Several red-flowered shrubs in the Acanthaceae (*Anisacanthus andersonii, Justicia candicans* [fig. 10.3], and *Tetramerium abditum*), the wild *jícama*, and the tree morning glory were important spring floral resources in TDF in Arroyo los Huérigos (650 m elevation, 28°25'48" N, 109°11'31" W) and at Rancho la Mula (900 m elevation, 28°28'50" N, 109°22'02" W). *Piojilla* was an important food

Figure 10.7. A male rufous hummingbird (*Selasphorus rufus*). This long-distance migrant is common in oak woodland and pine-oak forest along the Tónichi-Yécora transect in late summer.

plant in late July–early August at Rancho La Palmita (OW; 1460 m elevation, 28°22'18" N, 109°03'53" W; R. Coronado pers. comm.) on the north slopes of Mesa del Campanero. Maycoba sage was a locally important red-flowered shrub in September at Maycoba (OW; 1485 m elevation, 28°23'49" N, 108°40'00" W). Pineapple sage and the tree tobacco had flowers most of the year at El Aguajito (1640 m elevation, 28°22'18" N, 109°02'54" W), a steep mesic sycamore canyon with pine-oak forest on the slopes. On 3 separate days in early September 2001, a dense stand of the Texas betony attracted 11–12 hummingbird species in Barranca El Salto (POF, 1900–2100 m elevation, 28°19'31" N 109°02'00" W) on Mesa del Campanero.

A few anecdotal observations of visits to important nectar plants include a few berylline hummingbirds, recorded in TDF in spring (4 observations) feeding on limita and wild *jicama*. Violet-crowned hummingbirds were observed most often in both FTS and TDF either hawking insects or feeding from the introduced tree tobacco. In POF, blue-throated hummingbirds were most often observed feeding at tree tobacco. A few hardy white-eared hummingbirds were seen at pineapple sage flowers on December 12, 1999 and January 10, 2001 at El Aguajito (POF). On the first of those 2 days, the weather was very cold, with snow covering the ground.

Total Hummingbird Diversity along the Tónichi-Yécora Elevation Gradient

Of the 16 hummingbird species known from Sonora, 12 were observed by us along the Tónichi-Yécora elevation gradient. A 13th species, the plain-capped starthroat, was observed once visiting tree morning glory in TDF near Santa Ana during the December 22, 1999 Christmas Bird Count (J. Whetstone, pers. comm.). This species, which is evidently rare along the Tónichi-Yécora elevation gradient, is common in TDF near Alamos in southern Sonora from September through December (Russell and Monson 1998; S. A. Meyer, pers. comm.).

Three hummingbird species found in Sonora have not been documented along the Tónichi-Yécora elevation gradient: Allen's (*Selasphorus sasin*), Lucifer (*Calothorax lucifer*), and cinnamon (*Amazilia rutila*) hummingbirds. Of these 3 species, the Allen's hummingbird probably has the highest potential for occurring in the area. It breeds in high numbers in coastal California and winters chiefly in the Valley of Mexico (Phillips 1975; Howell and Webb 1995). In Sonora, it appears to be only a transient: it has been observed in January (Russell and Monson 1998) and September (S. A. Meyer, pers. comm.) at Alamos, in February at Puerto Peñasco (K. Kaufman and D. Stejskal pers. comm.), Punta Rosa (Russell and Lamm 1978), Hermosillo (E. Gómez L., pers. comm.), and San Carlos (H. and A. Etheridge, pers. comm.), and in March 59 km north of Hermosillo (E. López L., pers. comm.). The spring migration route for the species is along the Pacific slope and the northern half of the Baja California peninsula (Phillips 1975). During summer migration, Allen's hummingbirds move southward, chiefly through northern Baja California then eastward across the Sierra Madre Occidental (Howell and Webb 1995). Allen's hummingbirds also have been banded in southern Arizona in late summer (S. M. and R. O. Russell, pers. comm.), suggesting that some migration also occurs farther north through northwestern mainland Mexico. Russell and Monson (1998) suggested that Allen's hummingbirds may pass through the mountains of eastern Sonora in July and August, and this species was observed on July 29, 2001 in San Lázaro, just south of the Arizona border (E. López L., pers. comm.). Thus, some Allen's hummingbirds may have migrated through the study area in July or were misidentified as rufous hummingbirds, as the adult females and immature

Table 10.6. Plants visited by hummingbirds along the Tónichi-Yécora transect in eastern Sonora, arranged by vegetation type.

Localities and Dates	Flower Color	No of Hummingbird spp.	Hummingbird Species[a]											
			ANHU	BBLH	BCHU	BEHU	BLUH	BTLH	CAHU	COHU	MAHU	RUHU	VCHU	WEHU
Foothills thornscrub (*n* = 16)														
Anisacanthus thurberi (desert honeysuckle)	Orange	3		X						X		X		
Antigonon leptopus (queen's wreath)	Red	2		X								X		
*Bougainvilla spectabilis** (*bugambilia*)	Magenta, white	2								X			X	
Caesalpinia pulcherrima (red bird-of-paradise)	Red, yellow	2		X								X		
Callaeum macropterum (*gallinitas*)	Yellow	2		X						X				
Coursetia glandulosa (*sámota*)	Yellow, white	1		X										
Fouquieria macdougalii (tree ocotillo)	Red	4		X						X		X	X	
Guaiacum coulteri (*guayacán*)	Purple	1		X										
Havardia mexicana (*palo chino*)	White	4		X						X		X	X	
Ipomoea bracteata (wild *jícama*)	Purple	2		X								X		
Lycium andersonii (wolfberry)	White	1		X										
Merremia palmeri (morning glory vine)	White	2		X						X				
*Nicotiana glauca*** (tree tobacco)	Yellow	1								X				
Opuntia gosseliniana (prickly pear)	Yellow	1		X										
Stenocereus thurberi (organpipe cactus)	White	1		X										
Vitex mollis (*uvalama*)	Lavender	2		X						X				
Tropical deciduous forest (*n* = 14)														
Anisacanthus andersonii (*limita*)	Red	5		X		X						X	X	X
*Bauhinia variegata** (orchid tree)	Purple	1		X										
Caesalpinia pulcherrima (red bird-of-paradise)	Red, yellow	1		X										
Fouquieria macdougalii (tree ocotillo)	Red	3		X						X		X		
Ipomoea arborescens (tree morning glory)	White	5	X	X		X				X			X	
Ipomoea bracteata (wild *jícama*)	Purple	5		X		X		X				X	X	

(continued)

Table 10.6. Continued

Localities and Dates	Flower Color	No of Hummingbird spp.	Hummingbird Species[a]											
			ANHU	BBLH	BCHU	BEHU	BLUH	BTLH	CAHU	COHU	MAHU	RUHU	VCHU	WEHU
Justicia candicans (*rama del toro*)	Red	3		X						X				X
Lobelia laxiflora (bellflower)	Red	3		X				X				X		
*Nicotiana glauca*** (tree tobacco)	Yellow	5	X	X	X	X							X	
Operculina pteripes (morning glory vine)	Red	1		X										
*Opuntia ficus-indica** (prickly pear)	Yellow	1		X										
Opuntia karwinskiana (prickly pear)	Red	2		X									X	
Rosa sp. (rose)	Red	1		X										
Tetramerium abditum (*rama del toro*)	Red	1		X										

Oak woodland (*n* = 10)

Localities and Dates	Flower Color	No of Hummingbird spp.	ANHU	BBLH	BCHU	BEHU	BLUH	BTLH	CAHU	COHU	MAHU	RUHU	VCHU	WEHU
*Antirrhinum majus** (snapdragon)	Various	1				X								
Bouvardia ternifolia (scarlet bouvardia)	Red	2										X	X	
Cologania cf. *angustifolia* (pea vine)	Pink	1				X								
Cuphea llavea (bat-faced monkeyflower)	Red-purple	2				X							X	
Mandevilla foliosa (piojilla)	Yellow	6		X	X	X						X	X	X
*Nerium oleander** (oleander)	White	1											X	
*Nicotiana glauca*** (tree tobacco)	Yellow	3				X					X		X	
Salvia betulaefolia (Maycoba sage)	Red	5		X	X	X							X	X
Salvia elegans (pineapple sage)	Red	1			X									
Zinnia peruviana (wild zinnia)	Red	3		X		X							X	

Pine-oak forest (*n* = 16)														
*Antirrhinum majus** (snapdragon)	Various	1												X
Arbutus xalapensis (*madroño*)	Green-white	2						X						X
Arcostaphylos pungens (manzanita)	White, pink	1												X
Bouvardia ternifolia (scarlet bouvardia)	Red	1											X	
Castilleja patriotica (Indian paintbrush)	Orange-yellow	4			X							X	X	X
Castilleja tenuifolia (Indian paintbrush)	Red, yellow	1												X
Centaurea rothrockii (Madrean starthistle)	White, purple	3					X					X		X
*Dianthus caryophyllus** (carnation)	Red	1									X			
Gladiolus sp.* (gladiola)	Red	1												X
*Impatiens balsamina** (impatiens)	Various	1									X			
*Nicotiana glauca*** (tree tobacco)	Yellow	7		X	X	X	X					X	X	X
Ribes dugesii (gooseberry)	Purple	1												X
Rosa sp.* (rose)[b]	Red	1			X									
Salvia elegans (pineapple sage)	Red	6		X	X	X						X	X	X
Stachys coccinea (Texas betony)	Red	12	X	X	X	X	X	X	X	X	X	X	X	X
*Tropaeolium majus** (nasturtium)	Orange	1												

* = Cultivated species. ** = Non-native species.

[a]Hummingbird species: ANHU = Anna's hummingbird (*Calypte anna*); BBLH = broad-billed hummingbird (*Cynanthus latirostris*); BCHU = black-chinned hummingbird (*Archilochus alexandri*); BEHU = beryllline hummingbird (*Amazilia beryllina)*; BLUH = blue-throated hummingbird (*Lampornis clemenciae)*; BTLH = broad-tailed hummingbird (*Selasphorus platycercus*); CAHU = Calliope hummingbird (*Stellula calliope)*; COHU = Costa's hummingbird (*Calypte costae*); MAHU = magnificent hummingbird (*Eugenes fulgens*); RUHU = rufous hummingbird (*Selasphorus rufus*); VCHU = violet-crowned hummingbird (*Amazilia violiceps*); WEHU = white-eared hummingbird (*Hylocharis leucotis*).

[b]Same species as that described from tropical deciduous forest.

males of these two species have essentially identical plumages.

The lucifer hummingbird is a rare hummingbird in Sonora and is known from only the northeastern part of the state (Russell and Monson 1998). Cinnamon hummingbirds, which are common in tropical habitats from Sinaloa southward, have only been observed 3 times in southeastern Sonora from February 27 to March 1, 1948 (Russell and Monson 1998), although an individual observed in 1992 at a feeder in Patagonia, Arizona, likely passed through Sonora.

Hummingbird Seasonal Movements, Nectar Availability, and Habitat Association

In total, we documented nesting of 4 species (broad-billed, berylline, Costa's, and white-eared hummingbirds) along the Tónichi–Yécora transect. Our observations of berylline and white-eared hummingbird nests constitute first nesting records for the state for both species (Russell and Monson 1998; S. M. Russell, pers. comm.). The other 2 species observed nesting were the broad-billed and Costa's hummingbirds. Observation dates and patterns of abundance were suggestive of breeding for 4 additional species, the blue-throated, broad-tailed, magnificent, and violet-crowned hummingbirds. The blue-throated hummingbird is a summer resident in Sonora from mid-March to mid-September (Russell and Monson 1998). Our observation dates of February 13 at La Palmita and February 25 at El Aguajito may be early spring records for them in Sonora, although they could have been year-round residents (see below).

We documented the presence of 5 species during December and January. Of these 5 species, both the broad-billed and Costa's hummingbirds evidently winter in the area in sizable numbers. By comparison with broad-billed and Costa's hummingbirds, we observed Anna's, violet-crowned, and white-eared hummingbirds less frequently and/or in lower numbers during the winter. However, Christmas Bird Counts (CBCs) conducted in the area since 1997 (National Audubon Society 2003) confirm the existence of wintering populations of these 3 species along the Tónichi–Yécora transect. Anna's, violet-crowned, and white-eared hummingbirds have been observed every year during CBCs. Observed numbers of white-eared hummingbirds in particular reached 19 individuals in 1999, 15 in

2000, and 13 in 2001 (National Audubon Society 2003). At least some of the Anna's hummingbirds recorded during Yécora CBCs have been from TDF (J. Whetstone, pers. comm., 1999). Additionally, berylline, plain-capped starthroat, blue-throated, and rufous hummingbirds have been recorded during CBCs (National Audubon Society 2003). The berylline hummingbird was observed during all CBCs (e.g., 9 individuals in 2000) except in 2001 (in December 1999 individuals were observed feeding on tree morning glory and tree tobacco; J. Whetstone, pers. comm.). Thus, with the inclusion of the berylline species, at least 6 hummingbird species may have wintering populations along the Tónichi–Yécora elevation gradient. Observed numbers of blue-throated hummingbirds (1 individual in 2001 and 2002, constituting the first winter records of the species in Sonora), plain-capped starthroat (1 individual in 1999), and rufous hummingbird (1 individual in 1999) observed during CBCs (National Audubon Society 2003) suggest that these species are less regular in the area. The black-chinned, calliope, and magnificent hummingbird have not been recorded during the winter along the Tónichi-Yécora gradient (see also Russell and Monson 1998).

Notable differences seem to exist between hummingbird communities of low-elevation (FTS and TDF) and high-elevation (OW and POF) vegetation types. The broad-billed hummingbird was the only species apparently abundant year-round in FTS and TDF, although Costa's, rufous, and violet-crowned hummingbirds were also common in 1 or both of these vegetation types for part of the year. In OW and/or POF, likely seasonally dominant species are the berylline, black-chinned, blue-throated, rufous, and white-eared hummingbirds. Magnificent and blue-throated hummingbirds were seen only in OW and POF. Berylline and white-eared hummingbirds occurred at low elevation at least seasonally but apparently were much more common in OW and POF. In contrast, there were many more observations of broad-billed and Costa's hummingbirds in FTS and TDF than in the higher elevation vegetation types.

In FTS and TDF, most hummingbird observations were in the winter and spring, whereas in OW and POF they were during the summer and fall. Although greater hummingbird abundance and species diversity in the spring in FTS and TDF but in the summer-fall in OW and POF were readily apparent in the field, they cannot be established as facts with our data set. If true, however, these 2 patterns could be explained by any combination of the following mechanisms. First, some hummingbirds present in spring in FTS and TDF may remain in the general area year-round but somewhat shift their distribution upward along the elevation gradient in summer-fall. Conversely, some individuals of the higher-elevation resident species may range farther down in spring than in summer-fall. Second, migrants passing through the area in both spring and summer-fall may visit preferentially low-elevation habitats in spring and high-elevation habitats in summer-fall. Finally, migrants associated with high-elevation habitats may pass through the area only in the fall, while migrants associated with FTS and TDF may be present locally only in the spring. All 3 mechanisms might be at play and could reflect differences in the seasonal availability of nectar flowers, especially the lowland tree morning glory and tree ocotillo in spring and the montane red-flowered mints in summer-fall. The onset of cold temperatures in November above 1800 m elevation at locations such as Mesa del Campanero effectively reduces flowers available to hummingbirds in winter and early spring. This is in marked contrast to the migrant hummingbird wintering areas in the warmer montane forests in the highlands of Jalisco, Colima, and Michoacán farther south in Mexico, where mints and other nectar plants flower abundantly in winter.

A few of our findings are suggestive of local habitat shift or preferential use by hummingbirds. For most of the year, Costa's hummingbirds are the ultimate desert residents living in very dry, hot areas throughout the Sonoran Desert in Baja California, Baja California Sur, and Sonora (Wilbur 1987; Russell and Monson 1998). As shown with this study, Costa's hummingbirds are also common or abundant in the winter in both FTS and TDF and in the spring in FTS. Their nesting season in Sonora is from February to June (Russell and Monson 1998). From mid-July to mid-September, there is a paucity of Costa's hummingbird records in the state. Our results are generally congruent with Russell and Monson (1998), although we did observe Costa's hummingbirds in August and early September, including along the Río Yaqui (FTS, 220 m). Some observations are from POF, in the Barranca El Salto (1900–2100 m) on September 3–7, 2001. Observations of Costa's (and broad-billed) hummingbirds at Barranca El Salto suggest that these birds ascended Mesa del Campanero through the wet-

season TDF in Arroyo El Reparo, perhaps to exploit the seasonal opportunity provided by Texas betony. The species' occurrence at Barranca El Salto in the summer-fall suggests at least in part a seasonal shift in distribution along the elevation gradient. Similar patterns may characterize several other species observed during our study, including the violet-crowned and magnificent hummingbirds. The latter species may shift its distribution seasonally from OW in the spring to POF in the summer-fall, as it was recorded only in the spring in OW but seemed much more common in the summer-fall in POF.

Four species found along the Tónichi-Yécora transect are long-distance migrants: the black-chinned, broad-tailed, calliope, and rufous hummingbirds (Johnsgard 1997). More research is needed to evaluate patterns of local abundance of rufous and black-chinned hummingbirds in particular. However, our observed patterns of rufous hummingbird abundance in POF confirm the Sierra Madre Occidental as a late summer migration route for the species (moderate numbers of rufous hummingbirds were also found in the Río Magdalena and Río Sonora valleys in north-central Sonora in September). During spring migration, rufous hummingbirds travel along the coast of Sonora (Russell and Monson 1998) and through the central portion of the Sonoran Desert in central Sonora (Van Devender et al. 1994). Based on our study, a spring migration route also exists in eastern Sonora, but mostly in FTS and TDF, contrary to patterns observed in the summer-fall.

The black-chinned hummingbird has a wide breeding distribution that includes northern Sonora. Throughout the rest of the state, the species occurs only as a migrant, with numbers during the summer-fall exceeding those during the spring (Russell and Monson 1998). During our study there were only 2 observations of black-chinned hummingbirds in the spring. In the summer-fall, however, the species seemed much more common, although the lowlands west of the Tónichi-Yécora transect may be more important late-summer migration corridors, as large numbers of black-chinned hummingbirds were observed in the Río Magdalena and Río Sonora valleys in north-central Sonora in August–September 2001. Our results suggest that black-chinned hummingbird summer-fall migrants favored OW and POF at a time of the year during which nectar food plants are flowering.

In conclusion, we found evidence that the distribution and flowering phenology of nectar plants may strongly influence hummingbird seasonal movements (both long-distance migration and local shifts in distribution along the elevation gradient) along the Tónichi–Yécora transect. Worldwide, the protection of plant–pollinator interactions is an important conservation priority. Research needs in Sonora include further study of patterns of distribution and seasonal movements of hummingbirds and other pollinators (e.g., monarch butterflies, *Danaus plexippus*). In this study we identified a patch of Texas betony in Barranca El Salto attracting 12 hummingbird species. Ecologically important patches of nectar flowers such as the one we documented clearly deserve to be protected.

Acknowledgments

This chapter is dedicated to the loving memory of Bill Calder, *chuparrosero de corazón* extraordinaire. We gratefully acknowledge the keen observation skills of Steve and Ruth Russell, Susan Wethington, Lee Rogers, Joan Day Martin, Lorene Calder, and Eduardo Gómez, and their stimulating company on our field excursions. Jack Whetstone provided useful observations from the Yécora Christmas Bird Counts. We greatly appreciate the hospitality and companionship of the families at Rancho La Palmita (Don Rubén, Doña Ubelina, Sergio, Oscar, and Clarissa Coronado), Rancho La Mula (Don Carlos and Doña Amada Meléndrez), and Rancho La Palma Agujerada (Don Carlos and Doña Aurora Valenzuela).

Literature Cited

AOU. 1998. Check-list of North American Birds, 7th ed. American Ornithologists' Union, Washington, D.C.

Búrquez, A., A. Martínez Y., and P. S. Martin. 1992. From the high Sierra Madre to the coast: changes in vegetation along highway 16, Maycoba-Hermosillo. Pp. 239–252 *in* K. F. Clark, J. Roldán, and R. H. Schmidt (eds.), Geology and Mine Resources of the Northern Sierra Madre Occidental. Guide-book, El Paso Geological Survey Publication no. 24, El Paso, Texas.

Felger, R. S., M. B. Johnson, and M. F. Wilson. 2001. The Trees of Sonora, Mexico. Oxford University Press, New York.

Felger, R. S., and M. F. Wilson. 1995. Northern Sierra Madre Occidental and its Apachian outliers: a neglected center of biodiversity. Pp. 36–59 *in* Biodiversity and Management of the Madrean Archipelago: The Sky Islands of Southwestern United States and Northwestern Mexico. General Technical Report RM- GTR-264. U.S. Forest Service, Rocky Mountain Research Station, Fort Collins, Colorado.

Gentry, H. S. 1942. Río Mayo Plants. A Study of the Río Mayo, Sonora. Carnegie Institution of Washington Publication 527, Washington, D.C.

Howell, S. N. G., and S. Webb. 1995. A Guide to the Birds of Mexico and Northern Central America. Oxford University Press, New York.

Johnsgard, P. A. 1997. The hummingbirds of North America, 2nd ed. Smithsonian Institution Press, Washington, D.C.

Kearns, C. A., D. W. Inouye, and N. M. Waser. 1998. Endangered mutualisms: the conservation of plant-pollinator interactions. Annual Review of Ecology and Systematics 29: 83–112.

Lennartsson, T. 2002. Extinction thresholds and disrupted plant-pollinator interactions in fragmented plant populations. Ecology 83: 3060–3072.

Martin, P. S., D. Yetman, M. Fishbein, P. Jenkins, T. R. Van Devender, and R. K. Wilson. 1998. Gentry's Río Mayo Plants: The Tropical Deciduous Forest and Environs of Northwest Mexico. University of Arizona Press, Tucson.

National Audubon Society. 2003. Christmas Bird Count historical results. [Available: http://www.audubon.org/bird/cbc]

Phillips, A. R. 1975. The migrations of Allen's and other hummingbirds. Condor 77: 196–205.

Reina G., A. L., T. R. Van Devender, W. Trauba, and A. Búrquez M. 1999. Caminos de Yécora. Notes on the vegetation and flora of Yécora, Sonora. Pp. 137–144 *in* D. Vasquez del Castillo, M. Ortega N., and C. A. Yocupicio C. (eds.), Symposium Internacional sobre la Utilización y Aprovechamiento de la Flora Silvestre de Zonas Aridas, Universidad de Sonora, Hermosillo.

Russell, S. M., and D. W. Lamm. 1978. Notes on the distribution of birds in Sonora, Mexico. Wilson Bulletin 90: 123–131.

Russell, S. M., and G. Monson. 1998. The Birds of Sonora. University of Arizona Press, Tucson.

Van Devender, T. R., W. A. Calder, K. Krebbs, A. L. Reina G., S. M. Russell, and R. O. Russell. 2004. Hummingbird plants and potential nectar corridors of the rufous hummingbird in Sonora, Mexico. Pp. 96–121. *In* G. P. Nabhan, R. C. Brusca, and L. Holter (eds.), Conserving Migratory Pollinators and Nectar Corridors in Western North America. University of Arizona Press, Tucson.

Wilbur, S. R. 1987. Birds of Baja California. University of California Press, Berkeley.

11

Mammalian Phylogeography and Evolutionary History of Northern Mexico's Deserts

DAVID J. HAFNER

BRETT R. RIDDLE

Northern Mexico is dominated by the warm, southern deserts of North America: the Sonoran Desert west of the Sierra Madre Occidental and the Chihuahuan Desert of the Altiplano between the Sierra Madre Occidental and Oriental. These deserts contain the majority of arid-lands biodiversity of North America, and most of their surface area and biodiversity occur in Mexico. Our understanding of the evolutionary history of these deserts has advanced considerably since the seminal descriptive papers on the desert vegetation of North America by Forrest Shreve in 1942 and 1951. This new historical perspective has resulted from a large body of studies, including in-depth research on fossil floras and woodrat (*Neotoma*) middens; revolutionary advances in historical geology, historical biogeography, and genetics; and the development of new analytical techniques. Recent phylogeographic studies of the arid-adapted vertebrates of the Chihuahuan and Sonoran Deserts, particularly of a cadre of exemplar desert rodents, birds, and reptiles, have revealed a rich and complex Neogene and Pleistocene history. The system of southern deserts offers a superb opportunity and a model system in which to investigate a regional history complicated by temporally nested geological events.

Our purpose here is to discuss the current understanding of the geological and phylogeographic history of the southern deserts based mainly on mitochondrial (mt)DNA sequence analysis of desert-adapted, terrestrial mammals. We believe that this

historical perspective has general importance for evolutionary biology, ecology, and conservation biology. Identification of regional centers of distinct biodiversity is an important first step for conservation of desert areas, some of which already have been subjected to extreme, adverse human impact. We contend that sole reliance on floristic elements and on the species taxonomic level tend to obscure areas of unique biodiversity.

Historical Background

Sixty years ago, Shreve (1942) mapped the North American deserts, adding a fourth (the Mojave) to those previously recognized: the Great Basin, Sonoran, and Chihuahuan Deserts. Shreve (1942) differentiated these deserts based on physiognomy, community structure, and floristic composition. Although he recognized the dynamic nature of deserts, he stated that, "It is not yet possible to state, however, to what extent the differences in flora and floristic composition have been influenced by historical factors" (1942: 215). Shreve (1951) subsequently refined his description of the Sonoran Desert, defining its boundaries in more detail and recognizing 7 subdivisions. Turner and Brown (1982) further refined the boundaries of the subdivisions and referred 1 subdivision, the Foothills of Sonora, to thornscrub instead of desert (following Felger and Lowe 1976; Rzedowski 1978; Felger and Moser

1985; Búrquez et al. 1992). Schmidt (1989) reviewed 17 definitions of the Sonoran Desert (based on climate or flora, and in some cases including fauna), arriving at a compromise distribution based on the de Martonne (1926) Index of Aridity and the Dirección General de Geografía del Territorio Nacional (1983) climatic map for the northwestern boundary. There is continuing debate about the southern boundary of the Sonoran Desert (Búrquez et al. 1999).

Axelrod (1958) first considered and subsequently revealed (e.g., Axelrod 1983) the influence of history in the development of western North America's deserts. Hess (1962) and others then sparked a revolution in the earth sciences with the observation that the earth's crust moved laterally from long, volcanically active oceanic ridges. By 1970, plate tectonic theory was widely accepted, revealing a dynamic geological history of southwestern North America. Soon after, the science of biogeography underwent a related revolution and rejuvenation with the development of vicariance biogeography (e.g., Platnick and Nelson 1978), built on the groundbreaking theory of panbiogeography (Croizat 1952, 1958, 1960, 1964). Van Devender and his colleagues (e.g., Betancourt et al. 1990; Van Devender 1990a,b) provided detailed evidence from packrat middens for elevational and latitudinal shifts in desert vegetation associated with glacial–interglacial climates, emphasized the individualistic responses of species to climatic change (following Gleason 1926), and further suggested a fundamental difference in response between plants (e.g., greater sensitivity to climatic extremes such as winter freezing) and animals. With the advent of the polymerase chain reaction (Mullis et al. 1986), revolutionary techniques in genetic analysis provided the raw material for detailed reconstructions of the phylogeographic history of organisms. Foremost among these was the ability to sequence mtDNA, a rapidly evolving, nonrecombinant genetic molecule, and interpret sequence divergence among populations in a phylogeographic context (Avise et al. 1987; Avise 1994).

Analytical methods used to reconstruct the biogeographic history of regions (the "history of place" of Brown 1995) based on phyletic histories of regional taxa (Brown's "history of lineage") have advanced significantly since Brown (1995: 191) voiced skepticism that "it would be difficult or impossible to reconstruct the spatial pattern of the history of lineage." Efforts at historical reconstruction initially were hindered by a forced restriction to vicariant explanations and nonreticulate area cladograms. Areas in biogeographic analyses initially were treated like clades: just as clades, once diverged, were not "allowed" to join via hybridization (i.e., undergo reticulate evolution), no simple vicariant tree could depict sympatry following dispersal of formerly allopatric species, much less repeated cycles of vicariance and dispersal. Such episodes of dispersal would be analogous to homoplasies (independent evolutionary origins or losses of traits as opposed to evidence for common history) in a phylogenetic analysis of taxa. Simple vicariant explanations were regarded as the only testable hypotheses, and few algorithms existed to construct reticulate trees (or "reticulograms"), despite the common occurrence of nested geological events (a phenomenon familiar to geologists). The spatial history of many lineages undoubtedly involves repeated vicariant, dispersal, and extinction events of divergent ages in the same location.

Techniques recently developed from coalescent theory and population genetics provide a statistical approach to distinguish between the effects of population history (e.g., past fragmentation or recent range expansion) and the effects of recurrent evolutionary forces (e.g., gene flow, mutation, and drift) within a lineage. Coalescent theory is a framework for estimating a variety of demographic parameters from gene trees, including the estimation of divergence times. Combining robust phylogeographic reconstructions of lineages with these population-level techniques allows a more refined reconstruction of the spatial history of each lineage. Comparison of lineage reconstructions among co-occurring, ecologically similar taxa then allows construction of an area reticulogram representing common patterns (resulting from vicariance or biotic dispersal) and exceptions (resulting from idiosyncratic dispersal or extinction) among their regional histories. A variety of methods have been proposed to create reticulograms for biogeographic reconstruction (e.g., Brooks 1990; Hausdorf 1998; Legendre and Makarenkov 2002). For example, recent advances in the development of Brooks Parsimony Analysis (BPA; Brooks et al. 2001) fully and explicitly resolve deviations from the null hypothesis of a simple vicariance model of area relationships.

Since 1996, we have studied phylogeographic structure of mtDNA haplotypes within regional representatives of a restricted subset of desert-adapted rodents, all of which have low vagility. These include 4 species of white-footed mice (*Peromyscus*) of the subgenus *Haplomylomys*; 12 species of pocket mice (*Chaetodipus*); Merriam's kangaroo rat (*Dipodomys merriami*); and 3 species of antelope ground squirrels (*Ammospermophilus*). All are restricted to North American deserts and adjacent arid areas, and all are found in particular throughout the southern deserts. They are of sufficient antiquity to have been subjected to the same geological and paleoclimatic events during the development of the southern deserts and have sufficient variation in mtDNA haplotypes to be useful in examining phylogeographic questions of this temporal and geographic scale. The primary objective of our initial studies was to evaluate the role of deeper history in regional fragmentation of the southern deserts (Hafner and Riddle 1997; Riddle et al. 2000a,b). In the process, we noted strong genetic signals of embedded phylogeographic structure within each desert (Riddle et al. 2000c). Sequence divergences among haplotypes representing neighboring subregions ranged from shallow to deep, indicating a broad range of potentially causal vicariant events in the history of this structuring. We are currently working to elucidate phylogeographic structure at a considerably more refined spatial scale, involving 11 subregions depicted in figure 11.1.

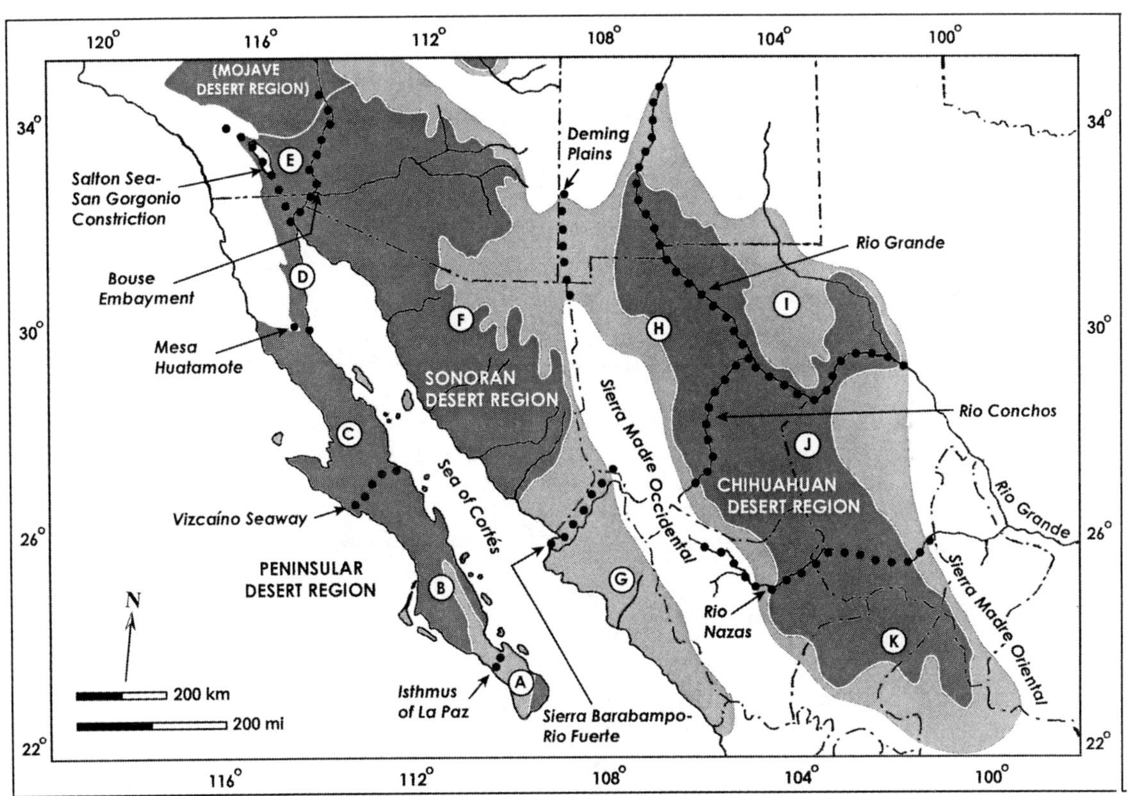

Figure 11.1. Southern desert regions of North America, as defined from an evolutionary perspective. Light shading indicates the combined distribution of 20 species of arid-adapted rodents; dark shading indicates southern deserts as defined by Shreve (1942, 1951), with the Sonoran Desert corresponding to 2 evolutionarily distinct regions: the Sonoran and Peninsular Desert Regions. Dotted lines indicate major (putative) vicariant events dividing subregions: *A*, San Lucan, exclusive of the Sierra Laguna; *B*, Magdalenan; *C*, Vizcaíno; *D*, San Felipe; *E*, Coloradan, *F*, Sonoran; *G*, Sinaloan, *H*, Chichuahuan, *I*, Trans-Pecos, *J*, Coahuilan; and *K*, Zacatecan.

The Southern Deserts from an Evolutionary Perspective

Three Evolutionarily Distinct Desert Regions?

Northern Mexico is dominated by the Sonoran and Chihuahuan deserts, which are distributed collectively in the shape of a south-pointing, three-tined fork (fig. 11.1). Floristically defined, these deserts make up nearly half (44%) of the land area of northern Mexico north of 22° N latitude.

Previously we emphasized the distinct evolutionary history of the Baja California peninsula portion of the Sonoran Desert based on the unique species of peninsular mammals, reptiles, birds, invertebrates (insects and scorpions), and flora (Hafner and Riddle 1997). This distinct history has been supported by phylogeographic studies of mammals, birds, amphibians, and reptiles (summarized in Riddle et al. 2000c) and more recent studies of scorpions (Gantenbein et al. 2001) and plants (Nason et al. 2002). In particular, some mammalian assemblages of the Baja California peninsula's Sonoran Desert, hereafter called the Peninsular Desert Region, are older and preserve a stronger historical signature than was previously suspected.

Mammalian phylogeography also indicates strong and ancient connections between the floristically defined deserts and other vegetation types. In discussing the history of our three evolutionarily distinct desert regions (Peninsular, Sonoran, and Chihuahuan), we are including some peripheral areas that support those other vegetation types. Most notably, our phylogeographically defined Sonoran Desert Region extends far to the south of the floristically defined Sonoran Desert, into thornscrub and tropical deciduous forest. The total area included in our analysis makes up more than two-thirds (67%) of northern Mexico.

We agree with Andersson (1996: 270) that "the central issue in . . . historical biogeography is the evolutionary history of biotas." We do not consider codistributed taxa to represent a discrete, highly integrated unit (*sensu* Clements 1916), but instead to represent taxa with similar vagility and ecological needs (*sensu* Gleason 1926). In this sense, the codistributed taxa may act as a faunal element (Udvardy 1969) that exhibits historical integrity (Armstrong 1972), such as that observed over ecological time by Frey (1992).

The Peninsular, Sonoran, and Chihuahuan Desert regions display clear subregional structure. We have identified 9 potential subregions, each with its own distinct biodiversity. There are also 2 transitional subregions. In most cases, the subregions correspond to those proposed based on flora (Shreve 1951), mammals (Burt 1938), or reptiles (Morafka 1977). These are (west to east, as labeled in fig. 11.1): San Lucan (A), Magdalenan (B), and Vizcaíno (C), subregions of our Peninsular Desert Region; San Felipe (D) and Coloradan (E), mixtures of Peninsular, Mojave, and Sonoran Desert regions; Sonoran (F) and Sinaloan (G), subregions of the Sonoran Desert Region; and Chihuahuan (H), Trans-Pecos (I), Coahuilan (J), and Zacatecan (K) subregions of the Chihuahuan Desert Region. This depiction should serve as a testable hypothesis in evaluating the possible existence of hidden biodiversity in other taxa (e.g., Gantenbein et al. 2001, for the scorpion, *Centruroides exilicauda*; Nason et al. 2002, for the columnar cactus, *Lophocereus*).

Systematists examining the evolutionary relationships among regional representatives of arid-adapted taxa may be misled by a strict assumption of uniformity within the southern deserts. Walpole et al. (1997), evaluating variation in mtDNA among populations of *Peromyscus eremicus* from the Chihuahuan and Sonoran deserts, included only samples from southeastern Arizona and extreme southwestern New Mexico (Subregion F in fig. 11.1) as representative of the Sonoran Desert, and New Mexico and Texas east of the Rio Grande (Subregion I) as representative of the Chihuahuan Desert. Walpole (1997: 397, 402) concluded that their data were "consistent with the hypothesis that the *P. eremicus* from the Sonoran and Chihuahuan deserts represent recently diverged species." However, they failed to detect the fact (revealed later by Riddle et al. 2000a) that populations of *P. eremicus* from the intermediate Chihuahuan Desert Subregion H possess the Sonoran (not Chihuahuan) mtDNA haplotype. If they instead had selected populations from Subregion H as representative of the Chihuahuan Desert, they would have concluded that there were no differences between populations of the 2 deserts. By including samples representing all subregions, Riddle et al. (2000a) corroborated the systematic conclusions of Walpole et al. (1997) while revealing incongruence between the traditional boundary of the 2 southern deserts and their respective forms of *P. eremicus*.

General Historical Patterns

Recent ecological research on the southern deserts and their vertebrate faunas has recognized, implicitly or explicitly, the importance of only a single historical event: the separation of the Sonoran and Chihuahuan deserts by the Sierra Madre Occidental. In addition, most studies of the historical zoogeography of these deserts have focused only on the Recent (late Pleistocene and Holocene) and ignored much geologic time beginning with the initial fragmentation of the Peninsular, Sonoran, and Chihuahuan Desert regions in Miocene or Pliocene times (Riddle et al. 2000c). Thus, despite a potentially great antiquity of the regional vertebrate lineages, it was generally assumed that all divergence of lineages was coincident with glacial–interglacial cycles of the late Pleistocene (e.g., Orr 1960; Savage 1960; Findley 1969; Hubbard 1974; Morafka 1974; Schmidly et al. 1993). Murphy (1983), Grismer (1994), and Upton and Murphy (1997) expanded their evolutionary models to include events throughout the Neogene and presented convincing evidence for the role of more ancient events in the history of the Peninsular Desert herpetofauna. We have presented evidence for the involvement of these Neogene events in lineage divergence in at least 2 rodent groups, *Peromyscus* (subgenus *Haplomylomys*) and *Chaetodipus* (Riddle 1995; Riddle et al. 2000a,b,c).

Evaluation of the ecology, phylogeny, and conservation of the southern deserts must be based on an appreciation of history dating as far back as the Neogene. The evolutionary distinctiveness of lineages (and distinct biodiversity) within each subregion may be causally related to identifiable geological and climatological events of 3 distinct time intervals: (1) Neogene dispersal events that followed the initial geomorphological division of the desert regions; (2) vicariance and dispersal that may have occurred during climatic oscillations (latitudinal and elevational) associated with pluvial–interpluvial cycles of the Pleistocene; or (3) other post-Wisconsinan filter-barriers, which would selectively have prevented or allowed dispersal of certain species. Finally, repeated and temporally nested geological and climatological events subsequent to the initial fragmentation of the 3 desert regions have resulted in a complex internal structure of each desert. We now have the analytical tools to tease apart the complicated histories of each region.

Historical Geology

Geological and climatic events involved in lineage divergence of arid-adapted mammals of the southern deserts are of 3 distinct time intervals: Neogene (Miocene to late Pliocene), early Pleistocene, and late Pleistocene. The Sierra Madre Oriental formed in the early Tertiary (along with the Rocky Mountains of the United States) from folding and thrusting of Cretaceous and Jurassic limestone related to Laramide (Hidalgoan) compressive orogeny (Ortega-Gutiérrez and Guerrero-Garcia 1982; chapter 1 this volume). At this time, a belt of volcanism related and parallel to a continuous subduction zone off the west coast represented the incipient Sierra Madre Occidental. During Oligocene-Miocene times (approximately 34–23 mya; McDowell and Keizer 1977, chapter 1 this volume), the subducting lithosphere underwent a main period of intensified activity, resulting in the largest ignimbritic (silicic volcanic rocks) province on earth (Ortega-Gutiérrez and Guerrero-Garcia 1982). Hundreds of isolated volcanic calderas and ash-flow aprons coalesced to form an elevated, relatively undisturbed plateau (the Sierra Madre Occidental), which extends from just south of the United States border for more than 1200 km to near Guadalajara, Mexico, where it passes beneath younger volcanic rocks of the Trans-Mexico Volcanic Belt (Swanson and McDowell 1984). Subsequent block faulting and erosion broke and wore away the eastern and western fringes of this elevated plateau, forming extensive basins in the elevated Mexican Plateau, which rises southward to the Altiplano of Zacatecas and San Luis Potosí.

Coincident with the initial orogeny of the Sierra Madres and Mexican Plateau, continental material was transferred to the west side of the Protogulf by plate movement (Spencer and Normark 1989), and the Cape Region and northern Peninsular Ranges of the Baja California peninsula were torn away from the Sinaloan and Sonoran mainland (respectively). During the late Miocene to early Pliocene (5.5–4 mya), the Gulf of California (or Sea of Cortez) began to form, due to separation of Baja California (Lonsdale 1989; Stock and Hodges 1989) and subsidence as a result of basin and range extension in North America (Gastil et al. 1983; summarized in Grismer 1994).

According to Murphy's (1983) transgulfian vicariance model, tropical-related reptiles entrapped in the Cape Region islands eventually gave rise to most

Peninsular Desert endemics. Traces of this pattern may be present in birds (Cody 1983), insects (Truxal 1960), spiders (Chamberlin 1924), and scorpions (Williams 1980; Gantenbein et al. 2001). Among peninsular terrestrial mammals, and in strong contrast to reptiles, there may be no relicts of more tropical forms in the Cape Region resulting from initial formation of the Gulf (Lawlor et al. 2002). Instead, relatively mesic-adapted species of terrestrial mammals occurring today in the Cape Region probably arrived by dispersal down the higher elevation spine of the peninsula during pluvial intervals of the Pleistocene, as recently as 10,000 years B.P. (e.g., *Sorex ornatus, Peromyscus truei*) or may have resulted from recent introduction by humans (*Oryzomys couesi*, Nelson 1921; Álvarez-Castañeda 1994; *Marmosa canescens* López-Forment and Urbano-V. 1977; Gardner and Cortés-Calva 1999). Whether tropical bats occurring in the Cape Region (e.g., *Balatiopteryx plicata, Pteronotus davyi, Mormoops megalophylla, Natalus stramineus*) represent ancient relicts or recent immigrants has not been investigated.

Continued plate-boundary expansion during the late Pliocene (ca. 3 mya) simultaneously enlarged and extended the Gulf of California (Buising 1990), elevated the Peninsular, Tehachapi, and Coast ranges of Baja California and California, and expelled marine waters from the Central Valley of California (Norris and Webb 1976). Northern extensions of the Gulf up the Salton Trough formed the San Gorgonio Constriction (Boehm 1984; Ingle 1987), and along the course of the Colorado River formed the Bouse Embayment (Blair 1978; Eberly and Stanley 1978; Buising 1990), effectively isolating the peninsula from continental regions. At the same time, the Cape Region was isolated from the rest of the peninsula by the Isthmus of La Paz (McCloy 1984). In mainland northern Mexico, volcanism was limited to isolated eruptions associated with continued block faulting of the remnant Sierra Madre Occidental, with the major geological processes being subsidence along faults, formation of basin and range topography, and erosion into interior basins of the Mexican Plateau or along coastal plains.

There is circumstantial evidence for a mid-peninsular seaway across the present-day Vizcaíno Desert during the early Pleistocene, about 1 mya (Upton and Murphy 1997; Riddle et al. 2000a,b). The biological evidence is of 2 types: disjunct distributions of marine fish and invertebrates on the Pacific and Gulf sides of the Vizcaíno (Present 1987;

Bernardi et al. 2003), and 17 species of mammals, birds, and reptiles that exhibit genetic discontinuities north and south of the Vizcaíno (Riddle et al. 2000c). The marine species occur in temperate and not in tropical waters and so do not have continuous distributions around the southern tip of the peninsula, yet they are genetically differentiated perhaps more than would be expected from a much more recent disruption in marine isotherms at the close of the Wisconsinan glaciation (Upton and Murphy 1997). In particular, 5 species of marine fish exhibit high levels of genetic divergence compatible with closure of a transpeninsular seaway approximately 1 mya (Bernardi et al. 2003).

Grismer (2002) correctly cautions that terrestrial discontinuities instead may have resulted from abrupt climatic or phytogeographic changes that also occur in this vicinity, and we agree that the 17 terrestrial discontinuities may include a mixture of recent and earlier (seaway-related) disjunctions. The most likely location of such a seaway or at least a severe constriction in the peninsula would have been east of San Ignacio at approximately 27°20' N, where the low-elevation flats surrounding the Laguna San Ignacio are separated from the Gulf of California by the Tres Virgenes volcanic field. Eruptions in the field (Sawlan 1986) date from 1.09 mya (Volcán La Reforma), through El Viejo (0.44 mya) to Las Tres Virgenes (possibly as recent as 1746 and 1857). These eruptions are thought to have been associated with a transform fault, and extensive uplift of marine sediments has been documented along the eastern edge of the volcanic field north of Santa Rosalía (Anderson 1950; Durham and Allison 1960). Moreover, marine coquina have been found beneath the oldest La Reforma flow deposits, indicating that marine waters of the Gulf occupied this region before eruption of La Reforma (B. Hausback, pers. comm.). However, the limestone deposits in the Vizcaíno Desert are virtually fossil-free (lacking even foraminiferans; J. Minch, pers. comm.), and older (several million years old) volcanic deposits surrounding San Ignacio are subaerial and do not appear to have been submerged for any length of time (i.e., no overlying marine deposits; B. Hausback, pers. comm.). Thus, available geological evidence would support a very narrow (e.g., 30 km) landbridge between a flooded Vizcaíno Desert and the Gulf of California about 1 mya, with extensive volcanic eruptions and uplift occurring within the narrow landbridge. Although an incomplete transpeninsular seaway plugged with an active volcanic field probably would have limited north–south dispersal of terrestrial forms (particularly lower-elevation, sand-dwelling species), it would not have permitted dispersal between Pacific and Gulf marine forms and thus would not account for the marine disjunct distributions.

Sea levels fluctuated and ecological zones shifted (in latitude and elevation) in response to repeated waxing and waning of the late Pleistocene glacial–interglacial climatic cycles, which became markedly longer and more extreme about 700,000 years B.P. (Webb and Bartlein 1992). During glacial intervals, the southern deserts were probably restricted down-slope and to the south, and riparian corridors and wetland habitat associated with pluvial lakes would have expanded, creating filter barriers to dispersal of arid-lands taxa. It is likely that a continuous band of more mesic vegetation in the vicinity of the Deming Plains blocked communication between Sonoran and Chihuahuan taxa, although continuous arid habitat around the northern Gulf of California apparently persisted throughout the Pleistocene (Van Devender et al. 1994). Along the Gulf, lowered sea levels (120 ± 60 m; Bloom 1983) connected landbridge islands to the adjacent mainland and exposed continental shelves. Freeze-intolerant or frost-sensitive desert plants were compressed to the south into isolated refugia (Betancourt et al. 1990), which probably included newly exposed shelves, emergent landbridges, and landbridge islands. Arid-adapted mammals were either restricted somewhat to these desert refugia or persisted in pockets of sclerophyllous woodland or marginal grassland habitats.

With the return of warmer, more seasonal interglacials, frost-sensitive plants spread from glacial refugia north and upslope into neighboring arid and sclerophyllous regions. Sonoran and Chihuahuan taxa regained contact across the Deming Plains, riparian and wetlands habitat was reduced, more extensive communication between the Peninsular and Sonoran Desert regions was reestablished around the northern lip of the Gulf (see Van Devender 1990b), and populations on land-bridge islands were isolated as ocean levels rose.

Pluvial–interpluvial cycles were repeated perhaps 15–20 times during the Pleistocene (Imbrie and Imbrie 1979) and intensified during the last 700,000 years B.P. (Webb and Bartlein 1992). The most recent pluvial conditions reached their maximum intensity between 10,000 and 20,000 years B.P. Warm, dry, and seasonal conditions returned and reached peak intensity approximately 6000 years B.P. (the "hypsithermal"; Pielou 1991), and spreading arid-lands fauna encountered filter barriers such as riparian corridors.

The geologic history of the southern deserts is complicated by recurring events at the same location, including fluctuations in sea level and climatic regimes that have resulted in alternating dispersal barriers and corridors occurring at the same site (fig. 11.2). As an example of temporally nested geologic events, the initial uplift of the Sierra Madre Occidental was an early vicariant event separating the Sonoran and Chihuahuan deserts, but expansion of deserts during the late Tertiary (Axelrod 1983) coupled with block faulting and erosion of the cordillera permitted dispersal and renewed contact between the 2 regions. More recently, climatic shifts during Pleistocene pluvial–interpluvial intervals repeatedly closed and reopened contact between the Chihuahuan and Sonoran deserts across the Deming Plains; the most recent corridor formed in the late Holocene, about 4500 years B.P. (Van Devender 1990a).

Historical Biogeography

A simple mapping of recognized species of arid-adapted rodents provides initial support for the existence of embedded substructure within the southern deserts. Such embedded structure has been dismissed as trivial (Morafka 1977) or included only in the context of late Pleistocene filter barriers (Petersen 1976; Schmidly 1977) without mention of deeper-history geologic or paleoclimatic events. Petersen (1976) and Schmidly (1977) listed species and subspecies of mammals that reach their geographic limits coincident with the Río Nazas and Rio Grande in the Chihuahuan Desert. In addition, at least 2 sets of species (*Chaetodipus intermedius* and *C. nelsoni*; and *Dipodomys spectabilis* and *D. nelsoni*) occupy alternate sides of the Río Conchos, and *Neotoma albigula* from east of the Rio Grande and south of the Río Conchos is now regarded as a separate species, *N. leucodon* (Edwards et al. 2001). In the Sonoran Desert Region, there is a marked turnover from 10 species of the arid-adapted rodents that occur north of the Río Yaqui (the southern margin of the Sonoran Desert as defined by Shreve 1942) to 4 species that occur only south of the Sierra Barabampo–Río Fuerte. Three other species occur throughout the area, and 11 of the

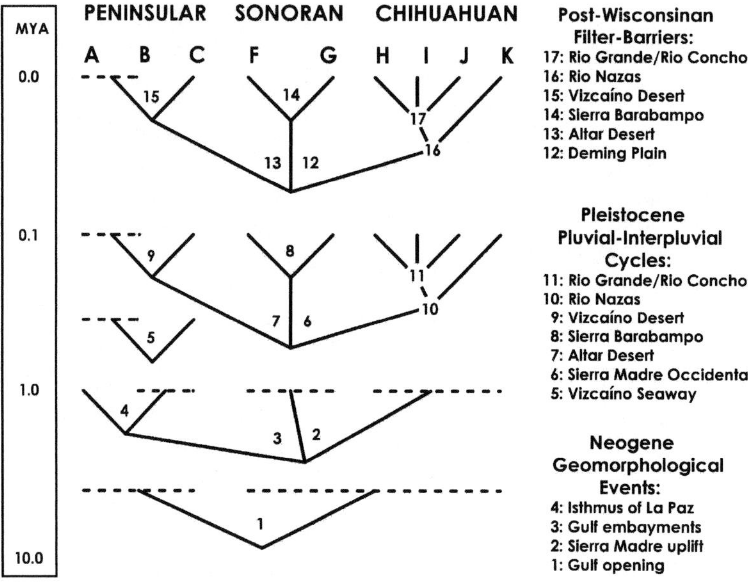

Figure 11.2. Vicariant events postulated as responsible for fragmentation of southern desert regions into embedded subregions. *A*, San Lucan, exclusive of the Sierra Laguna; *B*, Magdalenan; *C*, Vizcaíno; *D*, San Felipe; *E*, Coloradan, *F*, Sonoran; *G*, Sinaloan, *H*, Chichuahuan, *I*, Trans-Pecos, *J*, Coahuilan; and *K*, Zacatecan.

17 total species occupy a 200-km transition zone between the Río Yaqui and the Río Fuerte. On the Baja California peninsula, new locality records (Álvarez-Castañeda et al. 2001; Hafner and Riddle pers. obs.) indicate that 3 northern species (*Chaetodipus fallax*, *C. formosus*, and the silky pocket mouse, *Perognathus longimembris*) reach their southern limits at the Vizcaíno Desert of the central peninsula, which is also a zone of contact between *Peromyscus fraterculus* and *P. eva* (Lawlor 1971; Riddle et al. 2000a).

Heightened clarification of regional substructuring is provided by our preliminary analyses of mtDNA sequence data from some of the projected subregions (summarized in table 11.1). For example, the haplotype found in populations of *Peromyscus eremicus* north of the Río Conchos in the Chihuahuan Desert is more similar to that found in populations in the Sonoran Desert than to the haplotype characteristic of populations south of the Río Conchos (3.5% s.d.). Similar levels of divergence in some of our target taxa appear to be associated with the Sierra Barabampo–Río Fuerte region along the Sonoran–Sinaloan border (Riddle et al. 2000a; fig. 11.1). On the Baja California peninsula, possible relictual traces of the Isthmus of La Paz (Pliocene, 3 mya) are seen in a pocket mouse (*Chaetodipus arenarius*), the black-tailed brush lizard (*Urosaurus nigricaudus*), and a columnar cactus (*Lophocereus*; Riddle et al. 2000c; Nason et al. 2002). Meanwhile, the legacy of the Vizcaíno Seaway is retained in 5 of our 20 arid-adapted rodent species, 5 reptiles, and at least 1 bird species (Riddle et al. 2000c). Further, the zones of contact between neighboring desert regions appear to reflect a more complicated history than previously appreciated: the 2 subregions of the northern lip of the Gulf (D-E; San Felipe and Coloradan) host different mixtures of Peninsular, Mojave, and Sonoran elements, whereas the points of contact between Sonoran and Chihuahuan representatives of 2 species (*Peromyscus eremicus* and *Dipodomys merriami*) are along the Río Conchos rather than the Deming Plains, the traditionally recognized point of contact.

Riddle and Hafner (1999) have argued the advantages of using evolutionarily significant units (or ESUs; Moritz et al. 1995) instead of species as the fundamental unit of analysis in phylogeographic analyses. ESUs (as defined by Moritz et al. 1995) are geographically discrete evolutionary lineages

Table 1.11. Summary of the impact of vicariant events or filter barriers (A-B, B-C, etc.; see fig. 11.1) on selected arid-adapted mammals.

Taxon	SW	A-B	B-C	C-D	D-E	E-F	F-G	S	F-H	H-I	H-J	I-J	J-K	SE
Chaetodipus (*arenarius*)	X	12.8	2.2	—	X									
Chaetodipus (*spinatus*)	X	—	—	—	—	X								
Chaetodipus (*rudinoris, baileyi*)	**X**	—	**1.8**	—	—	8.7	X							
Ammospermophilus (*leucurus, harrisii, interpres*)	X	—	3.3	—	—	**1.2**	X		X	X	X	—	X	
Dipodomys (*merriami*)	X	—	4.1	—	—	—	X		—	2.6	—	—	1.5	X
Peromyscus (*eva, fraterculus, eremicus*)	X	—	**3.1**	—	**8.7**	—	—	X	—	3.5	3.5	—	—	X
Chaetodipus (*penicillatus, pernix, eremicus*)				X	—	3.3	**9.8**	X	**8.8**	—	—	—	—	X
Peromyscus (*merriami*)						X	5.2	X	X					
Chaetodipus (*intermedius*)						X	X		—	—	X			
Chaetodipus (*goldmani, artus, nelsoni*)						X	**5.1**	X	X	X	X	—	—	X

SW, S, SE = southwestern, southern, and southeastern margins of the desert regions; X = margin of species' geographic range; numbers = % s.d. values between haplotypes representing adjacent areas; boldface = parapatric sister taxa; dashes indicate missing comparisons.

revealed through analysis of molecular data, in our case mtDNA in animals, and therefore are operationally equivalent to a phylogroup (Avise 2000). This is only one of a number of alternative definitions of ESUs (Fraser and Bernatchez 2001). When used in this restricted sense of representing reciprocally monophyletic mtDNA lineages, these units are more sensitive indicators of historical separation and, due to the nonrecombinant nature of mtDNA, are more likely to retain the signature of such past events than are taxonomic species or subspecies, as has been demonstrated for the North American desert rodent biota (Riddle and Hafner 1999). Based on currently recognized and detected ESUs, each desert subregion may be characterized by the extent to which its arid-adapted rodent fauna is influenced by one desert region (including the neighboring Mojave) versus another (table 11.2). Quantifying the level of mtDNA sequence divergence for recognition of species boundaries has been the source of major criticism of ESUs. Recently, however, Bradley and Baker (2001) have attempted to determine the magnitude of genetic variation required to distinguish putative species under the genetic species concept (Simpson 1943; Dobzhansky 1950; Mayr 1969).

Simple comparisons of faunal similarities among areas, whether based on species or ESUs, fail to capture the temporally nested nature of recurrent geological events (fig. 11.2). Yet the variety of divergence levels within regional representatives of *Peromyscus* (subgenus *Haplomylomys*) and *Chaetodipus* (Riddle et al. 2000a,b,c) strongly supports the existence of temporally nested geological events. Recurrent branching patterns are particularly evi-

dent among species of *Chaetodipus* (fig. 11.3), in which area representatives appear repeatedly across a broad range of divergence values. A combination of new and powerful analytical tools, particularly the revised version of BPA (Brooks et al. 2001) and Nested Clade Analysis (NCA; Templeton et al. 1995), allow confident matching of specific branching patterns within each lineage with specific geological and paleoclimatic events of Pliocene, Pleistocene, or post-Wisconsinan age.

We based a preliminary BPA analysis of area relationships across the southern deserts (fig. 11.4) on 10 mammalian, 5 avian, and 4 reptilian species or species groups (Upton and Murphy 1997; Zink et al. 1997; Zink and Blackwell 1998a,b; Zink et al. 1998, 1999, 2000; Riddle et al. 2000a,b,c; Zink and Blackwell-Rago 2000; Serb et al. 2001; unpublished data from our lab). The primary BPA tree (traced in black in fig. 11.4) results from a parsimony analysis of the combined taxon relationships based on the original designated areas. Areas are then added to the primary tree (in gray; the secondary tree) to eliminate homoplasy. This combined tree represents a complex set of area relationships derived through a mixture of vicariance, postvicariance dispersal (with or without a subsequent round of divergence), or initial lack of response to vicariance (ancestrally widespread taxa). The BPA tree presents strong support for a Neogene to middle Pleistocene vicariant backbone, postulated as the following events (identified by lowercase letters along branches in fig. 11.4): (a) Neogene opening of the Gulf of California; (b) Neogene uplift of the Sierra Madre Occidental; (c) middle Pleistocene Vizcaíno Seaway; and (d) Neogene or Pleistocene barrier of the Sierra Madre Occidental.

Table 11.2. Percent influence of desert regions in the composition of the arid-adapted rodent fauna of the 11 subregions (labeled as in fig. 11.1).

Desert region	Subregion										
	A	B	C	D	E	F	G	H	I	J	K
Peninsular	**100**	**100**	67	38	14						
Mojave			33	25	36	10					
Sonoran				38	50	**70**	**100**	36	8		
Chihuahuan						20		**64**	**92**	**100**	**100**
Total species	6	8	12	16	14	20	7	11	13	14	13

Peninsular Desert Region (*sensu* Hafner and Riddle 1997): A = San Lucan, B = Magdalenan, C = Vizcaíno; transitional area between the Peninsular, Mojave, and Sonoran Desert regions: D = San Felipe, E = Coloradan; Sonoran Desert Region: F = Sonoran, G = Sinaloan; and Chihuahuan Desert Region: H = Chihuahuan, I = Trans-Pecos, J = Coahuilan, and K = Zacatecan. All blank values = 0. Percent influences of desert regions on their own subregions are boldfaced.

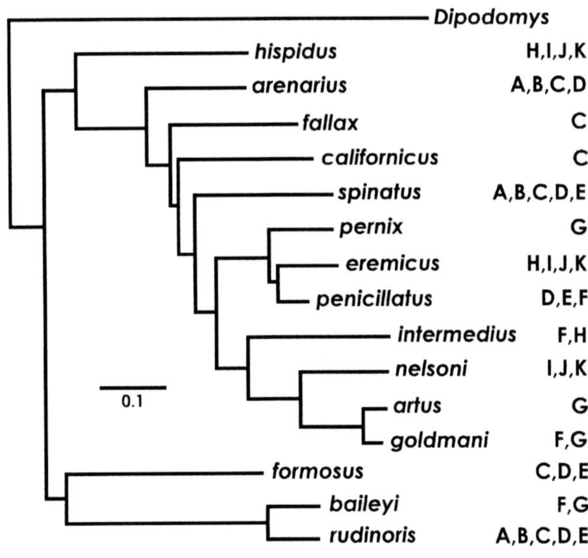

Figure 11.3. Preliminary phylogeny of species of Chaetodipus (Riddle et al. 2000b) and their highly redundant distribution among subregions of the 3 southern desert regions; subregions, right: *A*, San Lucan, exclusive of the Sierra Laguna; *B*, Magdalenan; *C*, Vizcaíno; *D*, San Felipe; *E*, Coloradan; *F*, Sonoran; *G*, Sinaloan; *H*, Chichuahuan; *I*, Trans-Pecos; *J*, Coahuilan; and *K*, Zacatecan.

The most significant difference between the primary (black) and secondary (gray) BPA trees derived from this analysis is due to a highly reticulate Sonoran area: duplications indicate a nearly equal split in area relationships to the west (Peninsular) and to the east (Chihuahuan). This preliminary analysis suggests the utility of the BPA approach for unraveling biogeographic pattern in biotas assembled through a complex history of vicariance and dispersal events.

For a phylogroup whose distribution encompasses 2 adjacent areas, 3 biogeographic scenarios are possible. First, the lineage may have been ancestrally widespread across both areas and has never been substantially isolated by a barrier. Second, the lineage may have expanded its range recently from one area to the other (e.g., post-Wisconsinan range expansion) after erosion of a previous barrier to dispersal. Third, the lineage may have been isolated only recently by a new barrier to dispersal such that separate phylogroups are not yet distinguishable but sorting of ancestral polymorphism and origination of new haplotypes

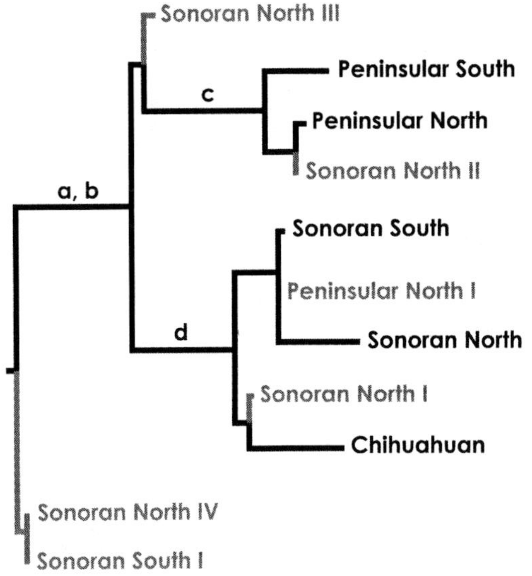

Figure 11.4. Area cladogram from secondary Brooks Parsimony Analysis (BPA) representing our recovery of vicariance-derived area relationships (letters *a–d*; see text for description of events) as well as a mixture of dispersal and additional divergence events throughout the tree. The single primary tree (in black) has 118 steps; confidence interval = 0.72, retention index = 0.63. Lighter lines represent area duplications (numbered).

have resulted in gene variation (Nei 1987) differing significantly between areas.

Several approaches from coalescent theory and population genetics may be used to distinguish between these alternatives, such as the maximum likelihood approach to estimating historical population sizes implemented in FLUCTUATE (Kuhner et al. 1998); and a more direct assessment of recent range expansion through examination of pairwise haplotype differences ("mismatch distributions"; Rogers and Harpending 1992). One approach, NCA (Templeton et al. 1995) provides a comprehensive, phylogenetically based framework to distinguish between historically equilibrial, expanding, or fragmented populations. NCA uses an intraspecific cladogram estimation procedure to help distinguish between the effects of population history, such as past fragmentation or recent range expansion, and the effects of recurrent evolutionary forces, such as current equilibrium between gene flow, mutation, and drift. NCA has an advantage over other meth-

ods in its explicit use of the phylogenetic and geographic information inherent in a given data set for differentiating between process explanations (but see Knowles and Maddison 2002, for a critique of NCA).

We have used NCA (Posada et al. 2000) to examine the population structure of one of our target taxa, *Dipodomys merriami* (fig. 11.5) based on mtDNA COXIII data (L. F. Alexander, pers. comm.). In the mtDNA genealogy for the entire species group, generated using the statistical parsimony procedure TCS (Clement et al. 2000), the clade from the Peninsular South area (not shown in fig. 11.5) is highly divergent from the remainder of *D. merriami* haplotypes (clade 5.1, including all haplotypes shown in fig. 11.5), postulated to result from isolation attributable to the middle Pleistocene Vizcaíno Seaway (fig. 11.1; Riddle et al. 2000c). In figure 11.5, a minimum-spanning haplotype network is shown for 93 individuals of *D. merriami* with connections having probabilities greater than 0.95. This network

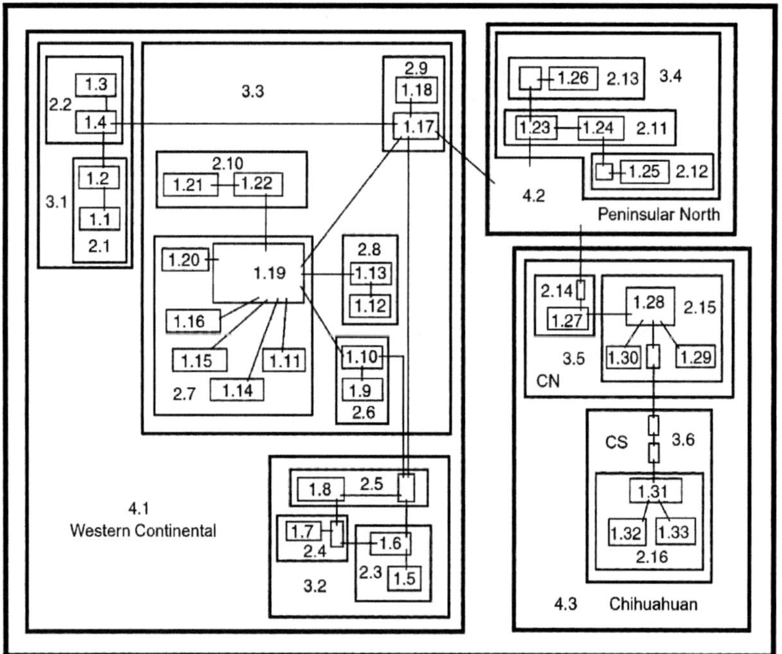

Figure 11.5. Statistical parsimony network and nested clade analysis using mtDNA COXIII sequences for 93 individuals in the *Dipodomys merriami* species group. The outer box encloses all of clade 5.1 (all samples except Peninsular South); haplotypes belonging to the same clade level are boxed together; clade level designations are given within each box that contains observed haplotypes.

was used to test the null hypothesis of a random geographical distribution of all clades within a nested clade. Four clades (3.3, 4.1, 4.3, and 5.1) had significant values with statistical inferences of past fragmentation (3.3), restricted gene flow with isolation by distance (4.1), past fragmentation (4.3), and range expansion with long-distance colonization (5.1). Of particular interest here is the inference of past fragmentation between clades 3.5 and 3.6 within the Chihuahuan clade 4.3. These clades correspond to populations occurring to the north and south of the Río Nazas in the Chihuahuan Desert, which is one of our postulated embedded barriers between areas J and K (fig. 11.1).

Desert Mammal Communities

Studies concerning the "macroecology" (*sensu* Brown 1995) of North American deserts may be hindered by a strict reliance on the original delineation of those deserts by Shreve (1942) without consideration of the major subdivisions within each (Shreve 1951; fig. 11.1). We have shown that an implicit assumption of homogeneity within the southern deserts obscured the evolutionarily distinct Peninsular Desert Region (*sensu* Hafner and Riddle 1997) and perhaps has led macroecologists to mistakenly assume that samples from the relatively depauperate periphery of a desert are representative of that entire desert (Riddle and Hafner 1999).

Brown and Kurzius (1987) developed a data set of desert mammal communities of the southwestern United States based on published and unpublished records that has served as the foundation for a variety of studies concerning community structure and assembly rules in a growing body of literature (Brown and Kurzius 1987; Patterson and Brown 1991; Fox and Brown 1993; Morton et al. 1994; Kelt et al. 1996, 1999; Kelt 1999; Brown et al. 2000). Brown and Kurzius (1987) developed this data set by drawing community samples from the literature that conformed to a reasonable set of criteria (≥100 trapnights in small, relatively uniform patches of habitat), and limited it to granivorous species of rodents. This data set was enlarged by Morton et al. (1994), who included more species of rodents, and it was further augmented by Kelt et al. (1996), who included samples from 2 biosphere reserves in Mexico (Mapimí, a 172,000-ha region in the central Chihuahuan Desert, and the 155,000-ha Pinacate section of the 714,556-ha Pinacate y Gran Desierto de Altar Reserve in extreme northwestern Sonora).

The focus of the body of literature cited above has been interregional or intercontinental comparisons of community structure. However, comments taken out of context might imply that the data set represents both regional and total arid-lands rodent diversity of North America. For example, Morton et al. (1994) describe the original data set as for granivorous rodents of the North American deserts instead of southwestern United States (Brown and Kurzius 1987). Kelt et al. (1996) applied this data set to an intercontinental comparison of community structure; they noted that due to extensive sharing of species across the Great Basin, Mojave, Sonoran (including our Peninsular Desert Region), and Chihuahuan deserts, they combined these as representing the North American desert region for comparison with desert communities in South America, Australia, and Asia. The data set was subsequently used to evaluate the degree of similarity among deserts for "the North American desert small mammal fauna" (Kelt 1999: 124–125) and to support certain community assembly rules "in local habitats throughout southwestern North America" (Brown et al. 2000: 314).

The community database developed by Brown and Kurzius (1987) and augmented by Morton et al. (1994) and Kelt et al. (1996) was developed to examine local community structure, not to capture the biodiversity of the North American deserts. It should not be used to compare faunal similarity among deserts (or desert regions) and should be used cautiously to generalize about macroecological patterns among the North American deserts. We have argued elsewhere (Riddle and Hafner 1999) that this database markedly underestimates the biodiversity within and among desert regions due to its reliance on the species as the selected unit of analysis. Although our commentary was taken as a criticism of Brown and Kurzius (1987) and Kelt et al. (1996; see Kelt and Brown 2000), our intent was to recommend broader inclusion of the recognized biodiversity if the database was to compare faunal composition of desert regions. In addition to reliance on the species unit, we note 5 other sources of bias that we believe render this data set inappropriate for interregional faunal comparisons: (1) inadequate sampling of Mexican desert regions (13 species left out); (2) unrecognized cryptic species (7 species, most of which were discovered subsequent to Kelt et al. 1996); (3) inclusion of species not restricted to arid regions (e.g., meadow mice, *Microtus*) and exclusion of other arid-adapted rodents (e.g., *Ammospermophilus*); (4) outdated location of the major shift from Sonoran to Chihuahuan communities (the sites in southeastern Arizona are included as "Chihuahuan," but exhibit higher Sonoran influence); and (5) inadequate trapping effort at each site to detect "rare or ephemeral species" (Brown and Kurzius 1987: 234), which might not be important for community analysis but are important for interregional faunal comparisons. Of the 45 species of arid-adapted rodents that we include as occurring in the southern desert regions, Brown and Kurzius (1987) included only 17, while Morton et al. (1994) and Kelt et al. (1996) included 21. The discrepancy between the actual arid-lands rodent biodiversity and that included within the community data set would be higher if ESUs were considered.

Conservation Issues

The southern deserts are disappearing from North America at an alarming rate, and Mexico should note the destruction of deserts north of the U.S.–Mexico border. The San Joaquin Valley of California, a unique desert area that includes 22 endemic species and subspecies of mammals, had suffered an estimated 98% conversion to agriculture by 1976 (Williams and Kilburn 1992). Explosive expansion of Los Angeles, California, Las Vegas, Nevada, and Phoenix, Arizona, has eliminated vast tracts of the Mojave and Sonoran deserts as desert is being lost due to urban expansion and related agricultural conversion, which also has decimated native desert around El Centro, California. Deserts in Mexico are being subjected to this same wholesale destruction due to explosive expansion of metropolitan areas, wholesale agricultural conversion, selective eradication of specific habitat associations, and introduction of exotics. Ceballos and Navarro-L. (1991) reported 24 arid-land–restricted species of mammals that are now extinct, endangered, threatened, or whose status is fragile in Mexico. Without an appreciation and delineation of the subregional diversity and structure within desert regions, whole biotic associations may be lost without any attempts at conservation. Future efforts to prioritize biodiversity hotspots in the North American deserts should incorporate quantitative measures of phylogenetic diversity (Moritz and Faith 1998) at the phylogeographic level of resolution presented in this chapter. Genetic diversity and phylogenetic structuring within species are important components of their evolutionary potential.

Agricultural conversion has eliminated much of the native habitat along the coastal plains of Sinaloa, the Altiplano of Zacatecas, the Magdalena Plains of Baja California Sur, Laguna Mayrán of Coahuila, and El Oriental basin in Puebla and Veracruz (Hafner and Riddle pers. obs.). Little remains of the peripheral desert habitat that previously existed in this latter region: all but the rocky hilltops have been plowed, and the soil blows in the wind except during the rain-dependent growing season of several months. Two species endemic to the basin, the Perote ground squirrel (*Spermophilus perotensis*) and the Perote mouse (*Peromyscus bullatus*) persist only in marginal habitat (Ceballos and Navarro-L. 1991; Valdéz and Ceballos 1997). In early 2002, we found 2 other Chihuahuan species, *Dipodomys phillipsi* and *Perognathus flavus*, surviving only in the thin margins between plowed fields and rocky hilltops.

Elsewhere, scattered and restricted vegetative associations have been targeted for destruction. Mesquite woodland in the Sonoran Desert, particularly between Tucson and Nogales, Arizona, has been cut for firewood and charcoal briquettes and to clear land for grazing, or supplanted by retirement housing developments, and the "fate of mesquite mice [*Peromyscus merriami*] in Arizona is precarious" (Hoffmeister 1986: 345). Our observations in northern Sonora indicate that these same processes are spreading to the south. The formerly expansive grassland distributions of the 2 living species of black-tailed prairie dogs (*Cynomys ludovicianus* and the Mexican endemic *C. mexicanus*) have been dramatically reduced in Mexico due to habitat destruction and deterioration (Ceballos et al. 1992). Relictual pockets of grassland throughout the Chihuahuan Desert Region have attracted first grazing then agricultural use, and native grassland no longer exists at many of the former locality records for *Chaetodipus hispidus*, a grassland indicator. As these relictual refugia disappear, so do the vertebrates that depend on them.

Recognizing the value of preserving the biodiversity of its desert regions, the Mexican federal government has established desert reserves across northern Mexico (for a description, go to http://www.conanp.gob.mx). These include 3 biosphere reserves (El Vizcaíno in northern Baja California Sur, El Pinacate y Gran Desierto de Altar in northwestern

Sonora, and Mapamí in northeastern Durango) and two protected areas (Islas del Golfo de California, and Valle de los Cirios on the Baja California peninsula).

Islands

The system of 25 larger and numerous smaller islands in the Gulf of California that has been called "the Galápagos of the northern hemisphere" has suffered often intense human impacts (Bahre 1983). Unfortunately, inadequate baseline data exist with which to evaluate continued change (either natural or due to human impact) in these islands and thus provide a better foundation to develop conservation strategies to mitigate continued adverse human impact.

Native mammals are known from 36 islands in the Gulf of California and along the Pacific Coast of the Baja California peninsula. The majority of these mammals are rodents, particularly of 3 genera: *Peromyscus*, *Chaetodipus*, and *Neotoma*. Most were detected during initial survey of the islands in the first half of the twentieth century, when sampling was done at usually 1 and rarely a few accessible sites on each island. With few exceptions, subsequent resampling of the islands during the last 50 years has taken place at the same sites. During the same period, extinction of at least 3 insular populations has been documented; 6 other insular forms are possibly extinct, and another 5 are considered vulnerable to or in imminent danger of extinction (Lawlor 1971; Mellink 1992a,b; Smith et al. 1993; Álvarez-Castañeda and Cortés-Calva 1996, 1999, 2002; Lawlor et al. 2002; Mellink et al. 2002). Causes of extinction on near-shore islands surrounding the peninsula include introduced predators and competitors (e.g., cats, goats, *Rattus*, *Mus*), habitat modification (e.g., clearing of ironwood; introduction of iceplant), and direct poisoning campaigns. Species may be disappearing that have never been detected. As a case in point, if Burt (1932) had not visited Isla Coronados and Isla San Pedro Nolasco in 1931 and collected the only specimens of *Neotoma bunkeri* and *Peromyscus pembertoni* (respectively), those species never would have been known to science (Lawlor 1971; Smith et al. 1993).

Similarly, there are no baseline data from which to evaluate the possible establishment of new species or genotypes on islands via either natural overwater dispersal (e.g., rafting) or human-related activities (e.g., via fishermen). Without a clear understanding of the taxonomic and genetic makeup of insular populations, interpretation of genetic relationships to mainland populations (established during submergence of land bridges; by ancient colonization; or by recent colonization, human-mitigated or not) are greatly weakened (Hafner et al. 2001).

The inventory of bats known from islands in the Gulf of California provides another indication of undersampling of island mammals: one night's netting on Isla Partida Sur (immediately north of Isla Espíritu Santo, and a frequently visited and surveyed island) yielded 2 new records for bat genera on the islands of the Gulf of California (*Nyctinomops* and *Lasiurus*; Lawlor et al. 2002; Riddle, Hafner, and Álvarez-Castañeda, pers. obs.).

Interest is rapidly building among private developers, state governments, and federal tourist agencies to develop Gulf of California islands and the adjacent Baja California coastline for tourist trade. Development has already begun on *Escalera Náutica*, a federally funded project that plans to increase yacht traffic in the Gulf of California by a factor of 10 over the next decade. *Escalera Náutica* will build a mid-peninsular highway to truck yachts on a 110-km shortcut between the Pacific Ocean and Gulf of California and will develop 22 service ports around the rim of the Gulf. Despite their inclusion in the Islas de Golfo de California natural protected area, 4 of the larger islands are currently privately or communally owned: Tiburón, del Carmen, Cerralvo, and San José; Espíritu Santo was only recently transferred to federal ownership. Proposed development of Isla San José includes 12 or 13 hotels, 3 residential developments, a golf course, airstrip, and a support town of some 2000 residents (*Boletín Oficial del Gobierno del Estado de Baja California Sur* 2000). Similarly ambitious plans are being proposed to develop the Loreto region (adjacent to Carmen, Coronados, and Danzante) and to expand tourist activity in the near-shore region between Loreto and La Paz.

Future Global Climate Change Scenarios

Peterson et al. (2002) used a genetic algorithm and museum specimen data to develop ecological niche models for 1870 species of birds, mammals, and butterflies occurring in Mexico. They then projected these niche models onto 2 climate surfaces modeled for 2055. They concluded that "although extinctions and drastic range reductions are predicted to be relatively few, species turnover in some local communities is predicted to be high (>40% of species), suggesting that severe ecological perturbations may result" (Peterson et al. 2002: 626). They further indicated that the "foci" of this turnover are the broad, open Chihuahuan Desert, the coastal plains of the Sonoran Desert, and the Baja California peninsula. This potential sensitivity to climatic change may enhance the vulnerability of the southern deserts to adverse human impact.

Acknowledgments

We thank L. Grismer and B. Hausback for helpful discussion. T. Van Devender, D. Kelt, G. Ceballos, and J.-L. Cartron provided useful comments on earlier drafts of the chapter. Financial support was provided by grants from the National Science Foundation to B.R.R. (DEB-9629787 and DEB-0237166) and D.J.H. (DEB-9629840 and DEB-0236957).

Literature Cited

Álvarez-Castañeda, S. T. 1994. Current status of the rice rat, *Oryzomys couesi peninsularis*. Southwestern Naturalist 39: 99–100.

Álvarez-Castañeda, S. T., and P. Cortés-Calva. 1996. Anthropogenic extinction of the endemic deer mouse, *Peromyscus maniculatus cineritius,* on San Roque island, Baja California Sur, Mexico. Southwestern Naturalist 41: 99–100.

Álvarez-Castañeda, S. T., and P. Cortés-Calva. 1999. Familia Muridae. Pp. 445–566 *in* S. T. Álvarez-Castañeda and J. Patton (eds.), Mamíferos del Noroeste de México. Centro de Investigaciones Biológicas del Noroeste, S. C., La Paz, Mexico.

Álvarez-Castañeda, S. T., and P. Cortés-Calva. 2002. Extirpation of Bailey's pocket mouse, *Chaetodipus baileyi fornicatus* (Heteromyidae: Mammalia), from Isla Montserrat, Baja California Sur, Mexico. Western North American Naturalist 62: 496–497.

Álvarez-Castañeda, S. T., E. Rios, and A. Gutierrez-Ramos. 2001. Noteworthy records of the little pocket mouse (Heteromyidae: *Perognathus longimembris*) on the Baja California Peninsula. Southwest Naturalist 46: 243–245.

Anderson, C. A. 1950. The 1940 E. W. Scripps cruise to the Gulf of California. Part I: Geology of the islands and neighboring land areas. Geological Society of American Memoirs 43: 1–53.

Andersson, L. 1996. An ontological dilemma: epistemology and methodology of historical biogeography. Journal of Biogeography 23: 269–277.

Armstrong, D. M. 1972. Distribution of mammals in Colorado. Monograph of the Museum of Natural History, University of Kansas 3: 1–415.

Avise, J. C. 1994. Molecular Markers, Natural History and Evolution. Chapman and Hall, New York.

Avise, J. C. 2000. Phylogeography: The History and Formation of Species. Harvard University Press, Cambridge, Massachusetts.

Avise, J. C., J. Arnold, R. M. Ball, E. Bermingham, T. Lamb, J. E. Neigel, C.A Reeb, and N. C. Saunders. 1987. Intraspecific phylogeography: the mitochondrial DNA bridge between population genetics and systematics. Annual Review of Ecology and Systematics 18: 489–522.

Axelrod, D. I. 1958. Evolution of the Madro-Tertiary geoflora. Botanical Review 24: 433–509.

Axelrod, D. I. 1983. Paleobotanical history of the western deserts. Pp. 113–129 *in* S. G. Wells and D. R. Haragan (eds.), Origin and Evolution of Deserts. University of New Mexico Press, Albuquerque.

Bahre, C. 1983. Human impact: the midriff islands. Pp. 290–306 *in* T. J. Case and M. L. Cody (eds.), Island Biogeography in the Sea of Cortéz. University of California Press, Berkeley.

Bernardi, G., L. Findley, and A. Rocha-Olivares. 2003. Vicariance and dispersal across Baja California in disjunct marine fish populations. Evolution 57: 1599–1609.

Betancourt, J. L., T. R. Van Devender, and P. S. Martin. 1990. Packrat Middens: The Last 40,000 Years of Biotic Change. University of Arizona Press, Tucson.

Blair, W. N. 1978. Gulf of California in Lake Mead area in Arizona and Nevada during the late Miocene time. Bulletin of the American Association of Petroleum Geologists 62: 1159–1170.

Bloom, A. L. 1983. Sea level and coastal changes. Pp. 42–51 *in* H. E. Wright, Jr. (ed.), Late Quaternary of the United States, vol. 2. The Holocene. University of Minnesota Press, Minneapolis.

Boehm, M. C. 1984. An overview of the lithostratigraphy, biostratigraphy, and paleoenvironments of the Late Neogene San Felipe marine sequence, Baja California, Mexico. Pp. 253–265 *in* V. A. Frizzell (ed.), Geology of the Baja California Peninsula. Field trip guidebook, vol. 39, Society of Economic Paleontologists and Mineralogists, Pacific Section, Los Angeles.

Boletín Oficial del Gobierno del Estado de Baja

California Sur. 2000. Plan maestro de desarollo y reglamento de usos del suelo, normas de ocupación, construcción, mantenimiento y control ambiental "Desarollo Ecoturístico Isla San José" vol 27, no. 38, November 30.

Bradley, R. D., and R. J. Baker. 2001. A test of the genetic species concept: cytochrome-*b* sequences in mammals. Journal of Mammalogy 82: 960–973.

Brooks, D. R. 1990. Parsimony analysis in historical biogeography and coevolution: methodological and theoretical update. Systematic Zoology 39: 14–30.

Brooks D. R., M. G. P. Van Veller, and D. A. McClennan. 2001. How to do BPA, really. Journal of Biogeography 28: 345–358.

Brown, J. H. 1995. Macroecology. The University of Chicago Press, Chicago, Illinois.

Brown, J. H., B. J. Fox, and D. A. Kelt. 2000. Assembly rules: desert rodent communities are structured at scales from local to continental. American Naturalist 156: 314–321.

Brown, J. H., and M. A. Kurzius. 1987. Composition of desert rodent faunas: combinations of coexisting species. Annales Zoologici Fennici 24: 227–237.

Buising, A. V. 1990. The Bouse Formation and bracketing units, southeastern California and western Arizona: implications for the evolution of the proto-Gulf of California and the lower Colorado River. Journal of Geophysical Research 95: 20, 111–20, 132.

Búrquez, A., A. Martínez-Yrízar, R. S. Felger, and D. Yetman. 1999. Vegetation and habitat diversity at the southern edge of the Sonoran Desert. Pp. 36–67 *in* R. H. Robichaux (ed.), Ecology of Sonoran Desert Plants and Plant Communities. University of Arizona Press, Tucson.

Búrquez, A., A. Martínez-Yrízar, and P. S. Martin. 1992. From the High Sierra Madre to the coast: changes in vegetation along Highway 16, Maycoba to Hermosillo, Sonora. Pp. 239–252 *in* K. F. Clark, J. Roldán-Quintana, and R. H. Schmidt (eds.), Geology and Mineral Resources of Northern Sierra Madre Occidental, Mexico. El Paso Geological Society, El Paso, Texas.

Burt, W. H. 1932. Descriptions of heretofore unknown mammals from islands in the Gulf of California, Mexico. Transactions of the San Diego Society of Natural History 16: 161–182.

Burt, W. H. 1938. Faunal relationships and geographic distribution of mammals of Sonora, Mexico. Miscellaneous Publications, Museum of Zoology, University of Michigan 39: 1–77.

Ceballos, G., and D. Navarro-L. 1991. Diversity and conservation of Mexican mammals. Pp. 167–198 *in* M. A. Mares and D. J. Schmidly (eds.), Latin American Mammalogy: History, Biodiversity, and Conservation. University of Oklahoma Press, Norman.

Ceballos, G., E. Mellink, and L. R. Hanebury. 1992. Distribution and conservation status of prairie dogs *Cynomys mexicanus* and *Cynomys ludovicianus* in Mexico. Biological Conservation 63: 105–112.

Chamberlin, R. V. 1924. Expedition of the California Academy of Sciences to the Gulf of California in 1921—the spider fauna of the shores and islands of the Gulf of California. Proceedings of the California Academy of Sciences 12: 561–694.

Clement, M., D. Posada, and K. A. Crandall. 2000. TCS: a computer program to estimate gene genealogies. Molecular Ecology 9: 1657–1660.

Clements, F. E. 1916. Plant succession: an analysis of the development of vegetation. Carnegie Institute Publication no. 242. Carnegie Institute, Washington, D.C.

Cody, M. L. 1983. The land birds. Pp. 210–245 *in* T. J. Case and M. L. Cody (eds.), Island biogeography in the Sea of Cortéz. University of California Press, Berkeley.

Croizat, L. 1952. Manual of Phytogeography. Dr. W. Junk, The Hague, Netherlands.

Croizat, L. 1958. Panbiogeography, 2 vols. Published by the author, Caracas, Venezuela.

Croizat, L. 1960. Principia Botanica. Published by the author, Caracas, Venezuela.

Croizat, L. 1964. Space, time, form: the biological synthesis. Published by the author, Caracas, Venezuela.

De Martonne, E. 1926. Aréisme et indice d'aridité, Compte-rendus de L'Académie des Sciences de Paris 182: 1393–1398. [In English: Regions of interior basin drainage. Geographical Review 17: 397–414.]

Dirección General de Geografía del Territorio Nacional. 1983. Carta de Climas: Estados Unidos Mexicanos. Secretaria de Programación y Presupuesto, INEGI, México.

Dobzhansky, T. 1950. Mendelian populations and their evolution. American Naturalist 74: 312–321.

Durham, J. W., and E. C. Allison. 1960. Geological history of Baja California and its marine environments. Symposium: the biogeography of Baja California and adjacent seas. Systematic Zoology 9: 47–91.

Eberly, L. D., and T. B. Stanley, Jr. 1978. Cenozoic stratigraphy and geologic history of southeastern Arizona. Bulletin of the Geological Society of America 89: 921–940.

Edwards, C. W., C. F. Fulhorst, and R. D. Bradley. 2001. Molecular phylogenetics of the *Neotoma albigula* species group: further evidence of a paraphyletic assemblage. Journal of Mammalogy 82: 267–279.

Felger, R. S., and C. H. Lowe. 1976. The island and coastal vegetation and flora of the northern part of the Gulf of California, Mexico. Natural History Museum of Los Angeles County Contributions in Science 285: 1–59.

Felger, R. S., and M. B. Moser. 1985. People of the Desert and Sea: Ethnobotany of the Seri Indians. University of Arizona Press, Tucson.

Findley, J. S. 1969. Biogeography of southwestern boreal and desert mammals. Pp. 113–128 *in* J. K. Jones (ed.), Contributions in Mammalogy. University of Kansas Miscellaneous Publications of the Museum of Natural History 51: 1–428.

Fox, B. J., and J. H. Brown. 1993. Assembly rules for functional groups in North American desert rodent communities. Oikos 67: 358–370.

Fraser, D. J., and L. Bernatchez. 2001. Adaptive evolutionary conservation: towards a unified concept for defining conservation units. Molecular Ecology 10: 2741–2752.

Frey, J. K. 1992. Response of a mammalian faunal element to climatic changes. Journal of Mammalogy 73: 43–50.

Gantenbein, B., V. Fet, and M. D. Barker. 2001. Mitochondrial DNA reveals a deep, divergent phylogeny in *Centruroides exilicauda* (Wood, 1863) (Scorpiones: Buthidae). Pp. 235–244 *in* V. Fet and P. A. Selden (eds.), Scorpions 2002. In memoriam Gary A. Polis. British Arachnological Society, Burnham Beeches, U.K.

Gardner, A. L., and P. Cortés-Calva. 1999. Didelphidae. Pp. 29–37 *in* S. T. Álvarez-Castañeda and J. Patton (eds.), Mamíferos del Noroeste de México. Centro de Investigaciones Biológicas del Noroeste (CIBNOR), S. C., La Paz, Mexico.

Gastil, G., J. Minch, and R. P. Phillips. 1983. The geology and ages of the islands. Pp. 13–25 *in* T. J. Case and M. L. Cody (eds.), Island Biogeography in the Sea of Cortéz. University of California Press, Berkeley.

Gleason, H. A. 1926. The individualistic concept of plant associations. Bulletin of the Torrey Botanical Club 53: 7–26.

Grismer, L. L. 1994. The origin and evolution of the Peninsular herpetofauna of Baja California, Mexico. Herpetological Natural History 2: 51–106.

Grismer, L. L. 2002. A re-evaluation of the evidence for a mid-Pleistocene mid-peninsular seaway in Baja California: a reply to Riddle et al. Herpetological Review 33: 15–16.

Hafner, D. J., and B. R. Riddle. 1997. Biogeography of Baja California Peninsular Desert mammals. Pp. 39–65 *in* T. L. Yates, W. L. Gannon, and D. E. Wilson (eds.), Life among the muses: papers in honor of James S. Findley. Special Publication, Museum of Southwestern Biology, University of New Mexico 3: 1–290.

Hafner, D. J., B. R. Riddle, and S. T. Álvarez-Castañeda. 2001. Evolutionary relationships of white-footed mice (*Peromyscus*) on islands in the Sea of Cortés, Mexico. Journal of Mammalogy 82: 775–790.

Hausdorf, B. 1998. Weighted ancestral area analysis and a solution of the redundant distribution problem. Systematic Biology 47: 445–456.

Hess, H. H. 1962. History of ocean basins. Pp. 599–620 *in* A. E. J. Engel, H. L. James, and B. F. Leonard (eds.), Petrological Studies: A Volume in Honor of A. F. Buddington. Geological Society of America, New York.

Hoffmeister, D. F. 1986. Mammals of Arizona. University of Arizona Press and Arizona Game and Fish Department, Tucson.

Hubbard, J. P. 1974. Avian evolution in the aridlands of North America. Living Bird 12: 155–196.

Imbrie, J., and K. P. Imbrie. 1979. Ice Ages: Solving the Mystery. Enslow Publishers, Short Hills, New Jersey.

Ingle, J. C., Jr. 1987. Paleooceanographic evolution of the Gulf of California: foraminiferal and lithofacies evidence. Abstracts with Programs, Geological Society of America 19: 721.

Kelt, D. A. 1999. On the relative importance of history and ecology in structuring communities of desert small animals. Ecography 22: 123–137.

Kelt, D. A., and J. H. Brown. 2000. Species as units of analysis in ecology and biogeography: are the blind leading the blind? Global Ecology and Biogeography 9: 213–217.

Kelt, D. A., J. H. Brown, E. J. Heske, P. A. Marquet, S. R. Morton, J. R. Reid, K. A. Rogouin, and G. Shenbrot. 1996. Community structure of desert small mammals: comparisons across four continents. Ecology 77: 746–761.

Kelt, D. A., K. Rogovin, G. Shenbrot, and J. H. Brown. 1999. Patterns in the structure of Asian and North American desert small mammal communities. Journal of Biogeography 26: 825–841.

Knowles, L. L., and W. P. Maddison. 2002. Statistical phylogeography. Molecular Ecology 11: 2623–2635.

Kuhner, M. K., J. Yamato, and J. Felsenstein. 1998. Maximum likelihood estimation of population growth rates based on the coalescent. Genetics 149: 429–434.

Lawlor, T. E. 1971. Distribution and relationships of six species of *Peromyscus* in Baja California and Sonora, Mexico. Occasional Papers, Museum of Zoology, University of Michigan 661: 1–22.

Lawlor, T. E., D. J. Hafner, P. Stapp, B. R. Riddle, and S. T. Álvarez-Castañeda. 2002. The mammals. Pp. 326–361 *in* T. J. Case, M. L. Cody, and E. Ezcurra (eds.), A New Island Biogeography of the Sea of Cortés. Oxford University Press, New York.

Legendre, P., and V. Makarenkov. 2002. Reconstruction of biogeographic and evolutionary networks using reticulograms. Systematic Biology 51:199–216.

Lonsdale, P. 1989. Geology and tectonic history of the Gulf of California. Pp. 499–521 *in* E. L. Winterer, D. M. Hussong, and R. W. Decker (eds.), The Eastern Pacific Ocean and Hawaii (The Geology of North America, vol. N), Geological Society of America, Boulder, Colorado.

López-Forment, C., and G. Urbano-V. 1977. Restos de pequeños mamíferos recuperados en regurgitaciones de lechuza, *Tyto alba*, en México. Anales del Instituto de Biología, Sería Zoología 1: 231–242.

Mayr, E. 1969. Principles of Systematic Zoology. McGraw-Hill, New York.

McCloy, C. 1984. Stratigraphy and depositional history of the San Jose del Cabo trough, Baja California Sur, Mexico. Pp. 267–273 *in* V. A. Frizzell (ed.), Geology of the Baja California Peninsula. Field trip guidebook, vol. 39, Society of Economic Paleontologists and Mineralogists, Pacific Section, Los Angeles.

McDowell, F. W., and R. P. Keizer. 1977. Timing of mid-Tertiary volcanism in the Sierra Madre Occidental between Durango City and Mazatlan, Mexico. Geological Society of America Bulletin 88: 1479–1487.

Mellink, E. 1992a. The status of *Neotoma anthonyi* (Rodentia, Muridae, Cricetinae) of Todos Santos Islands, Baja California, Mexico. Bulletin of the Southern California Academy of Science 91: 137–140.

Mellink, E. 1992b. Status de los heteromyidos y cricétidos endémicos del estado de Baja California. Comunicaciones académicas, serie ecología, Centro de Investigaciones Científicas y de Educación Superior de Ensenada. Ensenada, Mexico.

Mellink, E., G. Ceballos, and E. Luevano. 2002. Population demise and extinction threat of the Ángel de la Guarda deer mouse (*Peromyscus guardia*). Biological Conservation 108: 107–111.

Morafka, D. J. 1974. A biogeographical analysis of the Chihuahuan Desert through its herpetofauna. Ph.D. dissertation, University of Southern California, Los Angeles.

Morafka, D. J. 1977. Is there a Chihuahuan Desert? A quantitative evaluation through a herpetofaunal perspective. Pp. 437–454 *in* R. H. Wauer and D. H. Riskind (eds.), Transactions of the Symposium on the Biological Resources of the Chihuahuan Desert Region, United States and Mexico. National Park Service Transactions and Proceedings Series 3.

Moritz, C., and D. P. Faith. 1998. Comparative phylogeography and the identification of genetically divergent areas for conservation. Molecular Ecology 7: 419–429.

Moritz, C., S. Lavery, and R. Slade. 1995. Using allele frequency and phylogeny to define units for conservation and management. American Fisheries Society Symposium 17: 249–262.

Morton, S. R., J. H. Brown, D. A. Kelt, and J. R. W. Reid. 1994. Comparisons of community structure among small mammals of North American and Australian deserts. Australian Journal of Zoology 42: 501–525.

Mullis, K. B, F. A. Faloona, S. Scharf, R. K. Saiki, G. Horn, and H. A. Erlich. 1986. Specific enzymatic amplification of DNA in vitro: the polymerase chain reaction. Cold Spring Harbor Symposia on Quantitative Biology 1986.

Murphy, R. W. 1983. Paleobiogeography and patterns of genetic differentiation of the Baja California herpetofauna. Occasional Papers, California Academy of Sciences 137: 1–48.

Nason, J. D., J. L. Hamrick, and T. H. Fleming. 2002. Historical vicariance and postglacial colonization effects on the evolution of genetic structure in *Lophocereus*, a Sonoran Desert columnar cactus. Evolution 56: 2214–2226.

Nei, M. 1987. Molecular Evolutionary Genetics. Columbia University Press, New York.

Nelson, E. W. 1921. Lower California and its natural resources. Memoirs of the National Academy of Science 16: 1–194.

Norris, R. M., and R. W. Webb. 1976. Geology of California. John Wiley and Sons, New York.

Orr, R. T. 1960. An analysis of the Recent land mammals. Symposium: the biogeography of Baja California and adjacent seas. Systematic Zoology 9: 171–179.

Ortega-Gutiérrez, F., and J. C. Guerrero-Garcia.

1982. The geologic regions of Mexico. Pp. 99–104 *in* A. R. Palmer (ed.), Perspectives in Regional Geological Synthesis. D-NAG Special Publication 1, Geological Society of America, Boulder, Colorado.

Patterson, B. D., and J. H. Brown. 1991. Regionally nested patterns of species composition in granivorous rodent assemblages. Journal of Biogeography 18:395–402.

Petersen, M. K. 1976. The Rio Nazas as a factor in mammalian distribution in Durango, Mexico. Southwestern Naturalist 20: 495–502.

Peterson, A. T., M. A. Ortega-Huerta, J. Bartley, V. Sánchez-Cordero, J. Soberón, R. H. Buddemeier, and D. R. B. Stockwell. 2002. Future projections for Mexican faunas under global climate change scenarios. Nature 416: 626–629.

Pielou, E. C. 1991. After the Ice Age: The Return of Life to Glaciated North America. The University of Chicago Press, Chicago.

Platnick, N. I., and G. Nelson. 1978. A method of analysis for historical biogeography. Systematic Zoology 27: 1–16.

Posada, D., K. A. Crandall, and A. R. Templeton. 2000. GeoDis: a program for the Cladistic Nested Analysis of the geographical distribution of genetic haplotypes. Molecular Ecology 9: 487–488.

Present, T. M. C. 1987. Genetic differentiation of disjunct Gulf of California and Pacific outer coast populations of *Hypsoblennius jenkinsi*. Copeia 1987: 1010–1024.

Riddle, B. R. 1995. Molecular biogeography in the pocket mice (*Perognathus* and *Chaetodipus*) and grasshopper mice (*Onychomys*): the late Cenozoic development of a North America aridlands rodent guild. Journal of Mammalogy 76: 283–301.

Riddle, B. R., and D. J. Hafner. 1999. Species as units of analysis in ecology and biogeography: time to take the blinders off. Global Ecology and Biogeography 8: 433–441.

Riddle, B. R., D. J. Hafner, and L. F. Alexander. 2000a. Phylogeography and systematics of the *Peromyscus eremicus* species group and the historical biogeography of North American warm regional deserts. Molecular Phylogenetics and Evolution 17: 145–160.

Riddle, B. R., D. J. Hafner, and L. F. Alexander. 2000b. Comparative phylogeography of Baileys' pocket mouse (*Chaetodipus baileyi*) and the *Peromyscus eremicus* species group: historical vicariance of the Baja California Peninsular Desert. Molecular Phylogenetics and Evolution 17: 161–172.

Riddle, B. R., D. J. Hafner, L. F. Alexander, and J. R. Jaeger. 2000c. Cryptic vicariance in the historical assembly of a Baja California peninsular desert biota. Proceedings of the National Academy of Sciences, U.S.A. 97:14438–14443.

Rogers, A. R., and H. Harpending. 1992. Population growth makes waves in the distribution of pairwise genetic differences. Molecular Biology and Evolution 9: 552–569.

Rzedowski, J. 1978. La Vegetación de México. Limusa-Wiley, México D.F.

Savage, J. M. 1960. Evolution of a Peninsular herpetofauna. Symposium: the biogeography of Baja California and adjacent seas. Systematic Zoology 9: 184–212.

Sawlan, M. G. 1986. Petrogenesis of late Cenozoic volcanic rocks from Baja California Sur, Mexico. Ph.D. dissertation, University of California, Santa Cruz.

Schmidly, D. J. 1977. Factors governing the distribution of mammals in the Chihuahuan Desert region. Pp. 163–192 *in* R. H. Wauer and D. H. Riskind (eds.), Transactions of the Symposium on the Biological Resources of the Chihuahuan Desert Region, United States and Mexico. National Park Service Transactions and Proceedings Series 3: 1–658.

Schmidly, D. J., K. T. Wilkins, and J. N. Derr. 1993. Biogeography. Pp. 319–356 *in* H. H. Genoways and J. H. Brown (eds.), Biology of the Heteromyidae. Special Publication, American Society of Mammalogists 10: 1–719.

Schmidt, R. H., Jr. 1989. The arid zones of Mexico: climatic extremes and conceptualization of the Sonoran Desert. Journal of Arid Environments 16: 241–256.

Serb J. M., C. A. Phillips, and J. B. Iverson. 2001. Molecular phylogeny and biogeography of *Kinosternon flavescens* based on complete mitochondrial control region sequences. Molecular Phylogenetics and Evolution 18: 149–162.

Shreve, F. 1942. The desert vegetation of North America. Botanical Review 8:195–246.

Shreve, F. 1951. Vegetation of the Sonoran Desert. Carnegie Institution of Washington Publication no. 591, Washington, D.C.

Simpson, G. G. 1943. Criteria for genera, species, and subspecies in zoology and paleontology. Annals of the New York Academy of Science 44: 145–178.

Smith, F. A., B. T. Bestelmeyer, J. Biardi, and M. Strong. 1993. Anthropogenic extinction of the endemic woodrat, *Neotoma bunkeri* Burt. Biodiversity Letters 1: 149–155.

Spencer, J. E., and W. R. Normark. 1989. Neogene plate-tectonic evolution of the Baja California Sur continental margin and the southern Gulf of California, Mexico. Pp. 489–498 *in* E. L. Winterer, D. M.

Hussong, and R. W. Decker, (eds.), The Eastern Pacific Ocean and Hawaii (The Geology of North America, vol. N). Geological Society of America, Boulder, Colorado.

Stock, J. M., and K. V. Hodges. 1989. Pre-Pliocene extension around the Gulf of California and the transfer of Baja California to the Pacific Plate. Tectonics 8: 99–115.

Swanson, E. R., and F. W. McDowell. 1984. Calderas of the Sierra Madre Occidental volcanic field, western Mexico. Journal of Geophysical Research 89: 8787–8799.

Templeton, A. R., E. Routman, and C. A. Phillips. 1995. Separating population structure from population history: a cladistic analysis of the geographical distribution of mitochondrial-DNA haplotypes in the tiger salamander, *Ambystoma tigrinum*. Genetics 140: 767–782.

Truxal, F. S. 1960. The entomofauna with special reference to its origin and affinities. Symposium: the biogeography of Baja California and adjacent seas. Systematic Zoology 9: 165–170.

Turner, R. M., and D. E. Brown. 1982. 154.1 Sonoran desertscrub. Desert Plants 4: 181–221.

Udvardy, M. D. F. 1969. Dynamic Zoogeography with Special Reference to Land Animals. Van Nostrand Reinhold, New York.

Upton, D. E., and R. W. Murphy. 1997. Phylogeny of the side-blotched lizards (Phrynosomatidae: *Uta*) based on mtDNA sequences: support for a midpeninsular seaway in Baja California. Molecular Phylogenetics and Evolution 8: 104–113.

Valdéz, M., and G. Ceballos. 1997. Conservation of endemic mammals of Mexico: the Perote ground squirrel (*Spermophilus perotensis*). Journal of Mammalogy 78: 74–82.

Van Devender, T. R. 1990a. Late Quaternary vegetation and climate of the Chihuahuan Desert, United States and Mexico. Pp. 104–133 *in* J. L. Betancourt, T. R. Van Devender, and P. S. Martin (eds.), Packrat Middens: Late Quaternary Environments of the Arid West. University of Arizona Press, Tucson.

Van Devender, T. R. 1990b. Late Quaternary vegetation and climate of the Sonoran Desert, United States and Mexico. Pp. 134–165 *in* J. L. Betancourt, T. R. Van Devender, and P. S. Martin (eds.), Packrat Middens: Late Quaternary Environments of the Arid West. University of Arizona Press, Tucson.

Van Devender, T. R., T. L. Burgess, J. C. Piper, and R. M. Turner. 1994. Paleoclimatic implications of Holocene plant remains from the Sierra Bacha, Sonora, Mexico. Quaternary Research 41: 99–108.

Walpole, D. K., S. K. Davis, and I. F. Greenbaum. 1997. Variation in mitochondrial DNA in populations of *Peromyscus eremicus* from the Chihuahuan and Sonoran deserts. Journal of Mammalogy 78: 397–404.

Webb, T., III, and P. J. Bartlein. 1992. Global changes during the last 3 million years: climatic controls and biotic responses. Annual Review of Ecology and Systematics 23: 141–173.

Williams, D. F., and K. S. Kilburn. 1992. The conservation status of the endemic mammals of the San Joaquin Faunal Region, California. Pp. 329–345 *in* D. F. Williams, S. Byrne, and T. A. Rado (eds.), Endangered and Sensitive Species of the San Joaquin Valley, California. California Energy Commission, Sacramento.

Williams, S. C. 1980. Scorpions of Baja California, Mexico, and adjacent islands. Occasional Papers of the California Academy of Sciences 135: 1–127.

Zink R. M., and R. C. Blackwell. 1998a. Molecular systematics of the scaled quail complex (genus *Callipepla*). Auk 115: 394–403.

Zink R. M., and R. C. Blackwell. 1998b. Molecular systematics and biogeography of aridland gnatcatchers (genus *Polioptila*) and evidence supporting species status of the California gnatcatcher (*Polioptila californica*). Molecular Phylogenetics and Evolution 9: 26–32.

Zink R. M., and R. C. Blackwell-Rago. 2000. Species limits and recent population history in the curve-billed thrasher. Condor 102: 881–886.

Zink, R. M., G. F. Barrowclough, J. L. Atwood, and R. C. Blackwell-Rago. 2000. Genetics, taxonomy, and conservation of the threatened California gnatcatcher. Conservation Biology 14: 1394–1405.

Zink, R. M., R. C. Blackwell, and O. Rojas-Soto. 1997. Species limits in the Le Conte's thrasher. Condor 99: 132–138.

Zink R. M., D. L. Dittmann, J. Klicka, and R. C. Blackwell-Rago. 1999. Evolutionary patterns of morphometrics, allozymes, and mitochondrial DNA in thrashers (genus *Toxostoma*). Auk 116: 1021–1038.

Zink R. M., S. J. Weller, and R. C. Blackwell. 1998. Molecular phylogenetics of the avian genus *Pipilo* and a biogeographic argument for taxonomic uncertainty. Molecular Phylogenetics and Evolution 10: 191–201.

12

Extreme Succulent Plant Diversity on Cerro Colorado Near San Ignacio, Baja California Sur

MARK A. DIMMITT

JOHN F. WIENS

THOMAS R. VAN DEVENDER

The Baja California peninsula is rich in plant species, many of which are endemics. The first flora (Wiggins 1980) listed nearly 3000 taxa. Subsequent work has increased the estimate to 3676 taxa (R. Mitchel Beauchamp, compilation in progress), of which nearly one-third are endemic (Rebman 2001). The large flora obviously results from the extreme variability of habitats and climates, which in turn creates many ecological niches. Diversity is further augmented by proximity to the tropics. The peninsula spans 1270 km from the Mediterranean climate zone at 32.5° N southward through the horse latitudes and into the tropics at 23° N. The northern two-thirds receive winter rainfall, with the amount and dependability decreasing southward. The entire peninsula except the extreme northwest receives summer rain in at least some years; both the amount and dependability are greatest in the Cape Region and decrease northward. An east–west temperature and moisture gradient further increases habitat diversity. The west coast climate is ameliorated by cool, moist Pacific winds year-round. The mountains that extend nearly the entire length of the peninsula block these winds, so the atmosphere on most of the Gulf side tends to be extremely hot and arid in summer. The Gulf coastal strand and the Cape Region are usually humid, and this despite low rainfall in the case of the Gulf coastal strand. Finally, the Peninsular Ranges are composed of several rock types and provide both climatic (elevation and rainshadow effects) and edaphic diversity.

Areas of locally high diversity are more challenging to explain, especially in the case of small sites that differ markedly in diversity from nearby topographically and geologically similar locations. Over the past few decades botanists and amateur plant enthusiasts have noted several locations that have unusually high numbers of succulent species in the Baja California peninsula and elsewhere in the Sonoran Desert region. This chapter describes a locality that we think has more succulent species per unit area than any other small area in southwestern North America.

Description of Site and Survey Method

Cerro Colorado is a small volcanic hill located along Mexico Highway 1, 4 km east of San Ignacio, Baja California Sur (BCS, figs. 12.1 and 12.2). The hill and surrounding *bajadas* cover about 10 km² and range in elevation from 160–490 m. Cerro Colorado is on the southern border of the Vizcaíno Biosphere Reserve, which extends from the Pacific coast to the Gulf coast and from Guerrero Negro south to Santa Rosalía.

Cerro Colorado's rich succulent flora was first reported by Joseph (1985). Some of us revisited the site in 1992 and 1998 with other colleagues. In 2001 and 2002 we conducted a more thorough survey based on the establishment of 26 plots on the hill

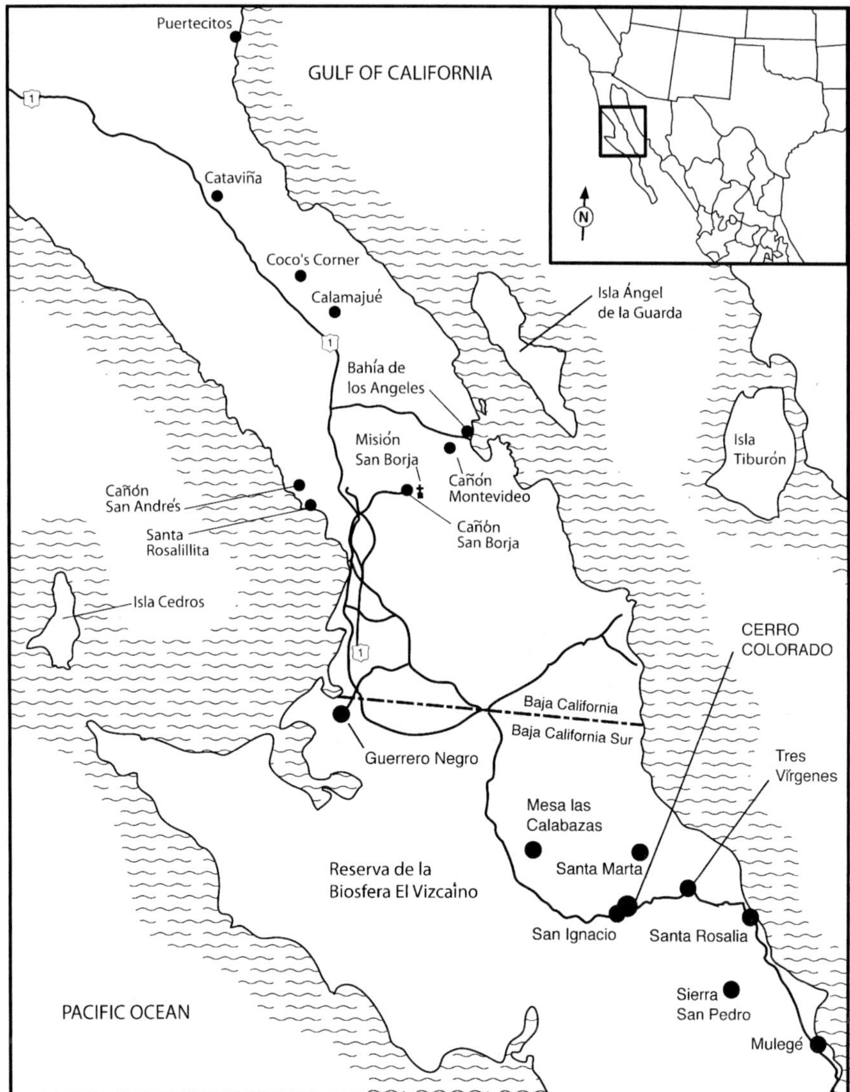

Figure 12.1. Central Baja California, with location of the study area.

and surrounding *bajadas* (of the same watershed) to sample the several habitats created by slope exposure, topography, elevation, and substrate. Most plots were approximately circular and covered about 1 ha. Those plots in canyons were elongated in order to stay within the same habitat type. A team of 5 people walked each plot until they determined with reasonable confidence that they had listed all of the identifiable plants (annuals were often in poor condition or absent). We estimated the frequency of each taxon on the list as abundant, common, uncommon or rare. This is a plotless *relevé* sampling method (Mueller-Dombois and Ellenberg 1974). We also recorded all species encountered while exploring the hill to characterize the habitats and select plot locations. Taxa of uncertain identity were vouchered and sent to U. S. and Mexican herbaria.

Definition of a Succulent

Classifying plants as succulent or nonsucculent was problematic for a number of taxa. Regional floras and popular books on succulents are all vague at

Figure 12.2. Cerro Colorado viewed from 4 km away across the oasis at San Ignacio, Baja California Sur. Deep rocky canyons are on the north (left) and east (far) sides of the hill. Mexico Highway 1 runs along the north base of the hill.

defining what makes a plant a succulent. For example, Rowley (1978) concluded only that many plants are difficult to categorize as to succulence. Popular publications on this topic often ignore clearly succulent plants such as many orchids and bromeliads simply because most succulent collectors do not grow them (e.g., Eggli 2001). Plant physiologists and systematists tend to be similarly noncommittal (D. Hunt, P. Nobel, R. Felger, and J. Weeks pers. comm.). Some authors use the term "semisucculent" for those plants with less obvious succulent characteristics, but this still leaves the separation between semisucculent and nonsucculent undefined.

A book by Von Willert et al. (1992) is the only source we found that attempted a concise description. These authors defined a succulent as any plant that possesses a succulent tissue, and further specified a succulent tissue as a tissue where water can be stored and made readily available for use by the rest of the plant, without any need for an external water supply in the short term. We use their definition here and restrict our discussion to their subcategory of xerophytic succulents, which excludes halophytic (salt-tolerant plants, often growing in saline wetlands) succulents and some geophytes (perennating organs below ground; e.g., potato, *Jatropha macrorhiza*, and most plants that are called "bulbs" in horticulture) with fleshy parts but which grow only when the soil is moist. The succulent tissues of halophytes and of most geophytes serve functions other than to support growth when soil moisture is unavailable. This definition of xerophytic succulent still leaves the status of a number of plants in question. We use the term "semisucculent" in tables 12.1 and 12.3 for those taxa that are not obviously succulent and thus subject to question. Elsewhere the term "succulent" includes semisucculents. We believe that all the plants listed as succulent fit the definition of xerophytic succulent in Von Willert et al. (1992).

Succulents may be classified by growth form or lifestyle. Growth-form classification schemes (e.g., trees, shrubs, vines) may include a succulent category consisting of plants with ample water storage tissues. Other schemes treat succulence as a subcategorical trait of other growth forms. For example, Felger (2000) classifies certain *Opuntia* species first as nanophanaerophytes (perennial shrubs with meristems 0.5 to 2.0 m above ground) and secondarily as succulents. The succulent lifestyle is a suite of characters that adapts a plant to xeric conditions. (Succulents also occur in some mesic habitats, such as saline wetlands. The function of succulence in these environments is poorly understood.) The primary lifestyle traits of xerophytic succulents are the abilities to (1) absorb and store water rapidly after light rains due to shallow root systems and often CAM idling metabolism (a benefit of crassulacean acid metabolism in which carbon dioxide and oxygen are recycled internally to support the complementary photosynthetic and respiratory cycles, enabling the plant to avoid becoming dormant when stomates are sealed to conserve internal water), and (2) use stored water efficiently and maintain significant levels of photosynthesis and other metabolic processes (even in species that lack CAM) when there is no available soil moisture. Although water storage requires a succulent tissue, this tissue may be thinly dispersed among other anatomical structures, so it is not always obvious. Thus, some plants lack phenotypic evidence of succulent tissue and would not be classified as having a succulent growth form but exhibit the succulent lifestyle.

Questionable succulent species in our study area are *candelilla* (giant cane milkweed, *Asclepias albicans*), *jumete* (desert milkweed, *A. subulata*), *torote prieto* (*Bursera hindsiana*), *calabacilla* (coyote gourd, *Cucurbita cordata*), *liga* (*Euphorbia xantii*), *tescalama* (rock fig, *Ficus palmeri*), *palo adán* (*Fouquieria diguetii*), *matacora* (*Jatropha cuneata*), *lomboy* (ashy limberbush, *J. cinerea*), and a mistletoe (*Psittacanthus sonorae*). We classified all except *Psittacanthus* as semisucculents (table 12.1).

Coyote gourd and many other cucurbits have large, tuberous roots that have considerable moisture as well as copious starch reserves. A sample of *Cucurbita foetidissima* root was 81% water (table 12.2). Coyote gourd and other cucurbits produce leafy shoots, flowers, and fruits well in advance of seasonal rains. This trait is itself insufficient to separate coyote gourd from clearly nonsucculent plants such as manzanita (*Arctostaphylos* spp.), which

sprout from woody crowns soon after dry-season fires. Metabolizing stored starch in manzanita crowns and roots generates enough water to support growth in the fall before the winter rains begin. Though tuberous-rooted cucurbits may also produce some of their water from starch breakdown, the high free-water content, along with their ability to produce growth even after a year without rain (M. Dimmitt, pers. obs.), leads us to classify them as xerophytic root semisucculents.

The growth form of rock fig (*Ficus palmeri*) is intermediate between a woody tree and a stem succulent. The fact that it can become established as a saxicole on exposed rock faces in the desert indicates that it has adaptations that typical woody trees do not have. The caudex of a young rock fig contained 68% water, more than stems of nonsucculent trees such as foothill palo verde (*Parkinsonia microphylla*, 53% water, table 12.2). Rock fig does not appear to have CAM. We tentatively classify it as a semisucculent based on its marginally elevated water content and lifestyle. (A closely related species, *F. petiolaris*, occurs in tropical deciduous forest, a community composed of so many similarly semisucculent tree species that the forest cannot support a fire.)

Most species of *Fouquieria* exhibit a woody shrub growth form, but they have a clearly succulent lifestyle: very shallow roots and the capacity to produce functioning leaves within 2 days after a light rainfall (ca. 7 mm; Dimmitt, unpublished data). The thin subcutaneous layer of moist tissue in these plants is succulent in nature (Henrickson, 1969a,b, 1972, pers. comm.). The rapid leaf production indicates the presence of an undescribed non-CAM idling metabolism (Dimmitt 2000).

Though the mistletoe (*Psittacanthus sonorae*) is slightly fleshy, it is probably not adapted to drought; it has a dependable supply of moisture as long as its host is alive.

Results and Discussion

In Joseph's (1985) initial survey of the northern slope and one of its flanking canyons, he found 26 taxa of succulent plants. Our surveys of the entire hill increased the number to 44 (table 12.1). The total vascular plant flora includes 151 perennial taxa (table 12.3). This is a fairly rich flora for a desert, but it is especially surprising that succulents make up 26% of the perennials. Both intensive

Table 12.1. Succulent and semisucculent taxa of Cerro Colorado.

Family/Species	Annual (A) or Perennial (P)	Succulence[a]	Storage Organs	Growth Season[b]	Distribution[c]	Distribution Status[d]
Agavaceae						
Agave cerulata ssp. *subcerulata*	P	XH	Leaves	W	C	N
Yucca valida	P	XH	Leaves	I	S	E
Aizoaceae						
Mesembryanthemum crystallinum	A	XS	Whole plant	W		W
Mesembryanthemum nodiflorum	A	XS	Whole plant	W		W
Trianthema portulacastrum	A	XH	Leaves	S	W	W
Asclepiadaceae						
Asclepias albicans	P	XH	Stems	I	N	W
Asclepias subulata	P	XH	Stems	I	C	W
Boraginaceae						
Heliotropium curassavicum	A/P	XH	Whole plant	I	W	W
Burseraceae						
Bursera hindsiana	P	XH	Stems	S	C	W
Bursera microphylla	P	XH	Stems	S	C	W
Cactaceae						
Cochemiea poselgeri	P	XS	Stems	S	S	E
Cochemiea setispina	P	XS	Stems	S	C	N
Cylindropuntia alcahes var. *alcahes*	P	XS	Stems	I	C	E
Cylindropuntia alcahes × *C. cholla*	P	XS	Stems	I		E
Cylindropuntia alcahes × *C. lindsayi*	P	XS	Stems	I		E
Cylindropuntia alcahes × *C. molesta*	P	XS	Stems	I		E
Cylindropuntia calmalliana	P	XS	Stems	I	C	N
Cylindropuntia cholla	P	XS	Stems, fruits	I	C	E
Cylindropuntia cholla × *C. lindsayi*	P	XS	Stems	I		E
Cylindropuntia molesta var. *molesta*	P	XS	Stems	I	C	E
Echinocereus brandegeei	P	XS	Stems	S	S	E
Ferocactus peninsulae var. *peninsulae*	P	XS	Stems	S	C	E
Ferocactus peninsulae × *F. rectispinus*	P	XS	Stems	S	C	E
Ferocactus rectispinus	P	XS	Stems	S	S	N
Grusonia invicta	P	XS	Stems	I	C	E
Lophocereus schottii	P	XS	Stems	S	S	W
Mammillaria dioica	P	XS	Stems	I	N	W
Mammillaria lewisiana	P	XS	Stems	?	C	E

Myrtillocactus cochal	P	XS	Stems	I	C	E
Opuntia cf. engelmannii	P	XS	Stems	I		W
Opuntia tapona	P	XS	Stems	I	S	E
Pachycereus pringlei	P	XS	Stems	S	C	W
Stenocereus gummosus	P	XS	Stems	S	C	E*
Stenocereus thurberi	P	XS	Stems	S	C	W
Crassulaceae						
Dudleya albiflora	P	XS	Leaves	W	S	N
Cucurbitaceae						
Cucurbita cordata	P	XH	Root	S	C	N
Ibervillea sonorae	P	XH	Caudex	S	S	W
Euphorbiaceae						
Euphorbia misera	P	XH	Stems	I	N	W
Euphorbia xantii	P	XH	Stems	I	S	E*
Jatropha cinerea	P	XH	Stems	S	C	W
Jatropha cuneata	P	XH	Stems	S	C	W
Pedilanthus macrocarpus	P	XS	Stems	S	C	W
Fouquieriaceae						
Fouquieria diguetii	P	XH	Stems	S	S	E*
Moraceae						
Ficus palmeri	P	XH	Caudex	S	S	W

* = non-native.

[a]XH = xerophytic semisucculent; XS = xerophytic succulent.

[b]Season of growth: W = winter; S = summer; I = indeterminate growth.

[c]Distribution relative to Cerro Colorado: C = Cerro Colorado site is near center of geographic range; S = most of distribution is to south of site; N = most of distribution is to north.

[d]Distribution status: N = narrow endemic; E = endemic to the Baja California peninsula; E* = endemic except for small populations on mainland of Mexico; W = widespread.

Table 12.2. Water content of cultivated, fully hydrated plants, expressed as percentage of fresh weight.

Species (Family)	Part	% Water	Source
Adelia virgata (Euphorbiaceae, nonsucculent)	Stem, 2 cm diameter	33	Dimmitt and Wiens (unpubl. data)
Agave aurea (Agavaceae)	Leaf	90	Dimmitt and Wiens (unpubl. data)
Aloe ramosissima (Aloaceae)	Leaf	82	Von Willert et al. (1992)
Asclepias subulata (Asclepiadaceae)	Stem	55	Dimmitt and Wiens (unpubl. data)
Bursera hindsiana (Burseraceae)	Caudex with bark	58	Dimmitt and Wiens (unpubl. data)
Bursera microphylla (Burseraceae)	Stem, 3 cm diameter	69	Dimmitt and Wiens (unpubl. data)
Bursera microphylla (Burseraceae)	Caudex with bark	83	Dimmitt and Wiens (unpubl. data)
Ceraria fruticulosa (Portulacaceae)	Leaf	80	Von Willert et al. (1992)
Ceraria fruticulosa (Portulacaceae)	Root	80	Von Willert et al. (1992)
Cheiridopsis robusta (Aizoaceae)	Leaf	84	Von Willert et al. (1992)
Cissus trifoliata (Vitaceae)	Leaf	92	Dimmitt and Wiens (unpubl. data)
Conophytum minutum (Aizoaceae)	Leaf	86	Von Willert et al. (1992)
Cotyledon orbiculata (Crassulaceae)	Leaf	84	Von Willert et al. (1992)
Cucurbita foetidissima (Cucurbitaceae)	Tuberous root	81	Dimmitt and Wiens (unpubl. data)
Eulophia petersii (Orchidaceae)	Leaf	92	Dimmitt and Wiens (unpubl. data)
Eulophia petersii (Orchidaceae)	Pseudobulb	94	Dimmitt and Wiens (unpubl. data)
Euphorbia xantii (Euphorbiaceae)	Stem, 0.5 cm diameter	76	Dimmitt and Wiens (unpubl. data)
Euphorbia xantii (Euphorbiaceae)	Stem, 2 cm thick	74	Dimmitt and Wiens (unpubl. data)
Ficus palmeri (Moraceae)	Caudex with bark	64	Dimmitt and Wiens (unpubl. data)
Hesperoyucca whipplei (Agavaceae)	Leaf	92	Dimmitt and Wiens (unpubl. data)
Ibervillea sonorae (Cucurbitaceae)	Caudex	85	Dimmitt and Wiens (unpubl. data)
Ipomoea arborescens (Convolvulaceae)	Caudex with bark	69	Dimmitt and Wiens (unpubl. data)
Ipomoea longifolia (Convolvulaceae)	Root, 5 cm diameter	88	Dimmitt and Wiens (unpubl. data)
Jatropha cuneata (Euphorbiaceae)	Stem, 2 cm diameter	69	Dimmitt and Wiens (unpubl. data)
Jatropha malacophylla (Euphorbiaceae)	Stem, 5 cm diameter	65	Dimmitt and Wiens (unpubl. data)
Pachypodium namaquanum (Asclepiadaceae)	Stem	87	Von Willert et al. (1992)
Parkinsonia microphylla (Fabaceae, nonsucculent)	Stem, 0.5 cm diameter	53	Dimmitt and Wiens (unpubl. data)
Pedilanthus macrocarpus (Euphorbiaceae)	Stem	84	Dimmitt and Wiens (unpubl. data)
Rhyncholaelia digbyana (Orchidaceae)	Leaf	90	Dimmitt and Wiens (unpubl. data)
Rhyncholaelia digbyana (Orchidaceae)	Pseudobulb	81	Dimmitt and Wiens (unpubl. data)
Senecio corymbiferus (Asteraceae)	Leaf	92	Von Willert et al. (1992)
Senecio corymbiferus (Asteraceae)	Stem	82	Von Willert et al. (1992)
Yucca baccata (Agavaceae)	Leaf	65	Dimmitt and Wiens (unpubl. data)
Yucca elata (Agavaceae)	Leaf	59	Dimmitt and Wiens (unpubl. data)

surveys were conducted in dry years, with the result that few annuals were found. Therefore our comparisons are based on perennial floras.

There are 6 general substrate types on Cerro Colorado (R. Scarborough, pers. comm.). A few species' distributions may be correlated with substrate, but not at the scale of our survey. For example, stands of cardón (*Pachycereus pringlei*) appeared visually to be denser on clayey soils, but they were uncommon on two of our four clay survey plots. *Lomboy* (*Jatropha cinerea*) is usually found on silty soils, but occasional stands occur on rocky slopes. The canyons had more succulents (and total species) than any other habitat type, but this is almost certainly due more to the multiple microhabitats on opposing canyon walls plus the concentration of runoff in the drainage bottoms than to substrate per se. Substrate presumably has some influence on plant occurrence, but it does not appear to be the major determining factor of the richness of these sites.

The biseasonal precipitation doubtlessly enhances diversity. In many areas of the central peninsula, 2 convergent elephant trees, copalquín (*Pachycor-*

Table 12.3. Flora of Cerro Colorado, Baja California Sur.

Family/Species	Annual (A) Perennial (P) or Indeterminate (X)	Succulent[a]
Acanthaceae		
Carlowrightia arizonica A. Gray	P	
Holographis virgata (Benth. & Hook.) T.F. Daniel ssp. *glandulifera* (Leonard & Morton) T.F. Daniel	P	
Justicia californica (Benth.) D.N. Gibson	P	
Justicia insolita Brandegee var. *insolita*	P	
Ruellia californica (Rose) I.M. Johnst.	P	
Agavaceae		
Agave cerulata Trel. ssp. *subcerulata* Gentry	P	XH
Yucca valida Brandegee	P	XH
Aizoaceae		
**Mesembryanthemum crystallinum* L.	A	XS
**Mesembryanthemum nodiflorum* L.	A	XS
Trianthema portulacastrum L.	A	XH
Amaranthaceae		
Amaranthus watsonii Standl.	A	
Iresine calea (Ibañez) Standl.	P	
Apocynaceae		
Vallesia glabra (Cav.) Link	P	
Asclepiadaceae		
Asclepias albicans S. Watson	P	XH
Asclepias subulata Decne.	P	XH
Matelea cordifolia (A. Gray) Woodson	P	
Matelea parvifolia (Torr.) Woodson	P	
Metastelma californicum Benth.	P	
Asteraceae		
Amauria brandegeana (Rose) Rydb.	P	
Amauria rotundifolia Benth.	P	
Ambrosia bryantii (Curran) W.W. Payne	P	
Ambrosia camphorata (Greene) W.W. Payne	P	
Ambrosia divaricata (Brandegee) W.W. Payne	P	
Ambrosia magdalenae (Brandegee) W.W. Payne	P	
Bebbia juncea (Benth.) Greene var. *juncea*	P	
Brickellia coulteri A. Gray var. *coulteri*	P	
**Centaurea melitensis* L.	A	
Coreocarpus dissectus (Benth.) S.F. Blake	P	
Encelia farinosa A. Gray var. farinosa	P	
Filago californica Nutt.	A	
Hymenoclea salsola A. Gray	P	
**Lactuca serriola* L.	A	
Malperia tenuis S. Watson	A	
Nicolletia trifida Rydb.	A	
Parthenice mollis A. Gray var. *peninsularis* Sauck	A	
Perityle californica Benth.	A	
Perityle emoryi Torr.	A	
Pleurocoronis laphamioides (Rose) R.M. King & H. Rob.	P	
Porophyllum gracile Benth.	P	
Trichoptilium incisum A. Gray	A	
Trixis californica Kellogg var. *californica*	P	
Verbesina encelioides (Cav.) Benth. & Hook.	A	
Viguiera deltoidea A. Gray s.l.	P	
Viguiera purissimae Brandegee	P	
Boraginaceae		
Cordia parvifolia A. DC.	P	
Cryptantha angelica I.M. Johnst.	A	

(*continued*)

Family/Species	Annual (A) Perennial (P) or Indeterminate (X)	Succulent[a]
Cryptantha angustifolia (Torr.) Greene	A	
Cryptantha grayi (Vasey & Rose) J.F. Macbr.	A	
Cryptantha maritima (Greene) Greene var. *pilosa* I.M. Johnst.	A	
Cryptantha sp.	A	
Heliotropium curassavicum L.	A/P	XH
Heliotropium procumbens Mill.	P	
Pectocarya recurvata I.M. Johnst.	A	
Brassicaceae		
Brassica tournefortii Gouan	A	
Descurainia pinnata (Walter) Britt.	A	
Draba cuneifolia Nutt.	A	
Lepidium lasiocarpum Torr. & A. Gray	A	
Lobularia maritima (L.) Desv.	A	
Lyrocarpa coulteri Harv.	P	
Sibara angelorum (S. Watson) Greene	A	
Sisymbrium irio L.	A	
Burseraceae		
Bursera hindsiana (Benth.) Engler	P	XH
Bursera microphylla A. Gray	P	XH
Cactaceae		
Cochemiea poselgeri (Hildmann) Britt. & Rose	P	XS
Cochemiea setispina (J.M. Coult.) Walton	P	XS
Cylindropuntia alcahes (F.A.C. Weber) F.M. Knuth var. *alcahes*	P	XS
Cylindropuntia alcahes × *C. cholla*	P	XS
Cylindropuntia alcahes × *C. lindsayi*	P	XS
Cylindropuntia alcahes × *C. molesta*	P	XS
Cylindropuntia calmalliana (J.M. Coult.) F.M. Knuth	P	XS
Cylindropuntia cholla (F.A.C. Weber) F.M. Knuth	P	XS
Cylindropuntia cholla × *C. lindsayi*	P	XS
Cylindropuntia molesta (Brandegee) F.M. Knuth var. *molesta*	P	XS
Echinocereus brandegeei (J.M. Coult.) Schumann	P	XS
Ferocactus peninsulae (F.A.C. Weber) Britt. & Rose var. *peninsulae*	P	XS
Ferocactus peninsulae × *F. rectispinus*	P	XS
Ferocactus rectispinus (Engelm.) Britt. & Rose	P	XS
Grusonia invicta (Brandegee) E.F. Anderson	P	XS
Lophocereus schottii (Engelm.) Britt. & Rose	P	XS
Mammillaria dioica Brandegee	P	XS
Mammillaria lewisiana H.E. Gates	P	XS
Myrtillocactus cochal (Orcutt) Britt. & Rose	P	XS
Opuntia cf. *engelmannii* Salm-Dyck	P	XS
Opuntia tapona Engelm.	P	XS
Pachycereus pringlei (S. Watson) Britt. & Rose	P	XS
Stenocereus gummosus (Engelm.) A. Gibson & Horak	P	XS
Stenocereus thurberi (Engelm.) A. Gibson & Horak	P	XS
Capparaceae		
Capparis atamisquea Kuntze	P	
Caryophyllaceae		
Drymaria holosteoides Benth.	A	
Celastraceae		
Schaefferia cuneifolia A. Gray	P	
Chenopodiaceae		
Atriplex barclayana (Benth.) D. Dietr.	P	
Atriplex elegans (Moq.) D. Dietr.	A	
Atriplex linearis S. Watson	P	
Atriplex pacifica A. Nels.	A	
Atriplex polycarpa (Torr.) S. Watson	P	

Table 12.3. Continued

Family/Species	Annual (A) Perennial (P) or Indeterminate (X)	Succulent[a]
Chenopodium murale L.	A	
Salsola tragus L.	A	
Convolvulaceae		
Evolvulus alsinoides L. var. *angustifolia* Torr.	P	
Jacquemontia abutiloides Benth.	P	
Crassulaceae		
Dudleya albiflora Rose	P	XS
Cucurbitaceae		
Cucurbita cordata S. Watson	P	XH
Ibervillea sonorae (S. Watson) Greene	P	XH
Ephedraceae		
Ephedra aspera S. Watson	P	
Euphorbiaceae		
Adelia virgata Brandegee	P	
Andrachne microphylla (Lam.) Baill.	A	
Bernardia viridis Millsp.	P	
Croton magdalenae Millsp.	P	
Ditaxis lanceolata (Benth.) Pax & K. Hoffm.	P	
Ditaxis neomexicana (Müll. Arg.) Heller	X	
Euphorbia arizonica Engelm.	X	
Euphorbia misera Benth.	P	XH
Euphorbia pediculifera Engelm. var. *pediculifera*	X	
Euphorbia tomentulosa S. Watson	P	
Euphorbia xantii Engelm.	P	XH
Jatropha cinerea (C.G. Ortega) Müll. Arg.	P	XH
Jatropha cuneata Wiggins & Roll.	P	XH
Pedilanthus macrocarpus Benth.	P	XS
Ricinus communis L.	P	
Stillingia linearifolia S. Watson	X	
Tragia jonesii Radcl.-Sm. & R. Govaerts	P	
Fabaceae		
Acacia farnesiana (L.) Willd.	P	
Acacia greggii A. Gray	P	
Acacia peninsularis (Britt. & Rose) Standl.	P	
Aeschynomene nivea Brandegee	P	
Calliandra californica Benth.	P	
Dalea bicolor Willd. var. *orcuttiana* Barneby	P	
Dalea mollis Benth.	X	
Desmanthus fruticosus Rose	P	
Ebanopsis confinis (Standl.) Britt. & Rose	P	
Lupinus arizonicus (S. Watson) S. Watson	A	
Lysiloma candidum Brandegee	P	
Macroptilium atropurpureum (A. DC.) Urban	P	
Marina parryi (Torr. & A. Gray) Barneby	X	
Olneya tesota A. Gray	P	
Parkinsonia aculeata L.	P	
Parkinsonia microphylla Torr.	P	
Parkinsonia praecox (Ruiz & Pav.) J. Hawkins	P	
Parkinsonia × *sonorae* (Rose) I.M. Johnst.	P	
Phaseolus filiformis Benth.	A	
Prosopis glandulosa Torr. var. *torreyana* (L.D. Benson) M.C. Johnst.	P	
Psorothamnus emoryi Rydb.	P	
Senna confinis (Greene) H.S. Irwin & Barneby	P	
Fouquieriaceae		
Fouquieria diguetii (Tiegh.) I.M. Johnst.	P	XH

(*continued*)

255

Table 12.3. Continued

Family/Species	Annual (A) Perennial (P) or Indeterminate (X)	Succulent[a]
Geraniaceae		
Erodium cicutarium (L.) Aiton	A	
Hydrophyllaceae		
Phacelia pauciflora S. Watson	A	
Phacelia scariosa Brandegee	A	
Phacelia sp.	A	
Krameriaceae		
Krameria erecta Shultes	P	
Krameria grayi Rose & Painter	P	
Lamiaceae		
Hyptis albida Kunth.	P	
Loasaceae		
Eucnide cordata (Curran) Kellogg	P	
Mentzelia adhaerens Benth.	A	
Petalonyx linearis Greene	P	
Loranthaceae		
Psittacanthus sonorae (S. Watson) Kuijt	P	
Malpighiaceae		
Callaeum macropterum (DC.) D.M. Johnson	P	
Janusia californica Benth.	P	
Malvaceae		
Abutilon incanum (Link) Sweet	P	
Abutilon palmeri A. Gray	P	
Herissantia crispa (L.) Brizicky	X	
Hibiscus biseptus S. Watson	P	
Hibiscus denudatus Benth.	P	
Horsfordia alata (S. Watson) A. Gray	P	
Horsfordia newberryi (S. Watson) A. Gray	P	
Horsfordia rotundifolia S. Watson	P	
Malva parviflora L.	A	
Sida alamosana S. Watson	P	
Sida xantii A. Gray	A	
Sphaeralcea axillaris S. Watson var. *axillaris*	P	
Sphaeralcea coulteri (S. Watson) A. Gray	A	
Sphaeralcea emoryi Torr.	X	
Moraceae		
Ficus palmeri S. Watson	P	XH
Nyctaginaceae		
Allionia incarnata L.	P	
Boerhavia coccinea Mill.	A	
Boerhavia erecta L.	A	
Mirabilis laevis (Benth.) Curran var. *crassifolia* (Choisy) Spellenberg	P	
Oleaceae		
Forestiera macrocarpa Brandegee	P	
Papaveraceae		
Argemone gracilenta Greene	A	
Passifloraceae		
Passiflora foetida L. var. *gossypifolia* (Ham.) Mast.	P	
Phytolaccaceae		
Stegnosperma halimifolium Benth.	P	
Plantaginaceae		
Plantago ovata Forssk.	A	
Poaceae		
Aristida adscensionis L.	A	
Aristida californica Thurb. var. *californica*	P	
Bouteloua barbata Lag.	A	

Table 12.3. Continued

Family/Species	Annual (A) Perennial (P) or Indeterminate (X)	Succulent[a]
Cenchrus palmeri Vasey	A	
Chloris chloridea (J. Presl) Hitchc.	P	
**Cynodon dactylon* (L.) Pers.	P	
Digitaria californica (Benth.) Henr.	P	
Enneapogon desvauxii P. Beauv.	P	
Eragrostis pectinacea (Michx.) Nees var. *pectinacea*	A	
Heteropogon contortus (L.) P. Beauv.	P	
Leptochloa fascicularis (Lam.) A. Gray	A	
Leptochloa panicea (Retz.) Ohwi ssp. *brachiata* (Steud.) N. Snow	A	
Muhlenbergia microsperma (A. DC.) Trin.	A	
**Pennisetum ciliare* (L.) Link	P	
Setaria macrostachya H.B.K.	P	
Pteridaceae		
Astrolepis sinuata (Sw.) Benham & Windham ssp. *sinuata*	P	
Notholaena californica D.C. Eaton	P	
Resedaceae		
Oligomeris linifolia (Vahl) J.F. Macbr.	A	
Rhamnaceae		
Colubrina viridis M.E. Jones	P	
Condalia globosa I.M. Johnst. var. *pubescens* I.M. Johnst.	P	
Sapindaceae		
Cardiospermum corindum L.	P	
Scrophulariaceae		
Antirrhinum cyathiferum Benth.	A	
Antirrhinum watsonii Vasey & Rose	A	
Simmondsiaceae		
Simmondsia chinensis (Link) Schneid.	P	
Solanaceae		
Datura discolor Bernh.	A	
Lycium californicum A. Gray	P	
Lycium fremontii A. Gray	P	
Lycium megacarpum Wiggins	P	
Lycium sp. #1	P	
Lycium sp. #2	P	
Nicotiana clevelandii A. Gray	A	
Physalis crassifolia Benth.	X	
Solanum hindsianum Benth.	P	
Sterculiaceae		
Ayenia filiformis S. Watson	P	
Melochia tomentosa L. var. *frutescens* A. DC.	P	
Ulmaceae		
Celtis pallida Torr. ssp. *pallida*	P	
Urticaceae		
Parietaria hespera Hinton	A	
Viscaceae		
Phoradendron californicum Nutt.	P	
Zygophyllaceae		
Fagonia californica Benth. ssp. *californica*	P	
Larrea divaricata Cav. ssp. *tridentata* (A. DC.) Felger & Lowe	P	
Viscainoa geniculata (Kellogg) Greene var. *geniculata*	P	

* = non-native.

[a]XS = xerophytic succulent; XH = xerophytic semisucculent.

mus discolor, Anacardiaceae) and *torote blanco* (*Bursera microphylla*, Burseraceae) grow side by side. The first species produces new leaves and stems only during the cool season; the second responds only to summer rains. The sympatric congeners boojum (*Fouquieria columnaris*) and *palo adán* (*F. diguetii*) exhibit the same seasonal dichotomy. Boojum bears leaves mainly in the cooler months and grows new stems only in early spring after wet winters. *Palo adán* will produce leaves after any rain that penetrates the soil more than about a centimeter, but long shoots grow only after hot-season rains. Four of the succulent taxa on Cerro Colorado are definitely attuned to the winter rains, and 20 respond primarily or exclusively to the summer rains (table 12.1). Most or all other succulents are indeterminate, responding opportunistically to rainfall at any time.

The Pacific side of the northern and central peninsula experiences frequent fog during the cooler months. The high humidity reduces evapotranspiration and thus considerably alleviates water stress in this region of low rainfall. But the fog banks are typically widespread, so it is difficult to cite fog as a major factor in explaining the much greater diversity on Cerro Colorado compared with nearby localities. Analysis of satellite images might reveal significant variations in local fog frequency and duration in this region.

The succulents of Cerro Colorado have diverse geographic distributions. Cerro Colorado is near the center of the latitudinal distribution for nearly half the succulent taxa (21). Three are northern taxa that are at or near their southern limits. These northern taxa respond opportunistically to rainfall. Another 11 taxa are near their northern limits. Fifty-seven percent of the succulents on Cerro Colorado (25 taxa) are endemic to the Baja California peninsula or have only small isolated populations on mainland Mexico. Six are narrow endemics.

Some of our plant records on Cerro Colorado are range extensions. *Cochemiea setispina* is significantly south and inland of its main distribution in the vicinity of Bahía de Los Angeles and the Sierra San Francisco. In contrast, *C. poselgeri* is 100 km north of its previously published northern limit near Mulegé. Cerro Colorado is probably the only locality where the 2 species are nearly sympatric (*C. setispina* occurs in deep arroyos on the north slope and *C. poselgeri* is restricted to the south slope and adjacent floodplain of Arroyo San Ignacio). *Opuntia tapona* is 160 km north of its published

distribution. *Ferocactus rectispinus* is at its northern limit. Most populations of non-introgressed individuals are restricted to within a few kilometers of the Gulf coast in the vicinity of Bahía Concepción; more inland plants are typically introgressed with *F. peninsulae*. Cerro Colorado supports an exceptionally dense hybrid swarm as well as a few apparently non-introgressed individuals of both parent species. Several nonsucculent taxa found on Cerro Colorado are also range extensions.

Four species were found only in the disturbed area along the highway (2 *Mesembryanthemum* spp., *Trianthema portulacastrum*, and *Heliotropium curassavicum*). All are at least facultative annuals; both *Mesembryanthemum* species are non-native. Several nonsucculent plant species were also found only along the road. Human activities thus almost certainly contribute to the diversity of the site.

Two species that are abundant in the midsection of the peninsula are conspicuously absent from the Cerro Colorado flora (table 12.4): boojum (*Fouquieria columnaris*) and *copalquín* (Baja elephant tree, *Pachycormus*). Boojum is near the southern edge of its range near San Ignacio. Unlike other *Fouquieria* species in Baja California, it is very succulent. It occurs 30 km north and east of Cerro Colorado at higher elevations (e.g., above 700 m on Volcán Tres Vírgenes). Perhaps Cerro Colorado is not high enough to catch sufficient winter rain or fog at this southerly latitude. This species grows almost to the Gulf coast at Bahía de Los Angeles (lat. 28.9°), where winter rainfall is nearly the same as at San Ignacio—60 and 55 mm, respectively (Hastings 1964; Warren 1979). However, the weather station at Bahía de Los Angeles is on the coast, where vegetation density is much lower than just a few kilometers inland, where the boojums grow. The vegetation and flora indicate that there is a steep rainfall gradient increasing inland from the Gulf coast.

Baja elephant tree does not seem to favor a specific soil type or texture. It is found on coarse basalt flows, fine silt, and other substrates. In the northern half of its range it is common from the Pacific coast almost to the Gulf coast, but in the southern half it grows mostly near the 2 coasts and is rare inland (Turner et al. 1995). San Ignacio is in the middle of the peninsula in the southern part of its habitat. However, it is common inland in the Sierra San Francisco only 20 km north of San Ignacio (J. Rebman, pers. comm.). This tree also occurs in the same elevational range as Cerro Colorado in the lava flows of Volcán Tres Vírgenes and on the slopes of Cerro Caguama adjacent to the lava. That population is 25 km southeast of Cerro Colorado and much more distant from its required winter rains and Pacific fogs. Why it is so patchily distributed in this region is unclear.

Additional succulents may be found on Cerro Colorado. Numerous species occur nearby, some of which are cryptic and may have been overlooked, or are so rare that we did not encounter one on the parts of the hill we surveyed (table 12.4).

Comparison with Nearby Sites

Cerro Colorado has the largest known number of succulent taxa of any small area of the Sonoran Desert Region. In places the species density exceeds 10 succulent taxa in a single hectare (fig. 12.3). No other locality less than 10 km² surveyed by us or reported in the literature has more than 31 taxa (table 12.5). The Sierra del Viejo, a limestone and granite range southwest of Caborca, Sonora, supports 40 taxa (Wiens, unpubl. data). However, its 400-km² area is 40 times that of Cerro Colorado. The Vizcaíno Biosphere Reserve covers 25,000 km² and has 78 known taxa of succulents (León de la Luz 1999; plus additions by us). It is remarkable that more than half of the succulents in a large region would be found on a single small hill.

Several sites have been visited and surveyed less intensively than Cerro Colorado. All appear to have a significantly lower number of succulent species (table 12.5). Cerro Colorado de Hilacha (hereafter "Hilacha") is 5 km northwest of Cerro Colorado. It is geologically similar to Cerro Colorado but is almost 200 m taller (680 m). An aerial survey in 1999 revealed a much lower density of vegetation than that of Cerro Colorado. The vegetation would not impede climbing this hill, whereas much of Cerro Colorado's slopes are clothed with nearly impenetrable thickets of shrubs and cacti. Hilacha is surrounded by rocky flats with dense thickets of agaves and chollas, making it nearly inaccessible by foot or vehicle. We were able to survey only the lower east slope. A survey of this hill is a high priority; its flora would presumably yield important information on the nature of the variability of biodiversity in the central Baja California peninsula.

Cerro Caguama is a volcanic hill contiguous with Volcán Tres Vírgenes 30 km east of Cerro Colorado. Though it lacks a deep canyon habitat, it is higher than Cerro Colorado and supports 31 taxa of succulents, the second highest number we have found among small sites anywhere in the Sonoran Desert Region (table 12.5). Being contiguous with a much more massive mountain may enrich its flora. The highway runs along the lower slope of this hill, as it does on Cerro Colorado. The disturbance-adapted annual succulents may also occur on Caguama; it was too dry for them when we surveyed the site.

Comparisons with Other Desert Areas

It would be valuable to compare our succulent lists with those of other deserts, but we have not found any for areas of similar size (up to 50 km²). The Chihuahuan Desert Region is recognized as a major

Table 12.4. Succulents not found on Cerro Colorado that occur nearby (distances are from Cerro Colorado).

Taxon	Nearest Known Location
Agave gigantensis	45 km SSE, San Sebastian
Bursera fagaroides	35 km SE, Volcán Tres Vírgenes
Cylindropuntia bigelovii	52 km E, Santa Rosalia
Cylindropuntia lindsayi	32 km E, Cerro Caguama
Cylindropuntia prolifera	3 km W, San Ignacio
Dudleya rubens	6 km NW, San Ignacio
Fouquieria columnaris	35 km SE, Volcán Tres Vírgenes
Fouquieria splendens	37 km ESE, Mentajo
Hesperoyucca whipplei	16 km NW, San Ignacio
Nolina palmeri	35 km SE, Volcán Tres Vírgenes
Opuntia chlorotica	40 km ENE, Cerro Azufre
Pachycormus discolor	24 km E, Mezquital
Peniocereus striatus	32 km E, Cerro Caguama

Figure 12.3. Detail of the north slope of Cerro Colorado. Twelve taxa of succulents, more than one-quarter of the succulent flora of the hill, can be seen in this single frame.

center of cactus diversity. Hernández et al. (2001) found 41 species of cacti in an area of San Luis Potosí covering 114 km² and 32 species on a single 3-km long transect. They did not list the other succulents in the area, but the number of Cactaceae suggests that this region may have equal or greater small-scale succulent diversity than in the central Baja California peninsula. Felger (2000) reported 18 species of cacti growing together in a small area of the Sonoran Desert. However, the other succulents in this area would not double the total.

The deserts of southern Africa unquestionably have the greatest succulent species count in the world. A large number of species are very narrow endemics. But whether a large number of species can be found in a small area cannot be determined without site surveys.

Conservation Status

Cerro Colorado's inclusion within the Vizcaíno Biosphere Reserve does not necessarily afford it suffi-

cient protection from degradation; most human activities are still permitted. Disturbance due to the highway does not appear to be a problem; in fact, it seems to have contributed several species to the flora. Trash from an unofficial dump site is abundant and unsightly along the highway, but it is probably not an ecological threat. Livestock are evident throughout the area, and the lower north slope is heavily impacted by their browsing. The site should be studied to determine whether cattle are threatening any of the plants. However, it does not appear that grazing has had a significant impact on the biodiversity of the site.

Although biodiversity is currently a major focus of ecological research, there seem to be few studies of small-scale diversity. This poorly understood phenomenon deserves more investigation in Mexico and elsewhere. Cerro Colorado is close to San Ignacio and is easy to access from the highway. Its location makes it a good candidate for an interpretive site to attract tourists and publicize the biodiversity of the reserve.

Table 12.5. Comparison of xerophytic succulent floras in the Sonoran Desert Region.

Locality and Region (Reference)[a]	Latitude (°N)	Elevation (m)	Area (km^2)	Total No. of Taxa	No. of Succulent Taxa	No. of Perennial Taxa	Succulents as % of Total No. of Taxa	Succulents as % of No. of Perennials	Survey Intensity/ Thoroughness
Vizcaíno Biosphere Reserve, BCS Sonoran Desert (Vizcaíno) (1)	26.5–28.0	0–1900	25,468	496	78	325	16	24	Moderate
Cerro Colorado, BCS Sonoran Desert (Vizcaíno)	27.0	200–500	10	221	44	151		26	Thorough except annuals
Sierra del Viejo (Caborca), SON Sonoran Desert (Central Gulf Coast) (2)	30.5	ca. 400–1000	400		40				Moderate; deficient in annuals
Tucson Mts, AZ Sonoran Desert (Arizona Upland) (3)	32.5	650–1430	750	646	38	350	6	11	Thorough
Cerro Caguama, BCS Sonoran Desert (Vizcaíno)	27.5	400–625	2	114	31	94		33	Moderate; deficient in annuals
San Andrés Canyon, BC Sonoran Desert (Vizcaíno)	29.0	50–150	2	71	28–29	68		41–43	Light
Sierra San Pedro, (S of Sta. Rosalia) BCS Sonoran Desert (Magdalena)	27.0	300–410	3	113	26	90		29	Light
Mesa Las Calabazas, BC Sonoran Desert (Vizcaíno)	27.5	300–400	2		26	49		53	Light
Montevideo Canyon, BC Sonoran Desert (Vizcaíno)	29.0	400–500	2	56	24	53		45	Light
Santa Marta, BCS Sonoran Desert (Vizcaíno)	27.5	525	2	98	23	82		28	Light
Calamajue Canyon, BC Sonoran Desert (Lower Colorado Valley)	29.5	500	1	70	22	60		37	Light
Table Top Mtn, AZ Sonoran Desert (Arizona Upland) (4)	33.0	500–1300	ca. 4		20				Moderate
Cerro Guatamote, BCS Sonoran Desert (Vizcaíno)	27.0	200–330	2	78	20	50		40	Low; deficient in annuals
San Borja Canyon, BC Sonoran Desert (Vizcaíno)	29.0	400	5	54	21	45		47	Light

(continued)

Table 12.5. Continued

Locality and Region (Reference)[a]	Latitude (°N)	Elevation (m)	Area (km²)	Total No. of Taxa	No. of Succulent Taxa	No. of Perennial Taxa	Succulents as % of Total No. of Taxa	Succulents as % of No. of Perennials	Survey Intensity/ Thoroughness
San Diego County, CA several communities (5)	32.5–33.5	0–2000	11,000	2,210	72	1385	3	5	Thorough
Cape Region, BCS Cape Thornscrub (6)	22.5–24.0	0–2200	6,800	1,116	76	779	7	10	Moderate?
Cataviña, BCN Sonoran Desert (Vizcaíno) (7)	30.0	500–800	20	296	29	180	6–9	16	Thorough
Chamela, JAL Tropical deciduous forest (8)	19.5	0–500	350	1,146	62+				Thorough
Las Bocas, SON Coastal thornscrub (9)	27.0	0–200	720	510	50–52	347	9	13	Thorough
Rio Cuchujaqui, SON Riparian, tropical deciduous forest (10)	27.0	200–360	46	740	36–39	470	5	8	Thorough
Sierra de Alamos, SON Tropical deciduous forest (11)	27.0	400–1730	929	963	59	786	6	8	Thorough
Sycamore Canyon (Santa Cruz Co.), AZ Riparian, Madrean Evergreen Woodland (12)	31.5	1100–1200	9	624	32–?	448	5	7	Thorough
Gran Desierto, SON Sonoran Desert (Lower Colorado Valley) (13)	31.0–32.0	0–1400	4,600	589	54	341	9	16	Thorough

Flora lists are by the authors unless otherwise noted. Percentages of succulent taxa among perennials are calculated unless the number of perennial taxa is not reported. Percentages of succulents among all taxa are not provided for light surveys or surveys deficient in annuals.

[a]References: (1) León de la Luz 1999, plus additions by authors; (2) Wiens unpubl. data; (3) Rondeau 1996, with additions by authors; (4) Felger pers. comm.; (5) Beauchamp 1986; (6) Lenz 1992; (7) Van Devender unpubl. data; (8) Lott 1993, updated in unpubl. ms; (9) Friedman 1996; (10) Van Devender et al. 2000; (11) Van Devender unpubl. data; (12) Toolin 1979; (13) Felger 2000.

Acknowledgments

This project was supported by a grant from the Cactus and Succulent Society of America. Our work in Mexico was conducted under a permit from José Luís León de la Luz, Instituto de Biología, Universidad Nacional Autónoma de México, La Paz, BCS. We owe thanks to several people who helped us characterize the succulent lifestyle: Richard S. Felger, Matthew B. Johnson, Gene Joseph, Jon Weeks, James Henrickson, Gordon Rowley, and David Hunt. Several people helped spot and identify plants in the field: Ana Lilia Reina G., David Seibert (1998), Anna Domitrovic (2001), Mark Sitter (2001), Robert Scarborough (2001), and Cory Martin (2002). This project is documented with additional images and graphics on the Arizona-Sonora Desert Museum's website at www.desertmuseum.org (Science Department page, section on research projects, biodiversity studies).

Literature Cited

Beauchamp, R. M. 1986. A flora of San Diego County, California. Sweetwater River Press, National City, California.

Dimmitt, M. A. 2000. Flowering plants of the Sonoran Desert. Pp. 153–264 *in* S. J. Phillips and P. W. Comus (eds.), A Natural History of the Sonoran Desert. University of California Press, Berkeley.

Eggli, U. (ed.). 2001. Illustrated Handbook of Succulent Plants: Monocotyledons. Springer Press, Heidelberg, Germany.

Felger, R. S. 2000. Flora of the Gran Desierto and Rio Colorado of Northwestern Mexico. University of Arizona Press, Tucson.

Friedman, S. L. 1996. Vegetation and flora of the costal plains of the Río Mayo region, southern Sonora, Mexico. Master's thesis, Arizona State University, Tempe.

Hastings, J. R. 1964. Climatological data for Sonora and northern Sinaloa. Technical Reports on the Meteorology and Climatology of Arid Regions 15. University of Arizona Institute of Atmospheric Physics, Tucson.

Henrickson, J. 1969a. An introduction to the Fouquieriaceae. Cactus and Succulent Journal (U.S.) 41: 97–106.

Henrickson, J. 1969b. The succulent *Fouquierias*. Cactus and Succulent Journal (U.S.) 41: 178–184.

Henrickson, J. 1972. A taxonomic revision of the Fouquieriaceae. Aliso 7: 439–537.

Hernández, H., C. Gómez-Hinostrosa, and R. T. Barcénas. 2001. Diversity, spatial arrangement, and endemism of Cactaceae in the Huizache area, a hot-spot in the Chihuahuan Desert. Biodiversity and Conservation 10: 1097–1112.

Joseph, G. 1985. "Succulent Hill", Baja California. Cactus and Succulent Journal (U.S.) 57: 212–213.

Lenz, L. W. 1992. An Annotated Catalogue of the Plants of the Cape Region, Baja California Sur, Mexico. Cape Press (Rancho Santa Ana Botanic Garden), Claremont, California.

León de la Luz, J. L. 1999. Listados Florísticos de México XVIII. Flora de la Región del Cabo de Baja California Sur. Instituto de Biología, Universidad Nacional Autónoma de México, Mexico D.F.

Lott, E. J. 1993. Annotated checklist of the vascular flora of the Chamela Bay region, Jalisco, Mexico. Occasional Papers of the California Academy of Sciences 148: 1–60.

Mueller-Dombois, D., and H. Ellenberg. 1974. Aims and Methods of Vegetation Ecology. John Wiley and Sons, New York.

Rebman, J. P. 2001. Succulent diversity in Lower California, Mexico. Cactus and Succulent Journal (U.S.) 73: 131–138.

Rondeau, R. 1996. Annotated flora and vegetation of the Tucson Mountains, Pima County, Arizona. Desert Plants 12: 3–46.

Rowley, G. 1978. The Illustrated Encyclopedia of Succulents. Leisure Books, London.

Toolin, L. J., T. R. Van Devender, and J. M. Kaiser. 1979. The flora of Sycamore Canyon, Pajarito Mountains, Santa Cruz County, Arizona. Arizona-Nevada Academy of Science Journal 14: 66–74.

Turner, R. M., J. E. Bowers, and T. L. Burgess. 1995. Sonoran Desert Plants. An Ecological Atlas. University of Arizona Press, Tucson.

Van Devender, T. R., A. C. Sanders, R. K. Wilson, and S. A. Meyer. 2000. Vegetation, flora, and seasons of the Río Cuchujaqui, a tropical deciduous forest near Alamos, Sonora. Pp. 36–102 *in* R. H. Robichaux and D. A. Yetman (eds.), The Tropical Deciduous Forest of Alamos: Biodiversity of a Threatened Ecosystem in Mexico. University of Arizona Press, Tucson.

Von Willert, D. J., B. M. Eller, M. J. A. Werger, E. Brinckmann, and H.-D. Ihlenfeldt. 1992. Life Strategies of Succulents in Deserts with Special Reference to the Namib Desert. Cambridge University Press, London.

Warren, D. K. 1979. Precipitation and temperature as climatic determinants of the distribution of *Fouquieria columnaris*. Ph.D. dissertation, University of Arizona, Tucson.

Wiggins, I. L. 1980. Flora of Baja California. Stanford University Press, Stanford, California.

13

Cactus Diversity and Endemism in the Chihuahuan Desert Region

HÉCTOR M. HERNÁNDEZ

CARLOS GÓMEZ-HINOSTROSA

Essentially confined to the New World, the family Cactaceae comprises about 100 genera and 1500 species (Barthlott and Hunt 1993). It is one of the most typical floristic elements of Mexico, which harbors 50 genera and about 550 species—the richest assemblage of cacti in the world. Most Mexican Cactaceae species occur in the arid and semiarid ecosystems and regions of the country, most notably the Chihuahuan Desert Region, the Sonoran Desert, and the Tehuacán Valley. Both nationwide and worldwide, the Chihuahuan Desert Region in particular stands out as having the highest diversity of Cactaceae (Hernández and Godínez 1994; Hernández and Bárcenas 1995).

Until recently, no complete catalogue of the cactus species in the Chihuahuan Desert Region (CDR) had been published. Hernández and Godínez (1994) had compiled a list of 115 species of endangered cacti in the Mexican portion of the CDR, including some marginal areas (see below). Some general patterns of diversity and distribution had also been described in the Cactaceae at a regional (Hernández and Bárcenas 1995, 1996) and local scale (Gómez-Hinostrosa and Hernández 2000; Hernández et al. 2001). Overall, however, the lack of a complete catalogue of cactus species in the CDR was regrettable, as it hindered efforts to characterize Cactaceae biogeographic patterns across the entire region.

This chapter is based on the first complete list of cactus species in the CDR, pending the discovery of new species. To compile this list, relevant bibliographic references were consulted (e.g., Sánchez-Mejorada 1978; Benson 1982; Pinkava 1984; Cornet 1985; Argüelles et al. 1991; Villarreal 1994; Bárcenas 1999; Gómez-Hinostrosa and Hernández 2000; Anderson 2001; Hernández et al. 2001). The preliminary list was verified and complemented with information contained in the Database of Cactus Collections from North and Central America, which we developed over the last decade. This database currently contains over 23,000 records of specimens from 35 national and international herbaria. The resulting checklist of the Cactaceae from the CDR (Hernández et al. 2004) includes species names and information on geographical distribution at the state level, conservation status, and herbarium vouchers. Entirely cultivated species are excluded. With a few exceptions, taxonomic nomenclature follows Hunt (1999). Recent work indicates that some of the smaller genera, such as some of the CDR endemics, are not well circumscribed, and some generic realignments may be expected as more information becomes available (Butterworth et al. 2002).

Based on Hernández et al.'s (2004) checklist, in this chapter we present patterns of cactus diversity and endemism in the CDR. In addition, we examine the conservation status of cacti in relation to both patterns of distribution and anthropogenic threats. Even with the completed checklist and our present analysis, knowledge of cactus biogeography remains fragmentary and at a coarse level of resolution.

The Chihuahuan Desert Region

The Chihuahuan Desert proper is difficult to define (Shreve 1942; Miranda 1955; Morafka 1977a,b; Schmidt 1979; Medellín-Leal 1982). It is heterogeneous, has large ecotones, and is interspersed with nondesert areas, typically higher-elevation mountains, subject to the strong influence of the nearby desert (Shreve 1942; Morafka 1977a,b; Medellín-Leal 1982). For these reasons, ecologists have often found it easier to include within one geographic unit the Chihuahuan Desert and, to a variable extent, its associated, less arid areas (Henrickson and Straw 1976; Morafka 1977a,b; Medellín-Leal 1982). This unit, often referred to as the Chihuahuan Desert Region (CDR), provides an ecologically relevant framework for the study of many taxa, in particular those with distributions that do not stop at the edge of the desert but extend largely into ecotonal areas or onto the lower slopes of associated mountains. This is true in particular of the Cactaceae.

The boundaries of the Chihuahuan Desert and the CDR have been disputed, particularly their southern and northern limits (Shreve 1942; Contreras 1955; Henrickson and Straw 1976; Johnston 1977; Morafka 1977a,b; Schmidt 1979; Medellín-Leal 1982). The Sierra Madre Occidental and the Sierra Madre Oriental in north-central Mexico represent the traditional western and eastern boundaries of the CDR. According to Henrickson and Straw (1976), the region has an estimated area of about 507,000 km^2 and extends from southern San Luis Potosí, at about 21° 40' N, into a small portion of Arizona and into the Rio Grande and Pecos River Basins in New Mexico and Texas, at about 34° 30' N.

Peripheral to the main body of the CDR, as outlined by Henrickson and Straw (1976), are several disjunct geographic units that we believe should be considered when the biota of the region is analyzed. These disjunct segments share climatic characteristics with Mexican desert and semidesert areas, although most of them lack the climatic extremes (e.g., prolonged winter frosts, extremely high summer temperatures, and annual precipitations well below 300 mm) prevalent farther north. They also support plant associations typical of the CDR, such as *Larrea*, *Larrea-Yucca*, and *Agave lechuguilla-Hechtia* scrub, and *Prosopis* thorn woodland.

Thus, our delineation of the CDR (fig. 13.1) includes many dry, intermontane valleys and canyons localized within the Sierra Madre Oriental (e.g., Rayones, Jaumave, and Aramberri valleys), where

the typical CDR floristic elements are visible. To the south of Henrickson and Straw's (1976) Chihuahuan Desert, we follow Medellín-Leal (1982) and also include several geographically and climatically isolated, dry depressions and valleys. These correspond in part to the region known as the Queretaroan-Hidalgoan Arid Zone (e.g., Barranca de Metztitlán, Valle del Mezquital, Valle de Actopan in Hidalgo, and the Extóraz River Basin in Querétaro), but also include all the dry portions of Guanajuato (e.g., Xichú, Atarjea).

To a large extent, our definition of the CDR is arbitrary, as have been all previous definitions of the CDR or the Chihuahuan Desert itself (Medellín-Leal 1982). However, beyond the climatic and botanical similarities noted above, our treatment of the CDR is also supported if, as we believe, the biotas of the main body of the CDR and disjunct fragments share a common evolutionary origin. In our opinion, endemic taxa in the disjunct areas are likely the result of a long history of isolation.

The hottest, driest, and lowest-elevation areas in the CDR are in the Rio Grande and Trans-Pecos region (600 m at the valley bottom), in the Bolsón de Cuatro Ciénegas (740 m), and in the Bolsón de Mapimí (1075 m). As elevation increases from these areas, there is increased precipitation (Secretaría de Programación y Presupuesto 1981; see also chapter 1). The northernmost areas of the CDR receive some snowfall during the winter months (Schmidt 1979).

By comparison with the northern portion of the CDR, several areas farther south have a more benign climate. The Huizache area, for example, has milder winters and summers than Cuatro Ciénegas or Big Bend. It is also characterized by high topographic heterogeneity, but most of this area (especially the lowland plains) is very dry, receiving less than 300 mm rain per year.

Some Definitions

Based on the above delineation of the CDR, in the rest of the chapter we refer to the CDR as an aggregation of 3 subregions (fig. 13.1). The Main Subregion corresponds to the large area defined by Henrickson and Straw (1976). Within this subregion are some nondesert areas (i.e., sky island mountains and ecotonal areas with xerophytic associations intermixing with grasslands or pine-oak-juniper woodlands), with La Paila and Parras, and

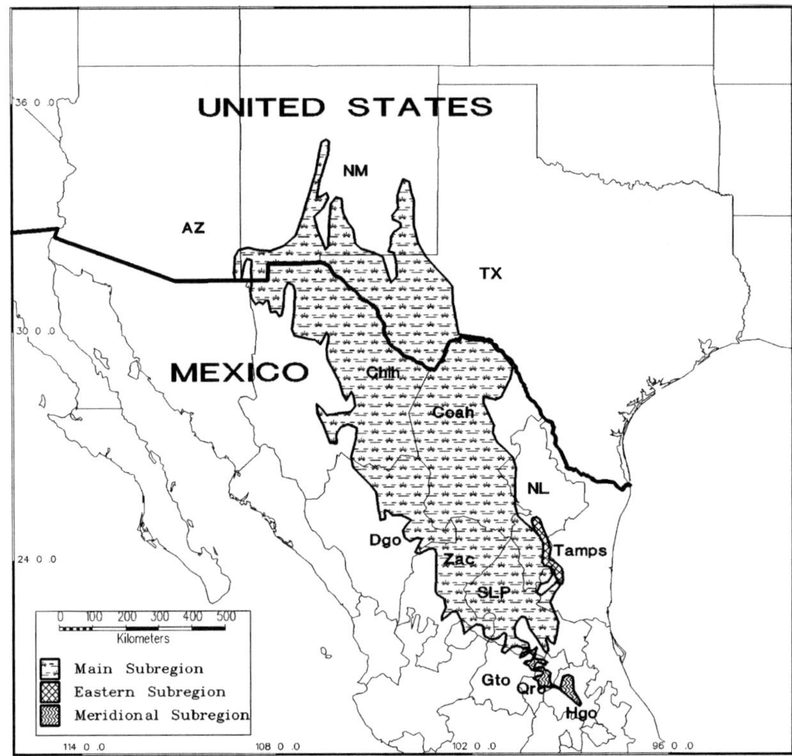

Figure 13.1. Approximate boundaries of the Chihuahuan Desert Region's subregions. The Main Subregion was delineated according to Medellín-Leal (1982). AZ, Arizona; NM, New Mexico; TX, Texas; Chih, Chihuahua; Coah, Coahuila; NL, Nuevo León; Tamps, Tamaulipas; Dgo, Durango; Zac, Zacatecas; SLP, San Luis Potosí; Gto, Guanajuato; Qro, Queretaro; Hgo, Hidalgo.

several other mountain ranges also included within the coverage polygon. Collectively, the southern, disjunct areas of Guanajuato, Querétaro, and Hidalgo are referred to as the Meridional Subregion of the CDR. Vegetation communities bordering desert areas of this subregion include tropical dry forest and oak forest. The valleys and depressions located in the Sierra Madre Oriental are called here the Eastern Subregion. Species richness and endemicity are examined at the regional, subregional, state, and local level.

Biogeographic Patterns

Species Diversity

We report here a total of 329 native cactus "species" (39 genera) in the CDR. Included within this total are 5 naturally occurring hybrids, all of them formally described and distinct, easily identifiable taxonomic entities. The taxonomy of cacti is complicated by the fact that cacti readily hybridize in nature. Even some taxonomic entities described as "normal" species, including many chollas, are suspected to have a hybrid origin.

The 329 species represent 59.8% of all the Cactaceae in Mexico (550 spp.). This outstanding diversity is all the more striking because the Mexican portion of the CDR covers only about 25% of the land area of this country. As shown in figure 13.2, 4 genera alone, *Mammillaria* (79 spp.), *Opuntia sensu stricto* (46 spp.), *Coryphantha* (36 spp.), and *Echinocereus* (30 spp.), account for 58% of the total diversity in the region. Conversely, 23 genera each provide only 1 or 2 species to the total cactus diversity of the region.

In the CDR most cactus species are small, inconspicuous globose plants. Only a few arborescent

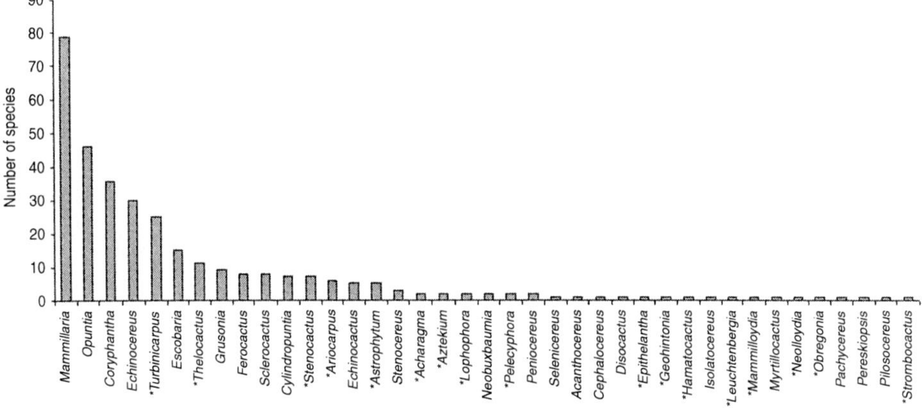

Figure 13.2. Generic composition of cacti from the Chihuahuan Desert Region, with number of species per genus. Endemic genera are marked with an asterisk.

cacti (e.g., *Myrtillocactus geometrizans*, *Isolatocereus dumortieri*, and *Stenocereus* spp.), some barrel-like species (e.g., *Echinocactus platyacanthus*, *Ferocactus pilosus*, and *F. histrix*), and a larger number of opuntioids (*Opuntia* and *Cylindropuntia*) are visible at a distance. Thus, the CDR contrasts with several other Mexican regions or ecosystems, where large, highly conspicuous cactus species dominate the landscape. The difference in growth forms is probably linked to the fact that through much of the CDR the relatively low winter temperatures inhibit the survival of columnar cactus species. In North America, the large arborescent, columnar or candelabriform growth forms have geographical ranges centered on areas influenced by tropical climatic regimens (e.g., the Tehuacán Valley, Balsas Basin, Tehuantepec Isthmus, and Sonoran Desert).

Species richness is not distributed evenly in the CDR (fig. 13.3). The states (Mexican or U.S.) with the lowest numbers of species are New Mexico, Arizona, and Chihuahua. The richest states are San Luis Potosí, Coahuila, Nuevo León, and Tamaulipas, all of which have more than 100 cactus species. The CDR (Main Subregion) portion of San Luis Potosí has 141 species—the highest diversity in the region (fig. 13.3). In its non-CDR portion, San Luis Potosí has several dozens of additional cactus species. Thus, it is the Mexican state with by far the highest number of cactus species.

Not surprisingly, it is in San Luis Potosí, Coahuila, Nuevo León, and Tamaulipas that most of the best-known, cactus-rich localities occur. Among them are the Huizache, Mier y Noriega, Doctor

Arroyo, Matehuala, Tula, Jaumave, and Cuatro Ciénegas areas (see fig. 3 in Hernández et al. 2001). Located in San Luis Potosí, at the southern extreme of the CDR Main Subregion, the Huizache area is characterized by an outstanding concentration of cactus species. Hernández et al. (2001) cataloged all the Cactaceae species in a square-shaped polygon measuring 2855 km². Most of the area supports plant associations typical of the CDR (e.g., *Larrea*, *Larrea-Yucca*, and *Agave lechuguilla-Hechtia* scrub), although relatively small fragments of more mesic vegetation types exist at the top of mountain ranges and in canyons. A total of 75 species were recorded—the highest density of cactus species in the world. A single portion of the Huizache area measuring 6' latitude by 6' longitude (ca. 114 km²) harbored 41 cactus species, only a few species less than in all of Cuatro Ciénegas (48 spp.; Pinkava 1984) or La Paila (44 spp.; Villarreal 1994). In this same area, 32 species were found along a single transect measuring only 3 km (Hernández et al. 2001).

Species richness in the CDR seems higher southeastward, in apparent correlation with higher precipitation and milder climatic conditions, plus very high habitat diversity. Hernández and Bárcenas (1995, 1996) mapped the geographical distributions of 93 species of cacti from the CDR. Although these were all endangered species, the results of the study may be applicable to the whole Cactaceae family. Highest endangered species richness was found in areas of moderate elevation, particularly toward the southeastern and, to a lesser extent, the eastern edge

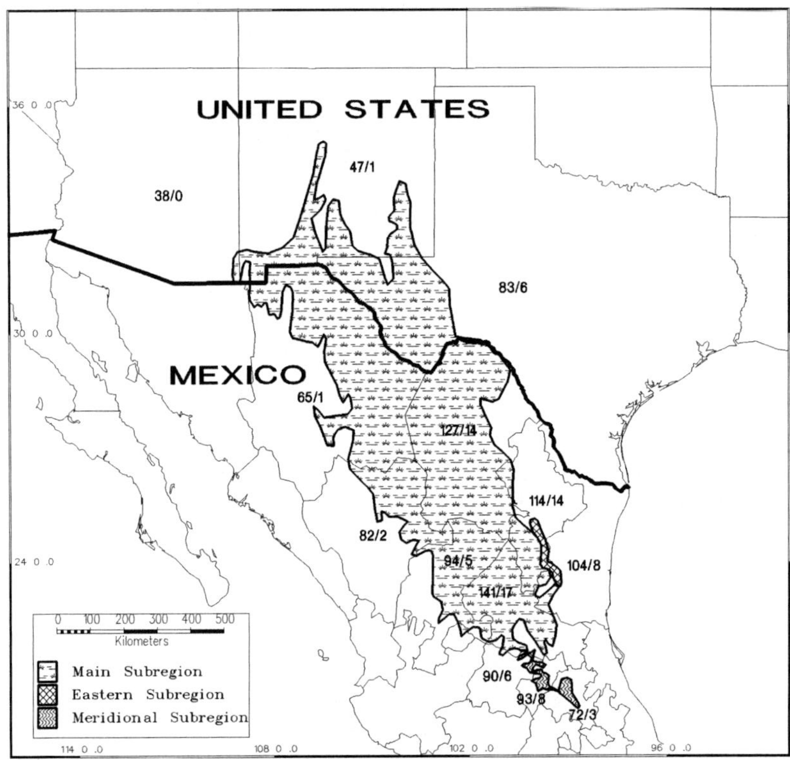

Figure 13.3. Total number of cactus species and number of endemic species by state in the Chihuahuan Desert Region. Only the species found within the limits of the Chihuahuan Desert Region are included.

of the CDR, in northern San Luis Potosí and in the southern portions of Coahuila, Nuevo León, and Tamaulipas. Endangered species richness decreases toward the western segment of the CDR and from the Cuatro Ciénegas region northward and northwestward. Another important area rich in endangered species corresponds to the Meridional Subregion, in Guanajuato (Xichú), Querétaro (Tolimán), and Hidalgo (Metztitlán).

Endemism

The CDR is known as a repository of endemic taxa (Johnston 1977; Minckley 1977; Pinkava 1984; Gómez-Hinostrosa and Hernández 2000; Hernández et al. 2001). Our analysis clearly supports this notion, as 17 (43.6%) out of the 39 cactus genera recorded in the CDR are endemic to this region (see also fig. 13.2). These 17 genera also represent 34% of Mexico's generic diversity of Cactaceae. Among the cactus species of the CDR, almost 70% are

strictly endemic, if the 3 subregions are considered together (fig. 13.1). A total of 152 cactus species, representing 46.2% of the total cactus diversity in the region, are restricted to the Main Subregion (fig. 13.4). An additional 41 species (12.5%) are endemic to the much smaller Meridional Subregion, with 13 more species found exclusively in the 2 subregions combined. Finally, there are 23 species (7%) endemic to the Eastern Subregion (fig. 13.4). In particular, this subregion is home to the endemic genera *Aztekium*, *Geohintonia*, and *Obregonia*.

The number of cactus species endemic to one state parallels patterns of total species richness (fig. 13.3). Among the 85 species restricted to 1 state, 17, 14, 14, and 8 are found in San Luis Potosí, Coahuila, Nuevo León, and Tamaulipas, respectively (table 13.1). As already seen, these 4 Mexican states also support the largest total numbers of Cactaceae (fig. 13.3). The disjunct Meridional Subregion is another important center of species endemism, especially when compared to the other end

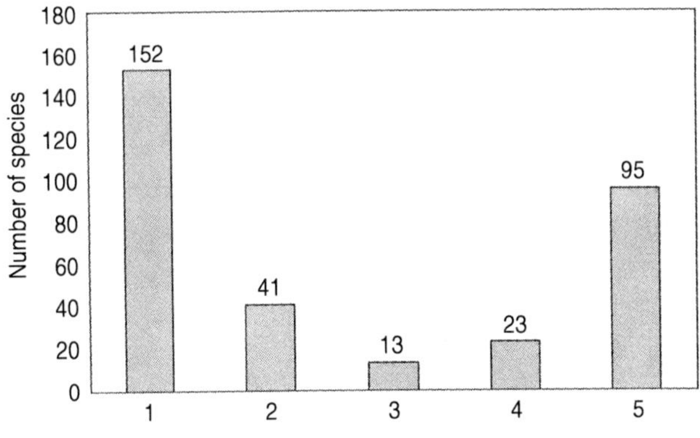

Figure 13.4. Geographic affinities and endemism of cacti from the Chihuahuan Desert Region. The numbers along the x-axis refer to: *1*, Main Subregion endemics; *2*, Meridional Subregion endemics; *3*, Main Subregion–Meridional Subregion endemics other than *1* or *2*; *4*, Eastern Subregion endemics; and *5*, Species not endemic to the Chihuahuan Desert Region.

of the CDR, in the north. The states of Querétaro and Guanajuato have 8 and 6 endemic species, respectively (fig. 13.3). These 2 states, together with Coahuila, San Luis Potosí, Nuevo León, and Tamaulipas, account for 67 species (78.8%) of the Cactaceae endemic to one state.

At a smaller spatial scale, the Huizache area again deserves to be singled out, with 10 locally endemic species. Sixty-three percent of the cactus species found in the Huizache are endemic to the CDR.

Rarity

Rarity is a concept highly relevant to conservation biology because, everything else being equal, rare species are more vulnerable to extinction than common ones. In the context of population and community biology, a species is considered rare if it has a very small distribution and/or low population density (Gaston 1994). These 2 attributes of rare species are observed frequently among Cactaceae in Mexico, and in particular in the CDR.

Only a few CDR cactus species occur in many Mexican or U.S. states (fig. 13.5). The most extreme examples of widespread species are *Opuntia engelmannii*, recorded in 23 states, and *Cylindropuntia imbricata* and *Acanthocereus tetragonus*, recorded in 21 states (Hernández et al., 2004; see also fig. 13.5). As with these 3 species, most of the widespread cacti are not endemic to the CDR. Conversely, the great majority of CDR cactus species are restricted to 1 or a few, usually contiguous, U.S. or Mexican states. More than one-third (41.3%) of the species occur in only 1 or 2 states, and two-thirds (69.6%) are in 5 or fewer states. A list of cactus species endemic to a single state is provided in table 13.1.

Some cactus species have extremely narrow distributions (fig. 13.6). The most extreme examples of narrow endemism are represented by species known from only 1 or a few localities (e.g., *Ariocarpus bravoanus*, *A. scaphirostris*, *Aztekium hintonii*, *Echinocereus waldeisii*, *Mammillaria carmenae*, *Opuntia chaffeyi*, *Turbinicarpus subterraneus*). Those species with distributions mapped in figure 13.6 represent only some of the known cases of stenoendemism in the CDR. Additional examples have been reported by Gómez-Hinostrosa and Hernández (2000) and Hernández et al. (2001), among others.

In our opinion, it is likely that the great majority of cactus narrow endemics are relict species. However, soil differentiation may also account for the existence of a number of geographically restricted species. This is suggested by the existence in the CDR of several obligate edaphic specialists, such as the gypsophilous *Aztekium ritteri*, *A. hintonii*, *Geohintonia mexicana*, and *Turbinicarpus zaragozae*. Johnston (1977) and Powell and Turner (1977) have both argued that atypical soil types play an important role in plant speciation in the CDR.

Rarity among CDR cacti is also expressed through low population densities. *Sclerocactus uncinatus* is a species widely distributed in the Main Subregion of the CDR, with numerous populations recorded in Chihuahua, Coahuila, Durango, Nuevo León, San Luis Potosí, Sonora, Tamaulipas, Zacatecas, and Texas. We have conducted surveys along more than one hundred 2–3 km transects, most of these yielding fewer than 10 individuals (Hernández and Gómez-Hinostrosa, unpubl. data). *Sclerocactus uncinatus* is a good example of a geographically widespread but locally rare species.

Low population densities and small distributions combine to produce small population sizes in a num-

Table 13.1. List of cactus taxa from the Chihuahuan Desert Region with ranges restricted to a single state.

Species	Chih	Coah	Dgo	Gto	Hgo	NM	NL	Qro	SLP	Tamps	TX	Zac
Acharagma aguirreana		Coah										
A. roseana		Coah										
Ariocarpus bravoanus									SLP			
A. scaphirostris							NL					
Astrophytum caput-medusae							NL					
Aztekium hintonii							NL					
A. ritteri							NL					
Coryphantha gracilis	Chih											
C. hintoniorum							NL					
C. maiz-tablasensis									SLP			
C. unicornis		Coah										
C. voghterriana									SLP			
C. wohlschlageri									SLP			
Cylindropuntia anteojoensis		Coah										
Echinocereus mapimiensis		Coah										
E. nivosus		Coah										
E. primolanatus		Coah										
E. rayonesensis							NL					
E. schmollii								Qro				
E. waldeisii									SLP			
Escobaria hesteri											TX	
E. lloydii												Zac
E. minima											TX	
Geohintonia mexicana							NL					
Grusonia bulbispina		Coah										
G. clavata						NM						
Hamatocactus crassihamatus				Gto								
Lophophora diffusa								Qro				
Mammillaria albiflora				Gto								
M. anniana										Tamps		
M. aureilanata									SLP			
M. baumii										Tamps		
M. carmenae										Tamps		
M. guelzowiana			Dgo									
M. herrerae								Qro				
M. humboldtii					Hgo							
M. laui										Tamps		
M. lenta		Coah										
M. luethyi		Coah										
M. marcosii				Gto								
M. mathildae								Qro				
M. melaleuca										Tamps		
M. microhelia								Qro				
M. sanchez-mejoradae							NL					
M. schwarzii				Gto								
M. tezontle									SLP			
M. theresae			Dgo									
M. wagneriana												Zac
M. weingartiana							NL					
M. wiesingeri					Hgo							
M. zeilmanniana				Gto								
M. sp.												Zac
Obregonia denegrii										Tamps		
Opuntia atrispina											TX	
O. chaffeyi												Zac

Table 13.1. Continued

Species	Chih	Coah	Dgo	Gto	Hgo	NM	NL	Qro	SLP	Tamps	TX	Zac
O. elizondoana								Qro				
O. ellisiana											TX	
O. megarrhiza									SLP			
O. pachyrrhiza									SLP			
O. pailana		Coah										
O. pyriformis												Zac
O. ×rooneyi											TX	
O. ×spinosibacca											TX	
Pelecyphora aselliformis									SLP			
Stenocactus vaupelianus					Hgo							
Thelocactus hastifer								Qro				
T. lausseri		Coah										
Turbinicarpus alonsoi				Gto								
T. bonatzii									SLP			
T. booleanus							NL					
T. hoferi							NL					
T. jauernigii									SLP			
T. knuthianus									SLP			
T. laui									SLP			
T. lophophoroides									SLP			
T. mandragora		Coah										
T. pailanus		Coah										
T. pseudomacrochele								Qro				
T. rioverdensis									SLP			
T. saueri										Tamps		
T. subterraneus							NL					
T. swobodae							NL					
T. ysabelae										Tamps		
T. zaragozae							NL					
T. ×mombergeri									SLP			
Number of taxa	1	14	2	6	3	1	14	8	17	8	6	5

States: Chih = Chihuahua, Coah = Coahuila, Dgo = Durango, Gto = Guanajuato, Hgo = Hidalgo, NM = New Mexico, NL = Nuevo León, Qro = Querétaro, SLP = San Luis Potosí, Tamps = Tamaulipas, TX = Texas, and Zac = Zacatecas.

ber of CDR cactus species. For example, *Turbinicarpus subterraneus* is a geographically and ecologically rare species known from only 2 populations in southern Nuevo León. One of these 2 populations apparently has only a handful of individuals, and no more than a few hundred individuals have been found in the second. An even more extreme example is that of *Opuntia chaffeyi*, known from only 2 disjunct localities in Zacatecas. Extensive searches where these 2 known populations occur have revealed no more than a combined total of 20 individual plants.

Beta Diversity

It has been hypothesized that high diversity in some desert environments can be explained, at least in part, by high beta diversity, or a high species turnover rate along geographical gradients. Goettsch (2001) sampled 23 cactus communities along a 250-km east–west transect near the southeastern end of the Main Subregion of the CDR. A total of 61 cactus species were found along the transect, and beta diversity values were calculated for every pairwise combination of contiguous and noncontiguous communities. In all cases beta diversity values were greater than 0, indicating that all communities had somewhat distinct cactus species compositions. In general, recorded beta diversity values were relatively high, even between contiguous communities. However, no total species turnover ($\beta = 1$) was recorded along the transect, the highest value being $\beta = 0.93$ between 2 noncontiguous sites. Perhaps the

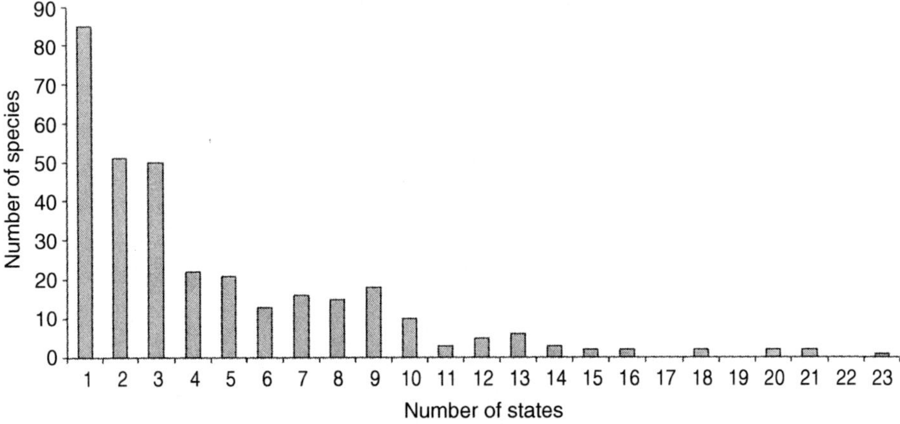

Figure 13.5. Frequency distribution of the number of states (Mexican and U.S. states included) represented in the range of cactus species occurring in the Chihuahuan Desert Region (CDR). Note that some species range outside the boundaries of the CDR. This figure is based on the complete ranges of species, rather than their ranges within the CDR.

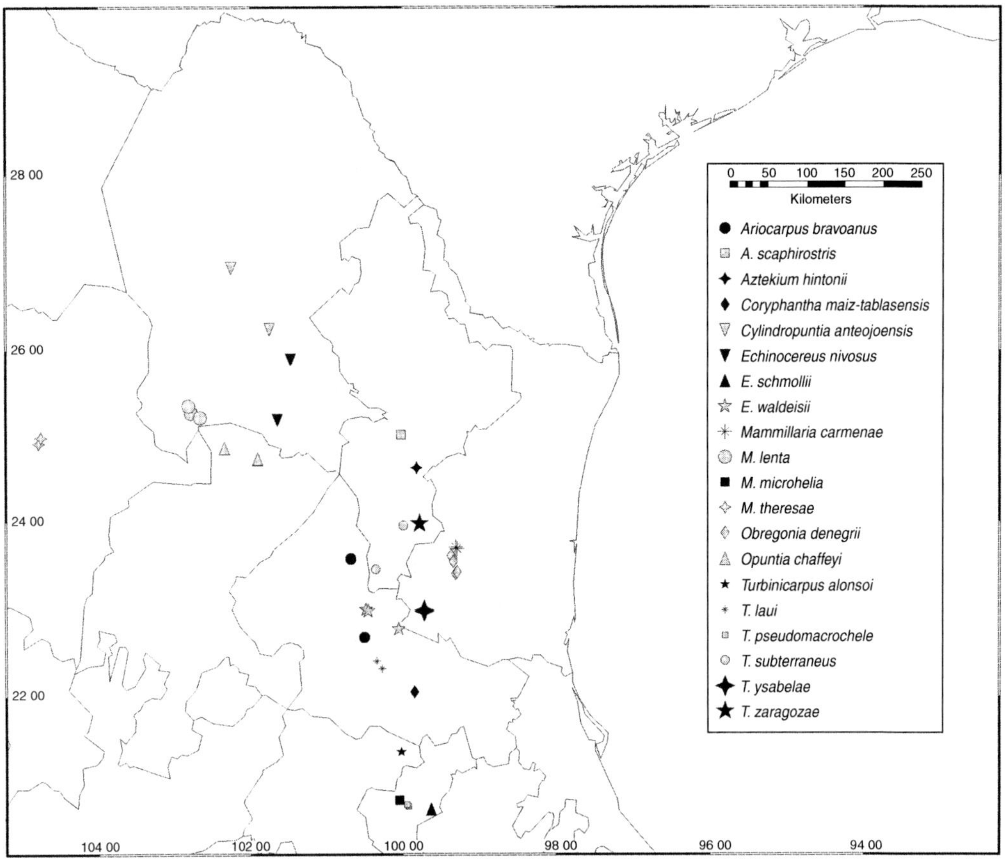

Figure 13.6. Geographic location of some narrowly distributed Chihuahuan Desert Region cactus species.

most important conclusion of this study was that the observed high beta diversity values were only partially a consequence of a real species turnover. Goettsch (2001) found that many of the cactus species recorded do not have a continuous distribution. Instead, they appeared on and off along the transect. This intermittent pattern of distribution, common in most widespread cactus species, causes a misleading perception of beta diversity results. A similar project to that conducted by Goettsch (2001) is currently in progress. This study is aimed at assessing cactus species turnover along a south–north transect across the entire CDR Main Subregion, from southern San Luis Potosí to the Big Bend area, Texas. It is expected that the data collected from this project will help us to better understand the nature of beta diversity in the CDR.

Conservation Status of Cactaceae in the CDR

The Cactaceae includes many endangered taxa. A total of 285 Mexican cacti are included in Mexico's official list of species At Risk (SEMARNAT 2002), of which 136 are found in the CDR. Many species of Cactaceae are also currently being reevaluated by IUCN (the World Conservation Union) for possible inclusion in the new Red List Categories (Hilton-Taylor 2000). It is expected that as the reevaluation nears completion, the Red List will represent the most objective assessment of the conservation status of cactus species.

Rarity is likely a natural phenomenon among most CDR cacti. At the same time, it is undeniable that people have amplified the phenomenon through overcollecting and habitat destruction. Natural populations of many CDR cactus species have been considerably affected by the collection of plants for use as ornamentals or as collectors' items. The entire cactus family is included in Appendix II of the Convention of International Trade of Endangered Species (CITES 1990), and, specifically, 40 species in the CDR currently appear in Appendix I. Although Mexican environmental authorities have established laws that preclude collecting of some of the flora and fauna, there is evidence that plants and seeds of Mexican cacti are still being collected illegally in the field. This is easily demonstrated by the fact that species like *Ariocarpus bravoanus*, *Aztekium hintonii*, *Geohintonia mexicana*, *Mammillaria luethyi*, *Turbinicarpus alonsoi*, and several other recently described taxa from the CDR, already exist in private collections in Europe, the United States, and Mexico, despite the fact that no permits to collect and export these species have been issued.

It has been proposed that controlled collection and distribution of seeds, particularly of rare cacti, would alleviate the pressure exerted by illegal collecting. However, the collection of fruits and seeds, even under theoretically controlled conditions would entail an important risk for various reasons. First, we know almost nothing about the demography and reproductive dynamics of most cactus species, and thus we lack the most fundamental data to regulate seed collection activities. Often, rare cacti are present in nature at extremely low densities. In many species (e.g., *Acharagma* spp., *Aztekium* spp., *Pelecyphora* spp., *Turbinicarpus* spp.) individual plants appear to produce relatively low numbers of fruits and seeds, and seedling recruitment is extremely rare and erratic. For these reasons, and pending more information on demographic attributes of every species, the collection of even a small number of seeds is inadvisable. A second issue relates to law enforcement. Mexican environmental authorities do not have the budget or manpower to exert a tight control on collecting, and many cactus populations are located in remote areas. Finally, human communities inhabiting arid regions in Mexico live in harsh and isolated settings, where education and culture are inappropriate for controlling in situ the collection of cactus seeds.

The most significant factor affecting the conservation status of Mexican cacti in general is the deterioration of their natural habitat. Large fragments of the CDR have been modified, in some cases dramatically, by agricultural development, goat raising, mining, road construction, dam building, and other human activities (Challenger 1998; chapter 3). These forms of disturbance have had a tremendous impact on cactus populations because these plants usually have slow growth rates, long life cycles, and low recruitment rates via seed germination; furthermore, as already mentioned, these plants frequently have narrow distribution ranges or occur at low densities. All these factors make CDR cacti extremely vulnerable to disturbance (Hernández and Godínez 1994).

Final Considerations

High diversity, high endemism, a tendency to be narrowly distributed and/or locally rare, and a high degree of endangerment are attributes defining the Cactaceae in the CDR. Under the continuing threat of habitat destruction and collecting, cacti should constitute one of the top priorities of conservation efforts in the CDR. Although not the only place deserving of special attention to conservation efforts, the Huizache area is a focal point of global importance for cactus conservation.

An essential prerequisite to develop sound conservation measures in a region is adequate knowledge of biodiversity. For this reason, we need to bring our understanding of species diversity and endemism among CDR cacti to a finer level of resolution. To accomplish this, more surveys are required throughout the CDR, but primarily in the numerous isolated valleys along the eastern and southeastern margins of the region and in other, less explored areas in northern Guanajuato and northern Zacatecas.

Exploratory surveys will improve our knowledge of total species richness in the Cactaceae. Several new species were described from the CDR during the last decade, and there is little doubt that more await to be discovered and described. In addition, we know almost nothing about the demography of cactus populations, and long-term population studies are necessary. Finally, we need to evaluate patterns of genetic variation in natural cactus populations. One important question is whether genetic variation is positively correlated with range size. This and other research goals are particularly important for effective conservation of the rarest, most vulnerable, cactus species.

Efforts have been made to increase the coverage of the natural protected areas in the CDR. In addition to the previously existing areas (Big Bend National Park, La Jornada Experimental Range, and the Mapimí Biosphere Reserve), several reserves were decreed in Mexico during the last decade: Cuatro Ciénegas, El Cielo, Metztitlán, Sierra Gorda, and the Real de Guadalcázar Natural Reserve, located in San Luis Potosí and comprising the Huizache area. However, many critical areas with high cactus species richness and endemism are not protected (e.g., Xichú, Guanajuato; Arramberi and Rayones, Nuevo León; and Tula, Tamaulipas). The number of areas under formal protection in the CDR needs to be increased, through the creation of a network of several additional, probably smaller protected areas. The selection of these should be based on a thorough analysis of variables such as species richness, degree of endemicity, taxonomic

uniqueness, and habitat diversity. Thus, conservation efforts in the CDR should be intensified through the consolidation of the already existing natural protected areas and the creation of new ones. Protected areas should be an effective tool for the conservation of the Cactaceae and their habitats.

Acknowledgments

We thank Fernando Chiang, Mark Olson, A. Michael Powell, and especially the editors of this book for constructive criticism of the manuscript. We also thank Barbara Goettsch for assisting us in the preparation of this chapter.

Literature Cited

Anderson, E. F. 2001. The Cactus Family. Timber Press, Portland, Oregon.

Argüelles, E., R. Fernández, and S. Zamudio. 1991. Listado florístico preliminar del estado de Querétaro. Flora del Bajío y regiones adyacentes. Fascículo complementario II. Instituto de Ecología, A.C., México D.F.

Bárcenas, R. T. 1999. Patrones de distribución de cactáceas en el estado de Guanajuato. Tesis de Licenciatura, Facultad de Ciencias, Universidad Nacional Autónoma de México, México D.F.

Barthlott, W., and D. Hunt. 1993. Cactaceae. Pp. 161–197 *in* K. Kubitzki, J. Rohwer, and V. Bittrich (eds.), The Families and Genera of Vascular Plants. II. Dicotyledons. Springer Verlag, New York.

Benson, L. 1982. The Cacti of the United States and Canada. Stanford University Press, Stanford, California.

Butterworth, C. A., J. H. Cota-Sanchez, and R. S. Wallace. 2002. Molecular systematics of tribe Cacteae (Cactaceae: Cactoideae): a phylogeny based on rpl16 intron sequence variation. Systematic Botany 27: 257–270.

Challenger, A. 1998. Utilización y Conservación de los Ecosistemas Terrestres de México. Pasado, Presente y Futuro. CONABIO, UNAM, and Agrupación Sierra Madre, México D.F.

CITES (Convention on International Trade in Endangered Species). 1990. Appendices I, II and III to the convention. U.S. Fish and Wildlife Service, Washington, D.C.

Contreras, A. 1955. Definición de las zonas áridas y su delimitación en la República Mexicana. Pp. 3–40 *in* E. Beltrán (ed.), Problemas de las Zonas Aridas de México. Instituto Mexicano de Recursos Naturales Renovables, México D.F.

Cornet, A. 1985. Las Cactáceas de la Reserva de

la Biosfera de Mapimí. Instituto de Ecología A.C., México D.F.

Gaston, K. J. 1994. Rarity. Chapman and Hall, London.

Goettsch, B. 2001. Diversidad beta e índices de similitud entre comunidades de cactáceas en el Desierto Chihuahuense. Tesis de Licenciatura, Facultad de Ciencias, Universidad Nacional Autónoma de México, México D.F.

Gómez-Hinostrosa, C., and H. M. Hernández. 2000. Diversity, geographical distribution, and conservation of Cactaceae in the Mier y Noriega region, Mexico. Biodiversity and Conservation 9: 403–418.

Henrickson, J., and R. Straw. 1976. A Gazetteer of the Chihuahuan Desert Region. A Supplement to the Chihuahuan Desert Flora. California State University, Los Angeles.

Hernández, H. M., and R. T. Bárcenas. 1995. Endangered cacti in the Chihuahuan Desert. I. Distribution patterns. Conservation Biology 9: 1176–1188.

Hernández, H. M., and R. T. Bárcenas. 1996. Endangered cacti in the Chihuahuan Desert. II. Biogeography and conservation. Conservation Biology 10: 1200–1209.

Hernández, H. M., and H. Godínez. 1994. Contribución al conocimiento de las cactáceas mexicanas amenazadas. Acta Botánica Mexicana 26: 33–52.

Hernández, H. M., C. Gómez-Hinostrosa, and R. T. Bárcenas. 2001. Diversity, spatial arrangement, and endemism of Cactaceae in the Huizache area, a hot-spot in the Chihuahuan Desert. Biodiversity and Conservation 10: 1097–1112.

Hernández, H. M., C. Gómez-Hinostrosa, and B. Goettsch. 2004. Checklist of Chihuahuan Desert Cactaceae. Harvard Papers in Botany 9: 11–26.

Hilton-Taylor, C. (comp.). 2000. 2000 IUCN Red List of Threatened Species. World Conservation Union, Gland, Switzerland.

Hunt, D. 1999. CITES Cactaceae Checklist. Royal Botanic Gardens/ International Organization for Succulent Plant Study, Kew, U.K.

Johnston, M. C. 1977. Brief resume of botanical, including vegetational, features of the Chihuahuan Desert Region with special emphasis on the uniqueness. Pp. 335–359 in R. H. Wauer and D. H. Riskind (eds.), Transactions of the Symposium on the Biological Resources of the Chihuahuan Desert Region, United States and Mexico. National Park Service, Washington, D.C.

Medellín-Leal, F. 1982. The Chihuahuan Desert. Pp. 321–372 in G. L. Bender (ed.), Reference Handbook on the Deserts of North America. Greenwood Press, Westport, Connecticut.

Minckley, W. L. 1977. Endemic fishes of the Cuatro Ciénegas basin, northern Coahuila, Mexico. Pp. 383–404 in R. H. Wauer and D. H. Riskind (eds.), Transactions of the Symposium on the Biological Resources of the Chihuahuan Desert Region, United States and Mexico. National Park Service, Washington, D.C.

Miranda, F. 1955. Formas de vida vegetales y el problema de la delimitación da las zonas áridas de México. Pp. 83–109 in E. Beltrán (ed.), Mesas Redondas sobre Problemas de las Zonas Áridas de México. Biblioteca Central de la Ciudad Universitaria, 24–28 de enero de 1955. IMRNRAC, Mexico.

Morafka, D. 1977a. Is there a Chihuahuan Desert? A quantitative evaluation through a herpetofaunal perspective. Pp. 437–454 in R. H. Wauer and D. H. Riskind (eds.), Transactions of the Symposium on the Biological Resources of the Chihuahuan Desert Region, United States and Mexico. National Park Service, Washington, D.C.

Morafka, D. 1977b. A biogeographical analysis of the Chihuahuan Desert through its herpetofauna. Biogeographica 9: 1–313.

Pinkava, D. J. 1984. Vegetation and flora of the Bolsón of Cuatro Ciénegas region, Coahuila, Mexico: summary, endemism and corrected catalogue. Journal of the Arizona-Nevada Academy of Science 19: 23–47.

Powell, A. M., and B. L. Turner. 1977. Aspects of the plant biology of the gypsum outcrops of the Chihuahuan Desert. Pp. 315–325 in R. H. Wauer and D. H. Riskind (eds.), Transactions of the Symposium on the Biological Resources of the Chihuahuan Desert Region, United States and Mexico. National Park Service, Washington, D.C.

Sánchez-Mejorada, H. 1978. Manual de Campo de las Cactáceas y Suculentas de la Barranca de Metztitlán. Sociedad Mexicana de Cactología, México D.F.

Schmidt, R. H., Jr. 1979. A climatic delineation of the "real" Chihuahuan Desert. Journal of Arid Environments 2: 243–250.

Secretaría de Programación y Presupuesto. 1981. Atlas Nacional del Medio Físico. Estados Unidos Mexicanos. Secretaría de Programación y Presupuesto, México D.F.

SEMARNAT. 2002. Norma Oficial Mexicana NOM–059–ECOL–2001, Protección Ambiental–Especies Nativas de México de Flora y Fauna Silvestres–Categorías de Riesgo y Especificaciones para su Inclusión, Exclusión o Cambio–Lista de Especies en Riesgo. Diario Oficial de la Federación, March 6, 2002.

Shreve, F. 1942. The desert vegetation of North America. Botanical Review 8: 195–246.

Villarreal, J. A. 1994. Flora vascular de la Sierra de la Paila, Coahuila, México. Sida 16: 109–138.

14

Cetacean Diversity and Conservation in the Gulf of California

JORGE URBÁN R.

LORENZO ROJAS-BRACHO

MERCEDES GUERRERO-RUÍZ

ARMANDO JARAMILLO-LEGORRETA

LLOYD T. FINDLEY

The Gulf of California, also called the Sea of Cortez, is an elongate marine basin with northwest-to-southeast orientation and is bounded on the west by the Baja California peninsula and on the east by mainland Mexico (fig. 14.1). It exceeds 1100 km in length and varies from 100 to 200 km in width. Altogether, it has an estimated surface area of nearly 260,000 km² (for a comprehensive description of the Gulf, see chapter 9).

The Gulf of California shelters a rich macrofauna of nearly 6000 known nominal species and subspecies (chapter 9) and is an especially important area for marine mammals, both pinnipeds and cetaceans. Pinnipeds (seals and sea lions) found in the Gulf include chiefly the California sea lion (*Zalophus californianus californianus*), with several stable colonies and an approximate population of 30,000 individuals (LeBoeuf et al. 1983). There are also sporadic records for elephant seal (*Mirounga angustirostris*), harbor seal (*Phoca vitulina*), and Guadalupe fur seal (*Arctocephalus townsendi*) (Aurioles et al. 1993; Urbán et al. 1997).

The notable diversity (species richness) and abundance of cetaceans in the Gulf, the focus of this chapter, can be ascribed to 3 main factors. First, the Gulf has exceptionally high rates of primary productivity (Zeitzchel 1969), supporting complex and productive food webs. Among toothed whales (Odontoceti), including fish-eating (ichthyophagi),

squid-eating (teuthophagi), and (otherwise) flesh-eating (sarcophagi) species, all find abundant and diverse prey to support generally large populations year-round. Among baleen whales (Mysticeti), the great abundance of euphausiid shrimps (krill) is the primary reason for the presence of the blue whale (*Balaenoptera musculus*) and the fin whale (*B. physalus*), among others. A second major factor is the great diversity of habitats reflecting the complex topography and oceanography of the Gulf (see chapter 9), allowing for the presence of both coastal and oceanic species, as well as species from both tropical and temperate waters. Third, the warm and relatively calm waters found in the Gulf during winter and spring are exploited by several migratory species. For example, humpback whales (*Megaptera novaeangliae*) give birth, nurse, and care for their newborn in the Gulf, and, until recently, so did part of the migrating population of the gray whale (*Eschrichtius robustus*) (Findley and Vidal 2002).

The Gulf of California is also an economically active region. Due to its ecological richness, high biological productivity, and proximity to the U.S. border, key economic activities have been steadily increasing along the Gulf's shores, contributing to uncontrolled coastal population growth. Major threats to the Gulf's biodiversity, including cetaceans, are uncontrolled or minimally controlled fisheries, tourism, aquaculture, and agriculture (see also

Figure 14.1. Gulf of California, with the Upper Gulf of California and Colorado River Delta Biosphere Reserve (arrow; see also fig. 14.3), and the approximate southern limit of the fin whale's distribution (dashed line).

chapter 3). The increasingly intensive fishing effort in the Gulf, along with relatively recent curtailment of freshwater inflow, has resulted in population declines, in some cases even commercial extinction, of several fishes, invertebrates, and marine turtles. Tourism alone attracts more than 12 million visitors per year and generates almost 2 billion dollars in revenue. Agricultural and aquacultural developments increasingly alter and pollute ecologically important coastal wetlands, releasing agrochemicals and destroying large swaths of the coastal zone (see also chapters 9 and 15).

Here we provide an overview of cetacean species diversity and patterns of abundance in the Gulf of California, with additional information on conservation status. Following some descriptions of local threats to whales and dolphins, and discus-

sion of current Mexican legislation relevant to cetacean conservation, we focus on 2 flagship species, the vaquita (*Phocoena sinus*) and the fin whale. For distinct reasons, both species represent conservation priorities. The vaquita is a critically endangered porpoise endemic to the northern reaches of the Gulf, while the population of the fin whale appears to be small and remarkably genetically isolated.

Cetacean Species Diversity and Population Status

Cetacean species diversity (or richness) in the Gulf of California is outstanding. With 31 species (21 genera) of whales and dolphins recorded, this sea has

37% of the world's 83 cetacean species (table 14.1). Three of the world's 4 families of baleen whales (Mysticeti) and 5 of the 10 families of toothed whales (Odontoceti) are represented.

Baleen Whales (Suborder Mysticeti)

The endangered North Pacific right whale (*Eubalaena japonica*) is the only local representative of the family Balaenidae (table 14.1), with only 1 confirmed sighting of this cold-water loving whale inside the Gulf, 10 miles east of San Jose del Cabo, near the tip of the Baja California peninsula (Gendron et al. 1999). In the monotypic Eschrichtiidae, the second baleen whale family represented in the region, the well-known gray whale is a frequent visitor to the southern Gulf during the winter and spring (December–April). Until recently, a small proportion of wintering pregnant females even gave birth and nursed their calves in coastal waters of southern Sonora and Sinaloa (Findley and Vidal 2002). Also, a few individuals (mainly juveniles) have been seen in summer and fall as far north as El Golfo de Santa Clara, Bahía San Jorge, Bahía de los Angeles, and Bahía San Luis Gonzaga (Vidal 1989; Vidal et al. 1993; Mellink and Orozco-Meyer 2002; Findley and Vidal unpubl. data), especially during cooler La Niña years (Urbán et al. 1999b; Sánchez-Pacheco et al. 2001). However, at that time of year most gray whales have migrated to boreal waters of the North Pacific (Urbán 2000a).

Also among the 8 baleen whale species recorded in the Gulf of California are 6 of the world's 7 species of rorquals (Balaenopteridae; table 14.1). The blue whale is a regular winter visitor, chiefly to the southwestern Gulf, especially off Loreto and Bahía de La Paz, but with sightings elsewhere and as far north as San Felipe in the northern Gulf (Tershy et al. 1990; Vidal et al. 1993; Thompson et al. 1996; Urbán 2000b). The remarkable fin whale, described in more detail further on, appears to be the only baleen whale with a resident population in the Gulf of California. The cold-water loving sei whale (*Balaenoptera borealis*) is the rarest rorqual, with less than 20 confirmed sightings, mainly in the central and southern Gulf (Gendron and Chávez 1996; Urbán 2000b). Bryde's whale (*B. edeni*) is one of the most common and widely distributed whales in the Gulf, with sightings year-round (Tershy et al. 1990; Urbán and Flores 1996). Based on genetic

studies, there may be 2 forms of this whale in the Gulf, one resident and the other seasonally immigrating from the closely related eastern tropical Pacific population (Dizon et al. 1995). The common minke whale (*B. acutorostrata*) has been recorded sporadically throughout the Gulf, from near the fishing town of El Golfo de Santa Clara (on the eastern side of the Colorado River delta) in the far north to Bahía de La Paz in the south (Tershy et al. 1990; Vidal et al. 1993; Urbán 2000b). The humpback whale is a regular visitor to the southern gulf, during winter, especially around the tip of the Baja California peninsula and in the southeasternmost Gulf off Nayarit and Jalisco, including the large Bahía Banderas (Urbán and Aguayo 1987; Urbán et al. 2000).

Toothed Whales (Suborder Odontoceti)

The monotypic family Physeteridae is represented in the Gulf by the sperm whale (*Physeter macrocephalus*), which is present year-round, especially in the deeper waters of the central Gulf south of the Midriff Islands, and along the peninsular coast off Santa Rosalía and Loreto (Vidal et al. 1993; Gendron 2000a; Jaquet and Gendron 2002).

Two kogiids occur in the Gulf, the dwarf sperm whale (*Kogia sima*) and the pygmy sperm whale (*K. breviceps*). The dwarf sperm whale is frequently sighted along the Baja California coast from Loreto to Isla Cerralvo, whereas the more rarely seen pygmy sperm whale has been recorded throughout the Gulf (Vidal et al. 1993; Gendron 2000b).

Among toothed whales, the beaked whales (Ziphiidae) are represented in the Gulf by 4 described species, Cuvier's beaked whale (*Ziphius cavirostris*), Baird's beaked whale (*Berardius bairdii*), Blainville's beaked whale (*Mesoplodon densirostris*), and pygmy beaked whale (*M. peruvianus*), and by 2 unidentified, potentially undescribed species, *Mesoplodon* sp. A (Pitman et al. 1987), and *Hyperoodon* sp. (Urbán et al. 1994; table 14.1). All of these species are rare or at least difficult to detect in the Gulf of California (Urbán and Pérez-Cortés 2000).

The family Delphinidae is represented by 13 (38%) of the world's 34 recognized species. Some, such as the melon-headed whale (*Peponocephala electra*), are of rare occurrence with only 1 or 2 recorded sightings. However, others, such as the common bottlenose dolphin (*Tursiops truncatus*),

Table 14.1. Cetaceans of the Gulf of California and their conservation status.

Scientific Name[a]	Common Name[a]	Conservation Status[b]		
		IUCN	CITES	NOM-59
Family Balaenidae				
Eubalaena japonica	North Pacific right whale	EN	I	IDE
Family Eschrichtiidae				
Eschrichtius robustus	Gray whale	LR	I	SP
Family Balaenopteridae				
Balaenoptera musculus	Blue whale	EN	I	SP
Balaenoptera physalus	Fin whale	EN	I	SP
Balaenoptera borealis	Sei whale	EN	I	SP
Balaenoptera edeni	Bryde's whale	DD	I	SP
Balaenoptera acutorostrata	Common minke whale	LR	I	SP
Megaptera novaeangliae	Humpback whale	VU	I	SP
Family Physeteridae				
Physeter macrocephalus	Sperm whale	VU	I	SP
Family Kogiidae				
Kogia breviceps	Pygmy sperm whale	DD	II	SP
Kogia sima	Dwarf sperm whale	DD	II	SP
Family Ziphiidae				
Ziphius cavirostris	Cuvier's beaked whale	DD	II	SP
Berardius bairdii	Baird's beaked whale	LR	I	SP
Hyperoodon sp.[c]	Bottlenose whale sp.	LR/—[d]	I	SP
Mesoplodon peruvianus	Pygmy beaked whale	DD	II	SP
Mesoplodon densirostris	Blainville's beaked whale	DD	II	SP
Mesoplodon sp. A[c]	Beaked whale sp. A	DD	II	SP
Family Delphinidae				
Steno bredanensis	Rough-toothed dolphin	DD	II	SP
Tursiops truncatus	Common bottlenose dolphin	DD	II	SP
Stenella attenuata	Pantropical spotted dolphin	LR	II	SP
Stenella longirostris	Spinner dolphin	LR	II	SP
Stenella coeruleoalba	Striped dolphin	LR	II	SP
Delphinus delphis	Short-beaked common dolphin	DD	II	SP
Delphinus capensis	Long-beaked common dolphin	DD	II	SP
Lagenorhynchus obliquidens	Pacific white-sided dolphin	DD	II	SP
Grampus griseus	Risso's dolphin	DD	II	SP
Peponocephala electra	Melon-headed whale	DD	II	SP
Pseudorca crassidens	False killer whale	DD	II	SP
Orcinus orca	Killer whale	LR	II	SP
Globicephala macrorhynchus	Short-finned pilot whale	LR	II	SP
Family Phocoenidae				
Phocoena sinus	Vaquita	CR	I	IDE

[a]Scientific and common names from Rice (1998) and International Whaling Commission (2001).

[b]IUCN: **CR** = critically endangered, **EN** = endangered, **VU** = vulnerable, **LR** = lower risk, **DD** = data deficient (from Hilton-Taylor, 2000). Note that the IUCN is in the process of reassessing all taxa using a revised system of criteria and categories. CITES: **I** = Appendix I (most endangered CITES-listed species), **II** = Appendix II (species not necessarily currently threatened with extinction but at risk of becoming so unless trade is strictly regulated) (from CITES Secretariat 2002). NOM-59-ECOL-2001: **IDE** = in danger of extinction (endangered), **SP** = subject to special protection (from NOM-59-ECOL-2001, DOF 2002a).

[c]Unidentified species, possibly new to science.

[d]The unknown, possibly new, bottlenose whale species of the Gulf does not appear in the current IUCN Red List of Threatened Species. The two known bottlenose whale species, *Hyperoodon ampullatus* and *H. planifrons*, are assigned to the category LR.

short-beaked common dolphin (*Delphinus delphis*), long-beaked common dolphin (*D. capensis*), and short-finned pilot whale (*Globicephala macrorhynchus*), have resident populations and are frequently sighted year-round. The rough-toothed dolphin (*Steno bredanensis*) appears to be only a summer visitor, whereas the Pacific white-sided dolphin (*Lagenorhynchus obliquidens*), more typical of the colder waters of the California Current, is present in the southwestern Gulf in winter. Some species, such as the pantropical spotted dolphin (*Stenella attenuata*) and the spinner dolphin (*S. longirostris*), are common in the southern Gulf but with no records from the northern part (Vidal et al. 1993; Urbán et al. 1997; Pérez-Cortés et al. 2000). Although seen infrequently, the striped dolphin (*S. coeruleoalba*) is widely distributed in the Gulf (Aguayo and Perdomo 1985; Gallo-Reynoso 1986; Pérez-Cortés et al. 2000). Killer whales (*Orcinus orca*) are not abundant. As they are present year-round, however, they presumably have a resident population in the Gulf (Guerrero-Ruíz et al. 1998).

The critically endangered vaquita (*Phocoena sinus*) is the only phocoenid species represented in the Gulf of California. Its population status is discussed later in this chapter.

Population Size Estimates

Little is known of population sizes of most Gulf cetaceans. Abundances have been estimated based on studies of varying duration and using different survey methodologies and observation platforms (from small boats to large oceanographic vessels and various aircraft). This lack of consistency introduces some uncertainty when comparing estimates of abundance among Gulf cetaceans. Nevertheless, the long-beaked common dolphin is clearly the most abundant species, with an estimated population size exceeding 60,000 individuals (Gerrodette and Palacios 1996; table 14.2). The common bottlenose dolphin, short-beaked common dolphin, pantropical spotted dolphin, and spinner dolphin populations are estimated at sizes varying from about 20,000 to 35,000 individuals (Gerrodette and Palacios 1996). The fin whale and Bryde's whale are the most abundant year-round resident baleen whales in the Gulf, with some population size estimates reaching about 800 and 600 individuals, respectively (table 14.2).

Conservation Status and Legal Protection

Conservation Status

The latest edition of the World Conservation Union's (IUCN) Red List of Threatened Species includes 1 species present in the Gulf of California as Critically Endangered (vaquita), 4 as Endangered (blue whale, fin whale, sei whale, and northern right whale), and 2 as Vulnerable (humpback whale and sperm whale) (Hilton-Taylor 2000; table 14.1).

The Convention on International Trade in Endangered Species of Wild Fauna and Flora (CITES 2002) includes all of the baleen whales, as well as the sperm whale, bottlenose whale, Baird's beaked whale, and vaquita in its Appendix I (species most endangered among CITES-listed animals and plants, with commercial international trade in specimens of these species prohibited). All the other toothed whales in the Gulf of California are included in its Appendix II (species for which commercial trade is strictly regulated with special permits delivered by the exporting country; table 14.1).

The Mexican government's Official Standard NOM-59-ECOL-2001 identifies several categories of species and subspecies of the country's terrestrial and aquatic flora and fauna classified as At Risk (*En Riesgo*). It includes all of the cetacean species entering Mexico's waters (DOF 2002a). The North Pacific right whale and the vaquita are listed as "Endangered" and all others as "Subject to Special Protection" (table 14.1).

Conservation Threats

The main human-caused mortalities of cetaceans in the Gulf of California are related to fisheries, especially coastal (inshore) fisheries. Zavala-González et al. (1994) mention more than 125 specimens recovered from artisanal fisheries and killed by gillnet entanglement, or even deliberately killed by harpoons or firearms. The cetaceans most frequently involved are the short-beaked common dolphin, long-beaked common dolphin, common bottlenose dolphin, pantropical spotted dolphin, and vaquita. Dolphins killed are used frequently as shark bait. Large cetaceans also have been affected by fisheries. At least 7 cases of gray whale entanglements in gillnets have been documented in the Gulf, with 5 of these whales being released (Vidal et al. 1994;

Table 14.2. Estimated population size and time-of-year presence of cetaceans in the Gulf of California.

Species	Population Size Estimate (95% Confidence Interval)	Survey Method	Time of Year Present	Reference[a]
North Pacific right whale	Few individuals	Field observations	Winter-spring	7
Gray whale	100–1000	Field observations	Winter-spring	7
Blue whale	100–1000	Line-transect	Winter-spring	7
Fin whale	820 (594–3229)	Line-transect	Year-round	1
	386 (282–488)	Mark-recapture	Year-round	2
Sei whale	> 100 (?)	Field observations	Winter-spring	7
Bryde's whale	564 (453–2085)	Line-transect	Year-round	1
	450	Mark-recapture	Year-round	3
Common minke whale	> 100 (?)	Field observations	Year-round	7
Humpback whale	1813 (918–2505)	Mark-recapture	Winter-spring	4
Sperm whale	417 (164–1144)	Line-transect	Year-round	1
Pygmy sperm whale	> 100 (?)	Field observations	Year-round	7
Dwarf sperm whale	100–500 (?)	Field observations	Year-round	7
Cuvier's beaked whale and *Mesoplodon* spp.	13,104 (4997–34,912)	Line-transect	Year-round	1
Baird's beaked whale	100 (?)	Field observations	Summer	7
Bottlenose whale sp.	100 (?)	Field observations	Year-round	7
Rough-toothed dolphin	6,341 (2853–14,757)	Line-transect	Summer	1
Common bottlenose dolphin	33,799 (20,500–58,358)	Line-transect	Year-round	1
Pantropical spotted dolphin	23,734 (14,419–40,913)	Line-transect	Year-round	1
Spinner dolphin	22,724 (12,411–43,572)	Line-transect	Year-round	1
Striped dolphin	8,642 (3314–23,603)	Line-transect	Year-round	1
Short-beaked common dolphin	28,681 (14,287–72,316)	Line-transect	Year-round	1
Long-beaked common dolphin	61,976 (31,295–154,153)	Line-transect	Year-round	1
Pacific white-sided dolphin	100–500	Field observations	Winter-spring	7
Risso's dolphin	16,918 (9027–33,205)	Line-transect	Year-round	1
Melon-headed whale	Few individuals	Field observations	?	7
False killer whale	> 100 (?)	Field observations	?	7
Killer whale	100–500	Field observations	Year-round	5
Short-finned pilot whale	3923 (1591–9829)	Line-transect	Year-round	1
Vaquita	567 (177–1073)	Line-transect	Year-round	6

Taxonomic order follows Rice (1998) and the International Whaling Commission (2001).

[a]References: (1) Gerrodette and Palacios 1996; (2) Enríquez 1996; (3) Urbán and Flores 1996; (4) Urbán et al. 1999b; (5) Guerrero-Ruíz 1997; (6) Jaramillo-L. et al. 1999; (7) Marine Mammal Research Program, Universidad Autónoma de Baja California Sur, La Paz.

Urbán et al. 2002). In February and March 2002, 2 humpback whales became entangled in gillnets, one in Bahía de La Paz and the other farther south, near Cabo Pulmo. Fortunately, both were released.

Another apparently human-caused mortality event occurred in the winter of 1995 in the upper Gulf of California, where 367 dolphins (including long-beaked common dolphins, common bottlenose dolphins, and striped dolphins), 8 baleen whales (including fin, common minke, and Bryde's whales), 51 California sea lions, and 215 sea birds (mostly Pacific loons [*Gavia pacifica*], eared grebes [*Podi-ceps nigricollis*], brown pelicans [*Pelecanus occidentalis*], and double-crested cormorants [*Phalacrocorax auritus*]) were found dead, possibly due to sea contamination by NK-19, a fluorescent cyanide compound used by narcotraffickers to mark drop-areas for unloading drugs (PROFEPA-SEMARNAP 1995; see also chapter 9). Although discounting that specific agent as the proximal cause of mortality, Vidal and Gallo-Reynoso (1996) agreed that the die-off was likely caused by an unknown toxic substance in the water or in prey ingested by the affected animals.

Formal Protection of Cetaceans in the Gulf of California

There is no single body of legislation enacted for the sole benefit of cetaceans in the Gulf. Instead, several different laws relevant to their conservation and management exist, and they apply to all of Mexico. The General Law of Ecological Balance and Environmental Protection (Ley General del Equilibrio Ecológico y la Protección al Ambiente), enacted in 1988, is currently the responsibility of the recently restructured Secretariat of the Environment and Natural Resources (SEMARNAT; see chapter 4). Articles 15 through 19 of the law provide SEMARNAT with a broad mandate to formulate policy and planning initiatives and to implement management actions for the protection of the nation's natural resources (Estados Unidos Mexicanos, 1993). The Fishing Law (Ley de Pesca) authorizes government agencies dealing with fisheries to "establish measures aimed at the protection of . . . marine mammals" (Secretaria de Pesca 1992: 10). Another piece of legislation, a 1991 addition to the Mexican Penal Code, Article 254 Bis, prohibits unauthorized capture of or injury to marine mammals and sea turtles. A prison term of 3 to 6 years is prescribed as the penalty (DOF 1991).

The General Law of Wildlife (Ley General de Vida Silvestre), under the responsibility of SEMARNAT, was approved on April 27, 2000 (DOF 2000a). This is the first pertinent Mexican law related to wildlife that confronts the challenges of balancing protection of the country's megadiversity with the need for socioeconomic development. On January 10, 2002, Article 60 Bis was added, stating that no specimen of any marine mammal can be the subject of subsistence or commercial use, with the exception of captures for scientific research and educational purposes, which still require prior approval of the authorities (DOF 2002b).

The Mexican government's Official Standard NOM-131-ECOL-1998 provides specific guidelines for whale-watching activities compatible with the conservation of whales and their habitat (DOF 2000b). In particular, the guidelines are species specific and define which areas and what period of the year whale watching is permitted, the number and type of boats allowed, and the distance to the whales and duration of observation.

Additionally, in May 2002, Mexico established the Mexican Whale Sanctuary (Santuario Ballenero Mexicano), encompassing its entire Exclusive Eco-nomic Zone (about 3 million km^2). The decree stipulates that environmental conditions required for biological functions of whales (e.g., breeding, calving, growth, migration, learning, and feeding) must be maintained. Species protected include all members of the families Balaenidae, Balaenopteridae, Eschrichtiidae, Physeteridae, Kogiidae, and Ziphiidae, in addition to killer whale, short-finned pilot whale, false killer whale (*Pseudorca crassidens*), pigmy killer whale (*Feresa attenuata*) and melon-headed whale in the family Delphinidae (DOF 2002c).

Marine Protected Areas in the Gulf of California

Upper Gulf of California and Colorado River Delta Biosphere Reserve

The Mexican government created the Upper Gulf of California and Colorado River Delta Biosphere Reserve on June 10, 1993, in part to protect the vaquita and the also endangered and endemic large corvinalike fish, the totoaba (*Totoaba macdonaldi*; see chapter 9). On June 29, 1994, the government published the Mexican Official Standard NOM-012-PEASC-1993 (DOF 1994) specifying the legal protection of the vaquita and the totoaba and providing a complement to the reserve's regulations. This standard stipulates specifically that any form of fishing is prohibited in the nuclear zone (core) of the reserve and prohibits the use of gillnets with mesh of 10 inches or more elsewhere in the reserve (the buffer zone). The most recent official standard, NOM-EM-139-ECOL-2002, lists actions to be taken to protect the marine and coastal ecosystems and those species under special protection in waters of the reserve (DOF 2002d; discussed in more detail further on).

Bahía de Loreto National Park

The greater Bahía de Loreto, on the central Gulf coast of the Baja California peninsula, harbors a high diversity of cetaceans. It is an important area for fin and blue whales in particular, as both species feed and, during the winter, also presumably mate and calve in the greater Bahía de Loreto. To define legally the management strategies necessary to preserve local natural resources and promote social development, the Bahía de Loreto National

Park (Parque Nacional Bahía de Loreto) was created on July 19, 1996 (DOF 1996).

Case Study: The Vaquita

Brief Description

The vaquita was described by Norris and McFarland (1958) from skull remains found on far-northern Gulf beaches. However, 29 years passed before fresh specimens became available to allow a full description of its external morphology and coloration (Brownell et al. 1987).

The vaquita is the smallest marine cetacean and smallest of all true porpoises (family Phocoenidae). Mean length of females is only 140.6 cm, and males are even slightly smaller (mean = 134.9 cm; Brownell 1983; Hohn et al. 1996). Although similar in external morphology to the harbor porpoise (*Phocoena phocoena*), its geographically nearest relative, it differs from that species and other phocoenids by its smaller size, proportionally larger flippers, and taller, more falcate dorsal fin.

The vaquita is built robustly (fig. 14.2). In profile, the head appears as a truncated cone with the posterior part of the melon (forehead) sloping slightly inward toward the blowhole. Anteriorly, the melon slopes abruptly to the snout tip. The pigmentation pattern is generally consistent on the body, with dark gray cape, pale gray lateral field, and white ventral field. The most conspicuous features are the relatively large black eye and lip patches (Brownell et al. 1987; Vidal et al. 1999).

Distribution and Abundance

The vaquita is the only marine mammal endemic to Mexico. Recent systematic surveys and other evidence support its historical and current geographical range as being restricted to the upper Gulf of California, especially its western portion (Barlow 1986; Brownell 1986; Silber 1990; Gerrodette et al. 1995; Barlow et al. 1997; Gallo-Reynoso 1998; Jaramillo et al. 1999; Vidal et al. 1999).

Even under the best weather conditions, vaquitas are difficult to detect and survey. In addition to their small body size, they tend to form only very small groups, live in perpetually murky-water habitat, and exhibit elusive diving/surfacing behaviors (Silber and Norris 1991; Barlow et al. 1993). Based on systematic surveys, rather than the limited survey

Figure 14.2. Vaquita killed by entanglement in a fishing net. Note the characteristic eye and lip pigment patches. (Photos by A. Jaramillo-Legoretta.)

data and educated guesswork of prior efforts (reviewed in Vidal et al. 1999), Barlow et al. (1997) estimated that the species' abundance ranged from a low of 224 (1993 ship survey) to a high of 885 individuals (1988–1989 aerial surveys). During August–September 1997, a more extensive population census, encompassing the entire upper Gulf of California and surveying all possible habitats, including shallow waters of the Colorado River delta, was cooperatively conducted by the U.S. Southwest Fisheries Science Center and Mexico's National Marine Mammal Program. The total population was estimated at 567 vaquitas (95% confidence interval, 177–1073; Jaramillo et al. 1999). This estimate is likely more accurate than previous attempts, which usually had a low number of sightings, relied on parameters adapted from surveys of other species and cruises, and did not cover all areas of potential occurrence.

Life History

Data on the life history of the vaquita were reported by Hohn et al. (1996), who analyzed the age distribution of 56 individuals recovered from gillnets (targeting various fish species), found as carcasses on beaches, or obtained from museums. The results of that analysis are striking. Most individuals (62%) were 0–2 years of age (immatures), and 31% were 11–16 years (total range of mature individuals examined was 7–21 years). There was a complete absence of individuals ages 3–6 years. Hohn et al. (1996) noted that the bimodal age structure, if not accurately reflecting the state of the current population (which it very well may do), could stem from a biased sample possibly due to age/sex segregation/distribution and/or differential susceptibility to entanglement in nets, as also pointed out by Rojas-Bracho and Jaramillo-Legorreta (2002).

Hohn et al. (1996) also reported that the vaquita is a highly seasonal (but nonprolific) reproducer, with most births occurring in early March. Based on data from the presumably similar harbor porpoise (*P. phocoena*) and a limited sample of mature vaquita ovaries (indicating nonannual ovulation), the gestation period was estimated to last 10–11 months, and it was concluded that mature females do not produce calves each year. The maximum lifespan recorded was 21 years. Age at sexual maturity was difficult to estimate because of the lack of juveniles in the sample, but all specimens of 3 or fewer years of age were immature, and all of 6 or more years were mature (Hohn et al. 1996).

Conservation and Management

The vaquita is 1 of the 3 most endangered cetacean species in the world (Jefferson et al. 1993). Classified as Critically Endangered by the World Conservation Union and as Endangered by the Mexican government (under NOM-59-ECOL-2001), the vaquita is also listed in Appendix I of CITES (table 14.1).

The vaquita is endangered in part because it is naturally rare, with only 1 small, narrowly distributed population (Rojas-Bracho and Taylor 1999; Taylor and Rojas-Bracho 1999). At the same time, however, there is no doubt that anthropogenic effects have greatly increased its risk of extinction. But which anthropogenic effect represents the most serious threat to the long-term persistence of the vaquita has been at the center of a long-standing debate (see below), which unfortunately has hindered efforts to protect the species (Rojas-Bracho and Taylor 1999). Additional specific conservation actions are needed beyond the simple designation of the Upper Gulf of California and Colorado River Delta Biosphere Reserve. The results of population surveys in 1993 and 1997 indicate that this (ostensibly) protected area does not adequately match the distribution of the vaquita. A large percentage (40%) of vaquita sightings were from outside the reserve's southern boundary (fig. 14.3). Further, there were no sightings from within the nuclear zone of the reserve, where all fishing is (ostensibly) prohibited (Gerrodette et al. 1995; Jaramillo et al. 1999).

Created by the Mexican government, the International Committee for the Recovery of Vaquita (CIRVA) is a more specific tool than the biosphere reserve for ensuring the preservation of the vaquita. CIRVA is composed of scientists from Mexico, Canada, the United Kingdom, and the United States. Its mandate is to propose a recovery plan for the species based on the best available scientific information. The plan must also evaluate the socioeconomic impacts of any promulgated regulations (Rojas-Bracho and Jaramillo-Legorreta 2002). To expedite decision making by CIRVA members, Rojas-Bracho and Taylor (1999) evaluated the impacts of 3 human-related factors on the population status of the vaquita: (1) habitat alteration from reduced inflow of the Colorado River; (2) high pollutant loads; and (3) increased mortality caused by gillnets. The debate mentioned above has centered largely on the respective impacts of water diversions and impoundments of the Colorado River and incidental mortality due to gillnets.

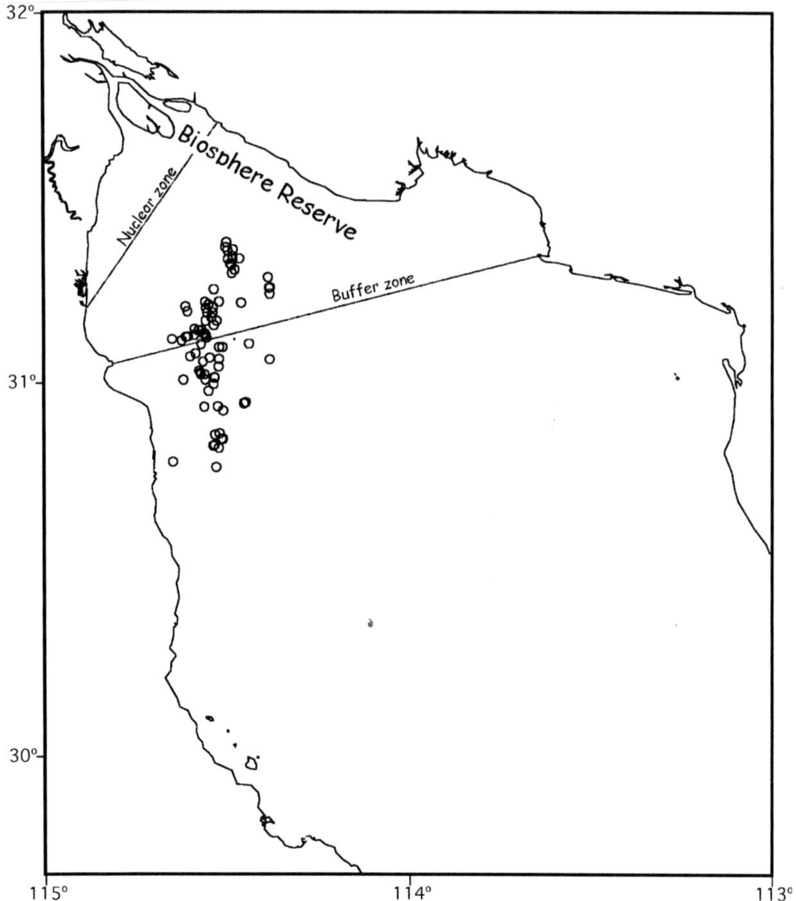

Figure 14.3. Sightings of the vaquita in relation to the Upper Gulf of California and Colorado River Delta Biosphere Reserve, August–September 1997. Based on data from Jaramillo L. et al. (1999).

Reduced Inflow of the Colorado River

The history of human impacts on the lower Colorado River and its delta region is reviewed in chapter 9. Since the early 1940s, the flow of the Colorado River to the upper Gulf of California has been greatly reduced by diversions for mainly agricultural purposes. Potential negative effects of reduced inflow of the Colorado River on vaquita survival first received attention during the 1970s, including during the 28th meeting of the International Whaling Commission (Brownell 1982; Gaskin 1982). Since that time, the position that flow reduction is the main threat to vaquitas has gained strength, particularly within the government sector of Mexico.

Different Mexican federal agencies, including the former National Marine Mammal Program, have maintained that this factor, not commercial fishing, was the reason for the population decline and current endangered status of the vaquita (Fleischer et al. 1994; Fleischer 1996).

Major rivers are important nutrient and sediment transportation agents. According to Villa-Ramirez (1993) and Fleischer (1996), reduced inflow of nutrients from the Colorado River has been responsible for degrading the vaquita's habitat. However, nutrient concentrations in the upper Gulf remain consistently well above those considered limiting to primary productivity. As measured through phytoplankton biomass and chlorophyll-a concentrations, primary productivity is also high. Zooplankton volumes exceed by a factor of 2 the values reported for strong upwelling regions such as those off the Pacific coasts of Costa Rica and Peru (for reviews of the oceanography of the Gulf of California, particularly its upper part, see Alvarez-Borrego 1983, 1992; Alvarez-Borrego and Lara-Lara 1991; Rojas-Bracho and Taylor 1999; and chapter 9).

Additionally, analyses of the stomach contents of 34 vaquitas showed them to be rather nonselective feeders on several species of marine benthic/demersal fishes and squids (Findley and Nava 1994; Findley et al. 1995; Pérez-Cortés 1996; Vidal et al. 1999), much like the harbor porpoise (Gaskin 1982) and Burmeister's porpoise (*P. spinipinnis*; Goodall et al. 1995). Vaquitas are thus unlikely to be strongly affected by the potential decline of prey species due solely to reduced freshwater input. To date, none of the vaquitas taken from gillnets has been reported to show signs of emaciation, including mothers, calves, and juveniles, those most likely to be stressed by food shortages (Vidal 1995; Hohn et al. 1996). In conclusion, habitat alteration resulting from the reduction in Colorado River inflow is currently a low-risk factor (Rojas-Bracho and Taylor 1999). It should also be pointed out that reduced inflow of freshwater is not the only possible mechanism through which the vaquita could be affected by lack of food. The continued high-level of commercial bottom trawling for shrimps in the upper Gulf undoubtedly has long-term negative impacts on fragile benthic communities and populations of species forming the vaquita's prey (Findley and Nava 1994; Nava-Romo 1994; chapter 9).

Reduced Fitness from High Pollutant Loads

Rojas-Bracho and Taylor (1999) reviewed the risk associated with pollutants for the vaquita in the northern Gulf of California. Chlorinated hydrocarbons, in particular, are always a source of concern: many are soluble in fat, and they can accumulate in living tissues and enter food webs. However, concentrations of chlorinated hydrocarbons measured in coastal waters of the Gulf are at least ten times below the U.S. Food and Drug Administration's acceptable limits for human consumption (Gutierrez-Galindo et al. 1992). Mora and Anderson (1991) found only low levels of organochlorine residues in seabirds from the northern Gulf. Calambokidis (1988) and Calambokidis et al. (1993) reported that

maximum values for total DDT and polychlorinated biphenyls (PCBs) in tissue samples from several vaquitas were relatively low, at 9.1 parts per million (ppm, wet weight) and 0.20 ppm (wet weight), respectively. V. Camacho (Department of Geochemistry, Instituto de Investigaciones Oceanológicas, Universidad Autónoma de Baja California, Ensenada, pers. comm. to Rojas-Bracho, April 1997) found DDE in concentrations of 1 ppm in tissues of 1 vaquita. By comparison, concentrations reported by Otterlind (1976) as possibly causing the decline of harbor porpoise in Swedish waters were 560 ppm (DDT) and 260 ppm (PCBs). According to Calambokidis (1988) and Calambokidis et al. (1993), the ratio of DDE to total DDT indicates that the source of DDT is not from recent applications of this pesticide, a conclusion congruent with Gutierrez-Galindo et al.'s (1988a,b) findings for the filter-feeding bivalves *Chione californiensis* and *Modiolus capax* in the northern Gulf. Thus, available data do not indicate that chlorinated hydrocarbon pesticides or PCBs are an immediate threat to the survival of the remaining vaquita population (Calambokidis et al. 1993). Because other pollutants appear less likely to compromise reproduction and/or induce mortality of marine mammals, pollutants per se currently present little or no risk to the vaquita (Rojas-Bracho and Taylor 1999).

Increased Mortality as Fisheries Bycatch

Although all marine mammals are susceptible to gillnet entanglement (Perrin et al. 1994), porpoises, including vaquitas, are particularly vulnerable to this type of fishing gear (Jefferson and Curry 1994). The first reports of incidental catch of vaquitas in gillnets designed for capture of totoaba and other fishes came from Norris and Prescott (1961) and W. E. Evans (in Brownell 1982). The highest incidental kills of vaquita have been in the large-mesh (15–30.5 cm) totoaba gillnets (Villa-Ramírez 1976; Brownell 1983; Vidal 1995; Vidal et al. 1999). Brownell (1982) suggested that the annual incidental kill in the upper Gulf during the early 1970s was in the range of tens to hundreds. Vidal (1995) documented the incidental mortality of 128 vaquitas (1985–1994) in gillnets with mesh sizes from 8.5 cm to 30.5 cm. Boyer and Silber (1990) and Vidal (1995) estimated roughly the bycatch at about 32 and 35 vaquitas per year, respectively.

Despite the above reports and numerous other studies (e.g., Mitchell 1975; Rojas-Bracho and Urbán-Ramírez 1993; D'Agrosa et al. 1995), the Mexican government fisheries agency continued to disregard gillnet mortality as a serious threat to vaquita survival in its communiqués and reports (Fleischer et al. 1994; Fleischer 1996). In 1995, however, the International Whaling Commission's subcommittee on small cetaceans recommended that immediate action be taken to eliminate incidental catches (International Whaling Commission 1995). The conclusion was based in part on demographic parameters drawn from harbor porpoise in the Gulf of Maine/Bay of Fundy. Later, D'Agrosa et al. (2000) reconfirmed the high incidental mortality caused by gillnets, monitoring those deployed by fishermen from just 1 of the region's 3 main fishing ports (El Golfo de Santa Clara). In that study, specifically addressing incidental vaquita mortality in gillnets, the authors documented the deaths of 11 individuals in 1113 fishing trips and estimated mortality as being between 39 and 84 vaquitas per year.

Rojas-Bracho and Taylor (1999) have confirmed that current mortality levels are high enough to drive the vaquita to extinction. These authors considered the most optimistic scenario using the lowest annual mortality estimate and the highest vaquita abundance estimate and obtained a human-caused annual mortality rate of 0.07 (39/567). To sustain or counterbalance such a high mortality rate, vaquitas would need high productivity, reflected in a high potential population growth rate. Demographic data for the vaquita are lacking, and thus it is currently not possible to estimate maximum population growth rate in this species. However, estimates of maximum annual population growth rate for other cetaceans typically vary from 0.02 to 0.04 (see Rojas-Bracho and Taylor 1999). Species with such maximum population growth rates could not sustain a mortality rate of 0.07. By extension, it is unlikely that vaquitas can sustain current mortality rates.

Given the current mortality rate, how long does the vaquita have before becoming extinct? Based on mean abundance estimates from the 4 rigorous vaquita surveys conducted to date (Barlow et al. 1997) and mortality estimates from D'Agrosa et al. (1995, 2000), Rojas-Bracho and Taylor (1999) calculated that current population growth rates likely ranged between $r = -0.05$ and, considering that D'Agrosa et al. (2000) estimated mortality for only 1 fishing town, $r = -0.15$. Although greater rates of decline are plausible, even a decline of -0.15 is sufficient for the population to decline over the next 15-year period to a critical level of about 50 individuals (fig. 14.4). Such a small population would then be likely to decline even faster due to stochastic factors and inbreeding depression (Rojas-Bracho and Taylor 1999).

Clearly, gillnets are the greatest threat to the survival of the vaquita. Annual mortality by gillnets represents 6–14% of the current population size estimate, again based on data from only 1 of the 3 regional fishing ports. Thus, short-term management actions should not be hindered by uncertainty in estimating the risk associated with other factors, and primary conservation efforts should be directed toward immediate elimination of incidental fishery mortality (Rojas-Bracho and Taylor 1999; Taylor and Rojas-Bracho 1999). However, CIRVA has also concluded that, in the long-term, changes in vaquita habitat due to reduction of Colorado River inflow are matters of concern and must be investigated further (Rojas-Bracho and Jaramillo-Legorreta 2002).

Published in September 2002, the Mexican Emergency Official Standard NOM-EM-139-ECOL-2002 (DOF 2002d) declared a ban on shrimp trawlers and large-mesh gillnets in the entire Upper Gulf of California and Colorado River Delta Biosphere Reserve. With help from the Mexican Navy, PROFEPA (Attorney General's Office for the Protection of the Environment; enforcement agency under SEMARNAT) began to enforce the new law, an action that was met with violent protest from the fishing industry. This protest led to a negotiated compromise allowing only the local fishing fleet to reenter the reserve and, among other conditions, prohibiting fishing in the main area of distribution of the vaquita. NOM-EM-139-ECOL-2002 is an important step toward the recognition that incidental mortality by gillnets and the destruction of vaquita habitat by shrimp trawlers should be eliminated completely, as recommended by CIRVA. Ideally, the southern boundary of the Biosphere Reserve should even be expanded to include the entire range of the vaquita (fig. 14.3), and gillnets and shrimp trawlers should be banned in the hopefully enlarged reserve (see Gallo-Reynoso 1998; Rojas-Bracho and Jaramillo-Legoretta 2002). Regrettably, this last protective measure would significantly impact the resource users of the upper Gulf of California, and full, immediate protection does not seem feasible.

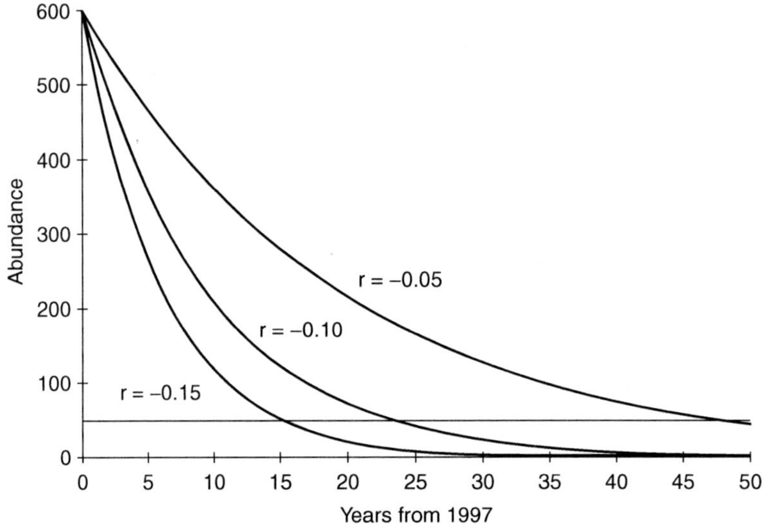

Figure 14.4. Projected population numbers of the vaquita over 50 years beginning in 1997, based on several scenarios. Abundance through time is estimated for population growth rates (*r*) of –0.05, –0.10, and –0.15. Each trajectory starts near the current best estimate of abundance. The horizontal line represents critical abundance of 50. From Rojas-Bracho and Taylor (1999); reprinted with permission.

An alternative proposal by CIRVA was that gillnet fishing within the range of the vaquita be removed in 3 stages, beginning with the largest-mesh gillnets. Meanwhile, strategies to offset economic hardship imposed by regulations on gillnet fishing should be promulgated with alacrity. Among other CIRVA recommendations were the need to gather data on habitat and seasonal movements of the vaquita, develop education and public awareness programs, and promote local community involvement (Rojas-Bracho and Jaramillo-Legoretta 2002).

Beginning in 2000, the World Wildlife Fund (WWF), CIRVA, and Conservation International (CI) have convened a series of stakeholder meetings on the recovery of the vaquita. The main outcome was the establishment of a working group, whose mandate is to develop a general strategy for recovery based on recommendations from CIRVA. In addition to WWF and CI, at least 30 other partner institutions and collaborators have committed to support the recovery strategy (Conservation and Sustainable Development Strategy for the Recovery of Vaquita [*Phocoena sinus*] and Its Habitat), consisting of 4 main elements: conservation, socioeconomic aspects, communications issues, and legal framework.

Case Study: The Fin Whale

Brief Description

The fin whale (or finback whale, fig. 14.5) is the second largest living animal after the blue whale, compared to which it is also more slightly built. At physical maturity, males and females average about 19 and 20 m in length, respectively. Adults have not been accurately weighed, but calculations suggest that a 25-m animal could weigh as much as 70,000 kg. Although showing considerable variation among populations, fin whales are generally brownish gray to blackish dorsally and whitish ventrally. They also often show streaks or crescents of lighter gray over and behind the head. The color pattern is asymmetrical in that the lower jaw usually is white or cream-colored on the right side and dark on the left. The dorsal fin is strongly curved and about 60 cm high, and an average of 85 throat grooves (pleats) extend to the navel.

Population Status and Movements

Fin whales are cosmopolitan baleen whales with an antitropical distribution. In the Northern Hemi-

Figure 14.5. Fin whale, *Balaenoptera physalus* (illustration by P. Folkens); reprinted with permission.

sphere, they can be observed as far south as the Mediterranean Sea or the Gulf of California, where they are relatively common. Several technical papers, field guides, and popular articles have mentioned that the Gulf of California population of fin whales appears to be resident (e.g., Gambell 1985; Vidal et al. 1993; Rice 1998). The earliest reference to a potentially resident population is by Gilmore (1957: 23), who stated that "finbacks inhabit the upper Gulf in numbers all year in what appears to be a land-locked population."

Until recently, Gilmore's suggestion of a resident and (presumably) isolated fin whale population proved difficult to substantiate. The hypothesis was supported by an analysis of vocalizations recorded from the Gulf population which indicated uniqueness in several acoustic characteristics and patterns compared to fin whales in other regions, thus implying isolation (Thompson et al. 1992). However, attempts to understand local and wider movements of Gulf fin whales through a modest mark (radio and visual tags) and recapture program in the 1980s

by investigators working out of Guaymas, Sonora, proved largely unsuccessful (i.e., almost no recaptures). And although a few photographic matches of identifiable individual fin whales were made between researchers working in different parts of the Gulf during the same time period, information on movements remained scant. Tershy et al. (1993) pointed out the necessity of genetic stock identification studies, comparison of photo-identifications of individuals from the Pacific with a comprehensive Gulf of California catalog, and satellite tagging of fin whales in the Gulf, especially during late spring.

Recent studies, particularly by researchers working out of La Paz, Baja California Sur, also support Gilmore's hypothesis. Of 519 fin whales photoidentified between 1981 and 2000 in different areas of the Gulf (La Paz, Loreto, Canal de Ballenas, San Luis Gonzaga, San Felipe, Puerto Peñasco, Puerto Libertad, Kino, Guaymas, and Agiabampo), 72 were resighted in more than 1 year (Urbán 1996; Urbán et al. 2001; fig. 14.6). The average time

Figure 14.6. Example of a distinctive fin whale rephotographed (resighted) in the Gulf of California during a period of 17 years; easily identified by the absence of a large section of its dorsal fin. It was first photographed in 1983 in Canal de Ballenas and last photographed (this photo) in 2000 in Bahía de Loreto. (Photo by A. Acevedo.)

between the first and last photographs of these whales was 6.8 years, with 2 resighted over an 18-year period. Irregular photographic effort among seasons and study sites makes it difficult to infer exact movement patterns of these animals. Fin whales did move around the Gulf, but interestingly, sightings and resightings tended to originate from those locations closest to one another (e.g., Loreto-Agiabampo; Canal de Ballenas-San Luis Gonzaga; San Luis Gonzaga-Puerto Libertad; and Guaymas-Kino). Also suggestive of year-round residency was that any regular spatial or seasonal movement patterns could not be detected with those data (Urbán et al. 2001).

Bérubé et al. (1998) conducted genetic analyses to evaluate the degree of isolation among fin whale populations across the North Atlantic, with samples collected in the Sea of Cortez also included. Results showed that the degree of genetic diversity among fin whales in the Gulf of California at nuclear and mitochondrial loci is highly reduced. Such reduced genetic variation signals a potentially small and isolated population (Bérubé et al. 1998).

To test if the reduced genetic variation detected in the Gulf of California is due to small population size or the result of a past bottleneck in an otherwise large eastern North Pacific population, Bérubé et al. (2002) analyzed differences in DNA sequence at 1 mitochondrial and 16 nuclear loci between samples collected from 8 fin whales off coastal California and 56 fin whales from the Gulf of California. For all 64 samples, sex was also determined. A highly significant difference in genetic sequence was observed between the samples from these 2 localities. The degree of divergence was estimated at $\theta = 0.24$ for the mitochondrial control region and $\theta = 0.22$ for all nuclear loci combined. This degree of genetic divergence is more than double that typically found for adjacent populations of the same cetacean species but is similar to that observed between populations occurring in different oceans.

The degree of isolation inferred by Bérubé et al. (2002) is surprising in light of the relatively unobstructed access to the Sea of Cortez, relatively short distance between coastal California and Gulf of California fin whale populations, and the potential for the species to migrate over great distances. In addition, fin whale abundance is lower in the summer and fall in Bahía de Los Angeles, Bahía de La Paz and off Guaymas, a pattern that can be interpreted in favor of seasonal movement out of the Gulf in the latter part of the spring or early sum-

mer (Tershy et al. 1993). During spring 2001, just before the seasonal decline in local abundance, 11 fin whales were tagged in the southern Gulf with satellite-monitored radio transmitters providing information on movements. All of the tagged whales remained in the Gulf, their movements suggesting seasonal migration to its northern portion rather than southward into the Pacific Ocean (Urbán et al. 2002; see also fig. 14.1). Thus, the genetic distinctiveness of Gulf of California fin whales (Bérubé et al. 2002) seems to be due to their year-round residency there. This is most unusual compared to populations of other species of baleen whales, which are almost invariably seasonal migrators over long distances. The Gulf of California is classified as a highly productive (>300 g C/m²–year) ecosystem (Alvarez-Borrego et al. 1978; Valdez-Holguin et al. 1995; chapter 9). This productivity likely explains why fin whales can, and apparently do, remain in the Gulf of California year-round (Tershy et al. 2002).

Feeding Ecology and Function of Low-Frequency Vocalizations

In 1999 and 2000, researchers from the University of California-Santa Cruz, Cornell University, and the Universidad Autónoma de Baja California Sur conducted a multifaceted study of the Gulf fin whale population. Systematic visual surveys from boats were combined with photo-identification every 5–7 days to independently estimate the density and distribution of whales within a 10 × 30-square-mile study area off Loreto. In addition, some whales were fitted with time-depth recorders. Vocalizing whales were tracked through sets of autonomous seafloor acoustic recorders, and biopsy samples were also taken from these individuals to determine their gender. Meanwhile, the abundance and distribution of krill (euphausiid shrimp) was also recorded through active acoustics (Clark et al. 2000).

Feeding was the primary activity of fin whales, as shown by their distribution relative to that of krill. Fin whales were feeding on krill (*Nyctiphanes simplex*) at depths of 97.9 ± 32.6 m, during only 6.3 ± 1.5 min of diving time (Croll et al. 2001). The short diving times were explained by the typical but energetically expensive feeding behavior, called "lunging," which apparently limits dive durations in this species (Acevedo-Gutierrez et al. 2002).

The long, patterned 15–30 Hz, or "20-Hz" vocal sequences characteristic of fin whale vocalizations can reach a sound intensity of 188 dB re 1μPa

(Watkins et al. 1987) and can be detected throughout the world's oceans (Richardson et al. 1995). Thompson et al. (1992) had earlier determined that Gulf fin whales exhibited unique vocalization characteristics, but during the recent study, these were shown to be only produced by males, despite a 1:1 overall sex ratio in the area (Croll et al. 2002). Thus, vocalizations are quite possibly male breeding displays used to attract females from great distances to concentrations of patchily distributed food. This finding lends support to growing concern over the impact of marine noise on *Balaenoptera* whales. Since at least the early 1960s, loud (190–250 dB re 1μPa at 1m) noise pollution from commercial ships, military sonar, seismic surveys, and oceanic acoustics research has increased in the frequency range used by large whales (Andrew et al. 2002). Noise pollution can mask other sounds and in particular reduces the distances over which receptive females might hear vocalizations of males. To the extent that *Balaenoptera* whale population growth is limited by the encounter rate of receptive females with singing males, the recovery of fin and other whale populations from past exploitation could be impeded by human-produced low-frequency sounds (Croll et al. 2002).

In conclusion, given the degree of genetic isolation, feeding ecology, singular vocalizations, movements, and residency status, the Gulf of California population of fin whales is unique. It is also an indicator of the regional ecosystem's health (productivity, level of noise pollution) and should be protected accordingly.

Conclusions

In this chapter we have described the diversity and estimated abundances of cetaceans in the Gulf of California. We have also identified and discussed some of the threats facing cetacean populations in the Gulf. Our coverage is necessarily incomplete: there are undoubtedly other threats to be documented, and even the most basic information on cetacean mortality caused by human actions is lacking for several areas of the Gulf. Moreover, total impact of the various threats cannot be predicted by simply summing their effects as though they were independent. At the same time, it may be difficult to evaluate synergistic effects. Wild populations are subject to pressures from both human activities and ecological variability.

As human use of the Gulf of California increases, a comprehensive, integrated management plan is sorely needed to ensure that economic development is compatible with conservation of cetaceans in this very productive and important sea. The plan should be based on solid evaluation of threats to cetaceans: incidental mortality caused by fisheries; habitat loss (e.g., mariculture installations, harbor/marina constructions) and degradation (e.g., debilitating noise levels, chemical pollution); and whale/dolphin watching activities (and other forms of ecotourism). More research is also needed to accurately assess population levels of cetaceans in the Gulf of California. Finally, a conservation plan cannot be successful without also considering as major goals a comprehensive educational program and the improvement of quality of life of local peoples, and we commend the several regional institutions and nongovernmental organizations that are now striving to realize such a plan.

Acknowledgments

We thank D. Bowen, editor of the journal *Marine Mammal Science*, and P. Folkens for permission to reprint some of the figures presented in this chapter. We dedicate this chapter to Anelio Aguayo Lobo for his pioneering studies on the biodiversity and conservation of marine mammals in the Gulf of California. Several of his former students have followed in his footsteps, focusing their work on the natural resources of the Gulf of California, whether through research, management, or education.

Literature Cited

Acevedo-Gutierrez, A., D. A. Croll, and B. R. Tershy. 2002. Unique feeding costs limit dive time in the largest whales. Journal of Experimental Biology 205: 1747–1753.

Aguayo L., A., and A. Perdomo. 1985. Range extension (*Stenella coeruleolaba*). Marine Mammal Science 1: 263.

Alvarez-Borrego, S. 1983. Gulf of California. Pp. 427–449 *in* B. H. Ketchum (ed.), Estuaries and Enclosed Seas. Elsevier, New York.

Alvarez-Borrego, S. 1992. Upper Gulf of California, the small habitat of the vaquita (*Phocoena sinus*). *In* Current Crisis in Marine Mammals Management: U.S. and Mexican Perspectives. Sixth Conference in the UC MEXUS Series, Critical Issues in U.S.–México Relations. University of California, Institute for Mexico and the United States, Riverside, California.

Alvarez-Borrego, S., and J. R. Lara-Lara. 1991. The physical environment and primary productivity of the Gulf of California. Pp. 555–567 *in* B. R. T. Simoneit and J. P. Dauphin (eds.), The Gulf and Peninsular Province of the Californias. American Association of Petroleum Geologists Memoir 47.

Alvarez-Borrego, S., J. A. Rivera, G. Gaxiola-Castro, M. J. Acosta-Ruíz, and R.A. Schwartzlose. 1978. Nutrientes en el Golfo de California. Ciencias Marinas 5: 53–71.

Andrew, R. K., B. M. Howe, and J. A. Mercer. 2002. Ocean ambient sound: comparing the 1960s with the 1990s for a receiver off the California coast. Acoustics Research Letters Online 3: 65–70.

Aurioles G., D., B. J. LeBoeuf, and L.T. Findley. 1993. Registros de pinnípedos poco comunes para el Golfo de California. Revista de Investigación Científica, Universidad Autonoma de Baja California Sur, La Paz, Mexico (no. especial. SOMEMMA) 1: 13–19.

Barlow, J. 1986. Factors affecting the recovery of *Phocoena sinus*, the vaquita or Gulf of California harbor porpoise. National Marine Fisheries Service, Southwest Fisheries Science Center Administrative Report no. LJ-86-37.

Barlow, J., L. Fleischer, K. A. Forney, and O. Maravilla-Chávez. 1993. An experimental aerial survey for vaquita (*Phocoena sinus*) in the northern Gulf of California, Mexico. Marine Mammal Science 9: 89–94.

Barlow, J., T. Gerrodette, and G. Silber. 1997. First estimates of vaquita abundance. Marine Mammal Science 13: 44–58.

Bérubé, M., A. Aguilar, D. Dedanto, F. Larsen, G. Notobartolo-di-Sciara, R. Sears, J. Sigurjonsson, J. Urbán R., and P. J. Palsbøll. 1998. Population genetic structure of North Atlantic, Mediterranean and Sea of Cortez fin whales, *Balaenoptera physalus* (Linnaeus, 1758): analysis of mitochondrial and nuclear loci. Molecular Ecology 7: 585–600.

Bérubé, M., J. Urbán R., A. Dizon, R. L. Brownell, and P. J. Palsbøll. 2002. Genetic identification of a small and highly isolated population of fin whales (*Balaenopetra physalus*) in the Sea of Cortez, Mexico. Conservation Genetics 3: 183–190.

Boyer, P., and G. K. Silber. 1990. An estimate of the mortality of vaquita, *Phocoena sinus*, in commercial fisheries. Unpublished paper presented at the International Whaling Commission Workshop on Mortality of Cetaceans in Passive Fishing, Nets and Traps, October 1990, La Jolla, California.

Brownell, R. L., Jr. 1982. Status of the cochito, *Phocoena sinus*, in the Gulf of California. FAO Advisory Committee on marine resources research. Mammals of the Seas: Small Cetaceans, Seals, Sirenians, and Otters. Food and Agriculture Organization, Fisheries Series 5: 85–90. Food and Agriculture Organization, Rome.

Brownell, R. L., Jr. 1983. *Phocoena sinus*. Mammalian Species 198: 1–3.

Brownell, R. L., Jr. 1986. Distribution of the vaquita, *Phocoena sinus*, in Mexican waters. Marine Mammal Science 2: 299–305.

Brownell, R. L., Jr., L. T. Findley, O. Vidal, A. Robles, and S. Manzanilla N. 1987. External morphology and pigmentation of the vaquita, *Phocoena sinus* (Cetacea: Mammalia). Marine Mammal Science 3: 22–30.

Calambokidis, J. 1988. Chlorinated hydrocarbons in the Gulf of California harbor porpoise (*Phocoena sinus*). Final Contract Report MM4465846-3 to U.S. Marine Mammal Commission. Available from Cascadia Research Collective, Olympia, Washington.

Calambokidis, J., L. T. Findley, R. L. Brownell, and J. Barlow. 1993. Chlorinated hydrocarbon concentrations in the Gulf of California harbor porpoise (*Phocoena sinus*). Pp. 13 *in* Abstracts, Tenth Biennial Conference on the Biology of Marine Mammals, November 11–15, Galveston, Texas.

CITES (Convention on International Trade in Endangered Species of Wild Fauna and Flora) Secretariat. 2002. Appendices I and II. [Available: www.cites.org]

Clark, C. W., D. Croll, A. Acevedo, and J. Urbán R. 2000. Multi-modal surveys of fin whales in the Sea of Cortez, Mexico. Journal of the Acoustical Society of America 108: 2539.

Croll, D. A., A. Acevedo, B. R. Tershy, and J. Urbán R. 2001. The diving behavior of blue and fin whales: is dive duration shorter than expected based on oxygen stores? Comparative Biochemistry and Physiology, Part A: Molecular and Integrative Physiology 129: 797–809.

Croll, D., C. W. Clark, A. Acevedo, B. Tershy, S. Flores, J. Gedamke, and J. Urbán. 2002. Only male fin whales sing loud songs. Nature 417: 809.

D'Agrosa, C., C. E. Lennert, and O. Vidal. 2000. Vaquita by-catch in Mexico's artisanal gillnet fisheries: driving a small population to extinction. Conservation Biology 15: 1110–1119.

D'Agrosa, C., O. Vidal, and W. C. Graham. 1995. Mortality of the vaquita (*Phocoena sinus*) in gillnet fisheries during 1993–94. Pp. 283–291 *in* A. Bjorge and G. Donovan (eds.), Biology of the Phocoenids. Reports of the International Whaling Commission, special issue 16.

Dizon, A. E., C. A. Lux, R. G. Leduc, J. Urbán R., M. Henshaw, and R. L. Brownell. 1995. An interim phylogenetic analysis of sei and Bryde's whale mitochondrial DNA control region sequences. Unpublished paper SC/47/NP23. International Whaling Commission 47th meeting, Dublin. IWC, Cambridge, U.K.

DOF. 1991. Articulo 254 bis al Codigo Penal. Diario Oficial de la Federación 30 Diciembre.

DOF. 1994. Norma Oficial Mexicana NOM-059-ECOL-1994. Que determina las especies y subespecies de flora y fauna silvestres terrestres y acuáticas en peligro de extinción, amenazadas, raras y las sujetas a protección especial y que establece especificaciones para su protección. Diario Oficial de la Federación 16 Mayo.

DOF. 1996. Decreto por el que se establece el Parque Marino Nacional Bahía de Loreto. Diario Oficial de la Federación 19 Julio.

DOF. 2000a. Ley General de Vida Silvestre. Diario Oficial de la Federación 4 Julio.

DOF. 2000b. Norma Oficial Mexicana NOM-131-ECOL-1998. Que establece lineamientos y especificaciones para el desarrollo de actividades de observación de ballenas, relativas a su protección y conservación de su hábitat. Diario Oficial de la Federación 10 Enero (Primera sección): 11–17.

DOF. 2002a. Norma Oficial Mexicana NOM-059-ECOL-2001. Protección ambiental—Especies nativas de México de flora y fauna silvestres—Categorías de riesgo y especificaciones para su inclusión, exclusión o cambio en la lista de especies en riesgo. Protección. Diario Oficial de la Federación 6 Marzo.

DOF. 2002b. Decreto por el que se reforman diversas disposiciones de la Ley General de Vida Silvestre. Diario Oficial de la Federación 10 Enero.

DOF. 2002c. Acuerdo por el que se establece como área de refugio para proteger a las especies de grandes ballenas de los sub-órdenes Mysticeti y Odontoceti, las zonas marinas que forman parte del territorio nacional y aquellas sobre las que la nación ejerce su soberanía y jurisdicción. Diario Oficial de la Federación 24 Mayo.

DOF. 2002d. Norma Oficial Mexicana NOM-EM-139-ECOL-2002. Establece medidas de protección para los ecosistemas marinos y costeros y de especies sujetas a protección especial de la Reserva de la Biosfera del Alto Golfo y delta del Río Colorado. Diario Oficial de la Federación 19 Septiembre.

Enríquez P., L. M. 1996. Ocurrencia, movimientos, estructura social y tamaño de las agregaciones de rorcual común *Balaenoptera physalus* (Linneaus, 1758) en el Golfo de California, México. Tesis profesional, Universidad Autónoma de Baja California Sur, La Paz, Mexico.

Estados Unidos Mexicanos. 1993. Ley General del Equilibrio Ecológico y la Protección al Ambiente. Editorial Porrúa, México, D. F.

Findley, L. T., and J. M. Nava. 1994. Food habits of the vaquita, *Phocoena sinus*. Project final report to Conservation International Mexico-Programa Golfo de California, Guaymas, Sonora.

Findley, L. T., J. M. Nava, and J. Torre. 1995. Food habits of *Phocoena sinus* (Cetacea: Phocoenidae). Abstracts, Eleventh Biennial Conference on the Biology of Marine Mammals, December 14–18, Orlando, Florida. [Also see Resumenes, XXI Reunión Internacional para el Estudio de los Mamíferos Marinos, Abril 8–12, 1996, Chetumal, Quintana Roo.]

Findley, L. T., and O. Vidal. 2002. Gray whale (*Eschrichtius robustus*) at calving sites in the Gulf of California, México. Journal of Cetacean Research and Management 4: 27–40.

Fleischer, L. A. 1996. Mexico. Progress report on cetacean research, April 1994 to March 1995. Report of the International Whaling Commission 46: 262–264.

Fleischer, L. A., R. Moncada, H. Pérez-Cortés, and A. Polanco. 1994. Análisis de la mortalidad incidental de la vaquita, *Phocoena sinus*, historia y actualidad. Informe Técnico para el Comité de la Preservación de la Vaquita y la Totoaba (CTPVT), Centro Regional de Investigación Pesquera, La Paz, Mexico.

Gallo-Reynoso, J. P. 1986. Notas acerca de un ejemplar del delfín listado *Stenella coeruleoalba* (Cetacea: Delphinidae) en San Blas, Nayarit, México. Anales del Instituto de Biología, Universidad Nacional Autónoma de México, Serie Zoología 56: 1035–1037.

Gallo-Reynoso, J. P. 1998. La vaquita marina y su hábitat crítico en el Alto Golfo de California. Gaceta Ecológica (INE-SEMARNAP) 47 [for 1997]: 29–44.

Gambell, R. 1985. Fin whale *Balaenoptera physalus* (Linnaeus, 1758). Pp. 171–192 *in* S. H. Ridgway and R. Harrison (eds.), Handbook of Marine Mammals, vol. 3. The Sirenians and Baleen Whales. Academic Press, London.

Gaskin, D. E. 1982. The Ecology of Whales and Dolphins. Heinemann Educational Books, London.

Gendron, D. 2000a. Familia Physeteridae. Pp. 635–637 *in* S. T. Alvarez-Castañeda and J. L. Patton (eds.), Mamíferos del Noroeste de México II. Centro de Investigaciones Biológicas del Noroeste, S.C, La Paz, Mexico.

Gendron, D. 2000b. Familia Kogiidae. Pp. 639–641 *in* S. T. Alvarez-Castañeda and J. L. Patton (eds.), Mamíferos del Noroeste de México II. Centro de Investigaciones Biológicas del Noroeste, S.C, La Paz, Mexico.

Gendron, D., and S. Chavez. 1996. Recent sei whale (*Balaenoptera borealis*) sightings in the Gulf of California, Mexico. Aquatic Mammals 22: 127–130.

Gendron, D., S. Lanham, and M. Carwardine. 1999. North Pacific right whale (*Eubalaena glacialis*) sighting south of Baja California, Mexico. Aquatic Mammals 25: 31–34.

Gerrodette, T., L. A. Fleischer, H. Pérez-Cortés and B. Villa-Ramírez. 1995. Distribution of the vaquita, *Phocoena sinus*, based on sightings from systematic surveys. Pp. 273–281 *in* A. Bjørge and G. Donovan (eds.), Biology of the Phocoenids. Reports of the International Whaling Commission, special issue 16, Cambridge, U.K.

Gerrodette, T., and D. M. Palacios. 1996. Estimates of cetacean abundance in EEZ waters of the eastern tropical Pacific. National Marine Fisheries Service, Southwest Fisheries Science Center Administrative Report no. LJ-96-10. La Jolla, California.

Gilmore, R. M. 1957. Whales aground in Cortes's Sea: tragic strandings in the Gulf of California. Pacific Discovery 10: 22–27.

Goodall, R. N. P., B. Würsig, M. W. Würsig, G. Harris, and K. S. Norris. 1995. Sightings of Burmeister's porpoise, *Phocoena spinipinnis*, off southern South America. Pp. 297–316 *in* A. Bjørge and G. Donovan (eds.), Biology of the Phocoenids. Reports of the International Whaling Commission, special issue 16, Cambridge, U.K.

Guerrero-Ruíz, M. 1997. Conocimiento actual de la orca *Orcinus orca* (Linnaeus, 1758) en el Golfo de California, México. Professional thesis, Universidad Autónoma de Baja California Sur, La Paz, Mexico.

Guerrero-Ruíz, M., D. Gendron, and J. Urbán R. 1998. Distribution, movements and communities of killer whales (*Orcinus orca*) in the Gulf of California, Mexico. Report of the International Whaling Commission 48: 537–543.

Gutierrez-Galindo, E. A., G. Flores-Muñoz, M. L. Ortega-García, and J. Villaescusa-Celaya. 1992. Pesticidas en aguas costeras del Golfo de California: programa de vigilancia con mejillón 1987–1988. Ciencias Marinas 18: 77–99.

Gutierrez-Galindo, E. A., G. Flores-Muñoz, M. F. Villa-Andrade, and J. Villaescusa-Celaya. 1988a. Organochlorine insecticides in fishes from the Mexicali Valley, Baja California, Mexico. Ciencias Marinas 14: 1–22.

Gutierrez-Galindo, E. A., G. Flores-Muñoz, and J. Villaescusa-Celaya. 1988b. Chlorinated hydrocarbons in molluscs of the Mexicali Valley and the upper Gulf of California. Ciencias Marinas 14: 91–113.

Hilton-Taylor, C. (comp.). 2000. 2000 IUCN Red List of Threatened Species. World Conservation Union, Gland, Switzerland.

Hohn, A. A., A. J. Read, S. Fernández, O. Vidal, and L. T. Findley. 1996. Life history of the vaquita, *Phocoena sinus* (Phocoenidae, Cetacea). Journal of Zoology, London 39: 235–251.

International Whaling Commission. 1995. Report of the Scientific Committee. Reports of the International Whaling Commission 45: 33.

International Whaling Commission. 2001. Appendix 3. Classification of the Order Cetacea (Whales, Dolphins and Porpoises). Journal of Cetacean Research and Management 3: v–xii.

Jaquet, N., and D. Gendron. 2002. Distribution and relative abundance of sperm whales in relation to key environmental features, squid landings and the distribution of other cetacean species in the Gulf of California, Mexico. Marine Biology 141: 591–601.

Jaramillo L., A., L. Rojas B., and T. Gerrodette. 1999. A new abundance estimate for vaquitas: first step for recovery. Marine Mammal Science 15: 957–973.

Jefferson, T.A., and B. Curry. 1994. A global review of porpoise (Cetacea: Phocoenidae) mortality in gillnets. Biological Conservation 67: 67–183.

Jefferson, T.A., S. Leatherwood, and M. A. Webber. 1993. FAO Species Identification Guide. Marine Mammals of the World. Food and Agriculture Organization, Rome.

Le Boeuf, B. J., D. Aurioles, R. Condit, C. Fox, R. Gisiner, R. Romero, and F. Sinsel. 1983. Size and distribution of the California sea lion population in Mexico. Proceedings of the California Academy of Sciences 43: 77–85.

Mellink, E., and A. Orozco-Meyer. 2002. A group of gray whales (*Eschrichtius robustus*) in the northeastern Gulf of California, Mexico. Southwestern Naturalist 47: 129–132.

Mitchell, E. (ed.). 1975. Report of the meeting on smaller cetaceans, Montreal, April 1–11, 1974. Journal of the Fisheries Research Board of Canada 32: 889–983.

Mora, M. A., and D. W. Anderson. 1991. Seasonal and geographical variation of organochlorine residues in birds from Northwest Mexico. Archives of Environmental Contamination and Toxicology 21: 541–548.

Nava-Romo, J. M. 1994. Impactos a corto y largo plazo en la diversidad y otras características ecológicas de la comunidad béntico-demersal capturada por la pesquería de camarón en el norte del alto Golfo de California, México. M.Sc. thesis, Instituto Tecnológico y de Estudios Superiores de Monterrey-Campus Guaymas, Sonora, Mexico.

Norris, K. S., and W. N. McFarland. 1958. A new harbor porpoise of the genus *Phocoena* from the Gulf of California. Journal of Mammalogy 42: 22–39.

Norris, K. S., and J. H. Prescott. 1961. Observations on Pacific cetaceans of California and Mexican waters. University of California Publications in Zoology 63: 291–402.

Otterlind, G. 1976. The harbour porpoise *Phocoena phocoena* endangered in Swedish waters. International Council for the Exploration of the Seas paper C.M. 1976/N:16. International Council for the Exploration of the Seas, Copenhagen, Denmark.

Pérez-Cortés M., H. 1996. Contribución al conocimiento de la biología de la vaquita, *Phocoena sinus*. M.Sc. thesis, Instituto de Ciencias del Mar y Limnología, Universidad Nacional Autónoma de México, México D.F.

Pérez-Cortés M., H., B. Villa R., A. Delgado E., and J. L. Patiño V. 2000. Familia Delphinidae. Pp. 597–626. *in* S. T. Alvarez-Castañeda and J. L. Patton (eds.), Mamíferos del Noroeste de México II. Centro de Investigaciones Biológicas del Noroeste, S.C., La Paz, Baja California Sur, Mexico.

Perrin, W. F., G. P. Donovan, and J. Barlow (eds.). 1994. Gillnets and cetaceans. Reports of the International Whaling Commission, special issue 15. Cambridge, U.K.

Pitman, R., A. Aguayo L., and J. Urbán R. 1987. Observations of an unidentified beaked whale in the eastern tropical Pacific. Marine Mammal Science 3: 345–352.

PROFEPA (Procuraduría Federal de Protección al Ambiente) -SEMARNAP (Secretaría de Medio Ambiente, Recursos Naturales y Pesca). 1995. Mortandad de mamíferos y aves marinas en el alto Golfo de California. Informe final. PROFEPA, México D.F.

Rice, D. W. 1998. Marine Mammals of the World: Systematics and Distribution. Society of Marine Mammalogy special publication 4. University of Central Florida, Orlando.

Richardson, W. J., C. R. Greene, C. I. Malme, and D. H. Thomson. 1995. Marine Mammals and Noise. Academic Press, New York.

Rojas-Bracho, L., and A. M. Jaramillo-Legorreta. 2002. Vaquita (*Phocoena sinus*). Pp. 1277–1280 *in* W. F. Perrin, B. Würsig, and J. G. M. Thewissen (eds.), Encyclopedia of Marine Mammals. Academic Press, San Diego, California.

Rojas-Bracho, L., and B. L. Taylor. 1999. Risk factors affecting the vaquita (*Phocoena sinus*). Marine Mammal Science 15: 974–989.

Rojas-Bracho, L., and J. Urbán-Ramírez. 1993. Vaquita: its environment, biology and problematic. Pp 48–72 *in* J. L. Ferman, L. Gómez-Morín, and D. W. Fisher (eds.), Coastal Management in Mexico: The Baja California Experience. Coastlines of the World Series, American Society of Civil Engineers, New York.

Sánchez-Pacheco, J., A. Vázquez-Hanckin, and R. de Silva-Dávila. 2001. Gray whales mid-spring feeding at Bahía de los Angeles, Gulf of California, Mexico. Marine Mammal Science 17: 186–191.

Secretaria de Pesca. 1992. Legal Framework for Fisheries 1992. Secretaria de Pesca, México D. F.

Silber, G. K. 1990. Occurrence and distribution of the vaquita (*Phocoena sinus*) in the northern Gulf of California. Fisheries Bulletin (U.S.) 88: 339–346.

Silber, G. K., and K. S. Norris. 1991. Geographic and seasonal distribution of the vaquita, *Phocoena sinus*. Anales del Instituto de Biología de la Universidad Nacional Autónoma de México, Serie Zoológica 62: 263–268.

Taylor, B. L., and L. Rojas-Bracho. 1999. Examining the risk of inbreeding depression in a naturally rare cetacean, the vaquita (*Phocoena sinus*). Marine Mammal Science 15: 1004–1028.

Tershy, B. R., D. Breese, and C. Strong. 1990. Abundance, seasonal distribution and population composition of balaenopterid whales in the Canal de Ballenas, Gulf of California, Mexico. Pp. 369–375 *in* P. S. Hammond, S. A. Mizroch, and G. P. Donovan (eds.), Individual recognition of cetaceans: use of photo-identification and other techniques to estimate population parameters. Reports of the International Whaling Commission, special issue 12, Cambridge, U.K.

Tershy, B. R., J. Urbán R., A. Acevedo-Gutiérrez, D. A. Croll, and B. Mate. 2002. Determination of fin whale residency patterns in order to prioritize conservation areas in the Gulf of California. Report to UC MEXUS-CONACyT 7/1/00-6/30/01. Institute for Mexico and the United States, University of California, Riverside.

Tershy, B. R., J. Urbán, D. Breese, L. Rojas, and L. T. Findley. 1993. Are fin whales resident to the Gulf of California? Revista de Investigación Científica (Universidad Autónoma de Baja California Sur) (no. especial. SOMEMMA 1) 1: 69–72.

Thompson, P. O., L. T. Findley, and O. Vidal. 1992. 20-Hz pulses and other vocalizations of fin whales, *Balaenoptera physalus*, in the Gulf of California, Mexico. Journal of the Acoustical Society of America 92: 3051–3057.

Thompson, P. O., L. T. Findley, and O. Vidal. 1996. Underwater sounds of blue whales, *Balaenoptera musculus*, in the Gulf of California, Mexico. Marine Mammal Science 12: 288–293.

Urbán R., J. 1996. La población del rorcual común *Balaenoptera physalus* en el Golfo de California. Final Report to the Comisión Nacional para la Bioversidad CONABIO-B040. CONABIO, México D.F.

Urbán R., J. 2000a. Familia Eschrichtidae [sic, Eschrichtiidae]. Pp. 655–659 *in* S. T. Alvarez-Castañeda and J. L. Patton (eds.), Mamíferos del Noroeste de México II. Centro de Investigaciones Biológicas del Noroeste, S.C., La Paz, Baja California Sur.

Urbán R., J. 2000b. Familia Balaenopteridae. Pp. 661–683 *in* S. T. Alvarez-Castañeda and J. L. Patton (eds.), Mamíferos del Noroeste de México II. Centro de Investigaciones Biológicas del Noroeste, S.C., La Paz, Baja California Sur, Mexico.

Urbán R., J., and A. Aguayo L. 1987. Spatial and seasonal distribution of the humpback whale, *Megaptera novaeangliae*, in the Mexican Pacific. Marine Mammal Science 3: 333–344.

Urbán R., J., C. Alvarez F., M. Salinas Z., J. Jacobsen, K. C. Balcomb III, A. Jaramillo L., P. Ladrón de Guevara P., and A. Aguayo L. 1999a. Population size of humpback whale, *Megaptera novaeangliae*, in waters off the Pacific coast of Mexico. Fishery Bulletin (U.S.) 97: 1017–1024.

Urbán R., J., and S. Flores R. 1996. A note on Bryde's whales (*Balaenoptera edeni*) in the Gulf of California, Mexico. Reports of the International Whaling Commission 46: 453–457.

Urbán R., J., A. Gómez-Gallardo U., V. Flores de Sahagún, M. Palmeros R., and S. Ludwig. 1999b. Changes in the abundance and

distribution of gray whales at Laguna San Ignacio, México, during the 1997–98 El Niño and the 1998–99 La Niña. Unpublished paper SC/51/AS31. International Whaling Commission 51st meeting, Grenada. IWC, Cambridge, U.K.

Urbán R., J., A. Gómez Gallardo U., M. Palmeros R., and G. Velazquez Ch. 1997. Los mamíferos marinos de la Bahía de La Paz. Pp. 193–217 *in* J. Urbán R. and M. Ramírez R. (eds.), La Bahía de La Paz: Investigación y Conservación. Universidad Autónoma de Baja California Sur, La Paz, Mexico.

Urbán R., J., A. Jaramillo L., A. Aguayo L., P. Ladrón de Guevara P., M. Salinas Z., C. Alvarez F., L. Medrano G., J. K. Jacobsen, K. C. Balcomb III, D. E. Claridge, J. Calambokidis, G. H. Steiger, J. Straley, O. Von Ziegesar, J. M. Waite, S. Miszroch, M. E. Dahlheim, J. D. Darling, and C. S. Baker. 2000. Migratory destinations of humpback whales wintering in the Mexican Pacific. Journal of Cetacean Research and Management 2: 101–110.

Urbán, J., S. Jaume, B. Tershy, J. Pettis, L. Findley, J. P. Gallo, A. Acevedo, D. Croll, and O. Vidal. 2001. Residency times and movement patterns of fin whales in the Gulf of California, Mexico. Pp. 220–221 *in* Abstracts, 14th Biennial Conference on the Biology on Marine Mammals, November 28–December 3, Vancouver, Canada.

Urbán R., J., B. Mate, M. Bérubé, A. Acevedo, S. Jaume, B. Tershy, and D. Croll. 2002. Los rorcuales comunes del Golfo de California: una población residente y aislada. Pp. 51 *in* Abstracts, 4th Congress of the Latin American Aquatic Mammals Society, 14–19 October 2002, Valdivia, Chile.

Urbán R., J., and H. Pérez-Cortés M. 2000. Familia Ziphiidae. Pp. 643–653 *in* S. T. Alvarez-Castañeda and J. L. Patton (eds.), Mamíferos del Noroeste de México II. Centro de Investigaciones Biológicas del Noroeste, S.C., La Paz, Baja California Sur, Mexico.

Urbán R., J., S. Ramírez, and J. C. Salinas V. 1994. First record of the bottlenose whale *Hyperoodon* sp. in the Gulf of California. Marine Mammal Science 10: 471–473.

Valdez-Holguin, J. E., G. Gaxiola-Castro, and R. Cervantes-Duarte R. 1995. Primary productivity in the Gulf of California, calculated from the relationship between superficial irradiance and chlorophyll in the euphotic zone. Ciencias Marinas 21: 311–329.

Vidal, O. 1989. La ballena gris, *Eschrichtius robustus*, en las áreas de crianza del Golfo de

California, México. M.Sc. thesis, Instituto Tecnológico y de Estudios Superiores de Monterrey-Campus Guaymas, Sonora, Mexico.

Vidal, O. 1995. Population biology and exploitation of the vaquita *Phocoena sinus*. Pp. 245–272 *in* A. Bjørge and G. Donovan (eds.), Biology of the Phocoenids. Reports of the International Whaling Commission, special issue 16, Cambridge, U.K.

Vidal, O., R. L. Brownell, Jr., and L. T. Findley. 1999. Vaquita, *Phocoena sinus* Norris and McFarland, 1958. Pp. 357–378 *in* S. H. Ridgway and S. R. Harrison (eds.), Handbook of Marine Mammals, vol. 6. The Second Book of Dolphins and the Porpoises. Academic Press, San Diego, California.

Vidal, O., L. T. Findley, and S. Leatherwood. 1993. Annotated checklist of marine mammals of the Gulf of California. Proceedings of the San Diego Society of Natural History 28: 1–16.

Vidal, O., and J.-P. Gallo-Reynoso. 1996. Die-offs of marine mammals and sea birds in the Gulf of California, México. Marine Mammal Science 12: 627–635.

Vidal, O., K. V. Waerebeek, and L. T. Findley. 1994. Cetaceans and gillnet fisheries in Mexico, Central America and the Wider Caribbean: a preliminary review. Pp. 221–233 *in* W. F. Perrin, G. P. Donovan, and J. Barlow (eds.), Gillnets and Cetaceans. Reports of the International Whaling Commission, special issue 15, Cambridge, U.K.

Villa-Ramirez, B. 1976. Report on the status of *Phocoena sinus*, Norris and McFarland, 1958, in the Gulf of California. Annales del Instituto de Biología, Universidad Nacional Autónoma de México, Serie Zoología 47: 203–208.

Villa-Ramírez, B. 1993. Recovery plan for the vaquita, *Phocoena sinus*. NTIS Report PB93-169415. Report to the U.S. Marine Mammal Commission, Washington, D.C.

Watkins, W. A., P. Tyack, K. E. Moore, and J. E. Bird. 1987. The 20-Hz signals of finback whales (*Balaenoptera physalus*). Journal of the Acoustical Society of America 82: 1901–1912.

Zavala-González, A., J. Urbán R. and C. Esquivel-Macías. 1994. A note on artisanal fisheries interactions with small cetaceans in Mexico. Pp. 235–237 *in* W. F. Perrin, G. P. Donovan, and J. Barlow (eds.), Gillnets and Cetaceans. Reports of the International Whaling Commission, special issue 15, Cambridge, U.K.

Zeitzchel, B. 1969. Primary productivity in the Gulf of California. Marine Biology 3: 201–207.

15

The Ecological Importance of Mangroves in Baja California Sur: Conservation Implications for an Endangered Ecosystem

ROBERT C. WHITMORE

RICHARD C. BRUSCA

JOSÉ LUIS LEÓN DE LA LUZ

PATRICIA GONZÁLEZ-ZAMORANO

RENATO MENDOZA-SALGADO

EDGAR S. AMADOR-SILVA

GINA HOLGUIN

FELIPE GALVÁN-MAGAÑA

PHILIP A. HASTINGS

JEAN-LUC E. CARTRON

RICHARD S. FELGER

JEFFREY A. SEMINOFF

CAROLE C. McIVOR

Conservation was made up by politicians who drive big cars and have never experienced hunger.
—a retired fisherman from Laguna San Ignacio
as quoted by Dedina (2000: 105)

Mangroves occur throughout the tropics, along shallow seashores protected from waves (Chapman 1976; Tomlinson 1986; Hogarth 1999). They grow in mud and other nonrocky substrates inundated during high tide. To cope with their environment (e.g., anoxic soil conditions and excess salt), they have developed a variety of morphological and physiological adaptations. For example, some mangroves have aerial roots for gas exchange, or they may actively secrete sodium chloride through salt glands in the leaves. Hydrological and edaphic conditions in mangrove ecosystems prevent all but a few other plant species from invading (Lugo 1998).

Mangroves do not constitute a discrete taxonomic group. Instead, the world's mangrove vegetation is the product of amazing convergence, probably due to biogeochemical and climatic factors and is hypothesized to have had at least 16 separate evolutionary origins (Duke 1995; Hogarth 1999). Based on mangrove species richness and composition, the world's tropics can be divided into 2 zones. The eastern zone (East Africa, India, southeastern Asia, Australia, and the western Pacific) shows a greater diversity of mangrove species than the western zone (West Africa, South, Central, and tropical North America; Tomlinson 1986).

In the tropics, characterized by an abundance of rainfall and fresh water, mangrove trees can reach a height of 30–40 m (Tomlinson 1986). They represent the dominant plant form in many coastal areas (e.g., Kunstadter et al. 1986) and often form forestlike communities several kilometers wide or more (Tomlinson 1986). Mangroves of the tropics are among the most productive ecosystems in the world (Farnsworth et al. 1996; Jennerjahn and Venugopalan 2002), providing not only habitat (Acosta and Butler 1997; Aliaume et al. 1997; Acosta 1999) but also nutrients—in the form of detritus—for a large number of organisms. Protozoa, diatoms, and phototrophic cyanobacteria, the latter forming dense mats up to 25 cm thick, all thrive in mangrove ecosystems (Lopez-Cortez 1991; Toledo et al. 1995; Sigueiros-Beltrones and Morzari 1999) and constitute the basis of highly complex food webs (Day and Yáñez-Arancibia 1985; Yáñez-Arancibia et al. 1993, 1994; Kaly and Jones 1998; Skilleter and Warren 2000; Holguin et al. 2001). Mangroves benefit human populations through their high productivity, and by also protecting coastal areas from storms and erosion (Menéndez et al. 1994).

In this chapter we discuss the ecological importance and conservation status of mangroves in Baja California Sur (hereafter BCS). In western North America, mangroves reach the northern edge of their distribution in coastal Sonora and along both sides of the Baja California peninsula (Turner et al. 1995). In this region, they grow under suboptimal conditions, and mangrove communities are far less extensive than in many parts of the tropics. They have been described for BCS or elsewhere in northwestern Mexico in a number of botanical works (e.g., Wiggins 1980; León de la Luz and Coria-Benet 1992; Turner et al. 1995; Felger et al. 2001). By comparison, however, little has been published on their associated fauna. Here, we place on record, as a basis for future research, lists of macroinvertebrate and vertebrate species inhabiting or regularly visiting BCS mangrove ecosystems. Despite being less extensive than in the tropics, BCS mangroves perform an important ecological role by sustaining a rich macrofauna, providing spawning/nursery habitat for many offshore species, and as a nutrient source for coastal ecosystems. Mangrove conservation is an important priority in Mexico. According to Herrera-Silveira and Ceballos-Cambranis (2000),

Mexico lost 65% of its mangrove communities between 1972 and 1992 due to direct exploitation and agricultural and urban development.

Distribution of Mangroves in Baja California Sur

Based on 1994 estimates, Mexico has 488,367 ha of mangrove vegetation, 12,120 ha (2.5%) of which are on the Baja California peninsula (Loza 1994). Stands of mangroves, sometimes referred to as "mangals" (Tomlinson 1986), or in northwestern Mexico as "*manglares*," are found in isolated coves, lagoons, and *esteros* of both sides of the peninsula (figs. 15.1–15.6). As in northwestern mainland Mexico, they occur in protected, shallow-water habitats that drain and fill daily. They do not tolerate stagnant water and soon perish if cut off from tidal circulation (e.g., Felger et al. 2001). Along the eastern (Gulf of California) side of the peninsula, mangroves are distributed from the Cape Region north to small islands in Bahía de Los Angeles (e.g., Isla Smith) in the state of Baja California. On the western (Pacific) side, mangroves have a more limited range. The northern limit of their distribution is near Laguna San Ignacio in BCS (Brusca 1975; Roberts 1989; Danemann and Carmona 1993; León de la Luz et al. 1995; Turner et al. 1995; Peterson 1998; Williams and Williams 1998).

Mangroves of BCS occur mainly in 5 coastal areas, referred to here as zones (fig. 15.1). The 3 most extensive mangrove ecosystems of the state are found in the Laguna San Ignacio complex (Zone I) and at Bahía Magdalena and adjacent shores (Zone II) along the Pacific coast; and along Bahía de la Paz (Zone IV) on the Gulf side.

Zone I (fig. 15.2) is centered on Laguna San Ignacio, located between 26°43' and 26°58' N, and 113°08' and 113°16' W. In the vicinity of Laguna San Ignacio are 3 other important locations: Estero la Bocana (also known as Pond Lagoon), Estero el Coyote (or Laguna la Escondida), and Estero San Juan. Traditionally, these 3 additional areas are grouped with Laguna San Ignacio to form what is referred to as the "San Ignacio complex."

Zone II (fig. 15.3) is centered on Bahía Magdalena. Together with adjacent coastal areas (e.g., Bahía Almejas), Bahía Magdalena forms a 240 km-long complex of bays and lagoons. It is often described as the "Chesapeake of the Pacific," due to its extensive size, beauty, and ecosystem dynamics (Dedina 2000: 125).

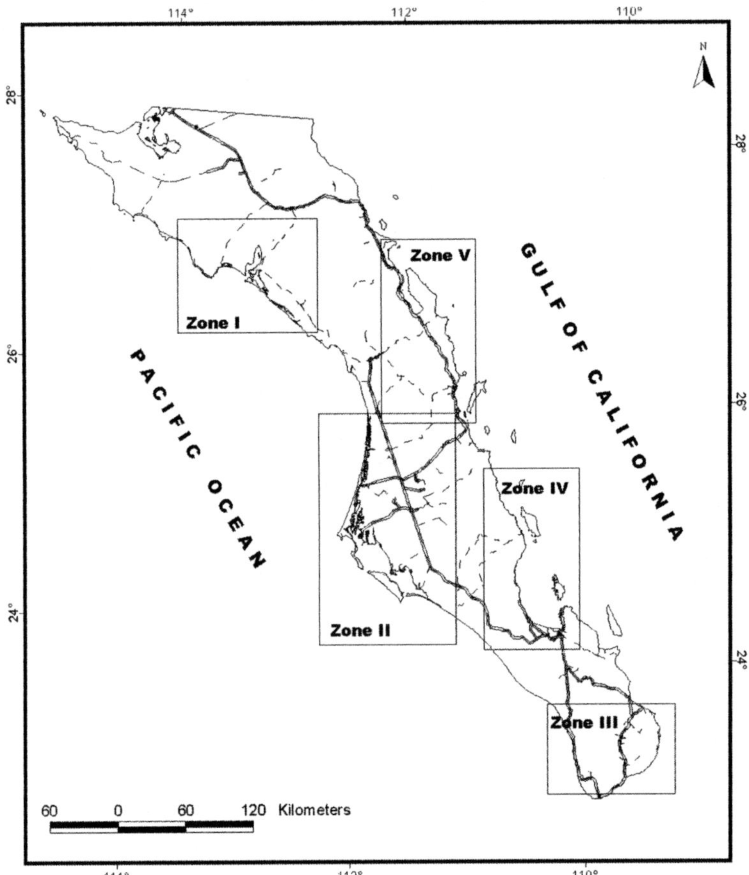

Figure 15.1. Baja California Sur, with the 5 coastal zones harboring notable mangrove stands.

The outer, ocean-facing shore is formed by 6 narrow islands, the largest of these being Isla Magdalena and Isla Margarita. Mangrove stands are spaced irregularly along almost 160 km of the coastline. Particularly important is the northern half of the Bahía Magdalena area (Boca Las Animas south to Boca La Soledad), where the stands occur almost continuously. Local natural resources support a fishing economy in the town of Puerto San Carlos (or San Carlos).

On the eastern coast of BCS, the largest mangrove areas are found in Zone IV (fig. 15.4), occupied by Bahía de la Paz. Mangroves are found along the southern end of the bay, including on the peninsula of El Mogote and along Ensenada de Aripes (also called Ensenada de La Paz). Some offshore islands have extensive and pristine mangrove stands, including at Bahía San Gabriel (24°27' N, 110°22' W) on Isla Espíritu Santo, and at Bahía

Amortajada (24°53' N, 110°35' W) on Isla San José. Across Ensenada de Aripes from El Mogote is the capital city of BCS La Paz, with its nearly 200,000 residents (INEGI 2001).

The Cape Region (Zone III, fig. 15.5) has less extensive, more discontinuous mangrove vegetation. Several isolated mangrove stands can be found between Todos Santos (23°26' N, 110°14' W) and Estero Migriño (23°00' N, 110°06' W) on the Pacific coast, and at Punta Colorada (23°30' N, 109°30' W).

Zone V (fig. 15.6) along the Gulf side of BCS, has only isolated mangrove pockets, most notably at Bahía Concepción and along Bahía de Loreto. Bahía Concepción (between 26°33' and 26°53' N and 111°42' and 111°56' W) has at least 8 small coves (e.g., Ensenada Morgán) with mangrove stands, none of which covers more than a few hectares. These mangrove stands are separated from one another

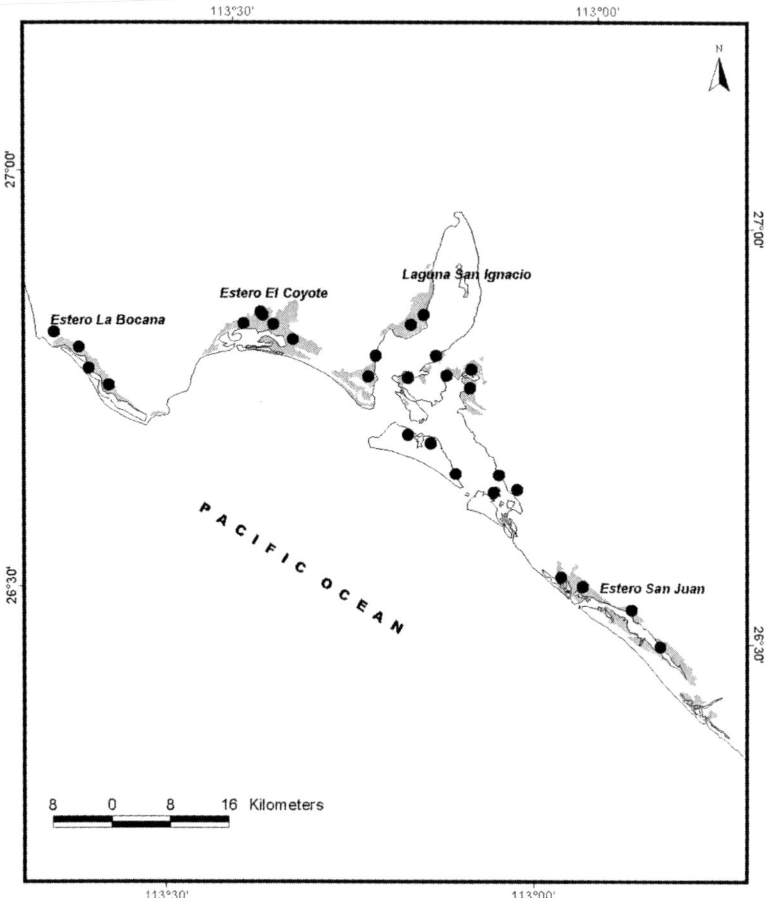

Figure 15.2. Mangrove distribution in Baja California Sur: Zone I, San Ignacio complex. Dots represent verified mangrove stands based either on on-site verification, photographic interpretation, or herbarium specimens. The shaded areas represent potential mangrove areas based on cartographic interpretation. Map generated using the Lambert Conformal Conic Projection. Cartographic source: INEGI 1:250,000 and 1:50,000; photographic source: INEGI 1:75,000. Map prepared by P. González and G. Arredondo, September 2002.

by a mean distance of 14 km, with 44 km representing the maximum distance between 2 stands (Whitmore et al. 2000). Several mangrove stands occur along Bahía de Loreto, including on Isla Danzante (25°47' N, 111°15' W) and Isla Monserrat (25°40' N, 111°03' W).

Vascular Plants

Worldwide, the northern distributional limit of mangroves is determined by reduced air and sea temperatures (Tomlinson 1986). In northwestern mainland Mexico, for example, this distributional limit coincides with occasional freezing weather (Felger and Moser 1985; Turner et al. 1995; Felger et al. 2001). Aridity is also believed to play an important role in limiting the establishment of mangroves, albeit only indirectly. Many mangroves are chiefly estuarine species growing in brackish waters. However, along arid coastal areas, freshwater runoff and river flow are minimal. The salt content of tidal waters is higher in arid regions than in wetter climates, and soils are typically poor and offer fewer

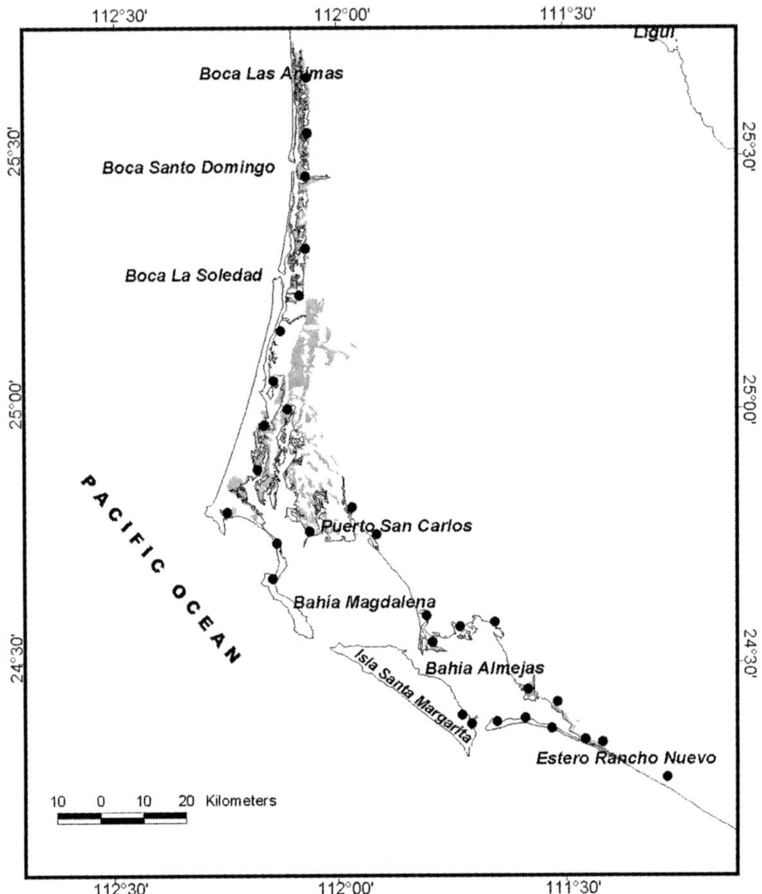

Figure 15.3. Mangrove distribution in Baja California Sur: Zone II, Bahía Magdalena area. Dots represent verified mangrove stands based either on on-site verification, photographic interpretation, or herbarium specimens. The shaded areas represent potential mangrove areas based on cartographic interpretation. Map generated using the Lambert Conformal Conic Projection. Cartographic source: INEGI 1:250,000 and 1:50,000; photographic source: INEGI 1:75,000. Map prepared by P. González and G. Arredondo, September 2002.

mineral nutrients for the growth of mangroves (Chapman 1976; Tomlinson 1986). In sum, mangroves grow under suboptimal conditions in areas with dry climates. As a result of the arid regional climate, mangroves in northwestern Mexico occur in strictly tidal saltwater, with the exception of 2 locations (Mulegé and Loreto) in BCS. Along *esteros* and some shallow bays (e.g., the inner coast of El Mogote in Bahía de la Paz), tidal waters are hypersaline.

At the landscape level, BCS mangrove communities constitute a relatively narrow and discontinuous band of desert-fringe vegetation. Individual plants consist typically of arborescent shrubs or small trees (León de la Luz and Coria-Benet 1992). Species diversity of mangroves in northwestern Mexico is low, as is the case for mangroves in arid regions elsewhere in the world. Three species dominate the mangrove vegetation nearly everywhere in BCS: red mangrove (*Rhizophora mangle*, Rhizophoraceae), black mangrove (*Avicennia germinans*, Avicenniaceae), and white mangrove (*Laguncularia racemosa*, Combretaceae). Overlapping zonation within the mangroves is pronounced and similar to that in Sonora (e.g., Felger et al. 2001): red mangrove extends into deep-

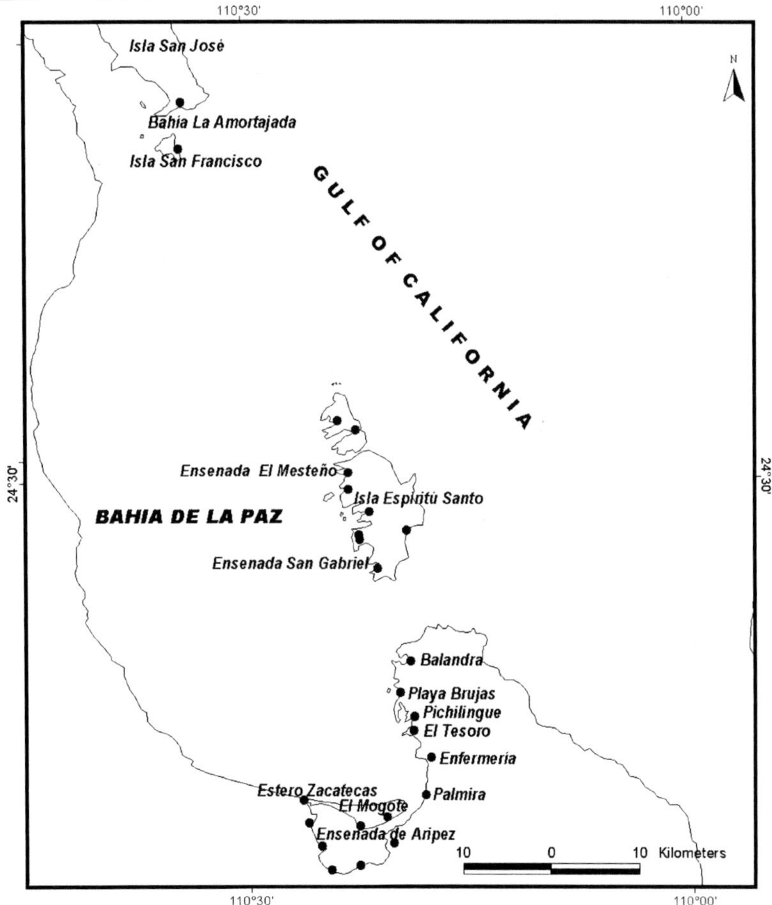

Figure 15.4. Mangrove distribution in Baja California Sur: Zone IV, Bahía de La Paz area. Dots represent verified mangrove stands based either on on-site verification, photographic interpretation, or herbarium specimens. Map generated using the Lambert Conformal Conic Projection. Cartographic source: INEGI 1:250,000 and 1:50,000; photographic source: INEGI 1:75,000. Map prepared by P. González and G. Arredondo, September 2002.

est water (the seaward zone), and black mangrove reaches maximum density in the shallowest water (landward zone); white mangrove reaches maximum density between the peak zones of the other 2. On the Pacific coast, the black mangrove extends its distribution north only to the Bahía Magdalena area. Farther north, such as at Laguna San Ignacio, stands of mangrove vegetation consists only of red and white mangrove (Centro de Investigaciónes Biologicas de Baja California Sur 1994).

A fourth mangrove species, buttonwood mangrove or *mangle botoncillo* (*Conocarpus erecta*; Combretaceae), occurs chiefly in the form of individual plants distributed sparsely at several locations of BCS. In the tropics, this species may grow as a tree, often reaching 15 m in height and forming thick stands. At Ensenada El Mezteño (24°31' N, 110°19' W) on Isla Espíritu Santo, buttonwood mangrove occurs as the only pure stand of this species on the peninsula. The buttonwood mangrove stand in this cove consists of 50–60 shrubby plants 2.5–3 m tall at the edge of a salt flat (J. León de la Luz, pers. obs.).

Red mangroves are characterized in part by arching "prop roots" descending from branches and stems, leathery leaves that are nearly opposite and decussate, wind-pollinated flowers, and viviparous

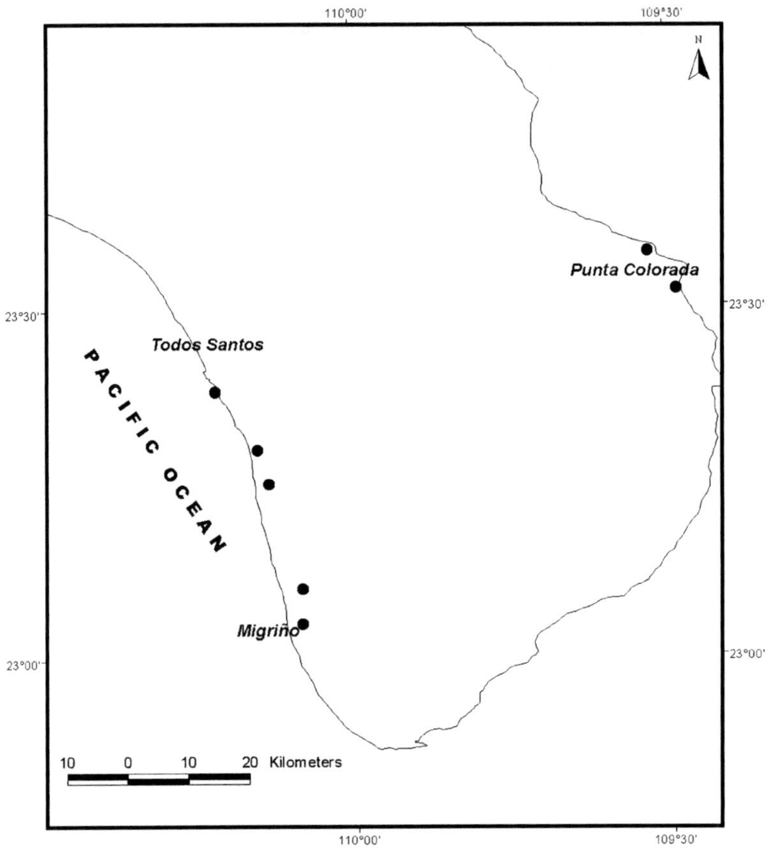

Figure 15.5. Mangrove distribution in Baja California Sur: Zone III, Cape Region. Dots represent verified mangrove stands based either on on-site verification, photographic interpretation, or herbarium specimens. Map generated using the Lambert Conformal Conic Projection. Cartographic source: INEGI 1:250,000 and 1:50,000; photographic source: INEGI 1:75,000. Map prepared by P. González and G. Arredondo, September 2002.

fruits (e.g., León de la Luz and Coria-Benet 1992; Felger et al. 2001). The bark, and to a lesser extent, the leaves, have a high tannin content. Red mangrove is used traditionally for tanning hides, but also as fuel wood and construction material for *covachas* (shelters). It is used to treat a number of ailments, including leprosy, fever, and sore throat (León de la Luz and Coria-Benet 1992). The Seri people, who live along the central Sonora coast, collected the driftwood as firewood and the roots to make a black dye. They also have used the fruits (enlarged embryos) in a variety of ways, for example to make a tea as a remedy for dysentery or, once roasted, to consume as food (Felger and Moser 1985).

Black mangrove is widespread in the western mangrove zone of the tropics, where it may reach a height of 25 m (Felger et al. 2001). In BCS—and elsewhere in northwestern Mexico—it reaches 6 m in height (León de la Luz and Coria-Benet 1992). Its root system includes subterranean cable roots, from which both anchoring roots and pneumatophores arise. The flowers produce nectar that is highly fragrant, especially at night (Felger et al. 2001). The fruits are viviparous. Black mangrove is used traditionally for tanning hides (León de la Luz and Coria-Benet 1992), and the Seris prized the wood for the curved ribs of their boats and used the driftwood for building fires (Felger and Moser 1985).

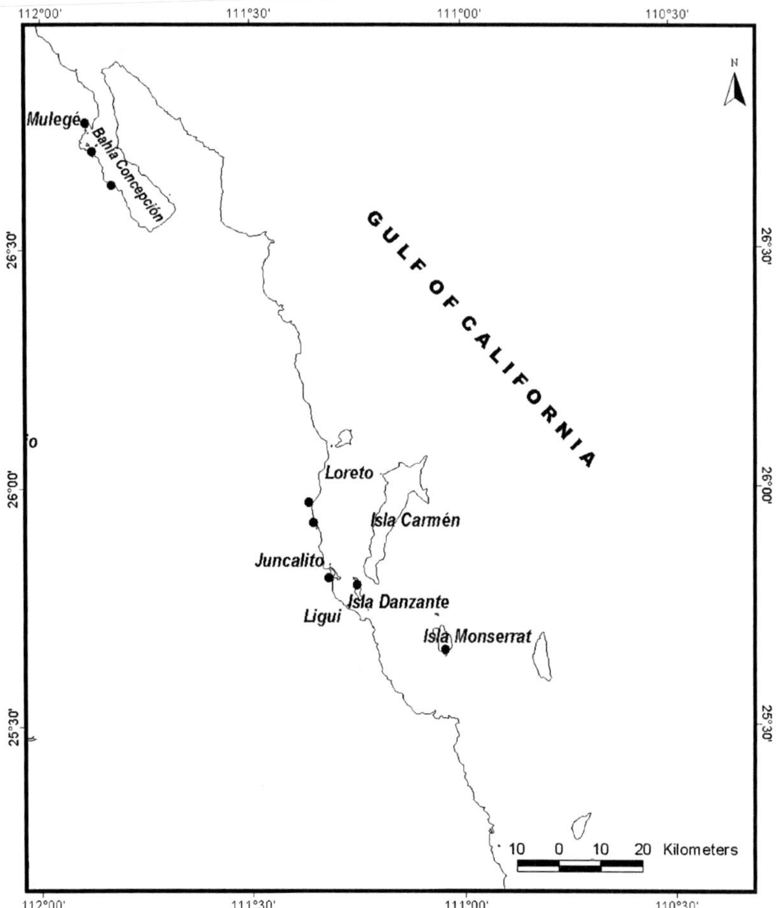

Figure 15.6. Mangrove distribution in Baja California Sur: Zone V, Bahía Concepción/ Bahía de Loreto area. Dots represent verified mangrove stands based either on on-site verification, photographic interpretation, or herbarium specimens. Map generated using the Lambert Conformal Conic Projection. Cartographic source: INEGI 1:250,000 and 1:50,000; photographic source: INEGI 1:75,000. Map prepared by P. González and G. Arredondo, September 2002.

White mangrove has shallow, horizontal roots and pneumatophores. The flowers, which appear from July to October in BCS (León de la Luz and Coria-Benet 1992), are visited by bees in the tropics (Felger et al. 2001). The Seris made use of the wood for boat paddles, harpoon shafts, and house posts and beams and used the leafy branches for roofing (Felger and Moser 1985).

A halophytic or salt-scrub vegetation is associated with mangroves, usually on the landward side. It is made up of highly predictable species of perennial saltgrasses (*Jouvea pilosa*, *Monanthochloe littoralis*, *Sporobolus virginicus*), perennial halophytic shrubs (e.g., *Allenrolfea occidentalis*, *Maytenus phyllanthoides*, *Salicornia subterminalis*, *Suaeda nigra*), and other halophytes (e.g., *Atriplex barclayana*, *Batis maritima*, *Heliotropium curassavicum*, *Salicornia bigelovii*, *S. virginicus*, *Sesuvium portulacastrum*). Extensive undersea meadows of *Zostera marina* occur near the mangrove stands in Bahía Magdalena (Ramírez-García and Lot 1994), and substantial quantities of this seagrass may seasonally wash into the mangroves. Stranded pieces of the seagrass often become entangled in the mangrove branches. Another seagrass, *Ruppia maritima* (Ruppiaceae), sometimes occurs in shallow water adjacent to mangrove stands—for example, at El Mogote (Bahía de La Paz) and Bahía Concepción (Ramírez-García and Lot 1994).

Macrofauna of Baja California Sur Mangroves

Baja California Sur mangroves and adjacent intertidal and subtidal waters provide habitat and nutrients for a large number of organisms, both terrestrial and marine. Habitats provided or strongly influenced by mangroves include their canopy and roots, soil surface, and tidal waters. Use of mangroves by birds and sea turtles has been fairly well documented. The inventory of macroinvertebrates, fish, and mammals remains incomplete. With the exception of some dipterans (biting midges and mangrove flies; Cheng and Hogue 1974), we are not aware of any published information on terrestrial or flying arthropods of BCS mangroves. Some of these are important, such as the insects pollinating black mangrove.

Biting Midges and Mangrove Flies (Order: Diptera)

Cheng and Hogue (1974) documented the presence of several dipterans associated with mangroves and adjacent mudflats in BCS. The mosquito *Deinocerites mcdonaldi* Belkin & Hogue (family Culicidae) breeds in the burrows of 2 land crab species, *Cardisoma crassum* and *Sesarma sulcatum* (see table 15.1). The infamous *jejene Culicoides furens* (Poey) is a common biting midge in the family Ceratopogonidae. It breeds around the aerial roots of black mangrove, in mud and burrows of *S. sulcatum*. Three other species, *Megaselia minutior* Borgmeier (Phoridae), *Dasyhelea* sp. (Ceratopogonidae), and *Smittia* sp. (Chironomidae), have been collected in emergence traps set on muddy flats around the aerial roots of black mangrove (Cheng and Hogue 1974).

Intertidal and Subtidal Macroinvertebrates

We present a list of all the (named) macroinvertebrates recorded from mangrove-lined lagoons, *esteros*, and coves on the Gulf side of BCS, including offshore islands (table 15.1). We are not aware of any other published compilation of mangrove-associated macroinvertebrates for any part of the

Baja California peninsula. In addition to the described (named) invertebrate fauna of the Gulf of California, there is a large undescribed fauna (Brusca et al. [chapter 9] estimate that less than half of the Gulf macroinvertebrates have been described). In particular, there are easily a dozen or more species of sponges (Porifera) and of tunicates (Urochordata) that are common in mangrove-lined waters of BCS. Of these, however, most are undescribed species.

The intertidal and subtidal macroinvertebrate fauna associated with BCS mangroves is diverse. Most species are intertidal, but some live in the permanent, subtidal channels of the mangrove lagoons. Our compilation lists 214 taxa, 15 of which occur only in association with rocky substrate near mangroves. Among the 214 taxa are 71 crustaceans, 63 bivalves (class Pelecypoda), 38 gastropods, 14 polychaete annelids, 6 echinoderms, 5 cnidarians (sea anemones), 5 sponges, 4 chordates (3 tunicates and 1 cephalochordate/amphioxus), 3 polyplacophorans (chitons), 3 ectoprocts (bryozoans), 1 nemertean, and 1 sipunculan (peanut worm). Tunicates and sponges dominate communities living on the roots of mangroves (i.e., the mangrove root microhabitat), whereas crustaceans and molluscs dominate the remaining habitats. Crustaceans and bivalves are not only diverse in these communities, but they also dominate the biomass. Virtually all of the bivalves are suspension feeders, attesting to the high productivity that characterizes these detritus-based ecosystems. Crustaceans are a mix of algal grazers, scavengers, and predators, as are the gastropods. The annelids are a mix of suspension-feeders and predators. Included in the list are 3 frequently seen visitors to coastal lagoons, species that live offshore in subtidal waters but occasionally break free in storm surge to be carried into mangrove embayments, where they can live for many weeks, rolling about with tides: *Zoobotryon verticillatum* (the gelatinous "spaghetti bryozoan," which may also grow on pier pilings in coastal lagoons), *Cliona* cf. *chilensis* (the "barrel sponge," also known as *Pseudosuberites pseudos*), and the bright orange *Aplidium* sp. (the colonial "ball ascidian").

The mangrove embayments of BCS also provide important refugia for young of the commercially valuable penaeid shrimps of the southern portion of the Gulf of California. Penaeid shrimps use these habitats as nursery grounds, migrating into them subsequent to their offshore planktonic larval phase. When they reach the juvenile or subadult stage, they migrate offshore once again. Loss of mangrove and other coastal lagoon habitats thus reduces the area

Table 15.1. Intertidal and subtidal invertebrate species documented from mangrove lagoons and *esteros* along the eastern (Gulf of California) coast of Baja California Sur.

Phylum	Subphylum	Class	Family	Scientific Name	Species Author
Porifera		Calcarea	Leucosoleniidae	*Leucosolenia cf. irregularis*	Jenkin, 1908
			Leucettidae	*Leucetta losangelensis*	(de Laubenfels, 1930)
		Demospongiae	Clionidae	*Cliona celata*	Grant, 1826
				Cliona cf. chilensis	Thiele, 1905
			Tetillidae	*Craniella crania*	(Müller, 1776)
Ectoprocta (Bryozoa)		Gymnolaemata	Bugulidae	*Bugula californica*	Robertson, 1905
			Thalamoporellidae	**Thalamoporella californica*	(Levinsen, 1909)
			Vesiculariidae	*Zoobotryon verticillatum*	(della Chiaje, 1828)
Cnidaria		Anthozoa	Cerianthidae	*Andvakia insignis*	(Carlgren, 1951)
			Hormathiidae	*Calliactis polypus*	(Verrill, 1869)
			Phyllactidae	*Phyllactis californica*	(McMurrich, 1893)
			Caryophylliidae	*Phyllangia consagensis*	(Durham & Barnard, 1952)
			Renillidae	*Renilla amethystina*	Verrill, 1866
Nemertea		Anopla	Lineidae	*Cerebratulus californiensis*	Coe, 1905
Annelida		Polychaeta	Capitellidae	*Capitella capitata*	(Fabricius, 1780)
				Notomastus magnus	Hartman, 1947
			Pilargidae	*Synelmis albini*	(Langerhans, 1881)
			Syllidae	*Branchiosyllis exilis*	(Gravier, 1900)
				Ehlersia cornuta	(Rathke, 1843)
				Typosyllis okadai	(Fauvel, 1934)
				Typosyllis regulata	Imajima, 1966
			Nereidae	*Perinereis bajacalifornica*	de León González & Solís-Weiss, 1998
			Amphinomidae	*Linopherus tripunctata*	(Kudenov, 1975)
			Onuphidae	*Diopatra farallonensis*	Fauchald, 1968
			Terebellidae	*Neoleprea spiralis*	(Johnson, 1901)
			Sabellidae	*Branchiomma cingulata*	(Grube, 1870)
				Branchiomma nigromaculata	(Baird, 1865)
				Sabella melanostigma	Schmarda, 1861
Sipuncula		Sipunculidea	Sipunculidae	*Sipunculus nudus*	Linnaeus, 1766
Arthropoda	Crustacea	Maxillopoda	Pollicipedidae	*Arcoscalpellum californicum*	(Pilsbry, 1907)
			Balanidae	*Balanus amphitrite*	Darwin, 1854
				Balanus eburneus	Gould, 1841
				Balanus improvisus	Darwin, 1854
				Balanus inexpectatus	Pilsbry, 1916
				Balanus trigonus	Darwin, 1854
				Chthamalus anisopoma	Pilsbry, 1916

(continued)

Table 15.1. Continued

Phylum	Subphylum	Class	Family	Scientific Name	Species Author
		Malacostraca	Amphilochidae	*Gitanopsis pusilloides*	Shoemaker, 1942
			Ampithoidae	*Amphitoe ramondi*	Audouin, 1826
			Corophiidae	*Gammaropsis thompsoni*	(Walker, 1898)
				Photis brevipes	Shoemaker, 1942
			Ischyroceridae	*Microjassa macrocoxa*	Shoemaker, 1942
			Lysianassidae	*Orchomene magdalenensis*	(Shoemaker, 1942)
			Cirolanidae	*Cirolana harfordi*	Lockington, 1877
			Corallanidae	*Excorallana tricornis*	(Hansen)
			Cymothoidae	*Ceratothoa gaudichaudii*	(Milne-Edwards, 1840)
				Ceratothoa gilberti	(Richardson, 1904)
				Cymothoa exigua	Schioedte & Meinert, 1884
				Elthusa menziesi	(Brusca, 1981)
				Elthusa vulgaris	(Stimpson, 1857)
				Enispa convexa	(Richardson, 1905)
				Livoneca bowmani	Brusca, 1981
				Mothocya gilli	Bruce, 1986
				Nerocila acuminata	Schioedte & Meinert, 1881
				Rocinela murilloi	Brusca & Iverson, 1985
			Sphaeromatidae	*Paracerceis sculpta*	(Holmes, 1904)
				Paracerceis richardsoni	Lombardo, 1988
			Ligiidae	*Ligia occidentalis*	Dana, 1853
			Leucosiidae	*Randallia ornate*	(Randall, 1839)
			Inachidae	*Stenorhynchus debilis*	(Smith, 1871)
			Xanthidae	*Cataleptodius occidentalis*	(Stimpson, 1871)
				Eurytium affine	(Streets & Kingsley, 1879)
				Eurytium albidigitum	Rathbun, 1933
				Hexapanopeus sinaloensis	Rathbun, 1930
				Panopeus purpureus	Lockington, 1877
				Pilumnus spinohirsutus	(Lockington, 1877)
				Pilumnus townsendi	Rathbun, 1923
			Pinnotheridae	*Pinnixa occidentalis*	Rathbun, 1893
				Raymondia clavapedata	(Glassell, 1935)
			Ocypodidae	*Ocypode occidentalis*	Stimpson, 1860
				Uca brevifrons	(Stimpson, 1860)
				Uca crenulata	(Lockington, 1877)
				Uca latimanus	(Rathbun, 1893)
				Uca musica	Rathbun, 1914

			Uca princeps	(Smith, 1870)
			Uca vocator	(Herbst, 1904)
			Uca zacae	Crane, 1941
			Ucides occidentalis	(Ortmann, 1897)
		Gecarcinidae	*Cardisoma crassum*	Smith, 1870
			Gecarcinus quadratus	de Saussure, 1853
		Grapsidae	*Aratus pisonii*	(H. Milne Edwards, 1837)
			Goniopsis pulchra	(Lockington, 1877)
			Grapsus grapsus	(Linnaeus, 1758)
			Pachygrapsus crassipes	Randall, 1839
			Pachygrapsus transversus	(Gibbes, 1850)
			Armases magdalenense	(Rathbun, 1918)
			Sesarma sulcatum	Smith, 1870
			**Tetragrapsus jouyi*	(Rathbun, 1893)
		Alphaeidae	*Alpheus normanni*	Kingsley, 1878
		Palaemonidae	**Periclimenes infraspinis*	(Rathbun, 1902)
		Penaeidae	*Farfantepenaeus brevirostris*	(Kingsley, 1878)
			Farfantepenaeus californiensis	(Holmes, 1900)
			Litopenaeus stylirostris	(Stimpson, 1874)
			Litopenaeus vannamei	(Boone, 1931)
			Trachysalambria brevisuturae	(Burkenroad, 1934)
		Diogenidae	*Clibanarius albidigitus*	Nobili, 1901
			Clibanarius digueti	Bouvier, 1898
			Clibanarius panamensis	Stimpson, 1859
			Petrochirus californiensis	Bouvier, 1895
		Galatheidae	*Munida hispida*	Benedict, 1902
		Porcellanidae	*Petrolisthes armatus*	(Gibbes, 1850)
Mollusca	Polyplacophora	Ischnochitonidae	**Lepidozona pectinulata*	(Pilsbry, 1893, ex Carpenter)
		Lepidopleuridae	**Leptochiton rugatus*	(Pilsbry, 1892)
		Mopaliidae	**Placiphorella velata*	Dall, 1879
	Gastropoda	Lottiidae	**Lottia atrata*	(Carpenter, 1864)
			**Colisella strongiana*	Hertlein, 1958
		Fissurellidae	*Diodora saturnalis*	(Carpenter, 1864)
		Columbellidae	*Parvanchis pygmaea*	(Sowerby, 1832)
		Neritidae	*Theodoxus luteofasciatus*	(Miller, 1879)
			**Nerita funiculate*	Menke, 1851
		Littorinidae	*Littoraria rosewateri*	Reid, 1999
			Littoraria variegata	(Souleyet, in Eydoux & Souleyet, 1852)
		Turritellidae	*Turitella gonostoma*	Valenciennes, 1832
		Modulidae	*Modiolus catenulatus*	(Philippi, 1849)

(continued)

Table 15.1. Continued

Phylum	Subphylum	Class	Family	Scientific Name	Species Author
			Cerithiidae	Cerithium stercusmuscarum	Valenciennes, 1833
			Potamididae	Cerithidea californica californica	Haldeman, 1840
				Cerithidea californica mazatlanica	Haldeman, 1840
				Cerithidea montagnei	(d'Orbigny, 1839)
				Cerithidea valida	(Adams, 1852)
			Strombidae	Strombus gracilior	Sowerby, 1825
			Calyptraeidae	Calyptraea mamillaris	Broderip, 1834
				Crepidula incurva	(Broderip, 1834)
				Crepidula striolata	Menke, 1851
				Crucibulum spinosum	(Sowerby, 1824)
			Triviidae	*Trivia californica	(Sowerby, 1832 ex Gray, MS)
			Trochidae	Tegula rugosa	(A. Adams, 1853)
			Muricidae	Hexaplex nigritus	(Philippi, 1845)
				Hexaplex erythrostomus	(Swainson, 1831)
				Muricopsis zeteki	Hertlein & Strong, 1951
				Ceratostoma unicorne	(Reeve, 1849)
				Thais kiosquiformis	(Duclos, 1832)
				Acanthina lugubris	(Sowerby, 1822)
			Buccinidae	Cantharus gatesi	(Berry, 1963)
				Melongena patula	(Broderip & Sowerby, 1829)
				Nassarius luteostomus	(Broderip & Sowerby, 1829)
			Olividae	Oliva incrassata	(Lightfoot, 1786)
			Turridae	Pyrgocythara scammoni	(Dall, 1919)
			Pyramidellidae	Turbonilla baegerti	Bartsch, 1917
			Melampidae	Melampus olivaceus	Carpenter, 1857
			Bullidae	Bulla gouldiana	Pilsbry, 1895
			Haminoeidae	Haminoea vesicula	(Gould, 1855)
			Aplysiidae	Stylocheilus longicauda	(Quoy & Gaimard, 1824)
		Pelecypoda	Arcidae	Arca pacifica	(Sowerby, 1833)
				Barbatia gradata	(Broderip & Sowerby, 1829)
				Barbatia reeveana	(d'Orbigny, 1846)
				Anadara adamsi	Olsson, 1961
				Anadara obesa	(Sowerby, 1833)
				Anadara tuberculosa	(Sowerby, 1833)
				Anadara nux	(Sowerby, 1833)
				Anadara emarginata	(Sowerby, 1833)
			Noetiidae	Noetia reversa	(Sowerby, 1833)

Family	Species	Author
Glycymerididae	*Glycymeris gigantea*	(Reeve, 1843)
	Glycymeris maculata	(Broderip, 1832)
	Glycymeris inaequalis	(Sowerby, 1833)
Mytilidae	*Lithophaga calyculata*	(Carpenter, 1857)
	Modiolus capax	(Conrad, 1837)
Pinnidae	*Pinna rugosa*	Sowerby, 1835
	Atrina maura	(Sowerby, 1835)
Pteriidae	*Pinctada mazatlanica*	(Hanley, 1856)
Ostreidae	*Saccostrea palmula*	(Carpenter, 1857)
Plicatulidae	*Plicatula penicillata*	Carpenter, 1857
Pectinidae	*Leptopecten velero*	(Hertlein, 1935)
Anomiidae	*Anomia peruviana*	d'Orbigny, 1846
Crassatellidae	*Crassinella varians*	(Carpenter, 1857)
Carditidae	*Carditamera affinis*	(Sowerby, 1833)
Ungulinidae	*Diplodonta subquadrata*	Carpenter, 1856
	Felaniella cornea	(Reeve, 1850)
Sportellidae	*Fabella stearnsii*	(Dall, 1899)
Chamidae	*Chama sordida*	Broderip, 1835
	Pseudochama saavedrai	Hertlein & Strong, 1946
Cardiidae	*Trachycardium consors*	(Sowerby, 1833)
	Trachycardium obovalis	(Sowerby, 1833)
	Trachycardium panamense	(Sowerby, 1833)
	Trachycardium procerum	(Sowerby, 1833)
	Papyridea aspersa	(Sowerby, 1833)
	Trigoniocardia granifera	(Broderip & Sowerby, 1829)
	Laevicardium elenense	(Sowerby, 1840)
Veneridae	*Globivenus isocardia*	(Verrill, 1870)
	Pitar lupanaria	(Lesson, 1830)
	Pitar concinnus	(Sowerby, 1835)
	Megapitaria squalida	(Sowerby, 1835)
	Dosinia dunkeri	(Philippi, 1844)
	Dosinia ponderosa	(Gray, 1838)
	Cyclinella singleyi	Dall, 1902
	Chione californiensis	(Broderip, 1835)
	Chione undatella	(Sowerby, 1835)
	Chione subrugosa	(Wood, 1828)
	Prototbaca asperrima	(Sowerby, 1835)
	Protothaca grata	(Say, 1831)
Mactridae	*Mulinia pallida*	(Broderip & Sowerby, 1829)
	Rangia mendica	(Gould, 1851)

(continued)

Table 15.1. Continued

Phylum	Subphylum	Class	Family	Scientific Name	Species Author
			Tellinidae	*Tellina mcneilii*	Dall, 1900
				Tellina simulans	C.B. Adams, 1852
				Tellina reclusa	Dall, 1900
				Tellina virgo	Hanley, 1844
				Macoma secta	(Conrad, 1837)
			Donacidae	*Donax carinatus*	Hanley, 1843
				Donax gracilis	Hanley, 1845
				Donax punctatostriatus	Hanley, 1843
				Donax transversus	Sowerby, 1825
			Psammobiidae	*Tagelus affinis*	(C.B. Adams, 1852)
				Tagelus politus	(Carpenter, 1857)
			Pholadidae	*Pholadidae melanura*	(Sowerby, 1834)
			Thraciidae	*Asthenothaerus diegensis*	(Dall, 1915)
			Corbiculidae	*Polymesoda mexicana*	(Broderip & Sowerby, 1829)
Echinodermata	Asterozoa	Asteroidea	Echinasteridae	*Echinaster parvispinus*	A.H. Clark, 1916
		Ophiuroidea	Ophiactidae	*Ophiactis savignyi*	(Muller & Troschel, 1842)
				Ophiactis simplex	(Le Conte, 1851)
			Ophiotrichidae	*Ophiothrix spiculata*	Le Conte, 1851
	Echinozoa	Echinoidea	Loveniidae	*Lovenia cordiformis*	A. Agassiz, 1872
		Holothuroidea	Stichopodidae	*Isostichopus fuscus*	(Ludwig, 1875)
Chordata	Urochordata	Ascidiacea	Molgulidae	*Molgula occidentalis*	Traustedt, 1883
			Polyclinidae	*Aplidium sp. ?*	
			Botryllidae	*Botrylloides diegensis*	Ritter & Forsyth, 1917
	Cephalochordata			*Branchiostoma californiense*	Andrews, 1893

Data are from the Macrofauna Golfo Database (Findley et al. in press), Holguin-Quiñones and García-Domínguez (1997), and the personal field notes of R. C. Brusca. *Occurrence restricted to rocks in mangrove waters.

that is critical to the life history of this commercially important invertebrate (Brusca 1980).

The grapsid crab *Goniopsis pulchra* has a wide distribution that includes both coasts of BCS and the eastern side of the Gulf (Brusca 1980). This bright-red semiterrestrial crab is distributed throughout the intertidal zone, and it is abundant in mangrove swamps of the middle and southern Gulf of California. At least in Sonora, *G. pulchra* is the primary herbivore of mangrove propagules that reach the swamp floor. In a study in 3 mangrove communities in Sonora, C. McIvor and A. I. Robertson (Charles Sturt University, Wagga Wagga, New South Wales, Australia, unpubl. data) determined that this crab fed preferentially on the propagules of black mangrove (*Avicennia*) when offered those in equal abundance to red mangrove (*Rhizophora*) propagules. Neverthe-less, when only red mangrove propagules were sea-sonally available, 8–40% of those propagules tethered were rendered nonviable for germination (growing tip removed, more than 50% of propagule consumed, or propagule pulled down a burrow) in less than 20 days. Preferential consumption of some as opposed to other species of mangrove propagules is believed to have ramifications for mangrove structure and zonation (e.g., Smith et al. 1989; Smith 1992).

In Sonora, *G. pulchra* took tethered propagules down their burrows for consumption when pos-sible; otherwise they grazed on this food source during ebb tides on the swamp floor. Mangrove herbivory by grapsid crabs of this and other genera has been repeatedly identified, especially in the Indo-West Pacific biogeographic realm, as an alternative pathway of carbon and organic matter processing to detritivory in mangrove ecosystems (e.g., Robertson 1991). In the infrequently flooded high intertidal zone (Robertson and Daniel 1989) and in poorly flushed basin forests (Twilley 1985), crab herbivory probably results in increased retention of mangrove organic matter within the swamp and thus lowered export. In lower intertidal zones subject to frequent tidal flooding, however, herbivorous crabs (or other invertebrates) are consumed by fish and turtles, and mangrove production is directly transferred out of the swamp to the adjacent coastal ecosystem.

Fish

Fishes are important inhabitants of mangrove waters around the world. Members of several groups such as gobies (Gobiidae) and mojarras (Gerreidae) occur in these areas for most or all of their lives, while others are temporary inhabitants as adults, juve-niles, or both. Mangrove ecosystems are thus im-portant for many tropical fishes (Blaber 1997), and this is true in particular with respect to the man-groves of BCS.

A detailed compilation of the fish fauna associ-ated with BCS mangrove waters has not been pub-lished, although the fishes of several of the larger bays of the state that include mangroves have been extensively surveyed and recently summarized by Galván-Magaña et al. (2000). The most extensively studied systems are Bahía de La Paz (Abitia-Cárdenas et al. 1994; Castro-Aguirre and Balart 1997; González-Acosta et al. 1999; Galván-Piña et al. 2003) and Bahía Concepción (Rodríguez-Romero et al. 1992, 1994, 1998), on the east coast of BCS, and Bahía Magdalena (de la Cruz-Agüero et al. 1994) and Laguna San Ignacio (Danemann and de la Cruz-Agüero 1993; de la Cruz-Agüero and Cota-Gómez 1998) on the Pacific side. These pub-lished surveys include all species of fishes recorded in these large and diverse lagoon systems and gen-erally do not indicate which species are associated with mangroves. Only 2 of these studies report on species from relatively restricted lagoons lined with mangroves within Bahía de La Paz. Castro-Aguirre and Balart (1997) report fishes recorded from Ensenada de Aripes, and González-Acosta et al. (1999) report fishes taken from a mangrove-lined swamp near the mouth of Ensenada de Aripes. Table 15.2 presents a list of fish species recorded in these 2 studies, as well as additional, unpublished records of fishes from that same area (Galván-Magaña, unpubl. data) and unpublished records of fishes collected adjacent to BCS mangrove stands and archived at the Scripps Institution of Ocean-ography Marine Vertebrates Collection. The latter collections are primarily from the lagoon systems along the Pacific coast of BCS, from Bahía Magdalena northward to near Punta Abreojos (26°49' N).

A total of 160 species of fishes have been re-corded from mangrove systems of BCS (table 15.2). This list includes a number of species that normally occur in mangrove waters as both juveniles and adults, as well as a number of others that also occur in other habitats both inside and outside of the larger lagoon systems. Dominant members include grunts (Haemulidae, 17 species), gobies and sleepers (Gobi-idae and Eleotridae, 15 species), drums (Sciaenidae, 11 species), jacks (Carangidae, 10 species), mojarras (Gerreidae, 9 species), anchovies (Engraulidae, 9 species), and seabasses (Serranidae, 8 species). Fishes

Table 15.2. Fish species recorded from mangrove waters of Baja California Sur.

Order	Family	Scientific Name
Carcharhiniformes	Triakidae	*Mustelus henlei* (Gill, 1863)
		Mustelus lunulatus Jordan & Gilbert, 1882
		Triakis semifasciata Girard, 1855
Rajiformes	Dasyatidae	*Dasyatis dipterura* (Jordan & Gilbert, 1880)
	Urolophidae	*Urobatis halleri* (Cooper, 1863)
		Urotrygon chilensis (Günther, 1871)
	Gymnuridae	*Gymnura marmorata* (Cooper, 1864)
	Rhinobatidae	*Rhinobatos productus* Ayers, 1854
Elopiformes	Elopidae	*Elops affinis* Regan, 1909
Albuliformes	Albulidae	*Albula nemoptera* (Fowler, 1911)
		Albula sp.
	Ophichthidae	*Myrichthys tigrinus* Girard, 1859
		Myrophis vafer Jordan & Gilbert, 1883
		Ophichthus zophochir Jordan & Gilbert, 1882
	Congridae	*Heteroconger digueti* (Pellegrin, 1923)
	Clupeidae	*Harengula thrissina* (Jordan & Gilbert, 1882)
		Lile stolifera (Jordan & Gilbert, 1882)
		Opisthonema libertate (Günther, 1867)
		Sardinops caeruleus (Girard, 1854)
	Engraulidae	*Anchoa argentivittata* (Regan, 1904)
		Anchoa exigua (Jordan & Gilbert, 1882)
		Anchoa ischana (Jordan & Gilbert, 1882)
		Anchoa lucida (Jordan & Gilbert, 1882)
		Anchoa mundeola (Gilbert & Pierson, 1898)
		Anchoa mundeoloides (Breder, 1928)
		Anchovia macrolepidota (Kner, 1863)
		Cetengraulis mysticetus (Günther, 1867)
		Engraulis mordax Girard, 1854
Gonorynchiformes	Chanidae	*Chanos chanos* (Forsskål, 1775)
Siluriformes	Ariidae	*Ariopsis planiceps* (Steindachner, 1877)
		Ariopsis seemanni (Günther, 1864)
		Bagre panamensis (Gill, 1863)
		Bagre pinnimaculatus (Steindachner, 1877)
		Galeichthys peruvianus Lütken, 1874
	Synodontidae	*Synodus sciptuliceps* Jordan & Gilbert, 1882
	Ophidiidae	*Ophidion galeoides* (Gilbert, 1890)
Batrachoidiformes	Batrachoididae	*Porichthys myriaster* Hubbs & Schultz, 1939
Mugliformes	Mugilidae	*Mugil cephalus* Linnaeus, 1758
		Mugil curema Valenciennes, 1836
		Mugil hospes Jordan & Culver, 1895
Atheriniformes	Atherinopsidae	*Atherinops affinis* (Ayers, 1860)
		Atherinopsis californiensis Girard, 1854
	Hemiramphidae	*Hyporhamphus naos* Banford & Collette, 2001
		Hyporhamphus rosae (Jordan & Gilbert, 1880)
Cyprinodontiformes	Fundulidae	*Fundulus parvipinnis* Girard, 1854
Gasterosteiformes	Fistulariidae	*Fistularia commersonii* Ruppell, 1838
	Syngnathidae	*Pseudophallus starksi* (Jordan & Culver, 1895)
		Syngnathus auliscus (Swain, 1882)
		Syngnathus euchrous Fritzsche, 1980
	Triglidae	*Prionotus stephanophrys* Lockington, 1881
Perciformes	Centropomidae	*Centropomus armatus* Gill, 1863
		Centropomus medius Günther, 1864
		Centropomus nigrescens Günther, 1864
		Centropomus robalito Jordan & Gilbert, 1882
		Centropomus viridis Lockington, 1877
	Serranidae	*Diplectrum euryplectrum* Jordan & Bollman, 1890

Order	Family	Scientific Name
		Diplectrum pacificum Meek & Hildebrand, 1925
		Epinephelus analogus Gill, 1863
		Epinephelus itajara (Lichtenstein, 1822)
		Epinephelus niphobles Gilbert & Starks, 1897
		Mycteroperca xenarcha Jordan, 1888
		Paralabrax maculatofasciatus (Steindachner, 1868)
		Paralabrax nebulifer (Girard, 1854)
	Carangidae	*Caranx caninus* Günther, 1867
		Caranx sexfasciatus Quoy & Gaimard, 1825
		Caranx vinctus Jordan & Gilbert, 1882
		Chloroscombrus orqueta Jordan & Gilbert, 1883
		Gnathanodon speciosus (Forsskål, 1775)
		Hemicaranx leucurus (Günther, 1864)
		Oligoplites altus (Günther, 1868)
		Selene brevoortii (Gill, 1863)
		Selene peruviana (Guichenot, 1866)
		Trachinotus paitensis Cuvier, 1832
	Lutjanidae	*Lutjanus aratus* (Günther, 1864)
		Lutjanus argentiventris (Peters, 1869)
		Lutjanus colorado Jordan & Gilbert, 1882
		Lutjanus novemfasciatus Gill, 1862
	Gerreidae	*Diapterus aureolus* (Jordan & Gilbert, 1882)
		Diapterus peruvianus (Cuvier, 1830)
		Eucinostomus currani Zahuranec, 1980
		Eucinostomus dowii (Gill, 1863)
		Eucinostomus entomelas Zahuranec, 1980
		Eucinostomus gracilis (Gill, 1862)
		Eugerres axillaris (Günther, 1864)
		Eugerres lineatus (Humboldt, 1821)
		Gerres cinereus (Walbaum, 1792)
	Haemulidae	*Conodon serrifer* Jordan & Gilbert, 1882
		Haemulon flaviguttatum Gill, 1862
		Haemulon maculicauda (Gill, 1862)
		Haemulon scudderii Gill, 1862
		Haemulon sexfasciatum Gill, 1862
		Haemulon steindachneri (Jordan & Gilbert, 1882)
		Haemulopsis axillaris (Steindachner, 1869)
		Haemulopsis elongatus (Steindachner, 1879)
		Haemulopsis leuciscus (Günther, 1864)
		Haemulopsis nitidus (Steindachner, 1869)
		Orthopristis chalceus (Günther, 1864)
		Orthopristis reddingi Jordan & Richardson, 1895
		Pomadasys bayanus Jordan & Evermann, 1898
		Pomadasys branicki (Steindachner, 1879)
		Pomadasys macracanthus (Günther, 1864)
		Pomadasys panamensis (Steindachner, 1876)
		Xenistius californiensis (Steindachner, 1876)
	Sparidae	*Calamus brachysomus* (Lockington, 1880)
	Polynemidae	*Polydactylus approximans* (Lay & Bennett, 1839)
	Sciaenidae	*Atractoscion nobilis* (Ayers, 1860)
		Bairdiella icistia (Jordan & Gilbert, 1882)
		Cynoscion parvipinnis Ayers, 1861
		Cynoscion xanthulus Jordan & Gilbert, 1882
		Menticirrhus nasus (Günther, 1868)
		Menticirrhus undulatus (Girard, 1854)

(continued)

Table 15.2. Continued

Order	Family	Scientific Name
		Micropogonias altipinnis (Günther, 1864)
		Ophioscion strabo Gilbert, 1897
		Umbrina roncador Jordan & Gilbert, 1882
		Umbrina wintersteeni Walker & Radford, 1992
		Umbrina xanti Gill, 1862
	Mullidae	*Pseudupeneus grandisquamis* (Gill, 1863)
	Chaetodontidae	*Chaetodon humeralis* Günther, 1860
	Pomacanthidae	*Pomacanthus zonipectus* (Gill, 1862)
	Kyphosidae	*Girella nigricans* (Ayers, 1860)
	Pomacentridae	*Abudefduf troschelii* (Gill, 1862)
	Labridae	*Halichoeres aestuaricola* Bussing, 1972
	Scaridae	*Nicholsina denticulata* (Evermann & Radcliffe, 1917)
	Labrisomidae	*Exerpes asper* (Jenkins & Evermann, 1889)
		Paraclinus sini Hubbs, 1952
	Blenniidae	*Hypsoblennius gentilis* (Girard, 1854)
	Eleotridae	*Dormitator latifrons* (Richardson, 1844)
		Gobiomorus maculatus (Günther, 1859)
	Gobiidae	*Bathygobius ramosus* Ginsburg, 1947
		Clevelandia ios (Jordan & Gilbert, 1882)
		Ctenogobius manglicola (Jordan & Starks, 1895)
		Ctenogobius sagittula (Günther, 1861)
		Evorthodus minutus Meek & Hildebr&, 1928
		Gillichthys mirabilis Cooper, 1864
		Gobionellus microdon (Gilbert, 1892)
		Gobiosoma chiquita (Jenkins & Evermann, 1889)
		Ilypnus gilberti (Eigenmann & Eigenmann, 1889)
		Microgobius brevispinis Ginsburg, 1939
		Microgobius cyclolepis Gilbert, 1890
		Microgobius tabogensis Meek & Hildebrand, 1928
		Quietula y-cauda (Jenkins & Evermann, 1889)
	Microdesmidae	*Microdesmus dorsipunctatus* Dawson, 1968
	Ephippididae	*Chaetodipterus zonatus* (Girard, 1858)
	Scombridae	*Auxis thazard* (Lacepède, 1800)
	Paralichthyidae	*Citharichthys gilberti* Jenkins & Evermann, 1889
		Cyclopsetta panamensis (Steindachner, 1876)
		Etropus crossotus Jordan & Gilbert, 1882
		Paralichthys californicus (Ayers, 1859)
		Paralichthys woolmani Jordan & Williams, 1897
		Syacium ovale (Günther, 1864)
	Pleuronectidae	*Hypsopsetta guttulata* (Girard, 1856)
	Achiridae	*Achirus mazatlanus* (Steindachner, 1869)
		Trinectes fonsecensis (Günther, 1862)
	Cynoglossidae	*Symphurus chabanaudi* Mahadeva & Munroe, 1990
Tetraodontiformes	Balistidae	*Balistes polylepis* Steindachner, 1876
		Pseudobalistes naufragium (Jordan & Starks, 1895)
	Tetraodontidae	*Sphoeroides annulatus* (Jenyns, 1842)
		Sphoeroides lobatus (Steindachner, 1870)
	Diodontidae	*Diodon holocanthus* Linnaeus, 1758
		Diodon hystrix Linnaeus, 1758

Data are from Castro-Aguirre and Balart 1997 (Ensenada de La Paz, Bahía de La Paz); González-Acosta et al. 1999 (a mangrove-lined swamp near the mouth of Ensenada de La Paz, Bahía de La Paz); the personal field notes of F. Galván-Magaña (Ensenada de Aripes, Bahía de La Paz); and unpublished records of fishes collected adjacent to mangrove stands in BCS and archived at the Scripps Institution of Oceanography Marine Vertebrates Collection. The latter collections are primarily from the lagoon systems along the Pacific coast of BCS, from Bahía Magdalena northward to near Punta Abreojos (26° 49' N). Taxonomy of species follows Eschmeyer (1998) and Findley et al. (in press).

from these families are common inhabitants of mangrove systems around the world (Blaber 1997).

Like the entire fish fauna of the Gulf of California (Walker 1960), the fishes of mangrove waters in BCS are dominated by tropical species. However, the lagoons along the Pacific coast of BCS also have several warm-temperate species, such as the barred sandbass (*Paralabrax nebulifer*), the white seabass (*Atractoscion nobilis*), and California killifish (*Fundulus parvipinnis*), all of them found along the Pacific coast from BCS northward to California. These complex lagoon systems are thus especially rich systems that lie in a zone of overlap between the tropical regions of the eastern Pacific and the more temperate areas to the north (Hubbs 1960; Galván-Magaña et al. 2000). There is no doubt that many other fishes, especially juveniles of species inhabiting other habitats as adults, will be recorded from BCS mangrove lagoon systems as their fish fauna is more thoroughly studied.

Sea Turtles

The coastal waters of Baja California Sur host 4 of the world's 7 species of sea turtles: the green turtle (*Chelonia mydas*), hawksbill (*Eretmochelys imbricata*), loggerhead (*Caretta caretta*), and olive ridley (*Lepidochelys olivacea*) (Cliffton et al. 1982; Nichols 2003; see also chapter 20). Within coastal mangrove systems the most common sea turtle species are the green turtle, and, to a lesser extent, the hawksbill (López-Mendilaharsu 2002; Seminoff et al. 2003; Brooks et al. 2003). Loggerhead turtles and olive ridleys typically prefer more offshore waters and are thus less common near mangroves (Nichols 2003).

The most important mangrove habitats for sea turtles in BCS include the Pacific Coast sites of Bahía Magdalena, Laguna San Ignacio, and Estero Coyote (Nichols 2003; J. Nichols, pers. comm.). Although a wide size-range of green turtles is typically seen, these areas appear to be most important as nursery grounds for small juveniles (Nichols 2003; Seminoff 2003). Seminoff (2003) has demonstrated that green turtles in mangrove systems of the Baja California peninsula are significantly smaller than turtles found in adjacent exposed, high-energy coastal areas and suggests the shallow and protected nature of these mangrove systems provides important predator-free habitat with abundant food for growing turtles.

Green turtles are primarily herbivorous throughout most of their global range, but in northwestern Mexico the species has been shown to consume a wide variety of both plants and invertebrates (Lopez-Mendilaharsu 2002; Seminoff et al. 2002a). Many potential food species are present near mangroves and, like elsewhere in the region (Seminoff et al. 2002b), green turtles likely maintain residency to these areas for extended periods while benefiting from the abundant local resources. In other parts of the world, it is the mangroves themselves that are consumed by green turtles. For example, the leaves of *Avicennia marina* make up a substantial portion of Australian green turtle diets (Pendoley and Fitzpatrick 1999; Limpus and Limpus 2000). The same can be said for green turtles foraging on red mangrove in the Galapagos (Pritchard 1971). However, although leaves and fruit of *A. germinans* are eaten on occasion in mangrove systems of BCS (Lopez-Mendilaharsu 2002), the primary value of these areas for sea turtles is the vast abundance of seagrass, marine algae, and invertebrate prey. Marine algae such as *Codium amplivesiculatum* and *Gracilaria textorii* are common near Pacific coast mangroves, and together with seagrasses they are the most important dietary components of local green turtles (Lopez-Mendilaharsu 2002). In addition, invertebrates such as molluscs and sponges have been found in the green turtle diet samples, the latter group being of primary importance also in the diet of local hawksbill turtles (Meylan 1988).

The hawksbill occurs on both coasts of the peninsula (Seminoff et al. 2003). Sightings of this now rare species are often in shallow, mangrove-lined bays, lagoons, and *esteros*. The hawksbill is seen occasionally in the mangrove *estero* at the mouth of the Río Santa Rosalía de Mulegé near the town of Mulegé (see below). Although it is principally a spongivore, it is known to eat the fruits, leaves, and bark of mangroves (Grismer 2002).

All sea turtles are threatened or endangered. Despite countrywide legal protection since 1990, illegal capture of sea turtles is still common, especially with green turtles in mangrove systems of the Baja California peninsula (Gardner and Nichols 2001; Nichols et al. 2002; chapter 20). Although some mangrove systems in BCS, particularly at Estero Coyote, continue to host a large number of turtles (Nichols 2003; J. Nichols, pers. comm.), it is clear that without stronger and immediate conservation action, sea turtle populations in and around these fragile ecosystems will continue to decline. For a discussion of sea turtle ecology and conservation in the Gulf of California, see chapter 20.

Birds

Much research has been conducted on birds inhabiting mangrove ecosystems in BCS. We present a list of species documented in mangrove stands and adjacent habitat (e.g., sand dunes, mudflats, and rocky outcroppings) of Bahía de La Paz, Bahía Magdalena, and the Laguna San Ignacio Complex (table 15.3). Our list is based on the published literature and the personal field notes (of E. S. A.-S., R. M.-S., and R. C. W.), plus those of Roberto Carmona and his students. The most detailed work was conducted in Bahía de La Paz. The list of winter migrants and transients, particularly songbirds (order Passeriformes) is likely incomplete. Elsewhere (e.g., the Caribbean; Sherry and Holmes 1996, Warkentin and Morton 2000, Reitsma et al. 2002), mangroves provide critical habitat for wintering species. Future research efforts in BCS should incorporate mist netting and banding outside the breeding season.

One hundred thirty-one species (representing 15 orders and 35 families) occur in association with the mangrove ecosystems of Bahía de La Paz, Bahía Magdalena, and the Laguna San Ignacio Complex. Sixty-six species have been documented at all 3 locations. The numbers of species documented at each of the 3 locations are not much different (Bahía de La Paz: 90 species; Bahía Magdalena: 104 species; Laguna San Ignacio Complex: 89 species). The list is dominated by the Scolopacidae (19 species), Anatidae (19 species), Laridae (16 species), and Ardeidae (12 species). Many of the species recorded also occur away from mangroves.

Of the 22 species using mangrove plants as nesting substrate, 10 are herons or egrets (family Ardeidae; see further on). With respect to ardeids, one location in particular stands out: Estero El Conchalito in Ensenada de Aripes, Bahía de La Paz (Carmona et al. 1994). During 1986–1991, 10 different species were documented nesting in the local black, red, and white mangroves. Yellow-crowned night-herons (*Nyctanassa violacea*), snowy egrets (*Egretta thula*), and cattle egrets (*Bubulcus ibis*) all nested colonially in black mangrove, while small colonies of black-crowned night-herons (*Nycticorax nycticorax*) and great blue herons (*Ardea herodias*) nested in red mangrove. Of all these species, only the great blue heron uses other nesting substrates besides mangroves (Carmona et al. 1994).

Other species recorded nesting in mangroves of BCS consist of the double-crested cormorant (*Phalacrocorax auritus*), magnificent frigatebird (*Fregata magnificens*), white ibis (*Eudocimus albus*), bald eagle (*Haliaeetus leucocephalus*), clapper rail (*Rallus longirostris*), Virginia rail (*Rallus limicola*), white-winged dove (*Zenaida asiatica*), Xantus's hummingbird (*Hylocharis xantusii*), western scrub-jay (*Aphelocoma californica*), verdin (*Auriparus flaviceps*), mangrove warbler (*Dendroica petechia castaneiceps*), and house finch (*Carpodacus mexicanus*). The Xantus's hummingbird is largely endemic to BCS. The bald eagle, recorded nesting on mangroves at Bahía Magdalena, is federally listed as Endangered (*En Peligro de Extinción*) by the Mexican government (DOF 2002). The clapper rail and the Virginia rail are federally listed as Subject to Special Protection (DOF 2002). The clapper rail nests in moderate numbers in red mangroves at El Mogote and along Ensenada Aripes (Massey and Palacios 1994; R. Carmona, pers. comm.). One Virginia rail nest was recorded in 1988 in a red mangrove in Bahía de La Paz. The mangrove warbler, a subspecies of the yellow warbler (*D. petechia*), is the only bird in the Baja California peninsula that is restricted to mangroves.

Several reasons have been postulated for the lack of a distinct mangrove avifauna worldwide, and they center on habitat structure considerations. Compared to terrestrial forests, such as those in the tropics, mangrove stands have low structural diversity, but they are extremely dense, consisting of "canopies of glossy, tough green leaves covering numerous gnarled to erect stems emerging from an inter-connected tangle of above-ground roots bedded in soft wet mud" (Duke 2001: 258). There is no understory vegetation, a principal component of terrestrial ecosystems to which many bird species are adapted (Maurer et al. 1980; Bell and Whitmore 1997). Although the world's mangroves do not have their own distinct avifauna, many birds use mangrove habitats for foraging on insects, as safe roosts, and for avoiding extreme temperatures during the day (e.g., Strong and Johnson 2001). Mangroves provide critical nesting habitat for many avian species (e.g., Hilton et al. 2000), most notably in BCS for the mangrove warbler, but also for most herons and egrets.

Mammals

The bottlenose dolphin (*Tursiops truncatus*) uses mangrove waters as feeding areas (Acevedo 1991; Ballance 1992; Felix 1994; J. Urbán-Ramírez, pers. comm.). North of Bahía Magdalena, in the Santo Domingo Channel (Puerto Lopez Mateos), bottle-

Table 15.3. Avian species documented in mangrove ecosystems of Baja California Sur.

Order	Family	Scientific Name	Common Name[a]	Area[b]
Gaviiformes	Gaviidae	*Gavia pacifica*	Pacific loon +	BM
		Gavia immer	Common loon +	BM
Podicipediformes	Podicipedidae	*Podilymbus podiceps*	Pied-billed grebe	BLP
		Podiceps nigricollis	Eared grebe +	BLP
		Aechmophorus clarkii	Clark's grebe +	BM
		Aechmophorus occidentalis	Western grebe +	BLP, BM, SI
Pelicaniformes	Sulidae	*Sula dactylatra*	Masked booby	BLP, BM
		Sula nebouxii	Blue-footed booby	BLP, BM
		Sula leucogaster	Brown booby	BLP
	Pelicanidae	*Pelacanus erythrorhynchos*	White pelican +	BLP, BM, SI
		Pelacanus occidentalis	Brown pelican	BLP, BM, SI
	Phalocrocoracidae	*Phalacrocorax auritus*	Double-crested cormorant	BLP, BM, SI*
		Phalacrocorax penicillatus	Brandt's cormorant	BLP, BM, SI
		Phalocrocorax pelagicus	Neotropic cormorant	BM
	Fregatidae	*Fregata magnificens*	Magnificent frigatebird	BLP, BM, SI*
Ciconiiformes	Ardeidae	*Botaurus lentiginosus*	American bittern	SI
		Ixobrychus exilis	Least bittern +	SI
		Ardea herodias	Great blue heron +, Pr	BLP*, BM*, SI*
		Ardea alba	Great egret	BLP*, BM*, SI*
		Egretta thula	Snowy egret	BLP*, BM*, SI*
		Egretta caerulea	Little Blue heron	BLP*, BM*, SI*
		Egretta tricolor	Tricolored heron	BLP*, BM*, SI*
		Egretta rufescens	Reddish egret Pr	BLP*, BM*, SI*
		Bubulcus ibis	Cattle egret	BLP*, BM
		Butorides virescens	Green heron	BLP*, BM*, SI*
		Nycticorax nycticorax	Black-crowned night-heron	BLP*, BM*, SI*
		Nyctanassa violacea	Yellow-crowned night-heron +	BLP*, BM*, SI*
	Threskiornithidae	*Eudocimus albus*	White ibis	BLP*, BM*, SI*
		Plegadis chihi	White-faced ibis +	BLP
		Ajaia ajaja	Roseate spoonbill +	BM
	Ciconiidae	*Mycteria americana*	Wood stork +, Pr	BLP
	Cathartidae	*Cathartes aura*	Turkey vulture	BLP, BM, SI
Anseriformes	Anatidae	*Anser albifrons*	Greater white-fronted goose +	SI
		Branta bernicula	Brant +	BM, SI
		Anas platyrhynchos	Mallard +	BLP
		Anas acuta	Northern pintail +	BLP, BM, SI

(*continued*)

Table 15.3. Continued

Order	Family	Scientific Name	Common Name[a]	Area[b]
		Anas cyanoptera	Cinnamon teal +	BLP, BM
		Anas clypeata	Northern shoveler +	BLP, BM
		Anas strepera	Gadwall +	BM
		Anas americana	American wigeon +	SI
		Aythya valisineria	Canvasback +	SI
		Aythya americana	Redhead +	BM, SI
		Aythya collaris	Ring-necked duck +	SI
		Aythya marila	Greater scaup +	BLP, SI
		Aythya affinis	Lesser scaup +	BLP, BM, SI
		Melanitta perspicillata	Surf scoter +	BM, SI
		Melanitta fusca	White-winged scoter +	BM
		Bucephala albeola	Bufflehead +	BLP, BM, SI
		Mergus merganser	Common merganser +	BM, SI
		Mergus serrator	Red-breasted merganser +	BLP, BM, SI
		Oxyura jamaicensis	Ruddy duck +	BLP, BM, SI
Falconiformes	Accipitridae	*Pandion haliaetus*	Osprey	BLP, BM, SI
		Elanus leucurus	White-tailed kite +	BM
		Haliaeetus leucocephalus	Bald eagle P	BM*
		Circus cyaneus	Northern harrier +	BM, SI
	Falconidae	*Falco sparverius*	American kestrel	BLP, BM
		Falco columbarius	Merlin +	BLP
		Falco peregrinus	Peregrine falcon Pr	BLP, BM, SI
Gruiformes	Rallidae	*Rallus longirostris*	Clapper rail Pr	BLP*, BM, SI
		Rallus limicola	Virginia rail Pr	BLP*, BM
		Porzana carolina	Sora +	BM
		Fulica americana	American coot	BLP
Charadriiformes	Charadriidae	*Pluvialis squatarola*	Black-bellied plover +	BLP, BM, SI
		Charadrius vociferus	Killdeer	BLP^, BM, SI
		Charadrius wilsonia	Wilson's plover	BLP^, BM^, SI^
		Charadrius semipalmatus	Semipalmated plover +	BLP, BM, SI
		Charadrius alexandrinus	Snowy plover	BLP^, BM, SI^
	Haematopodidae	*Haematopus bachmani*	Black oystercatcher	BM, SI
		Haematopus palliatus	American oystercatcher	BLP^, BM^, SI^
	Recurvostridae	*Himantopus mexicanus*	Black-necked stilt +	BLP, BM, SI
		Recurvirostra americana	American avocet +	BLP, BM, SI
	Scolopacidae	*Tringa melanoleuca*	Greater yellowlegs +	BLP, BM, SI
		Tringa flavipes	Lesser yellowlegs +	BLP, BM, SI

320

		Tringa solitaria	Solitary yellowlegs	BM
		Catoptrophorus semipalmatus	Willet +	BLP, BM, SI
		Heteroscelus incanus	Wandering tattler +	BLP, BM, SI
		Actitis macularia	Spotted sandpiper +	BLP, BM, SI
		Numenius phaeopus	Whimbrel +	BLP, BM, SI
		Numenius americanus	Long-billed curlew +	BLP, BM, SI
		Limosa fedoa	Marbled godwit +	BLP, BM, SI
		Arenaria interpres	Ruddy turnstone +	BLP, BM, SI
		Arenaria melanocephala	Black turnstone +	BLP, BM, SI
		Calidris alba	Sanderling +	BLP, BM, SI
		Calidris minutilla	Least sandpiper +	BLP, BM, SI
		Calidris mauri	Western sandpiper +	BLP, BM, SI
		Calidris canutus	Red knot +	BM
		Calidris alpina	Dunlin +	BLP, BM, SI
		Limnodromus griseus	Short-billed dowitcher +	BLP, BM, SI
		Limnodromus scolopaceus	Long-billed dowitcher +	BLP, BM, SI
		Phalaropus fulicaria	Red phalarope +	BLP, BM, SI
	Laridae	*Larus atricilla*	Laughing gull +	BLP, BM
		Larus philadelphia	Bonaparte's gull +	BLP, BM, SI
		Larus heermani	Heermann's gull Pr	BLP, BM, SI
		Larus delawarensis	Ring-billed gull +	BLP, BM, SI
		Larus californicus	California gull +	BLP, BM, SI
		Larus argentatus	Herring gull +	BM
		Larus livens	Yellow-footed gull Pr	BLP, BM, SI
		Larus occidentalis	Western gull	BM, SI
		Larus hyperboreus	Glaucous gull +	BM
		Xema sabini	Sabine's gull +	SI
		Sterna maxima	Royal tern	BLP, BM, SI
		Sterna elegans	Elegant tern Pr	BLP, BM, SI
		Sterna caspia	Caspian tern +	BLP, BM, SI
		Sterna forsteri	Forster's tern +	BLP, BM, SI
		Sterna antillarum	Least tern +, P	BLP^, BM, SI
		Rynchops niger	Black skimmer +	BLP, SI
Columbiformes	Columbidae	*Zenaida asiatica*	White-winged dove	BLP*
		Zenaida macroura	Mourning dove	BLP
Strigiformes	Strigidae	*Bubo virginianus*	Great-horned owl	SI
		Athene cunicularia	Burrowing owl	BLP, BM, SI
Caprimulgiformes	Caprimulgidae	*Chordeiles acutipennis*	Lesser nighthawk	BLP

(continued)

Table 15.3. Continued

Order	Family	Scientific Name	Common Name[a]	Area[b]
Apodiformes	Trochilidae	*Hylocharis xantusii*	Xantus' hummingbird	BLP*
		Archilocus alexandri	Black-chinned hummingbird +	SI
		Calypte costae	Costa´s hummingbird	BM
Coraciiformes	Alcedinidae	*Ceryle alcyon*	Belted kingfisher +	BLP, BM, SI
Piciformes	Picidae	*Melanerpes uropygialis*	Gila woodpecker	SI
Passeriformes	Laniidae	*Lanius ludovicianus*	Loggerhead shrike	BLP, BM, SI
	Corvidae	*Corvus corax*	Common raven	BM, SI
		Aphelocoma californica	Western scrub jay	BM*
	Alaudidae	*Eremophila alpestris*	Horned lark	BM, SI
	Remizidae	*Auriparus flaviceps*	Verdin	BLP*
	Mimidae	*Mimus polyglottos*	Northern mockingbird	BM, SI
	Motacillidae	*Anthus rubescens*	American pipit +	SI
	Parulidae	*Vermivora celata*	Orange-crowned warbler +	BM
		Dendroica petechia	Mangrove warbler	BLP*, BM*, SI*
		Seiurus aurocapillus	Ovenbird +	BM
		Seiurus novaboracensis	Northern waterthrush	BLP, BM, SI
	Emberizidae	*Amphispiza bilineata*	Black-throated sparrow	BM
		Passerculus sandwichensis	Savannah sparrow +	BLP, BM, SI
	Cardinalidae	*Passerina ciris*	Painted bunting +	BM
	Fringillidae	*Carpodacus mexicanus*	House finch	BLP*

Nomenclature follows the A.O.U. Checklist of North American Birds (American Ornithologists' Union 2000). Data are from Mendoza-Salgado (1983), Amador-Silva (1985), Wilbur (1987), Gutiérrez et al. (1989), Danemann and Guzmán-Poo (1992), Danemann and Carmona (1993), Carmona et al. (1994), Massey and Palacios (1994), Becerril and Carmona (1997), Page et al. (1997), and Howell et al. (2001), and the personal field notes of R. Carmona, R. Whitmore, R. Mendoza-Salgado, and E. Amador-Silva.

[a]+ migratory birds; P = species federally listed as in Danger of Extinction by the Mexican government (DOF 2002); Pr = species federally listed by the Mexican government as requiring Special Protection (DOF 2002).
[b]BLP = Bahía de La Paz; BM = Bahía Magdalena; SI = Laguna San Ignacio/Estero El Coyote/Estero La Bocana area; *species using mangroves as nesting substrate; ^ ground-nesting species.

nose dolphins are frequently seen in narrow channels among the mangroves. According to local fishermen, they search for mullets (*Mugil* spp.), moving in and out of mangrove waters with the tides. Bottlenose dolphins are also common near mangroves in Laguna San Ignacio (J. Urbán-Ramírez, pers. comm.). Numerous other marine mammal species are found in deeper waters that are energetically linked, but not adjacent to mangrove ecosystems, such as Bahía de La Paz or Laguna San Ignacio (Ballance 1992; Urbán-Ramírez et al. 1997).

On the terrestrial side, mangrove margins are visited by raccoons (*Procyon lotor*), coyotes (*Canis latrans*), badgers (*Taxidea taxus*), bobcats (*Lynx rufus*), and gray foxes (*Urocyon cinereoargenteus*) (Dedina 2000; R. Carmona, pers. comm.; R. Rodríguez-Estrella, pers. comm.). All of these species likely feed on mangrove-associated fauna, including crabs and birds, and perhaps also on the fruits of mangroves. Although their presence near mangroves has not been documented in BCS, the ring-tailed cat (*Bassariscus astutus*) and the spotted skunk (*Spilogale putorius*) might also visit these areas (R. Rodríguez-Estrella, pers. comm.).

Loss of Mangrove Vegetation in Baja California Sur

To date, the loss of mangroves on both coasts of BCS has been pronounced, but not as severe as that in other regions of the world, such as southeastern Asia or the Caribbean (Strong and Bancroft 1994; Colonnello and Medino 1998; Allen et al. 2001; Aube and Caron 2001; Mazda et al. 2002). As already mentioned, mangroves were used traditionally for tanning hides and for preparing remedies, and as fuelwood, charcoal, and construction material. Although not documented, the impact of some of these traditional uses was likely important in some areas (see Herrera-Silveira and Ceballos-Cambranis 2000). However, it pales in comparison with the wholesale clearing and destruction witnessed in recent decades and likely to continue in the future.

One probable cause of habitat loss, of historical interest, was the use of Bahía Magdalena as a gunnery range by the U.S. Navy during 1904–1910. The peak activity occurred when Theodore Roosevelt's "White Fleet," consisting of 28 coal-burning ships that included 16 of the largest battleships in the U.S. Navy, arrived in March 1908. "The fleet bombed the bay day and night" (Dedina 2000: 24–25). To this date, unexploded ordinance can still be found in the bay and its surrounding habitats. However, it is difficult to assess the level of mangrove destruction incurred at the time.

Before the completion of a paved Transpeninsular Highway linking Ensenada to the Cape Region, the Baja California peninsula remained fairly cut off from the U.S. and mainland Mexico. During parts of the year the gravel and dirt portions of the existing road were impassable due to flash flooding and lack of maintenance. The opening of the Transpeninsular Highway in 1973 transformed the situation of the peninsula. Immigration from other parts of Mexico and tourism both increased greatly, leading to the growth of cities such as La Paz, with associated clearing of mangrove stands (Carmona et al. 1994; see further on). Beach camping, once limited primarily to travelers arriving by air or sea, became very popular with American and Mexican tourists, and small recreational vehicle parks were created, among other places, within existing mangrove stands.

To a large extent, human impacts on BCS mangroves have been related to commercial and sport fishery and to tourism. In two towns, Puerto Aldofo López Mateos and Puerto San Carlos, several canneries with accompanying piers, and a large oil-fired power plant required taking out large portions of the original mangrove habitat in the Bahía Magdalena area. Although the local fishery is in decline (due to overfishing and probable reduced production from the mangrove stands; Holguín et al. 2001), human impacts are still seasonally important. Tourists descend on both Laguna San Ignacio and Bahía Magdalena each season (usually January through March) to watch chiefly gray whales (*Eschrichtius robustus*). As many as 10,000 people visit the area every year, spending an estimated 30 million dollars (Dedina 2000). Most people charter small pangas (flat-bottomed boats) to get an up-close look at adult whales and newborn calves. The result of this activity is localized fragmentation of mangrove stands to allow panga skippers access to the lagoon waters. Entryways, often 5–10 m wide, linking either dune or gravel areas to the bay, are cut into the fringe stands (the Puerto San Carlos area is notable in this regard). In addition, as part of a typical whale-watching tour, panga skippers take visitors on guided tours up the narrow inlets into the mangrove habitat for the purpose of bird and other wildlife observation. On occasion, new "inlets" are created by simply cutting away entire mangrove plants or harshly trimming overhanging branches.

An additional threat to mangrove habitat has begun to materialize. At the southeastern end of the Bahía Magdalena complex, a so-called experimental shrimp farm has been carved into the mangroves near Puerto Chale (24°25' N, 111°34' W), using the mouth of the Estero Grande Santa Rita. The declining native shrimp catch has, in other parts of the world, led to extensive shrimp aquaculture. Coastal mainland Mexico already has a large number of small to moderately sized shrimp farms, mostly in mangrove habitats (e.g., Cruz-Torres 2000; DeWalt et al. 2002). Some farms have been placed in mangrove habitats on the east coast of the Baja California peninsula and on the mainland across the Gulf from La Paz.

There is growing opposition to shrimp farming in coastal areas of Mexico. At the same time, shrimp farms are considered the country's "pink gold" and a key focus of Mexico's export-oriented fishing activity (World Rainforest Movement 2001). Shrimp aquaculture is a real threat to the mangrove ecosystems of the Baja California peninsula (Búrquez and Martínez-Yrízar 1997; Paez-Osuna 2001; chapter 3), especially since recent changes in land statutes have made it easier for foreign nationals to purchase property on the Baja California peninsula. *Ejidos* (see chapter 3) have recently been privatized, giving land title to the *ejido* leadership, allowing them to sell the land as needed (Dedina 2002). Recently, the Ejido Matancitas, which owns approximately 120 km of coastline in Bahía Magdalena, has offered to sell its land for the establishment of large shrimp farms (Dedina 2002). In addition to shrimp aquaculture, 2000 ha at Bahía Magdalena have been sold for construction of a planned resort (Dedina and Young 1995). The privatization of the *ejidos* has opened up BCS to unprecedented land grabs, placing coastal areas in jeopardy. Without strict conservation guidelines, mangrove fragmentation is likely to occur at an alarming rate in the coming years.

The growth of La Paz has already led to population declines of aquatic birds, in part due to the extensive loss of mangroves (Carmona et al. 1994; Becerril and Carmona 1997). Such development is likely to occur in other areas of BCS. In particular, the 1.6 billion dollar *Escalera Náutica* project will result in the construction of at least 22 new marinas and resort developments in coastal areas of northwestern Mexico (see also chapters 9, 11, and 16). To date, targeted areas on the peninsula include Punta Abreojos near the entrance to Laguna San Ignacio,

Bahía San Juanico (Scorpion Bay) near the town of San Juanico (26°15' N, 112°29' W) between Laguna San Ignacio and Bahía Magdalena, and Punta Canoas, north of Guerrero Negro (27°58' N, 114°05' W). Construction has already begun on a new marina at Santa Rosalillita also north of Guerrero Negro (Wildcoast 2002). The main plan behind this project is to connect the Pacific and Gulf of California coasts via a series of new marinas and improved roads. Santa Rosalillita is to be connected to Bahía Los Angeles, where a dredge permit has been awarded and plans drawn up for a new 175-m long marina (S. Dedina, pers. comm.).

Loss of mangroves has been better quantified in some areas of mainland Mexico (e.g., Ruíz-Luna and Berlanga-Robles 1999) than in BCS. One exception is the long-term study in Ensenada de Aripes just west of La Paz (table 15.4). Between 1973 and 1981 there was an overall 21% loss in mangrove area cover. El Mogote lost 18% and El Zacatal lost 28% due to tourism development and the construction of a marina, respectively. El Conchalito (the site supporting 10 nesting ardeid species) lost more than 37% due to the construction of two schools, a radio station, a highway, and a sewage conduit (Mendoza et al. 1984). More recently, Estero de Enfermería (adjacent to El Conchalito) was completely destroyed (loss of 5 ha) due to the construction of a highway that blocked seawater access to the mangrove system. Another notable loss was the construction of the Pichilingue port expansion, along the eastern side of Bahía de La Paz (north of La Paz), which destroyed several mangroves in the 1970s. And losses potentially occur anywhere urban sewage waste contamination is present (Machiwa 1999).

Mulegé

An example of step-by-step deterioration of mangrove habitat is that of Mulegé. This small town (26°53' N, 111°58' W) is located just north of Bahía Concepción. It sits at the mouth of the Río Santa Rosalía de Mulegé (or Río Mulegé), along an *estero*. The Río Santa Rosalía de Mulegé is fed by natural springs. It represents one of the few year-round large sources of fresh water on the peninsula. It also forms a beautiful oasis, with stagnant waters and lush vegetation along the banks. Associated with the fresh water marsh are introduced date palms (*Phoenix dactylifera*), native plants such as a Mexican fan palm (*Washingtonia robusta*), 2 rushes (*Juncus mexi-*

Table 15.4. Change in area coverage of mangrove stands at Ensenada de Aripes, Bahía de la Paz, between 1973 and 1981.

Location, with Plants Represented	Area Coverage in 1973 (ha)	Area Coverage in 1981 (ha)	Area Loss (ha)	% Area Loss
El Mogote *Avicennia germinans* *Rhizophora mangle* *Laguncularia racemosa*	149.62	122.42	27.22	18.18
Zacatecas *Avicennia germinans* *Rhizophora mangle* *Laguncularia racemosa*	15.93	15.93	0.00	0.00
El Comitán *Avicennia germinans* *Rhizophora mangle* *Laguncularia racemosa* *Maytenus phyllanthoides*	6.88	1.23	5.65	82.12
El Zacatal *Avicennia germinans* *Rhizophora mangle* *Laguncularia racemosa* *Maytenus phyllanthoides*	16.48	11.84	4.64	28.15
El Conchalito *Avicennia germinans* *Rhizophora mangle* *Laguncularia racemosa*	18.13	11.38	6.75	37.23
Total mangrove area	207.03	162.78	44.26	21.38

Of the plants represented, *Avicennia germinans*, *Rhizophora mangle*, and *Laguncularia racemosa* are true mangroves; *Maytenus phyllanthoides* is a mangrove-like shrub on the landward side of the stands. Data are from Mendoza et al. (1984).

cana and *J. acutus.*), a mesquite (*Prosopis articulata*), and Goodding's willow (*Salix gooddingii*), and limited cultivars including citrus and mango (Grismer and McGuire 1993). The marsh area upriver from the dam is highly impacted by domestic cattle and pigs, which run freely and have trampled significant portions of the edge habitat (Whitmore and Whitmore 1997). The river has also been dammed, not to regulate the flow of fresh water, but to prevent saltwater inundation of the oasis during high tides and storms.

Formerly, both sides of the *estero*, as well as numerous small islands within the waterway, were heavily vegetated with mangroves from the coast to the cement dam. Early disturbance was limited to access channels cut for local fishermen. Later, backfilled gravel and dirt roads were built on both sides of the *estero*, removing a large percentage of the mangroves vegetation.

The heaviest impacts did not occur until after the completion of the paved Transpeninsular Highway, when Mulegé began to blossom as a tourist destination. An abundance of mangroves on the south side were removed, and the terrain was backfilled to about 3 feet above the high-tide zone in order to construct recreational vehicle (RV) parks (fig. 15.7). In addition, boat docks, used by residents of the RV parks, were installed and a cement boat ramp constructed. At the eastern end of the *estero* a hotel with a small aircraft landing strip was built and today receives extensive use. The construction of houses along the *estero* has led to mangrove loss. Secluded portions of mangrove areas have been cut out for use as latrines by construction workers. One island

Figure 15.7. The south side of the *estero* at Mulegé, Baja California Sur, showing what was once a dense mangrove stand. The arrow points to a sea wall built during the construction of the original main north/south Transpeninsular Highway. The area has been backfilled and now that the paved version of highway has been completed and bypasses the area to the south, this location has been converted to a mobile home park with attendant permanent structures. Note that no mangroves appear in this photograph. (Photograph by Robert Whitmore, spring 2000.)

in the *estero* was clear-cut for the purpose of building a restaurant. Fortunately, the operation failed in the planning stages, and the island has partially revegetated (see below). Another island was similarly cut over to provide an area for pig grazing. Toward the western end of the *estero*, mangroves were severely pruned to allow residents of newly constructed homes to have a view of the water and boat access to it. These "single-trunk" red mangroves are destined to die since all prop roots have been removed. On the north side of the waterway similar removal has occurred. A small dirt road passes through the mangroves and leads to a restaurant on the sand beach at the eastern end of the *estero*. In addition, charter panga captains have cut passageways from the road to the water through the remaining fringe of mangroves. In sum, the remaining mangroves at Mulegé are only a small fraction of the original stands, and they are highly fragmented. Nonetheless, wildlife species, principally birds, use these fragments extensively. A small breeding population of mangrove warblers is present, and numerous species of wading birds nest or roost in the more isolated portions of the mangrove vegetation (Whitmore and Whitmore 1997; Whitmore et al. 1999). However, if development of the area is allowed to continue at its present rate, the future of the remaining mangroves is questionable.

Habitat Restoration

Historical data and recent studies suggest not only that mangrove fragmentation can be reversed, but also that new mangrove areas can be created where there were none before. In fact, it has been said that, "mangroves may be one of the easiest marine systems to reconstruct" (Kaly and Jones 1998: 656).

"Saline silviculture" has become the topic of an ever-widening body of literature (e.g., Teas 1982). Practiced since the eighteenth century, the planting, management, and harvesting of mangrove trees have become widespread, especially in some countries of the tropics such as Malaysia (Teas 1982). In Myanmar the planting of mangroves on previously clear-cut areas has been shown to be an economically feasible alternative to natural regeneration (Webb and Than 2000). A successful mangrove replanting program was initiated in Cuba in 1980, and by 1994 approximately 30,000 ha of mangroves had been restored (Menéndez et al. 1994). Mangrove areas were also established successfully in Hawaii (Cox and Allen 1999). Replanted mangrove habitat in the Philippines contained the highest shrimp densities of 4 microhabitats examined (Ronnback et al. 1999). And near La Paz in BCS, field-collected propagules of black mangrove have been successfully transplanted into a clear-cut zone at Laguna de Balandra, with 74% seedling survival after 2 years (Toledo et al. 2001).

Although mangrove stands may be easy to rehabilitate (Lugo 1998), they still require some attention, with a need also to actively manage the entire ecosystem (Christensen et al. 1996; Field 1999a,b; Ellison 2000), including biogeochemical pathways (McKee and Faulkner 2000). On the other hand, a uniform stand of replanted mangroves, especially if heavily managed, may actually provide less structural diversity and, hence, less species diversity than an unmanaged stand (Kumar 1999; Hsiang 2000). Fortunately, several nonprofit groups (e.g., Grupo Ecológico Manglar, Mangrove Replenishment Initiative) have taken an active role in mangrove conservation and creation in the United States and Mexico, with an emphasis on developing and defining a methodology for establishing self-sustaining mangrove stands.

Conclusions and Outlook

It should be noted that in Mexico, all mangrove species are protected by a variety of statutes including provisions under the General Law on Ecological Balance and Environmental Protection (see chapter 4). In the Mexican norm NOM-059-ECOL-2001, *Rhizophora mangle*, *Avicennia germinans*, *Conocarpus erecta*, and *Laguncularia racemosa* are all listed as Subject to Special Protection (DOF 2002). An environmental impact statement must be prepared for any proposed action requiring removal or alteration of mangrove habitat, and any outright destruction of mangroves is forbidden by law (Loza 1994; DOF 1999, 2002). [In 2004 the Mexican federal government substantially changed and relaxed the legal protection of mangroves, allowing clearing of mangroves against a fee or monetary compensation.—Ed.]

Also important is the fact that mangrove stands occur in at least 3 of the newly created biological reserves in BCS, Reserva de la Biósfera El Vizcaíno, Parque Nacional Bahía de Loreto, and Islas del Golfo de California. However, Mexican biological reserves suffer from a lack of law implementation and enforcement (see Suman 1994).

With the hope that tourism and shrimp export will boost the national economy, the government of Mexico will likely continue to sanction and devise plans that result in the destruction of mangrove ecosystems (World Rainforest Movement 1999), especially in areas that have strong drawing power (e.g., whale-watching sites). The *Escalera Náutica* project poses an enormous threat to mangrove ecosystems in the Baja California peninsula, although, to date, actual construction under this program has not impacted the key mangrove areas of La Paz, Bahía Magdalena, and the Laguna San Ignacio Complex.

Greater support for local conservation organizations is important. So is public education, with an emphasis on the role of mangroves in erosion prevention, land accretion, and ecosystem function. Mangrove systems should receive management priority and complete protection from disturbance. There has been some work on the Baja California peninsula on "reclaiming" disturbed habitats with native vegetation (Espejel and Ojeda 1995). Because of the ease with which disturbed mangrove habitats may be restored, a massive mangrove management and restoration program should even be initiated. It should be noted that regeneration of swamps and other habitats at the land–water interface is generally slow unless there is human assistance (Farnsworth et al. 1997).

Acknowledgments

We are very grateful to a large number of people: G. Arredondo for the production of the maps; Michael Whitmore, Craig Whitmore, and Wendy Niceler for assisting R. C. W in field data collection; José Urbán-Ramírez and Ricardo Rodríguez-Estrella for the communication of unpublished information on associations between mangroves and mammals; Lloyd Findley for kindly providing citations, especially in the Mexican literature; L. A. Abitia-Cárdenas, J. Rodríguez-Romero, and F. A. Gutiérrez-Sanchez for collection and identification of fishes from the coastal lagoons in BCS; Roberto Carmona for supplying copies of publications not readily available in the United States, and for sharing his field notes and those of his students; Jason Campbell for translating from Spanish into English; and Dara Whitmore for help with the bibliography. The late Annetta Mary Carter supplied the herbarium specimens used by P. G. Z. We thank Roberto Carmona and an anonymous reviewer for critical review of earlier drafts of this chapter. Funding for fieldwork (R. C. W.) and manuscript production was provided by West Virginia University.

Literature Cited

Abitia-Cárdenas, L. A., J. Rodríguez-Romero, F. Galván-Magaña, J. de la Cruz-Agüero, and H. Chávez-Ramos. 1994. Systematic list of the ichthyofauna of La Paz Bay, Baja California Sur, México. Ciencias Marinas 20: 159–181.

Acevedo, A. 1991. Behaviour and movements of bottlenose dolphins, *Tursiops truncatus*, in the entrance of Ensenada de La Paz, México. Aquatic Mammals 17: 137–147.

Acosta, C. A. 1999. Benthic dispersal of Caribbean spiny lobsters among insular habitats: implications for the conservation of exploited marine species. Conservation Biology 13: 603–612.

Acosta, C. A., and M. J. Butler. 1997. Role of mangrove habitat as nursery for juvenile spiney lobster, *Panulirus argus*, in Belize. Marine and Freshwater Research 48: 721–727.

Aliaume, C., A. Zerbi, and J. M. Miller. 1997. Nursery habitat and diet of juvenile *Centropomus* species in Puerto Rico estuaries. Gulf of Mexico Science 15: 77–87.

Allen, J. A., K. C. Ewel, and J. Jason. 2001. Patterns of natural and anthropogenic disturbance of the mangroves on the Pacific Island of Kosrae. Wetlands Ecology and Management 9: 279–289.

Amador-Silva, E. S. 1985. Avifauna de la Isla Santa Margarita. B.C.S. México. Memoria Profesional, Universidad Autónoma de Baja California Sur, La Paz, Mexico.

American Ornithologists' Union (AOU). 2000. Checklist of North American Birds, 7th ed. and supplements. American Ornithologists' Union, Washington, D.C.

Aube, M., and L. Caron. 2001. The mangroves of the north coast of Haiti. Wetlands Ecology and Management 9: 271–278.

Ballance, L. T. 1992. Habitat use patterns and ranges of bottlenose dolphin in the Gulf of California, Mexico. Marine Mammal Science 8(3): 262–274.

Becerril, M. F., and R. Carmona. 1997. Nesting of waterbirds in Ensenada de La Paz, Baja California Sur, Mexico (1992–1994). Ciencias Marinas 23: 265–271.

Bell, J. L., and R. C. Whitmore. 1997. Eastern towhee numbers increase following defoliation by gypsy moths. Auk 114: 708–716.

Blaber, S. J. M. 1997. Fish and Fisheries of Tropical Estuaries. Chapman and Hall, London.

Brooks, L. B., W. J. Nichols, and J. Harvey. 2003. Estero Banderitas Marine Protected Area: a critical component to the recovery of the east Pacific green turtle (*Chelonia mydas*). Pp. 63–64 *in* J. A. Seminoff (comp.), Proceedings of the Twenty-Second Annual Symposium on Sea Turtle Biology and Conservation, Miami, April 3–7, 2002. NOAA Technical Memorandum NMFS-SEFSC-503. National Oceanic and Atmospheric Administraton, Miami.

Brusca, R. C. 1975. Zoological classification of the Scripps Alpha Helix Baja California Expedition. Pp. 72–73 *in* Alpha Helix Research Program 1972–1974. Scripps Institution of Oceanography, University of California, San Diego, California.

Brusca, R. C. 1980. Common Intertidal Invertebrates of the Gulf of California. University of Arizona Press, Tucson.

Búrquez, A., and A. Martínez-Yrízar. 1997. Conservation and landscape transformation in Sonora, Mexico. Journal of the Southwest 39: 371–398.

Carmona, R., J. Guzmán, S. Ramírez, and G. Fernández. 1994. Breeding waterbirds of La Paz Bay, Baja California Sur, Mexico. Western Birds 25: 151–157.

Castro-Aguirre, J. L., and E. F. Balart. 1997. Contribución al conocimiento de la ictiofauna de fondos blandos y someros de la Ensenada y Bahía de La Paz, B.C.S. Pp. 139–149 *in* J. Urbán-Ramírez and M. Ramírez-Rodríguez (eds.), La Bahía de la Paz, Investigación y Conservación. Universidad Autónoma de Baja California Sur, Centro Interdisciplinario de Ciencias Marinas, and Scripps Institution of Oceanography, La Paz, Mexico.

Centro de Investigaciónes Biologicas de Baja California Sur. 1994. Manifestación de Impacto Ambiental Modalidad Intermedia:

Proyecto "Salitrales de San Ignacio". Segunda parte. Centro de Investigaciónes Biológicas de Baja California Sur, La Paz, Mexico.

Chapman, V. J. 1976. Mangrove Vegetation. Cramer, Vaduz, Liechtenstein.

Cheng, L., and C. L. Hogue. 1974. New distribution and habitat records of biting midges and mangrove flies from the coasts of southern Baja California, Mexico (Diptera : Ceratopogonidae, Culidae, Chironomidae, and Phoridae). Entomological News 85: 211–218.

Christensen, N. L., A. M. Bartuska, J. H. Brown, S. Carpenter, C. D'Antonio, R. Francis, J. F. Franklin, J. A. MacMahon, R. F. Noss, D. J. Parsons, C. H. Peterson, M. G. Turner, and R. G. Woodmansee. 1996. The report of the Ecological Society of America on the scientific basis for ecosystem management. Ecological Applications 6: 665–691.

Cliffton, K., D. O. Cornejo, and R. S. Felger. 1982. Sea turtles of the Pacific coast of Mexico. Pp. 199–209 *in* K. Bjorndal (ed.), Biology and Conservation of Sea Turtles. Smithsonian Institution Press, Washington, D.C.

Colonnello, G., and E. Medino. 1998. Vegetation changes induced by dam construction in a tropical estuary: The case of the Manamo river, Orinoco Delta (Venezuela). Plant Ecology 139: 145–154.

Cox, E. F., and J. A. Allen. 1999. Stand structure and productivity of the introduced *Rhizophora* mangle in Hawaii. Estuaries 22: 276–284.

Cruz-Torres, M. L. 2000. "Pink gold rush:" Shrimp aquaculture, sustainable development, and the environment in northwestern Mexico. Journal of Political Ecology 7: 63–90.

Danemann, G. D., and R. Carmona. 1993. Observations on esteros El Coyote and La Bocana, Baja California Sur, Mexico, in September 1991. Western Birds 24: 263–266.

Danemann, G. D., and J. de la Cruz-Agüero. 1993. Ichthyofauna of San Ignacio Lagoon, Baja California Sur, Mexico. Ciencias Marinas 19: 333–341.

Danemann, G. D., and J. R. Guzmán Poo. 1992. Notes on the birds of San Ignacio Lagoon, Baja California Sur, Mexico. Western Birds 23: 11–19.

Day, J. W., and A. Yáñez-Arancibia. 1985. Coastal lagoons and estuaries as an environment for nekton. Pp. 17–34 *in* A. Yáñez-Arancibia (ed.), Fish Community Ecology in Estuaries and Coastal Lagoons: Towards an Ecosystem Integration. Universidad Nacional Autónoma de México, México D.F.

Dedina, S. 2000. Saving the Gray Whale. University of Arizona Press, Tucson.

Dedina, S. 2002. Coastal conservation opportunity assessment of the Baja California peninsula. White paper, Wildcoast International Conservation Team, Imperial Beach, California.

Dedina, S., and E. Young. 1995. Conservation and development in the gray whale lagoons of Baja California Sur, Mexico. Report prepared for the U.S. Marine Mammal Commission, Washington, D.C. Available: http://scilib.ucsd.edu/sio/guide/z-serge.html.

de la Cruz-Agüero, J., and V. M. Cota-Gómez. 1998. Ichthyofauna of San Ignacio Lagoon, Baja California Sur, Mexico: new records and range extensions. Ciencias Marinas 24: 353–358.

de la Cruz-Agüero, J., F. Galván-Magaña, L. A. Abitia-Cárdenas, J. Rodríguez-Romero, and F. J. Gutiérrez-Sánchez. 1994. Systematic list of marine fishes from Bahía Magdalena, Baja California Sur (México). Ciencias Marinas 20: 17–31.

DeWalt, B. R., J. R. Ramírez Zavala, L. Noriega, and R. E. González. 2002. Shrimp aquaculture, the people and the environment in coastal Mexico. World Bank/Network of Aquaculture Centers in Asia-Pacific/World Wildlife Fund/Food and Agriculture Organization Consortium Program on Shrimp Farming and the Environment. Available: http:govdocs.aquake.org/cgi/collection/pacific_region.

DOF. 1999. Norma Oficial Mexicana de Emergencía NOM-EM-001–RECNAT-1999, que Establece las Especificaciones para la Preservación y Restauración del Manglar. Diario Oficial de la Federación 16 August.

DOF. 2002. Norma Oficial Mexicana NOM-059-ECOL-2001, Protección Ambiental-Especies Natives de México de Flora y Fauna Silvestres- Categoria de Riesgo y Especificaciones para su Inclusión, Extinction o Cambio- Lista de Especies en Riesgo. Diarío Oficial de la Federación 582 (no. 4, segunda sección), 6 March.

Duke, N. C. 1995. Genetic diversity, distributional barriers and rafting continents—more thoughts on the evolution of mangroves. Hydrobiologia 295: 167–181.

Duke, N. C. 2001. Gap creation and regenerative processes driving diversity and structure of mangrove ecosystems. Wetlands Ecology and Management 9: 257–269.

Ellison, A. M. 2000. Mangrove restoration: do we know enough? Restoration Ecology 8: 219–229.

Eschmeyer, W. N. (ed). 1998. Catalog of Fishes.

California Academy of Sciences Special Publications 1: 1–2905.

Espejel, I., and L. Ojeda. 1995. Native plants for recreation and conservation in Mexico. Restoration and Management Notes 13: 84–89.

Farnsworth, E. J., A. M. Ellison, and M. Fisher. 1997. The global conservation status of mangroves. Ambio 26: 328–334.

Farnsworth, E. J., A. M. Ellison, and W. K. Gong. 1996. Elevated CO_2 alters anatomy, physiology and reproduction of red mangroves. Oecologia 108: 599–609.

Felger, R. S., and M. B. Moser. 1985. People of the Desert and Sea: Ethnobotany of the Seri Indians. University of Arizona Press, Tucson.

Felger, R. S., M. B. Johnson, and M. F. Wilson. 2001. Trees of Sonora, Mexico. Oxford University Press, New York.

Felix, F. 1994. Ecology of the coastal bottlenose dolphin, Tursiops truncatus, in the Gulf of Guayaquil, Ecuador. Investigations on Cetacea 25: 235–256.

Field, C. D. 1999a. Rehabilitation of mangrove ecosystems: an overview. Marine Pollution Bulletin 37: 383–392.

Field, C. D. 1999b. Mangrove rehabilitation: choice and necessity. Hydrobiologia 413: 47–52.

Findley, L. T., M. E. Hendrickx, R. C. Brusca, A. M. van der Heiden, P. A. Hastings, and J. Torre. In press. Macrofauna del Golfo de California. CD-ROM, version 1.0. Macrofauna Golfo Project. Center for Applied Biodiversity Science, Conservation International, Washington, D.C., and Programa Golfo de California, Conservation International, Guaymas, Mexico.

Galván-Magaña, F., F. Gutiérrez-Sánchez, L. A. Abitia-Cárdenas, and J. Rodríguez-Romero. 2000. The distribution and affinities of the shore fishes of the Baja California Sur lagoons. Pp. 383–398 in M. Munawar, S. G. Lawrence, I. F. Munuwar, and D. F. Malley (eds.), Aquatic Ecosystems of Mexico: Status and Scope. Backhuys Publishers, Leiden, The Netherlands.

Galván-Piña, V. H., F. Galván-Magaña, L. A. Abitia-Cárdenas, F. J. Gutiérrez-Sánchez, and J. Rodriguez-Romero. 2003. Seasonal structure of fish assemblages in rocky and sandy habitats in Bahía de La Paz, Mexico. Bulletin of Marine Science 72: 19–35.

Gardner, S. C., and W. J. Nichols. 2001. Assessment of sea turtle mortality rates in the Bahía Magdalena region, Baja California Sur, Mexico. Chelonian Conservation and Biology 4: 197–199.

González-Acosta, G., J. de la Cruz-Argüero, and G. Ruíz-Campos. 1999. Ictiofauna asociada al manglar del estero El Conchalito, Ensenada de La Paz, Baja California Sur, México. Oceánides 14: 121–131.

Grismer, L. L. 2002. Amphibians and Reptiles of Baja California. University of California Press, Berkeley.

Grismer, L. L., and J. A. McGuire. 1993. The oases of central Baja California, Mexico. Part I. A preliminary account of the relic mesophilic herpetofauna and status of the oases. Bulletin of the Southern California Academy of Sciences 92: 2–24.

Gutiérrez, J. L., E. A. Silva, and R. M. Salgado. 1989. Avifauna costera de los esteros de la Bahía de La Paz, Baja California Sur, México. Revista de Investigaciones Marinas, Centro Interdisciplinario de Ciencias Marinas (CICIMAR) 4: 93–103.

Herrera-Silveira, J., and E. Ceballos-Cambranis. 2000. Manglares: ecosístemas valiosos. Biodiversitas 19: 1–10.

Hilton, G. M., T. Murray, T. Cleeves, B. Hughes, and E. G. Williams. 2000. Wetland birds in Turks and Caicos Islands II. Wetland Bird Communities. Wildfowl 51: 127–138.

Hogarth, P. J. 1999. The Biology of Mangroves. Oxford University Press, New York.

Holguin, G., P. Vazquez, and Y. Bashan. 2001. The role of sediment microorganisms in the productivity, conservation, and rehabilitation of mangrove ecosystems: an overview. Biology and Fertility of Soils 33: 265–278.

Holguin-Quiñones, O. E., and F. A. García-Domínguez. 1997. Lista anotada de las especies de moluscos recoletadas en La Bahía de La Paz, B.C.S. Pp. 93–117 in J. Urbán-Ramírez and M. Ramírez-Rodríguez (eds.), La Bahía de la Paz, Investigación y Conservación. Universidad Autónoma de Baja California Sur, Centro Interdisciplinario de Ciencias Marinas, and Scripps Institution of Oceanography, La Paz, Mexico.

Howell, S. N. G., R. A. Erickson, R. A. Hamilton and M. A. Patten. 2001. An annotated checklist of the birds of Baja California and Baja California Sur. Pp. 171–203 in R. A. Erickson and S. N. G. Howell (eds.), Birds of the Baja California Peninsula: Status, Distribution and, Taxonomy. Monographs in Field Ornithology no. 3. American Birding Association, Colorado Springs, Colorado.

Hsiang, L. L. 2000. Mangrove conservation in Singapore: a physical or psychological impossibility? Biodiversity and Conservation 9: 309–332.

Hubbs, C. L. 1960. The marine vertebrates of the outer coast. Systematic Zoology 9: 134–147.

INEGI. 2001. Tabulados Básicos Nacionales y por Entidad Federativa. Base de Datos y Tabulados de la Muestra Censal. XII Censo General de Población y Vivienda,

2000. México. Instituto Nacional de Estadística, Geografía, y Informática, México D.F.

Jennerjahn, T. C., and I. Venugopalan. 2002. Relevance of mangroves for the production and deposition of organic matter along tropical continental margins. Naturwissenschaften 89: 23–30.

Kaly, U. L., and G. P. Jones. 1998. Mangrove restoration: a potential tool for coastal management in tropical developing countries. Ambio 27: 656–661.

Kumar, R. 1999. Artificial regeneration of mangroves. Indian Forester 125: 760–769.

Kunstadter, P., E. C. F. Bird, and S. Sabhasri (eds.). 1986. Man in the Mangroves, The Socio-economic Situation of Human Settlements in Mangrove Forests. United Nations University, Tokyo.

León de la Luz, J. L. and R. Coria-Benet. 1992. Flora Iconográfica de Baja California Sur. Publicación no. 3 del Centro de Investigaciónes Biológicas de Baja California Sur, La Paz, Mexico.

León de la Luz, J. L., R. Coria-Benet, and J. Cansino 1995. XI Reserva de la Bíosfera El Vizcaíno, Baja California Sur. Serie Listados Florísticos de México, Instituto de Biología, Universidad Nacional Autónoma de México, México D.F.

Limpus, C. J., and D. J. Limpus. 2000. Mangroves in the diet of *Chelonia mydas* in Queensland, Australia. Marine Turtle Newsletter 89: 13–15.

López-Cortez, A. 1991. Microbial mats in tidal channels at San Carlos, Baja California Sur, Mexico. Geomicrobiology Journal 8: 69–86.

López-Mendilaharsu, M. 2002. Ecología alimenticia de *Chelonia mydas agassizii* en Bahía Magdalena, Baja California Sur, México. M.S. thesis, Centro de Investigaciones Biologicas del Noroeste, La Paz, Mexico.

Loza, E. L. 1994. Los manglares de México: synopsis general para su manejo. Pp. 144–151 *in* D. O. Suman (ed.), El ecosistema de Manglar en América Latina y la Cuenca del Caribe: Su Manejo y Conservación. Division of Marine Affairs and Policy, Rosenstiel School of Marine and Atmospheric Science, University of Miami, Miami.

Lugo, A. E. 1998. Mangrove forests: a tough system to invade but an easy one to rehabilitate. Marine Pollution Bulletin 37: 427–430.

Machiwa, J. F. 1999. Lateral fluxes of organic carbon in a mangrove forest partly contaminated with sewage wastes. Mangroves and Saltmarshes 3: 95–104.

Massey, B. W., and E. Palacios. 1994. Avifauna of the wetlands of Baja California, Mexico: current status. Studies in Avian Biology 15: 45–57.

Maurer, B. A., L. B. McArthur, and R. C. Whitmore. 1980. Habitat associations of breeding birds in clearcut deciduous forests in West Virginia. Pp. 167–172 *in* D. E. Capen (ed.), The Use of Multivariate Statistics in Studies of Wildlife Habitat. General Technical Report RM-87. U.S. Forest Service, Rocky Mountain Research Station, Fort Collins, Colorado.

Mazda, Y., M. Magi, H. Nanao, M. Kogo, T. Miyagi, N. Kanazawa and D. Kobashi. 2002. Coastal erosion due to long-term human impact on mangrove forests. Wetlands Ecology and Management 10: 1–9.

McKee, K. L., and P. L. Faulkner. 2000. Restoration of biogeochemical function in mangrove ecosystems. Restoration Ecology 8: 247–259.

Mendoza, R. E., E. S. Amador, J. Llinas, and J. Bustillos. 1984. Inventario de las áreas de Manglar en la Ensenada de Aripes, BCS. Pp. 43–52 *in* Memorias de la Primera Reunión sobre Ciencia y Sociedad "Presente y Futuro de la Ensenada de La Paz". Universidad Autónoma de Baja California Sur and Gobierno del Estado de Baja California Sur, La Paz, Mexico.

Mendoza-Salgado, R. E. 1983. Identificación, distribución y densidad de la avifauna marina en los manglares: Puerto Balandra, Enfermería y Zacatecas en la Bahía de La Paz, Baja California Sur, México. Tésis Profesional, Universidad Autónoma de Baja California Sur, La Paz, Mexico.

Menéndez, L., P. Alcolado, S. Oharriz, and C. Milián. 1994. Mangroves of Cuba: legislation and management. Pp. 74–84 *in* D. O. Suman (ed.), El Ecosistema de Manglar en América Latina y la Cuenca del Caribe: Su Manejo y Conservación. Division of Marine Affairs and Policy, Rosenstiel School of Marine and Atmospheric Science, University of Miami, Miami.

Meylan, A. B. 1988. Spongivory in hawksbill turtles: a diet of glass. Science 239: 393.

Nichols, W. J. 2003. Biology and conservation of sea turtles in Baja California, Mexico. Ph.D. dissertation. University of Arizona, Tucson.

Nichols, W. J., H. Aridjis, A. Hernandez, B. Machovina, and J. Villavicencios. 2002. Black market sea turtle trade in the Californias. Report to Wildcoast, San Diego, California.

Paez-Osuna, F. 2001. The environmental impact of shrimp aquaculture: causes, effects, and mitigating alternatives. Environmental Management 28: 131–140.

Page, G. W., E. Palacios, L. Alfaro, S. González, L. E. Stenzel, and M. Jungers. 1997. Numbers of wintering shorebirds in coastal wetlands of Baja California, Mexico. Journal of Field Ornithology 68: 562–574.

Pendoley, K., and J. Fitzpatrick. 1999. Browsing on mangroves by green turtles in Western Australia. Marine Turtle Newsletter 84: 10–11.

Peterson, W. 1998. The Baja Adventure Book, 3rd Ed. Wilderness Press, Berkeley, California.

Pritchard, P. C. H. 1971. Sea turtles in the Galapagos Islands. IUCN Publications, new series, supplementary papers 31: 34–37. IUCN (World Conservation Union), Gland, Switzerland.

Ramírez-García, P., and A. Lot H. 1994. La distribución del manglar y de los "pastos marinos" en el Golfo de California, México. Anales del Instituto de Biología, Universidad Nacional Autónoma de México, Serie Botánica 65: 63–72.

Reitsma, L., P. Hunt, S. L. Burnson, and B. B.Steele. 2002. Site fidelity and ephemeral habitat occupancy: northern waterthrush use of Puerto Rican black mangroves during the nonbreeding season. Wilson Bulletin 114: 99–105.

Roberts, N. C. 1989. Baja California Plant Field Guide. Natural History Publishing Company, La Jolla, California.

Robertson, A. I. 1991. Plant-animal interactions and the structure and function of mangrove forest ecosystems. Australian Journal of Ecology 16: 433–443.

Robertson, A. I., and P. A. Daniel. 1989. The influence of crabs on litter processing in high intertidal forests in tropical Australia. Oecologia 78: 191–198.

Rodríguez-Romero, J., L. A. Abitia-Cárdenas, J. de la Cruz-Agüero, and F. Galván-Magaña. 1992. Systematic list of marine fishes of Bahía Concepción, Baja California Sur, México. Ciencias Marinas 18: 85–95.

Rodríguez, R. J., L. A. Abitia C., F. Galván M., J. Arvizu, and B. Aguilar P. 1998. Ecology of fish communities from the soft bottoms of Bahía Concepción, Mexico. Archive of Fishery and Marine Research 46: 61–76.

Rodríguez, R. J., L. A. Abitia C., F. Galván M., and H. Chávez. 1994. Composition, abundance and specific richness of fishes from Bahía Concepción, Baja California Sur, México. Ciencias Marinas. 20: 321–350.

Ronnback, P., M. Troell, N. Kautsky, and J. H. Primavera. 1999. Distribution of shrimps and fish among Avicennia and Rhizophora microhabitats in the Pagbilao mangroves, Phillippines. Estuarine, Coastal and Shelf Science 48: 223–234.

Ruíz-Luna, A., and C. A. Berlanga-Robles. 1999. Modifications in coverage patterns and land use around the Huizache-Caimanero lagoon system, Sinaloa, Mexico: A multi-temporal analysis using Landsat images. Estuarine, Coastal and Shelf Science. 49: 37–44.

Seminoff, J. A. 2003. Ecology of Chelonia mydas at foraging areas in the Eastern Pacific Ocean: perspectives from Baja California. Pp. 77–78 in J.A. Seminoff (comp.), Proceedings of the Twenty-second Annual Symposium on Sea Turtle Biology and Conservation, Miami, April 3–7, 2002. NOAA Technical Memorandum NMFS-SEFSC-503. National Oceanic and Atmospheric Administration, Miami.

Seminoff, J. A., W. J. Nichols, A. Resendiz, and L. Brooks. 2003. Occurrence of hawksbill turtles, Eretmochelys imbricata (Reptilia: Cheloniidae), near the Baja California Peninsula, Mexico. Pacific Science 57: 9–19.

Seminoff, J. A., A. Resendiz, and W. J. Nichols. 2002a. Diet of the East Pacific green turtle, Chelonia mydas, in the central Gulf of California, Mexico. Journal of Herpetology 36: 447–453.

Seminoff, J. A., A. Resendiz, and W. J. Nichols. 2002b. Home range of the green turtle (Chelonia mydas) at a coastal foraging ground in the Gulf of California, Mexico. Marine Ecology Progress Series 242: 253–265.

Sherry, T. W., and R. T. Holmes. 1996. Habitat quality, population limitation, and conservation of Neotropical-Nearctic migrant birds. Ecology 77: 36–48.

Sigueiros-Beltrones, D. A., and L. H. Morzari. 1999. New records of marine benthic diatom species for the north-western Mexican region. Oceanides 14: 89–95.

Skilleter, G. A., and S. Warren. 2000. Effects of habitat modification in mangroves on the structure of mollusk and crab assemblages. Journal of Experimental Marine Biology and Ecology 244: 107–129.

Smith, T.J. I. 1992. Forest structure. Pp. 101–136 in A. I. Robertson and D. M. Alongi (eds.), Tropical Mangrove Ecosystems, Coastal and Estuarine Studies no. 4. American Geophysical Union, Washington, D.C.

Smith, T. J. S. III, H. T. Chan, C. C. McIvor, and M. B. Robblee. 1989. Comparisons of seed predation in tropical, tidal forests from three continents. Ecology 70: 146–151.

Strong, A. M., and T. G. Bancroft. 1994. Patterns of deforestation and fragmentation of mangrove and deciduous seasonal forests in the upper Florida Keys. Bulletin of Marine Science 54: 795–804.

Strong, A. M., and M. D. Johnson. 2001. Exploitations of a seasonal resource by nonbreeding plain and white-crowned pigeons: Implications for conservation of tropical dry forests. Wilson Bulletin 113: 73–77.

Suman, D. O. (ed.) 1994. El Ecósistema de Manglar en América Latina y la Cuenca del Caribe: Su Manejo y Conservación. Division of Marine Affairs and Policy, Rosenstiel School of Marine and Atmospheric Science, University of Miami, Miami.

Teas, H. J. 1982. Saline silviculture. Pp. 369–381 *in* A. San Pietro (ed.), Biosaline Research: A Look to the Future. Plenum Press, New York.

Toledo, G., Y. Bashan, and A. Soelder. 1995. Cyanobacteria and black mangroves in northwestern Mexico: colonization and diurnal and seasonal nitrogen fixation on aerial roots. Canadian Journal of Microbiology 41: 999–1011.

Toledo, G., A. Rojas, and Y. Bashan. 2001. Monitoring of black mangrove restoration with nursery-reared seedlings on an arid coastal lagoon. Hydrobiologia 444: 101–109.

Tomlinson, P. B. 1986. The Botany of Mangroves. Cambridge University Press, Cambridge.

Turner, R. M., J. E. Bowers, and T. L. Burgess. 1995. Sonoran Desert Plants: An Ecological Atlas. University of Arizona Press, Tucson.

Twilley, R. R. 1985. The exchange of organic carbon in basin mangrove forests in a southwest Florida estuary. Estuarine, Coastal and Shelf Science 20: 543–557.

Urbán-Ramírez, J., A. G. Gallardo-Unzueta, M. Palermos-Rodríguez, and G. Veláquez-Chávez. 1997. Los mamíferos marinos de la Bahía de La Paz, B.C.S. Pp. 201–236 *in* J. Urbán-Ramírez and M. Ramírez-Rodríguez (eds.), La Bahía de la Paz, Investigación y Conservación. Universidad Autónoma de Baja California Sur, Centro Interdisciplinario de Ciencias Marinas, and Scripps Institution of Oceanography, La Paz, Mexico.

Walker, B. W. 1960. The distribution and affinities of the marine fish fauna of the Gulf of California. Systematic Zoology 9: 123–133.

Warkentin, I. G., and E. S. Morton. 2000. Flocking and foraging behavior of wintering prothonotary warblers. Wilson Bulletin 112: 88–98.

Webb, E. L., and M. M. Than. 2000. Optimizing investment strategies for mangrove plantations by considering biological and economic paramaters. Journal of Coastal Conservation 6: 181–190.

Whitmore, R. C., and R. Craig Whitmore. 1997. Late fall and early spring bird observations for Mulegé, Baja California Sur, Mexico. Great Basin Naturalist 57: 131–141.

Whitmore, R. C., R. Craig Whitmore, and M. M. Whitmore. 1999. A previously unreported nesting colony of the yellow-crowned night-heron near Mulegé, Baja California Sur. Western Birds 30: 52–53.

Whitmore, R. C., R. Craig Whitmore, and M. M. Whitmore. 2000. Distributional notes on the Mangrove Warbler (*Dendroica petechia castaneiceps*) near the northern edge of its range in eastern Baja California Sur. Western North American Naturalist 60: 228–229.

Wiggins, L. L. 1980. Flora of Baja California. Stanford University Press, Stanford, California.

Wilbur, S. R. 1987. Birds of Baja California. University of California Press, Berkeley.

Wildcoast International Conservation Team. 2002. Nautical route marina construction underway in Baja California. [Available: http://www. wildcoast.net/ NewsDetails.asp?NewsId=45]

Williams, J., and B. Williams. 1998. The Magnificent Peninsula: Baja California, 6th ed. H. J. Williams Publications, Redding, California.

World Rainforest Movement. 1999. Mexico: mangrove destruction by tourism and shrimp farming. WRM Bulletin no. 22, April 1999. [Available: http://www.wrm.org.uy]

World Rainforest Movement. 2001. Mexico: growing opposition to industrial shrimp farming. WRM Bulletin no. 51, October 2001. [Available: http://www.wrm.org.uy]

Yáñez-Arancibia, A., A. L. Lara-Domínguez, and J. W. Day, Jr. 1993. Interactions between mangrove and seagrass habitats mediated by estuarine nekton assemblages: coupling of primary and secondary production. Hydrobiologia 264: 1–12.

Yáñez-Arancibia, A., D. Z. Lomelí, J. L. Rojas-Galavíz, and G. V. Zapata. 1994. Estudio de declaratoria como area ecológica de protección de flora and fauna silvestre de La Laguna de Términos, Campeche. Pp. 152–159 *in* D. O. Suman (ed.), El Ecosistema de Manglar en América Latina y la Cuenca del Caribe: Su Manejo y Conservacion. Division of Marine Affairs and Policy, Rosenstiel School of Marine and Atmosperic Science, University of Miami, Miami.

16

Avian Communities of Arroyos and Desert Oases in Baja California Sur: Implications for Conservation

RICARDO RODRÍGUEZ-ESTRELLA

MA. CARMEN BLÁZQUEZ

JUAN MANUEL LOBATO G.

The Baja California peninsula and its associated offshore islands have been the focus of much ornithological research since the late nineteenth century (e.g., Anthony 1893; Brewster 1902; Grinnell 1928; Bancroft 1930; Davis 1959; Stager 1960; Banks 1967; Cody 1983; Wilbur 1987; Erickson and Howell 2001). However, most studies have been descriptive, if not anecdotal. They typically have dealt with taxonomy and distribution, many of them focusing only on aquatic birds. In contrast, there have been few published studies on the ecology of land birds, and little is known in particular on the habitat preferences and movements (including migratory routes) of nonbreeders.

This lack of knowledge is all the more regrettable in view of the importance of the Baja California peninsula for the conservation of avian diversity at the scale of northwestern Mexico (Comisión para la Cooperacíon Ambiental 1999; Arizmendi and Marquez 2000). With a total of 4 endemic bird species (plus additional endemic subspecies, some of which may warrant recognition at the species level), the peninsula has even been designated as an Endemic Bird Area of the world (Stattersfield et al. 1998). It has been recognized as an important wintering area for a number of aquatic, wading, and land migratory bird species (Massey and Palacios 1994; Rodríguez-Estrella 1997).

The need to evaluate the importance of all habitats for land birds (and other wildlife) has become even more pressing recently due to the potential risk of losing biodiversity as regional anthropogenic impacts are increasing. Many of the peninsula's natural habitats are threatened by increasing urbanization and tourism development (e.g., chapter 15). Already, agriculture, grazing, and mining activity have been responsible for habitat loss. If implemented as planned, the Nautical Ladder Route (*Escalera Náutica*, FONATUR) project may soon produce sweeping changes in the regional landscape and prove a major threat to the conservation of biodiversity. The project calls for the development of buildings, roads, and small airports in many areas of the peninsula (chapter 15). This new infrastructure will increase human traffic and accessibility to many areas.

Two lowland habitat types of particular interest to us are the desert oasis and the arroyo, which we have both studied in Baja California Sur (BCS) since 1994. In lowlands of the southwestern United States, wetlands and riparian areas (i.e., terrestrial habitats associated with perennial, intermittent, ephemeral, or subsurface drainage systems; Dick-Peddie 1993; Patten 1998) support disproportionately large numbers of wildlife, and they serve as important corridors during bird migration (Hubbard 1977; Johnson et al. 1977; Naiman et al. 1993; Farley et al. 1994; Krueper 1996; Skagen et al. 1998; Cartron et al. 2000). Given the general aridity of BCS, the wetland and riparian vegetation of desert oases and also the xeroriparian vegetation along arroyos have the potential to serve a similar important function as that documented in the southwestern United States. Grismer and McGuire (1993) have demonstrated the ecological importance of desert oases on the Baja California peninsula (mostly in BCS) by documenting their relict mesophilic herpetofauna.

In this chapter we examine differences in landbird communities of oases, arroyos, and uplands (chiefly desertscrub, but also tropical deciduous forest in the Cape Region), during both the breeding and nonbreeding season. We also describe differences in species richness among oases as a function of seasons and biogeographic factors. Based on our research we recommend the development of a strategy for the protection of arroyos and oases, as currently there is no management plan in place for them.

Brief Overview of Baja California Sur and Its Avifauna

The vegetation and flora of BCS have been described in important botanical studies (e.g., Shreve 1937, 1951; Wiggins 1980; Turner and Brown 1982; León de la Luz and Domínguez 1989; Lenz 1992; Turner et al. 1995), and other works provide accounts of the oases and wetland plants (e.g., Lot et al. 1986; León de la Luz et al. 1997, 1999). At the lower elevations, BCS is dominated by desertscrub, but tropical deciduous forest is present in the Cape Region on both sides of the Sierra de la Laguna (e.g., León de la Luz and Domínguez 1989). Thornscrub is nearly absent in BCS. This vegetation type, which is transitional between desertscrub and tropical deciduous forest, is well represented in southern Sonora and Sinaloa. In BCS, however, it occurs only in a small area along the northern slope of the Sierra de la Giganta (J. L. León de la Luz, pers. comm.).

Structurally important plant species in desertscrub include copal (*Bursera* spp.), palo verde (*Cercidium* spp.), *ciruelo* (*Cyrtocarpa edulis*), cholla (*Cylindropuntia cholla*), palo ádan (*Fouquieria diguetii*), *matacora* (*Jatropha cuneata*), creosote bush (*Larrea divaricata*), the columnar giant cactus cardón or cardón pelón (*Pachycereus pringlei*), mesquite (*Prosopis* spp., mostly *P. articulata*), *pitahaya agria* (*Stenocereus gummosus*), ruellia (*Ruellia peninsularis*), and *datillo* (*Yucca valida*). Some important species found in dry, tropical deciduous forest include coralvine (*Antigonon leptopus*), palo eva (*Chloroleucon mangense* var. *leucospermum* [*Pithecellobium undulatum*]), chilicote (*Erythrina flabelliformis*), lomboy (*Jatropha vernicosa*), mauto (*Lysiloma divaricatum*), cardón-barbón (*Pachycereus pecten-aboriginum*), *palo de zorrillo* (*Senna atomaria*), and trumpet bush (*Tecoma stans*).

Arroyos (defined here as ephemeral washes, flowing only during stormwater discharge events) and desert oases (defined as isolated permanent or semipermanent bodies of spring-fed water and the vegetation associated with their edge; fig. 16.1) are embedded within the upland habitats described above. Arroyos are characterized by trees such as mesquite, sweet acacia (*Acacia farnesania*), *ciruelo*, and palo blanco (*Lysiloma candidum*; see table 16.1 for some other important species). Along some arroyos, the size of these trees may be notably larger than in surrounding desertscrub. The mesic vegetation of oases is variable, but often includes palms (*Washingtonia robusta, Phoenix dactylifera*), reed grass (*Phragmites australis*), and clumps of cattails (*Typha domingensis*) along the water edge or in associated freshwater marshes (Grismer and McGuire 1993; Arriaga et al. 1997; Rodríguez-Estrella et al. 1999; table 16.1). Approximately 88 typical oases (as defined in this chapter) have been located in BCS through aerial photography (Maya et al. 1997).

A total of 175 resident and migratory land-bird species have been recorded in BCS. The 4 land-bird species endemic to the Baja California peninsula are restricted largely, if not only, to BCS. They are the Belding's yellowthroat (*Geothlypis beldingi*), Xantus's hummingbird (*Hylocharis xantusii*), gray thrasher (*Toxostoma cinereum*), and Cape pygmy-owl (*Glaucidium hoskinsii*; endemic to Sierra de la Laguna following Howell and Webb 1995; Stattersfield et al. 1998). The Belding's yellowthroat in particular is listed as Endangered in the Mexican Official Standard NOM-059-ECOL-2001 (SEMARNAT 2002). Endemism is pronounced in the Sierra la Laguna, which in addition to the Cape pygmy-owl is home to the San Lucas robin (*Turdus migratorius confinis*), Baird's (yellow-eyed) junco (*Junco phaeonotus bairdi*), acorn woodpecker (*Melanerpes formicivorus angustifrons*), and band-tailed pigeon (*Columba fasciata vioscae*).

In addition to the endemic taxa mentioned above, breeding birds representative of the BCS avifauna include the crested caracara (*Caracara cheriway*), Harris's hawk (*Parabuteo unicinctus*), white-winged dove (*Zenaida asiatica*), Leconte's thrasher (*Toxostoma lecontei*), black-throated sparrow (*Amphispiza*

Figure 16.1. Oasis of San Ignacio.

bilineata), sage sparrow (*Amphispiza belli*), northern cardinal (*Cardinalis cardinalis*), pyrrhuloxia (*Cardinalis sinuatus*), and varied bunting (*Passerina versicolor*) (Wilbur 1987; Rodríguez-Estrella, pers. obs.). Breeding birds are typically year-round residents. One of the few exceptions is the purple martin (*Progne subis*), a summer migrant. Most of the BCS breeding population of that bird migrates south for the winter.

BCS also receives an influx of winter visitors, such as the northern harrier (*Circus cyaneus*), found in particular in the valley of La Laguna (Unitt et al. 1992; Rodríguez-Estrella et al. 1997). The Pacific-slope flycatcher (*Empidonax difficilis*) is another winter visitor that has been recorded in important numbers in the Sierra de la Laguna (Rodríguez-Estrella 1988; Rodríguez-Estrella et al. 1997). The Lincoln's sparrow (*Melospiza lincolnii*) becomes common during the winter from a latitude of 28° N southward (Rodríguez-Estrella et al. 1997). The white-crowned sparrow (*Zonotrichia leucophrys*), orange-crowned warbler (*Vermivora celata*), and

yellow-rumped warbler (*Dendroica coronata*) are all common winter visitors (Wilbur 1987). The peregrine falcon has presumably resident and wintering populations. It is fairly common as a breeder on Gulf of California islands. On the mainland, it does nest, but only in mountains, and increases in numbers during the winter (Rodríguez-Estrella et al. 1998, unpubl. data). Some of the winter visitors forage along coastal areas or among crops (e.g., Rodríguez-Estrella et al. 1998), but in this chapter we also document their distribution along arroyos, at oases, and in native upland vegetation.

Finally, some species occur (almost) exclusively during spring and fall migration, such as the willow flycatcher (*Empidonax traillii*), rufous hummingbird (*Selasphorus rufus*), bank swallow (*Riparia riparia*), Swainson's thrush (*Catharus ustulatus*), western tanager (*Piranga ludoviciana*), lazuli bunting (*Passerina amoena*), and Nashville warbler (*Vermivora ruficapilla*) (Wilbur 1987; Howell and Webb 1995; Unitt and Rodríguez-Estrella 1996; Erickson and Howell 2001). Breeding and nonbreeding birds

Table 16.1. Some important plant taxa present in desert oases and along arroyos of Baja California Sur (from Arriaga et al. 1997).

Oases	Arroyos
Phragmites australis (Poaceae)	*Prosopis articulata* (Fabaceae)
Typha domingensis (Typhaceae)	*Acacia farnesiana* (Fabaceae)
Arundo donax (Poaceae)	*Lysiloma candidum* (Fabaceae)
Washingtonia robusta (Arecaeae)	*Cyperus* (6 spp.) (Cyperaceae)
Phoenyx datilifera (Arecaeae)	*Juncus* (3 spp.) (Juncaceae)
Anemopsis californica (Saururaceae)	*Chloracantha spinosa* [*Aster spinosus*] (Asteraceae)
Cyperus (6 spp.) (Cyperaceae)	*Baccharis salicifolia* [*B. glutinosa*]
Scirpus americanus (Cyperaceae)	*B. sarothroides* (Asteraceae)
Juncus (3 spp.) (Juncaceae)	*Hymenoclea monogyra* (Asteraceae)
	Kosteletzkya depressa (Malvaceae)
	Scirpus americanus (Cyperaceae)

in mangrove areas of BCS are examined in chapter 15. Hereafter, all nonresident species are referred to as migrants.

Study Area and Methods

Some findings presented here have already been published (Rodríguez-Estrella et al. 1997, 1999). Most of the data on winter species assemblages in uplands, arroyos, and desert oases are original. They are based on extensive point-count surveys during the 1998–1999 winter, for which we are providing information on methodology. Relationships between oasis size and latitudinal position and species richness are also new and based on information collected since 1994 during both the breeding and nonbreeding seasons; they were analyzed using linear regressions.

Selection of Point Locations for the 1998–1999 Study

Oases, arroyos, and uplands surveyed in BCS during the 1998–1999 winter are shown in figure 16.2 (see also table 16.2 for more information on oases surveyed). To select points, we first delineated arroyos and oases on 1:50,000 and 1:250,000 INEGI maps of the region. We chose point locations to sample as many oases and arroyos as possible. Sampled oases were also selected based on latitude and size (Rodríguez-Estrella et al. 1997, 1999). For arroyo areas, we selected points that we considered a priori to represent the vegetation structure typical of this habitat type. Both small and large arroyos

were sampled (for the names and locations of the largest arroyos, see fig. 16.2). The same number of points were used as controls in adjacent uplands for each independent arroyo and oasis point. The remaining observation points were selected to cover most of the heterogeneity of upland habitats (but note that most survey points were in desertscrub) with variation in vegetation composition and structure, elevation, and topography. All these points, controls and natural remaining points, were surveyed from December 1998 through February 1999. We surveyed 421 points: 44 in oases, 65 in arroyos, and 312 in uplands.

Bird Point Surveys

We used single 15-minute standard point-count surveys (see Reynolds et al. 1980) at all locations. To optimize detection of land birds, we conducted all surveys between 0700 and 1000 hours and between 1600 and 1800 hours (Rodríguez-Estrella 1997; Rodríguez-Estrella et al. 1997). The observer remained stationary at the point location and recorded all species within a 50-m radius of his position. Recommended observation time for point-count surveys is in the range of 5–20 minutes (Hutto et al. 1986; Whitman et al. 1997). Both the 15-minute observation period and the 50-m radius design have proved effective for detecting most land-bird species in uplands, xeroriparian vegetation, and oases of the Baja California peninsula (Rodríguez-Estrella 1997, unpubl. data).

Species were identified by sight or by sound. We distinguished breeding and nonbreeding avian species based on previous regional field experience since

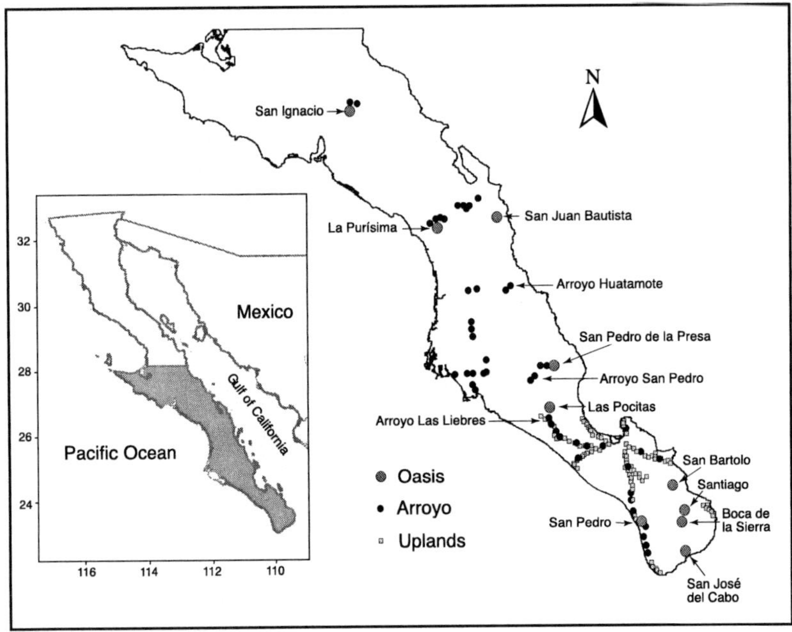

Figure 16.2. Baja California Sur point-count survey location in arroyos, oases, and scrub vegetation during December 1998 and January and February 1999.

1988 and also following several references (National Geographic Society 1987; Wilbur 1987; Howell and Webb 1995; American Ornithologists' Union 1998). The observer recorded to the minute the time of first detection of each species.

Measuring Community Similarity and Species Diversity and Abundance

To compare avian communities of the 3 habitat types, we relied on 5 indices or counts: (1) total species richness; (2) Shannon-Wiener diversity index (H') and evenness index (J') (Krebs 1989); (3) point diversity (average number of species recorded per survey point in each habitat type; (4) spatial heterogeneity (mean number of species detected per survey point /total number of species detected in the habitat); and (5) mean number of birds detected per survey point (Willson and Comet 1996). ANOVAs (with Bonferroni adjustments) were used to examine differences in species richness, species diversity, and evenness among upland, arroyo, and oasis survey points.

Community similarity of two habitat types was measured by Whittaker's (1975) community coefficient (CC) index:

$$CC = 2S_{ab}/(S_a + S_b)$$

where S_{ab} is the number of species shared by the two habitats a and b, and S_a and S_b are the numbers of species detected in each of these habitats. The maximum value the CC index can attain is 1, when the species communities of the two habitats are identical.

Land-Bird Communities of Oases, Arroyos, and Uplands

Nesting Land-Bird Communities

Since 1994 (1990 for uplands), we have documented nesting by 62 land-bird species in uplands, along arroyos, and in oases (table 16.3). A total of 51 species have been found nesting in uplands, compared to 22 species in arroyos and again 22 species in oases. Twenty-six (42%) species nest in at least

Table 16.2. Characteristics and location of oases included in point-count land-bird surveys during the 1998–1999 winter in Baja California Sur.

Oases	Area (km^2)	Elevation (m)	Latitude N	Longitude W
San Ignacio	2.695	220–230	27°17'–27°18'	112°54'–112°53'
La Purísima	2.254	130–140	26°11'	112°04'
San Juan Bautista	0.931	40–50	26°13'–26°14	111°29'–111°28'
San Pedro de la Presa	0.098	380	24°51'	110°02'
Las Pocitas	0.25	130–140	24°28'–24°30'	111°01'–111°00'
San Bartolo	0.588	510–520	23°44'–23°45'	109°51'–109°50'
Santiago	1.470	110–120	23°29'–23°30'	109°44'–109°42'
Boca Sierra	2.303	290–300	23°23'	109°50'–109°48'
San Pedro	0.196	0–10	23°22'	110°12'
San José Cabo	1.372	0–10	23°02'–23°03'	109°41'

2 of the 3 habitats. Twenty-five species may nest exclusively in uplands.

The 22 birds nesting in oases constitute a somewhat distinct assemblage, as they include 9 (41%) species apparently not shared with uplands and arroyos. Nesting species common to abundant in oases but not nesting in the other two habitat types include the black phoebe (*Sayornis nigricans*), vermilion flycatcher (*Pyrocephalus rubinus*), song sparrow (*Melospiza melodia*), and Belding's yellowthroat. In addition to these obligate oasis-nesters, based on our field experience there are nesting species shared by oases and uplands that appear more common in oases, such as Xantus's hummingbirds, hooded orioles (*Icterus cucullatus*), and Scott's orioles (*I. parisorum*). Xantus's hummingbirds occur also along arroyos, but in this habitat, too, they appear less abundant than in oases. Unlike Xantus's hummingbirds, hooded and Scott's orioles have not been found nesting along arroyos. In oases these 2 species nest in palm trees. Xantus's hummingbirds have been recorded usually nesting in smaller trees.

Only 2 nesting species have been found along arroyos but not in uplands: the white-throated swift (*Aeronautes saxatalis*) and the horned lark (*Eremophila alpestris*). The horned lark is also common during the breeding season in crop fields of BCS. We have documented nesting by white-throated swifts in several areas of BCS, in each case in cliffs along the bank of an arroyo. Despite those 2 species, however, the nesting land-bird community of arroyos appears largely to be a subset of that found in surrounding uplands. The endemic gray thrasher is an example of the greater resemblance of arroyos

with uplands compared to oases. This bird nests in uplands and along arroyos, but apparently not at oases. Altogether, we have found only 7 land-bird species nesting both along arroyos and at oases, and all of these are also found as breeders in the surrounding uplands.

Several species breed both in uplands and along arroyos but are found in greater densities in the latter habitat. Among them are the western screech owl (*Otus kennicottii*; a cavity nester), western scrub-jay (*Aphelocoma californica*), northern cardinal, and pyrrhuloxia. These species increase the distinctiveness of arroyos as nesting habitat. Birds nesting along arroyos in mesquites in particular include the white-winged dove, northern mockingbird (*Mimus polyglottos*), and phainopepla (*Phainopepla nitens*).

Avian Communities During the Winter

During the winter we have recorded 107 migrant and resident land-bird species in oases, arroyos, and uplands (table 16.4). More land-bird species have been detected in oases (92), followed by arroyos (83) and uplands (79). A total of 36 migrant passerines are known during the winter from oases, 33 along arroyos, but only 25 in the uplands. Arroyos seem characterized in particular by the highest number of *Empidonax* flycatchers and vireos. Ten species, including residents and migrants, have been found exclusively at oases; 5 only in arroyos and 5 only in uplands. We have detected 13 species at oases and along arroyos but not in uplands. Additionally, we have documented 30 species of aquatic

Table 16.3. Land-bird species known to nest in uplands (chiefly desertscrub), along arroyos, and in desert oases of Baja California Sur.

Common Name	Scientific Name	Uplands	Arroyos	Oases
Turkey vulture	*Cathartes aura*	X		
Osprey	*Pandion haliaetus*	X		
Harris's hawk	*Parabuteo unicinctus*	X		
Zone-tailed hawk	*Buteo albonotatus*			X
Red-tailed hawk	*Buteo jamaicensis*	X		
Crested caracara	*Caracara cheriway*	X		
American kestrel	*Falco sparverius*	X		
Peregrine falcon	*Falco peregrinus*	X		
California quail	*Callipepla californica*	X	X	
White-winged dove	*Zenaida asiatica*	X	X	X
Mourning dove	*Zenaida macroura*	X		
Common ground dove	*Columbina passerina*	X	X	X
Ruddy ground dove	*Columbina talpacoti*	X		X
Greater roadrunner	*Geococcyx californianus*	X		
Barn owl	*Tyto alba*	X	X	X
Western screech owl	*Otus kennicottii*	X	X	
Great horned owl	*Bubo virginianus*	X		X
Elf owl	*Micrathene whitneyi*	X		
Lesser nighthawk	*Chordeiles acutipennis*	X		
White-throated swift	*Aeronautes saxatalis*		X	
Xantus's hummingbird	*Hylocharis xantusii*	X	X	X
Black-chinned hummingbird	*Archilochus alexandri*	X	X	
Costa's hummingbird	*Calypte costae*	X	X	X
Belted kingfisher	*Ceryle alcyon*			X
Gila woodpecker	*Melanerpes uropygialis*	X		
Ladder-backed woodpecker	*Picoides scalaris*	X		
Northern flicker	*Colaptes auratus*	X		
Black phoebe	*Sayornis nigricans*			X
Vermillion flycatcher	*Pyrocephalus rubinus*			X
Ash-throated flycatcher	*Myiarchus cinerascens*	X		
Loggerhead shrike	*Lanius ludovicianus*	X	X	
Western scrub jay	*Aphelocoma californica*	X	X	
Common raven	*Corvus corax*	X		
Horned lark	*Eremophila alpestris*		X	
Purple martin	*Progne subis*	X		
Violet-green swallow	*Tachycineta thalassina*	X		
Verdin	*Auriparus flaviceps*	X	X	
Cactus wren	*Campylorhynchus brunneicapillus*	X	X	X
Rock wren	*Salpinctes obsoletus*	X	X	
Canyon wren	*Catherpes mexicanus*	X		
Bewick's wren	*Thryomanes bewickii*			X
Marsh wren	*Cistothorus palustris*			X
Blue-gray gnatcatcher	*Polioptila caerulea*	X	X	
California gnatcatcher	*Polioptila californica*	X	X	
Northern mockingbird	*Mimus polyglottos*	X	X	X
Gray thrasher	*Toxostoma cinereum*	X	X	
Phainopepla	*Phainopepla nitens*	X	X	
Belding's yellowthroat	*Geothlypis beldingi*			X
Yellow-breasted chat	*Icteria virens*	X		X
Canyon towhee	*Pipilo fuscus*	X		
Rufous-crowned sparrow	*Aimophila ruficeps*			X
Black-throated sparrow	*Amphispiza bilineata*	X		
Lark bunting	*Calamospiza melanocorys*	X		
Song sparrow	*Melospiza melodia*			X
Northern cardinal	*Cardinalis cardinalis*	X	X	

Table 16.3. Continued

Common Name	Scientific Name	Uplands	Arroyos	Oases
Pyrrhuloxia	*Cardinalis sinuatus*	X	X	
Western meadowlark	*Sturnella neglecta*	X		
Brown-headed cowbird	*Molothrus ater*	X		
Hooded oriole	*Icterus cucullatus*	X		X
Scott's oriole	*Icterus parisorum*	X		X
House finch	*Carpodacus mexicanus*	X		X
Lesser goldfinch	*Carduelis psaltria*	X		

birds at oases and/or along arroyos (listed in table 16.5 but not discussed in this chapter).

We recorded 90 resident and migrant land-bird species at the 421 survey points during the 1998–1999 winter. A total of 67 species were recorded in uplands, 62 along arroyos, and 49 in oases. The relatively high numbers of species found in oases and along arroyos are significant because most (74%) survey points were in uplands rather than those other two habitats. Additionally, average species richness per survey point was higher in arroyos and oases than in the uplands (Tukey tests: upland vs. arroyos, P = .001; uplands vs. oases, P < .001; oases vs. arroyos, P = .2). Average species diversity per survey point (H') was also higher in oases and arroyos than in uplands vegetation (Tukey tests: upland vs. arroyos, P < .001; uplands vs. oases, P < .001; oases vs. arroyos, P = .3). Survey points at oases and along arroyos showed the greatest observed degree of heterogeneity in terms of species richness (see appendix 16.1 for a summary of survey results).

Species assemblages recorded during the 1998–1999 surveys were fairly similar among the 3 habitats. The observed similarity in species composition was highest between uplands and arroyos when considering all species ($CC_{arroyo-oasis}$ = 0.68; $CC_{arroyo-uplands}$ = 0.74; $CC_{oasis-uplands}$ = 0.67) and just migrants ($CC_{arroyo-oasis}$ = 0.45; $CC_{arroyo-uplands}$ = 0.60; $CC_{oasis-uplands}$ = 0.50).

Observed mean abundance at survey points was highest at oases and lowest in uplands, but the difference among the 3 habitat types was not significant ($F_{2,418}$ = 0.29, P = .74). The same resident species tended to dominate in all 3 habitat types—for example, the verdin (*Auriparus flaviceps*), cactus wren (*Campylorhynchus brunneicapillus*), white-winged dove, and Costa's hummingbird (*Calypte costae*; table 16.6). None of the resident species frequently detected in 1 habitat type was absent else-where. Both Xantus's hummingbird (found at 9% and 8% of survey points along arroyos and in uplands, respectively) and the hooded oriole (11% and 7% of survey points along arroyos and in uplands, respectively) appeared substantially more common at oases.

Among migrants, only the orange-crowned warbler (*Vermivora celata*) and lark sparrow (*Chondestes grammacus*) tended to dominate in all 3 habitat types. In oases the most common migrant appeared to be the yellow-rumped warbler (*Dendroica coronata*), and this species was recorded at only 6% and 1% of survey points along arroyos and in uplands, respectively. The gray flycather (*Empidonax wrightii*), one of the most common migrants along arroyos and in the uplands, was not detected at oases.

Altogether, the observed frequency of detection of landbirds during the 1998–1999 surveys was notably higher in oases compared to uplands, even for the most common species of the uplands. This pattern was observed for birds such as the white-winged dove, verdin, and cactus wren. The blue-gray gnatcatcher (*Polioptila caerulea*) seemed much more common along arroyos than in uplands, but in general differences in abundance between these 2 habitats seemed smaller compared to those between oases and uplands.

Differences in Species Richness among Oases

At most oases, land-bird species richness is apparently higher during the winter, especially north of the Cape Region (fig. 16.3). The 2 oases where the highest numbers of land-bird species have been recorded during the winter are La Purísima (32 residents and 24 migrants) and San Ignacio (28 residents and 21 migrants), respectively. These 2 oases are both extensive, and they are located in the northern half of BCS. La Purísima lies at the mouth of Arroyo La

Table 16.4. Land-bird species recorded during the winter in uplands, along arroyos, and in desert oases of Baja California Sur.

Common Name	Scientific Name	Uplands	Arroyos	Oases
Turkey vulture	*Cathartes aura*	X	X	X
Osprey	*Pandion haliaetus*	X	X	X
Northern harrier*	*Circus cyaneus*	X	X	X
Sharp-shinned Hawk*	*Accipiter striatus*	X	X	X
Cooper's hawk*	*Accipiter cooperii*	X	X	X
Gray hawk*	*Asturina nitida*	X		
Harris's hawk	*Parabuteo unicinctus*	X	X	
Red-shouldered hawk*	*Buteo lineatus*	X		X
Zone-tailed hawk	*Buteo albonotatus*			X
Red-tailed hawk	*Buteo jamaicensis*	X	X	X
Crested caracara	*Caracara cheriway*	X	X	
American kestrel	*Falco sparverius*	X	X	X
Merlin*	*Falco columbarius*		X	X
Peregrine falcon	*Falco peregrinus*	X		X
California quail	*Callipepla californica*	X	X	X
White-winged dove	*Zenaida asiatica*	X	X	X
Mourning dove	*Zenaida macroura*	X	X	X
Common ground-dove	*Columbina passerina*	X	X	X
Ruddy ground-dove	*Columbina talpacoti*			X
Greater roadrunner	*Geococcyx californianus*	X	X	X
Great horned owl	*Bubo virginianus*	X	X	
Lesser nighthawk	*Chordeiles acutipennis*	X	X	X
White-throated swift	*Aeronautes saxatalis*		X	X
Xantus's hummingbird	*Hylocharis xantussi*	X	X	X
Black-chinned hummingbird	*Archilochus alexandri*	X	X	X
Costa's hummingbird	*Calypte costae*	X	X	X
Belted kingfisher	*Ceryle alcyon*		X	X
Gila woodpecker	*Melanerpes uropygialis*	X	X	X
Yellow-bellied sapsucker*	*Sphyrapicus varius*	X		
Ladder-backed woodpecker	*Picoides scalaris*	X	X	X
Northern flicker	*Colaptes auratus*	X	X	X
Hammond's flycatcher*	*Empidonax hammondii*		X	
Gray flycatcher*	*Empidonax wrightii*	X	X	
Dusky flycatcher*	*Empidonax oberholseri*		X	
Pacific-slope flycatcher*	*Empidonax difficilis*	X	X	X
Buff-breasted flycatcher*	*Empidonax fulvifrons*		X	
Black phoebe	*Sayornis nigricans*	X	X	X
Say's phoebe*	*Sayornis saya*	X	X	X
Vermillon flycatcher	*Pyrocephalus rubinus*	X	X	X
Ash-throated flycatcher	*Myiarchus cinerascens*	X	X	X
Cassin's kingbird*	*Tyrannus vociferans*	X	X	X
Loggerhead shrike	*Lanius ludovicianus*	X	X	X
Bell's vireo*	*Vireo bellii*	X	X	X
Gray vireo*	*Vireo vicinior*	X	X	X
Blue-headed vireo*	*Vireo solitarius*		X	
Warbling vireo*	*Vireo gilvus*		X	
Western scrub Jay	*Aphelocoma californica*	X	X	X
Common raven	*Corvux corax*	X	X	X
Horned lark	*Eremophila alpestris*		X	X
Purple martin	*Progne subis*	X		X
Tree swallow*	*Tachycineta bicolor*	X		X
Violet-green swallow*	*Tachycineta thalassina*	X	X	X
Barn swallow*	*Hirundo rustica*		X	X
Verdin	*Auriparus flaviceps*	X	X	X
Cactus wren	*Campylorhynchus brunneicapillus*	X	X	X

Table 16.4. Continued

Common Name	Scientific Name	Uplands	Arroyos	Oases
Rock wren*	Salpinctes obsoletus	X	X	
Canyon wren	Catherpes mexicanus	X	X	X
Bewick's wren*	Thryomanes bewickii	X		X
House wren*	Troglodytes aedon	X	X	X
Marsh wren	Cistothorus palustris			X
Ruby-crowned kinglet*	Regulus calendula		X	X
Blue-gray gnatcatcher	Polioptila caerulea	X	X	X
California gnatcatcher	Polioptila californica	X	X	X
Hermit thrush*	Catharus guttatus		X	X
Northern mockingbird	Mimus polyglottos	X	X	X
Gray thrasher	Toxostoma cinereum	X	X	X
Phainopepla	Phainopepla nitens	X	X	X
American pipit*	Anthus rubescens			X
Cedar waxwing*	Bombycilla cedrorum	X		X
Orange-crowned warbler*	Vermivora celata	X	X	X
Yellow warbler*	Dendroica petechia	X		X
Yellow-rumped warbler*	Dendroica coronata	X	X	X
Black-and-white warbler*	Mniotilta varia	X	X	X
Northern waterthrush*	Seiurus noveboracensis			X
MacGillivray's warbler*	Oporornis tolmiei		X	X
Common yellowthroat*	Geothlypis trichas	X	X	X
Belding's yellowthroat	Geothlypis beldingi			X
Wilson's warbler*	Wilsonia pusilla		X	X
Yellow-breasted chat	Icteria virens			X
Summer tanager*	Piranga rubra	X		
Western tanager*	Piranga ludoviciana			X
Green-tailed towhee*	Pipilo chlorurus	X	X	X
Canyon towhee	Pipilo fuscus	X	X	X
Rufous-crowned sparrow	Aimophila ruficeps	X	X	X
Chipping sparrow*	Spizella passerina		X	X
Clay-colored sparrow*	Spizella pallida	X	X	X
Brewer's sparrow*	Spizella breweri	X	X	X
Vesper sparrow*	Pooecetes gramineus			X
Lark sparrow*	Chondestes grammacus	X	X	X
Lark bunting	Calamospiza melanocorys	X		
Black-throated sparrow	Amphispiza bilineata	X	X	X
Song sparrow*	Melospiza melodia		X	X
Lincoln's sparrow*	Melospiza lincolnii		X	X
White-crowned sparrow*	Zonotrichia leucoprhys	X	X	X
Northern cardinal	Cardinalis cardinalis	X	X	X
Pyrrhuloxia	Cardinalis sinuatus	X	X	X
Rose-breasted grosbeak*	Pheucticus ludovicianus			X
Black-headed grosbeak*	Pheuticus melanocephalus	X	X	X
Blue grosbeak*	Guiraca caerulea	X		X
Lazuli bunting*	Passerina amoena		X	X
Western meadowlark	Sturnella neglecta	X		
Brown-headed cowbird	Molothrus ater	X	X	X
Hooded oriole	Icterus cucullatus	X	X	X
Scott's oriole	Icterus parisorum	X	X	X
House finch	Carpodacus mexicanus	X	X	X
Lesser goldfinch	Carduelis psaltria	X	X	X
House sparrow	Passer domesticus	X		X

*Migrant species.

Table 16.5. Aquatic birds recorded during the winter along arroyos and in desert oases of Baja California Sur.

Common Name	Scientific Name	Arroyos	Oases
Least grebe	*Tachybaptus dominicus*		X
Pied-billed grebe	*Podylimbus podiceps*		X
Eared grebe	*Podiceps nigricollis*		X
Brown pelican	*Pelecanus occidentalis*		X
Double-crested cormorant	*Phalacrocorax auritus*		X
Magnificent frigatebird	*Fregata magnificens*		X
Great blue heron	*Ardea herodias*	X	X
Great egret	*Ardea alba*		X
Snowy egret.	*Egretta thula*	X	X
Tricolored heron	*Egretta tricolor*		X
Reddish egret	*Egretta rufescens*		X
Cattle egret	*Bubulcus ibis*		X
Striated heron	*Butorides striatus*	X	
Black-crowned night-heron	*Nycticorax nycticorax*		X
White-faced ibis	*Plegadis chihii*		X
Northern shoveler	*Anas cyanoptera*	X	
Northern pintail	*Anas acuta*	X	
Redhead	*Aythya americana*	X	
Ruddy duck*	*Oxyura jamaicensis*	X	
Common moorhen	*Gallinula chloropus*	X	X
American coot	*Fulica americana*		X
Killdeer	*Charadrius vociferus*	X	X
Greater yellowlegs	*Tringa melanoleuca*	X	
Solitary sandpiper*	*Tringa solitaria*		X
Spotted sandpiper	*Actitis macularia*	X	X
Whimbrel	*Numenius phaeopus*		X
Bonaparte's gull	*Larus philadelphia*		X
Ring-billed gull	*Larus delawarensis*		X
Yellow-footed gull	*Larus livens*		X
Western gull	*Larus occidentalis*		X

*Migrant species.

Purísima, with lush vegetation that includes thick stands of palms (*Washingtonia robusta, Phoenix dactylifera*), reeds, and *Juncus acutus*. San Ignacio lies within a deep arroyo (Arroyo de San Ignacio); its vegetation is dominated by *Phragmites* and date palms (*P. dactylifera*), although there are also stands of *W. robusta*. *Arundo donax* is common along the water edge, mixed with *Phragmites*, whereas bamboo (*Bambusa vulgaris*) is mostly an ornamental species cultivated around houses. Another large oasis, Boca de la Sierra, has low land-bird species richness during the winter. It is located in the Cape Region. Using all data collected during the winter since 1994, there was no significant relationship between species richness and oasis size. However, land-bird species richness at oases increased significantly with higher latitude. This trend was not significant when we considered only migrants.

Based on data collected from 1994 to 1996, Rodríguez-Estrella et al. (1997) analyzed the community similarity of 7 oases during the winter. As shown in table 16.7, CC index values varied from 0.44 to 0.77. The highest degree of similarity was observed for the 2 oases of San Ignacio and La Purísima. Also similar in land-bird species composition were Santiago and San Bartolo. Santiago is nearly 3 times larger than San Bartolo, but both are situated in the eastern half of the Cape Region. Santiago also has been heavily impacted by anthropogenic effects (see further on). Geographic position is evidently important in determining land-bird communities in oases. The lowest CC index values observed by Rodríguez-Estrella et al. (1997) were between the oasis at San José del Cabo (near the southern tip of the peninsula) and three oases north of the Cape Region: San Ignacio, La Purísima, and Las Pocitas.

Table 16.6. Abundance of avian species expressed as the number (%) of points where a species was detected.

Arroyos		Oases		Uplands	
Species	Number (%)	Species	Number (%)	Species	Number (%)
Residents					
Auriparus flaviceps	49 (75)	*Campylorhynchus brunneicapillus*	38 (86)	*Cathartes aura*	212 (68)
Cathartes aura	41 (63)	*Melanerpes uropygialis*	36 (82)	*Auriparus flaviceps*	178 (57)
Melanerpes uropygialis	39 (60)	*Zenaida asiatica*	34 (77)	*Campylorhynchus brunneicapillus*	167 (54)
Myiarchus cinerascens	38 (58)	*Cathartes aura*	33 (75)	*Myiarchus cinerascens*	167 (54)
Campylorhynchus brunneicapillus	37 (57)	*Auriparus flaviceps*	31 (70)	*Melanerpes uropygialis*	153 (49)
Calypte coastae	35 (54)	*Icterus cucullatus*	30 (68)	*Calypte costae*	110 (35)
Polioptila caerulea	31 (48)	*Carpodacus mexicanus*	23 (52)	*Zenaida asiatica*	106 (34)
Colaptes auratus	24 (37)	*Calypte costae*	18 (41)	*Colaptes auratus*	97 (31)
Zenaida asiatica	22 (34)	*Myiarchus cinerascens*	13 (30)	*Falco sparverius*	81 (26)
Carpodacus mexicanus	20 (31)	*Hylocharis xantusii*	13 (30)	*Polioptila caerulea*	74 (24)
Migrants					
Chondestes grammacus	11 (17)	*Dendroica coronata*	32 (73)	*Vermivora celata*	26 (8)
Vermivora celata	8 (12)	*Vermivora celata*	15 (34)	*Chondestes grammacus*	23 (7)
Empidonax wrightii	7 (11)	*Chondestes grammacus*	9 (20)	*Tachycineta thalassina*	23 (7)
Empidonax difficilis	5 (8)	*Buteo lineatus*	7 (16)	*Empidonax wrightii*	21 (7)
Dendroica coronata	4 (6)	*Zonotrichia leucophrys*	7 (16)	*Empidonax difficilis*	11 (4)
Spizella pallida	4 (6)	*Melospiza melodia*	6 (13)	*Spizella pallida*	10 (3)
Tachycineta thalassina	4 (6)	*Melospiza lincolnii*	6 (13)	*Tyrannus vociferans*	8 (3)

The number of total survey points was 44 in oases, 65 in arroyos, and 312 in uplands (chiefly desertscrub).

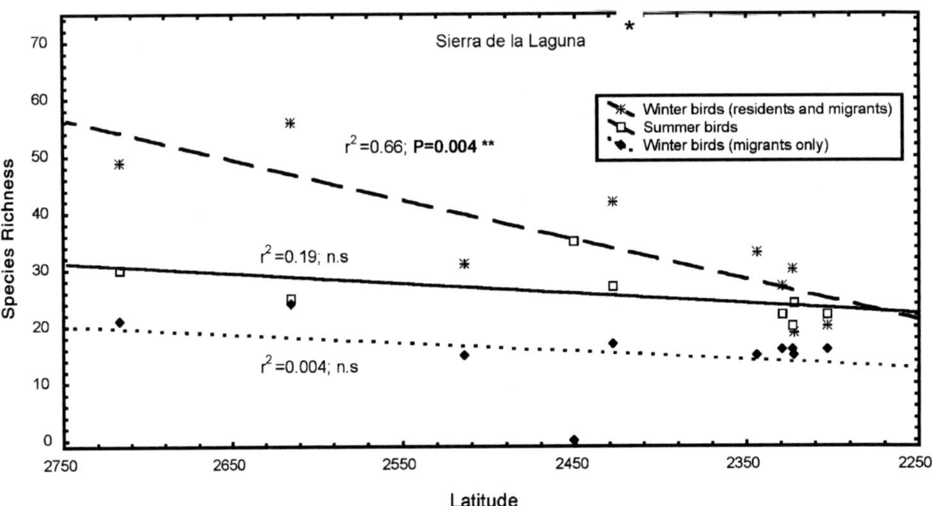

Figure 16.3. Bird species richness in oases plotted versus latitude. Note the Sierra de la Laguna oak–pine forest point. That point is provided for anecdotal comparisons only (it is not an oasis).

The small oasis near San Pedro de la Presa has the highest observed species richness of birds during the breeding season. A partial (but not complete) explanation of this finding may be that due to the small size of the oasis, we were more successful in finding secretive species. Using data collected during all breeding seasons, there was no significant relationship between species richness and oasis size or latitudinal position.

Importance of Oases and Arroyos in BCS

Most of our research on arroyos and oases has been conducted through comparisons within desertscrub, although a few areas dominated by tropical deciduous forest were included. More research is needed to study the influence of the surrounding vegetation on species richness and composition in oases and arroyos.

As strongly suggested by the patterns presented here, desert oases play an important role toward sustaining regional populations of many land-bird species. In BCS, several species seem to nest only in oases (at least at lower elevations). Oases appear to support most of the wintertime land-bird diversity of the lowlands. And 1 endemic species, Belding's yellowthroat, depends entirely on oases for its continued existence. This bird nests only in oases, where it forages mainly in reed grass and cattails and remains within 50 m of the water-edge vegetation

Table 16.7. Land-bird community similarity of seven oases of Baja California Sur.

	SI	PU	PO	SB	SA	SP	SJ
San Ignacio (SI)	1	0.77	0.6	0.64	0.58	0.64	0.54
La Purísima (PU)		1	0.66	0.64	0.60	0.65	0.47
Las Pocitas (PO)			1	0.68	0.55	0.68	0.44
San Bartolo (SB)				1	0.75	0.68	0.60
Santiago (SA)					1	0.62	0.65
San Pedro (SP)						1	0.65
San José Cabo (SJ)							1

Community similarity is expressed as Whitaker's (1975) community coefficient index. Modified from Rodríguez-Estrella et al. 1997.

(Rodríguez-Estrella et al. 1999). Besides land birds, oases are also associated with a relict or endemic mesophilic herpetofauna (e.g., *Hyla regilla*, *Trachemys nebulosa*, *Thamnophis hammondii*; Grismer and McGuire 1993; Grismer 2002). Based on our observations, raccoons (*Procyon lotor*) and ring-tail cats (*Bassariscus astutus*) are more common in oases than the surrounding uplands. The bobcat (*Lynx rufus*) and gray fox (*Urocyon cinereoargenteus*) visits oases associated with large amounts of water.

Oases represent small patches of rich habitat embedded in xerophythic vegetation. The availability of abundant food sources (e.g., insects, fruits) at oases is essential for migrating birds using them as stopovers (Jiménez et al. 1997; Rubio et al. 1997). Both resident and nonresident birds commonly forage at palms and edge vegetation where more insects are available. Also, birds can find protection against predators at oases (Rodríguez-Estrella and Arriaga 1997). Typically, the vegetation structure is more complex than in surrounding areas, providing suitable nest sites otherwise not available (Pineda et al. 1997).

The importance of arroyos for land birds is somewhat less pronounced. However, a number of species appear to nest or winter in greater densities along arroyos, and perhaps in a few cases are restricted to this habitat type (nesting: white-throated swift; wintering: several flycatchers and vireos; table 16.4). Species richness at a local scale was higher along arroyos than in uplands. During our research, resident and migrant birds were observed foraging mainly on mesquite and sweet acacia. Mesquite in particular is an important plant for birds in deserts, providing food (insects), shade, and protection (see Simpson 1977). Other notable plants visited by birds along arroyos included blue palo verde (*Parkinsonia florida*), palo ádan, and trumpet bush.

An important pattern we have found is that species nesting only or chiefly in arroyos, oases, or both are typically those described as obligate or preferential riparian/wetland breeding birds in lowlands of the southwestern United States (for a list, see Cartron et al. 1999, 2000). Among the 9 land-bird species found nesting only at oases, 8 occur in the southwestern United States, and of these 7 are considered (nearly) obligate riparian/wetland nesters at low and mid-elevations: the zone-tailed hawk (*Buteo albonotatus*), belted kingfisher (*Ceryle alcyon*), black phoebe, vermilion flycatcher, Bewick's wren (*Thryomanes bewickii*), marsh wren (*Cistothorus palustris*), and song sparrow (*Melospiza melodia*).

Among the birds apparently nesting in higher densities along arroyos than in uplands, the western screech owl, northern cardinal, and pyrrhuloxia are considered as preferential riparian/wetland species (Cartron et al. 1999).

Our research shows that during the winter land-bird species richness increases in northern oases of BCS. As already described, oases are visited during the winter by many migrants, but the increase in species richness seems largely the result of a habitat shift in resident birds from the surrounding desert (see figure 16.3).

The Sierra de la Laguna, located near the tip of the Baja California peninsula, appears to be the most important overwintering area for birds in BCS. The number of birds recorded at that location increases as the winter progresses. Migrants likely use oases of BCS not just as overwintering habitat but also as protracted stopovers during migration to and from the Sierra de la Laguna (Rodríguez-Estrella et al. 1997; Rubio et al. 1997). Skagen et al. (1998) suggested a "stepping stone" migration pattern involving oases in southeastern Arizona (see also Bairlein 1988; Yom-Tov 1993). Arroyos in Baja California may be used as corridors between habitats during migration and also as stopovers, but more studies are needed to confirm this possibility. We hypothesize that 2 general lowland migration north–south routes may exist along the length of the peninsula: 1 route using oases as stopovers, the other following arroyos and coastal areas (fig. 16.4).

Specific Threats to Oases and Arroyos

Based on our observations, the main anthropogenic impact on arroyos in BCS is currently mesquite wood extraction. The loss of old mesquite trees has been observed near many ranches. Local people in rural areas harvest mesquite trees for fuelwood, which is used for an increasing commerce with cities. Threats to oases are tourism, burning of reed grass and palm vegetation, cutting stems of reed grass for rural home construction, human water use, and cattle, horses, and burros feeding on reed grass (Rodríguez-Estrella et al. 1999).

Belding's yellowthroat no longer occurs at three historical breeding sites (Santiago, Miraflores, and El Triunfo). At Santiago, Brewster (1902) described the species as very abundant among reeds, with the water often 3–4 feet deep. Except for small reed patches, the oasis at Santiago no longer exists, as

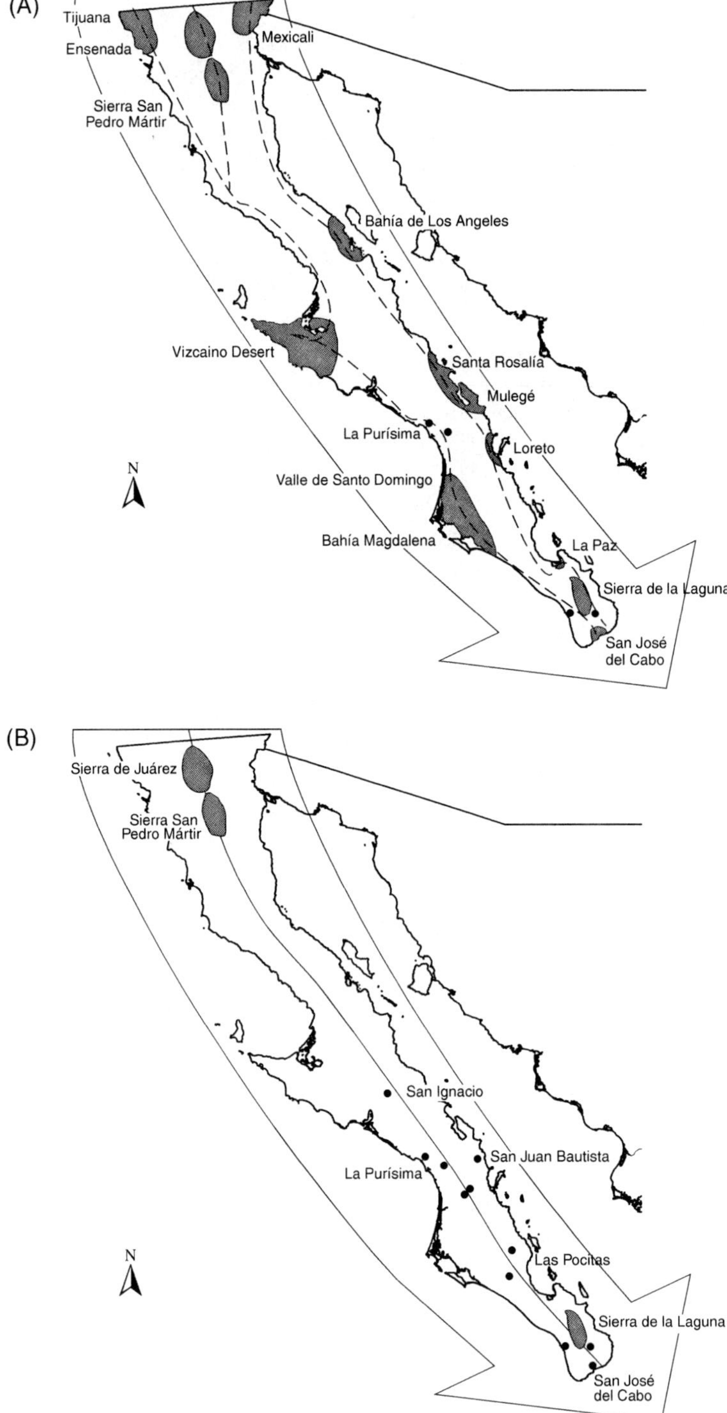

Figure 16.4. Proposed bird migratory routes along the length of the Baja California peninsula. (A) Route following coastal areas and arroyos; (B) route using oases as stopovers.

all of its water has been drained for agriculture. The oasis at El Triunfo has also vanished. Miraflores has only small reed patches and small water ponds (Rodríguez-Estrella et al. 1999).

General Implications for Conservation

Under Mexican law, only a few of the oases (i.e., San Ignacio and San José del Cabo) are protected on the Baja California peninsula, and there are no specific strategies designed to ensure this protection. All other oases lack any formal protection, in spite of their fragility and the threats they currently face (Rodríguez-Estrella et al. 1997, 1999). Conservation strategies for oases are urgently needed. We recommend conservation measures in particular for the oases of Mulegé, San Ignacio, La Purísima, San José, San Miguel de Comondú, San José del Cabo, San Pedrito, and Santiago. It might be tempting to give priority to large oases, but as seen in this chapter, size is not an accurate predictor of land-bird species richness. Along arroyos, cutting mature trees is a local threat to cavity nesters and other species. In our view, this practice should be regulated by the government. In addition, local arroyo conservation should be made a priority for environmental non-governmental organizations and grass-roots education (see chapter 20).

We propose that a landscape approach (emphasizing the arroyo, oasis, upland mosaic as a whole) be used to protect lowland avian species in BCS. Oases and arroyos may be both especially important and susceptible to anthropogenic effects (Rodríguez-Estrella and Arriaga 1997; Skagen et al. 1998), but uplands should be included, especially desertscrub. Many resident species of BCS show a high affinity for desertscrub, including for nesting, and they are also sensitive to alteration of this habitat (Rodríguez-Estrella 1997). A strategy for the protection of arroyos and oases is highly needed, as currently there is no management plan in place for them.

Acknowledgments

We thank J. Bustamante, M. de la Riva, A. Cota, L. Rubio, and E. Pineda for field assistance. Jean-Luc Cartron, Richard Felger, and R. Roy Johnson provided comments that greatly improved earlier drafts. Funding for our research was provided by CIBNOR and by Consejo Nacional de Ciencia y Tecnología CONACyT (Mexico)—CSIC (Spain), through a grant to R.R.E. This chapter was finished with support from project CONACyT (SEMARNAT-2002-C01-0318).

Appendix 16.1. Total species richness, average number of species recorded per survey point (ANS), Shannon-Wiener diversity index (H'), evenness index (J), and mean abundance per survey point in upland vegetation ($N = 312$ points), along arroyos ($N = 65$ points), and at oases ($N = 44$ points) in Baja California Sur during the winter season of 1998–1999.

Habitat Type	Total Species Richness	ANS	H'	J	Mean Abundance per Survey Point	Spatial Heterogeneity
Upland	67	7.43 ± 4.37	1.56 ± 0.64	0.86 ± 0.14	25.26 ± 66.86	8.09
Arroyo	62	9.55 ± 4.73	1.88 ± 0.54	0.89 ± 0.23	24.89 ± 30.01	9.49
Oasis	49	11.0 ± 3.80	2.06 ± 0.33	0.88 ± 0.06	32.96 ± 16.09	10.72

Statistical Tests

ANOVA		$F = 16.94$***	$F = 18.27$***	$F = 1.42$ (n.s.)	$F = 0.29$ (n.s.)	
Tukey tests						
Upland vs. arroyo		***	***			
Upland vs. oasis		***	***			
Arroyo vs. oasis		n.s.	n.s.			

***Statistical significance at the .005 level; n.s., no significant difference at the .05 level.

Literature Cited

American Ornithologists Union (AOU). 1998. Checklist of North American Birds, 7th ed. Allen Press, Lawrence, Kansas.

Anthony, A. W. 1893. Birds of San Pedro Mártir, Lower California. Zoe 4: 228–247.

Arizmendi, M. C., and L. V. Marquez (eds.). 2000. Areas de Importancia para la Conservación de las Aves en México. CIPAMEX, México D.F.

Arriaga, L., S. Díaz, R. Domínguez, and J. L. León de la Luz. 1997. Capítulo 6. Composición Florística y Vegetación. Pp. 69–106 *in* L. Arriaga and R. Rodríguez (eds.), Los Oasis de la Península de Baja California. Publication no. 13, CIBNOR, S.C., La Paz, Mexico.

Bairlein, F. 1988. How do migratory songbirds cross the Sahara? Trends in Ecology and Evolution 3: 191–194.

Bancroft, G. 1930. The breeding birds of central Lower California. Condor 32: 20–49.

Banks, R. C. 1967. Birds and mammals of La Laguna, Baja California. Transactions of the San Diego Society of Natural History 14: 205–232.

Brewster, W. 1902. Birds of the Cape region of Lower California. Bulletin of the Museum of Comparative Zoology 41: 1–241.

Cartron, J.-L. E., S. H. Stoleson, and R. R. Johnson. 1999. Riparian dependence, biogeographic status, and likelihood of endangerment in landbirds of the Southwest. Pp. 211–215 *in* D. M. Finch, J. C. Whitney, J. F. Kelly, and S. R. Loftin (eds.), Rio Grande Ecosystems: Linking land, Water, and People—Toward a Sustainable Future for the Middle Rio Grande Basin. Proceedings RMRS-P-7. USDA Forest Service, Rocky Mountain Research Station, Ogden, Utah.

Cartron, J.-L. E., S. H. Stoleson, P. L. L. Stoleson, and D. W. Shaw. 2000. Riparian areas. Pp. 281–327 *in* R. Jemison and C. Raish (eds.), Livestock Management in the American Southwest: Ecology, Society, and Economics. Elsevier Science, Amsterdam.

Cody, M. L. 1983. The land birds. Pp. 210–245 *in* T. J. Case and M. L. Cody (eds.), Island Biogeography of the Sea of Cortez. University of California Press, Berkeley.

Comisión para la Cooperación Ambiental (CCA). 1999. Areas Importantes para la Conservación de las Aves de América del Norte. Directorio de 150 Sitios Relevantes. CCA Publications, Quebec, Canada.

Davis, J. 1959. The Sierra Madrean element of the avifauna of the Cape district, Baja California, Mexico. Condor 61: 75–84.

Dick-Peddie, W. A. 1993. New Mexico Vegetation: Past, Present, and Future. University of New Mexico Press, Albuquerque.

Erickson, R. A., and S. N. G. Howell (eds.). 2001. Birds of Baja California: status, distribution, and taxonomy. Monographs in Field Ornithology 3.

Farley, G. H., L. M. Ellis, J. N. Stuart, and N. J. Scott. 1994. Avian species richness in different-aged stands of riparian forest along the Middle Rio Grande, New Mexico. Conservation Biology 8: 1098–1108.

Grinnell, J. 1928. A distributional summation of the ornithology of Lower California. University of California Publications of Zoology 32: 1–300.

Grismer, L. L. 2002. Amphibians and Reptiles of Baja California, Its Associated Pacific Islands, and Islands in the Sea of Cortez. University of California Press, Berkeley.

Grismer, L. L., and J. A. McGuire. 1993. The oases of Central Baja California, Mexico. Part I. A preliminary account of the relict mesofilic herpetofauna and the status of the oases. Bulletin of the Southern California Academy of Sciences 92: 2–24.

Howell, S. N. G., and S. Webb. 1995. A Guide to the Birds of Mexico and Northern Central America. Oxford University Press, New York.

Hubbard, J. P. 1977. Importance of riparian ecosystems: biotic considerations. Pp. 14–18 *in* R. R. Johnson and D. A. Jones (eds.), Importance, Preservation, and Management of Riparian Habitat: A Symposium. General Technical Report RM-43. USDA Forest Service, Rocky Mountain Forest and Range Experiment Station, Fort Collins, Colorado.

Hutto, R. L., S. M. Pletschet, and P. Hendricks. 1986. A fixed-radius point count method for non-breeding and breeding season use. Auk 103: 593–602.

Jiménez, M. L., C. Palacios, and A. Tejas. 1997. Los macroartrópodos. Pp. 107–124 *in* L. Arriaga and R. Rodríguez-Estrella (eds.), Los Oasis de la Península de Baja California. Publication no. 13, CIBNOR, S.C., La Paz, Mexico.

Johnson, R. R., L. T. Haight, and J. M. Simpson. 1977. Endangered species vs. endangered habitats: a concept. Pp. 68–79 *in* R. R. Johnson and D. A. Jones (eds.), Importance, Preservation, and Management of Riparian Habitat: A Symposium. General Technical Report RM-43. USDA Forest Service, Rocky Mountain Forest and Range Experiment Station, Fort Collins, Colorado.

Krebs, C. J. 1989. Ecological Methodology. Harper Collins Publisher, New York.

Krueper, D. J. 1996. Effects of livestock management on southwestern riparian systems.

Pp. 281–301 *in* D. W. Shaw and D. M. Finch (eds.), Desired Future Conditions for Southwestern Riparian Ecosystems: Bringing Interests and Concerns Together. General Technical Report RM-272. USDA Forest Service, Rocky Mountain Forest and Range Experiment Station, Fort Collins, Colorado.

Lenz, L. W. 1992. Plants of the Cape Region, Baja California Sur, Mexico. The Cape Press, Claremont, California.

León de la Luz, J. L., and R. Domínguez C. 1989. Flora of the Sierra de La Laguna, Baja California Sur, Mexico. Madroño 36: 61–83.

León de la Luz, J. L., R. Domínguez C., M. Domínguez L., and J. J. Pérez Navarro. 1997. Floristic composition of the San José del Cabo oasis, Baja California Sur, Mexico. Sida: 599–614.

León de la Luz, J. L., J. J. Pérez Navarro, M. Domínguez L., and R. Domínguez C. 1999. Listados Florísticos de México, XVIII, Flora de la Región del Cabo de Baja California Sur. Instituto de Biología, Universidad Nacional Autónoma de México, México D.F.

Lot, A., A. Novelo, and P. Ramírez. 1986. Listados Florísticos de México, V, Angiospermas Acuáticas Mexicanas. Instituto de Biología, Universidad Nacional Autónoma de México, México D.F.

Massey, B. W., and E. Palacios. 1994. Avifauna of the wetlands of Baja California, Mexico: current status. Studies in Avian Biology 15: 45–57.

Maya, Y., R. Coria, and R. Domínguez. 1997. Caracterización de los oasis. Pp. 5–25 *in* L. Arriaga and R. Rodríguez-Estrella (eds.), Los Oasis de la Península de Baja California. Publication no. 13, CIBNOR, S.C., La Paz, Mexico.

Naiman, R. J., H. Décamps and M. Pollock. 1993. The role of riparian corridors in maintaining regional biodiversity. Ecological Applications 3: 209–212.

National Geographic Society. 1987. Field Guide to the Birds of North America. 2nd ed. National Geographic Society, Washington, D.C.

Patten, D. T. 1998. Riparian ecosystems of semi-arid North America: diversity and human impacts. Wetlands 18: 498–512.

Pineda, E., R. Rodríguez-Estrella, L. Arriaga, and L. Rubio. 1997. Variaciones estacionales de la avifauna y estructura de la vegetación. Pp. 197–220 *in* L. Arriaga and R. Rodríguez-Estrella (eds.), Los Oasis de la Península de Baja California. Publication no. 13, CIBNOR, S.C., La Paz, Mexico.

Reynolds, R. T, J. M. Scott, and R. A. Nussbaum. 1980. A variable circular-plot method for estimating bird numbers. Condor 82: 309–313.

Rodríguez-Estrella, R. 1988. Avifauna. Pp. 185–208 *in* L. Arriaga (ed.), La Sierra de la Laguna de Baja California Sur. Publication no. 1, Centro de Investigaciones Biológicas de Baja california Sur, La Paz, Mexico.

Rodríguez-Estrella, R. 1997. Factores que condicionan la distribución y abundancia de las aves terrestres en Baja California Sur, México: El efecto de los cambios al hábitat por actividad humana. Ph.D. dissertation, Universidad Autónoma de Madrid, Spain.

Rodríguez-Estrella R., and L. Arriaga. 1997. Implicaciones ecológicas de las actividades humanas en la biota asociada a los oasis. Pp. 285–292 *in* L. Arriaga and R. Rodríguez-Estrella (eds.), Los Oasis de la Península de Baja California. Publication no. 13, CIBNOR, S.C., La Paz, Mexico.

Rodríguez-Estrella, R., J. A. Donázar, and F. Hiraldo. 1998. Raptors as indicators of environmental change in the scrub habitat of Baja California Sur, Mexico. Conservation Biology 12: 921–925.

Rodríguez-Estrella, R., L. Rubio, and E. Pineda. 1997. Los oasis como parches atractivos para las aves residentes e invernantes. Pp. 157–195 *in* L. Arriaga and R. Rodríguez-Estrella (eds.), Los Oasis de la Península de Baja California. Publication no. 13, CIBNOR, S.C., La Paz, Mexico.

Rodríguez-Estrella, R., L. Rubio, E. Pineda, and G. Blanco. 1999. The Belding's yellowthroat: current status, habitat preferences, and threats in oases of Baja California, Mexico. Animal Conservation 2: 77–84.

Rubio, L., R. Rodríguez-Estrella, and E. D. Pineda. 1997. Comparación del uso del hábitat por aves residentes e invernantes. Pp. 221–248 *in* L. Arriaga and R. Rodríguez-Estrella (eds.), Los Oasis de la Península de Baja California. Publication no. 13, CIBNOR, S.C., La Paz, Mexico.

SEMARNAT. 2002. Norma Oficial Mexicana NOM-059-ECOL-2001, Protección Ambiental-Especies Nativas de México de Flora y Fauna Silvestres-Categorías de Riesgo y Especificaciones para su Inclusión, Exclusión o Cambio-Lista de Especies en Riesgo. Diario Oficial de la Federación 582: 1–80.

Shreve, F. 1937. The vegetation of the Cape Region of Baja California. Madroño 4: 105–113.

Shreve, F. 1951. Vegetation of the Sonoran Desert. Carnegie Institution of Washington Publication no. 591, Washington, D.C.

Simpson, B. B. (ed.). 1977. Mesquite. Its Biology in Two Desert Scrub Ecosystems. U.S./IBP Series 4. Halsted Press, New York.

Skagen, S., C. P. Melcher, W. H. Howe, and F. L. Knopf. 1998. Comparative use of riparian corridors and oases by migrating birds in Southeast Arizona. Conservation Biology 12: 896–909.

Stager, K. E. 1960. Baja California symposium: The composition and origin of the avifauna. Systematic Zoology 9: 179–183.

Stattersfield, A. J., M. J. Crosby, A. L. Long, and D. C. Wege. 1998. Endemic bird areas of the world. Birdlife Conservation Series no. 7. Birdlife International, Cambridge, U.K.

Turner, R. M., J. E. Bowers, and T. L. Burgess. 1995. Sonoran Desert Plants: An Ecological Atlas. University of Arizona Press, Tucson.

Turner, R. M., and D. E. Brown. 1982. Tropical-subtropical desertlands. Desert Plants 4: 180–221.

Unitt, P., and R. Rodríguez-Estrella. 1996. Winter distribution of hermit thrush subspecies in the Sierra de la Laguna, Baja California Sur. Western Birds 27: 65–69.

Unitt, P., R. Rodríguez-Estrella, and A. Castellanos. 1992. Ferruginous hawk and pine siskin in the Sierra de La Laguna, Baja California Sur; subspecies of the pine siskin in Baja California. Western Birds 23: 171–172.

Whitman, A. A., J. M. Hagan III, and N. V. L. Brokaw. 1997. A comparison of two bird survey techniques used in subtropical forest. Condor 99: 955–965.

Whittaker, R. H. 1975. Communities and Ecosystems. McMillan, New York.

Wiggins I. L. 1980. Flora of Baja California. Stanford University Press, Stanford, California.

Wilbur, S. R. 1987. Birds of Baja California. University of California Press, Berkeley.

Willson, M.F., and T.A. Comet. 1996. Bird communities of northern forests: patterns of diversity and abundance. Condor 98: 337–349.

Yom-Tov, Y. 1993. The importance of stopover sites in deserts of Paleartic migratory birds. Israel Journal of Zoology 39: 271–273.

III

NATURAL RESOURCE IMPACTS AND CONSERVATION AT A POPULATION, SPECIES, AND LANDSCAPE LEVEL

17

Impact of Concrete Power Poles on Raptors and Ravens in Northwestern Chihuahua, Mexico

JEAN-LUC E. CARTRON

RICHARD E. HARNESS

ROBERT C. ROGERS

PATRICIA MANZANO-FISCHER

In the last 2 or 3 decades, much of rural northern Mexico has been electrified with three-phase power lines (i.e., power lines with 3 energized wires carrying voltages 120° apart in phase) mounted on poured concrete poles. These poles are reinforced inside by steel or "rebar" strands and are typically fitted with steel crossarms. Because steel and, to a lesser extent, concrete, are conductive, any bird that perches on the crossarm of a concrete pole becomes grounded. Whereas most electrocutions of raptors on nonconductive (e.g., wooden) poles and wooden crossarms involve simultaneous contact with two energized wires (APLIC 1996), on conductive poles the risk of electrocution is higher because a bird needs to touch only 1 energized conductor (Janss and Ferrer 1999).

To date, the biological impact of concrete power poles in Mexico has not been well documented. In this chapter, we assess the magnitude of this impact with a primary focus on the Janos–(Nuevo) Casas Grandes (JNCG) prairie-dog town complex, an area of northwestern Chihuahua with a recent history of bird electrocutions. We provide a list of birds with known or suspected mortality linked to concrete poles and a comparison of the incidence of electrocution in relation to power-pole configuration. We discuss the local and regional impact of concrete poles on frequently electrocuted species and describe specific measures to mitigate the problem. This chap-

ter is based mainly on the results of extensive monthly surveys conducted from December 2000 through November 2001 in the JNCG prairie-dog town complex and surrounding area, as well as more limited surveys elsewhere in northwestern Chihuahua.

Early Reports of Electrocuted Birds and the First Formal Surveys

In January 1999, a Christmas Bird Count (CBC) crew working in the JNCG prairie-dog town complex of northwestern Chihuahua noted dead birds bearing marks of electrocution at the base of recently installed concrete poles. Electrocuted birds found during this CBC survey as well as the next one, in December 1999, consisted of raptors, including a great horned owl (*Bubo virginianus*) and an immature bald eagle (*Haliaeetus leucocephalus*) (C. Melcher, pers. comm.).

CBC observations prompted Cartron et al. (2000) to conduct formal power line surveys in that same area in February and March 2000. During these surveys, 49 dead raptors and ravens were discovered, among them 11 golden eagles (*Aquila chrysaetos*) and 10 ferruginous hawks (*Buteo regalis*) (Cartron et al. 2000). Additional electrocuted birds were observed in late spring when the Comisión Federal de Electricidád (CFE) contracted EDM

International, Inc. to inspect local power lines and propose retrofitting solutions for poles. The contracted surveys established that electrocution events were not restricted to a small number of poles, nor to a specific pole configuration. Seventy-five percent of all observed bird remains were found beneath three-phase tangent units, but these units also accounted for 84% of all poles surveyed (R. E. Harness, unpubl. data). The remaining birds were found under a variety of other configurations, described later in the chapter.

The JNCG Prairie-Dog Town Complex and Surrounding Area

The JNCG prairie-dog town complex is located in a plain just east of the Sierra Madre Occidental in northwestern Chihuahua, Mexico (fig. 17.1). The area is included within the Chihuahuan Desert Region (Medellín-Leal 1982) and is characterized by a mosaic of habitats ranging from short-grass prairie to mesquite (*Prosopis* sp.) shrubland, xeroriparian vegetation, and agricultural fields (List and MacDonald 1998; Cartron et al. 2000). The short-grass prairie supports the largest complex of black-tailed prairie dog (*Cynomys ludovicianus*) towns remaining in North America (Ceballos et al. 1993; chapter 21). The area is also important for several wintering and nesting raptor species (Manzano-Fischer et al. 1999). Among them are the bald eagle and the golden eagle, 2 species listed in Mexico as Endangered and Threatened, respectively (SEMARNAT 2002), and the ferruginous hawk, which has declined recently in parts of its range and is now listed as Subject to

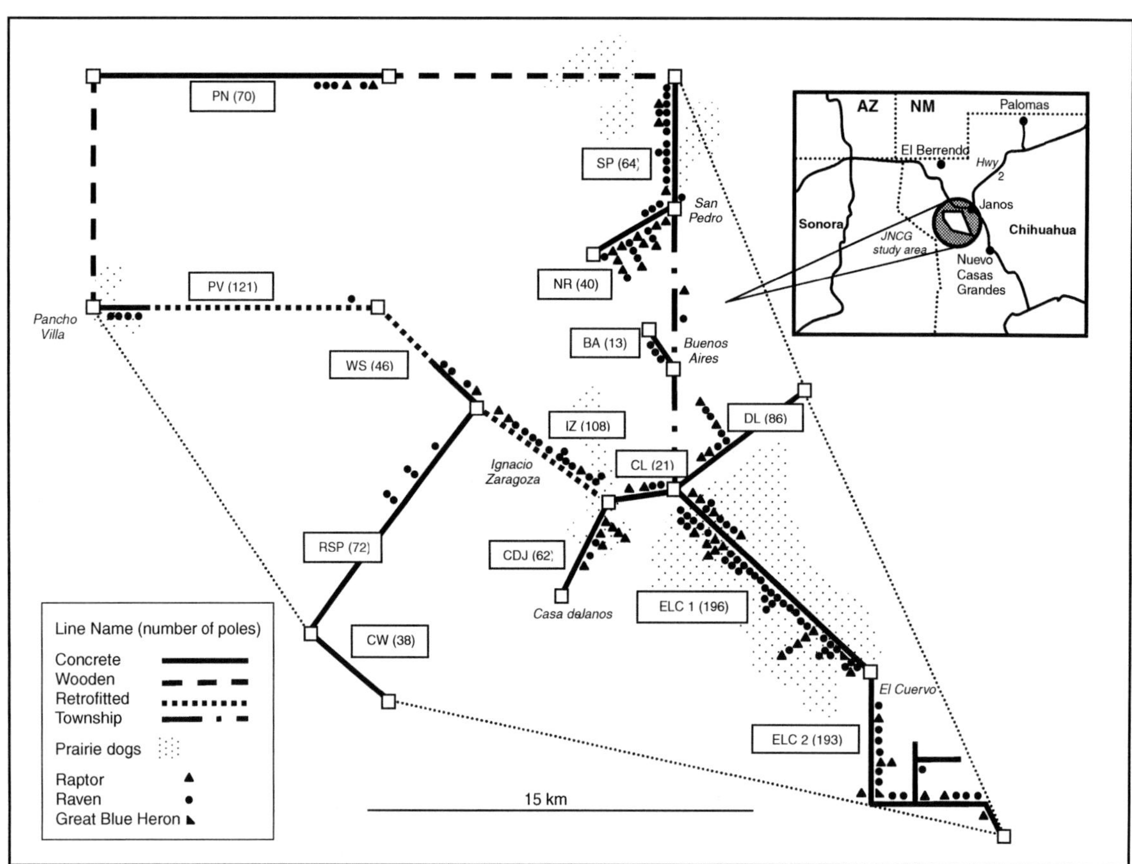

Figure 17.1. Map of the Janos–Nuevo Casas Grandes prairie-dog town complex and surrounding area. Note that 3 of the power lines were retrofitted in late spring. With the exception of 4 ravens, all casualties along those lines occurred before retrofitting.

Special Protection in Mexico (Bechard 1981; Houston and Bechard 1984; Schmutz 1984; Olendorff 1993; SEMARNAT 2002). According to CBC data and our own observations, the Chihuahuan raven (*Corvus cryptoleucus*) is one of the most common birds year-round in the JNCG complex area.

The area has also witnessed the development of ranches and rural communities of *ejidatarios* (peasants) and Mennonites. The first power lines to provide energy to local towns and private ranches were built in 1986. Over the next several years, the number of power lines with concrete poles increased steadily in the area, with some of these lines crossing prairie dog towns. The power line where the first electrocuted birds were discovered was built in October and November 1998. It connects the Mennonite colonies of Buenos Aires and El Cuervo and crosses the largest active prairie-dog town in North America.

Power Pole Configuration

The operating voltage of power lines in the JNCG area is 34.5kV (R. E. Harness unpubl. data). The most common type of pole, the three-phase tangent structure, consists of 2 energized conductors supported on a single steel crossarm with a third conductor on the pole top (fig. 17.2). The length of the crossarm is 2 m, and phase-to-phase (i.e. wire-to-wire) separation is approximately 1.5 m, as recommended for eagles (APLIC 1996). A variation of the typical configuration consists of all 3 energized conductors supported on a single steel crossarm. This results in greatly reduced phase-to-phase separation of approximately 0.8 m.

Tangent units are fitted with 2 types of insulators. The original lines were constructed with smaller pin-type insulators. In recent years, however, larger post-type insulators (fig. 17.2) have been used to increase the electrical separation from the wire to steel crossarm. Although this was done for lightning protection, it also affords additional protection for smaller birds.

Other types of concrete pole configurations occurring in the area consist of double deadend and deadend units. Both types of poles represent points along a power line where conducting wires terminate. Deadend units begin or terminate line sections. They are regularly used for branching of a primary line off the main circuit, with the last deadend unit on a lateral line tap usually including a three-phase transformer bank unit. Double deadend units occur where conductors from both directions terminate. They include an H-structure (i.e., 2 poles with a horizontal crossarm running between them) and angle poles (designed to accommodate directional changes in the distribution line), as well as double deadend units fitted with double crossarms (fig. 17.3). In the JNCG area, CFE uses noninsulated wires to connect transformers and to jumper between circuits on deadend and double deadend units. On double deadend units with double crossarms (hereafter referred to simply as double deadend units), exposed jumpers are routed under crossarms on the outer phases, but the center-phase jumper is allowed to go over the crossarms (fig. 17.3), a design that is typically safe on wooden poles where electrocutions are wire to wire. On a concrete double deadend pole, however, this same practice entails a much higher risk for perched birds. The increased risk may be compounded by the fact that double crossarms likely attract some bird species for their potential to support the construction of a nest.

Survey Methodology

The JNCG Area

From December 2000 to November 2001, we monitored 1146 poles in a 725-km^2 area centered on the JNCG prairie-dog town complex (fig. 17.1). Among these poles, 1130 (99%) were built with concrete and, at the onset of the study, were all fitted with steel crossarms. Sixteen double (H-structure) wooden poles with steel crossarms were interspersed with concrete units along the lines we surveyed and were included in our study. About 200 concrete poles also in the JNCG area were not formally included in our monitoring but were cursorily inspected from a moving car. They were located in the small towns of San Pedro, Ignacio Zaragoza, and Buenos Aires (fig. 17.1). Raptors and ravens seemed to avoid these areas, and no bird remains had been found there in earlier surveys (Cartron et al., unpubl data).

Among the poles we monitored, the same 297 (26%) poles along three power lines were inspected every month of the study period, while all others were searched less frequently. The average number of poles surveyed every month was 549 (SD = 189), and there was no significant difference in the number of poles surveyed per month among seasons (one-way ANOVA, *P* > .05). Along some power lines, bird remains were removed on November 25–26, 2000, 1 month before the beginning of the first

Figure 17.2. Tangent unit with post-type insulators and perched raven.

monthly survey. Elsewhere, based on the extent of rigor mortis, aspect of plumage, and carcass integrity, we estimated the age (< 1 week, < 1 month, > 1 month) of all remains newly encountered and recorded only those birds likely killed after November 2000. During our last survey in mid-November 2001, we inspected 1032 (91%) of the concrete poles in the JNCG area. Thus the results are for a period of approximately 1 year.

All surveys of power lines were conducted on foot or by car. Whenever possible, inspection routes between poles followed under the primary conductors to detect possible collisions or carcasses dragged away from poles by scavengers. The absence of dense vegetation through most of the year allowed easy detection of remains. Pole numbers and construction configurations were recorded. Bird remains were identified and examined for signs of electrocution. As dead birds were typically left in situ, remains > 1 week old were tallied only at poles with no mortality detected for the same species during the previous survey.

In late spring of 2001, CFE retrofitted about 210 concrete tangent units with 2.5-m-long wooden crossarms. The retrofitted poles were found along 3 power lines of the study area, including 1 of the power lines searched on a monthly basis (fig. 17.1).

Outside the JNCG Area

We conducted additional surveys along 7 power lines outside the JNCG area. However, none of these surveys was duplicated in time, and the total number of concrete poles we inspected was only 220. We also searched for obvious carcasses along power lines while traveling to and from the study area (i.e., along Highway 2 and the road to the border town of Palomas).

Results

Count of Electrocuted Birds, December 2000–November 2001

From December 2000 through November 2001, we found 178 sets of bird remains (i.e., whole carcasses, body parts, and clumps of feathers) in the JNCG prairie-dog town complex area. Among them, 72

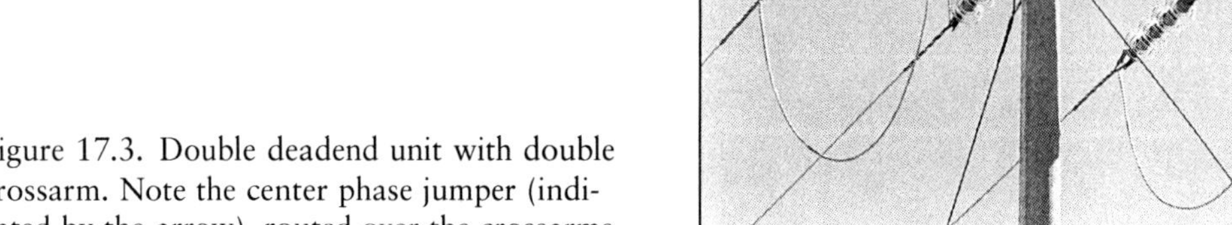

Figure 17.3. Double deadend unit with double crossarm. Note the center phase jumper (indicated by the arrow), routed over the crossarms.

(40%) showed definite signs of electrocution (i.e., singed feathers in nearly all cases; also detached legs). Remains that did not show any sign of electrocution typically consisted of only a few feathers or old, incomplete carcasses. With the exception of 2 dead birds found in Buenos Aires, all others were discovered during our formal surveys and amounted to an average of 1 bird killed every 6.5 concrete poles (1 dead bird was discovered at 1 of the double wooden poles). The incidence of mortality conclusively attributed to electrocution averaged 1 bird every 15.7 concrete poles.

The remains we found belonged to at least 11 different species (table 17.1). The Chihuahuan raven was the species most frequently found. There were 123 (69%) sets of raven remains found, among them 74 identified Chihuahuan ravens, but no common ravens (*C. corax*). The remaining 49 bird remains could not be identified conclusively to the species level, but based on the above finding, most were probably Chihuahuan ravens. Among all Chihuahuan raven remains, 41 (55%) showed singed feathers, but all remains without burn marks consisted of only clumps of feathers and bare bones.

Nine raptor species accounted for 52 (29%) of all remains. The red-tailed hawk (*Buteo jamaicensis*) was the second most frequently identified species, with a total of 27 (15%) dead birds, most (59%) of them immature individuals. Singed feathers were present on 17 (63%) of all red-tailed hawk remains. We also discovered 9 (5%) ferruginous hawks (2 adults, 3 immatures, 4 unknown) and 6 (3%) eagles. Among the eagles, only 3 (2 immatures) were identified as golden eagles. The other 3 eagles, not conclusively identified, could have been golden eagles or bald eagles. A juvenile golden eagle killed on a concrete power pole in the JNCG area on 25 November 2000, just before our baseline survey, was not included in our records of dead birds over the 1-year period. The remaining 6 raptor species and 2 great blue herons (*Ardea herodias*) amounted to 7% of all remains. Although recorded historically as electrocution casualties in the area (Cartron et al. 2000; C. Melcher, pers. comm.), bald eagles, American kestrels (*Falco sparverius*), and prairie falcons (*F. mexicanus*) were never found among remains during our monthly searches (table 17.1).

Outside the JNCG area, we discovered 12 remains including 6 (50%) red-tailed hawks (5 immatures and 1 adult) and 5 (42%) Chihuahuan ravens. The last set of remains belonged to an unidentified *Buteo* species. Seven (58%) of the remains showed signs of electrocution (i.e., singed feathers).

Table 17.1. Species identified among remains of birds killed on concrete power poles in northwestern Chihuahua, January 1999–November 2001.

Common Name	Scientific Name	Location	Signs of Electrocution
Great-blue heron[a]	*Ardea herodias*	Janos–Nuevo Casas Grandes complex area	None visible
Turkey vulture	*Cathartes aura*	Janos–Nuevo Casas Grandes complex area	Singed feathers
Osprey[a]	*Pandion haliaetus*	Janos–Nuevo Casas Grandes complex area	None visible
Bald eagle[a,b]	*Haliaeetus leucocephalus*	Janos–Nuevo Casas Grandes complex area	Detached legs
Swainson's hawk[a]	*Buteo swainsoni*	Janos–Nuevo Casas Grandes complex area	None visible
Red-tailed hawk	*Buteo jamaicensis*	El Berrendo	Singed feathers
		Highway 2	Singed feathers
		Janos–Nuevo Casas Grandes complex area	Singed feathers
		Road to Palomas	None visible
Ferruginous hawk	*Buteo regalis*	Janos–Nuevo Casas Grandes complex area	Singed feathers, detached legs
Golden eagle	*Aquila chrysaetos*	Janos–Nuevo Casas Grandes complex area	Detached legs
American kestrel[a,b]	*Falco sparverius*	Janos–Nuevo Casas Grandes complex area	Singed feathers
Prairie falcon[a,b]	*Falco mexicanus*	Janos–Nuevo Casas Grandes complex area	Singed feathers
Barn owl[a]	*Tyto alba*	Janos–Nuevo Casas Grandes complex area	None visible
Great-horned owl[a]	*Bubo virginianus*	Janos–Nuevo Casas Grandes complex area	Singed feathers
Short-eared owl[a]	*Asio flammeus*	Janos–Nuevo Casas Grandes complex area	Singed feathers
Chihuahuan raven	*Corvus cryptoleucus*	Highway 2	Singed feathers
		Janos–Nuevo Casas Grandes complex area	Singed feathers
		Road to Palomas	Singed feathers

[a]Less than four remains identified.
[b]Discovered only during earlier surveys (data from Cartron et al. 2000; C. Melcher pers. comm.).

Rate of Remains Disappearance

Eighteen (25%) of 72 raven remains monitored over time in the JNCG area disappeared completely during the first month after they were first discovered. By the following month, only 49% of all raven remains could still be observed. Twenty (95%) of 21 nonraven remains were still visible 1 month after being found. After 2 months, however, the proportion of remains still in place dropped to 63%.

Patterns of Variation in Observed Bird Mortality

Seasonal Variation in Observed Mortality

We found 105 (59%) of all observed bird remains in the study area in the last 4 months of surveys (fig. 17.4). Among these remains, 93 (89%) were either ravens (n = 69) or red-tailed hawks (n = 24). The number of raven remains we detected surged in August and September, whereas a sharp increase in the observed number of red-tailed hawk remains occurred in the fall (i.e., September–November; fig. 17.4). Where surveys had been conducted the month before along the same power line, or in the case of very recent carcasses, we were able to translate time of detection into approximate time of death. Thus, 38 (78%) of the 49 raven remains found in August–September were < 1 month old. Similarly, 19 (79%) of the 24 red-tailed hawks carcasses found in the fall resulted from deaths that had occurred less than 1 month before. Even so, because some power lines were surveyed less regularly than others, there is a potential bias associated with the use of survey results for estimating seasonal variation in mortality. However, disproportionately high mortality was detected from August–November in particular along the 2 non-retrofitted power lines we monitored monthly (fig. 17.1). Along the first power line (i.e., ELC1), no dead red-tailed hawk was ever found, but 24 raven remains discovered in August and September accounted for 55% and 41% of the annual mortality detected along that line for ravens and for all species combined, respectively. Along the second power line (CDJ), only 9 dead birds were found from December 2000–November 2001. Among them, 5 (4 red-tailed hawks and 1 raven) were found during the last 4 months. Dead red-tailed hawks and ravens found during that time represented 100% and 50% of the total number of

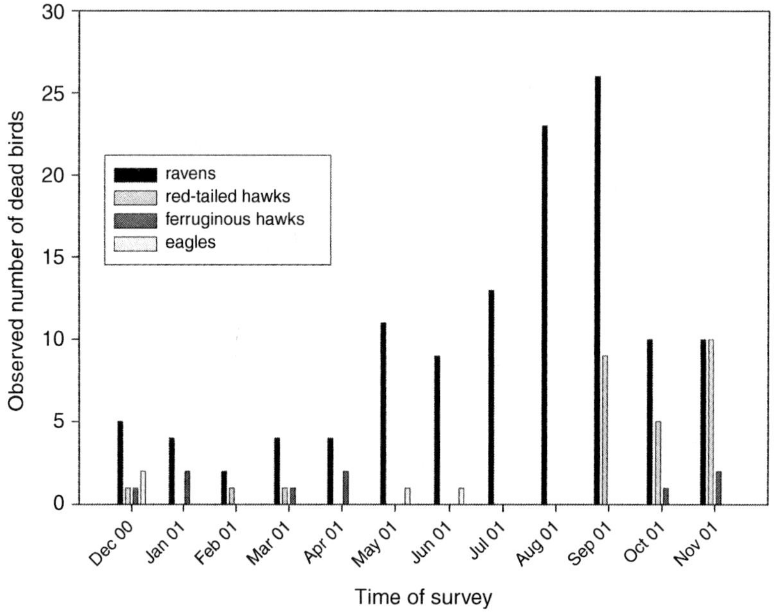

Figure 17.4. Observed number of dead birds from December 2000 to November 2001 in the Janos–Nuevo Casas Grandes prairie-dog town complex and surrounding area.

remains observed along the second power line for these species, respectively. Mortality also appeared to be disproportionately high in the fall for red-tailed hawks outside the JNCG area. Three (50%) of the 6 red-tailed hawk carcasses found outside the JNCG area were found in November and looked recent.

Ferruginous hawk remains were found only in small numbers but were observed during most months from October 2000 to April 2001. Most eagle deaths occurred in the winter, but 1 golden eagle was killed in late spring.

Differences in Observed Mortality among Power Lines

Along power line ELC1, we found 58 bird remains, or an observed average of 1 bird killed annually every 3.4 concrete poles. Along power line CDJ, annual mortality amounted to 1 dead bird every 4.9 concrete poles. The highest incidence of bird electrocution in the JNCG area appeared to be along a 40-pole power line just south and west of San Pedro, in shrubby vegetation dominated by *Ephedra* and characterized by the absence of prairie dogs. Twenty dead birds (i.e., 11 ravens, 7 red-tailed hawks, 1 unidentified eagle, and 1 osprey [*Pandion haliaetus*]) were recorded, of which 12 (60%) showed singed feathers. The observed incidence of mortality (i.e., 1 bird killed for every 2 poles) was only for a period of about 6 months: the power line was surveyed for the first time in July, and most (65%) of the dead birds were recorded in subsequent surveys. Conversely, some power lines in the area yielded a small number of dead birds. We did not detect any dead birds along a 38-pole power line that was surveyed twice (i.e., in April and November).

Outside the JNCG area, our one-time surveys of 220 poles yielded 5 remains, or an average of 1 dead bird for every 44 poles.

Observed Bird Mortality as a Function of Pole Configuration

We recorded 175 dead birds during formal surveys in the JNCG area at 134 concrete poles, with 1 additional bird found at the double wooden pole already mentioned. Most poles (73%) associated with bird mortality were tangent structures. However, dead birds were also found under three-phase transformer bank, H-structure, tap, angle, and double deadend units. The number of dead birds observed at double deadend units in particular was disproportionately high, given the small number of poles with this configuration (fig. 17.5). Along 2 power lines in particular, 1 in grassland with prairie dogs (n = 196 poles), and the other in shrubland (n = 40 poles), the incidence of electrocutions was 3.2 and 7.5 times higher for double deadend versus tangent poles. Overall, the observed average mortality per pole, compared along all 6 power lines with observed dead birds and both types of poles, was significantly higher for double deadend units than for tangent units (one-way ANOVA, F = 4.99, df = 1, P < .05). Thirteen (39%) of the 33 double deadend units monitored were associated with bird mortality. By comparison, we observed bird remains at 81 (13%) of the 631 non-retrofitted tangent poles in the study area. Three (25%) of the 12 dead birds observed outside the JNCG area were beneath double deadend units.

Among the 210 tangent poles that were retrofitted, only 3 (1%) yielded a total of 4 dead birds during monthly surveys. Three (75%) of these remains were observed during the July surveys, after heavy rains in the JNCG area. Surveys of 60 additional retrofitted concrete tangent poles in November 2001 did not reveal any bird remains.

Impact of Concrete Poles on a Local and Regional Scale

Electrocutions due to electric utility structures are an important cause of mortality among raptors in several parts of the world (e.g., Markus 1972; Haas 1980; Ledger and Annegarn 1981; Ferrer and Hiraldo 1991; LaRoe et al. 1995; Harness and Wilson 2001). The results of our surveys indicate that in the JNCG area, many raptors and also Chihuahuan ravens are electrocuted on concrete power poles. For at least the red-tailed hawk and the Chihuahuan raven, electrocutions on concrete power poles may represent a major cause of mortality beyond the limits of the JNCG area and through much of their distributions in northern Mexico.

The actual incidence of mortality on concrete poles in the JNCG area is likely higher than we documented from counts of remains. Our monthly survey coverage was <50% of the concrete poles in the area. Remains, especially those of ravens, vanished over time, and the observed rate of disappearance of remains does not reflect dead birds dragged

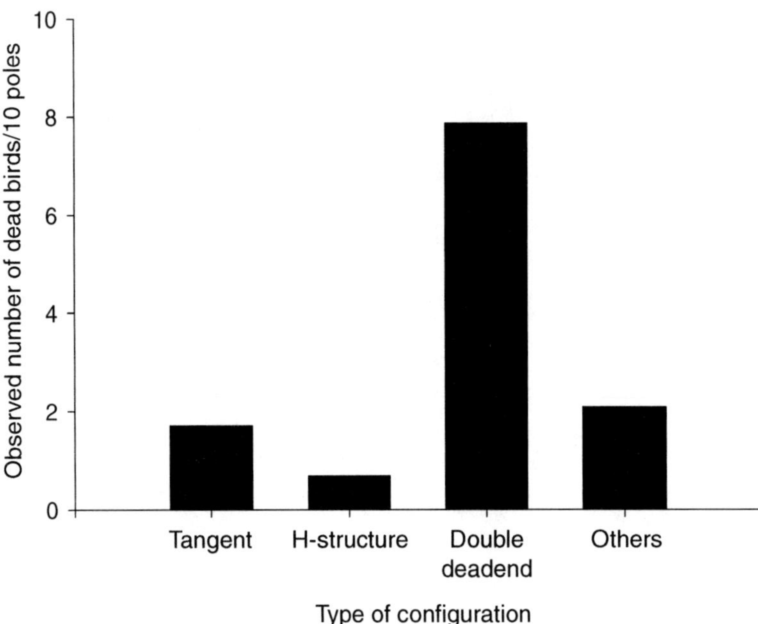

Figure 17.5. Observed number of dead birds per 10 poles as a function of pole configuration.

away by scavengers before they could be found. In several cases, we found single feathers under poles, which were not included in our tally of dead birds. One injured immature red-tailed hawk observed at the base of one pole in November 2001 disappeared without a trace during the same day.

The presence of dead birds under power poles often indicates the occurrence of electrocution, but death may also result from shooting or poisoning (Harness 1998). Shooting was not suspected as a major source of bird mortality in the JNCG area. Guns are tightly regulated in Mexico, and none of the remains we discovered exhibited feather shearing, a frequent finding on the carcasses of birds that have been shot (Olson 1999). The JNCG area does have a history of prairie dog poisoning (Ceballos et al. 1993). During our surveys, we also found evidence that some farmers used carbofuran (Furadan). This insecticide is known especially for the high toxicity of its granular form, but liquid carbofuran, the form presumably used in the JNCG area, can also be responsible for mortality of wildlife, including raptors (Allen et al. 1996).

Despite possible shooting and poisoning in the JNCG area, it is likely that electrocutions were by far the most common cause of mortality among those birds killed on concrete poles. Singed wing feathers were found on all (100%) complete Chihuahuan raven carcasses. A few fresh raptor carcasses did not exhibit burn marks, but bird electrocutions can be difficult to confirm without a necropsy (Harness 1998). Wildlife rehabilitators report that raptors surviving electrocutions may not display surface tissue damage (Yearout 1991).

Among all the poles present in the area, double deadend units with double crossarms were associated with the highest rate of mortality. Harness and Wilson (2001) pointed out that observed differences in mortality among pole-structure types can be biased by the nonrandom distribution of pole types across habitats. Raptors are often more strongly attracted to power poles in areas of low vegetation and/or prey concentrations (e.g., Benson 1981; Olendorff et al. 1981). Thus, a type of pole may be associated with higher observed mortality only because it occurs predominantly in areas of higher-than-average pole use. In our study, however, double deadend units were associated with a disproportionately high number of dead birds along several power lines that occurred in homogeneous habitat. Higher mortality at those poles thus most likely reflects an increased risk of electrocution.

The mortality data provided above pertain to particular types of construction units with all the necessary jumpering, mounting hardware, and protective devices. The data provide no specific information as to the actual energized component that may have been contacted by the bird. With the above cautionary note, the energized hardware contacted by ravens and raptors on double deadend units is probably the center phase jumper, often providing minimum clearance over the double steel crossarms (fig. 17.3).

The risk of electrocution on concrete poles appears to vary among species as a function of body size. Although American kestrels frequently perch on grounded crossarms in the JNCG area, they appear to be rarely electrocuted. Conversely, there are indications that for large raptors, the risk of electrocution is very high. During the first 6 monthly surveys combined, we observed only 4 hawks perched on concrete poles. In 2 of these 4 cases, during the next survey we found a carcass of the same species within three poles of the original sighting.

Rates and patterns of mortality on concrete poles seem also strongly conditioned by life cycle, including migration patterns. Our survey results suggest that most of the dead red-tailed hawks we found were fall migrants on their way south. Raven mortality increased at double deadend units at the beginning of the nesting season (April–May), suggesting that pairs were selecting these poles as platforms to support nests. Around fledging time, raven carcasses were typically found near nests active during the breeding season. Many of these carcasses were likely those of electrocuted fledglings. In the late summer, we observed many of the raven casualties in proximity to large flocks of this species (Cartron et al., unpubl. data).

The JNCG area is important for golden eagles and wintering ferruginous hawks (Manzano-Fischer et al. 1999). A stable-isotope analysis of feathers from golden eagle carcasses retrieved in February and March 2000 indicated that both winter and year-round residents occur in the JNCG area (Cartron and Kelly, unpubl. data). Among the dead ferruginous hawks we found an immature bird in December 2000 that had been banded the previous summer in Idaho. On a local scale, electrocutions by concrete power poles likely represent a substantial cause of mortality in both species. Including results from Cartron et al.'s (2000) surveys, 19 and 18 confirmed or suspected electrocution cases of ferruginous hawks and golden eagles,

respectively, have now been documented in the JNCG area.

Electrocutions by electric utility structures constitute the second greatest cause of mortality in golden eagles in North America north of Mexico (LaRoe et al. 1995). For example, Harness and Wilson (2001) reported 272 confirmed golden eagle electrocutions in the western United States from 1986 to 1996. By comparison, the number we report here for this species is low. However, power lines in northern Mexico have been surveyed for less than 2 years and in only a very small fraction of the species' regional distribution. In arid regions the incidence of raptor electrocutions is compounded by the scarcity of natural, high-above-ground perches (Benson 1981), and much of northern Mexico is characterized by low vegetation, although much of the region is also without power lines or has only wooden poles (J.-L. Cartron, pers. obs.). Our findings also have added significance because golden eagle numbers in Mexico are presumed to be lower than in the United States and Canada (G. Ceballos, pers. comm.).

Although power lines are generally not listed as a major cause of mortality among ferruginous hawks (see Bechard and Schmutz 1995), several confirmed cases of mortality by electrocution have been documented in this species in the western United States (Harness and Wilson 2001). Harmata et al. (2001) reported that 3 (20%) of 15 ferruginous hawks banded in Montana and later recovered were found dead in northern Mexico. Although the cause of mortality remained unexplained, the authors indicated the likelihood of mortality by collision with power lines or electrocutions in some areas of Mexico. In 1992, the U.S. Fish and Wildlife Service denied a petition to list the ferruginous hawk under the Endangered Species Act. At the time, however, there was no information on mortality of ferruginous hawks wintering in Mexico (Harmata et al. 2001).

Red-tailed hawks in the western United States die fairly frequently as a result of interactions with power lines (e.g., Harness and Wilson 2001). In contrast, there has been no published study on electrocutions among Chihuahuan ravens (Bednarz and Raitt 2002). The discovery of electrocuted red-tailed hawks and Chihuahuan ravens outside the JNCG area is important. Concrete poles are not just a localized threat in an area of unusually high prey densities attracting raptors. They may affect the population status of both the red-tailed hawk and the Chihuahuan raven regionwide or even on a larger scale.

The results of our surveys strongly suggest that every year, Chihuahuan raven electrocutions in Mexico range in the hundreds if not in the thousands. Estimates of mortality by electrocution in the red-tailed hawk are even more tenuous, but the possible impact of concrete poles is also great. This species is a year-round resident across all of northern Mexico (Preston and Beane 1993). Northern Mexico also receives an influx of migrants during the spring and fall, and our surveys suggest that mortality by electrocution on concrete poles occurs largely during fall migration. Although little information is available on migration routes in Mexico, migration occurs along a broad front in the United States (Preston and Beane 1993). Use of power poles by red-tailed hawks is probably high in all electrified areas with low vegetation. Based on area size alone, 27 red-tailed hawk electrocutions within the 725-km^2 JNCG area indicate the potential for as many as 13,407 red-tailed hawk electrocutions in an area the size of the Mexican Chihuahuan Desert (total surface area of the Chihuahuan Desert from Morafka 1977). Again, however, the JNCG area is not representative of the whole region. There are many large tracts of land free of development; other electrified areas only have wooden poles. Thus the figure above is almost certainly much too high. More surveys are needed on a regional scale.

Retrofitting of Poles and New Power Line Construction Practices

Currently in Mexico, concrete is preferred over wood for manufacturing power poles. Concrete is cheaper and, away from forested areas, also more readily available. Concrete poles are easier to install, as they do not need a large foundation. They also require less maintenance.

Conductive poles are commonly used in distribution line construction in Europe and other parts of the world, and they are associated with higher bird electrocution rates than nonconductive poles (Janss and Ferrer 1999). Concrete poles in Mexico pose a serious threat to raptors and ravens chiefly because they are fitted with steel crossarms. Depending on pole configuration, the risk of electrocution may be further compounded by insufficient phase-to-crossarm separation and/or the addition of non-insulated, conductive hardware (e.g., jumper wires). The recommendations we make here for new con-struction practices and for retrofitting are aimed at correcting at least 1 of the above factors. Our most recent traveling through the state of Chihuahua indicates that many concrete poles have already been retrofitted with wooden crossarms, perch guards, and polyvinyl chloride (PVC).

Retrofitting of double deadend poles should be considered a priority throughout northern Mexico, due to the seemingly very high incidence of electro-cution on these poles. As mentioned earlier, the placement of the 2 outer-phase jumpers under the crossarms is not sufficient to prevent electrocutions. The center-phase jumper, which is routed above the arm, should consist of insulated wire (600-V class insulation) with an insulated covering of Salisbury SALCOR to accommodate raptors and ravens.

At least in areas of high raptor or raven population numbers, three-phase tangent poles are another priority. In the JNCG area, they are associated with a high incidence of electrocution per pole, but also, due to the preponderance of these units along power lines, collectively they are also responsible for a large number of bird electrocutions. Neither pin-type nor post-type insulators appear adequate to protect ravens and larger raptors on three-phase tangent poles. The highest observed rate of mortality observed in this study was along a line where pin-type insulators were used. However, despite being fitted with post-type insulators, concrete tangent poles along the power line crossing the main prairie-dog town also exhibited a high rate of mortality, especially for ravens. These birds are smaller than the phase-to-crossarm separation. We assume that they are being electrocuted when they simply come close to the phase wire while perching on a grounded crossarm (i.e., through arcing).

In rural areas, three-phase tangent structures should continue to be framed to provide 1.5 m of phase-to-phase separation as recommended by APLIC (1996). The most serious problem is the use of con-ductive metal crossarms. A simple solution is to discontinue their use and replace them with non-conductive fiberglass crossarms. If metal crossarms continue to be used, the new lines can be made rap-tor friendly by adding insulating paint on the arms. An alternative solution is to suspend the outer con-ductors under the crossarms (instead of routing them over the crossarms). A pole-top cap must be used to discourage perching.

CFE has begun to retrofit tangent poles using wooden crossarms, PVC, or perch guards. Earlier attempts to curb electrocution rates consisted of using PVC around wires (this material is currently used around crossarms in Sonora; Cartron and Rogers pers. obs.). Although PVC is inexpensive, it deteriorates due to ultraviolet radiation and becomes brittle. Over time the retrofits begin to crack and fall off the line, requiring reapplication. Alternative materials with better UV-resistant properties such as "electrical" PVC should be investigated. CFE retrofit viewed by Harness (pers. obs.) in June 2000 included a PVC cap on the insulator. The cap was installed by simply compressing it over the tie wire attaching the conductor to the insulator. Approximately a month later this structure was reinspected, and the cap was found at the base of the pole (P. Manzano-Fischer, pers. obs.). Wooden crossarms have been used by CFE in the JNCG area and should drastically reduce the incidence of electrocution events, though they evidently do not provide maximum protection.

The JNCG area contains several irrigation loads that are inactive at least part of the year. A lateral power tap serving a well typically consists of a three-phase deadend tap structure, followed by several tangent structures, and terminating on a pole with a three-phase transformer bank. Any of these structures may be lethal to perching birds even when the well is idle. Among all the remains of birds observed in the JNCG area since the first formal power line surveys, 4 dead golden eagles and 2 dead ferruginous hawks were discovered during winter months along power taps serving wells. It may be feasible to simply open the cutouts (i.e., devices used to make a fused connection between primary conductors) at these taps when irrigation is not required. This simple solution may reduce overall raptor mortality, especially during the winter when no irrigation is required, and also decrease transformer losses at inactive services.

Recommendations for Future Research

In view of the massive retrofitting effort to be undertaken in northern Mexico, priority should be given to power lines known or suspected of causing very high bird mortality. Surveys are needed throughout the region, especially where large raptor or raven populations may exist, or in areas with a high incidence of power outages.

Our monthly surveys suggest that concrete poles retrofitted with wooden crossarms are associated with bird mortality almost exclusively during heavy rains, when wood becomes more conductive and the isolating properties of feathers are reduced. However, more surveys are needed to confirm that lower incidence of electrocutions. Likewise, the effectiveness (and durability) of PVC and perch guards used by CFE should be evaluated.

Nonwood poles are becoming more common not just in Mexico, but also in the United States, and they pose new risks to raptors (Harness 1998). Unfortunately, the high rate of bird electrocutions associated with these poles has been poorly documented in North America. In Mexico, mitigating the electrocution of birds at a large scale will likely require the long-term commitment of CFE, with collaboration from conservation organizations and wildlife scientists to monitor power lines.

In March 2002, a meeting was organized in Mexico City to address the issue of bird electrocutions in Mexico. The meeting was attended by representatives from Mexican government agencies, academic institutions, conservation organizations, and private consultants. Also present at this meeting were 25 engineers representing CFE from different states of the country. Two of us (Cartron and Manzano-Fischer) presented the results of our research. The meeting confirmed the lack of public information on bird electrocutions in Mexico, as no other data were presented. Legal aspects of the issue and viable solutions were among the topics discussed. A committee has been formed to establish communication channels among all the stakeholders and to ensure proper follow-up on the resolutions adopted at the meeting.

Acknowledgments

The research presented here was the result of collaboration with the Instituto de Ecología at UNAM. We thank S. Stoleson and G. Sadoti for help conducting the last set of surveys in November 2001. We are also grateful to C. Melcher for information on the earliest records of electrocuted birds in the JNCG area and to the Comisión Federal de Electricidád (CFE) for providing us with a history of power line construction and design. R. Dickerman and the Museum of Southwestern Biology assisted in species identification of some of the remains. Marco Restani, Eduardo Iñigo, Gerardo Ceballos, and

Richard Felger all provided helpful comments on an earlier draft of the chapter.

Literature Cited

Allen, G. T., J. K. Veatch, R. K. Stroud, C. G. Vendel, R. H. Poppenga, L. Thompson, J. A. Shafer, and E. Braselton. 1996. Winter poisoning of coyotes and raptors with Furadan-laced carcass baits. Journal of Wildlife Diseases 32: 385–389.

APLIC (Avian Power Line Interaction Committee). 1996. Suggested Practices for Raptor Protection on Power Lines: The State of the Art in 1996. Edison Electric Institute and Raptor Research Foundation, Washington, D.C.

Bechard, M. J. 1981. Historical nest records for the ferruginous hawk in Manitoba. Canadian Field-Naturalist 95: 467–469.

Bechard, M. J., and J. K. Schmutz. 1995. Ferruginous hawk (*Buteo regalis*). In A. Poole and F. Gill (eds.), The Birds of North America, no. 172. Academy of Natural Sciences, Philadelphia, and American Ornithologists' Union, Washington, D.C.

Bednarz, J. C., and R. Raitt. 2002. Chihuahuan raven (*Corvus cryptoleucus*). In A. Poole and F. Gill (eds.), The Birds of North America, no. 606. Academy of Natural Sciences, Philadelphia, and American Ornithologists' Union, Washington, D.C.

Benson, P. C. 1981. Large raptor electrocution and power pole utilization: a study in six western states. Ph.D. dissertation, Brigham Young University, Provo, Utah.

Cartron, J.-L. E., G. L. Garber, C. Findley, C. Rustay, R. Kellermueller, M. P. Day, P. Manzano-Fischer, and S. H. Stoleson. 2000. Power pole casualties among raptors and ravens in northwestern Chihuahua, Mexico. Western Birds 31: 255–257.

Ceballos, G., E. Mellink, and R. L. Hanebury. 1993. Distribution and conservation status of prairie dog Cynomys mexicanus and Cynomys ludovicianus in Mexico. Biological Conservation 63: 105–112.

Ferrer, M., and F. Hiraldo. 1991. Evaluation of management techniques for the Spanish imperial eagle. Wildlife Society Bulletin 19: 436–442.

Haas, D. 1980. Endangerment of our large birds by electrocution—a documentation. Okolgie der Vogel 2: 7–57.

Harmata, A. R., M. Restani, G. J. Montopoli, J. R. Zelenak, J. T. Ensign, and P. J. Harmata. 2001. Movements and mortality of ferruginous hawks banded in Montana. Journal of Field Ornithology 72: 389–398.

Harness, R E. 1998. Steel distribution poles—environmental implications. Pp. D1.1–D1.5 in Proceedings of the 1998 Rural Electric Power Conference. Institute of Electrical and Electronics Engineers, New York.

Harness, R. E., and K. R. Wilson. 2001. Electric-utility structures associated with raptor electrocutions in rural areas. Wildlife Society Bulletin 29: 612–623.

Houston, C. S., and M. J. Bechard. 1984. Decline of the ferruginous hawk in Saskatchewan. American Birds 38: 166–170.

Janss, G. F. E., and M. Ferrer. 1999. Mitigation of raptor electrocution on steel poles. Wildlife Society Bulletin 27: 263–273.

LaRoe, E. T., G. S. Farris, C. E. Puckett, P. D. Doran, and M. J. Mac (eds.). 1995. Our Living Resources: A Report to the Nation on the Distribution, Abundance and Health of U.S. Plants, Animals and Ecosystems. National Biological Service, U.S. Department of Interior, Washington, D.C.

Ledger, J. A., and H. J. Annegarn. 1981. Electrocution hazards to the Cape vulture Gyps coprotheres in South Africa. Biological Conservation 20: 15–24.

List, R., and D. W. MacDonald. 1998. Carnivora and their larger mammalian prey: species inventory and abundance in the Janos-Nuevo Casas Grandes prairie dog complex, Chihuahua. Revista Mexicana de Mastozoologia 3: 95–112.

Manzano-Fischer, P., R. List, and G. Ceballos. 1999. Grassland birds in prairie-dog towns in northwestern Chihuahua, Mexico. Studies in Avian Biology 19: 263–271.

Markus, M. B. 1972. Mortality of vultures caused by electrocution. Nature 238: 228.

Medellín-Leal, F. 1982. The Chihuahuan Desert. Pp. 321–372 in G. L. Bender (ed.), Reference Handbook on the Deserts of North America. Greenwood Press, Westport, Connecticut.

Morafka, D. J. 1977. A Biogeographical Analysis of the Chihuahuan Desert. W. Junk, The Hague, Netherlands.

Olendorff, R. R. 1993. Status, Biology, and Management of Ferruginous Hawks: A Review. Raptor Research and Technological Assistance Center Special Report. Bureau of Land Management, U.S. Department of Interior, Boise, Idaho.

Olendorff, R. R., A. D. Miller, and R. N. Lehman. 1981. Suggested practices for raptor protection on power lines: the state of the art in 1981. Raptor Research Report no. 4, Raptor Research Foundation, Inc., St. Paul, Minnesota.

Olson, C. V. 1999. Human-related causes of raptor mortality in western Montana: things are not always as they seem. Pp. 299–322 in R. G. Carlton (ed.), Avian Interactions with Utility and Communication Structures.

Product ID 1005180. Electric Power Research Institute, Palo Alto, California.

Preston, C. R., and R. D. Beane. 1993. Red-tailed hawk (*Buteo jamaicensis*). *In*: A. Poole and F. Gill (eds.), The Birds of North America, no. 52. Academy of Natural Sciences, Philadelphia and American Ornithologists' Union, Washington, D.C.

Schmutz, J. K. 1984. Ferruginous hawk and Swainson's hawk abundance and distribution in relation to land use in southeastern Alberta. Journal of Wildlife Management 48: 1180–1187.

SEMARNAT. 2002. Norma Oficial Mexicana NOM-059-ECOL-2001, Protección ambiental-Especies nativas de México de flora y fauna silvestres-Categorías de riesgo y especificaciones para su inclusión, exclusión o cambio-Lista de especies en riesgo. Diario Ofical de la Federación 582: 1–80.

Yearout, D. R. 1991. Electrocution and high wire accidents in birds. Pp. 43–54 *in* Proceedings of the 1990 International Wildlife Rehabilitation Council Conference. International Wildlife Rehabilitation Council, Oakland, California.

18

Baja California's Enduring Mediterranean Vegetation: Early Accounts, Human Impacts, and Conservation Status

RICHARD A. MINNICH

ERNESTO FRANCO-VIZCAÍNO

A unique and enduring traditional land-use system still exists in northern Baja California's Mediterranean grasslands, shrublands, oak woodlands, and conifer forests. This portion of the Californian floristic province has a pristine character rarely seen in Alta California, as the rural landscape remains little altered from the late eighteenth century, when Europeans first described it. Until recently, most of Baja California's biological environment was not as intensively exploited as Alta California's, despite the lack of effective formal protection for its wildlands. The region had experienced only a few brief gold-mining strikes. As late as 1880, agriculture essentially did not exist except around the Dominican missions. The only viable land-use was transhumance open-range cattle grazing. To this day, deliberate burning is still practiced by *vaqueros* and farmers, and wildland fires are largely uncontrolled.

To evaluate the effects of traditional land use and uncontrolled fire in northern Baja California, Minnich and Franco-Vizcaíno (1998) examined the diaries of the expeditions of Link, Crespi, Serra, Longinos-Martínez, and Arrillaga, written between 1766 and 1796. These diaries provide invaluable baseline information on the region's vegetation. Because of a mandate from the viceroy of Mexico to justify the construction of missions, the Spanish explorers were required to take daily observations of vegetation in expeditions that traversed some 1500 km throughout northern Baja California. Minnich and Franco-Vizcaíno (1998) also mapped the modern vegetation of northern Baja California using recent aerial photographs and compared it site-specifically with the Spanish accounts. They concluded that overall, the broadscale distribution, local patterning, and species composition recorded in the late-eighteenth century are still consistent with those of modern plant communities. A notable exception is the region's coastal sage scrub, significantly altered by urban and agricultural development, domestic livestock, and the spread of exotic annuals.

Based largely on the report of Minnich and Franco-Vizcaíno (1998), we summarize the extent of vegetation change in northwestern Baja California since the second half of the eighteenth century. We then review the regional history of traditional land uses, in particular grazing of domestic livestock, logging, and the lack of fire control. Additionally, we discuss the timing and impacts of the successive waves of invasions by exotic plants. In combination with grazing by domestic livestock, these have resulted in the widespread replacement by exotic annual grassland of coastal sage scrub and associated fields of native wildflowers. Finally, we provide an overview of current conservation efforts by Baja Californian governmental and nongovernmental agencies.

Climate and Physiography

Northwestern Baja California is a rugged landscape associated with 3 north–south mountain chains of the Peninsular Ranges (fig. 18.1). The Sierra Juárez comprises a discontinuous coastal chain of dissected mountains (peak elevations, 1200–1500 m) and an inland undissected tilted plateau (1600–2000 m). To the south is the Sierra San Pedro Mártir, a high mountain plateau with peaks reaching 2800–3000 m. Between these mountain chains are broad alluvial valleys and plateaus. Small coastal plains are found at Tijuana, Ensenada, and San Quintín.

The climate is Mediterranean, with winter frontal rains and summer drought (figs. 18.2 and 18.3). The mean annual precipitation ranges from 20 to 35 cm along the coast to 40 cm in the coastal Sierra Juárez, 40–50 cm in the interior Sierra Juárez, and 50–70 cm in the Sierra San Pedro Mártir (Minnich et al. 2000b). Snowfall occurs above 1600 m in the mountains. Mean winter temperatures decrease from 12–14°C in the coastal valleys to 0°C at 2200 m in the Sierra San Pedro Mártir. In summer, the North American monsoon causes afternoon thunderstorms over the inland sierras. Average summer rainfall (July–September) is locally as great as 5–10 cm along the crest of the Sierras Juárez and San Pedro Mártir, but is less than 1 cm along the Pacific coast (Minnich et al. 1993). Maximum temperatures near the coast average 20°C due to onshore flow of marine air from the upwelling Pacific Ocean, but land heating results in temperatures increasing to 30–40°C in the inland valleys and mountain uplands.

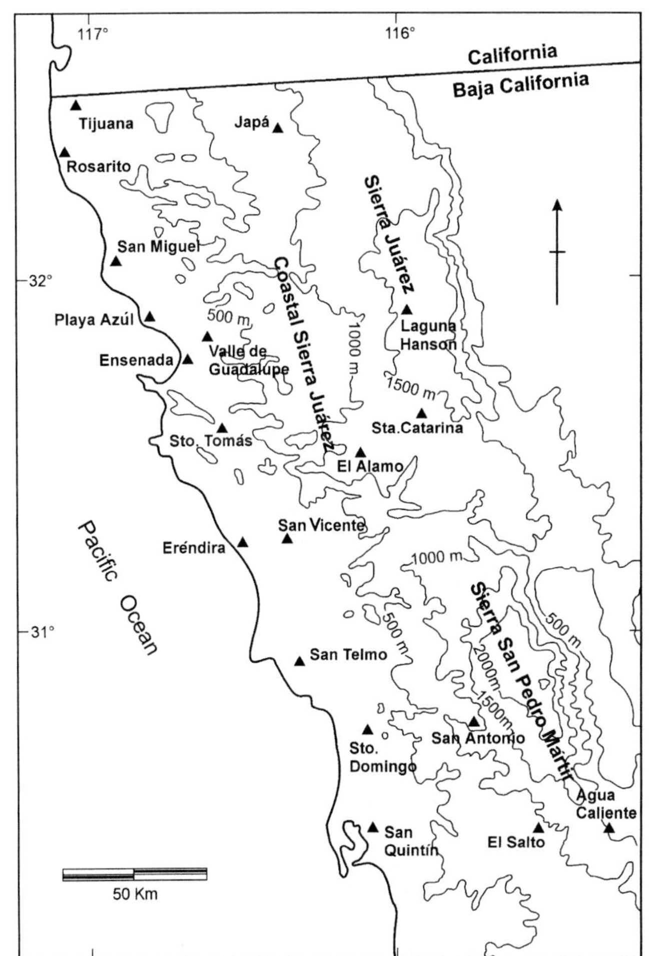

Figure 18.1. Topographic map and place names of northwestern Baja California.

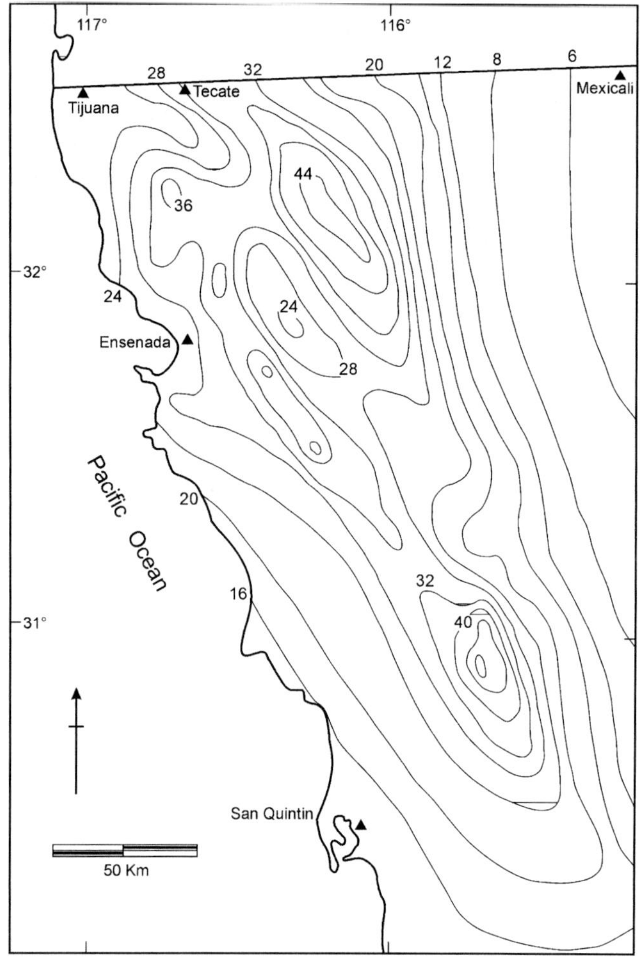

Figure 18.2. Mean annual precipitation (centimeters).

Vegetation

Plant communities show a broad altitudinal zonation similar to that in southern California (Minnich 1987; Hanes 1988; Thorne 1988; Passini et al. 1989; Peinado et al. 1994, 1995a,b). The description that follows and the vegetation map in figure 18.4 are generalized from detailed maps in Minnich and Franco-Vizcaíno (1998) and Minnich (2001). The northern coastal valleys are covered by exotic annual grasslands dominated by red brome (*Bromus rubens*), ripgut brome (*B. diandrus*), slender wild oat (*Avena barbata*), filaree (*Erodium cicutarium*), short-podded mustard (*Brassica geniculata* [= *Hirshfeldia incana*]), black mustard (*B. nigra*), and a few native herbs such as tarweed (*Hemizonia* spp).

The western foothills of the Sierras Juárez and San Pedro Mártir below 1000 m are covered by coastal sage scrub, or *matorral costero*, which consists of dense stands of drought-deciduous subshrubs 0.5–1.5 m tall mixed with woody deciduous shrubs and a few succulents. Important species include such subshrubs as California sagebrush (*Artemisia californica*), California buckwheat (*Eriogonum fasciculatum*), white sage (*Salvia apiana*), black sage (*Salvia mellifera*), and coastal brittle bush (*Encelia californica*). Maritime desertscrub, the southernmost phase of the coastal sage scrub, grows extensively in the lower foothills west of the Sierra San Pedro Mártir and northward along the coast to Eréndira. This plant community consists of a nearly continuous cover of subshrubs, but it is much richer in succulent taxa than coastal sage scrub farther north (Mooney 1988; Peinado et al. 1995b). Common species include members of coastal sage scrub as well as burbush (*Ambrosia*

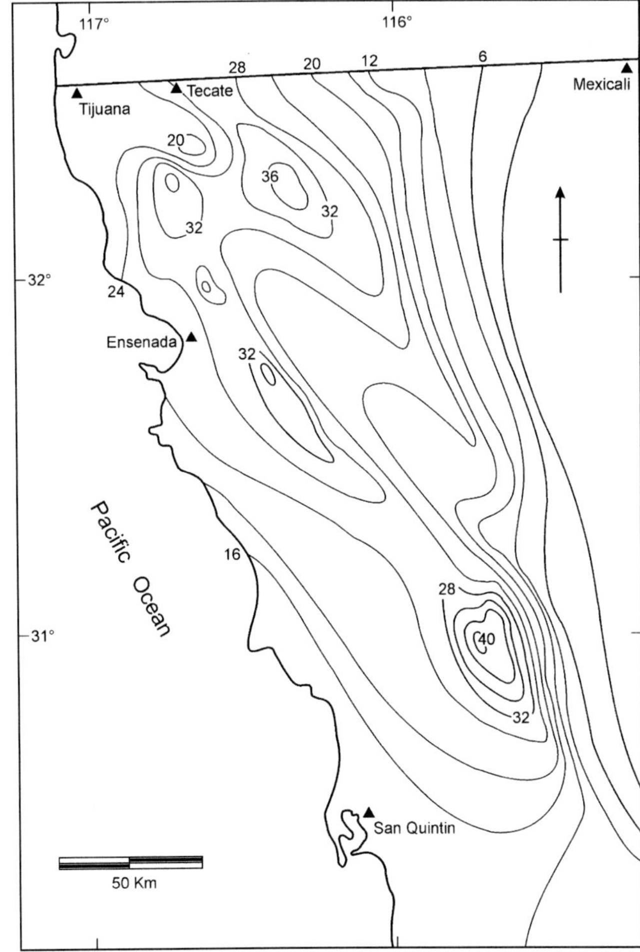

Figure 18.3. Mean winter precipitation (centimeters).

chenopodifolia), wild rose (*Rosa minutifolia*), San Diego sunflower (*Viguiera laciniata*), and desert almond (*Prunus fasciculata*). Both communities have woody deciduous shrubs of ash (*Fraxinus trifoliata*) and buckeye (*Aesculus parryi*) as well as evergreen sclerophyllous shrubs such as laurel sumac (*Malosma laurina* [= *Rhus laurina*]), lemonade berry (*Rhus integrifolia*), and jojoba (*Simmondsia chinensis*) (Peinado et al. 1995b).

Maritime desertscrub has abundant succulent taxa, including coastal agave (*Agave shawii*), velvet cactus (*Bergerocactus emoryi*), *pitaya agria* (*Stenocereus gummosus*), candelabra cactus (*Myrtillocactus cochal*), pincushion cactus (*Mammillaria dioica*), and chollas (*Cylindropuntia* spp.). Near the 30th parallel, this community grades into Sonoran desertscrub that contains the northernmost outposts of the bizarre boojum tree (*Fouquieria*

columnaris) and giant cardón cactus (*Pachycereus pringlei*).

Chaparral, which consists of evergreen sclerophyllous shrubs in carpetlike stands, grows on steep slopes and in shallow, rocky soils throughout the mountains from 400–800 m near the coast to as high as 2000–2400 m in the Sierra Juárez and the Sierra San Pedro Mártir. Chamise (*Adenostoma fasciculatum*) is the widespread dominant. Mixed chaparral, which comprises a mixture of species in the wild lilac (*Ceanothus*), manzanita (*Arctostaphylos*), and oak (*Quercus*) genera, is widespread on northern exposures of the near-coast ranges and in higher elevations of the Sierra Juárez. Other important genera are mountain mahogany (*Cercocarpus*), sumac (*Malosma, Rhus*), *Xylococcus*, and *Ornithostaphylos*. Red-shank chamise (*Adenostoma sparsifolium*) is widespread

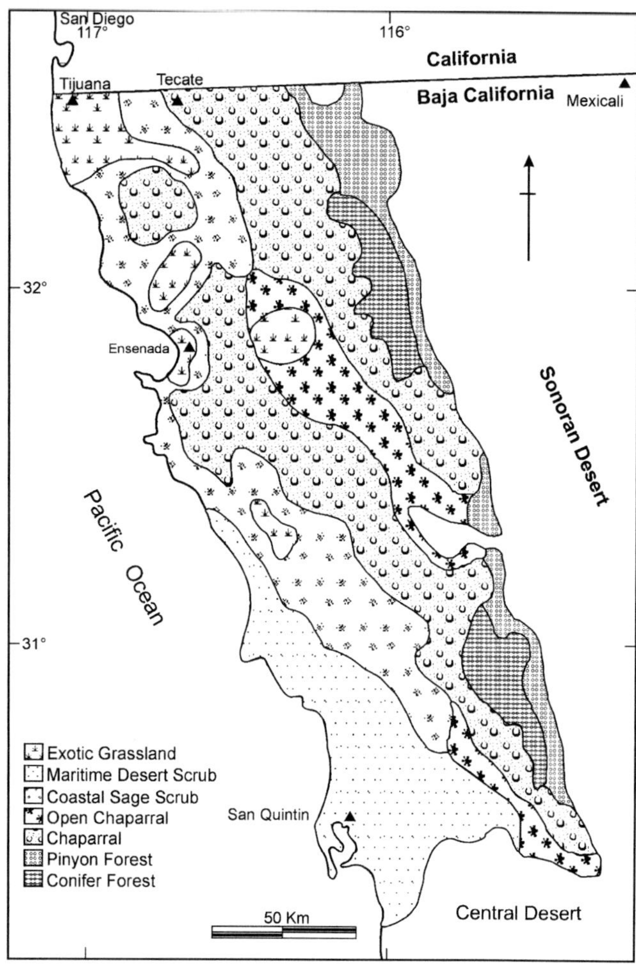

Figure 18.4. Generalized vegetation map of northwestern Baja California.

on the higher plateaus of Sierra Juárez and the west slope of Sierra San Pedro Mártir. Above 1800 m of both sierras, chamise and red shank chaparral is replaced by almost monotypic stands of peninsular manzanita (*Arctostaphylos peninsularis*).

The chaparral belt in the coastal Sierra Juárez and foothills of the Sierra San Pedro Mártir contains widely scattered closed-cone conifer forests, the dominant species having the serotinous or partially serotinous cone habit, including Tecate cypress (*Cupressus forbesii*), knobcone pine (*Pinus attenuata*), and Bishop pine (*P. muricata*). A few stands of Coulter pine (*Pinus coulteri*) occur locally in the inland Sierra Juárez and in the Sierra San Pedro Mártir, and small colonies of Cuyamaca cypress (*Cupressus arizonica* var. *arizonica*) grow in the southern Sierra Juárez. Riparian forests of western

cottonwood (*Populus fremontii*) and western sycamore (*Platanus racemosa*) occur along streams, and coast live oak (*Quercus agrifolia*) grows in open woodlands along arroyos and margins of basins and on scattered north-facing slopes.

The nonserotinous Parry pinyon (*Pinus quadrifolia*) forms patchy cover in the chaparral along the crest of the interior Sierra Juárez plateau and the western Sierra San Pedro Mártir. An extensive forest of Parry pinyon occurs in association with peninsular manzanita and canyon live oak (*Quercus chrysolepis*) on the upper eastern escarpment of the Sierra San Pedro Mártir. Canyon live oak grows on steep, north-facing exposures, along canyons and on cliffs in the upper margins of the chaparral belt, including the highest mountain tops of the near-coast ranges, the peaks rising above the

Sierra Juárez plateau, and the upper mesas of the southern Sierra Juárez. Canyon live oak is widespread in the conifer forests of the Sierra San Pedro Mártir. Monotypic forests of Jeffrey pine (*Pinus jeffreyi*) are widespread on the Sierras Juárez and San Pedro Mártir from 1500 to 2000 m. Stands of this pine occur mostly on basin floors, around the margins of meadows, and along arroyos. The Sierra San Pedro Mártir above 2000 m hosts extensive mixed-conifer forests (Minnich 2001). South-facing slopes are covered with Jeffrey pine mixed with white fir (*Abies concolor*) and sugar pine (*Pinus lambertiana*). White fir and sugar pine dominate steep northern exposures, including the upper eastern escarpment, where they grow with the endemic Sierra San Pedro Mártir cypress (*Cupressus montana*).

Lodgepole pine (*Pinus contorta*) is common along meadows and arroyos of the plateau above 2400 m, often in association with quaking aspen (*Populus tremuloides*). Incense cedar (*Calocedrus decurrens*) grows near watercourses and moist north-facing slopes. Important shrubs are manzanitas (*Arctostaphylos patula, A. pringlei, A. pungens*), snow bush (*Ceanothus cordulatus*), peninsular Emory oak (*Quercus peninsularis*), and canyon live oak. The forest belt contains local wet meadows that are the focal point of cattle grazing in summer, with several meadows reaching 500 ha. Dominant genera are *Juncus* and *Carex*, and common grasses and herbs include annual bluegrass (*Poa annua*), mat muhly (*Muhlenbergia richardsonis*), buttercup (*Ranunculus cymbalaria*), willow herb (*Epilobium adenocaulon*), locoweed (*Astragalus gruinus*), evening primrose (*Oenothera californica*), water parsnip (*Berula erecta*), and thistle (*Cirsium foliosum*). Herbaceous perennials such as yarrow (*Achillea millefolium*), cinquefoil (*Potentilla wheeleri*), rattleweed (*Astragalus circumdatus, A. palmeri*), and *Aster occidentalis* cover drier or overgrazed meadows.

Woodlands of single-needle pinyon (*Pinus monophylla*) grow below 1500 m on the eastern escarpments of both sierras. These communities are associated with widely scattered California juniper (*Juniperus californica*) and open stands of desert chaparral, in which oaks (*Quercus corneliusmulleri, Q. turbinella, Q. cedrosensis*), sugar bush (*Rhus ovata*), mountain mahogany (*Cercocarpus betuloides*), holly-leaf cherry (*Prunus ilicifolia*), and such leaf-succulents as Mojave yucca (*Yucca schidigera*), Parry nolina (*Nolina parryi*), and desert agave (*Agave deserti*) commonly occur.

European Land Use and Vegetation Change

The diaries of the Spanish missionaries show that the vegetation of much of northern Baja California was remarkably similar to that seen today (Minnich and Franco-Vizcaíno 1998). The diaries emphasize ethnobotanical plants (typically congeners of European species they readily recognized) and conspicuous trees such as palms. Examples of specific detail on desert species given in these manuscripts, along the explorer's routes, include the northern limits of copal (*Bursera hindsiana*) at San Felipe and coastal agave at Rosarito, as well as those of cardón and boojum tree in the southern Sierra San Pedro Mártir.

There are 2 striking examples of desert species with distributions virtually fixed in the region. With respect to copal, both *Bursera hindsiana* and *B. microphylla* grow near San Felipe. However, the stand of *B. hindsiana* located 4 km northwest of the town is apparently the same one identified in the 1906 Biological Survey (Nelson 1921) as *Elaphrium macdougalii*—a synonym for *B. hindsiana*. Nelson (1921:19) stated that this population is one of the most northerly representatives of the copals. José Joaquín Arrillaga apparently recorded the same colony on his second expedition of 1796 (Tiscareno and Robinson 1969). And in 1766, Fray Wenceslaus Linck recorded palms as far north as Agua Caliente (Burrus 1966) on the eastern escarpment of the southern Sierra San Pedro Mártir; this is the present northern limit in the range.

The diaries also indicate that the southern limits of several Mediterranean-climate species in the southern Sierra San Pedro Mártir were similar to the limits of modern ranges, including California juniper at El Salto and coast live oak at Rancho San Antonio, on the west face of the sierra, where Fray Juán Crespi made camp at a "large live oak" (*encino*) in 1769 (Bolton 1927). In particular, the diaries give a complete range of coast live oak along their routes, including sightings along the western Sierra San Pedro Mártir, in the coastal ranges from Santo Tómas to Valle Guadalupe, the west slope of the Sierra Juárez, and the transverse ranges northeast of San Vicente.

At Rancho San Antonio, Fray Junipero Serra made the remarkable observation of "two big pine trees among the rest" (Tibesar 1955:82–83). These were probably Jeffrey pines along the stream at an extraordinarily low elevation of 700 m. Jeffrey pines were seen there by Wiggins (1944), and specimens were collected by Reid Moran in 1967 (Minnich 1987).

José Joaquín Arrillaga, who left the most detailed accounts, described in 1796 the interdigitation of pine forest in basins and chaparral on ridges of the Sierra Juárez, which can be seen today throughout the range. Arrillaga wrote a detailed account of forests and chaparral on the west slope of the range near Laguna Hanson (Tiscareno and Robinson 1969:66–67).

[September 22]
[from La Matanza] I started out on the trail leading west. I went up a ravine and descended to an arroyo with many pines. . . . I climbed a hill covered with an abundance of chamizo, and descended to another spacious sink, formed by several low hills. . . . I continued along this sink, which was over a league long, and which had abundant pasture. Turning to the northwest [along this arroyo], I stopped . . . by some oaks.

[September 23]
Returning to the trail I left the previous day . . . I entered a short cañada and after leaving it on the right, I went up the hillside. I crossed a mesa full of chamizo, from which I descended to a sink with sufficient pasture [and] some oaks. . . . I proceeded to skirt [southward] along the flank of the mountain, descending to a narrow arroyo, where there are . . . a few pines. The [route that followed] was [covered by] chamizo and madroño and the entire view was the same. . . . From this mesa I climbed onto another mesa, and others followed. . . . Since I left this morning our route has been toward the southwest, but then we directed ourselves toward the south until we arrived at the place they call the Arroyo of San Rafael, where all the cañadas and arroyos I crossed . . . come together. . . . [At this site] are cottonwood trees, willows, and sycamores. From La Laguna [Laguna Hanson] to this arroyo [it] is all downhill.

Arrillaga had crossed a series of ridges and canyons that feed into Arroyo San Rafael. He described several trends in the vegetation that can still be observed today: (1) forests growing in the arroyos; (2) chaparral dominated by *chamizo* and occasionally by *madroño* [*Arctostaphylos* spp.] growing on the intervening hillslopes; and (3) with decreasing altitude, the forest composition shifting from pines (*pino*, Jeffrey pine) to oaks. The higher arroyos (Cañon La Bandeja, Los Barrancos) are covered by Jeffrey pine forest, whereas lower ones (La Casa Verde, La Rosa de Castilla) contain mostly woodlands of *Quercus agrifolia* with scattered *Pinus jeffreyi*. Chaparral covering the hillslopes is dominated by *Adenostoma fasciculatum*, *Arctostaphylos peninsularis*, *A. glauca*, and *A. pungens*. Chaparral is ubiquitous on the lower slopes, as suggested by Arrillaga's remark that "the entire view was the same."

In a final example, José Longinos-Martínez gave an extensive account of the pine forests of the Sierra San Pedro Mártir in 1792 (Simpson 1938, 1961). At Vallecitos he noted that "the range is thickly covered with pines (different species from those in the lower mountains)" (Simpson 1961:28), in apparent recognition of the richer mixed-conifer forests that cover the highest plateaus. In 1888, Colonel D. K. Allen (1888a,b, 1890) examined the forest at Vallecitos for prospective timber. Allen's data confirm that tree densities, species composition, and diameters were remarkably similar to those of today, apparently due to recurrent understory fires every 50 years (Minnich et al. 2000a, Minnich 2001). Photographs taken by Ford Carpenter during an expedition in 1903 (fig. 18.5) reveal open forest similar to that described by the Biological Survey (Nelson 1921) and at present (fig. 18.6).

Today, rural Baja California is a largely unfragmented, wild landscape, with scattered patches of agriculture, reminiscent of California in the nineteenth century. The vegetation has a largely pristine character in part because the region's isolation has been responsible for a centuries-long delay in the development of agriculture and ranching after the Spanish discovery of the peninsula (see Henderson 1964). Agriculture first expanded during the dictatorship of Porfirio Díaz in 1877–1911. During this period, much of the best land, mostly grasslands in the coastal valleys and desertscrub in the Mexicali Valley and the Colorado River Delta, was cleared for crops (Henderson 1964). Dry farming of wheat and barley was developed in the coastal plains and inland valleys. The total amount of land now under cultivation is 450,000 ha, about 10% of the state north of latitude 30°.

Domestic livestock have been grazed in the coastal valleys since the Dominican missions were established in the 1770s. Cattle were driven to the mountains for summer pasture after the mission system was extended inland to Santa Catarina and the Sierra San Pedro Mártir in the 1790s (the grazing history is summarized in Henderson 1964). The mountains

Figure 18.5. View of the forest at Vallecitos in the Sierra San Pedro Mártir in 1903. (Photograph by Ford Carpenter.) Reproduced with permission from the Mandeville Special Collections Library at the University of California, San Diego.

were said to be well stocked (i.e., "carrying capacities" were reached) by 1820. Cattle numbers increased after 1850 as a result of growing markets for meat during the California Gold Rush. Americans also drove cattle from California into the mountains of Baja California. However, the number of cattle in the nineteenth century was apparently never very large. By 1857 there were 43 ranches in Baja California, with 8260 head of cattle marketed to California (Henderson 1964). The number of cattle marketed domestically is unknown, but demand was apparently low until the gold strikes of the 1870s. It was estimated that cattle numbers in the Sierra San Pedro Mártir once reached 25,000 (Allen 1890). In 1911, after the Baja California gold strikes, there were reported to be 21,000 head in the Sierra Juárez and Sierra San Pedro Mártir (Henderson 1964).

In the late nineteenth century, foreign investors began organized sheep drives, led mostly by Basque shepherds. The herds were gathered in August near Tijuana and driven to pastures as far south as the Sierra San Pedro Mártir, returning 2 months later, in October. Sheep were driven through public do-

main and leased land, and the wool was sold to a mill at Ensenada. As many as 30,000 sheep per year were driven along this route between 1885 and 1905 to U.S. markets. Nelson (1921) reported that large numbers of sheep were grazed on the west side of the peninsula as far south as San Quintín. Sheep grazing may have intensified beginning in 1910, when this practice was prohibited by the U.S. government within the National Forest system in southern California (Lockmann 1981; Minnich 1988). Sheep grazing declined during the middle of the twentieth century because increasing agricultural settlement in the coastal valleys discouraged land leasing to shepherds. Still, there were as many as 8000 sheep in the Sierra San Pedro Mártir in 1956 (Henderson, 1964). Sheep were prohibited from the range beginning in 1964 (Meling-Pompa 1991a,b).

During the twentieth century cattle production became an important sector of the economy of Baja California. Cattle, which numbered 32,000 head at the time of the first agricultural census in 1930 (Secretaría de la Economía Nacional 1936) grew exponentially during the twentieth century, with

Figure 18.6. View of the forest at Vallecitos in the 1990s, likely within 1 km of Ford Carpenter's 1903 site. (Photograph by R. A. Minnich.)

180,000 head reported in 1990 (INEGI 1994). It is likely, however, that much of this growth reflects the development of feedlots in the irrigated Mexicali Valley rather than a significant increase in numbers of range cattle (Henderson 1964). The number of horses and mules in the state of Baja California peaked in the 1950s at 25,000 and 8000 respectively, then declined to 10,000 and 500 as a result of mechanization (Secretaría de Economía 1951, 1956; INEGI 1994).

Feral donkeys were numerous during the nineteenth and early twentieth centuries (Henderson 1964), but their numbers plummeted as they were used to provide dried and salted burro meat, which was still available in local markets through at least the 1950s. The impact of feral donkeys has likely been replaced by that of goats, which increased from about 5000 in the 1960s to 50,000 in 1990 (Secretaría de Economía 1951, 1956; INEGI 1994). Goats are typically grazed on the open range but return to corrals every night. However, significant numbers of goats have escaped and become naturalized,

and it is not known to what extent predation by coyotes and mountain lions can limit their numbers.

Goats are a potential threat because they can consume many of the coastal sage scrub plants and appear to be better adapted to them than are cattle. Genin and Badan-Dangon (1991) observed a goat herd that consumed 21 species of coastal sage scrub, of which 7 constituted 85% of the diet. These plants were deerweed (*Lotus scoparius*), California sagebrush, buckwheat (*Eriogonum fasciculatum and E. wrightii*), San Diego sunflower, lemonade berry, and bush mallow (*Malacothamnus fasciculatus*).

The impact of livestock grazing on the coastal sage scrub seems to depend on its intensity and frequency. In areas where grazing has been light to moderate, the coastal scrub can still be observed with various amounts of cover. The arrival of Old World exotic annual grasses and forb species may have made coastal sage scrub more attractive for livestock because some invasive species can produce more forage than the indigenous wildflowers, described in the region during the Spanish explorations and by botanists during the late nineteenth century (Minnich and Franco-Vizcaíno 1998).

With increases in carrying capacity, livestock may further augment browse pressure and physical removal of canopy. Currently, the coastal and inland basins show the unmistakable effects of grazing, including contoured livestock trails, thinning of the herbaceous layer, pruned shrubs, and fecal deposits. Browse preferences may result in selective removal of some shrub species (Genin and Badan-Dangon 1991). Although *Eriogonum fasciculatum* is preferred by cattle for fodder, this shrub apparently gains a selective edge over other coastal shrub species due to its ability to colonize disturbed ground. No historical baseline data exist to evaluate the species composition and structure of coastal sage scrub before grazing and the arrival of exotic annuals.

Cattle grazing is still practiced in seasonal transhumance in both the Sierras Juárez and San Pedro Mártir, and the same families have been in control of the highlands since the early nineteenth century (Meling-Pompa 1991a,b). During the rainy season, cattle graze on annual herbaceous cover in coastal sage scrub and in exotic grasslands, as well as on crop stubble in the agricultural zones. Once the annual grasslands have cured (i.e., died and dried), cattle are driven to the mountain meadows during May and June.

The mountain meadows are the primary source of feed, and estimated cattle-carrying capacities range from 2 to 15 ha per animal unit per year (Henderson 1964). Estimates of carrying capacity in chaparral, coast live oak, pinyon and Jeffrey pine forests at 4 *ejidos* in the northern Sierra Juárez vary from 20 to 50 ha per animal unit per year (summary in Minnich and Franco-Vizcaíno 1998). Estimated carrying capacities are lowest in chaparral because most shrub species are unpalatable to cattle. Along the international boundary east of Tecate, the carrying capacity is so low that livestock on both sides of the border depend almost entirely on irrigated pasture and/or supplemental feeding (Minnich and Bahre 1995). Estimated carrying capacities in pinyon forest range from 30 to 34 ha per animal unit per year.

Open range grazing has apparently had significant effects on the vegetation of the mountain meadows. Fragmentary descriptions in 1887–1889 indicate that meadows in both sierras were in excellent condition (Minnich and Franco-Vizcaíno 1998). The most notable impact of livestock grazing seems to have been the selection for species with low, prostrate growth forms such as *Achillea millefolium, Aster occidentalis, Potentilla wheeleri,* and *Trifolium wigginsii.* Today these plants typically reach heights of 3–5 cm during years of normal precipitation and reach 20–30 cm during El Niño years.

Cattle exclosure studies show that grazing by livestock reduces biomass in the meadows (Minnich et al. 1997), but results on consequent changes in species composition are inconclusive. During drought in 1989–1990, meadow biomass in exclosures was 50% greater than the unfenced controls. During years of normal or above-normal precipitation (e.g., 1991, 1992, 1993), meadow productivity increased by an order of magnitude, but there was little difference between exclosure and control biomass, perhaps because high productivity rates reduced cattle pressure.

In a similar finding, sapling counts and cattle exclosure studies in mixed-conifer forest (M. G. Barbour, pers. comm.) suggest that cattle grazing has no measurable effect on herbaceous and shrub cover, nor on conifer recruitment, in the forests. Although evidences of cattle such as fecal deposits are seen in most forests, animals spend most of their time in meadows because forests contain limited herbaceous cover. Most shrub species in the chaparral and mixed-conifer forest are unpalatable to livestock (Minnich and Bahre 1995).

Studies of pollen preserved in soils and sediments are needed to evaluate grazing impacts over the past 2 centuries. Four non-native species have been reported in the Sierra San Pedro Mártir, including filaree, cheat grass (*Bromus tectorum*), common dandelion (*Taraxacum officinale*), and *Koeleria macrantha* (Sosa-Ramírez and Franco-Vizcaíno 2001), but their distribution is limited and their abundance is low.

Outside the now extensive urban areas in northwestern Baja California, the most significant vegetation change in Baja California has been the invasion of Eurasian grasses and forbs across the coastal and inland valleys north of San Quintín. The timing of invasions may have been similar to that in California (Hendry 1931). In 1769, the Franciscan missionaries brought the first exotics that spread extensively in Baja California. Burr clover (*Medicago hispida*) was found at the Mission San Fernando Velicatá by 1769. Black mustard and filaree were recorded in mission bricks dating to

1780 at Santo Domingo and San Vicente. Duhaut-Cilly recorded that *B. nigra* became a terrible scourge in the plains of Los Angeles by 1820s (Carter 1929). It was probably widespread in Baja California by that time. In the late nineteenth century, Orcutt (1886a) saw houses made of mustard stalks at San Vicente, likely the alien black mustard. Other "Franciscan exotics" were abundant by that time. Orcutt (1886b: 41) states that the pasture at San Quintin "was exactly like that of Alta California: alfilleria [*Erodium cicutarium*], alfalfa [*Trifolium* spp. ?], burr [*Medicago polymorpha*] and red clovers [*Trifolium* spp.] make [up] the greater part of the forage plants." However, the alfilleria was "too scattered to be good feed" (Orcutt 1886b: 41). Wild oat (*Avena fatua*) was not recorded in mission bricks and may have been introduced some years after the construction of the missions.

Accounts from Alta California suggest that wild oat was widespread before the gold rush (e.g., Frémont 1848). In 1855, the U.S.–Mexico boundary survey reported that in coastal valleys of southern California,

> the wild oat *Avena fatua* is so extensively naturalized, that it gives every fertile tract the appearance of a cultivated field. The wide plains that border the sea in the neighborhood of Los Angeles are covered with the richest pasturage. The *Erodium cicutarium*, with several species of wild clover (*Trifolium* and *Medicago*) are mingled with a variety of other herbage, and thus serve to give a meadow-like aspect to this teeming land. (Parry 1859:18).

In 1861, José Matías Moreno wrote that in the pastures along the Pacific coast of northern Baja California, "the most abundant [species] are oats, alfilerillo, trifolium and grass" (Piñera-Ramírez and Martínez-Zepeda 1984:13). William M. Gabb, who traversed the entire Baja California peninsula in 1866–1867 stated that in Valle Guadalupe "the grass in the uncultivated parts . . . was unsurpassed by anything we saw on the whole journey" (Orcutt 1886b:38). Given the widespread extent of *Avena fatua* in California by that time, Gabb was likely describing oat fields in this valley.

Accounts by Davidson (1891), Abrams (1904), and Parish (1920) indicate that non-native brome grasses (*Bromus rubens, B. diandrus*) began increasing in coastal California only by 1900. The botanist Samuel Parish reported that bromes were com-

mon in southern California by 1920. The bromes, as well as slender wild oat and short-podded mustard, began to dominate grasslands and coastal sage scrub by the end of the century (Minnich and Dezzani 1998). Bromes were not described by Orcutt or the Biological Survey and thus must have expanded through the northern peninsula during the twentieth century.

Exotic annual grassland now covers areas described as pasture by early Spanish explorers. The ambiguities of late-eighteenth-century Spanish diaries lend little insight to pre-European herbaceous vegetation. The frequently used term *pasto* does not necessarily mean "grassland," as has been conventionally translated (Bolton 1927, 1930; Brown 2001), but rather "pasture" with an uncertain amount of grasses, if any (Minnich and Franco-Vizcaíno 1998). The "*Stipa*" bunch grass model, summarized in Heady (1988), is from Clements' (1920) "plant indicator" model of potential climax communities, but it is not based on historical evidence (Hamilton 1997).

Spanish diary entries in California during the spring months consistently describe fields of flowers, not bunch grasslands. The diary of Crespi, the only expedition to have crossed northern Baja California during the spring, did not describe flower fields in Baja California. Perhaps 1769 was a dry year. However, Orcutt (1886a) describes wildflowers at several locations such as Tijuana, Ensenada, San Telmo, and east of El Rosario. For example, he described the bay of Ensenada as being covered

> by magnificent fields of gold California poppy [*Eschscholzia californica*], phacelias, layia elegans [*Layia* sp.], orthocarpus, baerias [*Lasthenia* spp.], larkspurs [*Delphinium* spp.], *platystemon* and other delicate annuals and perennials which lent an added charm to the beauty of land and sea that was spread out before us on every hand. And thus for mile after mile we alternatively contemplate the rich garden of flowers and the beautiful scenery. (p. 54)

Invasive exotic grasses, especially bromes, have increased the flammability of herbaceous ecosystems. In the Central Valley and interior valleys of California, areas covered by forbs were described by early travelers as being barren in summer (Wester 1981). The journals of Crespi, Serra, and Arrillaga, which cover expeditions in late spring and early summer, indicate that barrenness was also characteristic of the valleys of Baja California in the dry season. Apparently the indigenous herbacous cover left little remnant fuels when they cured. Hence, herbaceous biomass and fire hazard may have increased with the invasion of exotic annuals, but fuel build up may have been reversed by livestock grazing. Grasslands only periodically support extensive fires, usually after wet years when inflammable biomass exceeds grazing pressure (Minnich 1983). Perhaps the greatest loss of biodiversity in Baja California has been the decline or extirpation of annual wildflowers due to invasions by exotic annuals. Spring wildflowers have become ever more infrequent along the Pacific coast in recent decades (E. Franco-Vizcaíno, pers. obs.).

Although most of the population of northern Baja California was rural during the nineteenth century, the impact of woodcutting for domestic fuel was probably local and of little regional significance. The introduction of natural gas and propane for heating and cooking predates the population explosion that occurred after World War II in the border cities. Gold discoveries between 1873 and 1889 resulted in the establishment of several boomtowns of 500–1650 people in the Sierras Juárez and San Pedro Mártir (Chaput et al. 1992). Because nearly all of the gold discoveries were placers, the demands for fuelwood for such processes as smelting, running stamps, pumps, and ore crushers were limited, and, moreover, the best ores played out within a few years of discovery (Chaput et al. 1992).

Cattle ranchers have used Jeffrey pine, California juniper, and red-shank chamise, as well as other chaparral species, for fence posts and corral construction since the early nineteenth century, but the impact from the construction and maintenance of such infrastructure is unknown. Wood utilization may have been light because most of the mountains have remained open range until recently. Most fencing was built to subdivide prime meadows and ranchsteads. Jeffrey pine has been used for construction of primitive summer shelters.

Low human population densities and the inaccessibility of the pine forests prevented significant removal of timber during the past 2 centuries. During the Dominican mission period, Arrillaga recorded in his journal that pine trees on the west slope of the central Sierra Júarez near San Salvador were cut for Misión Santo Tomás, but that likely involved only a few trees. Coulter pine forests were part of the territory under the jurisdiction of Misión San Miguel, but it is unclear whether that remote stand was ever exploited. Utilization of closed-cone

conifer forests was probably limited because most forests have been made inaccessible by impenetrable chaparral with which it normally grows. The trees also make poor fuelwood because of their small size.

Pine forests were logged during the mining strikes at Japá (1873–1874) and El Alamo (1889–1890). The International Colonization Company built a road from El Alamo to the forests at La Tableta (25 km south of Laguna Hanson), but the mill probably both lasted only a few years and had little impact on the forest there. The Ejido Sierra Juárez established a gasoline sawmill at Arroyo del Sauz, 5 km south of Laguna Hanson, in the 1930s and ran a limited logging operation in Jeffrey pine forests in the central part of the range. A federal census in 1940 states that 250 ha of a total of 23,000 ha of forest had been exploited (Minnich and Franco-Vizcaíno 1998). This *ejido* has not been actively logging since the 1970s, and we could not find any information on how much timber was removed between 1940 and 1970. The impact of the *ejido* logging operation on forest structure may have been negligible. Only selective logging of old-growth trees was practiced within a radius of 20 km of the Laguna Hanson (Haiman 1973), and understory fires (maps in Minnich and Chou 1997) have maintained open, mature stands in these cutover areas. The Ejido Bramadero obtained a permit in 1996 to log trees infected with dwarf mistletoe or infested with bark beetles in an 1800-ha parcel in *ejido* lands in the northwestern Sierra San Pedro Mártir outside the National Park, but logging operations were delayed several years and apparently stopped by the authorities in spring 2001 after only a few truckloads of logs had been exported to the United States. The continuation of logging remains under litigation.

Uncontrolled Fire and Dynamics of Baja California's Ecosystems

In Baja California's Mediterranean climate, wildland fire is strongly integrated into the ecological function, structure, and distribution of many plant communities. Fire regime properties (fire intervals, intensity, severity, removal of canopy) are outcomes of climate and the characteristics of the vegetation. Climate affects plant growth, productivity, and rates of fuel build up, and vegetation structure affects fire behavior and maintenance of the canopy. These factors exert selective forces on the distribution of fire-prone ecosystems (see Allen et al. 1991; Veblen et al. 1991; Christensen 1993; Barton 1994; Minnich 2001).

In Baja California's chaparral and mixed-conifer forests, fires result in fine-grained and self-organized patch mosaics of stands that range mostly from 500 to 5000 ha. Because a time lag exists between fuel accumulation and burning, the inflammability of communities increases with time. Fires preferentially burn old stands (> 40–50 years), whereas younger stands constrain the progress of burns (Minnich and Chou 1997; Minnich 2001). In both chaparral and mixed-conifer forests, fire occurs about twice per century. In Baja California, fires seem to occur at random in normal weather, and this results in slow spreading of flame lines and low fire intensities. In Alta California, the efficient suppression of small fires selects for extensive burning by relatively few, very intense fires. Because fire occurrence in Alta California is nonrandom and tends to coincide with drier, windier weather, fires have attained sizes as large as 60,000 ha. This enlargement in fire size in Alta California is also related to the homogenization of the patch mosaic (Minnich and Chou 1997).

Studies of post-fire succession in Alta California show that species composition and stand structure take on the characteristics of the surrounding mature communities within 20–40 years. Dominant species resprout or establish from soil seed banks within a few years after fire, and successions respond to variable fire intervals (summaries in Hanes 1988; Keeley 2000). Chaparral was degraded only when fire intervals were < 10 years because seeding species were burned before reaching reproductive maturity, and sprouters experienced increased mortality form carbohydrate depletion. Short fire sequences were encouraged by invasive Mediterranean annual grasses. Nutrient enrichment in chaparral after wildland fire is also less pronounced and more transitory in Baja California than is typically reported for burned shrublands in Alta California (Franco-Vizcaíno and Sosa-Ramírez 1997).

A chronosequence study along the U.S.–Mexico boundary that compared successional states of chaparral patches in coarse-grained mosaics on the U.S side with fine-grained mosaics on the Mexican side found that successional sequences were similar in both countries (Minnich and Bahre 1995). The authors concluded that the responses of chaparral species were independent of fire size due to sprouting and seedbank strategies. No dominant species required long-range seed dispersal to recolonize burns.

In contrast, grassland and coastal sage scrub have limited buildup of fuel and relatively high decomposition rates in winter; this results in more variable time intervals between fires than in chaparral (Minnich 1998). Stands have high levels of dead biomass during the dry season due to the drought dormancy of subshrubs and the curing of annuals. Decomposition rates are relatively high during winter because fine fuels have low lignin content. Thus, the levels of standing fuel primarily reflect annual productivity. The rapid accumulation of fuel makes entire landscapes inflammable, regardless of previous fire history. This results in a spatially random turnover of patches; that is, fires spread independently of patch structure. Consequently, fires in annual grasslands and coastal sage scrub in Baja California can grow as large as 10,000 ha, especially after a season of high rainfall and near the international boundary, where productivity is higher than farther south (Minnich 1983).

Conservation Issues

Protected by historical isolation, much of rural Baja California remains a showcase of unmanaged biota functioning with relatively little human exploitation or interference, in sharp contrast to wildlands in Alta California. Isolation continued into the late twentieth century, as urban and agricultural growth concentrated in the rich Mexicali Valley and the border cities. However, the opening of the trans-peninsular highway in 1973 contributed to rapid economic growth along the northern Pacific coast. The road to the National Astronomical Observatory opened the Sierra San Pedro Mártir to the outside world in the early 1970s. North of the 30th parallel, only two National Parks, Constitución de 1857 at Laguna Hanson and San Pedro Mártir, have Natural Protected Area status. Although the national parks were established by presidential decree in 1947, they have been actively administered (jointly by the state and federal governments) only since 1997. A new Natural Protected Area, Valle de los Cirios, was established in 2001 and covers nearly all lands from the 30th parallel to the southern boundary at the 28th parallel, including Isla Cedros.

The establishment of Natural Protected Areas (the Spanish acronym is ANP) is the beginning of a sustainable system for the protection of Baja California's extraordinary ecosystems, which are unique in Mexico. But because the current ANPs already represent such a large proportion of the state, it is unlikely that other areas deserving protection will be included in government plans in the near future. Although nongovernmental conservation organizations (NGOs) are weakly developed in Baja California compared to those in the United States, several NGOs including Pronatura, Pro-Esteros, and Terra Peninsular are currently working to conserve lands that do not enjoy official protection.

A critical priority for conservation is the coastal sage scrub in its natural setting adjacent to beaches. North of Ensenada, only a few kilometers of coastline at Playa Azúl still have relatively undisturbed coastal sage scrub. This area is currently threatened by the proposed development of one or more liquefied natural gas terminals that would off-load gas from ships to supply natural gas and electrical energy to southern California. Efforts are underway by NGOs to conserve this remaining patch of coastal sage scrub, possibly through a land swap that would locate the gas terminals at already developed sites. South of Ensenada, another important region for conservation of coastal sage scrub is the roadless area south of the Punta Banda peninsula.

A large portion of the maritime desertscrub adjacent to the ocean has been extirpated by agricultural and tourism development along the San Quintín coastal plain. Local inhabitants and NGOs have been working to protect San Quintín Bay and its spectacular volcanoes, covered with maritime desertscrub, from the development of a major tourist complex that would include a marina, hotel, golf course, and an airport. Other areas important for conservation of maritime desertscrub are south of Eréndira and the dune complex south of San Quintín.

Baja California's chaparral and conifer forests are still well protected by their resilience and isolation and may serve as a showcase for research in fire ecology for similar ecosystems throughout the southwestern United States and northwestern Mexico. In the long run, maintaining the current unmanaged fire regime will likely be more economical and less threatening to life and property. Establishment of a biosphere reserve in the Sierra San Pedro Mártir has been proposed as a way of involving the local population in the conservation and sustainable development of the chaparral and mixed-conifer ecosystems (Minnich et al. 1997). A biosphere reserve in the Sierras Juárez and San Pedro Mártir may be a more favorable model for resource management because it implies a more democratic and transparent decision-making process than that afforded by national parks and other ANPs, which are operated by a centralized authority.

The reintroduction of the endangered California condor (*Gymnogyps californianus*) in the Sierra San Pedro Mártir was an idea conceived independently and at about the same time by us and by Amadeo M. Rea (G. Ceballos, pers. comm.). It was proposed by Minnich et al. (1997) as a way of rallying regional, national, and international support for a biosphere reserve. The reintroduction effectively began with the release in October 2002 of the first California condors, in an area adjacent to the Sierra San Pedro Mártir National Park. The condor recovery team believes that the species has an excellent chance for successful naturalization in the Sierras Juárez and San Pedro Mártir (M. Wallace, pers. comm.). The eastern escarpment contains abundant cliff faces and ledges favorable to their roosting and nesting, and updrafts associated with the high relief are also favorable for soaring. Chaparral and other plant ecosystems have a diverse patch structure that allows location of carrion food sources as well as landing and takeoff. The reintroduction of the California condor to the sierras may serve as a management centerpiece for the national parks or the establishment of a biosphere reserve because this effort will bring international attention and require the integration of sustainable management of the region's natural resources, now under threat from accelerating development.

Acknowledgments

We thank Joaquín Sosa and Mario Salazar for help with botanical and census references. Michael Barbour provided helpful comments on an earlier draft of the chapter. Permission to reproduce the historical photograph by Ford Carpenter was granted by Lynda Claassen, director of the Mandeville Special Collections Library at the University of California, San Diego.

Literature Cited

Abrams, L. 1904. Flora of Los Angeles and Vicinity. Stanford University Press, Palo Alto, California.

Allen, D. K. 1888a. The San Pedro Mártir: The great pine region of the peninsula. Lower Californian, March 29, 1888. Lower Californian, Ensenada, Mexico [available on microfilm at the Museum of Natural History, Los Angeles, California].

Allen, D. K. 1888b. The San Pedro Mártir region. Lower Californian, April 19, 1888. Lower Californian, Ensenada, Mexico [available on microfilm at the Museum of Natural History, Los Angeles, California].

Allen, D. K. 1890. The big unexplored mountain on this peninsula. Lower Californian, December 26, 1890. Lower Californian, Ensenada, Mexico [available on microfilm at the Museum of Natural History, Los Angeles, California].

Allen, R. B., R. K. Peet, and W. L. Baker. 1991. Gradient analysis of latitudinal variation in southern Rocky Mountain forests. Journal of Biogeography 18: 123–129.

Barton, A. M. 1994. Gradient analysis of relationships among fire, environment and vegetation in a southwestern USA mountain range. Bulletin of the Torrey Botanical Club 121: 251–265.

Bolton, H. E. 1927. Fray Juan Crespi: Missionary Explorer on the Pacific Coast, 1769–1774. University of California Press, Berkeley.

Bolton, H. E. 1930. Anza's California Expeditions: Opening a Land Route to California. Diaries of Anza, Diaz, Garces, and Palou. University of California Press, Berkeley.

Brown, A. K. (ed., trans.). 2001. A Description of Distant Roads: Original Journals of the First Expedition into California, 1769–1770, by Juan Crespi. San Diego State University Press, San Diego, California.

Burrus, E. J. (trans.). 1966. Wenceslaus Linck's diary of his 1766 expedition to northern Baja California. Baja California Travel Series no. 5. Dawson's Book Shop, Los Angeles, California.

Carter, C. F. (trans.). 1929. Duhaut-Cilly's account of California in the years 1827–28. California Historical Society Quarterly 8 (2–4): 130–166, 214–250, 306–336.

Chaput, D., W. M. Mason, and D. Z. Loperena. 1992. Modest Fortunes: Mining in Northern Baja California. Natural History Museum of Los Angeles County, Los Angeles.

Christensen, N. L. 1993. Fire regimes and ecosystem dynamics. Pp. 233–244 in P. J. Crutzen and J. G. Goldammer (eds.), Fire in the Environment: The Ecological, Atmospheric, and Climatic Importance of Vegetation Fires. John Wiley and Sons, New York.

Clements, F. E. 1920. Plant indicators. Journal of Ecology 22: 39–68.

Davidson, A. 1891. British plants in southern California. Transactions and Journal of Proceedings of the Dumfriesshire and Galloway Natural History and Antiquarian Society, Session 1890–91, no. 7:112–115. Courier and Herald Offices, Dumfries, U.K. [Reprinted in Crossosoma (1997) 23:68–70.]

Franco-Vizcaíno, E., and J. Sosa-Ramírez. 1997. Soil properties and nutrient relations in burned and unburned Mediterranean-climate shrublands of Baja California, Mexico. Acta Oecologica 18: 503–517.

Frémont, J. C. 1848. Geographical memoir upon upper California. 30th Congress, 1st Sess., Senate Miscellaneous Documents 148. Washington, D.C.

Genin, D., and A. Badan-Dangon. 1991. Goat herbivory and plant phenology in a Mediterranean shrubland of northern Baja California. Journal of Arid Environments 21: 113–121.

Haiman, R. L. 1973. The biological environment and its modification by man in the Sierra de Juárez, Baja California. Ph.D. dissertation, University of California, Los Angeles.

Hamilton, J. G. 1997. Changing perceptions of pre-European grasslands in California. Madrono 44: 311–333.

Hanes, T. L. 1988. Chaparral. Pp. 417–469 in M. G. Barbour and J. Major (eds.), Terrestrial Vegetation of California. California Botanical Society, Davis.

Heady, H. F. 1988. Valley grassland. Pp. 491–514 in M. G. Barbour and J. Major (eds.), Terrestrial Vegetation of California. California Botanical Society, Davis.

Hendry, G. W. 1931. The adobe brick as a historical source. Agricultural History 5: 10–127.

Henderson, D. A. 1964. Agriculture and stock raising in the evolution of the economy and culture of the State of Baja California, Mexico. Ph.D. dissertation, University of California, Los Angeles.

INEGI. 1994. Resultados definitivos, tomo I. VII Censo Agrícola-Ganadero. Instituto Nacional de Estadística Geografía e Informática, Aguascalientes, Mexico.

Keeley, J. E. 2000. Chaparral. Pp. 203–253 in M. G. Barbour and W. D. Billings (eds.), Terrestrial Vegetation of North America. Cambridge University Press, Cambridge.

Lockmann, R. F. 1981. Guarding the Forest of Southern California: Evolving Attitudes toward Conservation of Watershed, Woodlands, and Wilderness. Arthur H. Clark Company, Glendale, California.

Meling-Pompa, D. 1991a. La ganaderia en la Sierra de San Pedro Mártir. Pp. 14–16 in E. Franco-Vizcaíno and J. Sosa-Ramírez (eds.), El Potencial de la Cordillera Peninsular de las Californias como Reserva de la Biosfera. Memorias de Congreso. ESE Report CIEC09101. Centro de Investigación Científica y Educación Superior de Ensenada, Ensenada, Mexico.

Meling-Pompa, D. 1991b. La ganaderia en San

Pedro Mártir. Pp. 17–19 *in* B. Bernaldez-Garza (ed.), Memoria de la III Semana de la Exploración y La Historia, Sierra San Pedro Mártir. Universidad Autónoma de Baja California, Ensenada, Mexico.

Minnich, R. A. 1983. Fire mosaics in southern California and northern Baja California. Science 219: 1287–1294.

Minnich, R. A. 1987. The distribution of forest trees in northern Baja California, Mexico. Madroño 34: 98–127.

Minnich, R. A. 1988. The biogeography of fire in the San Bernardino Mountains of California: a historical survey. University of California Publications in Geography 28: 1–121.

Minnich, R. A. 1998. Landscapes, landuse and fire policy: Where do large fires come from? Pp. 133–158 *in* J. M. Moreno (ed.), Large Forest Fires. Backhuys Publishers, Leiden, the Netherlands.

Minnich, R. A. 2001. Fire and elevational zonation of chaparral and conifer forests in the Peninsular Ranges of La Frontera. Pp. 120–142 *in* G. L. Webster and C. J. Bahre (eds.), Changing Plant Life of La Frontera: Observations on Vegetation in the United States/Mexico Borderlands. University of New Mexico Press, Albuquerque.

Minnich, R. A., and C. J. Bahre. 1995. Wildland fire and chaparral succession along the California-Baja California boundary. International Journal of Wildland Fire 5: 13–24.

Minnich, R. A., M. G. Barbour, J. H. Burk, and J. Sosa Ramírez. 2000a. California conifer forests under unmanaged fire regimes in the Sierra San Pedro Mártir, Baja California, Mexico. Journal of Biogeography 27: 105–129.

Minnich, R. A., and Y. H. Chou. 1997. Wildland fire patch dynamics in the Californian chaparral of southern California and northern Baja California. International Journal of Wildland Fire 7: 221–248.

Minnich, R. A., and R. J. Dezzani. 1998. Historical decline of coastal sage scrub in the Riverside-Perris Plain. Western Birds 29: 366–391.

Minnich, R. A., and E. Franco-Vizcaíno. 1998. Land of chamise and pines: historical accounts and current status of northern Baja California's vegetation. University of California Publications in Botany, vol. 80. University of California, Berkeley.

Minnich, R. A., E. Franco-Vizcaíno, and R. J. Dezzani. 2000b. The El Niño/Southern Oscillation and precipitation variability in Baja California, Mexico. Atmósfera 13: 1–20.

Minnich, R. A., E. Franco-Vizcaino, J. Sosa-Ramírez, J. H. Burk, W. J. Barry, M. G. Barbour, and H. de la Cueva-Salcedo. 1997. A land above: protecting Baja California's Sierra San Pedro Mártir within a biosphere reserve. Journal of the Southwest 39: 613–695.

Minnich, R. A., E. Franco-Vizcaino, J. Sosa-Ramírez, and Y. H. Chou. 1993. Lightning detection rates and wildland fire occurence in the mountains of northern Baja California. Atmósfera 6: 235–253.

Mooney, H. A. 1988. Southern coastal scrub. Pp. 472–487 *in* M. G. Barbour and W. D. Billings (eds.), North American Terrestrial Vegetation. Cambridge University Press, Cambridge.

Nelson, E. W. 1921. Lower California and its natural resources. Memoirs of the National Academy of Sciences 16: 1–194.

Orcutt, C. R. 1886a. A botanical trip. West American Scientist 2: 54–58.

Orcutt, C. R. 1886b. Northern Lower California. West American Scientist 2: 37–41.

Parish, S. B. 1920. The immigrant plants of southern California. Bulletin of the Southern California Academy of Sciences 19: 3–30.

Parry, C. C. 1859. Botany of the boundary, Introduction. *In* W. H. Emory (ed.), Report on the United States and Mexican Boundary Survey. 34th Congress, 1st Sess., Executive Document no. 135, vol. II, part I. Washington, D.C.

Passini, J. F., J. Delgadillo, and M. Salazar. 1989. L'ecosystème forestier de Basse-Californie: Composition floristique, variables écologiques principales, dynamique. Acta Oecologica 10: 275–293.

Peinado, M., F. Alcaraz, J. L. Aguirre, and J. Delgadillo. 1995a. Major plant communities of warm North American Deserts. Journal of Vegetation Science 6: 79–94.

Peinado, M., F. Alcaraz, J. L. Aguirre, J. Delgadillo, and I. Aguado. 1995b. Shrubland formations and associations in Mediterranean-desert transitional zones of northwestern Baja California. Vegetatio 117: 165–179.

Peinado, M., C. Bartolome, J. Delgadillo, and I. Aguado. 1994. Pisos de vegetación de Sierra de San Pedro Mártir, Baja California, Mexico. Acta Botánica 29: 1–30.

Piñera-Ramírez, D., and J. Martínez-Zepeda. 1984. Descripción del partido Norte de la Baja California, por José Matías Moreno, 1861. Fuentes Documentales para la História de Baja California, vol. 1, no. 2. Centro de Investigaciones Históricas UNAM-UABC, Tijuana, Mexico.

Secretaría de Economía. 1951. Segundo censo agrícola ganadero de los Estados Unidos Mexicanos 1940. Resumen General. Dirección General de Estadística, México D.F.

Secretaría de Economía. 1956. Tercer censo agrícola-ganadero y ejidal 1950. Resumen General. Dirección General de Estadística, México D.F.

Secretaría de la Economía Nacional. 1936. Primer censo agrícola-ganadero 1930. Resumen General. Estados Unidos Mexicanos. Dirección General de Estadística. Talleres Gráficos de la Nación, México D.F.

Simpson, L. B. (transl.). 1938. California in 1792. The Expedition of Jose Longinos Martinez. Huntington Library, San Marino, California.

Simpson, L. B. (transl. and ed.). 1961. Journal of José Longinos Martinez: Notes and Observations of the Naturalist of the Botanical Expedition in Old and New California and the South Coast, 1791–1792. John Howell Books, San Francisco, California.

Sosa-Ramírez, J., and E. Franco-Vizcaíno. 2001. Grazing impacts on mountain meadows of the Peninsular Ranges in La Frontera. Pp. 156–165 *in* G. Webster and C. Bahre (eds.), Changing Plant life of La Frontera: Observations of Vegetation in the United States-Mexico Borderlands. University of New Mexico Press, Albuquerque.

Thorne, R. F. 1988. Montane and subalpine forests of the Transverse and Peninsular Ranges. Pp. 537–557 *in* M. G. Barbour and J. Major (eds.), Terrestrial Vegetation of California. California Botanical Society, Davis.

Tibesar, A. (ed.). 1955. Writings of Junipero Serra [in English and Spanish]. Academy of American Franciscan History, Washington, D.C.

Tiscareno, F. (transl.), and J. W. Robinson (ed.). 1969. Jose Joaquin Arrillaga: Diaries of His Surveys of the Frontier, 1796. Dawson's Book Shop, Los Angeles, California.

Veblen, T. T., K. S. Hadley, and M. S. Reid. 1991. Disturbance and stand development of a Colorado subalpine forest. Journal of Biogeography 18: 707–716.

Wester, L. L. 1981. Composition of native grassland in the San Joaquin Valley, California. Madroño 28: 231–241.

Wiggins, I. L. 1944. Notes on the plants of northern Baja California. Contributions from the Dudley Herbarium 3: 289–305.

19

History, Ecology, and Conservation of the Pronghorn Antelope, Bighorn Sheep, and Black Bear in Mexico

RODRIGO A. MEDELLÍN

CARLOS MANTEROLA

MANUEL VALDÉZ

DAVID G. HEWITT

DIANA DOAN-CRIDER

TIMOTHY E. FULBRIGHT

Because the shape of Mexico resembles an upright funnel, most of the country's land surface area is in the north. A west-to-east cross-section of the country at around 25° N latitude would show the plains and mountains of the Baja California peninsula, coastal Sonoran plains, Sierra Madre Occidental, Mexican Plateau, Sierra Madre Oriental, and Gulf coastal plains (see chapter 1). This topographically intricate landscape encompasses a wide variety of habitats for wildlife, including montane oak and conifer woodlands and forests, grasslands, and extensive desert regions with plains and often sparsely vegetated peaks.

Among the wildlife present historically in these habitat types are several large, emblematic mammals. However, anthropogenic factors (e.g., hunting, predator-control programs, and habitat loss) in the region have already resulted in the extirpation of several of these species. The grizzly bear (*Ursus arctos*) and the Mexican wolf (*Canis lupus baileyi*) are 2 taxa now considered extinct in the wild in Mexico (Ceballos et al. 2002). Another large mammal, the elk (*Cervus elaphus*), occurs just in Coahuila, and only through successful reintroduction (Robles Gil et al. 1993).

Three additional large mammalian species are threatened in Mexico. The first species is the pronghorn antelope (*Antilocapra americana*), the sole remaining representative of the family Antilocapridae (order Artiodactyla), and a resident of the plains of temperate North America with extensions into desert plains. The second species is the bighorn sheep (Artiodactyla, Bovidae: *Ovis canadensis*), which inhabits the dry, mountainous areas of northwestern Mexico. The third species is the black bear (*Ursus americanus*), a large member of the order Carnivora (family Ursidae). In Mexico, the black bear inhabits the temperate coniferous and oak forests of the Sierra Madre Occidental and Oriental, as well as some of the outskirts of these mountain ranges. Between them, these 3 charismatic species once occurred over more than 90% of northern Mexico. Today, they have been identified by the Mexican government and various classification systems as species at risk. In this chapter we examine each of the 3 species in turn, reviewing their historical status and ecology and the conservation threats they face. We also discuss conservation and recovery programs specific to each of them.

Pronghorn Antelope

Traditional taxonomy indicates that there are 3 pronghorn subspecies with ranges limited to Mexico or extending into that country: the peninsular pronghorn (*A. a. peninsularis*) on the Baja California peninsula; the Sonoran pronghorn (*A. a. sonoriensis*) in extreme northwestern mainland Mexico and adjacent United States; and the Mexican pronghorn (*A. a. mexicana*) in the Chihuahuan Desert Region, from the central Mexican plateau to adjacent Texas and New Mexico (Hall 1981; fig. 19.1). On the basis of molecular genetic analyses, the taxonomic split may not be warranted (Conde 2000). Nevertheless, there is enough ecological information to indicate that pronghorn from grasslands and northern habitats may not survive under the harsh drought conditions of the Sonoran and Chihuahuan deserts. Although the pronghorn antelope is chiefly a resident of the North American Prairie, in Mexico it also inhabited an extensive area of the Sonoran and Chihuahuan deserts from the Baja California peninsula and northwestern Sonora eastward. In the Sonoran Desert, the species seems to occur naturally in lower densities than in the North American Prairie. All 3 Mexican subspecies are federally listed as Endangered in Mexico (Conde 2000; INE 2000a; SEMARNAT 2002).

History of the Decline

The pronghorn may be viewed as a relict species, as most of the North American Prairie, its chief historical habitat, has been altered or has disappeared. Before the arrival of European settlers, there may have been a total of 40–50 million pronghorn individuals. By 1929, however, pronghorn numbers had been drastically reduced to barely 30,000 individuals (Nelson 1925; Yoakum 1980; Cadieux 1987). Although uncontrolled harvesting was likely an additional important cause of decline, this 99.9% population reduction (Yoakum 1968) coincided with a loss in habitat of >75% (Christensen et al. 1995).

The original North American Prairie extended for several million square kilometers, from northern Mexico to southwestern Canada (Coupland 1992). This ecosystem was readily converted for human needs following the expansion of the European frontier, as colonists pushed north from the valley of Mexico City and west from the English colonies. Over time, the prairie receded, along with its indigenous human and nonhuman inhabitants (Coupland 1992). Today, most of the original prairie has been converted to some of the most strategically important agricultural fields, yielding grains such as corn and wheat. The remaining North

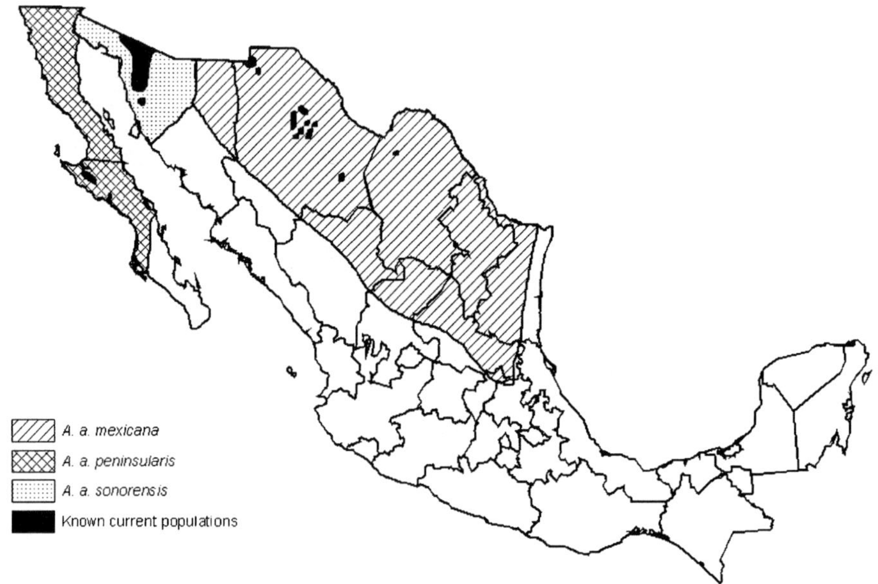

Figure 19.1. Historical distribution of the pronghorn antelope (*Antilocapra americana*) in Mexico, with locations of known current populations.

American Prairie is one of the most fragmented and endangered ecosystems of the continent.

Against this background of extensive habitat loss were other pronounced impacts of growing human populations on the resident wildlife. These impacts (overharvesting in the case of the pronghorn) caused the near-demise of several prairie dwellers in Mexico as well as in the United States. The black-footed ferret (*Mustela nigripes*) was considered extinct in the wild for a number of years in Mexico, but a recent reintroduction seems to have given the species a second chance (see chapter 21). Other species like the bison (*Bison bison*) are present in small pockets in national parks, biosphere reserves, and other protected areas and on some private ranches, through translocation operations and intensive management programs. The bison has all but disappeared entirely in the wild in Mexico (Meagher 1986), except for a relict population in the Janos-Casas Grandes region in northwestern Chihuahua (chapter 21).

In Mexico, the disappearance of the pronghorn from particular states has been poorly documented. In 1540, a hunt honoring the first viceroy of Mexico, Antonio de Mendoza, yielded some 600 animals, pronghorn and deer. The hunting area was less than 200 km northeast of Mexico City (Leopold 1965). In 1777, pronghorn were still abundant near the city of Durango, but toward the end of the nineteenth century they were already rare in northern Mexico. By 1922, the Mexican federal government banned pronghorn hunting, and the species was protected by a decree of President Obregón. In 1925, Nelson (1925) estimated that 2395 pronghorn occurred in northern Mexico: 600 animals in the state of Coahuila, 700 in Chihuahua, 595 in Sonora, and 500 in the Baja California peninsula. The last record from Zacatecas was in 1951 (Matson and Baker 1986), while the species was virtually extinct in San Luis Potosí by 1953 (Dalquest 1953). In 1955, a single herd was considered to be the last one in Chihuahua (Villa 1955).

All recent estimates indicate a pronghorn population of about 1000 animals or less in Mexico. By 1984, González-Romero and Lafón (1993) estimated that the species was present at only 11 separate locations, mostly private ranches in the states of Coahuila, Chihuahua, Sonora, and Baja California Sur, plus a small, reintroduced population in San Luis Potosí. Their methodology included a series of combined techniques that led those authors to estimate an 82% population decline in 60 years between 1925 and 1985 (González-Romero and Lafón 1993). Their very conservative estimate yielded a total of 446 pronghorn: 307 in Chihuahuan private ranches, 12 in Coahuila, 63 in Sonora, mostly in what is today the Pinacate Biosphere Reserve, and 64 in Baja California Sur. In 1993, the Baja California Sur population was estimated at 135 animals (Cancino et al. 1998). More recent estimates based primarily on fixed-wing airplane surveys, with also some ground searches, indicate between 282 and 564 in Chihuahua (Azuara et al. 2001), 200 in Baja California Sur (Cancino et al. 1995), and between 200 and 350 in Sonora (Valdés and Manterola 2001).

Ecology and Threats

Pronghorn are gregarious animals that maintain a well-defined social structure. Groups of pronghorn are largest during the winter and spring and are commonly led by old females (Byers 1997). Pronghorn have one of the highest reproductive rates per capita among ungulates. A large proportion of females produce twins in most reproductive events, and there is very close synchrony in the timing of birth, an important strategy to reduce relative mortality of fawns (Gregg et al. 2001). In this manner, the demographic strategy of the pronghorn is to rely on its numbers, satiating the predator guild by producing large numbers of young in a short period of time, in a relatively small area. For thousands of years, this strategy worked for the pronghorn. However, when the species' habitat became fragmented and converted to agriculture and hunting further reduced pronghorn numbers, the strategy became inadequate.

As already mentioned, the initial factors triggering the decline of pronghorn populations in Mexico and the rest of North America were likely habitat conversion and harvesting. Once the populations had dropped below a certain threshold, coyote predation and other sources of fawn mortality made it difficult for populations to recover, due to the high synchrony of female parturitions. Coyote populations have benefited from anthropogenic effects (e.g., garbage dumps, cattle ranching), and the extent and type of human activities can have important repercussions on coyote foraging behavior and predation pressure on prey species (Dumond et al. 2001). The number of pronghorn fawn surviving to recruitment into the population is correlated with the number of coyotes removed (Byers 1997). The peninsular pronghorn in particular faces very strong predation pressure: the impact of poaching is compounded by some of the greatest known coyote densities, as these predators receive supplemented food from the sea (marine carcasses and debris) precisely in the season when food resources should be limiting to them (Rose and Polis 1998).

Another likely negative factor has been the establishment of fences to divide pastures, properties, and right-of-way stretches near highways (Russell 1951; Gross et al. 1983). Some authors consider that pronghorn can negotiate most fences (Spillet 1964), while others believe fences are major obstacles and have invented devices to allow pronghorn to cross them (Mapston et al. 1970; Mapston 1972). However, we have witnessed pronghorn crossing under barbed wire fences. They can indeed cross these fences, but such action implies a virtual stop during which pronghorn kneel and crawl under the lower wire. This maneuver certainly makes them much more vulnerable to predation and poaching. All over Mexico, but especially in the grasslands of the north, fencing is widespread, with various materials and techniques used. Depending on land ownership, fences may extend over many miles, and several fences may occur along a 1-km transect. This pattern of fragmentation is likely to disrupt pronghorn movements, especially escape behaviors during encounters with predators.

Analogous to the effect of fences is that of roads with heavy traffic. Specifically, Highway 2 from Caborca to Sonoyta and Highway 45 from Chihuahua City to Juarez are 2 roads with increasingly heavy traffic, acting as effective barriers that prevent or severely curtail pronghorn movement. Highway 2 compounds the already dire situation of the Sonoran pronghorn, the most endangered subspecies.

Conservation and Recovery Efforts

After a dramatic population decline in the United States, the total number of pronghorn in that country has increased from an estimated 30,000 in 1924 to more than 1 million in 1983 (Cadieux 1987; see Flather et al. 1999, for a more conservative estimate). The species' recovery in the United States, achieved through active management, was all the more spectacular because more than 3.5 million pronghorn were legally harvested during this same period (Christensen et al. 1995).

Meanwhile, the Mexican pronghorn populations continued to decline, and the species disappeared from some Mexican states (see above). More recently, however, some measurable progress has been accomplished toward the recovery of the pronghorn in Mexico, through involvement of various nongovernmental organizations (NGOs), academic institutions, and government agencies. The peninsular pronghorn is undergoing an active reproductive program under semicaptivity inside the Vizcaíno Desert Biosphere Reserve (Cancino et al. 2001). From 10 males and 6 females captured as fawns in 1998, 1999, and 2000, today the program has more than 30 animals that continue to reproduce in large enclosures on site in the Vizcaíno Desert in Baja California Sur. These semicaptive individuals are in addition to 230 free-ranging animals occurring in Baja California Sur, and there is another effort in Nuevo León to reproduce the subspecies in captivity (J. Cancino pers. comm. to Medellín). The number of animals in captivity in Nuevo León is still small, but the goal is to continue breeding them until a reintroduction becomes feasible.

The Sonoran pronghorn is the focus of a recovery program spearheaded by local and federal government agencies and academic institutions. The Instituto del Medio Ambiente y Desarrollo Sustentable de Sonora (IMADES) and the Centro Cinegético Integral (CCI) have been conducting surveys, and they have a field station primarily for the study and protection of the Sonoran pronghorn (Valdés and Manterola 2001). Valdés and Manterola (2001) estimate a total of 40–60 individuals from central coastal Sonora, a region currently neither considered in recovery programs nor proposed for protection by a government agency. The Sonoran pronghorn also inhabits the Desierto del Pinacate Biosphere Reserve, where there are around 100 additional animals, but surveys and recovery programs in the reserve are urgently needed.

Some pronghorn recovery efforts rely on reintroductions, such as the one in Coahuila, where 2 groups (65 and 85 individuals) were reintroduced in 1996 and 1998, respectively. The reintroduced animals came from one of the few certified *Antilocapra americana mexicana* herds, in New Mexico, and the operation was carried out with the support of the New Mexico Game and Fish Department. Ongoing monitoring and protection of these groups occur in the Valle de Colombia, the release area. Monitoring indicates that, after an initial decline, presumably due to the acclimation and to dispersal outside the Valle de Colombia, there have been at least 2 reproductive events, and the population inside the valley is now calculated around 70 animals (Valdés and Manterola 2001). Unfortunately, this population, and probably others, too, face a serious challenge, one that is also potentially expensive for conservation. Although demographically this population may be healthy, with sufficient reproduction and recruitment and limited mortality, the habitat is rapidly being destroyed. Even if the population is viable in the long-term, it will not grow as a result of habitat fragmentation and alteration. As already mentioned, heavy traffic along Highway 2 is a threat to the Sonoran pronghorn. However, it is unlikely that the Mexican administration of highways will do much to facilitate safe passage of the Sonoran pronghorn across this highway. Thus, management programs should focus on minimizing pronghorn crossing.

Through the Federal Wildlife Division (Dirección General de Vida Silvestre), a division of the Minister of the Environment (Secretaría de Medio Ambiente y Recursos Naturales), the Mexican federal government established in 1997 a Committee for the Recovery of Priority Species. This committee comprises NGOs, academic institutions, and other organizations, and it is an advisory instrument for the decision-making process toward the recovery of the priority species (INE 1997), coordinated by the federal government. Committee members are the presidents of each of the subcommittees on particular species that have been established, plus members of the conservation community and government officials. The National Technical Advisory Subcommittee for the Conservation, Management, and Use of the Pronghorn in Mexico was established in 1999. It meets twice a year and is responsible for preparing guidelines, strategies, and projects to remove the 3 subspecies of pronghorn from the category of Endangered, through recovery (SEMARNAT 2002). The pronghorn subcommittee has established a web page to inform the public and update information about the species (http://www.berrendo.org.mx), and a book broadly describing the objectives and recovery program has been published (INE 2000a). Overall, the stage is set for the recovery effort, but there is much to do, especially for the Sonoran pronghorn.

Bighorn Sheep

The single representative of the subfamily Caprinae of the family Bovidae in Mexico, the bighorn sheep (*Ovis canadensis*) is one of the largest sheep in the world. It occurs in naturally fragmented populations throughout its distribution along the western mountain ranges of North America from northern Mexico to the northern Rocky Mountains in southwestern Canada (Hall 1981; Valdez and Krausman 1999). Six subspecies are recognized by taxonomists and are usually grouped in 2 morphotypes or ecotypes: the rocky mountain bighorn sheep, with 2 subspecies (*O. c. canadensis* and *O. c. californiana*), and the desert bighorn sheep, with 4 subspecies (*O. c. weemsi*, *O. c. cremnobates*, *O. c. nelsoni*, and *O. c. mexicana*). This classification is neither strict nor well defined (Valdez and Krausman 1999), and some studies have indicated that at least *O. c. nelsoni*, the California desert bighorn, belongs to the same subspecies as the Baja California desert bighorn, *O. c. cremnobates* (Wehausen and Ramey 1993). Of the 6 subspecies, 3 occur in Mexico: the southern Baja California bighorn sheep (*O. c. weemsi*), the Baja California bighorn sheep (*O. c. cremnobates*), and the Mexican bighorn sheep (*O. c. mexicana*). The bighorn sheep is federally listed as Subject to Special Protection in Mexico (SEMARNAT 2002).

History of the Decline

The historical distribution of the desert bighorn sheep in Mexico encompasses 3 distinct desert areas, 2 of them in the Sonoran Desert. It is found in the Vizcaíno Desert of the Baja California peninsula. It is also found in the desert areas of northern Baja California and Sonora. The third area is the Chihuahuan desert in central-northern Mexico, encompassing the xeric mountain ranges of Chihuahua, Coahuila, and a small portion of Nuevo León (fig. 19.2). Throughout this vast region, bighorn sheep populations occured in isolated patches of adequate habitat characterized by deep, steep canyons, rocky hills and mountains, and cliffs that conferred them a high degree of visibility (Risenhoover and Bailey 1985). This is typical of the sharp edges and abrupt slopes of desert mountains such as the Sierras Seri and Bacha in Sonora, the Sierras San Pedro Mártir and La Giganta in Baja California, and the Sierras Chupadero and El Soldado in Chihuahua.

In the first decade of the twentieth century, bighorn sheep populations began to suffer dramatic reductions, and their distribution has since then shrunk significantly (Buechner 1960; Bailey 1980; Hansen 1982). The exact severity of the decline is unclear, as is the original number of bighorn sheep in North America, before the arrival of European

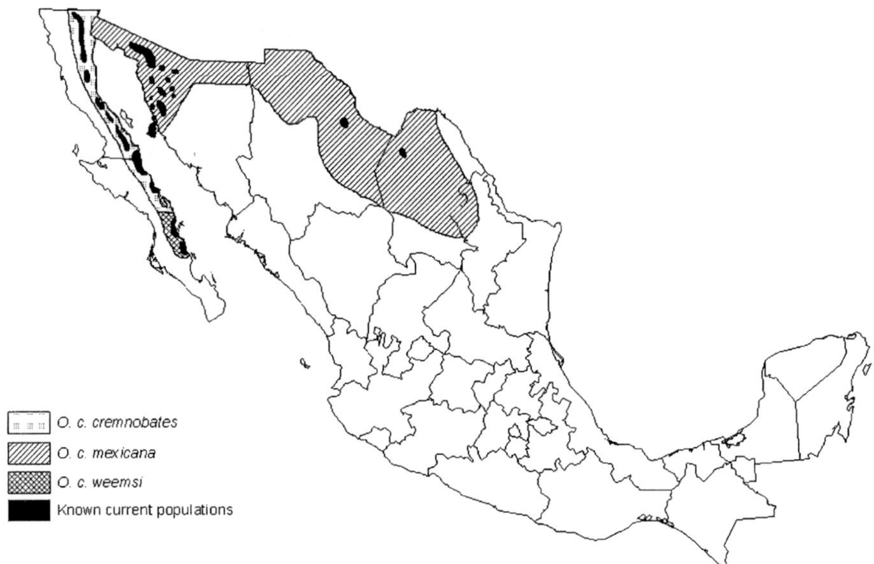

Figure 19.2. Historical distribution of the bighorn sheep (*Ovis canadensis*) in Mexico, with locations of known current populations.

settlers. There are estimates of 4 million sheep, including the Dall's and Stone sheep, *Ovis dalli* (Seton 1929), most of which in retrospect would have been bighorn (*O. canadensis*). However, Valdez (1988) estimated that there were never more than 500,000 wild sheep in all of North America. Buechner (1960) even suggests that in colonial times there were only between 15,000 and 20,000 wild sheep in the contiguous United States.

Sheldon (1925) indicates that, around 1860, there were bighorn sheep inhabiting the mountains surrounding the city of Chihuahua, and some rams were killed 20 miles north of that city. However, by the turn of the century (1898–1902), bighorn sheep were already rare and Sheldon reports only about 3 bands in the same region. Sheldon returned in 1925 and could not find signs of live sheep. They may have survived as isolated groups of individuals until later, as Anderson (1972) mentioned sightings in the Sierra del Diablo in Chihuahua shortly before 1954. He also reported occasional movement of sheep from the Big Hatchet Mountains in New Mexico south into Chihuahua through the Cañón de Santa Elena. Further anecdotal evidence suggests that there may have been some isolated sheep in Chihuahua as recently as the 1970s. Today, still, there may be transient animals crossing from the Black Gap or other surrounding areas. There is a

skull that seems to be at least 10 years old, in the possession of a rancher near the Sierra el Chupadero, south of Big Bend National Park. Overall, however, it is likely that the species is demographically extinct in Chihuahua.

Baker (1956) indicated that in the early 1950s, bighorn sheep were already very rare in Coahuila, and when Villa (1959) visited that state, he could not find a single verified live animal. Villa estimated that in all of Baja California there were between 1500 and 3000 sheep. He stated that he could not see a single live individual in Chihuahua in 4 visits in consecutive years. In Sonora, Villa (1959) estimated a population of about 200 animals, after flying in a small plane around Punta Cirios, Los Mochos, El Chino, and Cerro Viejo, all in central coastal Sonora. Villa stated that bighorn sheep had disappeared from El Pinacate, but this was an oversight, as sheep are known to have been there continuously until today. A land survey in the early 1970s was the basis for an estimate by Alvarez (1976) of between 4560 and 7800 animals for the entire Baja California peninsula. Mendoza (1976) estimated that in 1976 there were about 935 sheep in the state of Sonora. Monson (1980) examined the scant information published and suggested that there were between 4560 and 7800 sheep in all of Baja California, 900 in Sonora, 50 in Chihuahua, and 50 in Coahuila, for a grand total for Mexico of between 5560 and 8800 bighorn sheep in the late 1970s. By 1991, Valdez and Krausman (1999) estimated that there were around 23,000 desert bighorn sheep in Mexico and the United States.

Ecology and Threats

The bighorn sheep lives in naturally fragmented landscapes, habitat islands and archipelagos that are separated from one another by wide areas with unsuitable valleys and plateaus. This fragmented distribution is likely to bring about a metapopulation structure with reduced genetic flow between populations (Hansen 1980; Bleich et al. 1990) and thus increases the risk of extinction. Also relevant are the behavioral ecology and mating system (male-dominance polygyny) of bighorn sheep, which reduce the effective population size to less than the total population size (Fitzsimmons and Buskirk 1992).

The bighorn sheep's severe decline in recent times has been attributed to human-induced factors such as disease, overgrazing by cattle and feral burros, road and highway construction, urban development, irrigation channels, shrub cleaning or fire control, and recreational activities such as mountain hiking (Krausman et al. 1999). Many landowners have introduced barbary sheep (*Ammotragus lervia*) to the southwestern United States and northern Mexico (Ceballos et al. 2002). Wherever initially the barbary sheep coexisted with the bighorn sheep, the latter has disappeared. A similar situation has occurred with feral burros and domestic goats. Burros are abundant in many desert mountain ranges of northern Mexico, and barbary sheep are widespread in Chihuahua, Coahuila, and Nuevo León, 3 states that are devoid of bighorn sheep today. In addition to having a competitive advantage (wider niches, greater reproductive potential, and higher resilience against human activities), introduced species are more resistant to diseases than the bighorn sheep (Seegmiller and Simpson 1979; Sandoval 1980; Seegmiller and Ohmart 1981; Ginnett and Douglas 1982; Krausman et al. 1999). Contact with domestic livestock has repeatedly caused epizootic outbreaks (*Pasteurella*, pneumonia, scabies, mycobacterium, or other ecto- or endoparasitic diseases) in bighorn sheep populations resulting in local extinctions (Jessup 1985; Bunch et al. 1999).

Poaching is no small threat to bighorn sheep (see Kelly 1980a, for a summary of the origins of hunting of bighorn sheep and its implications). In Arizona and Nevada in the late 1960s, poaching accounted for most (41%) of the bighorn sheep mortality, with the second factor of mortality, road traffic, causing only 20% of bighorn sheep deaths (Welsh 1971). Geist (1971) includes poaching among the major causes of bighorn sheep population decline. In Mexico, federal agencies are overwhelmed and do not have the capacity to carry out realistic law enforcement operations. Thus, poaching continues to be a problem, particularly in areas where local landowners do not receive financial incentives.

Although bighorn sheep are affected by predators, they are not as susceptible to increased predation pressure as are the pronghorn, partly because of the low densities that predators can reach in bighorn sheep habitat (Kelly 1980b). The mountain lion seems to be the chief predator of sheep. Coyotes and bobcats may be important lamb predators, but only under certain circumstances such as around waterholes and areas with no escape terrain nearby (Krausman et al. 1999). Overall, however, predation rarely affects the persistence of a population (Krausman et al. 1999).

The likelihood that bighorn sheep will flee an area is highest in the presence of hikers, followed respectively by that of motor vehicles and that of mountain bicycles (Papouchis et al. 2001). Recreational activities in areas inhabited by bighorn sheep are a cause of concern. They may be fatal to sheep through stress-induced falls, runovers, or other accidents, and may also have caused sheep to abandon several areas, notably in California and Arizona (Geist 1971; Graham 1971, 1980; Etchberger et al. 1989; Harris 1992; Canfield et al. 1999). Recreational activities are affecting increasingly larger areas in bighorn sheep habitat in Mexico.

Conservation and Recovery Efforts

Population estimates since the 1990s suggest stability and in some cases limited recovery of the bighorn sheep in Mexico. In Baja California Sur, the number of sheep was estimated in 1992 at 500–700 animals (Jaramillo-Monroy and Castellanos-Vera 1992). In 1992, Lee and Mellink (1996) conservatively estimated population size in some mountain ranges in western Sonora at between 1488 and 2977 sheep. A subsequent survey in 1996 indicated that the populations in Sonora were "at least stable if not increasing." Also in 1992, Lee and Mellink estimated the population for the state of Baja

California (the northern half of the peninsula) at 780–1170. For this population they also concluded that between 1992 and 1995 the population was either stable or increasing. In continental Sonora, population size was estimated at a minimum of 2100 animals in 2000 (INE 2000b), with only some mountain ranges surveyed.

The population on Tiburón Island (see also below), off the coast of Sonora in the Sea of Cortez, was recently estimated at about 600 animals (Medellín unpubl. data; R. Lee, pers. comm. to Medellín). The population in the state of Baja California Sur was estimated at about 2000 animals, and in the state of Baja California the estimate is also about 2000 (INE 2000b), for a grand total for Mexico of 6700 sheep. Although the surveys are still incomplete and the historical record shows many gaps (Tarango and Krausman 1997), there is an apparent recovery or stability of the remaining populations after the sharp decline between 1900 and 1990. Clearly, however, 10 years of stability are not enough to indicate that the species is on its way to recovery. Also, where active conservation actions are not continuous and effective, poaching continues to be a significant threat.

In 1975, the more than 120,000 ha Tiburón Island was the object of a bighorn sheep introduction program. A group of bighorn sheep from continental Sonora, specifically the Pico Johnson in the Sierra Seri and Punta Cirios in Sierra Bacha, were placed on the island, where there are no previous records for the species (Montoya and Gates 1975; Felger and Moser 1985). Twenty sheep, 16 females and 4 males, were introduced. Of these, 16 lived long enough to leave descendants on the island. The first survey of this population was conducted in 1984 (DeForge et al. 1984), when the population was estimated at 120 animals. In 1993, Lee and López-Saavedra (1994) observed 293 sheep on Tiburón. As they indicated that only 30–60% of the sheep are seen during helicopter surveys (the method they used) their study would place the Tiburón population at 488–977 sheep. Annual surveys through 2001 (Medellín, unpubl. data; R. Lee, pers. comm. to Medellín) suggests that the population on Tiburón Island has been fluctuating between 360 and 500 (conservative estimate), or possibly between 500 and 800 animals (optimistic estimate).

In general, translocations of bighorn sheep do not succeed in establishing large populations (Douglas and Leslie 1999). Tiburón Island represents one of the most successful introductions of desert big-

horn sheep in the world, and it has been proposed also as an example of sustainable use of wildlife (Medellín et al. 1999). The growth rate of the bighorn sheep population on Tiburón Island is very high, possibly due to a series of factors, including the absence of mountain lions and cattle on the island, and the fact that Tiburón is the last great extension of the Sonoran Desert without major human disturbance. The program on this island has yielded a very significant income for the Seri Indian group, legal owners of the island but lacking adequate development programs. Today, there is a capable technical team trained to carry out surveys, participate in scientific studies, or guide hunts. Tiburón Island and Seri Indian involvement demonstrate that conservation, management, and economic development can all be part of the same strategy in a same area (Medellín et al. 1999; Medellín and Colchero 2002).

The Program for Conservation of Wildlife, launched by the federal government in 1996 (INE 1997), established the basis for local landowners to benefit from wildlife harvest through management and habitat improvement. As a result, local landowners started sheep conservation and recovery programs in Sonora and Baja California. There have been various translocations in Sonora and to Isla Carmen, another Island of the Gulf of California (DeForge et al. 1998). Today, the vast majority of occupied bighorn sheep habitat is registered as UMAs, or Units for the Conservation, Management, and Use of Wildlife (INE 1997). In the UMAs, wildlife harvesting is allowed, once a management plan with annual monitoring and habitat protection has been approved by the wildlife agency (Dirección General de Vida Silvestre). Translocations in Sonora have established at least 16 additional populations within the species' historical range (INE 2000b). These populations have received 157 animals from the Tiburon Island population. Most translocations have been private operations supervised by the federal government and, although recent (the first one received its founding stock in 1995), so far all of them have shown lamb recruitment and population growth (M. Valdéz, pers. obs.). Hunting is permitted only in the states of Sonora and Baja California Sur, where the number of permits per year has varied from 10 to 57 (both states) since 1995. An additional 22 sheep from Tiburon Island have been used to start a recovery program for the species in the state of Coahuila. This program, an initiative of the NGO Unidos para la Conservación with the participation of the Institute of Ecology of the National University of Mexico, is the first one in Mexico to attempt the species's recovery in a state where it had been extirpated.

Bighorn sheep populations are known to exist in at least 5 protected areas of Mexico: Reserva de la Biósfera El Vizcaíno and Parque Nacional Bahía de Loreto in the state of Baja California Sur, Parque Nacional Sierra de San Pedro Mártir, in the state of Baja California, Islas del Golfo de California, specifically Tiburón and Carmen islands in the Sea of Cortez, and Reserva de la Biosfera El Pinacate y Gran Desierto de Altar in Sonora. These areas have varying degrees of real protection, but Tiburón and Carmen islands, as well as the Reserva de la Biosfera El Vizcaíno, because of the continuous presence of personnel and intimate participation of local people, are clear examples of what adequate protection can achieve.

More time is needed to assess the bighorn sheep's recovery. As with the pronghorn, fragmentation and habitat alteration by roads, urban encroachment, and other factors all severely restrict bighorn movement and affect the probability of survival. In the future, the status of the bighorn sheep may be seriously negatively affected by the project *Escalera Náutica*, geared toward the development of urban infrastructure along both coasts of Baja California. Although some modification of the human infrastructure may benefit other wildlife such as deer, birds, or bats, sheep are typically very sensitive and do not tolerate human activities in their areas.

Within the Committee for the Recovery of Priority Species, a subcommittee for the bighorn sheep was recently established (E. Canales, pers. comm. to Medellín). The goals and operation are similar to those of the pronghorn subcommittee. Comparatively, the bighorn sheep is doing better than the pronghorn in Mexico. This is likely because the pronghorn once inhabited areas with much greater potential for human use (agriculture and cattle ranching).

Black Bear

History of the Decline

Like pronghorn antelope and bighorn sheep, black bears have large home ranges and normally avoid contact with people. Unlike pronghorn and bighorn, however, bears are primarily forest-dwelling animals. For this reason, black bears are less visible and more difficult to census, and less is known about their distribution and abundance. Constructing range maps is further complicated because young bears, particularly males, may travel up to 200 km from their natal area (Elowe and Dodge 1989; Rogers 1987), making reports of individuals difficult to interpret.

The distribution of black bears extends from the tree line in northern Alaska and Canada, southward into chaparral and forested areas of Mexico (Pelton et al. 1999). The best-documented range map, published by Leopold (1959), reported black bears in montane areas of the Sierra Madre Occidental from Mexico's northern border to Zacatecas, western Nayarit (Baker and Greer 1962), and northern Jalisco (Tinker 1978). Hall (1981) also included Aguascalientes in the range of the species. In the Sierra Madre Oriental, black bears were found in mountainous areas of Coahuila, Nuevo León, Tamaulipas, and San Luis Potosí (Dalquest 1953; Leopold 1959; Tinker 1978; Hall 1981; see fig. 19.3). Until the early 1900s, black bears shared their range in eastern Sonora, Chihuahua, and parts of Durango and Coahuila with grizzly bears (*Ursus arctos*), now considered extirpated from Mexico (Brown 1985; Ceballos et al. 2002).

How much of the black bear's historical distribution (fig. 19.3) is occupied today? A recent range map of the black bear (Pelton et al. 1999) showed known populations only in Chihuahua, Coahuila, and Nuevo León. However, reproductively active populations have now been documented in central and western Chihuahua (Doan-Crider, unpubl. data), northeastern Sonora, northern Coahuila (Doan-Crider and Hellgren 1996), central Tamaulipas (S. Ledezma Pineda, pers. comm. to Doan-Crider), and the Sierra Madre Oriental and Sierra Picachos in Nuevo León (Castro 1984; Nino Ramirez 1989; Zepeda-Gonzalez et al. 1997). The present-day status of populations in other areas of the bear's original range remains unverified (fig. 19.3).

The timing and extent of a decline of black bear populations is difficult to assess because there are so few data available on the species' abundance and distribution, past or current, in northern Mexico. Published accounts indicate that the species was decreasing in numbers before the mid-1980s, threatened by increased mortality rates caused by overhunting and encroachment on bear habitat (Baker 1956; Leopold 1959; Baker and Greer 1962). Apparent population declines were reported in particular

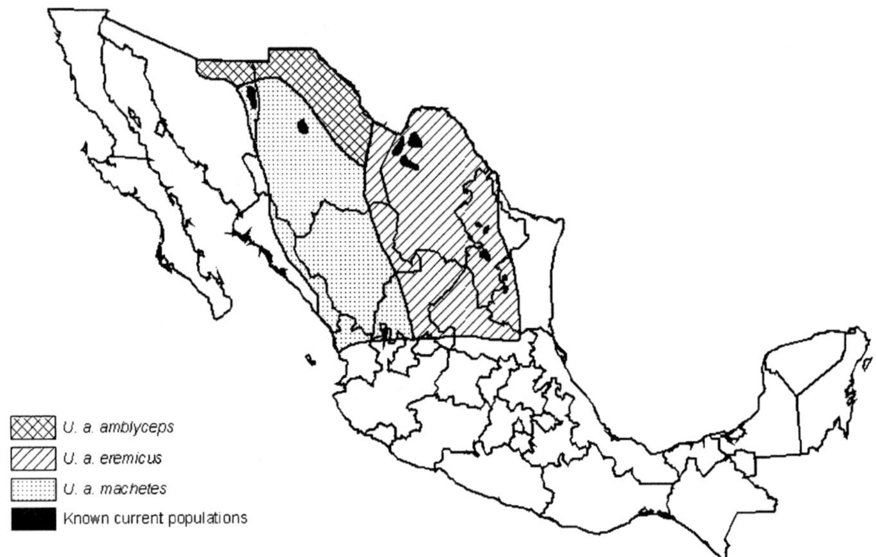

Figure 19.3. Historical distribution of the black bear (*Ursus americanus*) in Mexico, with locations of known current populations.

from both San Luis Potosí (Dalquest 1953) and Zacatecas (Matson and Baker 1986).

A population decline of the black bear occurred in the southwestern United States during the 1920s and 1930s, due to government trapping programs, use of poisons, and increased human use of the species' habitat (Brown 1985). In Mexico, and through most of the twentieth century, black bears were hunted, trapped, and poisoned for protection of property, for food, fat, and hides, and for sport (Burbridge 1908; Baker 1956; Leopold 1959). Land reform from 1915 through the 1940s brought many people into rural parts of northern Mexico (Whetten 1948), exposing bears to increased persecution because they were perceived as a threat to cattle and crops.

As populations of black bears declined, the Mexican government placed regulations on the harvest of these animals (Leopold 1959), and attempts were made at enforcement (Tinker 1978). Resources and personnel were insufficient, however, and bears were killed frequently. Baker (1956) and Leopold (1959) both felt that hunting in the 1940s and 1950s was no longer sustainable, and Baker (1956) recommended a ban on hunting black bear. Recent changes in attitudes toward black bears in the livestock industry and better law enforcement initiatives through the Mexican government have contributed to what appears to be increasing population trends for the black bear, at least in some areas (INE 1999). Currently, *Ursus americanus eremicus* is federally listed as Endangered in Mexico (SEMARNAT 2002).

Ecology and Threats

There are 3 primary habitat requirements for black bears. As members of the order Carnivora, black bears have simple digestive systems and narrow food requirements. They are unable to digest structural carbohydrates in plants (Pritchard and Robbins 1990), a physiological trait that makes the species dependent on high-quality foods such as animal material and plant parts that contain sugar, fat, or starch. Female bears in particular need a high-calorie, high-protein diet, as they give birth to their cubs during hibernation and lactate for at least 2 months before emerging from the den. While the female is in the den, her metabolism, gestation, and lactation are supported entirely by fat and protein stores acquired before hibernation (Hellgren 1998).

Even with the above requirements, the diet of black bears consists predominantly of plant material. In the Serranías del Burro, Coahuila, and Big Bend National Park (BBNP), Texas, plant material was 95% and 93%, respectively, of material found in bear feces (Doan-Crider 1995), and animal material was primarily insect remains. Oak (*Quercus* spp.)–shrub communities and oak forests of northern Mexico provide a wealth of bear foods. Oaks provide acorns, which are a dominant part of bear diets wherever they are found. Texas persimmon (*Diospyros texana*), agrito (*Berberis trifoliolata*), and manzanita (*Actostaphylos* spp.) are examples of shrubs providing fruits high in digestible carbohydrates.

The second requirement that bear habitat must provide is cover. Escape trees may be particularly important for female bears with cubs, for protection (Bartoskewitz 2001). Black bears require den sites in which females can secure their cubs during hibernation.

Water is the third requirement. It must be available in the landscape to sustain a bear population, as individuals outside of hibernation appear to depend on the presence of drinking water. In arid environments, bears and bear signs are frequently seen around water (Tinker 1978; Broyles 1995; D. Doan-Crider, pers. obs.). Livestock producers in Mexico have developed permanent water sources in many areas where surface water was lacking and thus may have influenced the local distribution of black bears.

In northern Mexico, habitat requirements of black bears are met in both oak–shrub communities and montane oak or oak–pine forests. Because such wooded habitats are limited to higher elevations, the species' habitat in Mexico is patchily distributed across the landscape. Although additional research is needed for conclusive evidence, black bear populations in northern Mexico likely function as meta-populations (Doan-Crider and Hellgren 1996; Onorato and Hellgren 2001). In northeastern Mexico, patches of oak–shrub communities and montane forest occur within a matrix of Chihuahuan desertscrub, grasslands, and Tamaulipan thornshrub. These low-elevation rangelands do not appear to be a geographic barrier to black bears during dispersal. Bears from northern Mexico colonized BBNP in Texas in the 1980s (Skiles 1995; LoBello 1989), an event that may have required them to traverse more than 25 km of Chihuahuan Desert. A small population of black bears has existed for at least 5 years in the Black Gap Wildlife Management Area, northeast of BBNP in Texas. As this area is dominated by lower elevation Chihuahuan Desert plants, we predict that some large mountain ranges with suitable habitat, such the Serranías del Burro, Coahuila (Doan-

Crider and Hellgren 1996), support bear populations acting as sources to help maintain adjacent ephemeral or sink populations.

Water is likely important in dispersal corridors. Bears dispersed from BBNP during a severe drought in 2000, and some died before they could reach Mexico (Onorato et al. 2002). Causes of death were not verified, but dehydration was suspected for 1 female and her cubs. It is possible that corridors, which correspond to lower elevation, drier lands, are more viable during good rain years than during drought. Ranchers have often reported increased problems with bears during drought, when these animals concentrate around cattle-drinking areas. In addition to water, landowners could play a critical role in dispersal corridors with their decisions of whether to tolerate the presence of bears on their properties.

Montane pine forests in northern Mexico have been impacted significantly by logging and other human activities (chapter 3). Although Leopold (1959) believed that deforestation was a factor in the decline of bear populations, its impact was likely not as prounounced as that of the more extensive loss of North American Prairie habitat on the pronghorn. The black bear is adapted to a wide range of wooded habitats. Overhunting and persecution rather than habitat loss have had the greatest impact on black bear populations. Black bears have many traits of a K-selected species: reproductive rates are low, survival rates are high (Doan-Crider and Hellgren 1996), and investment in offspring is high. Reproduction may vary among years (Rogers 1987), but after years of poor cub production, high adult survival rates ensure reproduction in subsequent years when conditions improve. As with most K-selected species, bear populations are highly susceptible to increased mortality (Miller 1990).

Human activities such as livestock grazing, agriculture, or even logging can have both positive and negative impacts on bears. These animals benefit from livestock water tanks, a fact that likely affects their movements, as they normally would have to wander to find water during dry times. Because black bears spend most of their time in search of food, they have learned to adapt to new, human-associated sources of food. Their foraging now includes predation upon vulnerable livestock such as sheep, pigs, and young calves, and feeding on bee hives, corn crops, and orchard trees such as pecans. Bears also learn to obtain easy food at garbage dumps, ranches, or townships. However, they sometimes cause extensive damage to human properties and, if they become no longer afraid of people, may even threaten human lives. Thus, increased food sources can augment reproductive rates in bears and help them stay fat for the winter, but they can ultimately result in further persecution and negative attitudes.

Little is known about the effects of logging on bears, but the removal of canopy cover may ultimately affect survival rates (LeCount et al. 1984; Bartoskewitz 2001). Alterations in vegetation communities that favor growth of food-producing woody species may benefit these animals, but unsustainable use that causes degradation of the watershed and soil may lead ultimately to a lower carrying capacity of the landscape. The effect of any human activity on the species depends on many factors, such as the intensity of the activity, the type of tools used, precipitation patterns, the vegetation community, and the attitudes of the people involved.

Conservation and Recovery Efforts

Although several efforts were made in the 1950s to protect bears in Mexico, the legislation was rarely enforced (INE 1999). In 1986, however, the Mexican government officially declared black bears an Endangered species, reemphasized previous bans on bear hunting, and followed up with some enforcement actions (INE 1999). Black bears are listed as an Appendix II species under the Convention on International Trade in Endangered Species (CITES), to which the Mexican government became a signatory in 1991 (Hemley 1994). This convention provides a mechanism to confront the illegal killing and export of black bears. The Mexican government also established the Technical Consulting Subcommittee for the Conservation and Management of the Black Bear in Mexico (INE 1999), with the intent of developing goals toward a nationwide conservation strategy for the species. Another important step in black bear conservation began during the 1970s and 1980s, when initiatives by groups of private landowners in several areas of northern Mexico began to actively promote black bear conservation on their lands. One example of their efforts is in the Serranías del Burro, Coahuila, where black bear density in the early 1990s was in the upper 25% of black bear population densities recorded in North America (Doan-Crider and Hellgren 1996). This population has maintained a high density through 2002 (Doan Crider and Hewitt, unpubl. data). It is

through dispersal from that population that black bears recently have recolonized areas of western Texas (see Taylor 1992).

The persistence of black bear populations in northern Mexico has been aided by many factors during the past 3 decades. First, actions by the federal government in the 1980s and 1990s, as described above, provided legal protection. Second, the attitude of the North American public toward bears changed during the twentieth century. Most North Americans now have positive views of bears and bear conservation, although rural residents are still more likely to have negative views (Kellert 1994). This change in attitude toward bears seems particularly prevalent in northern Mexico and has made people more tolerant of bear nuisance activity. A third factor contributing to the stabilization of black bear populations, despite a rapidly growing human population in the country, is a decline in the proportion of people that live in rural areas. In 1950, 57% of the Mexican population lived in rural communities (< 2499 inhabitants), whereas in 2000, only 25% of the population lived in rural areas (Whetten 1948; INEGI 2000). Although this may not represent a decline in the number of people living in rural areas, the fact that many of the northern states in which bears are found have some of the lowest human population densities in Mexico (INEGI 2000) means that interactions between bears and people may not be increasing as rapidly as they would elsewhere in Mexico.

A final change that has benefited black bears is an apparent change in land-use patterns. Although livestock production continues to be a prominent land use in much of northern Mexico, collaboration between ranchers and conservationists has introduced new perspectives. Black bear management programs such as in the Serranías del Burro, Coahuila, have demonstrated that black bears can successfully persist in high-intensity, livestock-producing areas. Many ranches are now owned in part for recreation. Their owners often derive aesthetic value from bears and manage their land, wildlife, and livestock in a way compatible with these animals. In northeastern Sonora, a similar situation perhaps prevails in Las Cuevas along the Bavispe River. During jaguar (*Pantera onca*) surveys, Valdez et al. (2002) recorded the local presence of black bears. They indicate that most of the surveyed area is private property, and cattle can be found throughout, with water available nearby.

Black bear recovery in Mexico will depend on determining the population status of the species throughout its potential range. As with the jaguar program (Sanderson et al. 2002), it will also require identifying priorities for conservation efforts. A first goal should be to ensure that established populations remain secure. The second goal should be to use abundance, location, and land-use patterns to identify those populations not yet secure but with a potential for long-term presistence. A third goal should be to understand the population structure of black bears and protect dispersing individuals.

In addition to management of the black bear, there is still a need for public education programs. The biology of the species, the true level of threat it presents to people and property, and means to benefit from and achieve peaceful coexistence with the black bear are important information for the public. To help improve the willingness of people to accept bears, programs should also be initiated to reduce bear–human conflicts by responding quickly and effectively when these animals cause problems. Sustainable use of montane areas supporting bear populations should be promoted. The species would benefit from maintaining a high diversity of food plants because total mast failures would become less likely, as would conflicts between people and bears. The ability of an area to support the species may be enhanced by the provision of supplemental water. Where the carrying capacity for black bears may have been exceeded, implementing management programs to assist landowners should be considered. Finally, laws regulating bear hunting should be actively enforced, but a flexible system must be in place to manage each bear population according to the needs of people and wildlife across the diverse Mexican landscape.

Conclusion

The pronghorn antelope, bighorn sheep, and black bear represent 3 species of large mammals that were quite widespread in North America. Conservation and recovery efforts for these species are the result of collaboration among academic institutions, NGOs, the government, and private landowners. They are also a good example of successful international collaboration. It is important to recognize, however, that the majority of breeding and reintroduction programs have been implemented through private or international funding, while government funds have been only sporadic and limited. Although in general the government has been supportive of conservation and recovery efforts, it is also true that some recovery actions have met with governmental opposition and diverse official obstacles.

Although much progress is needed, several populations of the pronghorn antelope, bighorn sheep, and black bear arguably have gained a foothold in northern Mexico, with promising signs of species recovery. Such recovery, however, can only be partial, as some regions within the historic range of each species are unlikely to be reclaimed. This is especially true for the pronghorn. This species is likely to recover the smallest proportion of its historical distribution, which now contains urban centers and other areas converted for the use of humans. For the pronghorn, then, we will have to be content with establishing populations in some relatively remote and well-conserved regions, especially in Baja California Sur, Sonora, Chihuahua, Coahuila, and maybe in Baja California Norte, Zacatecas, Durango, and San Luis Potosí. It will be an interesting challenge for the conservation community to break ground and successfully establish populations in states such as Nuevo León or Tamaulipas.

The black bear faces similar challenges, due primarily to conflicts with people but also to some loss of habitat. For these reasons, it would be economically and socially expensive and logistically and biologically difficult to recover populations in some areas. In the case of the bighorn sheep, the situation is somewhat easier and brighter. Because bighorn sheep inhabit areas that are inherently inhospitable to people, the probabilities of recovering populations in states where it became extirpated are better. It is not, however, directly feasible to do so. Many desert mountain ranges in central and western Mexico are populated by goats, barbary sheep, and other animals that outcompete bighorn sheep or are vectors of disease that kill them. Only with the adoption of an energetic policy to encourage landowners to remove exotic species will the bighorn sheep recover some ground.

In many regions of Latin America, the current economic trend of globalization pushes resources, monetary and otherwise, away from environmental issues. If governments and citizens react in time and open their eyes to the fact that only ecologically sound development is economically feasible, the future outlook for wildlife in Latin American countries will be positive. The time is fast approaching when the metaphoric coin that will decide the fate of both biological diversity and humanity will finally hit the ground. The pronghorn antelope, bighorn sheep, and black bear are 3 companions with whom we share this fate. It is up to us to recognize our responsibility as stewards and achieve success in conserving wildlife.

Acknowledgments

We thank J. L. Cartron, G. Ceballos, and R. Felger for their invitation to participate in this book. J. Cancino provided insight about the Baja California pronghorn. R. Lee has provided much guidance and logistic support to our efforts conserving bighorn sheep on Isla Tiburón. The Foundation for the North American Wild Sheep has also provided financial and organizational support. We thank P. Robles Gil, F. Ramírez-Ruiz de Velasco, D. Conde, G. O'Farrill, I. Cassaigne, F. Colchero, and especially F. Rodríguez and the Seri Indian Community for their enthusiastic commitment to conserve bighorn sheep. We also thank CONECO, AC, Sergio Ledezma P., Gildardo Castro, and Enrique Guadarrama for sharing their knowledge and providing support for our work with black bears. R. Felger, J. L. Cartron, G. Ceballos, and J. Arroyo dilligently provided useful comments that improved the manuscript. This chapter corresponds to publication 03-105 of the Caesar Kleberg Wildlife Research Institute and was finished while R. Medellín was on sabbatical at CERC, Columbia University.

Literature Cited

Alvarez, T. 1976. Status of the desert bighorn sheep in Baja California. Desert Bighorn Council Transactions 20: 18–21.

Anderson, S. 1972. Mammals of Chihuahua, taxonomy and distribution. Bulletin of the American Museum of Natural History 148: 149–410.

Azuara, D., R. A. Medellín, C. Manterola, and M. Valdés. 2001. Pronghorn (*Antilocapra americana mexicana*) population surveys in Chihuahua, estimated by aerial surveys. Pp. 95–106 *in* J. Cancino (ed.), Proceedings of the 19th Biennial Pronghorn Antelope Workshop. Centro de Investigaciones Biológicas del Noroeste, La Paz, Mexico.

Bailey, J. A. 1980. Desert bighorn forage, competition, and zoogeography. Wildlife Society Bulletin 8: 208–216.

Baker, R. H. 1956. Mammals of Coahuila. Publications of the Museum of Natural History, University of Kansas 9: 127–335.

Baker, R. H., and J. K. Greer. 1962. Mammals of the Mexican State of Durango. Michigan State University Museum Publication, Biological Series 2(2): 25–154.

Bartoskewitz, C. A. 2001. Spatial relationships related to reproductive status of female black bears in the Serranias del Burro, Coahuila. M.S. thesis, Texas A&M University, Kingsville.

Bleich, V. C., J. D. Wehausen, and S. A. Holl. 1990. Desert-dwelling mountain sheep: conservation implications of a naturally fragmented distribution. Conservation Biology 4: 383–390.

Brown, D. E. 1985. The Grizzly in the Southwest. University of Oklahoma Press, Norman.

Broyles, B. 1995. Desert wildlife water developments: questioning use in the Southwest. Wildlife Society Bulletin 23: 663–675.

Buechner, H. K. 1960. The bighorn sheep in the United States, its past, present, and future. Wildlife Monographs 4: 1–76.

Bunch, T. D., W. M. Boyce, C. P. Hibler, W. R. Lance, T. R. Spraker, and E. S. Williams. 1999. Diseases of North American wild sheep. Pp. 209–237 *in* R. Valdez and P. Krausman (eds.), Mountain Sheep of North America. University of Arizona Press, Tucson.

Burbridge, B. 1908. Bear and lion hunting in Mexico. Outdoor Life, January, 17–26.

Byers, J. A. 1997. Mortality risk to young pronghorns from handling. Journal of Mammalogy 78: 894–899.

Cadieux, C. L. 1987. Pronghorn antelope: Great Plains rebound. Pp. 132–143 *in* H. Kallman (ed.), Restoring America's Wildlife 1937–1987. U.S. Department of Interior, Fish and Wildlife Service. Supplemental Documents, U.S. Government Printing Office, Washington, D.C.

Cancino, J., R. Castellanos, and V. Sánchez. 2001. Capture, handrearing and captive management of peninsular pronghorn. P. 108 *in* J. Cancino (ed.), Proceedings of the 19th Biennial Pronghorn Antelope Workshop. Centro de Investigaciones Biológicas del Noroeste, La Paz, Mexico.

Cancino J., P. Miller, J. Bernal Stoopen, and J. Lewis. 1995. Population and habitat viability assessment for the peninsular pronghorn (*Antilocapra americana peninsularis*). IUCN/SSC Conservation Breeding Specialist Group, Apple Valley, Minnesota.

Cancino, J., A. Ortega-Rubio, and R. Rodríguez-Estrella. 1998. Population size of the peninsular pronghorn in Baja California Sur, Mexico. California Fish and Game 84: 25–30.

Canfield, J. E., L. J. Lyon, J. M. Hillis, and M. J. Thompson. 1999. Ungulates. Pp 6.1–6.25 *in* G. Joslin and H. Youmans (eds.), Effects of Recreation on Rocky Mountain Wildlife: A Review for Montana. Committee on Effects of Recreation on Wildlife, Montana Chapter of The Wildlife Society, Bozeman.

Castro, G. A. 1984. Algunas contribuciones al conocimiento de la biología del oso negro (*Ursus americanus*). Oficina de Fauna Silvestre, Monterrey, Nuevo León, Mexico.

Ceballos, G., J. Arroyo-Cabrales, and R. A. Medellín. 2002. The mammals of Mexico: composition, distribution, and conservation status. Occasional Papers, Museum of Texas Tech University 218: 1–27.

Christensen, A. G., L. J. Lyon, R. Pedersen, P. Harrington, W. L. Bodie, R. Johnson, and B. O'Gara. 1995. Ungulate assessment in the Columbia River Basin. Report to the Interior Columbia Basin Ecosystem Management Project. U.S. Department of Agriculture, Forest Service, and U.S. Department of Interior, Bureau of Land Management, Washington, D.C.

Conde, D. A. 2000. Filogenia y estructura genética del berrendo (*Antilocapra americana*) e implicaciones para su conservación. Thesis, Facultad de Ciencias, Universidad Nacional Autónoma de México, México D.F.

Coupland, R. T. (ed.). 1992. Ecosystems of the World: Natural Grasslands, vol. 8A. Elsevier Science, New York.

Dalquest, W. W. 1953. Mammals of the Mexican state of San Luis Potosí. Louisiana State University Studies, Biological Series 1: 1–229.

DeForge, J. R., S. Jiménez, S. D. Ostermann, E. M. Barrett, R. Valdez, and C. Hernández. 1998. Translocation and population modeling of Weems desert bighorn in Baja California Sur. Transactions of the Desert Bighorn Council 41: 51–73.

DeForge, J.R., R. Valdez, V. Suárez, and M. García. 1984. Status of investigations of desert bighorn sheep in Sonora, Mexico. Transactions of the Desert Bighorn Council 29: 41–43.

Doan-Crider, D. L. 1995. Food habits of the Mexican black bear in Big Bend National Park, Texas, and the Serranías del Burro, Coahuila, Mexico. Final Report to the U.S. Department of Interior. Big Bend National Park, Texas.

Doan-Crider, D. L., and E. C. Hellgren. 1996. Population characteristics and winter ecology of black bears in Coahuila, Mexico. Journal of Wildlife Management 60: 398–407.

Douglas, C. L., and D. M. Leslie, Jr. 1999. Management of bighorn sheep. Pp. 238–262 *in* R. Valdez and P. Krausman (eds.), Mountain Sheep of North America. University of Arizona Press, Tucson.

Dumond, M., M. A. Villard, and E. Tremblay. 2001. Does coyote diet vary seasonally between a protected and an unprotected forest landscaper? Ecoscience 8: 301–310.

Elowe, K. D., and W. E. Dodge. 1989. Factors affecting black bear reproductive success and cub survival. Journal of Wildlife Management 53: 962–968.

Etchberger, R. C., P. R. Krausman, and R. Mazaika. 1989. Mountain sheep habitat characteristics in the Pusch Ridge Wilderness, Arizona. Journal of Wildlife Management 53: 902–907.

Felger, R. S., and M. B. Moser. 1985. People of the Desert and Sea: Ethnobotany of the Seri Indians. University of Arizona, Tucson.

Fitzsimmons, N. N., and S. W. Buskirk. 1992. Effective population sizes for bighorn sheep. Proceedings of the Biennial Symposium of the North American Wild Sheep and Goat Council 8: 1–7.

Flather, C. H., S. J. Brady, and M. S. Knowles. 1999. Wildlife resource trends in the United States: A technical document supporting the 2000 Forest and Rangeland Renewable Resources Planning Act. U.S. Department of Agriculture, Forest Service, Rocky Mountain Research Station, Fort Collins, Colorado.

Geist, V. 1971. Mountain Sheep: A Study in Behavior and Evolution. University of Chicago Press, Chicago.

Ginnett, T. F., and C. L. Douglas. 1982. Food habits of feral burros and desert bighorn sheep in Death Valley National Monument. Transactions of the Desert Bighorn Council 26: 81–87.

González-Romero, A., and A. Lafón. 1993. Distribución y estado actual del berrendo (*Antilocapra americana*) en México. Pp. 409–420 *in* R.A. Medellín and G. Ceballos (eds.), Avances en el Estudio de los Mamíferos de México. Publicaciones especiales, vol. 1, Asociación Mexicana de Mastozoología, Mexico.

Graham, H. 1971. Environmental analysis procedures for bighorn in the San Gabriel Mountains. Transactions of the Desert Bighorn Council 15: 38–45.

Graham, H. 1980. The impact of modern man. Pp. 288–309 *in* G. Monson and L. Sumner (eds.), The Desert Bighorn, Its Life History, Ecology, and Management. University of Arizona Press, Tucson.

Gregg, M. A., M. Bray, K. M. Kilbride, and M. R. Dunbar. 2001. Birth synchrony and survival of pronghorn fawns. Journal of Wildlife Management 65: 19–24.

Gross, B. D., J. L. Holechek, D. Hallford, and R. D. Pieper. 1983. Effectiveness of antelope pass structures in restriction of livestock. Journal of Range Management 36: 22–24.

Hall, E. R. 1981. The Mammals of North America, 2nd ed. John Wiley and Sons, New York.

Hansen, M. C. 1980. Population dynamics. Pp. 217–25 *in* G. Monson and L. Sumner (eds.), The Desert Bighorn, Its Life History, Ecology, and Management. University of Arizona Press, Tucson.

Hansen, M. C. 1982. Status and habitat preference of California bighorn sheep on Sheldon National Wildlife Refuge, Nevada. M.S. thesis, Oregon State University, Corvallis.

Harris, L. K. 1992. Recreation in mountain sheep habitat. Ph. D. dissertation, University of Arizona, Tucson.

Hellgren, E. C. 1998. Physiology of hibernation in bears. Ursus 10: 467–477.

Hemley, G. (ed.). 1994. International Wildlife: A CITES Sourcebook. Island Press, Covelo, California.

INE. 1997. Programa de Conservación de la Vida Silvestre y Diversificación Productiva para el Sector Rural. Instituto Nacional de Ecología, SEMARNAP, México D.F.

INE. 1999. Proyecto para la Conservación y Manejo del Oso Negro (*Ursus americanus*) en México. Instituto Nacional de Ecología, SEMARNAP, México D.F.

INE. 2000a. Proyecto para la Conservación, Manejo y Aprovechamiento Sustentable del Berrendo (*Antilocapra americana*) en México. Instituto Nacional de Ecología, SEMARNAP, México D.F.

INE. 2000b. Proyecto para la Conservación, Manejo y Aprovechamiento Sustentable del Borrego Cimarrón (*Ovis canadensis*) en México. Instituto Nacional de Ecología, SEMARNAP, México D.F.

INEGI. 2000. Anuario Estadístico, Coahuila, Mexico. Instituto Nacional de Estadística Geografía e Informática, México D.F.

Jaramillo-Monroy, F., and A. Castellanos-Vera. 1992. Algunos aspectos de la población, manejo y conservación del borrego cimarrón en Baja California Sur, México. Ecologica 2(1): 25–30.

Jessup, D. A. 1985. Diseases of domestic livestock which threaten bighorn sheep populations. Desert Bighorn Council Transactions 29: 29–33.

Kellert, S. R. 1994. Public attitudes toward bears and their conservation. Pp. 43–50 *in* C. Swanson, D. McCollum, and M. Maj (eds.), Bears: Their Biology and Management. Proceeding of the 9th International Conference on Bear Research and Management. International Association for Bear Research and Management.

Kelly, W. E. 1980*a*. Predator relationships. Pp. 186–196 *in* G. Monson and L. Sumner (eds.), The Desert Bighorn, Its Life History, Ecology, and Management. University of Arizona Press, Tucson.

Kelly, W. E. 1980*b*. Hunting. Pp. 336–342 *in* G. Monson and L. Sumner (eds.), The Desert Bighorn, Its Life History, Ecology, and Management. University of Arizona Press, Tucson.

Krausman, P. R., A. V. Sandoval, and R. C. Etchberger. 1999. Natural history of desert bighorn sheep. Pp. 139–191 *in* R. Valdez and P. Krausman (eds.), Mountain Sheep of North America. University of Arizona Press, Tucson.

LeCount, A. L., R. H. Smith, and J. R. Wegge. 1984. Black bear habitat requirements in central Arizona. Special Report no. 14. Arizona Game and Fish Department, Phoenix.

Lee, R. M., and E. E. López-Saavedra. 1994. A second helicopter survey of desert bighorn sheep in Sonora, Mexico. Transactions of the Desert Bighorn Council 38: 12–13.

Lee, R. M., and E. Mellink. 1996. Status of bighorn sheep in Mexico–1995. Transactions of the Desert Bighorn Council 39: 35–39.

Leopold, A. S. 1959. Wildlife of Mexico: The Game Birds and Mammals. University of California Press, Berkeley.

Leopold, A. S. 1965. Fauna silvestre de México. Instituto Mexicano de Recursos Naturales Renovables, México D.F.

LoBello, R. L. 1989. Status of the black bear in Big Bend National Park, Texas. Proceedings of the Third Regional Conference of the US–México Border States on Parks and Wildlife. Big Bend National Park, Texas.

Mapston, R. D. 1972. Guidelines for fencing on antelope ranges. Pp. 167–170 *in* H. O. Compton and D. Pyrah (eds.), Proceedings of the 5th Biennial Antelope States Workshop, June 19–22, 1972, Billings, Montana.

Mapston, R. D., R. S. Zobell, K. R. Winter, W. D. Dooley. 1970. A pass for antelope in sheep-tight fences. Journal of Range Management 23: 457–459.

Matson, J. O., and R. H. Baker. 1986. Mammals of Zacatecas. Special Publications, Museum of Texas Tech University 24: 1–88.

Meagher, M. J. 1986. *Bison bison*. Mammalian Species 266: 1–8.

Medellín, R. A., and F. Colchero. 2002. Recuadro XVII.3. Los borregos cimarrones de la Isla Tiburón: conservación y desarrollo sustentable. Pp. 510–512 *in* R. Primack, R. Rozzi, P. Feinsinger, R. Dirzo, and F. Massardo (eds.), Fundamentos de Conservación Biológica, Perspectives Latinoamericanas. Fondo de Cultura Económica, México D.F.

Medellín, R. A., F. Colchero, C. Manterola, F. Ramírez, and G. Ceballos. 1999. The Tiburon Island Bighorn Sheep Program: an example of binational, interinstitutional collaboration for conservation and sustainable development in a Mexican Indian and protected area. Wild Sheep (spring 1999): 71–72.

Mendoza, J. 1976. The bighorn sheep of the State of Sonora. Transactions of the Desert Bighorn Council 20: 25–26.

Miller, S. D. 1990. Population management of bears in North America. International Conference on Bear Research and Management 8: 357–373.

Monson, G. 1980. Distribution and abundance. Pp. 40–51 *in* G. Monson and L. Sumner (eds.), The Desert Bighorn, Its Life History, Ecology, and Management. University of Arizona Press, Tucson.

Montoya, B., and G. Gates. 1975. Bighorn capture and transplant in Mexico. Transactions of the Desert Bighorn Council 19: 28–32.

Nelson, E. W. 1925. Status of the pronghorned antelope, 1922–1924. U. S. Department of Agriculture Bulletin 1346: 1–64.

Nino Ramirez, J. A. 1989. Análisis preliminar de la dieta de verano del oso negro (*Ursus americanus*) en la Sierra los Picachos, Higueras, Nuevo León, México. Undergraduate thesis, Universidad Autónoma de Nuevo León, Monterrey, Nuevo León, Mexico.

Onorato, D. P., and E. C. Hellgren. 2001. Black bear at the border: Natural recolonization of the Trans-Pecos. Pp. 245–259 *in* D. S. Maehr, R. Noss, and J. Larkin (eds.), Large Mammal Restoration: Ecological and Sociological Challenges in the 21st Century. Island Press, Covello, California.

Onorato, D. P., E. C. Hellgren, D. M. Leslie, Jr., and R. A. Van Den Bussche. 2002. Conservation ecology of an isolated population of black bears in the Big Bend ecosystem. Final report, Natural Resources Preservation Program–Project 97-08. Oklahoma Cooperative Fish and Wildlife Research Unit, Stillwater, Oklahoma.

Papouchis, C. M., F. J. Singer, and W. B. Sloan. 2001. Responses of desert bighorn sheep to increased human recreation. Journal of Wildlife Management 65: 573–582.

Pelton, M. R., A. B. Coley, T. H. Eason, D. L.

Doan Martínez, J. A. Pederson, F. T. van Manen, and K. M. Weaver. 1999. American black bear conservation action plan. Pp. 144–156 *in* C. Servheen, S. Herrero, and B. Peyton (eds.), Status Survey and Conservation Action Plan: Bears. World Conservation Union, Gland, Switzerland.

Pritchard, G. T., and C. T. Robbins. 1990. Digestive and metabolic efficiencies of grizzly and black bears. Canadian Journal of Zoology 68: 1645–1651.

Risenhoover, K. L., and J. A. Bailey. 1985. Foraging ecology of mountain sheep: implications for habitat management. Journal of Wildlife Management 49: 797–804.

Robles Gil, P., G. Ceballos, and F. Eccardi. 1993. Diversidad de Fauna Mexicana. CEMEX, Monterrey, Mexico.

Rogers, L. L. 1987. Effects of food supply and kinship on social behavior, movements, and population growth of black bears in northeastern Minnesota. Wildlife Monographs 97.

Rose, M. D., and G. A. Polis. 1998. The distribution and abundance of coyotes: the effects of allochthonous food subsidies from the sea. Ecology 79: 998–1007.

Russell, P. T. 1951. Crisis of antelope management. Wyoming Wildlife 15 (6): 4–9.

Sanderson, E. W., K. H. Redford, C. B. Chetkiewicz, R. A. Medellín, A. R. Rabinowitz, J. G. Robinson, and A. B. Taber. 2002. Planning to save a species: the jaguar as a model. Conservation Biology 16: 58–72.

Sandoval, A. V. 1980. Management of a psoroptic scabies epizootic in bighorn sheep (*Ovis canadensis mexicana*) in New Mexico. Transactions of the Desert Bighorn Council 24: 21–28.

Seegmiller, R. F., and R. D. Ohmart. 1981. Ecological relationships of feral burros and desert bighorn sheep. Wildlife Monographs 78: 1–58.

Seegmiller, R. F., and C. D. Simpson. 1979. The barbary sheep: some conceptual implications of competition with desert bighorn. Transactions of the Desert Bighorn Council 23: 47–49.

SEMARNAT (Secretaría de Medio Ambiente y Recursos Naturales). 2002. Norma Oficial Mexicana NOM-059-ECOL-2001, Protección Ambiental-Especies Nativas de México de Flora y Fauna Silvestres-Categorías de Riesgo y Especificaciones para su Inclusión, Exclusión o Cambio-Lista de Especies en Riesgo. Diario Oficial, 6 March: 1–56.

Seton, E. T. 1929. The bighorn. Pp. 519–573 *in* E. T. Seton (ed.), Lives of the Game Animals, Vol. 3. Doubleday, Garden City, New York.

Sheldon, C. 1925. The big game of Chihuahua,

Mexico. Pp. 138–181 *in* G. B. Grinnell and C. Sheldon (eds.), Hunting and Conservation, the Book of the Boone and Crockett Club. Yale University Press, New Haven, Connecticut.

Skiles, J. R. 1995. Black bears in Big Bend National Park—The Tex-Mex connection. Pp. 67–73 *in* J. Auger and H. L. Black (eds.), Proceedings of the 5th Western Black Bear Workshop. Brigham Young University Press, Provo, Utah.

Spillet, J. J. 1964. Effects of livestock fencing on pronghorn antelope movements. M.S. thesis, Utah State University, Logan.

Tarango, L. A., and P. R. Krausman. 1997. Desert bighorn sheep in Mexico. Transactions of the Desert Bighorn Council 41: 1–7.

Taylor, R. B. 1992. Black bear status. Federal Aid Project No. W-125-R-3, Job 68. Texas Parks and Wildlife Department, Austin.

Tinker, B. 1978. Mexican Wilderness and Wildlife. University of Texas Press, Austin.

Valdés, M., and C. Manterola. 2001. La conservación del berrendo (*Antilocapra americana*) en México. Biodiversitas 7(9): 1–6.

Valdez, R. 1988. Wild sheep and wild sheep hunters of the New World. Wild Sheep and Goat International, Mesilla, New Mexico.

Valdez, R., and P. R. Krausman. 1999. Description, distribution, and abundance of mountain sheep in North America. Pp. 3–22 *in* R. Valdez and P. Krausman (eds.), Mountain Sheep of North America. University of Arizona Press, Tucson.

Valdez, R., A. Martínez-Mendoza, and O. Rosas-Rosas. 2002. Componentes históricos y actuales del habitat del jaguar en el noreste de Sonora, México. Pp. 367–378 *in* R. A. Medellín et al. (eds.), El Jaguar en el Nuevo Milenio. Fondo de Cultura Económica, Instituto de Ecología, Universidad Nacional Autónoma de México, and Wildlife Conservation Society, México D.F.

Villa, R. 1955. Observaciones acerca de la última manada de berrendos (*Antilocapra americana mexicana*) en el estado de Chihuahua, México. Anales del Instituto de Biología, Universidad Nacional Autónoma de México 26: 229–236.

Villa R., B. 1959. Estado que guarda actualmente la población de borregos silvestres en el territorio de México. Pp. 422–434 *in* Memoria de la Segunda Convención Nacional Forestal. Secretaría de Agricultura, México D.F.

Wehausen, J. D., and R. R. Ramey II. 1993. A morphometric reevaluation of the Peninsular bighorn subspecies. Transactions of the Desert Bighorn Council 37: 1–10.

Welsh, G. W. 1971. What's happening to our sheep? Transactions of the Desert Bighorn Council 15: 63–73.

Whetten, N. L. 1948. Rural Mexico. University of Chicago Press, Chicago.

Yoakum, J. D. 1968. A review of distribution and abundance of American pronghorn antelope. Pp. 4–14 *in* Third Biennial Antelope States Workshop, February 5–6, 1968, Casper, Wyoming.

Yoakum, J. D. 1980. Habitat management guides for the American pronghorn. Technical Note 347. U.S. Department of Interior, Bureau of Land Management, Washington, D.C.

Zepeda-González, J., J. Arroyo-Cabrales, O. J. Polaco, and A. Jiménez-Guzmán. 1997. Notas acerca de la distribución de algunos mamíferos del sur de Nuevo León, México. Revista Mexicana de Mastozoología 2: 101–112.

20

Sea Turtles in Northwestern Mexico: Conservation, Ethnobiology, and Desperation

RICHARD S. FELGER

WALLACE J. NICHOLS

JEFFREY A. SEMINOFF

The waters of northwestern Mexico (fig. 20.1) have been among the most important feeding and developmental grounds in the eastern Pacific for 5 of the world's 7 species of sea turtles. These are, in order of abundance: the green turtle (*Chelonia mydas*), known locally as the "black turtle," the loggerhead (*Caretta caretta*), the olive ridley (*Lepidochelys olivacea*), the leatherback (*Dermochelys coriacea*), and the hawksbill (*Eretmochelys imbricata*). Most depend on shallow costal habitats for the abundant food resources, but the leatherback, a pelagic species, instead cruises offshore. Due to exploitation of eggs and turtles as food and, to a lesser extent, incidental mortality relating to marine fisheries and degradation of marine and nesting habitats, sea turtle populations have declined throughout the region. Worldwide, sea turtle populations continue to plummet, even as research and conservation efforts escalate (Limpus 1995; Mosier et al. 2002).

All sea turtle species in northwestern Mexico are listed either as Endangered (Cm, Cc, Lo) or Critically Endangered (Dc, Ei) by IUCN–World Conservation Union (Hilton-Taylor 2000). Since joining the Convention on International Trade of Endangered Species (CITES) in 1991, Mexican authorities and nongovernmental organizations have acted to limit illegal trade and reduce incidental capture of sea turtles. Despite strong policies, shortcomings in our understanding of sea turtle life history as well as the inability to root out corruption and conflicts of interest have impeded progress.

Sea turtles are migratory and use a wide range of broadly separated localities and habitats during their lifetime. Upon departing nesting beaches as small hatchlings, sea turtles begin an oceanic phase during which many believe they associate with flotsam and jetsam and drift passively with these "floating islands" for several years, a life stage referred to by Archie Carr (1973) as the "lost years." Eventually, all species except the leatherback settle out of the pelagic environment and become tied to shallow coastal waters, where they commonly reside in coral reef, rocky reef, marine algae, and seagrass habitats. Leatherbacks are less frequently encountered in near-shore habitats, instead preferring life in the high seas. At sexual maturity, all sea turtles begin a migratory phase that brings them back to their natal nesting beaches for mating and egg laying. Incredibly, a single individual may use habitats separated by massive expanses of open water, often traveling thousands of miles and traversing the world's oceans.

Recent conservation initiatives have paved the way for collaboration between scientists and local peoples throughout northwestern Mexico (Nichols et al. 2002a; Nichols 2003a). However, we must realize that through sea turtle migrations, the coastal waters of northwestern Mexico are connected to other regions of the Pacific Ocean, some near and some very far away. Therefore, effective conservation must involve international partnerships and information exchange, thereby enabling the

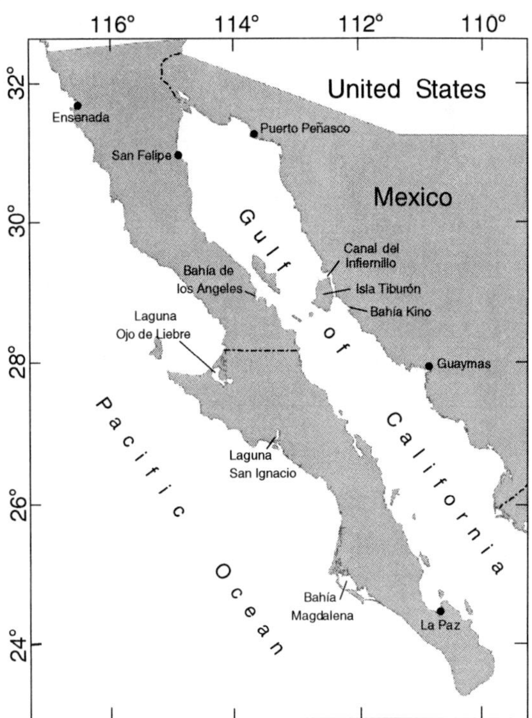

Figure 20.1. Northwestern Mexico.

development of protection strategies that encompass the entire life history of these "ocean connectors" (Pesenti and Nichols 2002).

Sea turtles are unique in that they are an important component of coastal marine ecosystems as well as vital to the culture and traditions of many indigenous peoples around the world. As sea turtle populations have diminished globally, there has been a corresponding decrease in the indigenous knowledge and practices associated with them. In this chapter we discuss the conservation of sea turtles in northwestern Mexico, their biology and ecology as revealed by western science and traditional knowledge, and the role of these animals in indigenous and modern cultures of the region.

Indigenous Knowledge of Sea Turtles

Native peoples of northwest Mexico have been strongly tied to sea turtles, benefiting from their abundance for both food and nonfood uses since prehistoric times. Travelers to the region have described this connection for more than four centuries. For example, during his voyage to the Colorado River delta region in 1539, Francisco de Ulloa com-

mented that the indigenous people near Bahía San Luiz Gonzaga along the east coast of Baja California used "thick hooks made of tortoise shell bent in fire" to catch fish (Wagner 1929: 36). Similarly, O'Donnell (1974) suggests indigenous people in Bahía de los Angeles likely used sea turtles, and Jesuit missionary records indicate that turtles were abundant in both the Gulf of California and the Pacific Ocean (Aschmann 1966). Although various indigenous peoples in northwestern Mexico used sea turtles, some, such as the Cochimí of Gulf coast areas of the Baja California peninsula, have long been culturally extinct, and their knowledge, too, has perished (Aschmann 1966). However, no indigenous group has been as culturally linked to sea turtles as the Seris, or the Comcáac, as they call themselves. Their cultural identity remains strong, and during the past several decades community elders have willingly shared their rich knowledge of sea turtles (e.g., Felger and Moser 1985; Felger 2000).

The Seris are well known for their traditional reliance on sea turtles for subsistence and material culture and for the vigorous role of sea turtles in many other aspects of their cultural heritage (e.g., Felger and Moser 1985; Nabhan 2003). Original Seri territory coincides with the Sonoran portion of

Shreve's (1951) Gulf Coast phytographic subdivision of the Sonoran Desert, from Puerto Lobos southward to Guaymas, including Isla Tiburón and Isla San Esteban (Felger and Moser 1985). Coastal waters of this region hosted an abundance of sea turtles and other marine resources now in serious decline. Although Seri culture remains strong, there has been serious loss of knowledge associated with vanquished biological richness and diversity.

Species Accounts Based on Seri and Scientific Knowledge

The rich knowledge of sea turtles in Seri culture provides a unique opportunity to compare indigenous information with that generated from western scientific research. The following species accounts describe the biology and status of sea turtles based on both the Seri perspective and that of western biologists. Information on Seri knowledge of sea turtles summarized here is derived in part from Felger's fieldwork in collaboration with E. Moser and M. B. Moser, beginning in the 1960s (see Felger 2000). The "ethnographic present" is used when appropriate, and past tense employed when it can be applied with reasonable accuracy. Seri sea turtle knowledge is remarkably consistent with scientific information. Seris distinguished 16 kinds or variants of sea turtles: 10 green turtles, 2 loggerheads, 2 hawksbills, the olive ridley, and the leatherback (Felger and Moser 1985). Each genus recognized by biologists is clearly distinguished as an ethnogenus. Some of the ethnotaxa may represent micro-races or populations from different nesting places converging on feeding grounds.

Leatherback, Dermochelys coriacea

Perhaps the most intriguing of the Gulf's macrovertebrates is the leatherback, the world's largest living turtle, reaching more than 2 m long and as much as 700 to even more than 850 kg (Grismer 2002). Common names in Mexico are *siete filos* (seven ridges) and *tortuga laud* (lute turtle). It is unique among the living sea turtles with its flexible, smooth, and streamlined leathery shell with 7 dorsal keels, lack of epidermal scales, and enormous front flippers; a body plan ideally adapted for a life at sea.

The usual Seri name for the leatherback is *mosnípol* ("black sea turtle"). Traditional beliefs and oral history surrounding the leatherback are extensive, and the Seris continue to hold these giant reptiles in special spiritual esteem. The leatherback is the only sea turtle for which a special ceremony is held, a 4-day celebration similar to the girl's puberty ceremony. The Seris say that the leatherback was once one of the legendary Giants mourning the death of a loved one at the time of its transformation during the Great Flood (Felger and Moser 1985; also see Bowen 2000). Now, as a leatherback, the face is still black from much crying. And the leatherback does indeed have a black head. The association with death makes it dangerous and requires it to be handled with great care. It is also dangerous because spirit power resides in any animal or artifact of exceptionally great size. The purpose of the ceremony is to appease or cheer up the dangerous spirit of the turtle and assure no harm will come to the harpooner and his family. It is said that all will go well for the people who take part in the ceremony and conform to the prescribed rules and that there will be an abundance of green turtles (*moosni*) in the days ahead. For more details see Felger and Moser (1985) and Burckhalter (1999).

Although the Seris have no knowledge of the leatherback ever being common, they report that it has become even rarer than in the past, and in recent years there have been fewer and fewer leatherback ceremonies (M. B. Moser, pers. comm.). This is consistent with the fact that the leatherback is in serious decline throughout the Pacific, and it is predicted to become extinct in the Pacific in less than 2 decades (Spotila et al. 2000). These declines are particularly evident in the western Pacific Ocean, where the world's largest nesting colony, at Terengganu, Malaysia, has nearly vanished, although as recently as the 1960s there were more than 5000 nests per year (Chan and Liew 1996). In the eastern Pacific the situation is tragically similar. At Llano Grande, Guerrero, once the largest nesting colony in Mexico, there were more than 2000 nests annually in the early 1990s, but only 7 females nested during the 2001–2002 season (A. Barragan, pers. comm.). At Mexiquillo, Michoacán, once the second largest rookery in Mexico, more than 1000 females nested per season in the 1980s, but there were fewer than 10 during the 2001–2002 season (L. Sarti, pers. comm.). Incidental mortality from pelagic long-line and drift-net fisheries is the primary reason for the continued mortality (Eckert 1997), although egg poaching and earlier illegal slaughter for leather contributed heavily to the devastation (Cliffton et al. 1982).

Drastic conservation measures are required to avert imminent ecological extinction, including total beach and nest protection, elimination of egg poaching, and closure of long-line and drift-net fisheries in leatherback migratory corridors (e.g., Morreale et al. 1996). In addition, conservation efforts must address the problem of marine pollution. Around the world, substantial numbers of leatherbacks die each year because of choking and intestinal blockage caused by eating clear plastic bags, perhaps consumed because of their resemblance to certain jellyfish on which these turtles normally feed (Fritts 1982; Carr 1987; Grismer 2002).

Loggerhead, Caretta caretta

Common names for the loggerhead in Mexico are *tortuga perica* (parrot turtle), *javelina*, and *amarilla*. The Seris distinguish two kinds of loggerheads, *xpeeyo* and *moosni ilítcoj caacöl*, both described as weighing a maximum of 55–65 kg and having large heads. The *xpeeyo* is said to have the largest head of any sea turtle except the leatherback. The oil (fat) of the *xpeeyo* has a bad smell and taste, and the meat smells differently from that of other turtles. The carapace is distinguished by color and pattern and is rounded and relatively small, reaching about 50 cm long, and is wider than that of the hawksbill and not as elongated. Like the hawksbill, it snaps or attempts to bite when captured. Its local distribution was largely restricted to an area in the Canal del Infiernillo (El Infiernillo; fig. 20.1) with abundant oyster beds.

The *moosni ilítcoj caacöl* ("large-headed sea turtle") differs from the *xpeeyo* by being greener, having a smaller head, and does not snap or attempt to bite. It looks similar to the *moosniáa* (*Chelonia*, see further on), except for its larger head, and has green fat similar to that of the common *Chelonia*. It was very rare after the mid-twentieth century, and was known only from Bahía Kino and southward. It seems to be a name for loggerhead populations to the south of the Canal del Infiernillo.

The Seris explain that both forms of loggerheads eat the soft parts (meat) of certain large mollusks, which is confirmed by well-known scientific information (Dodd 1988). The Seris state that loggerhead meat does not taste good, and they never heard of anyone finding eggs in these turtles. That they have not seen eggs is consistent with the fact that the closest nesting beaches are on the opposite side of the Pacific Ocean. In the Pacific, loggerheads carry out an extensive developmental migration, traveling from nesting areas in Japan and Australia to developmental and foraging habitats in the eastern Pacific (Bowen et al. 1995). After spending years foraging in the eastern Pacific, upon sexual maturity loggerheads return to their natal nesting beaches for reproduction (Resendiz et al. 1998; Nichols et al. 2000a) and remain in the western Pacific for the remainder of their life cycle (Kamezaki et al. 1997). Thus, any turtles encountered in the Gulf of California are likely immature animals: the 55–65 kg maximum size observed by the Seris is consistent with the size of a large but immature loggerhead.

Loggerheads worldwide can be found in both shallow coastal waters and offshore pelagic zones; they consume a variety of marine invertebrates (Dodd 1988). Once common in northwestern Mexico, the loggerhead is now scarce. It has declined tremendously and vanished from many of the foraging habitats throughout its range, largely due to illegal poaching and incidental capture in long-line and drift-net fisheries (Wetherall 1996; Kamezaki 1997). Although loggerhead meat was considered inferior due to its oily and odiferous qualities, there was an increased harvesting of loggerheads in offshore waters along the Pacific coast as *Chelonia* populations in northwestern Mexico crashed during the late twentieth century.

Hawksbill, Eretmochelys imbricata

The Seris recognize 2 kinds of hawksbill (*tortuga carey*), *moosni quipáacalc* and *moosni siipoj*. In describing the *moosni quipáacalc* ("sea-turtle overlaps") they refer to the overlapping, brown- and yellow-mottled carapace plates, the strongly developed beak, and the head, which viewed from above is more pointed than in other sea turtles. This hawksbill feeds on certain large bivalves. It was most often found at a specific place in the Canal del Infiernillo and was rarely seen elsewhere. Although once common, by the mid-twentieth century it had become very rare.

The other hawksbill, *moosni siipoj* ("osprey sea turtle"), is considered to have a smooth carapace and a more pointed beak than the *moosni quipáacalc*, hooking downward and backward like that of an osprey. It is said to eat eelgrass (*Zostera marina*) and be more common south of Bahía Kino, and it was found during the warm months. By the 1970s or earlier it had become very rare.

The hawksbill has been heavily exploited for its shell rather than the meat, and tortoiseshell has been a cornerstone of international artisan markets. The hawksbill is nearly extirpated from northwestern Mexico. Despite occasional glimpses of immature and adult hawksbills (Seminoff et al. 2003b), large-scale nesting rookeries no longer exist on the Pacific coast of the Americas (National Marine Fisheries Service and U.S. Fish and Wildlife Service 1998). Sporadic nesting still occurs in Jalisco and Nayarit (Chavez 1989; R. Briseño, pers. comm.) and perhaps at the Tres Marias Islands (Márquez 1990).

Despite tremendous declines in the Gulf of California and elsewhere, hunting and incidental mortality continue (Gardner and Nichols 2001; Seminoff et al. 2003b). For example, fishermen unanimously report that if a hawksbill is captured, it is never returned to the sea. In addition to the value of the shell for the tortoiseshell industry, a stuffed hawksbill or carapace is hung above a doorway to bring good luck to a home.

Olive Ridley, Lepidochelys olivacea

The olive ridley (*golfina, tortuga golfina, mestiza*) is the smallest sea turtle in the Pacific Ocean, measuring only up to 70 cm in straight carapace length and weighing a maximum of nearly 45 kg or less—slightly larger than its Atlantic counterpart, the Kemp's ridley (*Lepidochelys kempii*). The Seris refer to the olive ridley as *moosni otác* ("toad turtle"). It is characterized as roundish and flattish, and as not having a pointed beak, and local non–Seri fishermen agreed on this description for the *golfina*. Seris explain that the *moosni otác* differs from the regular green turtle or *moosni* as follows: the shape of the carapace is somewhat narrower at the anterior end, flared at the sides (middle), and flattish, the head is the same size, and the flippers are shorter. The carapace color is like that of the *xpeeyo* (loggerhead), somewhat yellowish with gray. The plastron is whitish with a bit of yellow. It was eaten out of necessity, although generally preferred to the loggerhead because it was more docile and did not snap or bite. The *moosni otác* was once relatively common in the Seri region, but has become increasingly rare since sometime before the mid-twentieth century.

The Seris describe the olive ridley as the only sea turtle that contains shelled eggs, an observation consistent with the fact that olive ridleys and leatherbacks are the only 2 species that nest, albeit uncommonly, on Gulf of California beaches. In addition, there is a substantial nesting beach along the tip of the Baja California peninsula near Cabo Pulmo (Fritts et al. 1982).

The olive ridley is currently the most common sea turtle in the eastern Pacific. Outside of northwestern Mexico, in more tropical regions, this species exhibits the unique group-nesting behavior known as *arribada* nesting: huge congregations of olive ridleys, sometimes more than 100,000 females, storm the beaches en masse to lay eggs. These arrivals are usually timed with specific lunar and tidal cycles and may be an antipredation strategy.

The entire pre-1950 western Mexico population of olive ridleys is estimated to have been on the order of 10,000,000 (Cliffton et al. 1982). In the late 1960s and 1970s, the olive ridley population crashed following unrestrained industrialized harvesting, poaching, and habitat destruction (e.g., Felger 2003). The population in the seas of western Mexico was about 3,185,000 in the mid-1960s, but an estimated 2,000,000 were killed in coastal waters during 5 years in the late 1960s (Cliffton et al. 1982). Most of the great as well as minor *arribadas* of the eastern Pacific nearly vanished, but recent conservation initiatives have enabled the recovery of 3 of them, 2 in Costa Rica and 1 in Mexico. Harvest closures and vigorous protection of nests and nesting sites have resulted in the spectacular recovery of the *arribada* at Escobilla, Oaxaca (Instituto Nacional de Ecología 2000) where on the order of 650,000 nests were deposited during the summer 2001 (J. Vasconcelos pers. comm.).

Although significant increases have been observed at the majority of *arribada* sites since 1990, especially along the Pacific coast of southern Mexico (García et al. 2003; ridleys are smaller than other species and quicker to mature), recovery of former abundances has yet to be achieved. Farther north, at El Verde north of Mazatlán, Sinaloa, and despite the heroic conservation efforts spearheaded by Daniel Rios over the past 3 decades, fewer than 800 female olive ridleys have nested annually in recent years (D. Rios pers. comm.), out of more than 5000 that would do so before the mid-1960s (Márquez 1976). Although the reasons for lack of greater success are not wholly understood, we can point to excessive fishing pressure (resulting in higher bycatch) and greater human activity in the north.

Green Turtle, Chelonia mydas

The green turtle, *Chelonia mydas* (also known locally as the black turtle or *caguama, tortuga prieta,* or *negra*) may have been the most abundant large marine vertebrate in the Pacific Ocean. These turtles provided most Seri groups with their single most important food resource, and there was vast traditional knowledge of hunting, ecology, food and nonfood uses, ritual, and other cultural aspects centered on *Chelonia* (e.g., McGee 1898; Felger and Moser 1985; Burckhalter 1999; figs. 20.2 and 20.3). Since the sixteenth century Spanish chronicles, the Seris have been noted for their tall stature (commonly 6 ft, about 1.8 m), undoubtedly largely due to sustained high-protein diet, often of fish and sea turtle meat, beginning in early childhood (e.g., Felger and Moser 1985; M. B. Moser, pers. comm.). The Seris recognized 10 different kinds of *Chelonia*, briefly characterized below.

(1) *Moosni* is the generic term for all sea turtles. It is also used to designate specifically the most common or "regular" kind of *Chelonia*. The *moosni* generally has been taken any time of year in the Canal del Infiernillo on the east side of Isla Tiburón, and in the warmer months also along deeper waters of the other shores of Isla Tiburón, Isla San Esteban, and the mainland coast north and south of the Canal del Infiernillo.

(2) *Mosnáapa* refers to the "true" or "ultimate" *moosni*. This was the kind most commonly eaten many years ago by the "ancestors." Therefore, it was the "true" turtle, the one most important to the Seris. By the 1970s it was either very rare or extinct in the region. It resembled the ridley in size and color, but had a different beak. *Mosnáapa* had a brownish-gray hue to its carapace and flippers, whereas the regular *moosni* is dark with a greenish hue and has a greenish-gray carapace.

(3) *Mosníil* ("blue sea turtle") was exceptionally large and long, with a smooth carapace. It was common but became rare, and by the mid-twentieth century had not been seen for many years. It was generally found along the northwestern coast of Isla Tiburón and westward into the open sea, and apparently elsewhere. It was hunted in summertime, generally in open water away from land. These huge turtles were strong, and the surfaces of the carapace and flippers were said to be bluish. The blue carapace was thought to have "painted" the harpoon and line when rubbed against it. One of the Seri names for *Asparagopsis taxiformis* (a red alga) is

Figure 20.2. Seri turtle hunter, Guadalupe López, harpooning a green turtle in the Canal de Infiernillo, Sonora; early spring 1976. Note monofilament line. (Photo by Kim Cliffton.)

Figure 20.3. Guadalupe López bringing a harpooned green turtle onboard in the Canal de Infiernillo, Sonora; early spring 1976. Note the 2 harpoon heads and monofilament line. (Photo by Kim Cliffton.)

mosníil ihaquéepe or "what the blue turtle eats" (Felger and Moser 1985). *Asparagopsis* contains unique halogenated compounds (bromines; e.g., McConnell and Fenical 1976) that may account for some of the information for the mysterious blue turtle.

(4) The *mosníctoj* ("pink sea turtle") is the rarest modern turtle known to the Seris, and the information and name suggest an albino. It was mostly seen at the south end of the Canal del Infiernillo near a place called *Mosníctoj Iime* ("pink sea turtle's home"). The occurrence of such a rare genotype at a specific place hints at a degree of territoriality, and recent studies indicate green turtles do indeed maintain fidelity to specific home ranges (Seminoff et al. 2002c).

(5) The *quiquíi* is found in various sizes and noted as being thin with sunken eyes and thin flippers that it can hardly move. There is much fat and little meat. Although rather rare, it was still occasionally caught in the 1980s. The Seris said it is not sick, but had died and its spirit was reborn. It may occur anywhere during the warmer months when turtles are moving and is easy to harpoon.

(6–8) The Seris recognize 3 phases of the green turtle which, based on detailed behavioral and morphological descriptions, indicate migrating animals. They have been scarce since the mid-twentieth century. *Cooyam* are the young of the turtle called *ipxom haaco iima*. Turtles of intermediate size are called *cooyam caacöl*. All 3 contain relatively large amounts of fat and may have empty stomachs and narrow intestines.

(6) The *cooyam* were young turtles entering the region from the south for the first time. Two moons of the Seri calendar, approximating February and March, are named for the seasons when these turtles passed through from the south and made their way north.

The *cooyam* is noted for the large amount of fat around the stomach. The fat is grayish, whereas fat of the *moosni* is greenish. The stomach contains little or no food, and the intestines are likewise empty and thin. The liver is whitish on the exterior, and full of fat, or oil. In contrast, the liver of the *moosni* is dark, and said to be really oily only in the case of a female carrying (noncalcified) eggs.

Cooyam front flipper bones are smaller than those of the *moosni*. The capillary bed or network in the mesentery surrounding the stomach is also smaller than that of the *moosni*. It seems reasonable that in a nonfeeding migratory animal, the blood vascularization associated with digestion would not be as prominent as in one that is actively feeding. When clubbed on the head to kill it, (prior to being butchered) an oily fluid with a skunk-like smell called *isa* is ejected from pores of Rathke's glands (the glands are described by Ehrenfeld and Ehrenfeld 1973). The Seris were careful to remove these glands and point out that the *moosni* (the common *Chelonia*) also have these glands, but they contain very little fluid and do not spray when the animals are killed.

The *cooyam* has a white plastron, dark head with a reddish hue, and the carapace has definite patches, streaks, and tinges of dull reddish or amber color. In contrast, the *moosni* is noted as generally having a darker plastron and a greener carapace than the *cooyam* or *ipxom haaco iima*. Some Seris said that the green will rub off, apparently due at least in part to a thin algal growth on the *moosni* carapace. The concept of the *cooyam* involves about 15 features: seasonality, schooling, northward migration, body size, carapace color, plastron color, color of head and neck or throat region, reduced bone size in front flippers, fluid from Rathke's glands, fat content, color of fat, liver color, empty stomach and intestines, slender intestines, and reduced vascularization of stomach mesentery.

(7) As an adolescent sea turtle growing up in the region, the *cooyam* was reclassified as *cooyam caacöl* ("large *cooyam*"). It occurred most often off the west coast of Isla Tiburón. Otherwise it was similar to the *cooyam*.

(8) *Ipxom haaco iima* ("the one whose fat is spiritually poweful") is said to be the adult *cooyam*. It is found in the more remote and deeper waters between Isla San Esteban and Isla Tiburón and off the north and west sides of the latter island (a region of cooler water in summertime and without eelgrass). Some Seris thought it stays between San Esteban and Tiburón, and others believed it moved north. It is solitary rather than gregarious like the *cooyam* and is seen only during the warm months. It does not swim straight north like the *cooyam* in the Infiernillo, but is seen moving in all directions, and not on the surface but rather swims 2–3 m below the surface. The carapace is about 1 m long, and there is a very thick layer of fat, perhaps 10–12 cm thick, thicker than that of the *cooyam*. As

with the *cooyam*, the liver differs from that of other sea turtles in being white, and it turns into yellowish grease when cooked with water.

Some Seris said this turtle eats large jellyfish (undoubtedly the Portuguese man-o-war; *Physalia utriculus*), called *ipxom haaco iimj quih oiitoj* ("what the *ipxom haaco iimj* eats" and a similar kind of organism called the *copsíij* (unidentified jellyfish). These turtles were easily harpooned when feeding on these jellyfish (Cnidaria) because they closed their eyes to avoid the stinging tentacles. Some people, however, said that these turtles had empty stomachs (when butchered).

(9) The *moosni quimoja* is simply an oversized or giant *moosni*. The term *quimoja* is an archaic word for which the meaning is lost. It may weigh 100 kg or more, and the carapace may be about 1 m long. It is rare, but until about the 1970s was occasionally taken in the Canal del Infiernillo.

(10) The *moosni ctam hax ima* ("male sea turtle that isn't brave") has a shorter tail (about 18–20 cm long) than the regular male *moosni* (the tail reaches about 40 cm long). It is never seen mating and reminds the Seris of a kind of adult mule deer in which the testicles are not developed.

The green turtle is generally herbivorous, although in the Gulf of California and adjacent areas in the eastern Pacific, it consumes larger portions of animal matter than do *Chelonia* in other regions (Bjorndal 1997; Seminoff et al. 2002b). In the gulf as well as elsewhere in the world, food selection is habitat specific, and turtles may be primarily seagrass consumers (Felger and Moser 1973) or marine-algae consumers (Seminoff 2000; Seminoff et al. 2002b). The Canal del Infiernillo between Mexico's largest island, Isla Tiburón, and mainland Sonora, is a particularly important area for green turtles in northwestern Mexico. This region hosts the most extensive annual beds of eelgrass, *Zostera marina*, in the eastern Pacific, covering 6000 ha (Torre-Cosío 2002), and green turtles have been shown to depend on this food source (Felger and Moser 1973). They also feed on the other 2 seagrasses in the Infiernillo, ditch grass, *Ruppia maritima* (Felger and Moser 1985) and shoal grass, *Halodule wrightii* (R. Felger unpubl. data), which become extensive in hot weather as eelgrass seasonally disappears (e.g., Felger et al. 1980; Meling-López and Ibarra-Obando 1999). Although it is clear that significant threats are present from illegal poaching of green turtles and incidental mortality in commercial fisheries (Seminoff

2000; Gardner and Nichols 2001), a new threat is emerging along the central Gulf coast of the Baja California peninsula. In the areas of Bahía las Animas and Bahía de los Angeles, a commercial algae harvest operation (PHYCOS, Ensenada) has been extracting the red alga *Gracilariopsis lemaneiformis* from shallow water benthic habitats. This alga has been shown to be the primary diet constituent of green turtles in this region (Seminoff et al. 2002b). Green turtles are opportunistic foragers and, as such, may be able to shift diets to compensate for the diminished abundance of this algal species. However, the potential for such a shift in behavior and the consequences of this large-scale algae extraction are not known.

Green Turtle Overwintering

During winter months, waters of the northern and central Gulf of California and Pacific coast of the Baja California peninsula may drop below 15°C (Robinson 1973). In response, green turtles become sluggish and may exhibit overwintering torpor (Felger et al. 1976), a behavior that has not been described in other green turtle populations. The Seris had extensive knowledge of overwintering *Chelonia*. These torpid or "buried" turtles were sought at specific places and conditions (e.g., clear, calm water) from the muddy-sandy sea floor of the Canal del Infiernillo, involving special techniques employing harpoons as long as 10 m. The harpooners were young men with exceptional vision who could discern the faint outline of the exposed portion of the carapace (Felger et al. 1976).

Mexican fishermen from Bahía Kino, in the course of diving for spiny lobsters during the early 1970s, discovered torpid, overwintering turtles on rocky ledges surrounding the Midriff Islands. Using a hookah arrangement with compressor-diving equipment, they harvested these torpid turtles until they had become economically extinct within only several years (Felger et al. 1976). In the decades since then, overwintering *Chelonia* have been discovered on rocky ledges and caves throughout the central and northern Gulf of California (Nichols and Seminoff unpubl. data). In addition, divers harvesting sea cucumbers (*Isostichopus fuscus*; see Castro 1995) opportunistically take torpid sea turtles when they find them, but nowhere approaching the levels of the Midriff Island harvests of the 1970s. Nichols and Seminoff (unpubl. data) have also documented torpid, overwintering *Chelonia* in San Ignacio Lagoon

on the Pacific coast of the Baja California peninsula. Overwintering behavior might not be confined to the green turtle. K. Cliffton encountered torpid, probably overwintering loggerheads at about 15-m depth, one in the Canal del Infiernillo in March 1978, and another in Magdalena Bay in February 1979 (Cliffton et al. 1982).

Green versus Black: The Taxonomic Debate

Eastern Pacific and Gulf of California green turtles are part of the *Chelonia* complex that has sparked a long-standing and unresolved taxonomic debate (Bowen and Karl 1999; Karl and Bowen 1999; Pritchard 1999). The eastern Pacific and Gulf of California green turtle is sometimes recognized as a full species, *Chelonia agassizii*, originally described by Bocourt (1868). Carr (1952) changed its rank to a subspecies, *C. mydas agassizii*, based on unique morphology, including the dark coloration and high-domed carapace (Carr 1961). After several years of work in the Gulf of California, Caldwell and Caldwell (1962) described the northeastern Pacific population as the subspecies *C. mydas carrinegra* (named for their colleague, Archie Carr, and the turtles' dark pigmentation). It is an apparent synonym of *C. mydas agassizii*. Caldwell and Caldwell (1962) described *C. mydas carrinegra* as a permanent, stable population, with small as well as large, breeding-size turtles of both sexes (coupled with reports of nesting), and suggested genetic isolation in Baja California waters.

It is now known that northwestern Mexico green turtles originate from at least 2 distant rookeries: Michoacán and the Revillagigedo Islands (Nichols et al. 2000b; P. Dutton, pers. comm.). By the 1970s fishermen and biologists realized that Gulf of California green turtles migrate to breed and nest at Colola Beach and Maruata Bay, Michoacán (Cliffton et al. 1982; fig. 20.4). For example, Cliffton et al. (1981: 203) note that "our tag returns indicate that the turtles traverse the Mexican coastline in their migrations, as well as travel to distant feeding grounds in Central and South America (Instituto de Pesca, unpublished data)." An additional scientific link came from an adult female green turtle tagged at the primary nesting beach in Michoacán on April 12, 1984 and captured in 1985 near Piedra San Bernabe in the northwestern Gulf, a distance of 1750 km (Alvarado and Figueroa 1992; Figueroa et al. 1993). Further connections between areas in

Figure 20.4. Green turtle nesting at Colola Beach, Michoacán, December 1982. (Photo by Kim Cliffton.)

southern Mexico and foraging sites in the Gulf of California have been established by satellite telemetry (Nichols et al. 1999; Nichols 2003a), genetics (Nichols et al. 2000a; P. Dutton, pers. comm.), and flipper-tag recoveries (Seminoff et al. 2002a).

Early Exploitation

As in other parts of the world, seventeenth- and eighteenth-century pirates and explorers caught "turtles that swarmed about," and used turtle meat while sailing Mexican waters (Gerhard 1963: 5–6). Sea turtles were taken from beaches and at sea (O'Donnell 1974). Sailing near Cabo San Lucas in October 1793, Captain James Colnett reported the sea "almost covered with turtles, and other tropical fish" (Colnett 1798: 94). Sea turtles were reported by early missionaries and explorers to be so abundant that navigators were sometimes slowed by massive flotillas of migrating turtles (Cliffton 1990).

Nineteenth-century whalers and sealers relied on the ever-available turtle meat, stocking up on an abundant, high-quality, and easy-to-keep food source (O'Donnell 1974). In his journal of winter 1858–1859, Scammon (1970) reported that the provisioning with sea turtles before leaving the great lagoons of the Pacific coast of Baja California was commonplace and easy.

Green turtles in northwest Mexico were considered an inexhaustible resource in the early twentieth century (e.g., Nelson 1921). Turtles were shipped from northwest Mexico to canneries in California as well as restaurants in California and Arizona and on to inland cities such as Chicago. Up to 1000 turtles were shipped weekly to San Diego, and Los Angeles and San Diego markets received 35,000 kg in 1920 (O'Donnell 1974).

On the Pacific coast of the Baja California peninsula, the development of canneries, large turtle-fishing vessels, and improved capture techniques resulted in a surge in turtle captures in the early twentieth century. Townsend (1916: 445) reported that "a single haul of a seine 600 feet long brought to shore 162 green turtles . . . probably half as many more escaped." In some places nets 100–400 m long were spread across the mouths of lagoons after flood tides. One common practice involved the use of rowboats to retrieve turtles from massive nets, then unloading them to launches that shuttled turtles to a schooner, which after being loaded with about 350 turtles, returned to San Diego (Averett 1920). Turtles were often held in ponds in San Diego until slaughtered. Up to 2 tons of turtle products were produced per day; canned meat and rendered oil were the main products. Craig (1926) described similar harvesting at San Felipe in the northern Gulf of California in a report on the new "totuava" (*Cynoscion macdonaldi*) fishery (see also chapter 3).

Modern Sea Turtle Exploitation

Turtle exploitation escalated after World War II due to demands from growing northwestern Mexico communities. By the 1960s the harvests reached a new production peak, primarily for domestic consumption. Between 1955 and 1961, landings from the Baja California peninsula made up 39% of the total for all Mexico. Turtle harvesting boomed by the mid-1960s, due in part to government promotion because of a simultaneous decline in crocodile resources, resulting in great demand for turtle hides. Between 1962 and 1967 the sea turtle catch in Mexico increased 633%, and Mexico produced more turtle products than any other nation (O'Donnell 1974).

Fishermen at Laguna San Ignacio recalled year-round sea turtle hunting from the 1940s through the 1970s. During summer they rowed or sailed small boats by lamplight at night in the lagoon, harpooning turtles when the sea was calm. Standing in the bow, the harpooner looked for distinctive bioluminescence, or during daylight the hunter looked for the turtle's gasping head at the surface (fig. 20.5). The men partially filled their boats with water in order to heave huge turtles over the gunwales (F. Savin, pers. comm. 1996).

By the 1960s entanglement nets and ever-larger outboard motors became common throughout northwestern Mexico, replacing harpoons, sails, and oars and allowing easier capture of more turtles. Harvest rates in northwestern Mexico in the 1960s and 1970s were greater than at any other time. In addition to harvests through the more conventional techniques of harpooning and netting, during the 1970s overwintering green turtles were collected in large numbers (see discussion on green turtle overwintering above). At the same time, large-scale harvest of nesting females and eggs occurred at nesting beaches farther south on the Pacific coast. Cliffton et al. (1982) estimated 375,000 green turtles harvested on the Pacific coast of Mexico from 1966 to 1970. Construction of the Baja California trans-peninsular highway in the early 1970s, better roads in Sonora, and the new coastal highway in Michoacán brought increased numbers of people and expanded sea turtle markets. Yet even as hunting efforts intensified as boats became more secure and larger outboard motors permitted longer forays, catches declined. By 1988, nesting female green turtles on Colola Beach, Michoacán, reached an all time low of 79 animals, down from an estimated 25,000 per year in the early 1970s (Cliffton et al. 1982). Government control effectively began in 1978 on the main green-turtle nesting beaches in Michoacán (Cliffton et al. 1982). Due to these conservation efforts, by the year 2000 there were about 500 females nesting annually (Alvarado-Díaz et al. 2001).

Despite increasing but mixed protection on the nesting beaches, in-water harvest continued in northwestern Mexico (fig. 20.6). Fishermen evaded closed regions, seasons, and size limits by holding out-of-season turtles in pens or tying them to mangrove roots until the legal season started, by misrepresenting species, and by transporting turtles between regions during closures. Illegal harvest continued by fishermen who lacked permits, and quotas were routinely exceeded, perhaps at 3 times the authorized levels (Cliffton et al. 1982). A regional population crash occurred at Bahía de los Angeles between 1961 and 1971 (A. Resendiz, pers. comm.). Local fishermen indicate that by the early 1980s it had become commercially infeasible to hunt sea turtles, although they continued to opportunistically take turtles. Nevertheless, despite reduced capture

Figure 20.5. Green turtle surfacing, vicinity of Bahía de Los Angeles, July 2000. (Photo by David Barron.)

Figure 20.6. Bound green turtles, in a covered, beachside corral to await slaughter or consignment to northern markets, Bahía de los Angeles, Baja California, about 1977. (Photo by Kim Cliffton.)

rates and the disappearance of an organized, legal turtle market, sea turtle hunting continues (Gardner and Nichols 2001; Nichols et al. 2002a; Nichols 2003a; Seminoff et al. 2003a).

Sea Turtle Conservation in Mexico

Management techniques were initiated in Mexico in 1966 (Instituto Nacional de Ecología 2000). The goal of the Ministry of Fisheries (Secretaría de Pesca) was to manage for maximum yield by allowing populations to recover while simultaneously protecting nesting beaches. The temporary seasonal ban on turtle hunting, between May and August (typically the peak of the season) proved useless due to inadequate enforcement. Tagging programs, one of the first steps in conservation research, were initiated in the 1960s throughout the country, including the Baja California peninsula (Márquez 1976; A. Resendiz, pers. comm.) and Sonora (K. Cliffton, pers. comm.). In addition, attention was focused on

a network of conservation camps and research programs at nesting beaches. The study of sea turtles in the water, however, proved to be difficult and expensive. In 1968, the Ministry of Commerce elaborated regulations related to the capture, use, and commercialization of sea turtles, emphasizing and obliging use of the entire animal, rather than only the hide, in order to halt the standard wastage (Cantú and Sánchez 2000; table 20.1).

A total ban of turtle taking was implemented in 1971 as an interim measure and response to diminishing populations. The unenforceable ban was lifted in March 1973, ostensibly due to incorrect perception of population increase (Luis García Cacho, Chief Fisheries Officer, Ensenada, Baja California, May 1973; cited in O'Donnell 1974). Between 1973 and 1976, annual quotas were established and exclusive permits granted to fishing cooperatives (table 20.1). In theory, the turtle cooperatives would manage their local fisheries, and quotas would ensure the effectiveness of the new program. O'Donnell (1974) predicted the difficulty in imposing unenforceable

Table 20.1. Conservation measures and legislation affecting sea turtles in Mexico (Cantú and Sánchez 2000).

Year	Action
1968	Commerce ministry develops rules for capture, use, and commerce in sea turtles
1971	Ban on all species for 2 years
1972	Ban on capture, except for fishing cooperatives
1979	Total ban for leatherback and hawksbill turtles
1982	Creation of first environmental ministry (SEDUE, *Secretaria de Desarrollo Urbano y Ecología*)
1986	Seventeen nesting beaches declared reserves
1986	Enactment of Federal Law of Fisheries
1990	Agreement establishes a total ban on all species of sea turtle (May)
1991	Mexico becomes a member of CITES (July)
1991	Creation of Article 254 bis on the penal code that establishes jail penalties for taking, killing, or commercializing sea turtles and their products
1992	Environmental issues become the responsibility of Social Development Ministry (SEDESOL, *Secretaría de Desarollo Social*)
1993	Agreement creates the Intersecretarial Commission for the Protection and Conservation of Sea Turtles and the National Committee for the Protection and Conservation of Sea Turtles
1994	Creation of a ministry (SEMARNAP, *Secretaría de Medio Ambiente, Recursos Naturales y Pesca*; now SEMARNAT, *Secretaría de Medio Ambiente y Recursos Naturales*) dedicated only to environmental issues (December)
1994	Creation of the Subattomey Office of Natural Resources in the Federal Attorney General's Office for the Protection of Environment (PROFEPA, *Procuraduría Federal de Protección al Ambiente* (Decémber)
1996	First Mexican Official Norm of Emergency that establishes the use of turtle excluder devices (TEDs) in the Pacific (March)
1998	Mexico signs the Inter-American Convention for the Protection and Conservation of Sea Turtles (December)
2001	Maximum penalty under penal code increased to 12 years imprisonment for harming sea turtles

regulations on economically disadvantaged people depending on sea turtles, echoing earlier concerns of Caldwell (1963).

Despite attempts to slow the take of sea turtles in national waters during the 1970s, a number of slaughterhouses continued to process unprecedented numbers of sea turtles. None was more notorious and deadly to sea turtles than the one operated in San Isidro, Oaxaca by Antonio Suarez, an influential businessman living in Mexico City (Cahill 1978; Felger 2003). In the late 1970s, as many as 40,000 turtles (mostly olive ridleys) were processed annually at Suarez's San Isidro factory-style slaughterhouse. As a result, the amazing group-nesting phenomenon, or *arribada*, in nearby Escobilla, Oaxaca all but vanished. Subsequent vigorous conservation efforts have resulted in a return of the Escobilla *arribada*.

In 1991 Mexico joined CITES, having declared a complete ban on the use of all sea turtles and eggs in 1990 (Aridjis 1990; table 20.1). This ban resulted largely from international and national pressure on fisheries officials. Stiff fines and jail terms up to 3 years were implemented to punish those who con-

tinued trade in turtle products. The following list of sanctions and penalties was excerpted from a flyer distributed in 1990 to fishermen in Mexico:

- In accordance with the Fisheries laws, the extraction, capture, possession, transport, or commercialization of sea turtles is considered an infraction according to Article 24 Section XIX.
- Consistent with the same law, this infraction establishes fines of 1,001 to 2,000 times the minimum daily salary in force in the capital (D.F.), as well as the confiscation of equipment, fishing tackle, or that with which the infraction was committed.
- Independent of this administrative proceeding, Article 254 bis of the penal code considers the prior conduct a crime.
- Article 254 bis.—Whoever intentionally captures, gravely harms, or deprives the life of marine mammals or turtles, or collects or commercializes their products in whatever form without authorization, will receive from the

appropriate [authorities] a penalty of between six months and three years of prison.

- The same penalty as in the preceding paragraph will be imposed by the authorities on those who intentionally, without authorization, capture aquatic species, declared prohibited (*en veda*).

Despite the 1990 ban,

in some fishing camps and principally during the period from March to July, the traffic of *Caretta* and *Chelonia* is the most important activity. And while capture of sea turtles is a federal offense, the majority of turtles are slaughtered on islands or beaches far from the fishing camps. Nevertheless, this isn't the primary problem, if the exploitation fell only on the adults, it would be possible to think about recovery in the long term. Disgracefully, as much in *Caretta* as in *Chelonia*, the species most frequently captured, the fishery is primarily directed at the immature animals (on beaches 90% of the carapaces found have been subadults). (Olguin-Mena 1990: 62)

By 1997 it became clear that in addition to protecting nesting beaches (García et al. 2003) and reducing illegal consumption, there was a need to address commercial marine fisheries activities impacting sea turtles. Each year thousands of shrimp trawlers scour the sea floor in coastal habitats all along Mexico's Pacific coast, and through accidental capture, thousands of sea turtles are drowned in the process. In response to this tremendous sea turtle by-catch mortality, Mexico mandated the use of turtle excluder devices (TEDs; DOF 1999), a contraption fitted into the shrimp net that enables turtles to escape before being trapped in the net (fig. 20.7). The use of TEDs has been effective in reducing sea turtle mortality in many shrimping areas, but regrettably TEDs have not yet been fully accepted by the local shrimping fleets.

The proposed *Escalera Náutica* mega-project calls for a system of 22 marinas and tourism facilities along the coasts of northwest Mexico to attract tourists from the United States (see chapters 9, 15, and 16). The potential negative environmental impacts are staggering; negative impacts on sea turtles and their habitats include contamination of critical nursery and feeding areas, habitat destruction, boat traffic and turtle strikes within marine protected areas, and increased poaching concomitant with an increase in the human population (e.g., Nichols 2003b). We propose instead an *Escalera Ecologica* to strengthen existing and proposed marine, coastal wetland, and terrestrial reserves and alternative economic initiatives (Broyles et al. 2001; Sala et al. 2002; see also introduction of this volume).

Cultural Context of Sea Turtle Exploitation

Loss of sea turtles in northwestern Mexico represents far more than a threat to a few commercial species or loss of biological diversity. Many coastal-area families relied on sea turtles to earn a living. Turtle meat was—and still is for some—an important source of protein. Turtle meat was equated with health, vitality and stamina, virility, and celebrations. Sea turtle blood and oil were used medicinally to treat ailments such as anemia and bronchitis. These traditions have not abated, and most coastal inhabitants still include sea turtle meat as part of their diet and culture (Garcia-Martínez and Nichols 2000).

The use of sea turtles is associated with neither poverty nor hunger. Turtles are eaten by fishermen, government employees, teachers, military personnel, and virtually anyone who grew up with the tradition of turtle feasts. Sea turtle meat remains the premier *plato tradicional* at special events, holiday celebrations, and to honor visiting dignitaries. The cultural use of turtle meat, the deep traditions surrounding its use, and its perceived benefits, completely override adherence to laws protecting endangered sea turtles. Although some have tried, there is no culinary replacement. A turtle feast is among the highest honors and displays of trust.

Sea turtle consumption peaks during Holy Week, or *semana santa*—the week before Easter. During this holiday Catholics consume vast quantities of sea turtle due to an errant belief that it qualifies as fish under Lenten rules. Recently a group of conservationists and spiritual leaders have petitioned the Vatican to correct this phylogenetic error and save the lives of many thousands of endangered sea turtles (Carleton 2002; Nichols et al. 2002b).

Despite more than a decade of complete protection by law (DOF 1990), sea turtle bycatch and harvest in northwestern Mexico continue at a rate of 7,800 to 35,000 turtles annually (Cantú and Sánchez 2000; Gardner and Nichols 2001; Nichols et al.

Figure 20.7. Turtle excluder device (TED), Laguna Ojo de Liebre, Baja California Sur. The metal grill pictured is in front of the collection bag at the end of the trawl net. Turtles captured hit these bars and are pushed out a trap door just in front of the collection bag. (Photo by J. Seminoff.)

2002b). Based on a survey of the entire Mexico coastline during 1993 and 1994 by Nichols and Seminoff (Seminoff 1994), we concluded that northwest Mexico has the highest sea turtle mortality per capita in Mexico and perhaps in all of North America. Many accidentally captured turtles enter the black market, are traded locally, or are consumed domestically. The problems are typical of conservation-law enforcement in developing countries (Gomez 1982): misunderstanding and circumvention of the law, limited knowledge of the resource by local residents, lack of adequate enforcement, conflict of interest, and tradition of overharvest (tragedy of the commons).

Recommendations and Potential Solutions

Many researchers in Mexico have embraced education and bringing people together rather than enforcement alone (Tennesen 1999). Since 1999 the Sea Turtle Conservation Network of the Californias (*Grupo Tortuguero de las Californias*), a grassroots organization composed of fishermen, local residents, researchers, government resource managers, and conservationists, has initiated programs designed to decrease local pressures on sea turtles (Instituto Nacional de Ecología 2000; Nichols 2003a). Some of the goals and conservation efforts in northwest Mexico include:

- Development of alternative income sources. NGOs, fishing cooperatives, *ejidos*, and government agencies are working together to create local wildlife refuges and marine reserves, sustainable fisheries, and to develop community-based ecotourism, adventure tourism, and catch–release sport fishing businesses. Nascent sea turtle ecotourism is functioning in the Cape Region of Baja California Sur. In some areas aquaculture is promoted as a viable option, although there are serious inherent problems.

- Education. By sharing information on the status of marine resources, sea turtle life histories, and current research results, local groups have made tremendous progress. Some residents unaware of the endangered status of sea turtles and the consequences of cumulative use have changed their habits and joined the recovery effort. Such efforts need to be extended and include development of educational curricula describing sea turtle natural history and ecology.
- Research. At the base of any recovery plan are studies of the resources and the community of users. Long-term efforts to monitor trends in sea turtle populations as well as the perspective of the human communities are underway. Application of social sciences to solutions of resource exploitation is particularly warranted (Piper 1992).
- Community involvement in research. Involving fishermen in research is a form of education, for both the researcher and the fisher. For some it is an additional source of income. Moreover, because local residents typically have more intimate knowledge of local resources than do visiting researchers, projects that attempt to integrate local people into field activities are more likely to succeed with both research and conservation goals (Nichols et al. 2002a).
- Enforcement. There is no substitute for strong enforcement.

Much of the biological diversity of sea turtles as well as the cultural diversity related to human–sea turtle interactions is irretrievably lost. Most sea turtle species in northwest Mexico are commercially extinct and teeter on the brink of ecological extinction. Restoration may still be possible, but the opportunity is rapidly running out.

Acknowledgments

We thank first and foremost the legions of researchers, students, fishers, conservationists, volunteers, soldiers, NGOs, and government and law enforcement personnel who are devoted to the protection and recovery of Mexico's sea turtles. Without their tireless work, now spanning more than four decades, there can be no possibility for sea turtle recovery. We also thank members of the Seri communities who provided hundreds of hours of consultation on sea turtle information. Financial and logistical support was provided by the Center for Field Research at Earthwatch Institute, Wallace Research Foundation, the PADI Foundation, National Geographic Society, and the Lerner-Grey Foundation. We thank the following individuals for their generous assistance: Fernando Arcas, Betty Aridjis, Homero Aridjis, Ana Barragan, Luis Bourillón, Louise Brooks, Kim Cliffton, Dennis Cornejo, Serge Dedina, Mahina Drees, Barbara Edwards, Jennifer Gilmore, Linda Wallace-Gray, Adan Hernandez, the late Roberto Herrera T., Gabriel Hoeffer, Tod Jones, Arturo Juarez, the late Guadalupe López, Jose Luis López, Cathy Moser Marlett, Stephen Marlett, Humberto Morales, the late Edward Moser, Mary Beck Moser, Angelica Narvaez, Dana Nichols, Hoyt Peckham, Rodrigo Rangel Acevedo, Phillip J. Regal, Antonio Resendiz, Bety Resendiz, the late Alexander Russell, Jean Russell, Laura Sarti, Jorge Torre-Cosío, and Michael Wilson. Orthography of Seri terms was checked and revised by Stephen Marlett and Mary Beck Moser. Todd Jones provided helpful comments on a draft of the chapter. Research described herein was authorized by Secretaría del Medio Ambiente, Recursos Naturales, y Pesca, México (permit nos. 150496-213-03, 190698-213-03, 280499-213-03, and 280597-213-03).

Literature Cited

Alvarado, J., and A. Figueroa. 1992. Recapturas post-anidatorias de hembras de tortuga marina negra (*Chelonia agassizi*) marcadas en Michoacán, México. Biotropica 24: 560–566.

Alvarado-Díaz, J., C. Delgado-Trejo, and I. Suazo-Ortuño. 2001. Evaluation of black turtle project in Michoacán, México. Marine Turtle Newsletter 92: 4–7.

Aridjis, H. 1990. Mexico proclaims total ban on harvest of turtles and eggs. Marine Turtle Newsletter 50: 1–3.

Aschmann, H. 1966. The Natural and Human History of Baja California. Dawson's Book Shop, Los Angeles, California.

Averett, W. E. 1920. Lower California green turtle fishery. Pacific Fisherman 18(7): 24–25.

Bjorndal, K. A. 1997. Foraging ecology and nutrition of sea turtles. Pp. 199–232 *in* P. L. Lutz and J. A. Musick (eds), The Biology of Sea Turtles. CRC Press, Boca Raton, Florida.

Bocourt, M. 1868. Description de quelques chéloniens nouveaux appartenant à la faune mexicaine. Annales des Sciences Naturelles—Zoologie et Biologie Animale 10: 121–122.

Bowen, B. W., and S. A. Karl. 1999. In war, truth is the first casualty. Conservation Biology 13: 1013–1016.

Bowen, B. W., F. A. Abreu-Grobois, G. H. Balazs, N. Kamezaki, C. J. Limpus, and R. J. Ferl. 1995. Trans-Pacific migration of the loggerhead sea turtle demonstrated with mitochondrial DNA markers. Proceedings of the National Academy of Sciences U.S.A. 92: 3731–3734.

Bowen, T. 2000. Unknown Island: Seri Indians, Europeans, and San Esteban Island in the Gulf of California. University of New Mexico Press, Albuquerque.

Broyles, B., R. Felger, and C. Bowden. 2001. El Corredero Sonorense. Pp. 245–265 in P. Robles Gil, E. Ezcurra, and E. Mellink (eds.), El Golfo de California, un Mundo Aparte. Agrupación Sierra Madre, México D.F.

Burckhalter, D. 1999. Among Turtle Hunters and Basket Makers: Adventures with the Seri Indians. Treasure Chest Books, Tucson, Arizona.

Cahill, T. 1978. The shame of Escobilla. Outside Magazine February 1978, 22–64.

Caldwell, D. K. 1963. The sea turtle fishery of Baja California, México. California Fish and Game 49: 140–151.

Caldwell, D. K., and M. C. Caldwell. 1962. Sea turtles in Baja California waters (with special reference to those of the Gulf of California), and the description of a new subspecies of northeastern Pacific green turtle. Los Angeles County Museum of Natural History, Contributions in Science 61: 1–31.

Cantú, J. C., and M. E. Sánchez. 2000. Tráfico Ilegal de Tortugas Marinas en México, Situación Histórica y Actual. Investigación de Teyeliz, A.C., México D.F.

Carleton, J. 2002. Environmentalists struggle to rescue Mexican sea turtles from poachers. Wall Street Journal, March 29, 2002, 1B.

Carr, A. 1952. Handbook of Turtles of the United States, Canada, and Baja California. Cornell University Press, Ithaca, New York.

Carr, A. F. 1961. Pacific turtle problem. Natural History 70: 64–71.

Carr, A. F. 1973. So Excellent a Fishe: A Natural History of Sea Turtles. Anchor Press/Doubleday, Garden City, New York.

Carr, A. 1987. Impact of nondegradable marine debris on the ecology and survival outlook of sea turtles. Marine Pollution Bulletin 18 (6 part B): 352–356.

Castro, L. R. S. 1995. Management options of the commercial dive fisheries for sea cucumbers in Baja California, Mexico. SPC Beche-de-Mer Information Bulletin 7: 20.

Chan, E-H., and H-C. Liew. 1996. Decline of the leatherback population in Terengganu, Malaysia, 1956–1995. Chelonian Conservation and Biology 2: 196–203.

Chavez, M. 1989. Presencia de tortuga carey, Eretmochelys imbricata, en Playa Platanitos, Nayarit, México. Pp. 28–29 in L. Sarti (ed.), Memorias del VI Encuentro Interuniversitario Mexicano sobre Tortugas Marinas. Universidad Automoma de México, July 7–10, 1989, México D.F.

Cliffton, K. 1990. Leatherback turtle slaughter in Mexico. Tucson Herpetological Society Newsletter 3(5): 44–46.

Cliffton, K., D. O. Cornejo, and R. S. Felger. 1982. Sea turtles of the Pacific coast of Mexico. Pp. 199–209 in K. Bjorndal (ed.), Biology and Conservation of Sea Turtles. Smithsonian Institution Press, Washington, D.C.

Colnett, J. 1798. A Voyage to the South Atlantic and Round Cape Horn into the Pacific Ocean. W. Bennett, London.

Craig, J. A. 1926. A new fishery in Mexico. California Fish and Game 12: 166–169.

Dodd Jr., C. K. 1988. Synopsis of the biological data on the loggerhead sea turtle, Caretta caretta (Linnaeus 1758). U.S. Fish and Wildlife Service Biological Report 88–14, Washington, D.C.

DOF. 1990. Acuerdo por el que se establece veda para las especies y subespecies de tortuga marina en aguas de jurisdicción Federal del Golfo de México y Mar Caribe, así como en las costas del Océano Pacífico, incluyendo el Golfo de California. Diario Official de la Federación, May 28.

DOF. 1999. Modificación a la Norma Oficial Mexicana 002-PESC-1993, Para ordenar aprovechameinto de las especies de camarón en aguas de jurisdicción federal de los Estados Unidos Mexicanos, publicada el 31 de diciembre de 1993. Diario Oficial, Primera Sección, December 29.

Eckert, S. A. 1997. Distant fisheries implicated in the loss of the world's largest leatherback nesting population. Marine Turtle Newsletter 78: 2–7.

Ehrenfeld, J. G., and D. W. Ehrenfeld. 1973. Externally secreting glands of fresh water and sea turtles. Copeia 1973: 305–314.

Felger, R. S. 2000. The Seris and the Guy Who Cuts the Tops Off Plants. Seri hands, a special issue. Journal of the Southwest 42: 521–543.

Felger, R.S. 2003. Sea turtles, Seri Indians, and conservation in the Gulf of California. Pp. 82–83 in J.A. Seminoff (comp.), Proceedings of the Twenty-Second Annual Symposium on Sea Turtle Biology and Conservation. NOAA Technical Memorandum NMFS-SEFSC-503. National Oceanic and Atmospheric Administration, Miami, Florida.

Felger, R. S., K. Cliffton, and P. J. Regal. 1976. Winter dormancy in sea turtles: independent discovery and exploitation in the Gulf of California, Mexico, by two local cultures. Science 191: 283–285.

Felger, R. S., and M. B. Moser. 1973. Eelgrass (*Zostera marina* L.) in the Gulf of California: Discovery of its nutritional value by the Seri Indians. Science 181: 355–356.

Felger, R. S., and M. B. Moser. 1985. People of the Desert and Sea: Ethnobotany of the Seri Indians. University of Arizona Press, Tucson.

Felger, R. S., M. B. Moser, and E. W. Moser. 1980. Seagrasses in Seri Indian culture. Pp. 260–276 *in* R. C. Phillips and C. P. McRoy (eds.), Handbook of Seagrass Biology, an Ecosystem Perspective. Garland STPM Press, New York.

Figueroa, A., J. Alvarado, F. Hernández, Gerardo Rodríguez, and J. Robles. 1993. The ecological recovery of sea turtles of Michoacán, México. Special attention to the black turtle (*Chelonia agassizi*). Final report to WWF-USFWS. World Wildlife Fund, Albuquerque, New Mexico.

Fritts, T. 1982. Plastic bags in the intestinal tracks of leatherback marine turtles. Herpetological Review 13: 72–73.

Fritts, T. H., M. L. Stinson, and R. Marquez M. 1982. Status of sea turtle nesting in southern Baja California, Mexico. Bulletin of the Southern California Academy of Sciences 81: 51–60.

García, A., G. Ceballos, and R. Adaya. 2003. Intensive beach management as an improved sea turtle conservation stategy in Mexico. Biological Conservation 111: 253–261.

Garcia-Martínez, S., and W. J. Nichols. 2000. Sea turtles of Bahía Magdalena, Baja California Sur, Mexico: demand and supply of an endangered species [Abstract]. *In* R. S. Johnston and A. L. Shriver (compil.), Proceedings of the Tenth Biennial Conference of the International Institute of Fisheries Economics and Trade: Microbehavior and Macroresults. Oregon State University, Corvallis, July 11–14, 2000 (CD).

Gardner, S. C., and W. J. Nichols. 2001. Assessment of sea turtle mortality rates in the Bahía Magdalena region, Baja California Sur, México. Chelonia Conservation and Biology 4: 197–199.

Gerhard, P. 1963. Pirates in Baja California. Editorial Tlilan, Tlaplan, Mexico.

Gomez, E. D. 1982. Problems of enforcing sea turtle conservation laws in developing countries. Pp. 537–539 *in* K. A. Bjorndal (ed.), Biology and Conservation of Sea Turtles. Smithsonian Institution Press, Washington, D.C.

Grismer, L. L. 2002. Amphibians and Reptiles of Baja California, Including Its Pacific Islands and the Islands in the Sea of Cortés. University of California Press, Berkeley.

Hilton-Taylor, C. 2000. 2000 IUCN Red List of Threatened Species. IUCN—World Conservation Union, Gland, Switzerland.

Instituto Nacional de Ecología. 2000. Programa Nacional de Protección, Conservación, Investigación y Manejo de Tortugas Marinas. Secretaría de Medio Ambiente, Recursos Naturales, y Pesca, México D.F.

Kamezaki, N. 1997. Effects of global warming on sea turtles. Pp. 254–272 *in* A. Domoto and K. Iwakuni (eds.), Threats of Global Warming to Biological Diversity. Tsukiji Shokan, Tokyo.

Kamezaki, N., I. Miyakawa, H. Suganuma, K. Omuta, Y. Nakajima, K. Goto, K. Sato, Y. Matsuzawa, M. Samejima, M. Ishii, and T. Iwamoto. 1997. Post-nesting migration of Japanese loggerhead turtles, *Caretta caretta*. Wildlife Conservation Japan 3: 29–39.

Karl, S. A., and B. W. Bowen. 1999. Evolutionary significant units versus geopolitical taxonomy: Molecular systematics of an endangered sea turtle genus. Conservation Biology 13: 990–999.

Limpus, C. J. 1995. Global overview of the status of marine turtles: a 1995 viewpoint. Pp. 605–610 *in* K. A. Bjorndal (ed.), Biology and Conservation of Sea Turtles, 2nd ed. Smithsonian Institution Press, Washington, D.C.

Márquez M., R. 1976. Turtle programme in Baja California, Mexico. Marine Turtle Newsletter 1: 5.

Márquez, M. R. 1990. FAO species catalog. Sea turtles of the world. An annotated and illustrated catalog of sea turtle species known to date. FAO Fisheries Synopsis vol. 11, no. 125. Food and Agricultural Organization, Rome. 81 pp.

McConnell, O. J., and W. Fenical. 1976. Halogen chemistry. Phytochemistry 16: 367–368.

McGee, W. J. 1898. The Seri Indians. Seventeenth Annual Report of the Bureau of American Ethnology, part I. Smithsonian Institution, Washington, D.C.

Meling-López, A. E., and S. E. Ibarra-Obando. 1999. Annual life cycles of two *Zostera marina* L. populations in the Gulf of California: contrasts in seasonality and reproductive effort. Aquatic Botany 65: 59–69.

Morreale, S. J., E. A. Standora, J. R. Spotila, and F. V. Paladino. 1996. Migration corridor for sea turtles. Nature 384: 319–320.

Mosier A, A. Foley, and B. Brost (eds). 2002. Proceedings of the Twentieth Annual Symposium on Sea Turtle Biology and Conservation. NOAA Technical Memorandum NMFS-SEFSC-477. National Oceanic and Atmospheric Administration, Miami, Florida.

Nabhan, G. P. 2003. Singing the Turtles to Sea. University of California Press, Berkeley.

National Marine Fisheries Service and U.S. Fish and Wildlife Service. 1998. Recovery plan for U.S. Pacific populations of the hawksbill turtle (*Eretmochelys imbricata*). National Marine Fisheries Service, Silver Spring, Maryland.

Nelson, E. W. 1921. Lower California and its natural resources. Memoirs of the National Academy of Sciences 16: 1–194.

Nichols, W. J. 2003a. Biology and conservation of sea turtles in Baja California. Ph.D. dissertation, University of Arizona, Tucson.

Nichols, W. J. 2003b. Sinks, sewers and speed bumps: the impact of marina development on sea turtles in Baja California, Mexico. Pp. 17–18 *in* J.A. Seminoff (comp.), Proceedings of the Twenty-second Annual Symposium on Sea Turtle Biology and Conservation. NOAA Technical Memorandum NMFS-SEFSC-503. National Oceanic and Atmospheric Administration, Miami, Florida.

Nichols, W. J., K. E. Bird, and S. Garcia. 2002a. Community-based research and its application to sea turtle conservation in Bahía Magdalena, BCS, Mexico. Marine Turtle Newsletter 89: 4–7.

Nichols, W. J., P. H. Dutton, J. A. Seminoff, E. Bixby, F. A. Abreu, and A. Resendiz. 2000b. Poi or papas: do Hawaiian and Mexican green turtles feed together in Baja California waters? Pp. 14–15 *in* H. J. Kalb and T. Wibbels (comp.), Proceedings of the Nineteenth Annual Symposium on Sea Turtle Biology and Conservation. NOAA Technical Memorandum NMFS-SEFSC-443. National Oceanic and Atmospheric Administration, Miami, Florida.

Nichols, W. J., A. Resendiz, J. A. Seminoff, and B. Resendiz. 2000a. Transpacific migration of a loggerhead turtle monitored by satellite telemetry. Bulletin of Marine Science 67: 937–947.

Nichols, W. J., C. Safina, and L. Grossman. 2002b. Divine intervention: lobbying the Vatican to save sea turtles. Marine Turtle Newsletter 99: 29.

Nichols, W. J., J. A. Seminoff, A. Resendiz, P. Dutton, and F. A. Abreu-Grobois. 1999. Using molecular genetics and biotelemetry to study life history and long distance movement: a tale of two turtles. Pp. 102–103 *in* F. A. Abreu-Grobois, R. Briseño, R. Márquez, and L. Sarti (comps.), Proceedings of the Eighteenth Annual Symposium on Sea Turtle Biology and Conservation. NOAA Technical Memorandum NMFS-SEFC-436. National Oceanic and Atmospheric Administration, Miami, Florida.

O'Donnell, J. 1974. Green turtle fishery in Baja California waters: history and prospect. Master's thesis, California State University, Northridge.

Olguin-Mena, M. 1990. Las tortugas marinas en la costa oriental de Baja California y costa occidental de Baja California Sur, Mexico. Thesis, Universidad Autonoma de Baja California Sur, La Paz.

Pesenti, C., and W. J. Nichols. 2002. Signs of success: Fourth Annual Meeting of the Sea Turtle Conservation Network of the Californias (Grupo Tortuguero de las Californias). Marine Turtle Newsletter 97: 14–16.

Piper, J. C. 1992. Anthropology, sustainability and the case of Mexico's sea turtles. Master's thesis, University of Arizona, Tucson.

Pritchard, P. C. H. 1999. Status of the black turtle. Conservation Biology 13: 1000–1003.

Resendiz, A., B. Resendiz, W. J. Nichols, J. A. Seminoff, and N. Kamezaki. 1998. First confirmation of a trans-Pacific migration of a tagged loggerhead sea turtle (*Caretta caretta*), released in Baja California. Pacific Science 52: 151–153.

Robinson, M. K. 1973. Atlas of monthly mean sea surface and subsurface temperatures in the Gulf of California, México. San Diego Society of Natural History Memoir 5: 1–97.

Sala, E., O. Aburto-Oropeza, G. Paredes, I. Parra, J. C. Barrera, and P. K. Dayton. 2002. A general model for designing networks of marine reserves. Science 298: 1991–1993.

Scammon, C. M. 1970. Journal aboard the Bark Ocean Bird on a Whaling Voyage to Scammon's Lagoon, Winter of 1858–1859, D. A. Henderson (ed.). Dawson's Book Shop, Los Angeles, California.

Seminoff, J. A. 1994. Conservation of the marine turtles of Mexico: a survey of nesting beach conservation projects. M.S. thesis, University of Arizona, Tucson.

Seminoff, J. A. 2000. The biology of the East Pacific green turtle (*Chelonia mydas agassizii*) at a warm temperate foraging area in the Gulf of California, Mexico. Ph.D. dissertation, University of Arizona, Tucson.

Seminoff, J. A., J. Alvarado, C. Delgado, J. L. Lopez, and G. Hoeffer. 2002a. First direct evidence of migration by an East Pacific green sea turtle from Michoacán, México, to a foraging ground on the Sonoran Coast of the Gulf of California. Southwestern Naturalist 47: 314–316.

Seminoff, J. A., T. T. Jones, A. Resendiz, W. J. Nichols, and M. Y. Chaloupka. 2003a. Monitoring green turtles (*Chelonia mydas*) at a coastal foraging area in Baja California, Mexico: multiple indices to describe population status. Journal of the Marine Biological Association of the United Kingdom 83: 1355–1362.

Seminoff, J. A., W. J. Nichols, A. Resendiz, and L. Brooks. 2003b. Occurrence of hawksbill turtles, *Eretmochelys imbricata* (Reptilia: Cheloniidae), near the Baja California Peninsula, México. Pacific Science 57: 31–38.

Seminoff, J. A., A. Resendiz, S. Hidalgo, and W. J. Nichols. 2002b. Diet of the East Pacific green turtle, *Chelonia mydas*, in the central Gulf of California, México. Journal of Herpetology 36: 447–453.

Seminoff, J. A., A. Resendiz, and W. J. Nichols. 2002c. Home range of the green turtle (*Chelonia mydas*) at a coastal foraging ground in the Gulf of California, México. Marine Ecology Progress Series 242: 253–265.

Shreve, F. 1951. Vegetation of the Sonoran Desert. Carnegie Institution of Washington Publication 591: 1–192.

Spotila, J. R., R. D. Reina, A. C. Steyermark, P. T. Plotkin, and F. V. Paladino. 2000. Pacific leatherback turtles face extinction. Nature 405: 529–530.

Tennesen, M. 1999. Mexico's turtle wars. International Wildlife Magazine. November/December 1999, 44–51.

Torre-Cosío, J. 2002. Inventory, monitoring and impact assessment of marine biodiversity in the Seri Indian territory, Gulf of California, Mexico. Ph.D. dissertation, University of Arizona, Tucson.

Townsend, C. H. 1916. Voyage of the Albatross to the Gulf of California in 1911. Bulletin of the American Museum of Natural History 31: 117–130.

Wagner, H. R. 1929. Spanish Voyages to the Northwest Coast of America in the Sixteenth Century. California Historical Society, San Francisco.

Wetherall, J. A. 1996. Assessing impacts of Hawaiian longline fishing on Japanese loggerheads and Malaysian leatherbacks: some exploratory studies using TURTSIM. Pp. 57–75 in A. B. Bolten, J. A. Wetherall, G. H. Balazs, and S. G. Pooley (eds.), Status of Marine Turtles in the Pacific Ocean Relevant to Incidental Take in the Hawaii-based Pelagic Longline Fishery. NOAA Technical Memorandum NMFS-SWFSC-230. National Oceanic and Atmospheric Administration, La Jolla, CA.

21

Prairie Dogs, Cattle, and Crops: Diversity and Conservation of the Grassland–Shrubland Habitat Mosaic in Northwestern Chihuahua

GERARDO CEBALLOS

RURIK LIST

JESÚS PACHECO

PATRICIA MANZANO-FISCHER

GEORGINA SANTOS

MARIO ROYO

One of the greatest surprises for the Spanish conquistadors, priests, and explorers during their discovery of what is now known as the Chihuahuan Desert Region was a harsh but highly diverse landscape, teeming with unusual wildlife and strange plants. The Camino Real, the historical route that linked Mexico City to Santa Fe in New Mexico, crossed a large part of the Chihuahuan Desert Region. Along that road, the Spaniards marvelled at the sight of apparently endless grasslands, which in fact extended practically uninterrupted from northern Mexico north beyond the Camino Real and the edge of the Chihuahuan Desert Region to southern Canada.

Did the pristine grasslands of the Chihuahuan Desert Region also support large populations of the black-tailed prairie dog (*Cynomys ludovicianus*)? According to Bailey (1932), a prairie-dog town in 1908 covered an estimated 1000 square mile area in the Animas Valley of New Mexico, near the border with Chihuahua. Prairie dogs might initially have increased their range after the introduction of cattle (see Hubbard and Schmitt 1984), but arguably they were always a common feature of the Chihuahuan Desert Region grasslands, as they were always a common feature of the Great Plains.

Unfortunately, 5 centuries after the discovery of the New World, grasslands have been massively converted to croplands, rangelands, and urban environments. They are among the most threatened ecosystems in North America. The rate and severity of their disappearance represents a major conservation concern at a landscape, species, and population level (Samson and Knopf 1996; Henwood 1998).

The disappearance of grasslands has been compounded by overexploitation of some species and attempts at eradicating others, including prairie dogs (see chapter 19). Drastic population declines of prairie dogs, in turn, have negatively affected the black-footed ferret (*Mustela nigripes*), mountain plover (*Charadrius montanus*), and other species depending on prairie dogs and the grassland special habitat type they create (Clark 1989; Ceballos et al. 1993; Knopf 1994; Miller et al. 1994, 2000; Knap et al. 1999). As discussed in this chapter, prairie dogs are both keystone species and "ecosystem engineers" in North American grasslands because of their influence on the structure and functioning of these ecosystems (Weltzin et al. 1997a; Ceballos et al. 1999; Kotliar et al. 1999). Prairie dogs profoundly impact both the abiotic and biotic characteristics of the environment, and they can modify many environmental features at regional and local scales (Uresk 1985; Archer et al. 1987; Whicker and Detling 1988; Cid et al. 1991; Weltzin et al. 1997b).

In 1987, our group rediscovered extensive grasslands occupied by prairie dogs in the Janos and Casas Grandes Municipalities (counties, hereafter referred to as the Janos–Casas Grandes Region) of northwestern Chihuahua (fig. 21.1). This discovery led us to begin a long-term ecological study and conservation project on the prairie dog ecosystem and its surrounding area. We present here a summary of our results so far. We have divided the chapter into 3 parts, covering (1) the unique regional biodiversity, with an emphasis on species of conservation concern, (2) the threats associated with human activities, and (3) the prospects to improve land-use practices for the biological conservation of the region. We also review the role of prairie dogs in grassland ecosystems.

Due in part to the large number of threatened species, the vertebrate fauna of the Janos–Casas Grandes Region ranks among the most important in northwestern Mexico (List et al. 1998). For the conservation of mammalian diversity, it even ranks among the first 3 sites in Mexico (Ceballos 1999). It is an important breeding and wintering site for grassland birds at a national scale (Manzano-Fischer et al. 2000). The biological wealth and importance of the Janos–Casas Grandes Region for the conservation of North America's biodiversity has been recognized in both Mexico and the United States (Dinerstein et al. 1999; CONABIO 2000).

Biological Diversity: Ecosystems and Species

Vegetation

At the lower elevations of the Janos–Casas Grandes Region, the vegetation consists of a mosaic of grasslands and shrublands, with also some riparian woodlands (fig. 21.2). Southward and westward, in the

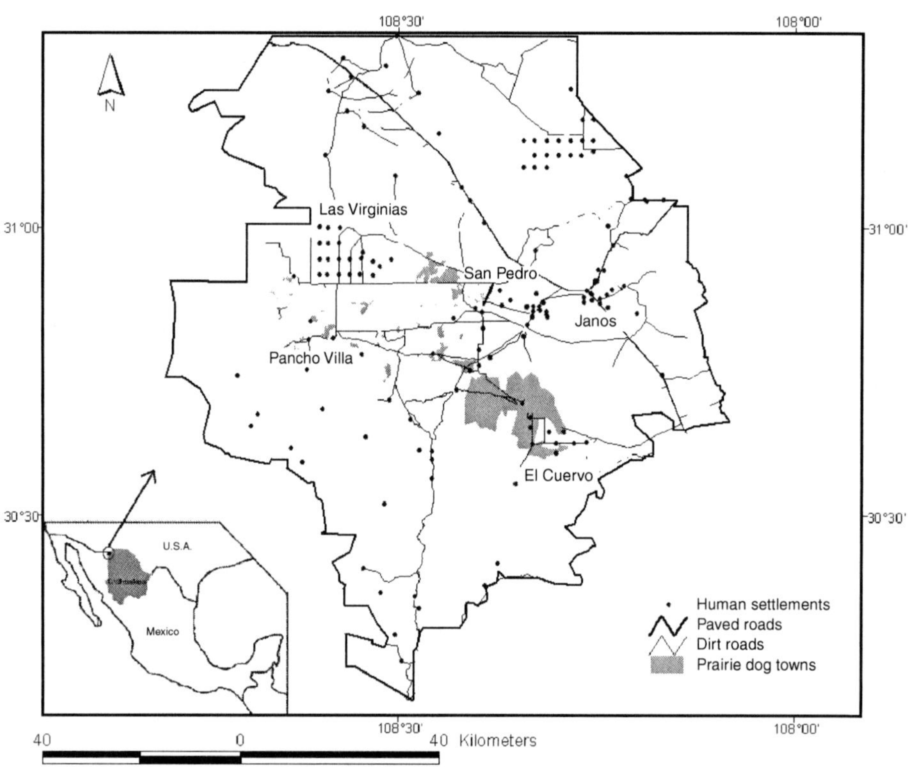

Figure 21.1. Janos–Casas Grandes Region, with distribution of associated prairie dog towns.

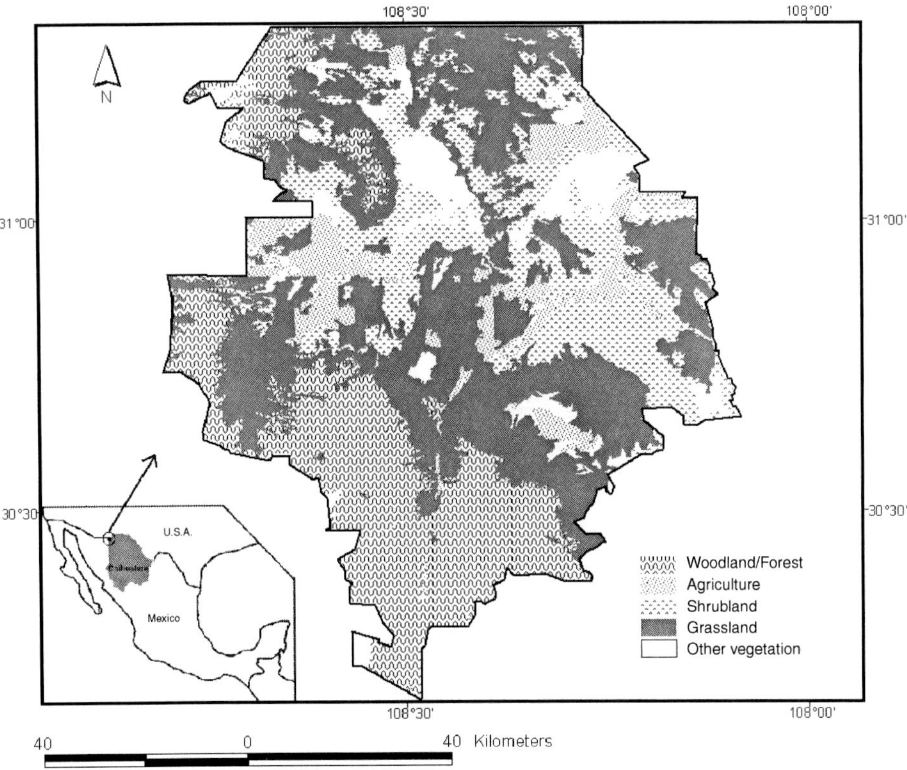

Figure 21.2. Distribution of vegetation types and agricultural areas in the Janos–Casas Grandes Region (Janos and Casas Grandes Counties). Note that westward and southward, in the foothills of the Sierra Madre Occidental, the vegetation changes to mainly oak and pine woodlands.

foothills of the Sierra Madre Occidental, the vegetation changes to piñon and oak woodlands.

Based on our fieldwork and the Comisión Técnico Consultiva de Coeficientes de Agostadero (COTECOCA 1978), various types of grasslands occur in the region. (1) Shortgrass prairie, with blue grama (*Bouteloua gracilis*) and black grama (*B. eriopoda*), is found in areas with flat topography or low, smooth hills (slope of 0–8%) at an altitude of 1450–1500 m. In higher areas (1500–1600 m) and along steeper slopes (7–20%), the shortgrass prairie is dominated by blue grama. As a result of overgrazing and prairie dog poisoning, some of the original grassland has been replaced by shrubs, mainly mesquite (*Prosopis glandulosa*) but also Mormon tea (*Ephedra trifurca*), cholla (*Opuntia imbricata*) and broom snakeweed (*Gutierrezia sarothrae*). (2) Annual grassland with *Aristida adscensionis* and six-weeks grama (*Bouteloua barbata*) has been promoted by cattle overgrazing and by the defoliating activity of

prairie dogs, whose towns are found in this vegetation type as well as shortgrass prairie. (3) Bunch grassland with red grama (*Bouteloua trifida*) and side-oats grama (*B. curtipendula*) is found on the slopes of mountains and hills, with 12–30% slopes, in the 1500 to 1600 m altitudinal range. Along seasonal drainages, this grassland is invaded by mesquite, yucca (*Yucca carnerosana, Y. filifera*), and sotol (*Dasylirion wheeleri*). (4) Shrub and halophytic grassland with alkali sacaton (*Sporobolus airoides*), mesquite, and four-wing saltbush (*Atriplex canescens*) is found in the sandbanks and margins of small washes in closed watersheds with slopes up to 4%, and ranges in altitude from 1300 to 1450 m. (5) Halophytic grassland with alkali sacaton and tobosa (*Hilaria mutica*) occupies small areas with slow internal drainage and slopes of 0–4%, at an altitude of 1300–1350 m.

Shrublands include (1) thornscrub with whitethorn acacia (*Acacia constricta*), mesquite, and catclaw (*Mimosa biuncifera*), found in isolated patches in the northern part of the region, in areas of 0–5% slope, and at altitudes of 1350–1400 m. This vegetation type has a grassland component (i.e., graminoid species), which is restricted to the immediate vicinity of bushes. (2) Desertscrub with mesquite and tarbush (*Flourensia cernua*) is found near the town of Janos, in areas with less than 5% slope, and at altitudes of 1350–1450 m.

Riparian wooded communities are found along the seasonal streams, washes, and perennial ponds and are characterized by a tree canopy, which includes sycamore (*Platanus wrightii*), Arizona walnut (*Juglans major*) and willows (*Salix* spp.), with an understory of *Muhlenbergia repens* and *Sporobolus giganteus* (List et al. 1999).

In many areas, the natural vegetation intermingles with croplands. Grasslands, which have been both converted into crops and invaded by shrubs, still occupy 34% of the region (fig. 21.2). Currently, shrublands cover 16% of the natural vegetated land. The oak and pine forests are beyond the scope of this chapter.

Terrestrial Vertebrates

A total of 333 terrestrial vertebrate species have been documented in the mosaic of grasslands, shrublands, and riparian wooded communities of the Janos–Casas Grandes Region (table 21.1). Although the inventories are relatively complete, it is likely that there will be records of additional species, mainly of bats, reptiles, and amphibians. Many of the regional taxa depend on grasslands, prairie dogs, or both. Among these are the bison (*Bison bison*), pronghorn antelope (*Antilocapra americana*), kit fox (*Vulpes macrotis*), golden eagle (*Aquila chry-*

saetos), ferruginous hawk (*Buteo regalis*), mountain plover, and green toad (*Bufo debilis*).

Even without considering the higher elevations, there are 77 species of mammals in the region (table 21.1), making it one of the most diverse temperate regions in Mexico (Ceballos 1999). The Pennsylvania vole (*Microtus pennsylvanicus*) is not included in our species count. Anderson (1972) found this species at the Ojo de Galeana marsh (the only location in Mexico where the vole has been recorded), in the neighboring county of Galeana. If the mammals from the oak and pine forests are included, the region supports 100 species, representing around 21% of all the mammals from Mexico. There are no endemic mammals. However, several species, such as the bison and black-tailed prairie dog, are at the southern edge of their distribution. Thus, they have very restricted ranges in Mexico. Conversely, species such as the white-sided jackrabbit (*Lepus callotis*) and jaguarundi (*Herpailurus yaguarondi*) reach the northern end of their distributions in the Janos–Casas Grandes Region, marginally entering the United States (see also Cook 1986; Walter et al. 1990).

The region is unusually rich in terms of avian diversity, with 218 species (Manzano-Fischer et al. 1999, 2000). Although species composition and abundance can change dramatically depending on precipitation and vegetation cover, most of the species can be described as rare (136); some are common (54), and only a few are abundant (28). Approximately 61% of the birds are migratory. They include 80 winter residents, such as the bald eagle (*Haliaeetus leucocephalus*) and several falcon species that take advantage of the high density of prey; 31 summer residents such as the Swainson's hawk

Table 21.1. Terrestrial vertebrate diversity in the Janos–Casas Grandes Region of northwestern Chihuahua (lower elevations only).

Order	Families	Genera	Species	Endemic	Threatened
Mammals	19	45	77[a]	0	11
Birds	47	129	218	0	18
Reptiles	8	20	31[b]	0	13
Amphibians	3	3	7[b]	0	1
Total	77	197	333	0	43

Threatened species are those listed by the IUCN (Hilton-Taylor 2000) or by the Mexican federal government (SEMARNAT 2002).

[a]Includes the recently reintroduced black-footed ferret (*Mustela nigripes*).
[b]The inventory of the regional herpetofauna has not yet been completed.

(*Buteo swainsoni*) and the American avocet (*Recurvirostra americana*); and 22 transients such as the osprey (*Pandion haliaetus*) and willow flycatcher (*Empidonax traillii*). Among the (year-round) resident species (39% of the total avifauna) are the Chihuahuan Raven (*Corvus cryptoleucus*) and the scaled quail (*Callipepla squamata*).

Some of the species with resident populations also receive an influx of migratory individuals. For example, the horned lark (*Eremophila alpestris*) has a small resident population that increases during the autumn and winter due to migratory individuals. Another species in this category is the burrowing owl (*Athene cunicularia*), whose population increases during the breeding season, mostly in the summer. Yet another species that apparently has both resident and migratory populations is the mountain plover. However, further studies are needed to gather more information on this important species.

The strong seasonality of rainfall, scarcity of permanent bodies of water, low winter temperatures, and relative habitat homogeneity are reflected in the fairly low diversity of reptiles and amphibians. There are only 31 reptile and 7 amphibian species represented in the grassland–shrubland mosaic of the region, although additional species are expected (Domínguez et al. 1974; table 21.1). Frogs and toads are the only amphibians represented. One species, the plains spadefoot (*Scaphiopus bombifrons*), has been recorded in mesquite shrubland. However, the Great Plains toad (*Bufo cognatus*), green toad, Woodhouse's toad (*B. woodhousii*), Couch's spadefoot toad (*Scaphiopus couchii*), and western spadefoot toad (*Spea hammondii*) all occur in grassland. The last species, the bullfrog (*Rana catesbeiana*), occurs in irrigation canals and ponds within grassland habitat.

Reptiles are more diverse than amphibians, in a pattern similar to most arid regions in northern Mexico and in the southwestern United States. The reptiles most largely represented are snakes (15 spp.), followed by lizards (14 spp.) and turtles (2 spp.). Four rattlesnakes are conspicuous and abundant in the prairie dog colonies, due to the abundance of burrows and prey. The 4 rattlesnakes are the Mojave rattlesnake (*Crotalus scutulatus*), black-tailed rattlesnake (*C. molossus*), western diamondback rattlesnake (*C. atrox*), and western rattlesnake (*C. viridis*). Reptiles and amphibians have markedly seasonal patterns of activity. Their activity peak coincides with the rainy season, and during the colder months of the winter and spring they are inactive, burrowed underground.

Prairie Dogs and Other Priority Species

The region has a large number of flagship and endangered species, which is important for conservation. A total of 43 vertebrate species (13% of the total) are classified as At Risk by the Mexican federal government (SEMARNAT 2002) or as Threatened by the World Conservation Union (Hilton-Taylor 2000; table 21.1). Of these 43 species, 32 (74%) have been recorded in grassland habitat, and most of the others occur only in riparian vegetation (table 21.2).

Taxonomically, the regional list of threatened species consists of 11 mammals, 18 birds, 13 reptiles, and 1 amphibian (table 21.2). Among the mammals, threatened species are predominantly medium-sized or even large, like the pronghorn antelope and the black bear (*Ursus americanus*), and all of them are threatened by hunting, poisoning, and habitat fragmentation.

Most of the birds listed are raptors, including the golden and bald eagles and the ferruginous hawk. Currently, the main threat to these species is the high incidence of electrocutions on concrete power poles (chapter 17, J.-L. Cartron pers. comm.). However, some listed raptors and other species, such as the mountain plover, also face the threat of habitat loss. Hunting is another potential problem, but its magnitude remains uncertain.

A disproportionately high number of the reptiles in the region are considered of conservation concern. Rattlesnakes in particular are heavily exploited and are listed as Subject to Special Protection. Only 1 amphibian, the green toad, is listed. The small representation of amphibians may reflect the wide distribution of most species, the lack of information about their population status, or both.

When we began studying the prairie dog colonies in the Janos–Casas Grandes Region in 1987, they occupied an area of more than 55,000 ha (Ceballos et al. 1993). The prairie-dog town complex was the largest one left in North America, (fig. 21.3). This was an important discovery at a time when the whole remaining black-footed ferret population had just been taken into captivity, following its near extinction due largely to the decline of prairie dog towns in the United States and Canada (Miller et al. 1996). The finding of such a large prairie-dog town complex brought hope for potential future reintroductions.

The black-footed ferret is considered the most endangered North American mammal. It became extinct in the wild after the last 18 remaining

Table 21.2. Habitat associations and conservation status of threatened vertebrate species in the Janos–Casas Grandes Region (lower elevations only).

Species	Habitat Associations[a]	Conservation Status[b]
Mammals		
Notiosorex crawfordi	G	T
Tadarida brasiliensis	G	LR/NT
Vulpes macrotis	G, S, R?	T
Herpailurus yaguaroundi	S	T
Taxidea taxus	G, S	T
Ursus americanus	R	E
Antilocapra americana	G	E
Bison bison	G, S	E, LR/CD
Cynomys ludovicianus	G	T, LR/NT
Erethizon dorsatum	G, S, R	E
Lepus callotis	G	LR/NT
Birds		
Aquila chrysaetos	G, S, R	T
Haliaeetus leucocephalus	G	E
Accipiter striatus	R	Pr
Accipiter cooperii	R	Pr
Buteo swainsoni	G, S, R	Pr
Buteo regalis	G, S	Pr, LR/NT
Parabuteo unicinctus	S, R	Pr
Falco femoralis	G, S	Pr
Falco mexicanus	G	T
Falco peregrinus	G	Pr
Rallus limicola	R	Pr
Grus canadensis	R	Pr
Charadrius montanus	G	VU
Numenius americanus	G	LR/NT
Asio flammeus	G, R	Pr
Myadestes townsendi	R	Pr
Anthus spragueii	G	VU
Oporonis tolmiei	S, R	T
Reptiles		
Crotaphytus collaris	G	T
Phrynosoma cornutum	S, R	T
Heterodon nasicus	G, S	Pr
Lampropeltis getula	G, S, R	T
Masticophis flagellum	G, R	T
Thamnophis cyrtopsis	R	T
Thamnophis eques	G	T
Thamnophis marcianus	G	T
Crotalus atrox	G, S, R	Pr
Crotalus molossus	G, S, R	Pr
Crotalus scutulatus	G, S, R?	Pr
Crotalus viridis	G	Pr
Terrapene ornata	G	Pr, LR/NT
Amphibians		
Bufo debilis	G	Pr

Species included here are those listed by the IUCN (Hilton-Taylor 2000) or by the Mexican federal government (SEMARNAT 2002).

[a]Habitat associations: G = grassland; S = shrubland; R = riparian communities.

[b]Conservation status in Mexico according to SEMARNAT: T = Threatened; E = Endangered; Pr = Special Protection (i.e., species not currently endangered or threatened but declining or subject to heavy exploitation); and according to IUCN: LR/NT = Low Risk, Near Threatened; LR/CD = Low Risk, Conservation Dependent; VU = Vulnerable.

Figure 21.3. Example of grassland found in the Janos–Casas Grandes Region. Note the extensive prairie dog town.

individuals were captured in Wyoming in 1987 in order to begin a captive breeding program (Miller et al. 1996). In Mexico, there are no recent museum records of the species. However, the presence of ferrets in Mexico is confirmed by a prehistoric record, which was associated with prairie dogs, from Jiménez in the Chihuahuan Desert, 400 km to the south of Janos (Messing 1986).

Black-footed ferrets depend on prairie dogs for food and shelter. They have been reintroduced at 7 sites in the United States. However, at only 1 site has reintroduction been truly successful: Conata Basin in South Dakota. The other sites have only small prairie-dog town complexes, often decimated by plague. To date, the Janos–Casas Grandes prairie-dog town complex has remained plague-free and is considered one of very few sites where long-term viable populations of ferrets can be established (Vargas et al. 1999).

The first 94 ferrets were released in the Janos–Casas Grandes complex between September and November 2001 (releases of ferrets occur in the fall, at the time when females and young normally dis-perse), in cooperation with the U.S. Fish and Wildlife Service. The release of these ferrets represented the first time that a globally threatened species was reintroduced in Mexico after it had become extirpated in that country (Ceballos and Pacheco 2000). In our survey of September 2002, we found at least 17 ferrets, of which 9 were survivors from 2001 and 3 were wild-born (5 were not identified), suggesting that the reintroduction would prove successful.

The bison was one of the most abundant large mammals in North American grasslands, and it was once distributed from Canada to northern Mexico (Hall 1981). Bison were hunted almost to extinction and were presumed to have been extirpated from Mexico (Ceballos and Navarro 1991). However, in 1988 we found that a herd, earlier reported in the region (Di Peso et al. 1973), was still roaming free in the Janos–Casas Grandes Region. This is the only wild bison herd in Mexico, and the species is listed as Endangered according to the Mexican federal government (SEMARNAT 2002). The Janos herd moves extensively along the border between New Mexico and Chihuahua, making the bison one of the migratory species shared by Mexico and the United States. There were approximately 120 bison, concentrated mainly on the El Berrendo and Las Palmas ranches to the north of Janos, and also several ranches in southern Hidalgo County, New Mexico. In 1998, however, part of the herd, mainly cows and calves, were fenced inside the Hurt ranch in New Mexico, stopping their movements to Mexico. We are working on an agreement with the U.S. Fish and Wildlife Service to let these bison rejoin the remaining free-roaming herd. We are also evaluating the genetic makeup of the Mexican herd and its ecology. If it is not descended directly from the Yellowstone herd, it could provide additional genetic variability to the North American bison population.

The pronghorn antelope, which is associated with the North American prairies, once occupied vast areas in northern Mexico. Currently, about 1000 individuals are found in scattered populations in the Baja California Peninsula, Sonora, and Chihuahua (chapter 19). A major problem facing this species is the conversion of grasslands to agriculture or mesquite shrubland, thereby favoring coyote (*Canis latrans*) predation on young antelope. Additionally, pronghorn antelope are still intensively hunted. In the Janos–Casas Grandes Region, there are 2 or 3 groups of only 4 or 5 individuals, but on the U.S. side, adjacent to the international border, there are larger herds (Cook 1986; Ceballos pers. obs.).

The North American porcupine (*Erethizon dorsatum*) is a common species in the United States, including the Southwest (Woods 1973). However, in northern Mexico, there are only a few records from a large geographic area, and the species is considered At Risk (SEMARNAT 2002). The only resident population reported for Mexico is found in the Janos–Casas Grandes Region, where it occurs mainly in riparian vegetation (List et al. 1999).

The kit fox is considered Threatened in Mexico (SEMARNAT 2002). The grasslands this fox inhabits are being lost to agriculture and rural development. In the Janos–Casas Grandes Region, however, the kit fox is the most abundant carnivore. Kit foxes depend heavily on prairie dogs, their main prey (List et al. 2003), and they also use prairie dog burrows for dens and as escape holes from predators (Moehrenschlager and List 1996).

The occurrence of several species in the United States depends on the conservation of their source populations in Mexico. For example, the jaguarundi is rare in extreme southeastern Texas (Davis and Schmidly 1994), while in Arizona there have been only unconfirmed sightings of the species (Girmendonk 1994). One of the known populations of jaguarundis closest to the United States is found in the Janos–Casas Grandes Region, less than 75 km to the south of the Arizona–New Mexico border. In this area the species has been sighted in riparian vegetation (J. Harris, pers. comm.; R. List, pers. obs.).

The aplomado falcon (*Falco femoralis*) is listed as Endangered by the U.S. Fish and Wildlife Service. Its reintroduction at several locations in Texas was made possible by a captive breeding program started with nestlings from Mexico (e.g., Perez et al. 1996). Much like for the jaguarundi, the status of the aplomado falcon may ultimately depend on populations of this species in Mexico. The aplomado falcon is common in grasslands of Chihuahua (see chapter 22). Since 1997, there have been sightings of aplomado falcons year-round in the Janos–Casas Grandes Region, including nesting birds (Dieni et al. 2003; J. Harris, pers. comm.).

Other grassland birds, such as the mountain plover, ferruginous hawk, long-billed curlew (*Numenius americanus*), Sprague's pipit (*Anthus spragueii*), and Baird sparrow (*Ammodramus bairdii*), are all present in the region. They are among species showing population declines across some of their ranges due to habitat loss on both breeding and wintering grounds (McNicholl 1988; Knopf 1994). Overgrazing and agriculture in Mexico could well be playing an important role in the decline of grassland birds on wintering grounds (chapter 22). Finally, bald eagles and especially golden eagles are common in the area, as they prey heavily on prairie dogs (Manzano-Fischer et al. 1999). Golden eagles have both a year-round and winter resident population.

Cattle and Crops: Anthropogenic Threats

The natural ecosystems of northwestern Chihuahua and their biodiversity are disappearing due to habitat loss and degradation, overgrazing, hunting, and other human activities (see chapters 17 and 22). The first documented anthropogenic impact on Chihuahuan grasslands dates back to 1598. At that time, the Spanish conquistador Juan de Oñate and his party followed the Camino Real to take 7000 head of cattle from Chihuahua City to Santa Fe (Gehlbach 1993). In the following centuries millions of livestock were to devour the regional grasses to the roots, trample springs and soils, and cause massive erosion and desertification, thus changing biological communities and permanently altering the landscape. In recent decades, the massive impact of human activities has worsened because human presence has soared from a few scattered ranches and small settlements to large towns such as Janos, Casas Grandes, Ascención, and Nuevo Casas Grandes. If present trends continue, we believe that grasslands destruction and fragmentation will cause the disappearance of the prairie dog towns as a functional ecosystem within 2 decades (see further on).

Hunting and predator-control activities long ago caused the disappearance in Mexico of species such as the grizzly bear (*Ursus arctos*), Mexican wolf (*Canis lupus*), and elk (*Cervus elaphus*), and the near extirpation of the pronghorn antelope, bighorn sheep (*Ovis canadensis*), bison, and other species (Villa 1955; Leopold 1959; Baker 1977; Brown 1983; Ceballos and Navarro 1991; Pacheco et al. 2000; chapter 19). Populations of other species, such as the mountain lion (*Puma concolor*), black bear, and mule deer (*Odocoileus hemionus*), have also suffered from overkill (Leopold 1959; Baker 1977; Challenger 1998). The bighorn sheep was last recorded in the region in 1960 (F. Perez-Higareda, pers. comm.), the grizzly bear in 1974 (G. Ceballos, unpubl. data), and the wolf in 1977 (Brown 1983).

As early as 1955, overgrazing was reported as a conservation problem in the Janos–Casas Grandes Region (Villa 1955). Overgrazing continues today, although the effects of droughts have worsened the situation to the extent that parts of the landscape are completely barren, with gullies formed by rain-caused erosion. Reduced amounts of vegetation available as food, coupled with competition with cattle, have likely contributed to the decline of prairie dogs. Another impact of cattle ranching on the prairie dog has been through poisoning. Poisoning was the leading cause of prairie dog town loss until 1996 (List 1997). Then, a combination of factors, such as the expansion of the utility network to supply the Mennonite communities with electricity (chapter 17) and the market demand for certain crops, particularly potatoes, caused the expansion of agricultural areas in various parts of the prairie dog complex (Marcé 2001).

The conversion of grasslands to agriculture is an ongoing process. In the central part of the complex, 1169 ha of prairie dog towns were lost between 1996 and 2000, 50% due to agriculture, with 4 prairie dog towns disappearing altogether, and the rest becoming highly fragmented (Marcé 2001).

Currently, irrigated crops occupy only 5% of the land of the region, and most of them are adjacent to the larger human settlements. Meanwhile, irrigated pastures are found in more remote grassland areas. With crops beginning to appear farther from settlements, agriculture may soon have a significant negative effect on wildlife not only through conversion of grassland to crop, but also through pesticide, prairie dog and jack-rabbit control around the crops, and the increased human activities of planting and tending crops.

The invasion of grasslands by mesquite leads to quantitative changes in species composition and abundances of plants and vertebrates (e.g., Ceballos et al. 1999; Royo and Báez 2001; for a review, see also Miller et al. 1994, 2000; Kotliar et al. 1999; Kotliar 2000). In Janos, the abundance and regional distribution of many species, ranging from small mammals and reptiles to large birds of prey and mammals, have likely paralleled changes in the habitat mosaic. Where prairie dogs have been eliminated, and overgrazing occurs, mesquite begins to invade grassland, promoting complex changes in plant and animal communities, which in turn further facilitate the establishment and spread of mesquite. The positive feedback loop is particularly obvious with heteromyid rodents, which become more abundant in mesquite-dominated communities, and facilitate the dominance of mesquite with their foraging activities. Eventually, return to grassland conditions may become impossible due to changes in soil properties (Weltzin et al. 1997a; Marcé 2001).

In 1988, we calculated that there were approximately 55,000 ha of prairie dog towns dispersed in more than 150,000 ha of grasslands (Ceballos et al. 1993). By then, prairie dogs had been eliminated from most areas in the United States and Canada, leaving only scattered and isolated small towns across the range and leaving the Janos–Casas Grandes complex as the largest prairie dog town complex in North America (Ceballos et al. 1993). Ten years later, however, the area occupied by the Janos–Casas Grandes prairie dog towns had been reduced by 45%, to about 30,000 ha. Meanwhile, the largest prairie dog town on the North American continent had decreased from 35,000 to 15,000 ha (Marcé 2001). Although the towns in the central part of the complex suffered the greatest reduction in size, the northern part of the complex may have been even more severely affected, as only a few small towns survived, completely isolated and highly prone to extirpation.

The loss of prairie dogs has important local and regional ecological impacts because their abundance, burrowing habits, and foraging activities largely shape the structure and functioning of their grassland habitats (fig. 21.4). Prairie dogs significantly affect vegetation structure, productivity, and ecosystem processes such as nutrient cycling (Coppock et al. 1983; Detling and Whicker 1988; Whicker and Detling 1988, 1993; Detling 1998). Because nitrogen levels are higher in prairie dog towns than in grasslands without prairie dogs, prairie dogs increase the nutritional value and digestibility of the vegetation for herbivores, including cattle (Whicker and Detling 1993). To build their burrow systems, prairie dogs turn over 5930 kg of soil per hectare in the Janos–Casas Grandes grasslands, whereas other rodents outside the prairie dog colonies only remove from 371 to 1867 kg/ha (Ceballos et al. 1999). To detect predators, prairie dogs keep the vegetation of their towns low, creating favorable habitat for certain grassland species, including the mountain plover (Samson and Knoff 1996; Manzano-Fischer et al. 1999). Prairie dog burrows are used by 21 species of vertebrates in the area (List 1997; Ceballos et al. 1999). The high densities of prairie dogs attract unusually high concentrations of mammalian but also avian predators, such as the golden eagle (Threatened in Mexico) and the fer-ruginous hawk (Subject to Special Protection; table 21.2). Perhaps the greatest effect of prairie dogs at a regional scale is the result of the suppression of mesquite and woody plants that now cover many natural types of grassland and homogenize the landscape through much of the southwestern United States and northern Mexico (Weltzin et al. 1997a,b; Kotliar 2000; Miller et al. 2000; Royo and Baéz 2001). Overall, the disappearance of prairie dogs from the region would likely result in a net reduction of regional biodiversity. For some species, such as the black-footed ferret and the mountain plover, chances of survival either in Mexico, the United States, or in both countries would be seriously jeopardized.

Conservation Strategy

A major challenge and responsibility is to halt the deterioration of the natural ecosystems of the region and to devise a strategy that addresses the development needs at local and regional scales and the long-term conservation of grasslands, prairie dogs, and regional biodiversity. To achieve these goals, we are studying patterns of land use in the region to help plan agricultural development in productive soils that are not in biologically important places and also to restrict cattle from riparian and other sensi-

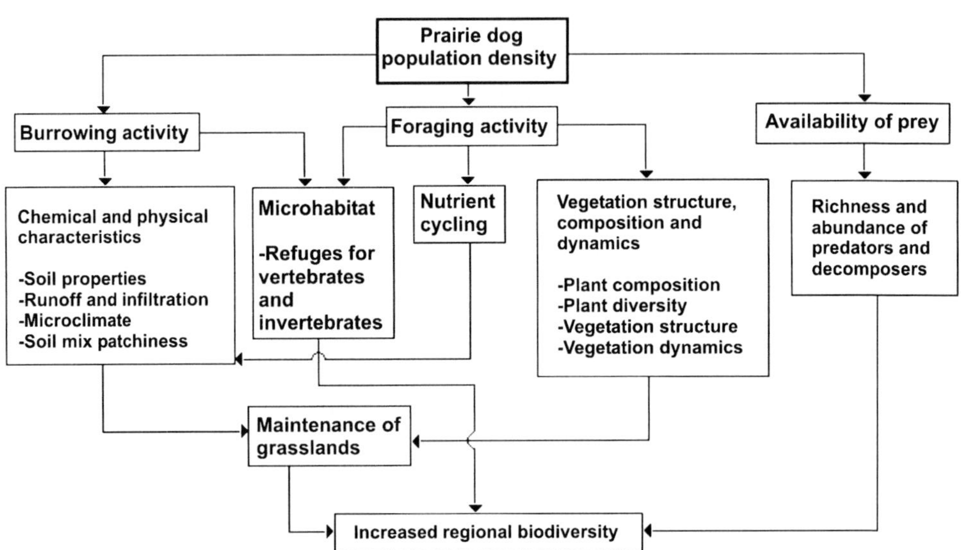

Figure 21.4. Role of prairie dogs in grassland ecosystems. Prairie dogs are considered both keystone species and ecosystem engineers, and they are critical for the maintenance of regional biodiversity (modified from Ceballos et al. 1999).

tive areas. To improve the condition of the grassland, to halt or abate soil erosion, and to restore degraded areas, we are also working with ranchers to improve management techniques, which currently are unproductive and highly detrimental to the grassland. Additional objectives are to (1) develop economic alternatives for the inhabitants of the area, ranging from regulated hunting to increases in the value of organic produce; (2) create a biosphere reserve in the region to ensure the long-term protection of the biodiversity; and (3) purchase land to further enhance the protection of the reserve's core area.

In the 1940s Janos and Ascención counties became a wildlife refuge by presidential decree, an early recognition of the region's biodiversity. We are proposing to modify that decree to create a biosphere reserve. The proposed status of biosphere reserve would identify conservation as a priority, while at the same time permitting compatible social development. Under the Mexican environmental law, biosphere reserves afford maximum protection in the core area, while allowing for productive, sustainable activities by local people in the buffer zone. No other type of protected area provides such an integrated and flexible conservation and management regime. We have already identified the boundaries of the proposed reserve, which would extend as far north as the international border. We have a biological and social justification for these boundaries and have also tentatively delineated the core and buffer zones. It is our expectation that the reserve will be decreed no later than 2005.

The potential of the area to make a binational conservation area is remarkable. Two organizations in the United States (Sky Islands Alliance and New Mexico Wilderness Alliance) are working to create a large conservation network in southeastern Arizona and southwestern New Mexico. Together with these and other organizations, we recognize that the long-term survival of many species depends on adequate protection on both sides of the U.S.–Mexico border and on habitat connectivity between the countries. The proposed conservation area's limit to the north is the Gray Ranch, which is a very large property (owned by Las Animas Foundation) with conservation-friendly management. The Gray Ranch is adjacent to other conservation-minded ranches belonging to members of the Malpai Borderlands Group. Therefore, the effective size of the reserve from a biological standpoint would expand beyond its legal boundaries.

Concluding Remarks

The Janos–Casas Grandes Region is important for the conservation of biodiversity in North America. As such, it deserves the attention of both Mexico and the United States. Our work has contributed to the recognition of the region as a conservation priority at a national and continental scale (CONABIO 2000; List et al. 2000; Manzano-Fischer et al. 2000). More important, however, we believe that our work shows the importance of linking basic and applied research to solve practical environmental problems, an issue crucial especially for developing countries, where conservation needs are often neglected to meet pressing social and economic demands.

At the turn to the twenty-first century, we firmly believe that we deserve the opportunity to preserve representative tracts of grasslands and other ecosystems for future generations. Looking at the immensity of the Janos–Casas Grandes grasslands, we wonder about the future. Only time will tell if we were successful. Only time will prove if the correct answers were provided.

Acknowledgments

We are extremely grateful to our friends and colleagues who have helped us carry out our research in Chihuahua through the years: Guadalupe Mondragón, Beatriz Hernández, Yolanda Domínguez, Hugo Rivas, Gisselle Oliva, Juan Cruzado, Bejamín Vieyra, Erika Marcé, David Macdonald, Cynthia Melcher, Eric List, the Janos Christmas Bird Count crews, Sandy Lanham, and Eduardo Medinilla, as well as dozens of volunteers. Conn Nugent and Alberto Szekely have shared our dream of creating a biosphere reserve and provided guidance toward achieving that goal. The research and conservation work has been possible thanks to the generous support of J.M. Kaplan Fund, DGAPA (UNAM), CONACyT, CONABIO, Canadian Wildlife Service, Defenders of Wildlife, the Dutch Embassy in Mexico, Environmental Flying Services, Ecociencia, Foundation for Deep Ecology, National Black-footed Ferret Recovery Foundation, National Fish and Wildlife Foundation, National Wildlife Federation, Naturalia A.C., People's Trust for Endangered Species, Phoenix Zoo, Research Ranch Foundation, U.S.A.I.D., U.S.F.W.S., Sky Island Alliance, The British Council, The Wildlands Project, and the Turner Foundation. The chapter was greatly improved through comments from Jean-Luc Cartron, Richard Felger, and Jorge Salazar.

Literature Cited

Anderson, S. 1972. Mammals of Chihuahua: taxonomy and distribution. Bulletin of the American Museum of Natural History 148: 1–410.

Archer, S., M. G. Garrett, and J. K. Detling. 1987. Rates of vegetation change associated with prairie dog (*Cynomys ludovicianus*) grazing in North American mixed-grass prairie. Vegatatio 72: 159–166.

Bailey, V. 1932. Mammals of New Mexico. USDA Bureau of Biological Survey, North American Fauna, vol. 53. U.S. Department of Agriculture, Washington, DC.

Baker, R. H. 1977. Mammals of the Chihuahuan Desert region future prospects. Pp. 221–225 *in* R. H. Wauer and D. H. Riskind (eds.), Transactions of the Symposium on the Biological Resources of the Chihuahuan Desert Region, United States and Mexico. U.S.D.I. National Park Service Transactions and Proceedings Series no. 3, Washington, D.C.

Brown, D. E. 1983. The Wolf in the South-west. University of Arizona Press, Tucson.

Ceballos, G. 1999. Conservación de los mamíferos de México. Biodiversitas 27: 1–8.

Ceballos, G., and D. Navarro. 1991. Diversity and conservation of Mexican mammals. Pp. 167–198 *in* M. A. Mares and D. J. Schmidly (eds.), Topics in Latin American Mammals: Ecology, Evolution and Education. University of Oklahoma Press, Norman.

Ceballos, G., and J. Pacheco. 2000. Los perros llaneros de Chihuahua: importancia biológica y conservación. Biodiversitas 32: 1–5.

Ceballos, G., E. Mellink, and L. R. Hanebury. 1993. Distribution and conservation status of prairie dogs *Cynomys mexicanus* and *Cynomys ludovicianus* in Mexico. Biological Conservation 63: 105–112.

Ceballos, G., J. Pacheco, and R. List. 1999. Influence of prairie dogs (*Cynomys ludovicianus*) on habitat heterogeneity and mammalian diversity in Mexico. Journal of Aridlands 41: 161–172.

Challenger, A. 1998. Utilización y Conservación de los Ecosistemas Terrestres de México Pasado Presente y Futuro. CONABIO, México D.F.

Cid, M. S., J. K. Dietling, A. D. Whicker, and M. A. Brizuel. 1991. Vegetational responses of mixed-grass prairie site following exclusion of prairie dog and bison. Journal of Range Management 44: 100–105.

Clark, T. W. 1989. Conservation biology of the black-footed ferret *Mustela nigripes*. Special Scientific Report no. 3. Wildlife Preservation Trust, Philadelphia.

CONABIO. 2000. Regiones Prioritarias Terrestres de México. Comisión Nacional para el Conocimiento y Uso de la Biodiversidad, México D.F.

Cook, J. A. 1986. The mammals of Animas Mountains and adjacent areas, Hidalgo County, New Mexico. Occasional Papers of the Museum of Southwestern Biology 4: 1–45.

Coppock, D. L., J. E. Detling, J. E. Ellis, and M. I. Dwyer. 1983. Plant-herbivore interactions in North American mixed-grass prairie—I. Effects of black-tailed prairie dogs on intraseasonal aboveground plant biomass and nutrient dynamics and plant species diversity. Oecologia 56: 1–9.

COTECOCA. 1978. Comisión Técnico Consultiva de Coeficientes de Agostadero. Secretaría de Agricultura y Recursos Hydráulicos, Chihuahua, Mexico.

Davis, W. B., and D. J. Schmidly. 1994. Mammals of Texas, revised ed. Texas Parks and Wildlife Press, Austin.

Detling, J. K. 1998. Mammalian herbivores: ecosystem-level effects in two grassland national parks. Wildlife Society Bulletin 26: 438–448.

Detling, J. K., and A. D. Whicker. 1988. Control of ecosystem processes by prairie dogs and other grassland herbivores. Pp. 23–29 *in* Eighth Great Plains Wildlife Damage Control Workshop Proceedings, April 28–30, 1987, Rapid City, South Dakota. General Technical Report RM-154. Rocky Mountain Research Station, USDA Forest Service, Fort Collins, Colorado.

Dieni, J. S., W. H. Howe, S. L. Jones, P. Manzano-Fischer, and C. P. Melcher. 2003. New information on wintering birds of northwestern Chihuahua. Pp. 26–31 *in* American Birds: 103rd Christmas Bird Count. National Audubon Society, New York.

Dinerstein, E., D. Olson, J. Atchley, C. Loucks, S. Contreras-Balderas, R. Abell, E. Iñigo, E. Enkerlin, C. E. Williams and G. Castilleja. 1999. Ecoregional-based Conservation in the Chihuahuan Desert: A Biological Assessment and Biodiversity Vision. World Wildlife Fund, Washington, D.C.

Di Peso, C. C., J. B. Rinaldo, and G. J. Fenner. 1973. Casas Grandes, a Fallen Trading Center of the Grand Chichimeca. Northland Press, Flagstaff, Arizona.

Domínguez, P., T. Álvarez, and P. Huerta. 1974. Colección de anfibios y reptiles del noroeste de Chihuahua, México. Revista de la

Sociedad Mexicana de Historia Natural 35: 117–142.

Gehlbach, F. 1993. Mountain Islands and Desert Seas: A Natural History of the U.S.–Mexico Borderlands. Texas A & M University Press, College Station.

Girmendonk, A. L. 1994. Ocelot, jaguar, and jaguarundi sighting reports: Arizona and Sonora, Mexico. Technical Report 25, Nongame and Endangered Wildlife Program, Arizona Game and Fish Department, Phoenix.

Hall, E. R. 1981. The Mammals of North America, 2nd ed. John Wiley and Sons, New York.

Henwood, W. D. 1998. An overview of protected areas in the temperate grasslands biome. Parks 8(3): 3–8.

Hilton-Taylor, C. 2000. 2000 IUCN Red List of Threatened Species. World Conservation Union, Gland, Switzerland.

Hubbard, J. P., and C. G. Schmitt. 1984. The black-footed ferret in New Mexico. Final report to the Bureau of Land Management. Contract no. NM-910-CT1-7. Endangered Species Program, New Mexico Department of Game and Fish, Santa Fe.

Knap, A. K., J. M. Blair, J. M. Briggs, S. L. Collins, D. C. Barnett, L. C. Johnson, and E. G. Towne. 1999. The keystone role of bison in the North American tallgrass prairie. Bioscience 49: 39–50.

Knopf, F. L. 1994. Avian assemblages on altered grasslands. Studies in Avian Biology 15: 247–257.

Kotliar, N. B. 2000. Application of the new keystone-species concept to prairie dogs: how well does it work? Conservation Biology 14: 1715–1721.

Kotliar, N. B., B. W. Baker, A. D. Whicker, and G. Plumb. 1999. A critical review of assumptions about the prairie dog as a keystone species. Environmental Management 24: 177–192.

Leopold, A. S. 1959. Wildlife of Mexico: The Game Birds and Mammals. University of California Press, Berkeley.

List, R. 1997. Ecology of the kit fox (*Vulpes macrotis*) and coyote (*Canis latrans*) and the conservation of the prairie dog ecosystem in northern Mexico. Ph.D. dissertation, Oxford University, Oxford.

List, R., P. Manzano, and D. W. Macdonald. 2003. Coyote and kit fox diets in a prairie dog complex in Mexico. Pp. 183–188 *in* M. Sovada and L. Carbyn (eds.), The Swift Fox: Ecology and Conservation of Swift Foxes in a Changing World. Canadian Plains Research Center, University of Regina, Regina, Canada.

List, R., O. Moctezuma, and C. Martínez del Río. 2000. Cooperative conservation: Wildlands Project efforts in the Sierra Madre. Wild Earth 10: 51–54.

List, R., O. Moctezuma, and M. Miller. 1998. What to protect in Chihuahua and Sonora. Pp. 82–89 *in* Proceedings of the 9th US/Mexico Border States Conference on Recreation, Parks and Wildlife, Tucson, June 3–6, 1998. Rocky Mountain Research Station, USDA Forest Service, Fort Collins, Colorado.

List, R., J. Pacheco, and G. Ceballos. 1999. Status of the North American porcupine (*Eerethizon dorsatum*) in Mexico. Southwestern Naturalist 44: 400–404.

Manzano-Fischer, P., G. Ceballos, R. List, O. Moctezuma, and J. Pacheco. 2000. Janos-Nuevo Casas Grandes. Pp. 171–172 *in* C. Arizmendi and L. Márquez Valdemar (eds.), Áreas de Importancia para la Conservación de las Aves en México. CONABIO, México D.F.

Manzano–Fischer, P., R. List, and G. Ceballos. 1999. Grassland birds in prairie dog towns in Northwestern Mexico. Studies in Avian Biology 19: 263–271.

Marcé, E. 2001. Distribución y fragmentación de las colonias de perros llaneros de cola negra (*Cynomys ludovicianus*) en el noroeste de Chihuahua, México. Bachelor's thesis, Universidad Nacional Autónoma de México, México D.F.

McNicholl, M. K. 1988. Ecological and human influences on Canadian populations of grassland birds. Pp. 1–25 *in* P. D. Goriup (ed.), Ecology and Conservation of Grassland Birds. Technical Publication no. 7. International Council for Bird Preservation, Cambridge, U.K.

Messing, J. H. 1986. A late Pleistocene-Holocene fauna from Chihuahua, Mexico. Southwestern Naturalist 31: 277–288.

Miller, B., G. Ceballos, and R. Reading. 1994. The prairie dog and biotic diversity. Conservation Biology 8: 677–681.

Miller, B., R. Reading, and S. Forrest. 1996. Prairie Night: Conservation of the Black-footed Ferret. Smithsonian Institution Press, Washington, D.C.

Miller, B., R. Reading, J. Hoogland, T. Clark, G. Ceballos, R. List, S. Forrest, L. Hanebury, P. Manzano, J. Pacheco, and D. Uresk. 2000. The role of prairie dogs as keystone species: a response to Stapp. Conservation Biology 14: 318–321.

Moehrenschlager, A., and R. List. 1996. Comparative ecology of North American prairie foxes—conservation through collaboration. Pp. 22–28 *in* D. W. Macdonald and F. H.

Tattersall (eds.), The WildCRU Review. Wildlife Conservation and Research Unit, Oxford, U.K.

Pacheco, J., G. Ceballos, and R. List. 2000. Los mamíferos de la región de Janos-Casas Grandes, Chihuahua, México. Revista Mexicana de Mastozoología 4: 71–85.

Perez, C. J., P. J. Zwank, and D. W. Smith. 1996. Survival, movements and habitat use of aplomado falcons released in southern Texas. Journal of Raptor Research 30(4): 175–182.

Royo, M., and A. D. Báez. 2001. Descripción del hábitat de áreas colonizadas y sin colonizar por el perrito llanero (*Cynomys ludovicianus*) en el noroeste de Chihuahua. Técnica Pecuaria Mexicana 39: 89–104.

Samson, F. B., and F. L. Knopf. 1996. Prairie Conservation: Preserving North America's Most Endangered Ecosystem. Island Press, Washington, D.C.

SEMARNAT. 2002. Norma Oficial Mexicana NOM-059-ECOL-2001, Protección ambiental-Especies nativas de México de flora y fauna silvestres-Categorías de riesgo y especificaciones para su inclusión, exclusión o cambio-Lista de especies en riesgo. Diario Oficial de la Federación 582: 1–80.

Uresk, D. W. 1985. Effects of controlling black-tailed prairie dogs on plant production. Journal of Range Management 38: 466–468.

Vargas, A., P. Gober, M. Lockhart, and P. Marinari. 1999. Black-footed ferrets: recovering an endangered species in an endangered habitat. *In* H. Griffith (ed.), Mustelids in a Modern World: Conservation Aspects of Human-Carnivore Interactions. Hull University Press, Hull, U.K.

Villa R., B. 1955. Observaciones acerca de la última manada de berrendos (*Antilocapra americana*) en el Estado de Chihuahua, México. Anales del Instituto de Biología 26: 229–236.

Walter, H., D. S. Pahlavan, and J. B. Forke. 1990. Terrestrial vertebrates of the U.S.-Mexican borderlands: results of a preliminary zoogeographic transect. Pp. 407–432 *in* P. Ganster and H. Walter (eds.), Environmental Hazards and Bioresource Management in the United States–Mexico Borderlands. Latin American Center, University of California, Los Angeles.

Weltzin, J. F., S. Archer, and R. K. Heitschmidt. 1997a. Small-mammal regulation of vegetation structure in a temperate savanna. Ecology 78: 751–763.

Weltzin, J. F., S. L. Dowhower, and R. K. Heitschmidt. 1997b. Prairie dog effects on plant community structure in southern mixed-grass prairie. Southwestern Naturalist 42: 251–258.

Whicker, A. D., and J. K. Detling. 1988. Ecological consequences of prairie dog disturbances. Bioscience 38: 778–785.

Whicker, A. D., and J. K. Detling. 1993. Control of grassland ecosystem processes by prairie dogs. Pp. 18-27 *in* J. L. Oldemeyer, D. E. Biggins, B. J. Miller, and R. Crete (eds.), Proceedings of the Symposium on the Management of Prairie Dog Complexes for the Reintroduction of the Black-footed Ferret. Biological Report 13. U.S. Department of the Interior, Fish and Wildlife Service, Washington, D.C.

Woods, C. A. 1973. *Erethizon dorsatum.* Mammalian Species 29: 1–6.

22

Habitat Associations and Conservation of Grassland Birds in the Chihuahuan Desert Region: Two Case Studies in Chihuahua

MARTHA J. DESMOND

KENDAL E. YOUNG

BRUCE C. THOMPSON

RAUL VALDEZ

ALBERTO LAFÓN TERRAZAS

Across much of northern North America, the last 30 years have witnessed alarming population declines of one-third of all grassland-restricted and half of all grassland-adapted bird species (Knopf 1994; Vickery and Herkert 2001). Causal factors include conversion of native grasslands to agricultural and urban areas, removal of native herbivores, and widespread fire suppression (Knopf 1994; Saab et al. 1995; Vickery and Herkert 2001). Despite much concern over these population declines, few grassland areas receive legal protection across North America.

Threats to Mexico's grassland birds, many of which have migratory populations shared with the United States, have not been well documented. The vast grasslands of northern Mexico's Chihuahuan Desert Region are finally receiving recognition as important breeding, wintering, and migratory stop-over grounds for many bird species (Manzano-Fischer et al. 1999; Mendez-Gonzalez 2000; Dinerstein et al. 2000; Desmond 2004; chapter 21). However, the extent to which human activities and habitat loss in northern Mexico are affecting the regional grassland avifauna or even possibly contributing to the population declines observed farther north is not known.

Within the Chihuahuan Desert Region, the largest remaining expanses of grassland are found in the state of Chihuahua (Dinerstein et al. 2000). Dominant grasses include tobosa (*Pleuraphis mutica*), alakali sacaton (*Sporobolus airoides*), black grama (*Bouteloua eriopoda*), side-oats grama (*B. curtipendula*), hairy grama (*B. hirsuta*), blue grama (*B. gracilis*), and burrograss (*Scleropogon brevifolius*). Shrubs include creosote bush (*Larrea tridentata*), tarbush (*Flourensia cernua*), honey mesquite (*Prosopis glandulosa*), acacia (*Acacia* spp.), soaptree yucca (*Yucca elata*), and Torrey yucca (*Y. torreyi*) (Johnson 1974; Dinerstein et al. 2000; chapter 21). The result at a landscape level is a mosaic of open grasslands, grasslands with shrub associations, and shrublands.

In the northern Chihuahuan Desert Region in the southwestern United States, grassland bird distribution and abundance vary dramatically among winters and are linked to monsoonal summer rains that influence annual seed production (Pulliam and Brand 1975; Pulliam and Parker 1979; Pulliam and Dunning 1987; Niemela 2002). In southern New Mexico, Niemela (2002) found that seed production and sparrow abundance in the winter were influenced by the pattern as well as the amount of rainfall in the preceeding summer. Dunning and Brown (1982) linked 20 years of Christmas Bird Count data in southeastern Arizona with the amount of summer precipitation.

The rural landscape in Chihuahua is dominated by large private ranches, community-owned properties known as *ejidos* (chapter 3), and Mennonite colonies. Land is used primarily for livestock grazing and agriculture. Croplands are expanding in northern Chihuahua as private ranches are being sold and *ejido* lands are leased to farming communities, including Mennonite colonies. Shrub encroachment is also occurring, resulting in the expansion of shrublands at the expense of grasslands. Here we present case studies focusing chiefly on the habitat associations (with an emphasis on vegetation structure) of raptors and wintering passerines in Chihuahua's grasslands. We discuss conservation and management needs for these avian assemblages. Several threats to grassland birds not addressed here are discussed in other chapters of this volume. These threats include electrocution on concrete power poles, pesticide use, and fragmentation of prairie dog colonies (see chapters 3, 17, and 21).

Case Study:
Wintering Grassland Passerines

We first present information on wintering grassland birds in northwestern Chihuahua, with a special focus on passerine species. We collected data during winter months from 1998 to 2000 in grassland areas centered around the towns and cities of Oscar Soto Maynez, Santa Clara, Buenaventura, Chihuahua, and Janos (fig. 22.1). At the lower elevations, typical semidesert grasslands were present, often dominated by tobosa, black grama, and alkali sacaton. Blue grama-dominated plains grasslands occurred at elevations up to 2000 m in the Sierra Madre Occidental.

The first of 3 unpublished data sets discussed in this section contains information collected throughout northwestern Chihuahua, and it was designed to test specific habitat associations of grassland passerines. This data set was based on 3 surveys of 88 3-ha plots in January–March 1998, a year with normal rainfall and overall good grassland condition. We excluded areas with moderate to high shrub abundance from the study design. On each plot, birds were surveyed using the area search method (Desmond 2004), and characteristics of the vegetation (e.g., shrub abundance, grass height, grass cover) were also recorded. We then tested relationships between bird occupancy patterns (i.e., presence or absence) and attributes of the vegetation through logistic regressions.

Within the highly agricultural area surrounding Oscar Soto Maynez in the Sierra del Nido region (fig. 22.1), we also recorded avian assemblages along agricultural field borders, which in some places in Chihuahua represent the last remnants of grassland habitat. The specific objective of this effort was to determine whether birds of surrounding open grasslands used agricultural field borders. Fifteen such borders were selected and surveyed twice in January–February of 2000 along 1000 × 12-m transects. Agricultural field borders were characterized by variable vegetation composition and structure. Some were dominated by grass but were heavily grazed. Others had dense weedy cover. Dominant grass species included blue grama, dropseed (*Sporobolus* spp.), three-awns (*Aristida* spp.), and fluffgrass (*Dasyochloa pulchella*). Also present were forbs and some shrubs, including mesquite and ephedra (*Ephedra* spp.). Agricultural fields bordering transects consisted of stubble from grain crops or were barren, having been recently plowed.

In the same area of Oscar Soto Maynez, we studied differences in passerine assemblages between a ranch owned and managed by the Universidad Autónoma de Chihuahua (Teseáchic) and an adjacent *ejido* (Francisco Villa). The objective of this effort was to examine avian use of grasslands in relation to grazing intensity. Whereas the private ranch was dominated by blue grama grasslands, the adjacent *ejido* lands were dominated by fluff grass, false buffalograss (*Munroa squarrosa*), and various annuals. Continuous overgrazing had resulted in the loss of perennial blue grama grasslands from these sites. *Ejido* lands were grazed to 1–2 cm or less in height (fig. 22.2a), and the only cover available for passerines on these sites consisted of small clusters of shrubs associated with arroyos. The methodology for comparing species assemblages on *ejido* (overgrazed) and ranch (not overgrazed) lands involved a total of 16 3-ha plots: 8 on the *ejido* and 8 on the private ranch. Plots were separated by a minimum of 250 m. We surveyed them using the area search method 3 times in January and February 1998. Total avian abundance and abundance of individual species were averaged on a per plot basis and analyzed with t-tests. A similar study conducted in 1999 in the black-tailed prairie dog (*Cynomys ludovicianus*) colony complex near Janos (fig. 22.1; see also chapters 17 and 21) has already

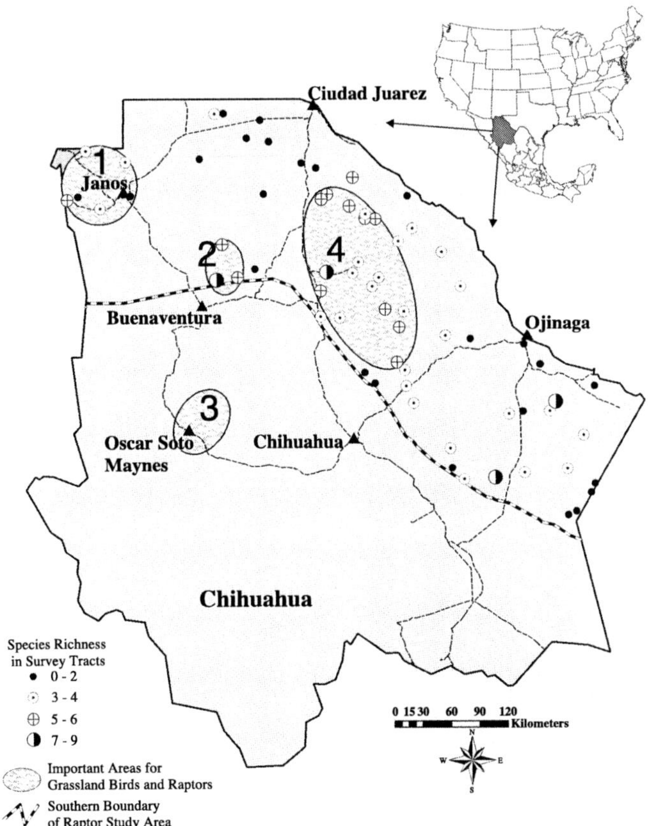

Figure 22.1. Study area locations and species richness (number of raptorial bird species) distribution in tracts surveyed for raptors and ravens in 1998–1999 in northern Chihuahua, Mexico. Gray circles represent important conservation areas for grassland birds (including raptors and wintering passerines), based on our research. *1*, Janos–Casas Grandes prairie-dog town complex area; *2*, Buenaventura area; *3*, Sierra del Nido Region; *4*, north-central Chihuahuan grasslands.

been published (Desmond 2004). Its results are summarized further on.

Habitat Associations of Wintering Passerines

We detected 49 (33 passerine) species on all study plots combined (table 22.1). Eight (6 passerine) species were recorded only along agricultural field borders. Twenty-five (15 passerine) species were detected only in grasslands. We observed 16 (12 passerine) species in both habitat types.

The most common passerine birds observed were savannah sparrows (*Passerculus sandwichensis*), Baird's sparrows (*Ammodramus bairdii*), grasshopper sparrows (*A. savannarum*), Cassin's sparrows (*Aimophila cassinii*), vesper sparrows (*Pooecetes gramineus*), Brewer's sparrows (*Spizella breweri*), white-crowned sparrows (*Zonotrichia leucophrys*), horned larks (*Eremophila alpestris*), chestnut-collared longspurs (*Calcarius ornatus*), meadowlarks (*Sturnella* spp.), and Sprague's pipits (*Anthus spragueii*). Together with meadowlarks, savannah, vesper, and Brewer's sparrows were abundant in

Figure 22.2. Examples of grassland condition during the winters of 1998 and 1999 in Soto Maynez, Chihuahua, Mexico: (a) different grazing practices common in Chihuahua with *ejido* property in the foreground and a private ranch on the opposite side of the fenceline; (b) dense vegetation occupied by *Ammodramus* sparrows during the winter of 1998; (c, d) photographs from the same pasture taken in January 1998 (c) and January 1999 (d).

grassland and field borders. White-crowned sparrows were found using only field borders. All remaining species were detected only on grassland plots. Surveys of grasslands and agricultural fields conducted during the same year, over several years, are needed to confirm the absence of some species in these habitat types.

Baird's and grasshopper sparrows, Sprague's pipit, and chestnut-collared longspur showed strong habitat preferences that were similar with respect to shrub density but otherwise quite distinct. *Ammodramus* sparrows (Baird's and grasshopper sparrows) exhibited the strongest association with open grasslands and dense cover. Specifically, their presence was both negatively correlated with shrub abundance and positively correlated with percentage of grass cover (Wald's Chi-Square, $\chi^2 = 13.78$, df = 2, P =.002). These two species occurred as solitary individuals. They were most abundant in blue grama grasslands (in the Sierra del Nido region), where they used tall, dense vegetation for cover (fig. 22.2b). Sprague's pipits also occurred in open grasslands, and their presence was negatively associated with shrub abundance ($\chi^2 = 9.63$, df = 2, P = .002). However, in contrast with Baird's and grasshopper sparrows, Sprague's pipits used both dense and sparsely vegetated grassland. They were usually solitary but were also frequently observed in groups of 2 or 3 individuals. The chestnut-collared longspur used open grasslands, and its presence was negatively associated with shrub abundance ($\chi^2 = 4.29$, df = 2, P = .04). As is usually the case with this flocking species, the chestnut-collared longspur was most common in open areas of short and sparse vegetation, in flocks often as large as 50–100 individuals. Yet it was also frequently observed in smaller flocks of 5–25 individuals in tall grass. The reason for their use of taller grasslands was unclear.

Table 22.1. Avian species recorded on study plots in grasslands and agricultural field borders in northwestern Chihuahua, Mexico.

Species	Grasslands	Field Borders
Turkey vulture (*Cathartes aura*)	X	
White-tailed kite (*Elanus leucurus*)	X	X
Northern harrier (*Circus cyaneus*)	X	X
Golden eagle (*Aquila chrysaetos*)	X	
Red-tailed hawk (*Buteo jamaicensis*)		X
Ferruginous hawk (*Buteo regalis*)	X	
American kestrel (*Falco sparverius*)	X	X
Merlin (*Falco columbarius*)	X	X
Prairie falcon (*Falco mexicanus*)	X	
Scaled quail (*Callipepla squamata*)	X	
Long-billed curlew (*Numenius americanus*)	X	
Mourning dove (*Zenaida macroura*)		X
Greater roadrunner (*Geococcyx californianus*)	X	
Short-eared owl (*Asio flammeus*)	X	
Burrowing owl (*Athene cunicularia*)	X	
Ladder-backed woodpecker (*Picoides scalaris*)	X	
Say's phoebe (*Sayornis saya*)	X	X
Loggerhead shrike (*Lanius ludovicianus*)	X	X
Chihuahuan raven (*Corvus cryptoleucus*)	X	X
Common raven (*Corvus corax*)	X	
Horned lark (*Eremophila alpestris*)	X	
Tree swallow (*Tachycineta bicolor*)	X	
Cactus wren (*Campylorhynchus brunneicapillus*)	X	
Rock wren (*Salpinctes obsoletus*)	X	
Mountain blue bird (*Sialia currucoides*)	X	X
Curve-billed thrasher (*Toxostoma curvirostre*)	X	
Sprague's pipit (*Anthus spragueii*)	X	
Green-tailed towhee (*Pipilo chlorurus*)		X
Canyon towhee (*Pipilo fuscus*)	X	X
Cassin's sparrow (*Aimophila cassinii*)	X	
Chipping sparrow (*Spizella passerina*)	X	X
Clay-colored sparrow (*Spizella pallida*)	X	X
Brewer's sparrow (*Spizella breweri*)	X	X
Lark sparrow (*Chondestes grammacus*)	X	
Black-throated sparrow (*Amphispiza bilineata*)	X	
Grasshopper sparrow (*Ammodramus savannarum*)	X	
Baird's sparrow (*Ammodramus bairdii*)	X	
Lark bunting (*Calamospiza melanocorys*)	X	
Savannah sparrow (*Passerculus sandwichensis*)	X	X
Vesper sparrow (*Pooecetes gramineus*)	X	X
White-crowned sparrow (*Zonotrichia leucophrys*)		X
Chestnut-collared longspur (*Calcarius ornatus*)	X	
McCown's longspur (*Calcarius mccownii*)	X	
Eastern meadowlark (*Sturnella magna*)	X	X
Western meadowlark (*Sturnella neglecta*)	X	X
Brewer's blackbird (*Euphagus cyanocephalus*)		X
Brown-headed cowbird (*Molothrus ater*)		X
House finch (*Carpodacus mexicanus*)		X
Pine siskin (*Carduelis pinus*)		X

Based on surveys conducted in 1998 (grasslands) and 2000 (field borders).

This may point to different use of the vegetation by roosting and foraging individuals, or it may indicate some flexibility of the species to forage in habitat consisting of denser vegetation.

No other species showed negative relationships with shrub abundance. Savannah sparrow presence in grassland plots was positively associated with the percentage of grass cover but was not related to shrubs (χ^2 = 15.52, df = 2, P < .001). This species was present on plots with scattered shrubs. It often co-occurred on plots with *Ammodramus* sparrows and was most abundant in blue grama grasslands. Unlike *Ammodramus* sparrows, savannah sparrows often occurred in small, loose flocks and were found in large numbers using agricultural field borders. Meadowlarks were common in grassland plots, and, in addition to the percentage of grass cover, their presence was positively associated with both forb production and grass height but not with shrub abundance (χ^2 = 14.05, df = 2, P = .001). Meadowlarks exhibited a preference for dense grasslands and were observed roosting in loose aggregations on grassland plots in groups of up to 20 individuals. Meadowlarks were also abundant along some agricultural field borders—namely, those dominated by grasses or those interspersed with shrubs—suggesting a flexibility to occupy a variety of habitat types during the nonbreeding period.

Cassin's sparrows were absent on the majority of plots and where present used grasslands interspersed with shrubs. Although their abundance has been positively related to the abundance of shrubs (M. Desmond unpubl. data), the only relationship we found was a negative association with forbs (χ^2 = 4.86, df = 2, P = .028). This may indicate a preference for undisturbed areas. The species is both resident and migratory in the Chihuahuan Desert Region, and distributional patterns are not well understood.

Horned larks, which often co-occurred with chestnut-collared longspurs in mixed flocks, showed no negative association with shrub abundance. This was likely related to the presence of small numbers of horned larks on plots with sparse vegetation and some shrub encroachment. The presence of this species was negatively related to the percentage of grass cover (χ^2 = 8.68, df = 2, P = .003). This is in agreement with studies that associate horned larks with heavily grazed areas on both breeding and wintering grounds (Beason 1995; Desmond 2004). For many of the above species, the lack of a negative relationship with shrub abundance likely reflected the absence of study plots with heavy shrub encroachment. Although these species exhibited tolerance for scattered shrubs, they may be absent from sites with higher shrub density.

It was difficult to characterize the presence of species in plots consisting of agricultural field borders. Although Brewer's, savannah, white-crowned and vesper sparrows, and meadowlarks were all abundant on these plots, we could not model their presence. White-crowned sparrows are an edge species and, although abundant in agricultural field borders, were never observed using grassland plots. Brewer's sparrows were abundant along agricultural field borders but were also found in grassland plots with shrubs, and other wintering studies within the northern Chihuahuan Desert Region have linked the abundance of Brewer's sparrows with the abundance of mesquite (Niemela 2002). Savannah and vesper sparrows and meadowlarks were present in large numbers in both grasslands and agricultural borders. Savannah sparrows and meadowlarks selected sites with higher canopy cover, and many of the agricultural field borders were heavily vegetated. Vesper sparrows appeared the most versatile. They were present on plots with different grass cover and height and were also abundant in agricultural field borders. A larger study examining agricultural field borders would be necessary to elucidate some of these relationships.

Patterns of detection during surveys could be used to assess the degree of dependency of some species (those with sufficient numbers of records) on open grasslands versus all grassy areas including agricultural field borders. Grassland obligates included Baird's and grasshopper sparrows, chestnut-collared longspur, and Sprague's pipit. Grassland associates (birds exhibiting a strong preference for grassland but occurring also along agricultural field borders) included savannah and vesper sparrows, horned lark, and western meadowlark (*S. neglecta*). Species that were associated with grass-dominated habitats but had no (strong) association with grasslands were the white-crowned and Brewer's sparrows. The Cassin's sparrow was difficult to classify. This species was clearly associated with grasslands interspersed with shrubs; however, the level of shrub encroachment suitable for occupancy by this species needs to be defined. Finally, there were some passerines that appeared to be strongly associated with grasslands but for which we did not have sufficient data on our plots. These included the Say's phoebe (*Sayornis saya*), loggerhead shrike (*Lanius ludovicianus*), lark bunting (*Calamospiza melanocorys*), and lark sparrow (*Chondestes grammacus*). However, both lark buntings and loggerhead shrikes appear to be associated with a shrub component.

Influence of Grazing Regime

In Oscar Soto Maynez in 1998, there were pronounced differences in grassland condition between the private ranch and the overgrazed, adjacent *ejido* (fig. 22.2a). We also observed differences in avian species assemblages between plots on the private ranch and adjacent *ejido* lands. *Ejido* plots had a lower total number of species (7 compared to 18 on the adjacent private ranch) and in particular did not support any Baird's, grasshopper and savannah sparrows, meadowlarks, or Sprague's pipits. Abundances of vesper sparrows and chestnut-collared longspurs were similar on plots of the private ranch and the adjacent *ejido* (both *P* values > .05). Horned larks were found only on *ejido* plots.

As already mentioned, the distribution and abundance of wintering passerines are also influenced by rainfall, and grasslands that are usually in good condition may be overgrazed in dry years. A good example of this is presented in figure 22.2. The same pasture had high abundance of Baird's sparrows in 1998 (fig. 22.2c) but very few individuals of this species the following winter (fig. 22.2d). The poorer grassland condition as a result of below-normal rainfall resulted in heavier grazing pressure, leaving little suitable habitat for species such as Baird's and grasshopper sparrows.

Near Janos, patterns of abundance and species diversity were similar to those observed in Oscar Soto Maynez. Surveys within the El Cuervo prairie dog colony (the largest colony in the Janos–Casas Grandes complex) revealed a total of 19 wintering species on plots located on a private ranch. Plots on the overgrazed *ejido* lands within the El Cuervo prairie dog colony only supported 8 wintering species (Desmond 2004). Prairie dog colonies are important habitat for wintering grassland birds, but the overgrazed portion of the El Cuervo colony (on *ejido* lands) provided little value to grassland birds with the exception of horned larks.

Case Study: Raptors

Published information is limited regarding raptor–habitat associations in northern Mexico. The litera-ture indicates that several species, including the Swainson's (*Buteo swainsoni*) and ferruginous hawks (*B. regalis*), peregrine (*Falco peregrinus*) and aplomado (*F. femoralis*) falcons, and burrowing owl (*Athene cunicularia*) are primarily associated with open grassland and *Yucca*–grassland communities (Hunt et al. 1988; Johnsgard 1990; Lanning and Hitchcock 1991; Rodríguez-Estrella and Ortega 1993; Bednarz 1995; Howell and Webb 1995; Miller 1996; Montoya et al. 1997; Rodríguez-Estrella 2000; Bak et al. 2001; Young et al. 2004). Habitat associations with mesquite- and acacia-dominated communities have also been reported but overall are less substantiated.

We surveyed for raptors (and ravens) in northern Chihuahua in 1998–1999 to investigate raptor habitat associations as part of a study to predict potential habitat for aplomado falcons in the northern Chihuahuan Desert (Young et al. 2002).

Field Surveys

Raptor and raven surveys were conducted in 68 tracts of variable size within a 160-km belt south of the U.S.–Mexico border in northern Chihuahua (Young et al. 2004; fig. 22.1). Tracts were placed randomly in proportion to dominant vegetation communities derived from a vegetation map and from a sample of grasslands identified from an aerial flight (Young et al. 2004). We surveyed each tract systematically by placing point stations at 0.3- to 1.0-km intervals along roads, depending on vegetation and topography. Three to 5 minutes were spent at each point station where observers recorded all raptors and ravens detected using 8× binoculars and a 20× spotting scope. Between point stations, observers drove < 6 km/hour and recorded all raptors or ravens observed. In areas with limited road access, observers walked and established point stations to ensure complete coverage of the tract. Because differentiating Chihuahuan ravens (*Corvus cryptoleucus*) and common ravens (*C. corax*) at long distances in the field requires highly trained personnel and additional survey time, we grouped the 2 species for analyses. Actual surveyed area in tracts was estimated by buffering survey point stations 600 m (approximate maximum observable distance from most point stations) in ArcView 3.2. As such, tract size and configuration depended on juxtaposition and number of survey stations established, resulting in all areas being similarly covered (Young et al. 2004). Habitats were broadly classified into

4 grass-dominated and 3 shrub-dominated communities, grading from structurally simple types to more structurally diverse types with various density, canopy, and interspersion (table 22.2; Buffington and Herbel 1965; Johnson 1974; Rzedowski 1990; Dinerstein et al. 2000).

We surveyed 1242 km² between May and October 1998 and an additional 1182 km² between February and October 1999. Survey tracts averaged 31.1 ± 1.5 km², but 2 tracts were < 10 km², and 6 tracts were > 50 km². Tracts surveyed in 1998 were primarily shrub-dominated communities (83%), composed of creosote bush-tarbush, mesquite-acacia, or creosote bush-tarbush/mesquite-acacia. Conversely, in 1999 tracts surveyed were primarily grassland communities (78%), represented by grasslands with or without complexes of mesquite-acacia, creosote bush-tarbush, or yucca (table 22.2). There were no differences in raptor or raven detections in grassland or shrubland between years ($\chi^2 = 0.87$, df = 1, $P = .351$), so data were pooled across years for analyses. We examined habitat associations through frequency of occurrence of raptors and ravens observed in the 7 dominant vegetation communities estimated from the surveyed tracts.

Species–Habitat Associations

Most of the taxa observed during our study were resident breeders. Swainson's hawks, red-tailed hawks (*Buteo jamaicensis*), ravens, and American kestrels (*Falco sparverius*) were widely distributed throughout the study area and across the 7 vegetation types examined (table 22.3). Aplomado falcons, burrowing owls, and northern harriers (*Circus cyaneus*) were detected in most vegetation types. Merlins (*Falco columbarius*), ferruginous hawks, and bald eagles (*Haliaeetus leucocephalus*) were observed only occasionally on tracts. These birds are known to breed only in the United States and Canada, so individuals observed were likely migrants (Sodhi et al. 1993; Bechard and Schmutz 1995). Although prairie falcons (*Falco mexicanus*) are resident breeders in the Chihuahuan Desert, they were also only occasionally detected on survey tracts. This likely was due to the absence of nesting sites (i.e., cliffs or bluffs) near the survey tracts (Steenhof 1998).

Associations with grassland-dominated sites were statistically significant for the northern harrier ($\chi^2 = 6.24$, df = 1, $P = .022$), aplomado falcon ($\chi^2 = 10.60$, df = 1, $P = .002$), and American kestrel ($\chi^2 = 10.38$, df = 1, $P = .002$). The burrowing owl's association with grasslands approached significance ($\chi^2 = 3.58$, df = 1, $P = .085$); however, this species was also found in shrublands (table 22.3). A marginally significant ($\chi^2 = 4.17$, df = 1, $P = .054$) association was also found between Swainson's hawk and grassland-dominated sites. Our observed distribution and detection of species among grassland and shrub-dominated tracts were generally consistent with the technical literature on habitat associations (table 22.3).

Grasslands with soaptree/Torrey yucca had the greatest number (11) of raptor species observed, while desert shrubland with acacia/honey mesquite had the lowest number (4) of raptor species observed (table 22.3). On most survey tracts (51, 75%), 4 or fewer taxa were detected. Only on 2 tracts were 8 or more taxa detected. Areas that were consistently higher in species richness occurred in the north-central part of the study area, north of Chihuahua City and south of Ciudad Juarez (fig. 22.1).

Table 22.2. Area of 7 dominant vegetation communities surveyed for raptors and ravens in northern Chihuahua, Mexico, 1998–1999.

Dominant Community Vegetation	1998 Area (km²)	%	1999 Area (km²)	%
Creosote bush-tarbush (CT)	312	25	160	14
Creosote bush-tarbush/mesquite-acacia (CT/MA)	492	40	60	5
Mesquite-acacia (MA)	220	18	29	2
Grass/creosote bush-tarbush (G/CT)	38	3	112	9
Grass/mesquite-acacia (G/MA)	90	7	421	36
Grass/yucca (G/Y)	43	4	170	14
Grassland (G)	47	4	230	19
Total	1242		1182	

Table 22.3. Raptor species and raven–habitat associations and relative detection among 7 vegetation community categories observed during surveys of 68 sample tracts conducted in northern Chihuahua, Mexico, May 1998–October 1999.

Species	Shrub Dominated[a]				Grass Dominated[a]				
	Lit. Assoc.	CT (12)	CT/MA (17)	MA (10)	Lit. Assoc.	G/CT (5)	G/MA (12)	G/Y (5)	G (7)
White-tailed kite (*Elanus leucurus*)			L					M	L
Bald eagle (*Haliaeetus leucocephalus*)					*			M	
Northern harrier (*Circus cyaneus*)[b]		L	L		*	M	M	H	M
Accipiter spp. (*Accipiter* spp.)						M			L
Harris's hawk (*Parabuteo unicinctus*)	*	L			*	L			
Swainson's hawk (*Buteo swainsoni*)[b]	*	H	M	H	*	H	H	H	H
Red-tailed hawk (*Buteo jamaicensis*)[b]		M	H	M	*	H	M	M	H
Ferruginous hawk (*Buteo regalis*)								M	
Golden eagle (*Aquila chrysaetos*)			L		*	M	L	M	
American kestrel (*Falco sparverius*)[b]		L	L	L		M	M	H	H
Merlin (*Falco columbarius*)									L
Aplomado falcon (*Falco femoralis*)[b]	*		L	M	*	H	M	H	M
Peregrine falcon (*Falco peregrinus*)[b]	*	L	L		*		L	M	L
Prairie falcon (*Falco mexicanus*)	*				*	L			
Burrowing owl (*Athene cunicularia*)[b]	*	L	L		*	M	L	H	L
Raven spp. (*Corvus* spp.)[b]	*	H	H	H	*	H	H	H	H

H = ≥ 60% detection frequency; M = 21–59% detection frequency; L = ≤ 20% detection frequency; blank cells indicate no detection.
* = association with shrub- or grassland-dominated communities based on literature.
[a]Vegetation communities (number of sample tracts in parentheses): CT = desert shrubland with creosote bush/tarbush, CT/MA= desert shrubland with acacia/honey mesquite and creosote bush/tarbush, MA= desert shrubland with acacia/honey mesquite, G/CT= grassland with creosote bush/tarbush, G/MA= grassland with acacia/honey mesquite, G/Y= grassland with soaptree/Torrey yucca, and G= grassland.
[b]Species that were detected in ≥ 5 survey tracts.

This area coincides with the north-central grasslands of Chihuahua identified as a conservation priority by Dinerstein et al. (2000; see further on).

Of the raptor species found in northern Chihuahua primarily in grasslands, several are Subject to Special Protection in Mexico (DOF 2002). Subject to Special Protection status is given to species or populations that could become threatened from factors that negatively affect their viability (DOF 2002). This is true of the aplomado falcon, which is also federally listed as Endangered in the United States. In the Chihuahuan Desert Region, aplomado falcons occupy desert grama or tobosa grasslands with a woody vegetation component, as seen in this study (table 22.3; see also Young et al. 2004). They prey upon small birds, insects, rodents, and reptiles

(Montoya et al. 1997; Keddy-Hector 2000). In northern Chihuahua, 7 bird species accounted for 68% (frequency of occurrence) of the avian diet of aplomado falcons (Montoya et al. 1997). Meadowlarks were the most frequent avian prey item, followed by the common nighthawk (*Chordeiles minor*), northern mockingbird (*Mimus polyglottos*), western kingbird (*Tyrannus verticalis*), brown-headed cowbird (*Molothrus ater*), Scott's oriole (*Icterus parisorum*), and mourning dove (*Zenaida macroura*) (Montoya et al. 1997).

Aplomado falcons are secondary nesters that rely on nests previously constructed by other raptors or ravens (Keddy-Hector 2000). Aplomado falcon nests in the Chihuahuan Desert have been located typically in soaptree yucca, Torrey yucca, honey mesquite, and netleaf hackberry (*Celtis reticulata*), and on power poles (Montoya et al. 1997; Young et al. 2002). Plant structure seems important for nesting, and most aplomado falcons have been found in large and complex soaptree yuccas. Primary nest builders in northern Chihuahua include ravens, Swainson's hawks, and red-tailed hawks (Young et al. 2004). Although these birds are common throughout northern Chihuahua, they were detected more frequently in tracts with aplomado falcons than in tracts without aplomado falcons (Young et al. 2004). Thus, the presence of aplomado falcons in a grassland area might be strongly influenced by both the vegetation and local densities of nesting ravens and raptors.

Conservation

Grasslands occupy only 7% of the Chihuahuan Desert Region, and yet they are vital to many species (including wintering and breeding grassland birds) and to the overall biological diversity of the region (Biodiversity Support Program et al. 1995; Beck and Gibbens 1999; see also chapter 21). The grasslands of Chihuahua are the most extensive in the region and appear to harbor sizable populations of sensitive species. Common or abundant wintering passerines found in the first case study included the savannah, Baird's, grasshopper, Cassin's, Brewer's, and white-crowned sparrows, chestnut-collared longspur, eastern (*Sturnella magna*) and western meadowlarks, and Sprague's pipit. All of these species are associated with documented widespread (and statistically significant) declines according to Breeding Bird Surveys from 1966 through 2001 (Sauer et al. 2001).

In chapter 3, Stoleson et al. described the ecological impact of buffelgrass (*Pennisetum ciliare*), an exotic species that often escapes planted pastures and threatens Sonoran desertscrub through changes in the fire regime. Overtime, Sonoran desertscrub shifts to monotypic grasslands dominated by buffelgrass. Ironically, in the Chihuahuan Desert Region the situation is somewhat reversed. Native grasslands are not only being converted to crops but are subject to encroachment of shrubs such as mesquite, creosote, and tarbush (see chapter 21). Anthropogenically driven changes from perennial grasslands to shrublands and associated shifts in avian species assemblages have been documented especially in the northern part of the Chihuahuan Desert Region, in New Mexico, and in southeastern Arizona (Buffington and Herbel 1965; Raitt and Pimm 1976, 1977; Allred 1996; Lloyd et al. 1998; Beck and Gibbens 1999; Desmond 2004).

According to our research, 4 passerine species in particular may be highly vulnerable to anthropogenic effects in the Chihuahuan Desert Region: the Baird's and grasshopper sparrow, Sprague's pipit, and chestnut-collared longspur. These 4 species were strongly associated with open grasslands, and their presence was negatively related to shrub abundance. Avian use and diversity along agricultural field border transects were high, but species composition was different from nearby grasslands. Baird's, grasshopper, and Cassin's sparrows, chestnut-collared longspur, and Sprague's pipit were not found using these linear habitats, suggesting that, in addition to overgrazed grasslands, narrow, linear strips of habitat are not adequate for some wintering grassland sparrows. More extensive studies are needed to confirm this last finding.

Based on our research and personal observations, Chihuahuan grasslands are generally in better condition on private ranches than on property owned by *ejidos*. Uncertainty in property rights and the lack of land management incentives in the *ejido* system lead to greater resource exploitation compared to private rangelands (Molinar et al. 1998). Further, private ranches often institute a rotational grazing system, whereas grasslands located on *ejidos* usually have a continuous grazing regime, resulting in overuse (fig. 22.2a). Large private ranches protect expanses of grasslands from human encroachment and development.

The degradation of the Chihuahuan Desert Region has led to a cooperative effort of various agencies and programs in the United States and Mexico to protect and conserve this fragile ecoregion. This has led to actions such as protecting and conserving threatened and endangered species, designating priority sites that sustain biodiversity and endemism, and focusing management practices on the restoration of severely impacted areas. Current conservation efforts through organizations including World Wildlife Fund, The Nature Conservancy, Commission for Environmental Cooperation, Universidad Nacional Autónoma de México, Instituto Tecnológico y de Estudios Superiores de Monterrey (ITESM), PROFAUNA, PRONATURA Noreste, Comisión Nacional para la Conocimiento y Uso de la Biodiversidad (CONABIO), Universidad Autónoma de Chihuahua, Rio Grande/Río Bravo Basin Coalition, Agrupación Pradera, Forgotten River Action Committee, Alliance for Rio Grande Heritage, Southwest Environmental Center, El Paso Municipal Utility District, International Boundary and Water Commission, and Rio Grande Restoration are making the public aware of conservation issues in Chihuahua and throughout northern Mexico.

The Chihuahuan Desert Region contains few protected areas designed primarily for conservation of biological diversity. In fact, only 2.5% of its total area is under formal protection (Dinerstein et al. 2000), and although 75% of the region is in Mexico, all scientific reserves (IUCN level I protection) are in the United States (Dinerstein et al. 2000). Our two studies indicate the need for conserving a wide range of grassland habitats instead of just open grasslands. Yucca grasslands appear to support a high diversity of raptors and seem important especially for the aplomado falcon (Young et al. 2004). Priority sites for conservation throughout the Chihuahuan Desert Region, including large tracts of grasslands, have been identified by the World Wildlife Fund and others (e.g., Miller 1996; Dinerstein et al. 2000; chapter 21). Based on previous work and our research, 4 areas of northern Chihuahua are important as conservation sites: the north central Chihuahuan grasslands, the Sierra del Nido region, the Janos–Casas Grandes prairie dog complex (which is part of the Chiricahua-Sierra Madre complex), and the Buenaventura area (not previously identified; fig. 22.1). The high-elevation plains grasslands of the Sierra Madre Occidental in western Chihuahua (Sierra del Nido region) are critical for wintering Baird's and grasshopper sparrows and are home to a disjunct breeding population of savannah sparrows. The higher and more predictable rainfall in this area supports a grassland dominated by blue grama that is typically taller and denser than semidesert grasslands. Because of its fertile soils, much of this area has already been converted to agriculture, especially in the Barbicora and Namiquipa valleys. Farther north, the black-tailed prairie dog complex in the Janos-Nuevo Casas Grandes area provides important habitat for many species, but most notably for large wintering and migratory flocks of chestnut-collared longspurs, horned larks, and lark buntings, as well as mountain plovers (*Charadrius montanus*), ferruginous hawks, and burrowing owls. The north-central Chihuahuan grasslands represent important breeding, migratory and wintering grassland bird (including raptor) habitat (Mendez-Gonzalez 2000; Young et al. 2004). This area in particular has the largest remaining tract of Chihuahuan Desert grassland throughout the region and is critical for maintaining biological diversity within the Chihuahuan Desert (Dinerstein et al. 2000). Grasslands north of Buenaventura are important for a diversity of breeding raptors, including the aplomado falcon.

Acknowledgments

Enrique Carreon, Eduardo Carrillo, Claudia Chacon, Martha Leon, Cesar Méndez, Carlos Morales, Suzana Peluca, Mauro Ramos, Roberto Rodríguez, Laurie Sager, Carlos Sánchez, and Elsa Zamarrón devoted long hours to field work. Julie Cummings at New Mexico State University assisted with reviewing literature, summarizing raptor–habitat associations, and identifying conservation implications. Scott Schrader assisted with graphics. Janet Ruth provided very insightful comments on an earlier version of the chapter. We especially thank the numerous private landowners in Chihuahua, Mexico, for allowing access to land. Research was accomplished with support from the Bureau of Land Management, New Mexico State University, the National Science Foundation, Texas A&M University-Kingsville, T&E Inc., White Sands Missile Range, the Wray Trust, Universidad Autónoma de Chihuahua, and the U.S. Fish and Wildlife Service.

Literature Cited

Allred, K. 1996. Vegetative changes in New Mexico rangelands. New Mexico Journal of Science 36: 169–231.

Bak, J. M., K. G. Boykin, B. C. Thompson, and D. L. Daniel. 2001. Distribution of wintering ferruginous hawks (*Buteo regalis*) in relation to black-tailed prairie dog (*Cynomys ludovicianus*) colonies in southern New Mexico and northern Mexico. Journal of Raptor Research 35: 124–129.

Beason, R. C. 1995. Horned lark (*Eremophila alpestris*). *In* A. Poole and F. Gill (eds.), The Birds of North America, no. 195. Academy of Natural Sciences, Philadelphia, and American Ornithologists' Union, Washington, D.C.

Bechard, M. J., and J. K. Schmutz. 1995. Ferruginous hawk (*Buteo regalis*). *In* A. Poole and F. Gill (eds.), The Birds of North America, no. 172. Academy of Natural Sciences, Philadelphia, and American Ornithologists' Union, Washington, D.C.

Beck, R. F., and R. P. Gibbens. 1999. The Chihuahuan Desert ecosystem. New Mexico Journal of Science 39: 45–85.

Bednarz, J. C. 1995. Harris' hawk (*Parabuteo unicinctus*). *In* A. Poole and F. Gill (eds.), The Birds of North America, no. 146. Academy of Natural Sciences, Philadelphia, and American Ornithologists' Union, Washington, D.C.

Biodiversity Support Program, Conservation International, The Nature Conservancy, Wildlife Conservation Society, World Resources Institute, and World Wildlife Fund. 1995. A regional analysis of geographic priorities for biodiversity conservation in Latin America and the Caribbean. Biodiversity Support Program, Washington, D.C.

Buffington, L. C., and C. H. Herbel. 1965. Vegetational changes on a semiarid desert grassland range from 1858 to 1963. Ecological Monographs 35: 139–164.

Desmond, M. J. 2004. Effects of grazing practices and fossorial rodents on a winter avian community in Chihuahua, Mexico. Biological Conservation 116: 235–242.

Dinerstein, E., D. Olson, J. Atchley, C. Loucks, S. Contreras-Balderas, R. Abell, E. Inigo, E. Enkerlin, C. Williams, and G. Castilleja. 2000. Ecoregion-based conservation in the Chihuahuan Desert—A biological assessment. World Wildlife Fund, Comisión Nacional para el Conocimiento y Uso de la Biodiversidad (CONABIO), The Nature Conservancy, PRONATURA Noreste, and Instituto Tecnológico y de Estudios Superiores de Monterrey. Washington, D.C.

DOF. 2002. Norma Oficial Mexicana NOM-059-ECOL-2001, Protección Ambiental—Especies Nativas de México de Flora y Fauna Silvestres—Categorías de Riesgo y Especificaciones para su Inclusión, Exclusión o Cambio—Lista de Especies en Riesgo. Diario Oficial de la Federación, Segunda Sección.

Dunning, J. B., and J. H. Brown. 1982. Summer rainfall and winter sparrow densities: a test of the food limitation hypothesis. Auk 99: 123–129.

Howell, S. N. G., and S. Webb. 1995. A Guide to the Birds of Mexico and Northern Central America. Oxford University Press, New York.

Hunt, W. G., J. H. Enderson, D. V. Lannning, M. A. Hitchcock, and B. S Johnson. 1988. Nesting peregrines in Texas and northern Mexico. Pp. 115–121 *in* T. J. Cade, J. H. Enderson, C. G. Thelander, and C. M. White (eds.), Peregrine Falcon Populations: Their Management and Recovery. University of Wisconsin Press, Madison.

Johnsgard, P. A. 1990. Hawks, Eagles and Falcons of North America: Biology and Natural History. Smithsonian Institution Press, Washington, D.C.

Johnson, M. C. 1974. Brief resume of botanical, including vegetational features of the Chihuahuan Desert region with special emphasis on their uniqueness. Pp. 335–359 *in* R. H Wauer and D. H Riskind (eds.), Transactions of the Symposium on the Biological Resources of the Chihuahuan Desert Region: United States and Mexico. National Park Service Transactions and Proceedings, series no. 3. U.S. Department of the Interior, Washington, D.C.

Keddy-Hector, D. P. 2000. Aplomado falcon (*Falco femoralis*). *In* A. Poole and F. Gill (eds.), The Birds of North America, no. 549. Academy of Natural Sciences, Philadelphia, and American Ornithologists' Union, Washington, D.C.

Knopf, F. L. 1994. Avian assemblages on altered grasslands. Pp. 232–246 *in* J. R. Jehl, Jr. and N. K. Johnson (eds.), A Century of Avifaunal Change in Western North America. Studies in Avian Biology no 15.

Lanning, D. V., and M. A Hitchcock. 1991. Breeding distribution and habitat of prairie falcons in northern Mexico. Condor 93: 762–765.

Lloyd, J., R. W. Mannan, S. Destefano, and C. Kirkpatrick. 1998. The effects of mesquite invasion on a southeastern Arizona grassland bird community. Wilson Bulletin 110: 403–408.

Manzano-Fischer, P., R. List, and G. Ceballos. 1999. Grassland birds in prairie dog towns in northwestern Chihuahua, Mexico. Studies in Avian Biology 19: 263–271.

Méndez-Gonzalez, C. E. 2000. Abundancia relativa y biomasa de aves de pastizal dentro de territorios de halcones aplomados (*Falco femoralis*) en Chihuahua, México. M.S. thesis, Universidad Autónoma de Chihuahua, Chihuahua, Mexico.

Miller, B. 1996. Delineating a new protected area in northern Chihuahua. Wild Earth 6: 14–16.

Molinar, F., H. Souza Gomes, J. L. Holechek, and R. Valdez. 1998. Mexico, macro-economics, and range management. Rangelands 20: 16–23.

Montoya, A. B., P. J. Zwank, and M. Cardenas. 1997. Breeding biology of aplomado falcons in desert grasslands of Chihuahua, Mexico. Journal of Field Ornithology 68: 135–143.

Niemela, S. A. 2002. Influences of habitat heterogeneity and seed distribution on a wintering Chihuahuan Desert avifauna. M.S. thesis, New Mexico State University, Las Cruces.

Pulliam, H. R., and M. R. Brand. 1975. The production and utilization of seed in plains grassland of southeastern Arizona. Ecology 56: 1158–1166.

Pulliam, H. R., and J. B. Dunning. 1987. The influence of food supply on local density and diversity of sparrows. Ecology 68: 1009–1014.

Pulliam, H. R., and T. A. Parker III. 1979. Population regulation of sparrows. Fortshritte der Zoologie 25: 137–147.

Raitt, R. J., and S. L. Pimm. 1976. Dynamics of bird communities in the Chihuahuan Desert, New Mexico. Condor 78: 427–442.

Raitt, R. J., and S. L. Pimm. 1977. Temporal changes in northern Chihuahuan Desert bird communities. Pages 579–590 *in* R. H. Wauer and D. H. Riskind (eds.), Transactions of the Symposium on the Biological Resources of the Chihuahuan Desert Region: United States and Mexico. National Park Service Transactions and Proceedings, series no. 3. U.S. Department of the Interior, Washington, D.C.

Rodríguez-Estrella, R. 2000. Breeding success, nest-site characteristics, and diet of Swainson's hawk (*Buteo swainsoni*) in a stable population in northern Mexico. Canadian Journal of Zoology 78: 1052–1059.

Rodríguez-Estrella, R., and R. A Ortega. 1993. Nest site characteristics and reproductive success of burrowing owls (Strigiformes, Strigidae) in Durango, Mexico. Revista de Biología Tropical 41: 143–148.

Rzedowski, J. 1990. Vegetación Potencial. IV.8.2. Atlas Nacional de México, vol. II. Escala 1:4,000,000. Instituto de Geografía, Universidad Nacional Autónoma de México, México D.F.

Saab, V. A., C. E. Bock, T. D. Rich, and D. S. Dobkin. 1995. Livestock grazing effects in western North America. Pp. 311–353 *in* T. E. Martin and D. M. Finch (eds.), Ecology and Management of Neotropical Migratory Birds. Oxford University Press, New York.

Sauer, J. R., J. E. Hines, and J. Fallon. 2001. The North American Breeding Bird Survey, Results and Analysis 1966–2000. Version 2001.2, U.S.G.S Patuxent Wildlife Research Center, Laurel, Maryland.

Sodhi, N. S., L. W. Oliphant, P. C. James, and I. G. Warkentin. 1993. Merlin (*Falco columbarius*). *In* A. Poole and F. Gill (eds.), The Birds of North America, no. 44. Academy of Natural Sciences, Philadelphia, and American Ornithologists' Union, Washington, D.C.

Steenhof, K. 1998. Prairie falcon (*Falco mexicanus*). *In* A. Poole and F. Gill (eds.), The Birds of North America, no. 346. Academy of Natural Sciences, Philadelphia, and American Ornithologists' Union, Washington, D.C.

Vickery, P. D., and J. R. Herkert. 2001. Recent advances in grassland bird research: where do we go from here? Auk 118: 11–15.

Young, K. E., B. C. Thompson, D. M. Browning, Q. H. Hodgson, J. L. Lanser, A. Láfon Terrazas, W. R. Gound, and R. Valdez. 2002. Characterizing and predicting suitable aplomado falcon habitat for conservation planning in the northern Chihuahuan Desert. New Mexico Cooperative Fish and Wildlife Research Unit. Las Cruces, New Mexico.

Young, K. E., B. C. Thompson, A. Lafón Terrazas, A. B. Montoya, and R. Valdez. 2004. Aplomado falcon abundance and distribution in the northern Chihuahuan Desert of Mexico. Journal of Raptor Research 38: 107–117.

23

Nesting Seabirds of the Gulf of California's Offshore Islands: Diversity, Ecology, and Conservation

ENRIQUETA VELARDE

JEAN-LUC E. CARTRON

HUGH DRUMMOND

DANIEL W. ANDERSON

FANNY REBÓN GALLARDO

EDUARDO PALACIOS

CRISTINA RODRÍGUEZ

The Gulf of California (fig. 23.1; described in detail in chapter 9) is an elongated, semi-enclosed sea separating the Baja California peninsula from northwestern mainland Mexico. Its southern boundary is somewhat subjective, but based on geological and bathymetric parameters, it should be extended south to a line joining Cabo San Lucas at the tip of the Baja California peninsula to Cabo Corrientes, Jalisco. Following this view, the southern Gulf corresponds here to a triangular area lying between the coast of the Mexican mainland from Mazatlán, Sinaloa south to Cabo Corrientes; the Gulf's southern boundary, as defined above; and another imaginary line joining Cabo San Lucas to Mazatlán (fig. 23.1).

The Gulf is dotted with a large number of offshore islands and islets, most of which are concentrated in the Midriff Islands Region and along the eastern coast of Baja California Sur. In comparison, the Gulf south of Isla Cerralvo (the southernmost island off Baja California Sur) harbors only a small number of offshore islands. The southern Gulf along the coast of Nayarit has 1 island (Isla Isabel) and 2 archipelagos, Islas Marías (4 islands) and Islas Marietas (2 main islands). Isla Tiburón and Isla

Angel de la Guarda, both in the Midriff Region, are the Gulf's 2 largest islands, with areas of 1208 km^2 and 895 km^2, respectively (Gastil et al. 1983). Except in the southern Gulf, the climate is arid, and the vegetation is characteristic of the Sonoran Desert (Felger and Lowe 1976; Cody et al. 1983), although some islands also have sheltered lagoons and bays fringed with mangroves (see chapter 15). Floristically, the islands of the southern Gulf are very different. There are virtually no trees on Islas Marietas, but Isla Isabel and Islas Marías have tropical deciduous forest vegetation (see further on).

The Gulf's waters are characterized by high primary productivity (Alvarez-Borrego 1983), which supports a diverse macrofauna, including 891 fish, 7 marine reptiles (1 sea snake, 1 crocodile, and now-endangered populations of 5 sea turtles), 4 pinnipeds (only 1 common species, the California sea lion [*Zalophus californianus*]), and 31 cetaceans (see chapters 9, 14, and 20). In addition, the offshore islands harbor numerous colonies of nesting seabirds, whose diversity, ecology, and conservation status are examined here. A number of shorebirds and wading birds nest on the Gulf's islands, and they often have been studied together with seabirds

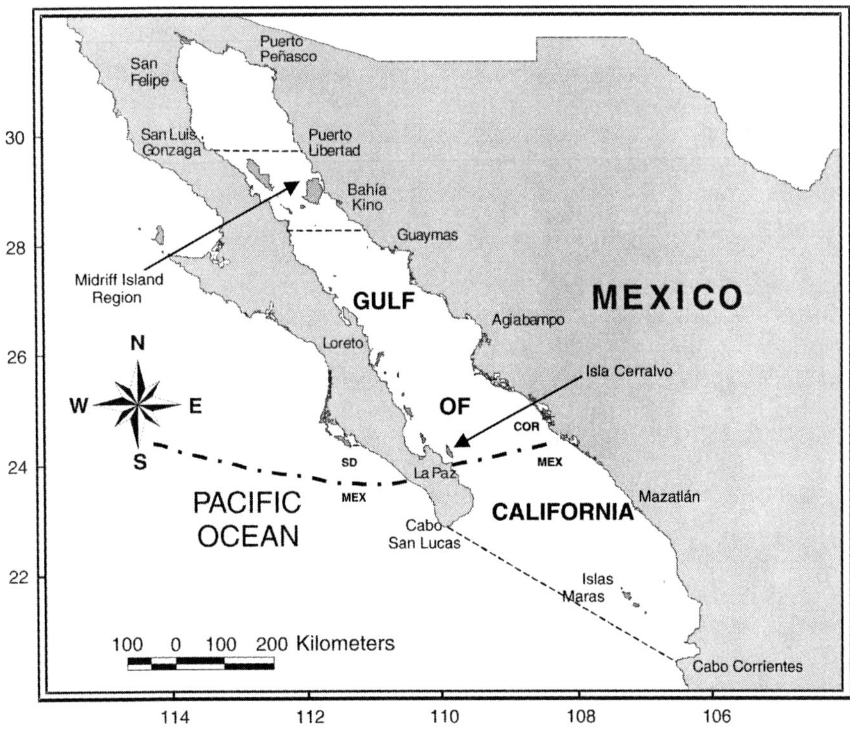

Figure 23.1. Gulf of California. Note the dashed line between Cabo San Lucas and Cabo Corrientes, indicating the southern boundary of the Gulf. Briggs' (1974) marine provinces: COR = Cortez Province; MEX = Mexican Province; SD = San Diego Province. Boundaries between provinces are from Anderson (1983).

(e.g., Anderson et al. 1976; Anderson 1983). However, they are excluded from this chapter, as are inshore colonies of seabirds, typically on sandbars, mudflats, and mangrove islets inside estuaries.

As noted by Anderson (1983), extensive mangrove estuaries occur on the mainland toward the southern end of the Gulf, providing important nesting grounds for aquatic birds. The Neotropic cormorant (*Phalacrocorax brasilianus*), for example, has huge colonies near Culiacán in Sinaloa (D. Anderson pers. obs.). At the other end of the Gulf, in the Río Colorado Delta, Isla Montague has nesting colonies of 5 seabird species, including the gull-billed tern (*Sterna nilotica*) and the least tern (*Sterna antillarum*), both of which do not breed on offshore islands (Palacios and Mellink 1992, 1993, 1996). The community of seabirds nesting offshore constitutes a somewhat distinct assemblage. Perhaps more important, their conservation in the Gulf centers on a distinct set of threats, especially introduced ground predators.

Specific information on seabird distribution and conservation needs in the Gulf of California has been published in 3 major summaries: Anderson (1983), Everett and Anderson (1991), and Velarde and Anderson (1994). Also important are 2 books that cover the region (Wilbur 1987; Howell and Webb 1995), as well as works focusing on select species or islands (e.g., Velarde 1999; Rebón 2000; Erickson and Howell 2001; Mellink 2001). This chapter provides detailed information from the less well-known southern Gulf.

Overview of the Nesting Seabird Fauna

To date, research on marine faunal assemblages continues to be based on Briggs' (1974) worldwide delineation of marine biogeographic zones, regions, and provinces. This classification scheme reflects patterns of overlapping distribution in invertebrates and fish, but it may be also applied to breeding seabirds (Anderson 1983).

Using Briggs' (1974), the Gulf of California south to Topolobampo (Briggs 1974) or Mazatlán (Walker 1960) on the Mexican mainland side constitutes its own distinct Cortez Province (fig. 23.1). Together with the San Diego Province (along the Pacific coast from Conception Point, California south to Bahía Magdalena, Baja California Sur), the Cortez Province makes up the larger California Region, situated north of the tropical convergence. The southern Gulf lies in the Mexican Province of the Eastern Pacific Region. Of importance here is the fact that the Gulf of California spans the boundary between temperate and tropical zones. To a large extent, this is reflected in the assemblages of seabird species in the Gulf. For example, several species of tropical affinity nest only on the southernmost islands (see further on).

A total of 20 nesting seabirds have been recorded nesting on the Gulf's offshore islands. Of these, only the yellow-footed gull (*Larus livens*) is (nearly) endemic to the Gulf. However, 4 species, the least storm-petrel (*Oceanodroma microsoma*), Heermann's gull (*Larus heermanni*), elegant tern (*Sterna elegans*), and Craveri's murrelet (*Synthliboramphus craveri*), have ≥ 95% of their total populations concentrated in the Gulf during the nesting season (Anderson 1983; Velarde and Anderson 1994). The Heermann's gull, in particular, nests almost exclusively inside the Gulf (Mellink 2001), with only a few small colonies documented off the west coast of the Baja California peninsula, on Isla Benito del Centro (Islas San Benitos) and Isla San Roque (Anthony 1925; Huey 1927; Jehl 1976). The least storm-petrel breeds outside the Gulf on Islas San Benitos (Velarde 2000a). The elegant tern has two small nesting colonies along the Pacific coast of southern California (Schaffner 1982; Velarde 2000b). The Craveri's murrelet breeds on Islas San Benitos (Velarde 2000c). After breeding, these 4 species become more widely distributed. They disperse southward, northward, or both, along the Pacific coast of North America. The elegant tern winters chiefly along the west coast of South America (Harrison 1983).

Based on the above information, the Gulf of California is not a major center of seabird endemism. Endemism becomes somewhat more pronounced at a larger scale, that of the California Region (Anderson 1983). Compared to the 5 species above, the black storm-petrel (*Oceanodroma melania*) has a larger portion of its total breeding population outside of the Gulf of California, but still ranges within the California Region during the breeding season (Velarde 2000d). The black-vented shearwater (*Puffinus opisthomelas*), a seabird that once nested in the Gulf of California, has its breeding range restricted to Pacific islands off the west coast of the Baja California peninsula, with 95% of its breeding population nesting on Isla Natividad (Velarde and Keitt 2000).

Two additional species, the double-crested cormorant (*Phalacrocorax auritus*) and Brandt's cormorant (*P. penicillatus*), have temperate affinities, although the first of these 2 species breeds also along the eastern coast of North America as far south as Belize (Howell and Webb 1995). Along the western coast of North America (the Brandt's cormorant is restricted to that coast), both cormorants reach the southern end of their distribution in the Gulf. Northward, their breeding distributions extend beyond the San Diego Province (California Region).

Eight species are seabirds of tropical waters and reach the northern end of their breeding distribution in the Gulf of California (and on or off the Pacific coast of the Baja California peninsula). They are the red-billed tropicbird (*Phaethon aethereus*), magnificent frigatebird (*Fregata magnificens*), brown booby (*Sula leucogaster*), blue-footed booby (*S. nebouxii*), red-footed booby (*S. sula*), bridled tern (*Sterna anaethetus*), sooty tern (*S. fuscata*), and brown noddy (*Anous stolidus*) (Anderson 1983). Five of these species nest in the Gulf only on southern islands: the magnificent frigatebird, red-footed booby, bridled tern, sooty tern, and brown noddy.

The laughing gull (*Larus atricilla*) breeds along the eastern and western coasts of North America, occupying both temperate and tropical waters. In the west, it breeds from Isla Montague in the Río Colorado Delta south to Colima (Palacios and Mellink 1992; Howell and Webb 1995). The breeding distributions of 2 more species, the brown pelican (*Pelecanus occidentalis*) and the royal tern (*Sterna maxima*), extend well into temperate latitudes but are centered in Central America (Harrison 1983). The breeding distribution of the California brown pelican (*P. o. californicus*) is wide, from southern and central California south to at least the state of Guerrero, Mexico (Gress and Anderson 1983). Even here, the genetic status of brown pelicans from Guerrero south and including Panama is largely unknown, and *P. o. californicus* could extend even farther south (this is an area ripe for research). Within the known breeding range of *P. o. californicus*, Gress and Anderson (1983) have roughly defined 4 major breeding populations, based mainly on distributional characteristics (these have been mapped in Anderson et al. 1996): (1) a Southern California Bight population, (2) a southwestern Baja California coastal population, (3) a desert-island nesting population in the Midriff Region of the Gulf of California (accounting for at least 70% of the total subspecies' numbers), and (4) a Mexican mainland and southern Gulf insular population, nesting mainly in vegetation. These were delineated, not on a genetic basis (genetic studies have never been done), but geographically and to help conservationists define more convenient management units (more details can be found in Gress and Anderson 1983).

Some Characteristic Seabirds of the Gulf

Least Storm-Petrel

The least storm-petrel has a very narrow nesting distribution (Anderson et al. 1976; Velarde and Anderson 1994). Only 4 islands or groups of islands have been recorded as nesting sites, 3 in the Gulf of California (Isla Partida in the Midriff Region [fig. 23.2], Los Islotes, and Roca Consag) and 1 off the Pacific coast of the Baja California peninsula (Islas San Benitos; Velarde 2000a). The species nests in crevices along cliffs and among boulders and lays only 1 egg. The incubation period is long, due to the size of the egg, which is large in relation to the adult's body size (Ainley 1984). The chick-rearing period is also long. The chick is fed with a concentrated oil that the adult produces from a diet of plankton. Natural predators of the least storm-petrel include the magnificent frigatebird, peregrine falcon (*Falco peregrinus*), and barn owl (*Tyto alba*) (Velarde 2000a). If an egg or chick is lost to predation, there is no production of a new egg until the following nesting season (Ainley 1984). For this reason, introduced mammals are a major threat to the species (Jehl 1984; Everett and Anderson 1991). The only island free of introduced predators on which the least storm-petrel nests is Isla Partida. This island has nearly 95% of the species' breeding population (Velarde 2000a; table 23.1).

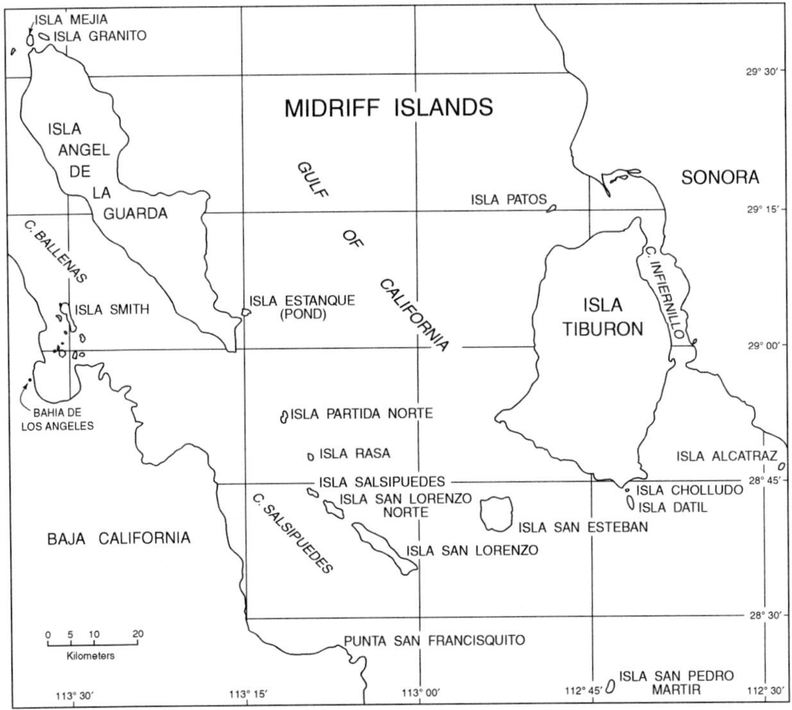

Figure 23.2. Midriff Region. Reproduced with permission from Lindsay (1983).

Table 23.1. Nesting seabirds of the Gulf of California, with locations of known insular colonies of about 300 breeding individuals or more.

Species	Islands	Largest Colony Size[a] (Island)	References[b]
Black storm-petrel (*Oceanodroma melania*)	Partida	50,000	1
Least storm-petrel (*Oceanodroma microsoma*)	Partida	500,000	1
Red-billed tropicbird (*Phaethon aethereus*)	Isabel, San Pedro Mártir	300 (San Pedro Mátir)	1
Blue-footed booby (*Sula nebouxii*)	Cholluda, Isabel, Islas Marietas, San Ildefonso, San Jorge, San Pedro Mártir, Tortuga	220,000 (San Pedro Mártir)	1, 2
Brown booby (*Sula leucogaster*)	Isabel, Islas Marietas, San Jorge, San Ildefonso, San Pedro Mártir	148,000 (San Pedro Mártir)	1, 2
Brown pelican (*Pelecanus occidentalis*)	Angel de la Guarda, Cerralvo, Cholluda, Espíritu Santo, Las Ánimas, Salsipuedes, San Ildefonso, San Lorenzo, San Luis, San Pedro Mártir, Santa Catalina, Tortuga	20,000 (Las Ánimas)	1
Double-crested cormorant(*Phalacrocorax auritus*)	Alcatraz, San Jorge, San Luis	2,000 (Alcatraz)	1, 3, 4
Brandt's cormorant (*Phalacrocorax penicillatus*)	Islote Partido, San Esteban	300 (Partido, San Esteban)	1
Magnificent frigatebird (*Fregata magnificens*)	Isabel, Islas Marietas	6,000–7,000 (Isabel)	5
Heermann's gull (*Larus heermanni*)	Cholluda, Cardonosa, Isabel, Islas Marietas, Rasa, San Jorge, San Ildefonso	260,000 (Rasa)	1, 6, 7
Yellow-footed gull (*Larus livens*)	Angel de la Guarda, Carmen, Espíritu Santo, Partida, San Esteban, San Ildefonso, San José, San Luis, San Marcos, Tortuga	1,000 (San Esteban)	1
Laughing gull (*Larus* atricilla)	Islas Marietas	6,000 (Islas Marietas)	8
Royal tern (*Sterna maxima*)	Rasa	17,000 (Rasa)	1, 9
Bridled tern (*Sterna anaethetus*)	Islas Marietas	290 (Islas Marietas)	10, 11
Sooty tern (*Sterna fuscata*)	Isabel	3,000–5,000 (Isabel)	5
Elegant tern (*Sterna elegans*)	Rasa	180,000 (Rasa)	12, 13
Brown noddy (*Anous stolidus*)	Islas Marietas	560 (Islas Marietas)	8, 14

[a]Expressed as number of breeding individuals.
[b]References: (1) Velarde and Anderson 1994; (2) Tershy and Breese 1997; (3) Anderson et al. 1976; (4) Jímenez Serranía et al. 2002; (5) H. Drummond unpubl. data; (6) Velarde 1999; (7) Mellink 2001; (8) F. Rebón unpubl. data; (9) Velarde 1989; (10) Rebón 1997; (11) Rodríguez 1997; (12) Tobón 1992; (13) E. Velarde unpubl. data; (14) Mora 1998.

Brown Pelican

The California brown pelican is ubiquitous in the Gulf and easily wins the title "King of the Sea of Cortez." Major nesting colonies in the Gulf of California, in order of decreasing size, include the San Lorenzo Archipelago (Islas Animas, San Lorenzo, and Salsipuedes), the San Luis Archipelago (Islas San Luis and Cholluda), Isla San Pedro Mártir, Puerto Refugio (Islas Angel de la Guarda, Pelicano, and Granito), Isla Piojo in the Bahía de los Angeles area, Isla San Ildefonso, plus many other, smaller (100–400 nests) breeding colonies. In the San Lorenzo and San Luis Archipelagos, in non–El Niño years, there may be some 20,000 and 10,000 active nesting pairs, respectively. Colonies in close proximity to one another (e.g., Islas San Luis and Cholluda, Islas San Lorenzo and Animas) are generally considered of the same population, and individuals breeding among nearby colonies tend to share similar demographic features (see also Anderson and Gress 1983). After breeding, Gulf of California brown pelicans, as do all other populations, travel widely and commonly intermix as far north as southern British Columbia, Canada (rarely a few individuals even reach Alaska) and south to Central America (Anderson and Anderson 1976; Briggs et al. 1981; see also Shields 2002; D. Anderson in prep.).

The female pelican typically lays 3 eggs, which are then incubated under the webbed feet of the parents (Schreiber 1977). In years of intense El Niño events, there may be very little or no breeding, nesting success is severely reduced, and breeding populations tend to be found in areas other than the Midriff Region of the Gulf because they travel more widely (Anderson et al. 1996; D. Anderson in prep.). Compared to the Southern California Bight, food items for brown pelicans in the Gulf are of greater size and are much more diverse. Nonetheless, breeding effort and success are both closely related to biomass of proximate food sources (see Anderson et al. 1982).

Heermann's Gull

The closest relative of the Heermann's gull is the gray gull (*Larus modestus*), distributed along or near the South American Pacific coast. The gray gull's nesting distribution encompasses the barren Atacama Desert, 50–100 km inland (Howell et al. 1974). All known colonies of the Heermann's gull are on islands. It is interesting that the island with nearly the entire world's population of this species (see further on) is largely devoid of vegetation, as are most of the other islands where the species nests (i.e., Islas Cardonosa, San Ildefonso, Cholluda). Adaptations to barren conditions on islands of the Gulf include regurgitation of water by adults for the chicks.

The Heermann's gull breeds in high densities, with up to 110 nests per 100 m^2 (Velarde 1999). Breeding synchrony is high, and chicks may be defended communally against predators. The species is a food specialist; its diet is restricted largely to small shoaling pelagic fish, mainly sardines, anchovies, and mackerels (Velarde et al. 1994). On Islas Marietas, the Heermann's gull also feeds on the eggs and chicks of both blue-footed boobies and black-bellied whistling-ducks (*Dendrocygna autumnalis*) (Rebón 1997).

During the spring and early summer of each year, most of the Heermann's gull population concentrates in the Midriff Islands Region, when the fish that make up its diet also concentrate in that same area, due to the low water temperatures and high productivity at these times. There is plenty of food for rearing chicks, even during mild El Niño years, and infanticide was never observed from 1979 to 2002 (E. Velarde, pers. obs.). During strong El Niño years, such as 1998 and 2003, Heermann's gulls congregate as usual, but only a small proportion of all pairs actually lays eggs, many of which are abandoned before incubation is completed (E. Velarde, pers. obs.).

After breeding, and for a period of 8 months each year, Heermann's gull disperses along Pacific coastal waters mainly from southern Canada to central Mexico, with some individuals remaining within the Gulf of California.

Yellow-footed Gull

The nearly endemic yellow-footed gull has its closest relative and ecological counterpart in the kelp gull (*Larus dominicanus*). Compared to some of the other seabird species in the Gulf, the yellow-footed gull tends to be widely but also sparingly distributed. It may nest individually or in colonies, the largest of which may reach only about 500 pairs (Spear and Anderson 1989; Velarde and Anderson 1994; table 23.1).

In contrast to Heermann's gull, the yellow-footed gull has a very diverse diet. It may scavenge or take fish, shellfish, marine arthropods, fishermen's offal, the eggs and chicks of other seabirds, and even the adults of small seabird species such as black and least storm-petrels and eared grebes (*Podiceps nigricollis*) (Burger and Gochfeld 1996; E. Velarde, pers. obs.). Nests must be near the shoreline (boulder beaches and inaccessible cliffs are the most typical terrain) to allow the chicks to drink at the edge of the water within the first 2 days after hatching.

Elegant Tern

The elegant tern used to nest on several islands of the Gulf of California and near Guaymas on the continental mainland. It also nested along the Pacific side of the Baja California peninsula, both on the mainland and on Isla San Roque. However, direct human disturbance, mainly egg collection, as well as the introduction of exotic species, have reduced the number of nesting colonies (Saunders 1896; Brewster 1902; Ridgway 1919; Bent 1921; Mailliard 1923; Bancroft 1927a; AOU 1998; Velarde 2000b). At present there is only 1 known colony in the Gulf of California, on Isla Rasa (about 180,000 individuals, or 95% of the species' total population; table 23.1), with 2 other colonies near San Diego, California (Schaffner 1982). Much of what is known about the breeding and feeding ecology of the species is derived from the studies of those 3 colonies (Schaffner 1982; Tobón 1992; Velarde et al. 1994, in press).

The elegant tern generally nests in flat, open areas and lays 1 egg. On Isla Rasa it starts nesting during the first week of April, and breeding success is about 1 chick per every 2 nests during normal years. The elegant tern feeds on small pelagic fish, mainly the northern anchovy (*Engraulis mordax*) in the Pacific and the Pacific sardine (*Sardinops caeruleus*) and northern anchovy in the Gulf of California. Based on catch volume, the Pacific sardine is the most important species for Mexican fisheries. In the Gulf, the elegant tern has been found to be an excellent predictor of Pacific sardine total catch and catch per unit effort for the commercial fleet operating out of the state of Sonora (Velarde et al., in press).

Craveri's Murrelet

Craveri's murrelet is the only alcid nesting in the Gulf of California, where it is has a widespread distribution (DeWeese and Anderson 1976). Outside of the Gulf, Craveri's murrelet also breeds on Islas San Benitos, off the Pacific coast of the Baja California peninsula. Its breeding behavior and ecology have not been studied in detail. Craveri's murrelets begin to arrive in the Gulf of California in January, at which time they are probably already paired. Nesting occurs on rocky islands, in burrows and crevices just above the high-tide mark (Bancroft 1927b; DeWeese and Anderson 1976). The species lays usually 2 eggs (Bent 1919). After only 2–4 days in the nest, the flightless chicks are taken by their parents to sea, where they are fed larval fishes and small adult pelagic fishes (Bent 1919; Bancroft 1927b; DeWeese and Anderson 1976). After breeding, the species disperses along the Pacific coast from southern California to Nayarit. Due to the presence of introduced predators, these birds have been extirpated from some of the islands where they used to nest, such as Isla Rasa (Bent 1919; Bancroft 1927b; Banks 1963; DeWeese and Anderson 1976; AOU 1998; BirdLife International 2000; Velarde 2000c).

Distribution and Conservation of Nesting Colonies

The distribution of nesting seabirds in the Gulf is influenced by many factors (Anderson 1983). Besides faunal affinity, already mentioned, the other 2 most important factors likely consist of predation pressure and food availability. In the Gulf, seabird breeding population size seems negatively correlated with island size and positively correlated with distance from the mainland, presumably because small, isolated islands are the least likely to have terrestrial predators (Anderson 1983). Most of the largest seabird colonies are located in the very productive waters of the Midriff Region (fig. 23.2). Several small islands in the southern Gulf have large colonies of seabird species at the edge of their nesting distribution.

There are currently 3 main threats to seabirds nesting offshore in the Gulf of California (Anderson and Keith 1980; Jehl 1984; Anderson 1988; Anderson et al. 1989; Everett and Anderson 1991; Velarde and Anderson 1994; Howell and Webb 1995). Human disturbance is widespread and common, due to the uncontrolled access of people to the islands. Cats, rodents, goats, and other mammals have been introduced on many of the Gulf's islands (see Lawlor et al. 2002). Introduced mammals almost certainly have been responsible for the apparent extirpation or extinction of several native small mammals (e.g., Lawlor 1983; Mellink et al. 2002), and likewise their impact on nesting colonies of seabirds has been important. The last threat consists of overfishing. Commercial fisheries have already had severe impacts on shrimp populations, as well as on the Pacific sardine. The small, pelagic Pacific sardine serves as the food base of many larger fish (some of them also important for commercial fisheries), marine mammals, and seabirds (Velarde and Anderson 1994).

Conservation efforts in the Gulf of California islands began in 1963 when Isla Tiburón was declared a Natural Reserve Zone and Wildlife Refuge. The islands of the Gulf south to Isla Cerralvo on the peninsular side and to Mazatlán on the Mexican mainland side have been designated collectively as the Area de Protección de Flora y Fauna Islas del Golfo de California. In 2002, however, the Mexican federal government issued a decree designating Isla San Pedro Mártir and adjacent waters as a Biosphere Reserve (DOF 2002). The southernmost Gulf islands are not included in the Area de Protección de Flora y Fauna Reserva Islas del Golfo de California, but Isla Isabel is a National Park and the Mexican government may be preparing to designate the Islas Marías archipelago as a Biosphere Reserve and relocate all the resident prisoners (Vicenteño 2003).

As discussed later in the chapter, a few eradication programs have been successfully carried out, but much work is still required (McChesney and Tershy 1998; Tershy et al. 2002; Ramírez et al. ms.). Isla Rasa was the first Gulf island to benefit from a successful rodent eradication campaign (McChesney and Tershy 1998; Ramírez et al. ms). Isla Rasa and Isla Isabel are models of on-site protection incorporating research and public education.

Midriff Region

Isla Partida

Isla Partida is 17 km from the eastern shore of the Baja California peninsula. It has a surface of approximately 1.2 km², and is formed by 2 mountain masses of volcanic origin, joined by a lower and narrower land area. Its coasts form 2 coves of small, round boulders, and the bases of the 2 mountain areas have cliffs. The vegetation of the island is scant, mainly composed of cardón cactus (*Pachycereus pringlei*) (Bourillón et al. 1988). The island has one of the largest yellow-footed gull nesting

colonies in the Gulf (table 23.1). The 2 mountains that form the island are made of fragmented volcanic rock, among which both black storm-petrels and least storm-petrels nest. These 2 species have their largest Gulf colonies on Isla Partida, and their huge numbers are the most spectacular feature of the island (Velarde and Anderson 1994; table 23.1). The Craveri's murrelet has also been recorded breeding on Isla Partida, although no estimates of the nesting population are available (DeWeese and Anderson 1976).

Approximately 12,000 to 15,000 fishing bats (*Myotis vivesi*, a Gulf endemic species) roost among the 2 storm-petrels (Villa-Ramírez 1979). There is no other mammalian species on the island, and there are no snakes. Undoubtedly, the lack of land predators has favored the use of ground crevices by several species. Both storm-petrels and the fishing bat are preyed upon by yellow-footed gulls and barn owls nesting on the island (E. Velarde, unpubl. data).

Isla Rasa

Isla Rasa lies 18 km from the coast of the Baja California peninsula (Gastil et al. 1983). This small (0.6 km²) island rises to a maximum elevation of only 35 m (hence the name *Rasa*, meaning flat in Spanish). Most of Isla Rasa consists of low hills of volcanic rock, with also valleys and 3 tidal lagoons (Boswall and Barrett 1978; Velarde 1999). There is little vegetation, chiefly cardón cactus, 2 cholla cacti species (*Opuntia* spp.), and the saltbush *Atriplex barclayana* (Barreto 1973). The island is almost totally covered by guano.

Isla Rasa is most important for its colony of Heermann's gulls, estimated at 260,000 individuals, or 95% of the species' world breeding population (Velarde 1999). However, it also has very large colonies of 2 other larid species, the elegant and royal terns (table 23.1). The yellow-footed gull nests on Isla Rasa in small numbers; between 5 and 9 nests are established every year (E. Velarde, pers. obs.). Fewer than 10 landbirds and shorebirds nest on the island. In total, however, nearly 90 species of land, shore, and seabirds have been documented on the island as nonbreeders (Velarde 1989; Cody and Velarde 2002).

Guano mining on an industrial scale began in the 1870s on Isla Rasa and continued into the 1910s (Willard 1890, 1891; Muñoz Lumbier 1919; Villa Ramírez 1976; Bowen 2000). Commercial harvesting of gull and tern eggs also occurred from the late nineteenth century until the 1960s. The number of eggs collected and shipped to coastal Gulf towns eventually reached half a million a year, resulting in the drastic reduction of nesting colonies on the island (Walker 1965; Barreto 1973).

In 1964, the island was declared a Zone for Refuge and Reproduction of Migratory Birds and Wildlife by the Mexican federal government (Velarde and Anderson 1994). Legal protection was the result of lobbying efforts by many Mexican and U.S. institutions and individuals, including the California Academy of Sciences (George Lindsay and Robert Orr), Arizona-Sonora Desert Museum (Louis W. Walker), Audubon Society, and Universidad Nacional Autónoma de México (UNAM, Bernardo Villa), and some local residents (notably Antero Díaz). At present it has been recategorized as an Area for the Protection of Flora and Fauna. Since the 1970s, it has benefited from on-the-ground protection by Mexican biologists, first Bernardo Villa Ramírez, then one of us (E.V.). Protection has been permanent since 1979, with biologists staying on the island during the nesting season of seabirds.

At some point during the history of commercial exploitation of the island, both black rats (*Rattus rattus*) and house mice (*Mus musculus*) were introduced, and black rats in particular preyed on eggs and newly hatched young of Heermann's gulls. However, due to a rodent eradication program carried out in 1995 and coordinated by Jesús Ramírez, black rats and house mice are no longer found on the island. Since 1995, the breeding success of Heermann's gulls has increased 5 fold in previously rat-infested areas, and the elegant tern nesting population (77,000 individuals in 1995) has more than doubled (E. Velarde, unpubl. data).

On-the-ground protection is currently guarding against disturbance by a large number of ecotourists who visit the island every year and against egg collecting. A group of students and some researchers, coordinated by E. Velarde, have gathered data on seabird populations, behavior, and ecology since 1979. Thus, while visitors are guided and educated, students learn and acquire first-hand experience about seabirds and their conservation. Important information is also collected for the environmental authorities. Financial support for on-site protection and research on Isla Rasa has come from many organizations, both Mexican and international (mainly United States), as well as from some private donors.

Isla San Pedro Mártir

Isla San Pedro Mártir is located nearly halfway between the coasts of the Baja California peninsula and Sonora. Rising to an elevation of 320 m, this small (1.9 km^2) volcanic island is characterized by steep terrain covered on the lower slopes by guano and on the upper slopes by stands of cardón cactus and globe mallow (*Sphaeralcea hainesii*). Currently, San Pedro Mártir has only temporary fishing and, in some years, research camps (H. Drummond, pers. obs.) Historically, it was mined intensively for guano (Bowen 2000), and the mining operations were likely responsible for the introduction of the black rat on the island.

Tershy and Breese (1997) lived on San Pedro Mártir for 13 months in the early 1990s, documenting nesting colonies of 7 seabird species: red-billed tropicbird, blue-footed booby, brown booby, Brandt's cormorant, brown pelican, Heermann's gull, and yellow-footed gull. The island is ecologically important, especially for its large colonies of blue-footed and brown boobies. The size of these colonies appears to fluctuate widely through time (Nelson 1978; B. Tershy, pers. comm.), but in 1990 they reached 110,000 and 74,000 pairs, respectively (Tershy and Breese 1997; table 23.1). With about 150 pairs, the red-billed tropicbird colony on San Pedro Mártir is the largest for this species in the Gulf (Velarde and Anderson 1994; table 23.1). Black and least storm-petrels may have nested on the island before rats were introduced. The same may be true for Craveri's murrelet. One pair of this species was found nesting in 1991 on 1 of the island's satellite offshore rocks (Tershy and Breese 1997).

Other Islands

Unfortunately, human disturbance of seabird nesting colonies occurs throughout the Midriff Region. A single visitor strolling through a colony may cause the loss of several hundred eggs and/or chicks. Visitors may be tourists, fishermen, photographers, or even researchers. Several ecotourism operators are now well aware of these threats and have specific codes of conduct to prevent disturbance. However, people who travel in small private boats, such as sailboats, yachts, or skiffs, may be ignorant of the dangers they pose to nesting seabirds.

Brown pelican breeding numbers reflect oceanographic conditions and changes in the distribution of fish resulting from El Niño events as well as commercial fisheries activities (see Anderson et al. 1980, 1982; Ainley et al. 1988; Velarde and Ezcurra 2002). Brown pelicans are also highly sensitive to disturbance from tourists, fishermen, and others (Anderson and Keith 1980; Anderson 1988). On Isla San Lorenzo Norte, young but mobile brown pelicans occasionally become impaled on chollas (Anderson and Keith 1980). This type of mortality can occur naturally but increases greatly as a result of disturbance, which often flushes the young into areas with chollas. The eggs of brown pelicans can be depredated by yellow-footed gulls or common ravens (*Corvus corax*), especially when the parents have left the nest, again, due to human disturbance (Anderson and Keith 1980).

Southern Islands

At least 13 species of seabirds breed on the islands of the southern Gulf. This area corresponds to the southern limit of the breeding range of Nearctic species such as the Brant's cormorant and Heermann's gull. Southern islands also support the northernmost breeding colonies of Neotropical species such as the red-footed booby, bridled tern, and brown noddy (Rebón 1997, 2000).

The southern Gulf harbors several of the Mexican Pacific's largest colonies of the laughing gull, bridled tern, and brown noddy and the second largest colony of the brown booby (after Isla San Pedro Mártir; Rebón 1997, 2000). Bailey (1906) published the first ornithological annotated checklist from Islas Marías and Isla Isabel. Friedmann et al. (1950) and Blake (1953) reported important information on the distribution of seabirds in the southern Gulf. Grant (1964, 1965) and Grant and Cowan (1964) contributed extensively to the knowledge of the birds (including seabirds) of Islas Marías and Marietas. Here again, large colonies (such as those on Isla Isabel and Islas Marietas) tend to be on small islands. Most seabirds in the southern Gulf nest on the ground, but magnificent frigatebirds and some brown pelicans nest on the upper branches of small trees and bushes (H. Drummond pers. obs.).

Conservation threats include human disturbance of seabird colonies by the growing ecotourism industry in Puerto Vallarta, Punta Mita, and Nuevo Vallarta, as well as introduced mammals. On Islas Marietas, egg theft and fires started intentionally represent additional threats (Rebón 1997).

Isla Isabel

Twenty-eight kilometers from the coast of Nayarit lies a small (0.8 km^2) volcanic island known as Isla Isabel or Isabella. Standing fresh water is limited to a single tiny pond (1 m diameter), and most of the island's surface is vegetated, principally with a deciduous tropical forest comprising 2 tree species: *Crataeva tapia* and *Euphorbia schlechtendalii*. Seabirds that nest in the forest, grassland, and cliffs of Isla Isabel are listed in table 23.2.

Biologists at UNAM's Instituto de Ecología have been studying the seabirds on Isla Isabel continuously since 1980. Their research has included long-term behavioral and ecological studies of the blue-footed boobies and magnificent frigatebirds and occasional studies of the brown boobies, Heermann's gulls, and brown pelicans. Every year these biologists camp on

Table 23.2. Seabirds nesting on Isla Isabel.

Species	Estimated Nesting Pairs
Magnificent frigatebird, *Fregata magnificens*	3000–3500
Blue-footed booby, *Sula nebouxii*	1000–1500
Brown booby, *Sula leucogaster*	1500–2500
Red-footed booby, *Sula sula*	2–8
Brown pelican, *Pelecanus occidentalis*	50–100
Red-billed tropicbird, *Phaethon aethereus*	100–150
Heermann's gull, *Larus heermanni*	150–200
Sooty tern, *Sterna fuscata*	3000
Brown noddy, *Anous stolidus*	60–70

Sources: Gaviño and Uribe 1978; Rodríguez et al., unpubl. data; J. L. Osorno, pers. comm.

the island for 5 months or more, with the support of the Mexican Navy and fishermen from coastal villages.

In 1980 the island was declared a National Park, and in 1994 the administration of the park was ceded to the Instituto de Ecología, which has managed the island since that date with varying degrees of co-participation by the environmental authority (currently CONANP, the Comisión Nacional de Areas Naturales Protegidas, of the SEMARNAP). At the time of writing, the institute and CONANP are putting the finishing touches on the official management program for the island and negotiating a new contract for co-administration. The personnel of the island's administration limit the fishermen and tourists to camping in the island's main bay and control access to other parts of the island, mostly keeping them well away from seabird colonies. Thus, if anything, the disturbance caused by visiting yachts, tourists from San Blas, and the fishermen has tended to diminish since the island was designated a National Park and the UNAM biologists established their long-term research program. In addition, those biologists have carried out programs to limit the impact of introduced mammals on the island's seabirds and inform local fishermen about conservation issues.

Although the populations of frigatebirds, boobies, pelicans, tropicbirds, and brown noddies appear to have been stable over the last 25 years, and this stability has been confirmed in one case by a population study of the blue-footed booby (e.g., Drummond et al. 2003), the sooty tern population has undergone a progressive, drastic decline that may recently have been arrested. In 1978, Gaviño and Uribe (1978) estimated (with an unstated methodology) that there were nearly 300,000 individuals in the nesting population, but just 13 years later Osorio and Torres (1991) estimated that the population had declined to a mere 3000 sooty terns, and suggested that predation by feral domestic cats (*Felis catus*), introduced 50 years earlier, was the main cause. Osorio and Torres (1991) further estimated that cats killed 25% of nesting adults in 1990 when, with the annual arrival of the terns at the island, some cats shifted their diet from fish (discarded by fishermen) to terns. According to Osorio and Torres (1991), Isla Isabel had at that time one of the densest island populations of cats in the world (113 cats/ km^2; Fitzgerald and Veitch 1985), and a model by Macias Garcia (Osorio and Torres 1991) showed that the tern population could go extinct in 15 years.

In 1991 and 1994 the sooty terns abandoned their clutches and broods, possibly in response to cat predation, apparently resulting in complete reproductive failure of the colony. The cats never seemed to prey upon booby chicks, but in 1994 they were observed occasionally killing magnificent frigatebird and Heermann's gull chicks (H. Drummond, pers. obs.).

In response to this situation, Rodríguez and Drummond carried out a cat eradication campaign, using the methodology of Veitch (1985), based on poisoning, trapping, and hunting (Rodríguez 1998) and funded by the Comisión Nacional para el Uso y la Conservación de la Biodiversidad (CONABIO). After 4 years of intermittent work, involving numerous fishermen, professional hunters, and a professional trapper, the last cat was killed in 1998, year in which only 4% of reproductive pairs of terns suffered predation (probably by peregrine falcons). Since then, the number of sooty terns nesting on the island has approximately doubled to roughly 6000, indicating that population recovery may be underway.

Just before eliminating the cats, Rodríguez, Drummond, and Ibarra Contreras also attempted to eliminate black rats from the island, using the methodology of Veitch and Bell (1989). Although intensive poisoning was sustained for an unusually long period of more than 40 days, at times involving up to 26 technicians, fishermen, and volunteers, the population of rats was never seriously depleted (Rodríguez 1998). Since then the island administration has resorted to simply controlling the number of rats in the zones where fishermen and researchers camp, by periodic poisoning.

In a further attempt at conserving the island ecosystem and reconciling conservation with the interests and needs of the local community, between February 1997 and July 1999, the Instituto de Ecología implemented a program of environmental education with the hundreds of fishermen who camp on the island but live in villages along the mainland coast. The aim was to promote changes in their knowledge, attitudes, and behavior by installing latrines and systems for handling waste and by stimulating reflection and discussion through workshops, mural newspapers (newspapers posted on walls rather than hand distributed), leaflets, and so on (Ibarra Contreras 2002). Financed by the Fondo Mexicano para la Conservación de la Naturaleza, the program included a novel component of objective "before and after" attitude measurement, which showed significant gains in the fishermens' attitudes to the island environment, in addition to substantial improvements in their living conditions.

In 2000 a novel source of perturbation threatened the island's reefs and ecosystem: the company Thunnus Acuícola de Nayarit installed underwater cages beside the island for the commercial fattening of yellowfin tuna fish (*Thunnus albacares*), by permit from the Subsecretaría de Pesca. However, the company had not complied with legislation requiring analysis of environmental impact, and the island's administration objected that reefs and seabird populations could be harmed. A legal battle followed. Eventually, impressed by the risk to long-term studies of seabird biology involving many years of financial investment by the federal government and the national university, the courts and environmental authorities ruled in favor of the island administration, and in 2002 the cages (which had recently suffered hurricane damage) were towed away by Thunnus Acuícola under legal duress.

Islas Marietas

The small archipelago known as Islas Marietas (fig. 23.3) lies 9.5 km to the southwest of Punta Mita along the coast of Nayarit, inside Bahía de Banderas, a submerged valley established 18,000 years ago after the glacial high of the Wisconsinian (Ordoñez 1946). It consists of 2 islands, Isla Redonda (Round Island) and Isla Larga (Long Island); 2 islets, El Morro and La Corbeteña; and a couple of rocks, La Ampolla (Bladder) and Los Morros Cuates. All are within the boundaries of the continental platform near the southern limit of the California Cold Current. The islands lack standing water. They are virtually treeless and are mostly covered by grassland (*Pennisetum setosum*) and small thickets of bromelias (Rebón 1997, 2000).

Isla Redonda (20°42' N, 105°35' W) is 1 km wide and 0.5 km long, with an area of 0.2 km² (Rebón 2000). Maximum elevation is 40 m, in the southeastern section of the island. The natural arrangement of rocks creates several caves and hollows used by birds. The sea has eroded cliffs along some portions of the island, and one can walk over the roof of marine caves. Two small, fine-sand beaches are visible only at low tide. Off the northwestern end of the island are 2 rocks separated from it by a distance of 15 m. Both rocks are used for nesting by some seabirds (Rebón 1997, 2000). Dominant plants include *Jouvea pilosa* (Gramineae), *Cyperus ligularis* (Cy-

peraceae), *Bromelia pinguin* (Bromeliaceae), and a cactus (*Stenocereus* sp.). A metallic lighthouse was built on the island.

Isla Larga (20°41' N, 105°36' W) is separated from Isla Redonda by a distance of 1 km (Rebón 2000). It is 1.1 km wide and 0.9 km long, with an area of about 0.4 km². Maximum elevation is 25 m. The island has 16 small rocky or sandy beaches. It consists of a large plateau, which has 6 rocky hills 10–20 m high, with associated caves. The vegetation is fairly varied. It is dominated by Gramineae and Cyperaceae, but other vegetation is present, including Cactaceae and *Orbignya* palms. Isla Larga also has a metallic lighthouse (Rebón 1997, 2000).

The irregularly shaped islet La Corbeteña (20°43' N, 105°51' W) is 32 km off Punta Mita. The islet El Morro (20°41' N, 105°40' W) lies 7.4 km west of Isla Larga; it is 80 m long and 50 m wide, with a maximum elevation of 30 m. Both islets have no vegetation, and they are covered in guano, giving them their whitish color (SEGO and SEMAR 1987). There are no nesting colonies on La Corbeteña and El Morro, presumably due to strong local winds. Seabirds use the 2 islets only for resting (F. Rebón, pers. obs.).

The bird life of the 2 main islands was studied between 1987 and 1999 by one of us (F.R.G.), along with students. There are 84 species, 60 on Isla Redonda and 69 on Isla Larga (Rebón 1997, 2000). Ten species nesting on the archipelago are seabirds (table 23.3). Five of them nest on both islands, 3 only on Isla Redonda, and 2 only on Isla Larga. The largest nesting population is that of brown boobies. Islas Redonda and Larga collectively have the largest nesting populations of bridled terns and brown noddies in the Gulf. They harbor the largest nesting population of laughing gulls in the Gulf and Pacific Ocean Region (Rebón 1997, 2000). The red-billed tropicbird, blue-footed and brown boobies, Brandt's cormorant, and magnificent frigatebird are year-round residents. The 5 larid species are summer residents.

Grant (1964) documented breeding colonies of blue-footed and brown boobies, magnificent frigatebirds, and Heermann's gulls on the archipelago. However, his list of breeding species did not include the other 6 species. The laughing gull, royal tern, and Brandt's cormorant, in particular, have recently begun to breed on Islas Marietas. Conversely, the brown pelican nested on Islas Marietas in Grant's time. From 1987 to 1999 brown pelicans were not observed breeding on Islas Marietas (breeding was

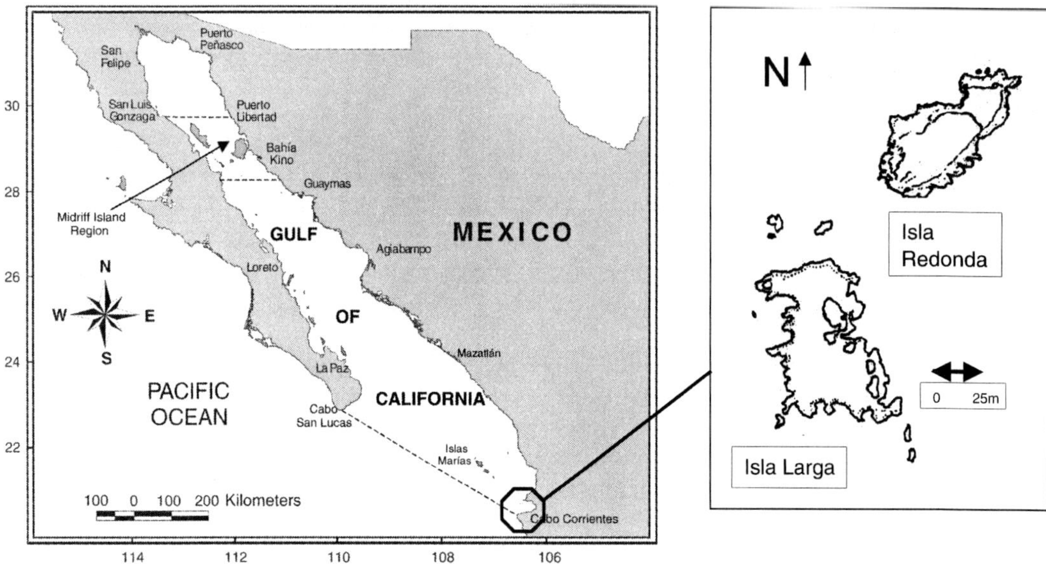

Figure 23.3. Islas Marietas, Nayarit.

documented on the islet Los Arcos near the town of Mismaloya in Bahía de Banderas). Up to 361 resting pelicans were noted on Isla Redonda, and the species could well nest on this island in the near future, provided it is not subjected to disturbance from tourism (Rebón 1997, 2000).

Additional seabirds occur on Islas Marietas, but nesting by these species has not been documented. Yet, based on their behavior the possibility exists that they do breed on the archipelago, in places difficult to observe. If they do not breed locally now, they may do so in the future (like the brown peli-

can). Those species are the Townsend's shearwater (*Puffinus auricularis*), Caspian tern (*Sterna caspia*), and Sandwich tern (*S. sandvicensis*) (Rebón 1997).

Although not a seabird as defined in the chapter, the American oystercatcher (*Haematopus palliatus*) deserves to be mentioned here. The species was common in Bahía de Banderas in the 1980s (C. Esquivel, pers. comm.) and perhaps bred on Islas Marietas. This species has disappeared locally, presumably in relation to intertidal fishing and disturbance caused by tourism. Massey and Palacios (1994) mention that the population of the Ameri-

Table 23.3. Seabirds nesting on the two main islands of Islas Marietas, with estimates of total colony size (number of breeding individuals).

Common Name	Scientific Name	Isla Redonda	Isla Larga	Status	Estimated Colony Size
Red-billed tropicbird	*Phaethon aethereus*	B		YRR	4
Blue-footed booby	*Sula nebouxii*	B (1,074)	B (1,526)	YRR	2,600
Brown booby	*Sula leucogaster*	B (20,132)	B (10,368)	YRR	30,500
Brandt's cormorant	*Phalacrocorax penicillatus*	B		YRR	40
Magnificent frigatebird	*Fregata magnificens*	B		YRR	396
Laughing gull	*Larus atricilla*		B	SR	6,000
Heermann's gull	*Larus heermanni*	B (4,100)	B (900)	SR	5,000
Royal tern	*Sterna maxima*		B	SR	150
Bridled tern	*Sterna anaethetus*	B (172)	B (118)	SR	290
Brown noddy	*Anous stolidus*	B (398)	B (162)	SR	560

B = breeding; YRR = year-round resident; SR = summer resident.

can oystercatcher from the El Vizcaíno region (Baja California peninsula) has substantially declined, also due to human disturbance.

The decline of the American oystercatcher offers insight on potential threats to seabirds of Islas Marietas. There is no permanent human presence on the islands, but there are temporary fishing camps. The nearest towns on the mainland (Corral del Risco, Punta Mita, La Cruz de Huanacaxtle, and Bucerías) depend mainly on tourism, fishing, and agriculture, and Islas Marietas are visited by growing numbers of people, including tourists.

Ecotourism has boomed in recent years around Bahía de Banderas (a whale-watching area) and the major resort of Puerto Vallarta. In particular, ever larger tourboats anchor briefly beside the Islas Marietas to allow tourists to snorkel, although these tourists rarely land on the islands, and the major impact of the tourboats is surely on the surrounding reefs and shallows. On February 3 and 4, 1997, a major fire caused by a Canadian tourist devastated 93% of the island. Six days later, F. Rebón coordinated an effort involving UNAM, SEMARNAP, and the Mexican Navy to investigate the impact of the fire and incinerate bird remains (Rebón et al. 1997). They found the charred remains of 136 blue-footed boobies (including 17 chicks) and 194 brown boobies (including 7 chicks), as well as scores of eggs that were burned or destroyed by gulls and scores of abandoned nests of both booby species (Rebón et al. 1997).

Currently the islands are not protected by any specific legislation, but local authorities manage to largely discourage landing or camping on the islands, and CONANP (the federal agency managing Natural Protected Areas) is preparing a conservation plan for them. Given the increasing expansion of tourism in the bay and demand for ecotourism activities, implementation is urgent.

Islas Marías

The Islas Marías Archipelago (185 km^2) consists of 4 islands, María Madre, María Magdalena, María Cleofas, and San Juanito. Located 120 km off San Blas, along the coast of Nayarit, they were purchased in 1905 by the Mexican federal government. The largest island, Isla María Madre, is used as a penitentiary, where inmates live in 11 villages (linked by a circular road) and carry out agricultural activities (the archipelago as a whole is a federal penal zone). Isla María Cleofas is in more pristine condition, and access is strictly limited by the

Secretaría de Gobernación and the Mexican Navy to prevent escapes from Isla María Madre. Cattle, goats, dogs, cats, pigs, rats, and fruit trees have been introduced to the islands of the archipelago (Arizmendi and Márquez 2000; Vicenteño 2003).

There is little information on the avifauna of the Islas Marías archipelago. Grant and Cowan's (1964) bird checklist indicated that red-billed tropicbirds, blue-footed boobies, brown boobies, brown pelicans, and magnificent frigatebirds were local breeders. When Drummond visited María Cleofas in the 1990s, he found an extensive and diverse tropical deciduous forest. According to local fishermen (unauthorized visitors) there is fresh water and medium-sized native mammals on this island. Drummond saw numerous brown pelicans, boobies, and magnificent frigatebirds but suspected that colonies may not be large if preyed upon by local mammals, although the island is enormous in comparison with nearby Isla Isabel. As already mentioned, there is a plan to designate Islas Marías as a Biosphere Reserve and relocate prisoners.

Conclusions and Outlook

Seabird conservation is mostly a matter of island protection. About 4% of the marine area and almost all the islands in northwestern Mexico, including those harboring major seabird colonies, are federally protected under different categories of natural protected areas, such as Biosphere Reserve, National Park, and Area for Protection of Flora and Fauna. There is a proposal to protect the Pacific islands off Baja California (B. Tershy, pers. comm.).

Island and seabird conservation in the region is complex, and all developed conservation planning needs to consider the following aspects: (1) Local, state, and federal governments must ensure the protection and restoration of island ecosystems by enforcing environmental laws, using data from academia for informed decision-making, and supporting applied conservation-oriented research, such as monitoring indicator species and fisheries resources and eradication of exotic species. (2) Conservation of natural habitat must continue to be an important approach, but should be accompanied by completion and implementation of management plans. (3) Cooperative programs between Canada, the United States, and Mexican wildlife agencies should expand, but solutions to the problems must be site specific and operational.

The shortage of baseline data makes it impossible to quantify changes in numbers in most seabird species over time. Numerical baseline data are sketchy, as early ornithologists reported species' presence and breeding information, but seldom reported numbers (Massey and Palacios 1994). Furthermore, when baseline data are available and recommendations based on them are provided by researchers, managers and users seldom, if ever, take them into account for resource administration decisions. Therefore, a collaborative attitude needs to be developed among governmental agencies so that sound administation measures are developed and implemented. We also need to plan and promote seabird monitoring on an international basis and to strive for cooperation of seabird specialists throughout the region. Ideally, a few widespread species should be monitored throughout their range in the Pacific, an objective that will require an internationally coordinated effort. Finally, financial resources need to be secured for at least long-term on-site protection and environmental education among local people, both of which are essential for achieving basic conservation.

Literature Cited

Ainley, D. 1984. Storm-petrels. Pp. 58–63 *in* D. Haley (ed.), Seabirds of Eastern North Pacific and Arctic Waters. Pacific Search Press, Seattle, Washington.

Ainley, D. G., H. R. Carter, D. W. Anderson, K. T. Briggs, M. C. Coulter, F. Cruz, J. B. Cruz, C. A. Valle, S. I. Fefer, S. A. Hatch, S. A. Hatch, E. A. Schreiber, R. W. Schreiber, and N. G. Smith. 1988. Effects of the 1982–1983 El Niño-Southern Oscillation on Pacific Ocean bird populations. Proceedings of the International Ornithological Congress 19: 1747–1758.

Alvarez-Borrego, S. 1983. Gulf of California. Pp. 427–449 *in* C. B. H. Ketchum (ed.), Estuaries and Enclosed Seas. Elsevier, Amsterdam.

Anderson, D. W. 1983. The seabirds. Pp. 246–264 *in* T. J. Case and M. L. Cody (eds.), Island Biogeography in the Sea of Cortez. University of California Press, Berkeley.

Anderson, D. W. 1988. Dose-response relationship between human disturbance and brown pelican breeding success. Wildlife Society Bulletin 16: 339–345.

Anderson, D. W., and I. T. Anderson. 1976. Distribution and status of brown pelicans in the California Current. American Birds 30: 3–12.

Anderson, D. W., and F. Gress. 1983. Status of a northern population of California brown pelicans. Condor 85: 79–88.

Anderson, D. W., F. Gress, and D. M. Fry. 1996. Survival and dispersal of oiled brown pelicans after rehabilitation and release. Marine Pollution Bulletin 32: 711–718.

Anderson, D. W., F. Gress, K. F. Mais, and P. R. Kelly. 1980. Brown pelicans as anchovy stock indicators and their relationships to commercial fishing. California Cooperative Oceanic Fisheries Investigations Report 21: 54–61.

Anderson, D. W., F. Gress, K. F. Mais, and P. R. Kelly. 1982. Brown pelicans: influence of food supply on reproduction. Oikos 39: 23–31.

Anderson, D. W., and J. O. Keith. 1980. The human influence on seabird nesting success: conservation implication. Biological Conservations 18: 65–80.

Anderson, D. W., J. O. Keith, G. R. Trapp, F. Gress, and L. A. Moreno. 1989. Introduced small ground predators in California brown pelican colonies. Colonial Waterbirds 12: 98–103.

Anderson, D. W., J. E. Mendoza, and J. O. Keith. 1976. Seabirds in the Gulf of California: a vulnerable, international resource. Natural Resources 16: 483–505.

Anthony, A. W. 1925. Expedition to Guadalupe Island, Mexico, in 1922. The birds and mammals. Proceedings of the California Academy of Sciences 14: 277–320.

AOU. 1998. Check-list of North American Birds, 7th ed. American Ornithologists' Union, Washington, D.C.

Arizmendi, M. del C., and L. Márquez (eds.) 2000. Areas de Importancia para la Conservación de las Aves en México. CIPAMEX, A.C., México D.F.

Bailey, H. H. 1906. Ornithological notes from western Mexico and the Tres Marias and Isabella Islands. Auk 23: 369–391.

Bancroft, G. 1927a. Breeding birds of Scammon's Lagoon, Lower California. Condor 30: 29–57.

Bancroft, G. 1927b. Notes on the breeding coastal and insular birds of central Lower California. Condor 29: 188–195.

Banks, R. C. 1963. Birds of the Belvedere Expedition to the Gulf of California. Transactions of the San Diego Natural History Society 13: 49–60.

Barreto, J. A. 1973. Isla Raza, B.C., refugio de gaviotas y gallitos de mar. Bosques y Fauna 10 (4): 3–8.

Bent, A. C. 1919. Life histories of North American diving birds. U.S. National Museum Bulletin 107, Washington, D.C.

Bent, A. C. 1921. Life histories of North American gulls and terns. U.S. National Museum Bulletin 113, Washington, D.C.

BirdLife International. 2000. Threatened Birds of the World. Lynx Edicions and Birdlife International, Barcelona and Cambridge.

Blake, E. R. 1953. Birds of Mexico: A Guide for Field Identification. University of Chicago Press, Chicago.

Boswall, J., and M. Barrett. 1978. Notes on the breeding birds of Isla Raza, Baja California. Western Birds 9: 93–108.

Bourillón, L., A. Cantú, F. Eccardi, J. Ramírez, E. Velarde, and A. Zavala. 1988. Islas del Golfo de California, Secretaría de Gobernación and Universidad Nacional Autónoma de México, México D.F.

Bowen, T. 2000. Unknown Island: Seri Indians, Europeans, and San Esteban Island in the Gulf of California. University of New Mexico Press, Albuquerque.

Brewster, W. 1902. Birds of the Cape Region, Lower California. Bulletin of the Museum of Comparative Zoology 41: 21–22.

Briggs, J. C. 1974. Marine Zoogeography. McGraw-Hill, New York.

Briggs, K. T., D. B. Lewis, W. B. Tyler, and G. L. Hunt, Jr. 1981. Brown pelicans in southern California: habitat use and environmental fluctuations. Condor 83: 1–15.

Burger, J., and M. Gochfeld. 1996. Family Laridae (Gulls). Pp. 572–623 *in* J. del Hoyo, A. Elliott, and J. Sargatal (eds.), Handbook of the Birds of the World, vol. 1. Lynx Edicions, Barcelona.

Cody, M., R. Moran, and H. Thompson. 1983. The plants. Pp. 49–97 *in* T. J. Case and M. L. Cody (eds.), Island Biogeography in the Sea of Cortez. University of California Press, Berkeley.

Cody, M. L., and E. Velarde. 2002. The landbirds. Pp. 271–312 *in* T. J. Case, M. L. Cody, and E. Ezcurra (eds.), A New Island Biogeography of the Sea of Cortés. Oxford University Press, New York.

DeWeese, L. R., and D. W. Anderson. 1976. Distribution and breeding biology of Craveri's murrelet. Transactions of the San Diego Society of Natural History 18: 155–168.

DOF. 2002. Decreto por el que se Declara Área Natural Protegida con la Categoría de Reserva de la Biosfera, la Región Denominada Isla San Pedro Mártir, Ubicada en el Golfo de California, Frente a las Costas del Municipio de Hermosillo, Estado de Sonora, con una Superficie Total de 30,165-23-76.165 hectáreas. Diario Oficial de la Federación, June 13, primera sección, Pp. 6–14.

Drummond, H., R. Torres, and V. V. Krishnan. 2003. Buffered development: resilience after aggressive subordination in infancy. American Naturalist 161: 794–807.

Erickson, R. A., and S. N. G. Howell (eds.). 2001. Birds of Baja California: status, distribution, and taxonomy. Monographs in Field Ornithology 3: 29–63.

Everett, W. T., and E. W. Anderson. 1991. Status and conservation of the breeding seabirds on offshore Pacific islands of Baja California and the Gulf of California. Pp.115–139 *in* J. P. Croxall (ed.), Seabird Status and Conservation: A Supplement. ICBP Technical Publication 11. International Council for Bird Preservation, Cambridge, U.K.

Felger, R. S., and C. H. Lowe. 1976. The island and coastal vegetation and flora of the northern part of the Gulf of California. Los Angeles County Natural History Museum Contributions in Science 285.

Fitzgerald, B. M., and C. R. Veitch. 1985. The cats of Herekopare Island, New Zealand: their history, ecology and effects on birdlife. New Zealand Journal Zoology 12: 319–330.

Friedmann, H., L. Griscom, and R. T. Moore. 1950. Distributional checklist of the birds of Mexico. Pacific Coast Avifauna 29: 1–202.

Gastil, G., J. Minch, and R. P. Phillips. 1983. The geology and ages of the islands. Pp. 13–25 *in* T. J. Case and M. L. Cody (eds.), Island Biogeography in the Sea of Cortez. University of California Press, Berkeley.

Gaviño, G., and Z. Uribe. 1978. Algunas observaciones ecológicas en Isla Isabel, Nayarit, con sugerencias para el establecimiento de un Parque Nacional. Unpublished manuscript, Instituto de Biología, Universidad Nacional Autónoma de México, México D.F.

Grant, P. R. 1964. The birds of the Tres Marietas Islands, Nayarit, Mexico. Auk 81: 514–519.

Grant, P. R. 1965. A systematic study of the birds of the Tres Marias Islands, Mexico. Postilla 90: 106.

Grant, P. R., and I. McT. Cowan. 1964. A review of the avifauna of the Tres Marias Islands, Nayarit, Mexico. Condor 66: 221–228.

Gress, F., and D. W. Anderson. 1983. California brown pelican Recovery Plan. U.S. Fish and Wildlife Service, Portland, Oregon.

Harrison, P. 1983. Seabirds: An Identification Guide. Houghton Mifflin, Boston.

Howell, S. N. G., and S. Webb. 1995. A Guide to the Birds of Mexico and Northern Central America. Oxford University Press, New York.

Howell, T. R., B. Araya, and W. R. Millie. 1974. Breeding biology of the gray gull, *Larus modestus*. University of California Publications in Zoology 104: 1–57.

Huey, L. 1927. The bird life of San Ignacio and Pond lagoons on the western coast of Lower California. Condor 29: 239–243.

Ibarra Contreras, A. 2002. Diseño y evaluación del programa de educación ambiental aplicado a los pescadores del parque nacional Isla Isabel. Master's thesis, Universidad Nacional Autónoma de México, México D.F.

Jehl, J. R., Jr. 1976. The northernmost colony of Heermann's gull. Western Birds 7: 25–26.

Jehl, J. R., Jr. 1984. Conservation problems of seabirds in Baja California and the Pacific Northwest. Pp. 41–48 *in* J. P. Croxall, P. G. H. Evans, and R. W. Schreiber (eds.), Status and Conservation of the World's Seabirds. ICBP Technical Publication 2. International Council for Bird Preservation, Cambridge, U.K.

Jiménez Serranía, V. T. Pfister, and G. Suárez Gracida. 2002. Monitoreo de Cormorán de doble cresta (*Phalacrocorax auritus*) en Isla Alcatraz, Bahía de Kino, Sonora, México. Unpublished manuscript, Prescott College, Prescott, Arizona.

Lawlor, T. 1983. The mammals. Pp. 265–289 *in* T. J. Case and M. L. Cody (eds.), Island Biogeography in the Sea of Cortez. University of California Press, Berkeley.

Lawlor, T. E., D. J. Hafner, P. Stapp, B. R. Riddle, and S. T. Álvarez-Castañeda. 2002. The mammals. Pp. 326–361 *in* T. J. Case, M. L. Cody, and E. Ezcurra (eds.), A New Island Biogeography of the Sea of Cortés. Oxford University Press, New York.

Lindsay, G. E. 1983. History of scientific exploration in the Sea of Cortéz. Pp. 3–12 *in* T. J. Case and M. L. Cody (eds.), Island Biogeography in the Sea of Cortéz. University of California Press, Berkeley.

Mailliard, J. 1923. Expedition of the California Academy of Sciences to the Gulf of California in 1921: the birds. Proceedings of the California Academy of Sciences 12: 443–456.

Massey, B. W., and E. Palacios. 1994. Avifauna of the wetlands of Baja California, Mexico: current status. Studies in Avian Biology 15: 45–57.

McChesney, G. J., and B. R. Tershy. 1998. History and status of introduced mammals and impacts to breeding seabirds on the California Channel and northwestern Baja California islands. Colonial Waterbirds 21: 335–347.

Mellink, E. 2001. History and status of colonies of Heermann's gull in Mexico. Waterbirds 24: 188–194.

Mellink, E., G. Ceballos, and J. Luévano. 2002. Population demise and extinction threat of the Angel de la Guarda deer mouse (*Peromyscus guardia*). Biological Conservation 108: 107–111.

Mora A., L. 1998. Reproducción de la golondrina marina gorriblanca *Anous stolidus* (Aves: Sterninae) en las Islas Marietas, Nayarit, colonia de anidación más grande registrada para México. Tesis de Licenciatura, Facultad de Ciencias, Universidad Nacional Autónoma de México, México D.F.

Muñoz Lumbier, M. 1919. Algunos datos sobre las islas Mexicanas para contribuir al estudio de sus recursos naturales. Anales del Instituto Geológico de México 7. Secretaría de Industria, Comercio y Trabajo, Departamento de Exploraciones y Estudios Geológicos, México D.F.

Nelson, J. B. 1978. The Sulidae: Gannets and Boobies. Oxford University Press, Oxford, U.K.

Ordoñez, E. 1946. Principales provincias geográficas y geológicas de la República Mexicana. Sobretiro de la Guía del Explorador Minero. Universidad Nacional Autónoma de México, Institutó de Geología, México D.F.

Osorio, B. M., and L. R. Torres A. 1991. Depredación por mamíferos introducidos *Felis catus* y *Rattus rattus*: riesgo de extinción local de *Sterna fuscata* en la Isla Isabel, Nayarit. Reporte de la biología de campo, Facultad de Ciencias, Universidad Nacional Autónoma de México, México D.F.

Palacios, E., and E. Mellink. 1992. Breeding bird records from Montague Island, northern Gulf of California. Western Birds 23: 41–44.

Palacios, E., and E. Mellink. 1993. Additional breeding records from Montague Island, northern Gulf of California. Western Birds 24: 259–262.

Palacios, E., and E. Mellink. 1996. Status of the least tern in the Gulf of California. Journal of Field Ornithology 67: 48–58.

Ramírez, J., E. Velarde, and G. Ceballos. Unpublished ms. Eradication of the introduced rodents and its effects on the breeding success of nesting seabirds in Isla Rasa, Gulf of California, Mexico. Available upon request from the authors.

Rebón G., F. 1997. Análisis de la avifauna presente en el Archipiélago de las Islas Marietas y sus aguas adyacentes, Nayarit, México. Tesis de Maestría en Ciencias (Biología), Facultad de Ciencias, Universidad Nacional Autónoma de México, México D.F.

Rebón G., F. 2000. Distribución, abundancia y conservación de la avifauna de las Islas

Marietas, Nayarit, México. Anales del Instituto de Biología Universidad Nacional Autónoma de México. Serie Zoología 71 (1): 59–88.

Rebón, F., E. Cornejo L., and E. Rodríguez A. 1997. Informe de la salida realizada a las Islas Marietas, Nayarit, durante los días 10–12 de Febrero de 1997, seis días después del incendio provocado en Isla Larga. Unpublished manuscript, Universidad Nacional Autónoma de México, México D.F.

Ridgway, R. 1919. The Birds of North and Middle America. U.S. National Museum Bulletin, part VIII 16: 1–852.

Rodríguez A., E. 1997. Reproducción del gallito de mar bridado *Sterna anaethetus Nelson Ridgway 1919* (aves: Sterninae) en las Islas Marietas, Nayarit, México. Tesis de Licenciatura, Facultad de Ciencias, Universidad Nacional Autónoma de México, México D.F.

Rodríguez, N., Ma. C. 1998. Erradicación de gatos y ratas en una isla tropical del Pacífico de México. Master's thesis, Facultad de Ciencias, Universidad Nacional Autónoma de México, México D.F.

Saunders, H. 1896. Gaviae. Catalogue of Birds at the British Museum 25: 1–339.

Schaffner Jr., F. C. 1982. Aspects of the reproductive ecology of the elegant tern (*Sterna elegans*) at San Diego Bay. Master's thesis, San Diego State University, San Diego, California.

Schreiber, R. W. 1977. Maintenance behavior and communication in the brown pelican. Ornithological Monographs no. 22.

SEGO (Secretaría de Gobernación) and SEMAR (Secretaría de Marina). 1987. Islas Mexicanas, Régimen jurídico y catálogo. SEGO, México D.F.

Shields, M. 2002. Brown pelican (*Pelecanus occidentalis*). *In* A. Poole and F. Gill (eds.), The Birds of North America, no. 609. Academy of Natural Sciences, Philadelphia, and American Ornithologists' Union, Washington, D.C.

Spear, L. B., and D. W. Anderson. 1989. Nest-site selection by yellow-footed gulls. Condor 91: 91–99.

Tershy, B. R., and D. Breese. 1997. The birds of San Pedro Mártir Island, Gulf of California, Mexico. Western Birds 28: 96–107.

Tershy, B. R., C. J. Donlan, B. S. Keitt, D. A. Croll, J. A. Sanchez, W. Wood, M. A. Hermosillo, G. R. Howald, and N. Bianvaschi. 2002. Island conservation in northwest Mexico: a conservation model integrating research, education and exotic mammal eradication. Pp. 293–300 *in* C. R. Veitch and M. N. Clout (eds.), Turning the Tide: The Eradication of Invasive Species. Invasive Species Specialist Group of the World Conservation Union, Gland, Switzerland.

Tobón G., E. D. 1992. Ecología reproductiva de la golondrina marina elegante (*Sterna elegans*) con énfasis en la conducta dentro de las guarderías. B.S. thesis, Facultad de Ciencias, Universidad Nacional Autónoma de México, México D.F.

Veitch, C. R. 1985. Methods of eradicating feral cats from offshore islands in New Zealand. Pp. 125–141 *in* ICBP Technical Publication no 3. International Council for Bird Preservation (now BirdLife International), Cambridge, U.K.

Veitch, C. R., and D. B. Bell. 1989. Eradication of introduced animals from the islands of New Zealand. Pp. 137–146 *in* D. R. Towns, H. C. Daugherty, and A. I. Atkinson (eds.), Ecological Restoration of New Zealand Islands. Conservation Science Publication 2, Department of Conservation, Wellington, New Zealand.

Velarde, E. 1989. Conducta y ecología de la reproducción de la gaviota parda (*Larus heermanni*) en la Isla Rasa, Baja California. Tesis de Doctorado, Facultad de Ciencias, Universidad Nacional Autónoma de México, México D.F.

Velarde, E. 1999. Breeding biology of the Heermann's gull (*Larus heermanni*) in Isla Rasa, Gulf of California, Mexico. Auk 116: 513–519.

Velarde, E. 2000a. Paíño mínimo (*Oceanodroma microsoma*). Pp. 84–85 *in* G. Ceballos and L. Márquez Valdelamar (eds.), Las Aves de México en Peligro de Extinción. Fondo de Cultura Económica/UNAM/CONABIO, México D.F.

Velarde, E. 2000b. Charrán elegante (*Sterna elegans*). Pp. 164–166 *in* G. Ceballos and L. Márquez Valdelamar (eds.), Las Aves de México en Peligro de Extinción. Fondo de Cultura Económica/UNAM/CONABIO, México D.F.

Velarde, E. 2000c. Mérgulo de Craveri (*Synthliboramphus craveri*). Pp. 169–171 *in* G. Ceballos and L. Márquez Valdelamar (eds.), Las Aves de México en Peligro de Extinción. Fondo de Cultura Económica/UNAM/CONABIO, México D.F.

Velarde, E. 2000d. Paíño negro (*Oceanodroma melania*). Pp. 80–81 *in* G. Ceballos and L. Márquez Valdelamar (eds.), Las Aves de México en Peligro de Extinción. Fondo de Cultura Económica/UNAM/CONABIO, México D.F.

Velarde, E., and D. W. Anderson. 1994. Conservation and management of seabird islands in the Gulf of California: setbacks and successes. BirdLife Conservation Series no. 1: 229–243.

Velarde, E., and E. Ezcurra. 2002. Breeding dynamics of Heermann's gulls. Pp. 313–325 *in* T. J. Case, M. L. Cody, and E. Ezcurra (eds.), A New Island Biogeography of the Sea of Cortés. Oxford University Press, New York.

Velarde, E., E. Ezcurra, M. A. Cisneros-Mata, and M. F. Lavin. In press. Seabird ecology, El Niño anomalies, and prediction of sardine fisheries in the Gulf of California. Ecological Applications.

Velarde, E., and B. Keitt. 2000. Pardela mexicana (*Puffinus opisthomelas*). Pp. 76–78 *in* G. Ceballos and L. Márquez Valdelamar (eds.), Las aves de México en Peligro de Extinción. Fondo de Cultura Económica/ UNAM/CONABIO, México D.F.

Velarde, E., M. S. Tordesillas, R. Esquivel, and L. Vieyra. 1994. Seabirds as indicators of important fish populations in the Gulf of California. California Cooperative Oceanic Fisheries Investigations Reports 35: 137–143.

Vicenteño, D. 2003. Serán Islas Marias reserva ecológica. El Archipiélago dejará de ser un centro de reclusión federal, antes de que concluya la administración foxista, para convertirse en un área natural protegida. Reforma Newspaper Group. September 7, 2003, 1A, 8A–11A.

Villa Ramírez, B. 1976. Isla Rasa, Baja California, Enigma y Paradigma. Supervivencia 2: 17–29.

Villa-Ramírez, B. 1979. Algunas aves y la rata noruega, *Rattus norvegicus*, "versus" el murciélago insulano *Pizonyx vivesi* en las islas del Mar de Cortés, México. Anales del Instituto de Biología, Universidad Nacional Autónoma de México, Serie Zoología 50: 729–736.

Walker, B. W. 1960. The distribution and affinities of the marine fish fauna of the Gulf of California. Systematic Zoology 9: 123–133.

Walker, L. W. 1965. Baja's island of birds. Pacific Discovery 18: 27–31.

Wilbur, S. R. 1987. Birds of Baja California. University of California Press, Berkeley.

Willard, A. 1890. Annual Report 1890. Addressed to Hon. W. F. Wharton, Assistant Secretary of State, December 31, 1890. Microfilm, RG 59, General Records of the Department of State, Consular Dispatches, Guaymas, vol. 8. U.S. National Archives, Washington, D.C.

Willard, A. 1891. Annual Report 1891. Addressed to Hon. W. F. Wharton, Assistant Secretary of State, December 31, 1891. Microfilm, RG 59, General Records of the Department of State, Consular Dispatches, Guaymas, vol. 8. U.S. National Archives, Washington, D.C.

Index

Abies, 64; *A. concolor*, 174, 375; *A. durangensis*, 206

Acacia, 46, 347, 439, 446; *A. cochliacantha*, 205; *A. constricta*, 427; *A. farnesiana*, 335; as fuelwood, 64; boatthorn, 205; sweet, 335; whitethorn, 427

Acanthaceae, 217

Acanthocereus tetragonus, 269

Achillea millefolium, 375, 379

Acipenser oxyrinchus, 143, 144–145

Acorns, as black bear food, 397; as food, 66

Adenostoma fasciculatum, 173, 373, 376; *A. sparsifolium*, 173, 373

Aegopogon cenchroides, 112, 116; *A. tenellus*, 112, 116

Aeronautes saxatalis, 339, 340, 342, 347

Aesculus parryi, 373

Agave, 46; *A. deserti*, 175, 375; *A. lechuguilla*, 65, 265, 267; *A. shawii*, 173, 373; *A. vivipera*, 65; coastal, 173, 373; desert, 175, 375

Agiabampo, 289, 290

Agriculture, 334; conversion of native habitats for, 60, 238; decline of grassland bird species, 432; diversion of Colorado River water, 189; effect on prairie dogs, 433; effect on pronghorn antelope, 388; field borders, 440, 442, 444, 448; impact on bears, 397; irrigated, 59; shift to agro-industry, 59; threat to fish, 159

Agrito, 397

Agrupación Pradera, 449

Agua Amarilla, 110

Aguaceros, 171

Aguascalientes, 395

Aimophila cassinii, 441, 443, 444, 448

Ajos-Bavispe Protected Natural Area, 58

Alacran, 130

Alamillo, 206

Alamos, 134, 218

Alder, 206

Alfalfa, 118, 380

Alfilleria, 380

Algae, marine, 317, 412; red, 410, 413

Aliso, 206

Allenrolfea occidentalis, 305

Alliance for Rio Grande Heritage, 449

Almond, desert, 373

Alnus oblongifolia, 206

Aloysia wrightii, 65

Altiplano, loss of habitat, 238

Amazilia beryllina, 208, 209, 210, 213, 214, 215, 216, 218, 222; *A. rutila*, 218, 222; *A. violiceps*, 207, 209, 210, 218, 222, 223

Amazon, green-cheeked, 60; parrots, trade in, 69

Amazona, pet trade of, 69; *A. viridigenalis*, 60

Ambrosia chenopodifolia, 372; *A. dumosa*, 175

Ameiurus melas, 143, 146–147

Ammodramus bairdii, 432, 441, 442, 443, 444, 445, 448, 449; *A. savannarum*, 441, 442, 443, 444, 445, 448, 449

Leptotyphlops humilis, 171, 177
Lepus callotis, 62, 428, 430
Ley de Aguas Nacionales (LAN), *see* National
    Waters Act; Ley de Pesca (LP), *see* Fisheries
    Act; Ley Federal sobre Metrología y Normali-
    zación, *see* Federal Law on Metrology and
    Normalization; Ley General de Desarrollo
    Forestal Sustenable (LGDFS), *see* General Act
    for Sustainable Forestry Development; Ley
    General de Vida Silvestre (LGVS), *see* General
    Wildlife Act; Ley General del Equilibrio
    Ecológico y la Protección al Ambiente
    (LGEEPA), *see* General Act for Ecological
    Balance and the Protection of the Environment
Lichanura trivirgata, 69, 167, 171, 177
Liga, 249
Ligusticum porteri, 67
Lilac, hoary-leaf, 173; wild, 373
Limberbush, ashy, 249
Limita, 214, 218, 219
Linares, 160
Linck, Fray Wenceslaus, 375
Lindsay, George, 460
Lippia berlandieri, 65
Lithophile(s), among scorpions, 124, 131
Lizard, Baja California collared, 169, 177; Baja
    California legless, 171, 175; Baja California
    leopard, 169, 177; Baja California spiny, 169,
    176, 177; banded rock, 69, 167, 168, 177;
    black-tailed brush, 169, 177, 232; California
    legless, 171, 175; coast horned, 170, 177;
    Colorado Desert fringe-toed, 170, 176; desert
    horned, 170, 175; desert night, 171, 176; desert
    spiny, 169, 176, 177; flat-tailed horned, 170,
    175; granite night, 171, 175; granite spiny, 169,
    177; Great Basin collared, 169; Islas Coronado
    alligator, 168, 175; long-nosed leopard, 169,
    176, 177; long-tailed brush, 170, 176; side-
    blotched, 171, 177; Sierra los Cucapás collared,
    168, 169, 175; southern alligator, 170, 175;
    southern sagebrush, 170, 175; tree, 170, 176;
    western fence, 170, 175; zebra-tailed, 170, 172,
    176
Llano Grande (Guerrero), 407
Llanos de San Juan Bautista, 60
Locoweed, 375
Logging, 63–65; and loss of old-growth forest, 63;
    impact on black bears, 397, 398; in northern
    Baja California, 381
Loliolopsis chiroctes, 187
Lomboy, 249, 252, 335
Longinos-Martínez, José, 376
Longspur, chestnut-collared, 441, 442, 443, 444,
    445, 448, 449
Loon, Pacific, 281, 319
López, Guadalupe, 410, 411
Lophocereus, 228, 232

Loreto, 278, 289, 290, 302; development of, 239;
    scorpion diversity, 134
Los Barrancos, 376
Lotus scoparius, 378
Lovegrass, Lehmann's, 118–119; plains, 116
Lower Colorado Valley Region (phytographic
    division), 175, 176
Lucania parva, 148–149, 155, 156
Luziola gracillima, 113, 116
Lynx rufus, 323, 347, 393
Lysiloma, 43; *L. candidum*, 335; *L. divaricatum*,
    205, 335; *L. watsonii*, 205

Macaws, trade in, 69
Macrhybopsis aestivalis, 143, 154, 155
Macrocystis, 180
Macrofauna, defined, 180; Golfo Project, 180, 184
Madera, 119
Madrone, 205; madroño, 64, 205, 206, 376
Magdalena, 176; Plain Province (biogeographic
    subdivision) and scorpion diversity, 132; Plains,
    loss of habitat, 238
Magdalenan subregion, 228
Mahogany, mountain, 373, 375
Maíz, 118; *blanco*, 119; *cócono*, 119 *reventador*, 119
Maizillo, 116
Malacothamnus fasciculatus, 378
Mallow, bush, 378; globe, 460
Malosma, 373; *M. laurina*, 173, 373
Malpai Borderlands Group, 435
Malpighiaceae, 46
Mammal(s), impacted by vicariant events, 233;
    in the Janos-Casas Grandes area, 430;
    phylogeography of, 228
Mammillaria, diversity in the Chihuahuan
    Desert Region, 266; *M. carmenae*, 269, 272;
    M. dioica, 373; *M. lenta*, 272; *M. luethyi*, 273;
    M. microhelia, 272; *M. theresae*, 272
Management Units for the Conservation of Wildlife
    (UMAS), 94, 394
Mandevilla foliosa, 205, 214, 217
Mangals, as a synonym of mangrove stands, 299
Manglares, 299
Mangle botoncillo, 303
Mango, 325
Mangrove(s), 70; birds recorded from, 319–322;
    black, 302, 303, 304, 306, 313, 318;
    buttonwood, 303; change in stand coverage,
    325; characteristics of, 299; ecosystems of,
    298–327; estuaries, 183; fish recorded from,
    314–316; invertebrates recorded from, 307–
    312; red, 30, 313, 318, 326; restoration of,
    326–327; white, 302, 303, 304, 304, 318
Mangrove Replenishment Initiative, 327
Manifestación de impacto ambiental, *see*
    Environmental Impact Assessment legal
    requirement